DISTURBANCE ANALYSIS FOR POWER SYSTEMS

DISTURBANCE ANALYSIS FOR POWER SYSTEMS

Mohamed A. Ibrahim
New York Power Authority
Director of Protection and Control (Retired)

IEEE
IEEE PRESS

WILEY
A JOHN WILEY & SONS, INC., PUBLICATION

Published by John Wiley & Sons, Inc., Hoboken, New Jersey
Published simultaneously in Canada

Library of Congress Cataloging-in-Publication Data:
Ibrahim, Mohamed A., 1943-
Disturbance analysis for power systems / Mohamed A. Ibrahim.
p. cm.
Includes index.
ISBN 978-0-470-91681-0 (cloth)
1. Electric power system stability. 2. Transients (Electricity) 3. Electric power failures. 4. Electric network analysis. I. Title.
TK1010.I27 2011
621.319–dc22

2010048274

ePDF ISBN: 978-1-118-17211-7
oBook ISBN: 978-1-118-17209-4
ePub ISBN: 978-1-118-17210-0

10 9 8 7 6 5 4 3 2 1

To my mother, who taught me without knowing how to read or write; my father; my wife; and my family

CONTENTS

PREFACE

The fault recording equipment used in monitoring power systems evolved from a wet trace and light beams writing on special photo-sensitive paper or film oscillograms to digital, microprocessor-based technology. Some of the old records took days to develop, as in the case of the wet trace and recurring problems with sensitive papers. As a result, some key records were lost, making the analysis of power system disturbances extremely difficult. In addition, starting recording equipment was a hassle, causing unreliable oscillograph operations. A digital fault recorder (DFR) is considered an intelligent electronic device that can be accessed via communication links to send fault records automatically to remote operating centers and engineering offices immediately following a disturbance. This allowed a rapid analysis to make it possible to restore the system. Accurate root-mean-square measurements as well as a host of software packages can be executed to verify the system model and to assess the impact of disturbances on power system equipment.

Analysis of power system disturbances is an important function that monitors the performance of a protection system. It can also provide a wealth of valuable information regarding correct behavior of the system. Understanding power system phenomena can be simplified, and adoption of safe operating limits and protective relaying practices can be enhanced. Review of DFR and numerical relay fault records for system operations can help to isolate incipient problems so that corrections can be implemented before the problems become serious. Understanding power system oscillations and system relaying response during a power swing condition can be enhanced, thus avoiding system blackouts. In addition, understanding power system engineering concepts and the use of symmetrical components in the analysis of power system faults can be enforced and enhanced through DFR analysis.

A bulk power system is normally protected by two redundant relaying systems. The performance of these systems can be monitored through an analysis of system disturbances. Restoration of a power system requires correct analysis of the disturbance that caused the outage to confirm that it is safe to reenergize the system. Correct analysis can contribute to safe restoration without the fear of energizing faulty power system equipment. In addition, through proper system disturbance analysis confidence can be gained in the philosophy behind relaying application.

To facilitate the reader's review process, the DFR records are accompanied by unique functional system diagrams that show the voltages and currents monitored, using designation labels that match the records. A section is devoted to documenting power system phenomena as they appear in actual case studies. This will provide engineers who have limited experience with such problems the necessary background to perform their own analyses of their systems.

The book serves as a forum to document and present my 40+ years of experience in the area of power system disturbance analysis. Many colleagues from the American Electric Power Service Corp., the New York Power Authority, and several utilities have contributed to the book directly or indirectly, and I am grateful for their input. It has been my intention to simplify the topics presented and provide clear guidance as well as basic education to relay engineers. In this new format, the theory and basic fundamentals of relay applications are first briefly explained. This is then followed by real case studies involving system disturbances, to enforce these basics. The studies are based on actual occurrences collected through my years of involvement in the protection of utility systems. The real names of utility plants, substations, and lines have been replaced by generic labels.

In the old vertical integration environment, training and education were essential to most utilities. In the highly competitive new environment, exchange of experience and technical information is hampered, as is passing useful experience to young engineers. At this point in the history of protective relaying, the fundamentals that have been handed down from generation to generation are in danger of becoming lost. This has given me the impetus to document my experience in a useful format that can benefit engineers, since little training is now available for engineers entering the protection and control field in the area of system disturbance analysis.

In the book I present in detail how power system disturbance analysis is used as an important tool to judge the performance of protection systems. Actual DFR records, oscillograms, and numerical relay fault records are analyzed to demonstrate how to deduce the sequence of events. Topics such as the information needed for analysis, fault incident angle, and power system phenomena and their impact on relay system performance are covered. Power system phenomena derived from an analysis of system disturbances are described. In addition, case studies of actual system disturbances involving the performance of protection systems for generators, transformers, overhead transmission lines, cable feeders, and breaker failures are included. Several chapters are devoted to system disturbance analysis as a tool for optimizing the performance of relaying schemes. In addition, the book can serve as a tool for validating power system models and provides a wealth of technical information about the behavior of power systems.

The book is intended primarily for engineers and technicians working in the areas of protection and control, power system operation, and electrical power system equipment. It is also intended for operators and support staff at energy control centers to enhance their technical background in the safe restoration of a power system following a disturbance. The book will provide engineers with a basic background in most power system phenomena and their impact on the behavior of protection systems. The book can also be used as a textbook for undergraduate and graduate students seeking to enhance their power backgrounds. A chapter is devoted to problems, to enhance understanding of the system disturbance analysis function. The book can thus provide an incentive to colleges to offer the system disturbance analysis topic in either an undergraduate or graduate course.

Mohamed A. Ibrahim

1

POWER SYSTEM DISTURBANCE ANALYSIS FUNCTION

An analysis of system disturbances provides a wealth of valuable information regarding power system phenomena and the behavior of protection systems. Experience can be enhanced and knowledge can be gained from the analysis function. This book is organized, first, to cover the analysis function and how it can be implemented. Then, in the following sections, phenomena related to system faults and the clearing process of faults from the power system are described. Power system phenomena derived from an analysis of system disturbances are stated. In addition, case studies of actual system disturbances involving the performance of protection systems for generators, transformers, overhead transmission lines, cable feeders, and breaker failures are provided. A section is devoted to problems that enhance an understanding of the system disturbance analysis function.

Analysis of system disturbance is based on 60-Hz phenomena associated with power system faults. Therefore, sampling rates of digital fault recorders (DFRs) are designed to fulfill this requirement. High-frequency power system transient analysis requires special devices other than conventional DFRs and numerical relays, with unique requirements different from those of a traditional power system disturbance analysis function.

Disturbance Analysis for Power Systems, First Edition. Mohamed A. Ibrahim.

To analyze the performance of protective relaying systems, high-speed digital fault and disturbance recording devices need to be employed properly. Equipment can be used for continuous monitoring of the behavior of relaying installed on a power system during the occurrence of either faults or power swing or switching operations. The equipment can be used to explain undesired operations and to assess system performance during correct operation. Analysis of fault records will help in adapting operating and protection practices and in assuring the reliability of a bulk power system. The analysis will also help to isolate problems and incipient failures. In addition, the strategic placement of DFR equipment should provide adequate coverage of the overall system response to any type of system fault or wide-area system disturbance. For this reason, DFR applications and implementation on a bulk power system are mandated by industry standards and regulations.

A review of DFR records for every operation in a system will help to isolate incipient difficulties so that corrections can be provided before a serious problem develops and to provide basic useful information about the performance of the relaying system. A review of all fault records for disturbances on a system can enhance the reliability of a relay system. Systematic analysis of disturbances can play an important role in system blackout avoidance. When they occur during the early stage of analysis, flagging relay and system problems should be addressed before they precipitate into wider-area interruption and system blackouts. This can be accomplished by analyzing correct operations and finding the causes of incorrect operations. In addition, it can provide a better assessment of the validity of relay setting calculations, correct current transformer (CT) and voltage transformer (PT) ratios, and correct breaker operations. It can also enhance the system restoration process by providing fault types and locations and a better measure of power quality.

The proposed NERC Reliability Standard PRC-002-02, "Disturbance Monitoring and Reporting Requirements," is noted here as a document which ensures that regional reliability organizations establish requirements for the installation of disturbance-monitoring equipment and reporting of disturbance data to facilitate analyses of system events and verification of system models.

1.1 ANALYSIS FUNCTION OF POWER SYSTEM DISTURBANCES

Analysis of power system disturbances can be summarized on the basis of the following primary functions:

1. The need to view fault data as soon as possible after a fault or disturbance occurs so as to restore the system safely.
2. The need to design the DFR with a reasonable pre-fault time (5 to 10 cycles) to capture incipient initiating conditions (e.g., surge arrester spillover).
3. The need to design the DFR with a long post-fault time, adjustable from 0 to 5 s, to be able to analyze backup protection clearing times (60 cycles or more) and

limited power system swings (several seconds) following the occurrence of system disturbances.

4. The need to manipulate the data time base on the DFR record to analyze the effect of faults.
5. The need, finally, to manipulate the DFR data channels and view only those selected.

Ideally, the analysis function should be carried out for all relay operations in a system. The normally cleared events can lead to the discovery of equipment problems and can also be used as a teaching example for power system behavior and phenomena. From the analysis function, monthly disturbance analysis reports can be prepared. In addition, other reports can be generated. The analysis function will focus primarily on providing answers to the following basic questions:

1. What happened?
2. Why did it happen?
3. What is going to be done about it?

In essence, a sequence-of-events report, or time line, needs to be developed. Traditionally, a DFR monitors power system voltages and currents, whereas a sequence-of-events recorder (SER) monitors relay outputs, breaker and disconnect switch positions, alarms, relay targets, and relay communication channels. A DFR can integrate both functions by monitoring events and analog quantities. The following are some of the functions that analysis of DFR records, in conjunction with SER records, can provide:

1. Sequence of operation
2. Fault types
3. Clearing times
4. Reclosing times
5. Relay problems such as:
   (a) Failure to trip
   (b) Failure to target
   (c) Failure to reset
   (d) Delayed clearing
6. Communication problems such as:
   (a) False operation of blocking schemes during carrier transmission holes
   (b) Failure to operate for permissive overreaching transfer trip schemes during signal loss
7. Circuit breaker problems such as:
   (a) Contact arcing
   (b) Unequal pole closing

    (c) Unequal pole opening
    (d) Re-strike
    (e) Reignition
8. Fault current and voltage magnitudes to confirm a short-circuit model
9. CT saturation
10. Asymmetrical current caused by dc (direct current) offset
11. Fault locations, currently provided by numerically based distance relaying, can also be provided by DFRs when sufficient analog signals per line are monitored

1.2 OBJECTIVE OF DFR DISTURBANCE ANALYSIS

Data obtained from DFRs and numerical relaying can be used for continuous monitoring of the behavior of the relay system and assist in setting operating margins on critical control and protective apparatus in an electric power system during system disturbance events such as faults, power swings, and switching operations. Analysis of the data can have the dual role of explaining undesired operations and assessing system performance during correct operation.

The primary objective of obtaining and analyzing DFR data is for the purpose of adapting operating and protection practices as well as control strategies to assure the security and dependability of the bulk power protection system. The secondary objective is for the purpose of helping to isolate problems and incipient failures. This requires a review of all DFR data for every operation, to detect and correct incipient troubles before they become a serious problem. The ability should exist for remote interrogation for data analysis and manipulation. Data need to be viewed as soon as possible after a fault or disturbance occurs. The data time base for the DFR record should be manipulated for analysis. The ability should exist to manipulate data channels and view only those of importance. This will ensure that other channels will not obscure vital data.

It is a good idea to analyze all disturbances in a system, but this may require additional personnel who may not be available within the utility's environment. Indeed, it should be realized that the knowledge gained from analyzing mundane operations may prove to be very valuable. Following are some of the benefits that may be gained from an analysis of system disturbances:

1. Knowledge of the performance of the relaying system and associated inputs, outputs, communication system and circuit breakers
2. Root-mean-square (RMS) ground current calculations confirming the power system model
3. Development of statistics summarizing a fault

4. Optimization of the performance of the relaying system by optimizing the design process through analysis feedback
5. Identification of power system phenomena of interest to be used as teaching tools for engineers to enhance their basic technical backgrounds
6. Review of mundane operations that result in successful fault clearing to reveal valuable power system phenomena and correction of system design and modeling errors

1.3 DETERMINATION OF POWER SYSTEM EQUIPMENT HEALTH THROUGH SYSTEM DISTURBANCE ANALYSIS

As mentioned earlier, an analysis of system disturbances can provide feedback regarding the integrity of power system equipment and associated protection systems. The following are examples of some of the feedback of analysis results that can be used to assess equipment health:

1. Detection of excessive capacitor bank outrush currents into close-in faults requires assessment of current transformer (CT) secondary-connected burdens to reduce overvoltage stress across CT secondary circuits.
2. Detection of circuit breaker (CB) re-striking current during the CB fault current interruption process requires CB inspection and examination for possible testing and maintenance.
3. Detection of unequal CB pole closing or opening requires inspection and examination for possible testing and maintenance of the circuit breaker.
4. Disappearance of third-harmonic current flow in generator neutrals requires assessment of generator neutrals for the possibility of either an open neutral or a stator ground fault near the neutral.
5. Determination of undesired relay operation and follow-up analysis can help in the detection of misapplications of relay settings.
6. Detection and follow-up analysis of undesired relay operation can lead to the discovery of certain hidden relay failures before the undesired operation can precipitate into a serious event that can stress the system.
7. Detection of mutual coupling phenomena can help in fine-tuning ground distance relay settings.
8. Detection of magnetic flux cancellation for CB tripping functions can help in identifying single failure criteria that can have a serious impact on clearing future occurrences of system faults.
9. Detection of excessive capacitative voltage transformer (CVT) transients upon the occurrence or clearing of close-in faults can lead to fine-tuning of the zone 1 distance relay setting reach.

10. Thorough system disturbance analysis can lead to optimization of protective relaying dc schematics.
11. Thorough system disturbance analysis can lead to optimization of protective relay settings.
12. Thorough system disturbance analysis can lead to detection of surge arrester spillover that can lead to mitigation of the spill prior to failure.
13. Thorough system disturbance analysis can lead to detection of power system oscillation, which may require running stability studies to determine the need to add either out-of-step blocking or tripping relay schemes.
14. Thorough system disturbance analysis can lead to detection of CT saturation, which may require either reduction of secondary fault currents (by raising the CT ratio) or reduction of CT-connected burden.
15. Simulation and running of actual system faults, based on disturbance analysis, using computer-based test sets with the help of the COMTRADE fault current format can lead to a determination of failed relays and their associated auxiliary relays.
16. Analysis of multiphase faults caused by lightning strikes can lead to optimization of transmission tower footing resistance.
17. Determination of flashover at voltage peaks, leading to insulation failure as the main cause, can provide feedback to correct any insulation design deficiency.

1.4 DESCRIPTION OF DFR EQUIPMENT

Figure 1.1 illustrates the basic subsystem blocks in a digital fault recorder. The analog input signals are first interfaced to a surge suppression package and sampling filters. The input current flows through a shunt and is converted to a voltage that is sampled, converted to digital form by an analog-to-digital (A/D) converter, and then read and processed by the microprocessor. Similarly, the input voltage is scaled down to a range compatible with the A/D range to be converted and then read and processed by the microprocessor. The A/D has to be checked periodically with sufficient accuracy and an acceptable A/D conversion resolution of a true 16 bits. Delta–sigma A/D converters implemented on a commercial single-chip design, with built-in autocalibration capabilities and built-in linear-phase multistage digital decimation and filtering capability are used for some commercial DFRs to guarantee no aliasing in analog input-sampled signals. Binary inputs representing various functions within the substation are also sampled to give a time resolution of about 1 ms. The basic concepts of a DFR function of sampling and storing data whenever a trigger threshold is exceeded is executed inside the device memory by instruction steps within specific firmware. RAM memory is used for data and is normally checked on startup of the DFR device. ROM and PROM are used in the DFR algorithm and software analysis package and checked periodically by memory check sum routines. EPROM is used to store trigger and parameter settings. A programmable digital signal processing

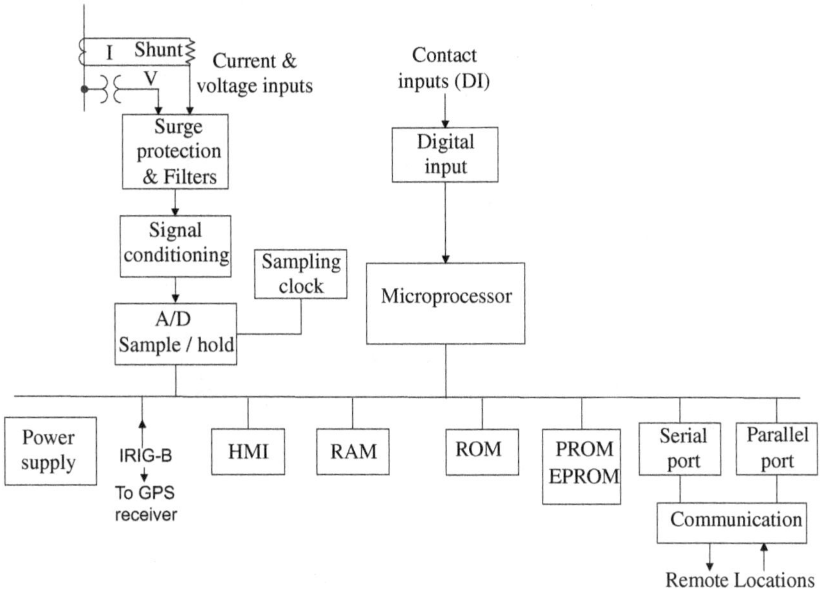

Fig. 1.1 Subsystems of a DFR device.

microprocessor is used to perform serial–parallel conversions and extended-precision adder functions, triggering of recording via various algorithms, and trigger timing management. The DFR-captured data can be retrieved from a remote location via an acquisition computer called the master station. The DFR system should be time-synchronized using an IRIG-B signal from global positioning satellite (GPS) receivers. DFR equipment offers normal communication capability to allow for remote retrieval of fault and event records, making for immediate disturbance analysis and reducing the time and cost needed to perform the analysis task.

1.5 INFORMATION REQUIRED FOR THE ANALYSIS OF SYSTEM DISTURBANCES

The sequence of events can be derived from an analysis of the fault information that may be available from several devices. Presently, the problem is that too many data are available from every intelligent electronic device (IED) and the challenge is for relay and operating engineers to select the most vital data, which need to be analyzed quickly to restore the affected system safely. A sequence-of-events report may be developed using some of the following data:

1. Digital fault recorder records and/or oscillograms (if applicable)
2. Sequence-of-event recorder records

3. Relay targets
4. Numerically based protection oscillograph fault records (if applicable)
5. Phasor measurement records
6. System operation logs
7. Event story as created by field personnel
8. SCADA record, indicating system configurations and loading
9. PC-based short-circuit study simulations
10. As-built one-line, ac three-line, elementary, wiring, and logic diagrams
11. Operating procedures
12. Computer logs and customer information
13. Description of system clearances in the event of an operating or technician error
14. Strip/chart recording or smart IED meters of power system quantities (active power, reactive power, frequency, voltage, and current)

1.6 SIGNALS TO BE MONITORED BY A FAULT RECORDER

1.6.1 Analog Signals

A DFR will monitor voltages and currents as well as digital inputs from the electrical power system. Channel assignments to the DFR should consider monitoring sufficient information to implement the fault location option. This requires the monitoring of three phase-neutral voltages and three phase currents with an option to either monitor or calculate the neutral current (I_n) for each transmission line. In addition, the DFR should monitor all neutral currents and ground sources at the substation to be able to validate the short-circuit model for ground faults. Validation of a short-circuit model for phase faults is difficult to accomplish, due to the effect of loading, which is normally not factored in a steady-state quasi-short-circuit study simulation.

The analog channels are normally configurable as voltage or current inputs. The phase-to-neutral voltage inputs may be scaled for about 66.4 V, with a range of 0 to 250 V RMS, allowing a margin of more than 2 pu (per unit) overvoltage. Current inputs may be scaled for 5 A RMS (nominal load current) and at least 100 A full-scale input using calibrated shunts. The thermal duty can be rated at least 10 A RMS continuous and at least 200 A RMS short time for 2s. Monitoring a generator dc field current can provide valuable educational information about negative-sequence double-frequency-induced rotor current during unbalanced system faults. In addition, monitoring a generator dc field will reveal the 60-Hz induced rotor current during inadvertent energization of generator incidents. Both phenomena are illustrated herein through applicable generator case studies.

Dedicated sensors with over, under, and rate-of-change value settings were used for traditional (conventional) oscillographs. DFRs can also be programmed for each analog channel for over, under, or rate-of-change settings. Additional sensors may include positive-sequence current or voltage, negative-sequence current or voltage, zero-sequence current or voltage, frequency transducers rate of change of impedance during a power system swing (long-term rate of change), and total harmonic distortion.

Following is a list of typical analog channels monitored at the substation level:

- Phase-to-neutral voltages
- Line phase and neutral currents
- Transformer neutral currents
- Transformer tertiary currents
- Transformer polarizing currents (sum of more than one current)
- Capacitor currents (phase and neutral)
- Shunt reactor currents
- Transformer high- and low-side currents
- Zero-sequence voltages
- Bus voltages
- Generator neutral voltages
- Generator fields
- Generator currents
- Generator phase-to-neutral voltages

Monitoring of tertiary ($3I_0$) current by a DFR may help in the classification of ground faults. The CTs for all the phases are paralleled to collect ground current ($3I_0$) and filter out any loading currents (the sum of balanced positive-sequence currents $= 0$). For breaker-and-one-half substation configuration, monitoring of the middle breaker ground current can provide valuable information for circuit breaker maintenance by showing the last breaker of the two that will interrupt the fault current. In addition, determination of which of the two line breakers is exhibiting a re-strike during the fault-clearing process can be accomplished.

1.6.2 Event (Digital or Binary) Inputs and Outputs

Most DFR systems provide means for event recording. This may be status change (closing or opening) of an auxiliary contact associated with a circuit breaker or a disconnect switch operation or the presence of voltage at a control circuit node, which would indicate that a certain control logic function was performed. Examples of events are positions for circuit breakers; disconnect switches, dc presence for control circuits, relays, auxiliary relays, lockout relays, and protection communication signals. Event recording can also be performed by dedicated SERs in the form of stand-alone packages or as part of other systems, such as remote terminals for SCADA systems. Most SER systems are designed with a typical 1-ms resolution time.

1.7 DFR TRIGGER SETTINGS OF MONITORED VOLTAGES AND CURRENTS

In older oscillograph equipment, recording was generally begun using dedicated start sensors to capture fault records. Delta tertiary zero-sequence currents and transformer neutral currents were commonly used to sense ground faults. Undervoltage sensors were also used at key voltage points within the substation, together with an operation limiter, to sense phase faults. Dedicated negative-sequence sensors were also used to trigger the device for unbalanced faults.

The present state-of-the-art DFR is designed with trigger algorithms that are capable of detecting over, under, rate-of-change, and swing conditions for each analog input channel. The trigger algorithm provides concurrent user selectivity for step change, ramp change, and oscillatory conditions. The DFR is normally triggered to capture a record by all analog channels and selected binary inputs. The DFR monitors for line faults three phase-to-neutral voltages and three phase and neutral currents for each line connected at the substation. Phase undervoltage and phase overcurrents will trigger the DFR for phase line faults, and neutral currents will trigger for ground line faults. One analog trigger is sufficient to capture a DFR record.

In addition, triggering can be initiated using positive-, negative-, and zero-sequence symmetrical components as a supplement for shunt faults and as a main trigger for series imbalance, such as open phase. A frequency computation from a bus voltage can also trigger a frequency deviation. Total harmonic distortion and individual harmonic distortion for a specific frequency can also be programmed to trigger a DFR to provide an analysis of power quality. Impedance can also be calculated and used to trigger a DFR. Power swing amplitude for voltage, current, and active and reactive power, as well as oscillation frequency and rate of change of impedance, can also be used to trigger a DFR. In addition, selected digital inputs can be used to capture a record: for example, emergency shutdown lockout relays, which can be energized by many abnormal conditions at a generating plant.

Manual triggering is also provided to test the data capture and output function of a DFR. The manual trigger may be hardwired or software based, with an option for remote acquisition from a master station location. The DFR can also be configured to have a very slow scan to capture long-term events such as power system oscillations or out-of-step conditions. Each trigger function is user programmable with an individual dual-mode limiter function. This function prevents excessive recording both in case a trigger condition persists for an extended period of time and in case a "chattering" trigger should occur. The operation limiter feature will restrict data recording to a selectable length in the event of a continuous long-term trigger condition. An example is the use of undervoltage to trigger the capture of a record for phase faults on a system. Since all analog-monitored channels will be used as triggers, this voltage may be associated with a transmission line. When the line is removed from service during a scheduled outage, the undervoltage sensor will trigger the DFR to capture a record. However, a mean must be established to limit the length of the record since triggering will continue as long as the line is out of service. It should be noted that if phase overcurrent is used to trigger a DFR for faults, the operation limiter feature is not required.

The pre-trigger is normally set at 5 cycles, with an adjustable range from 1 to 10 cycles of the power system frequency. The post-fault range is normally set at 1 s, with an adjustable range from 1 to 5 s. A DFR sensor setting must be established carefully, similar to a relay setting. Every channel should trigger and the DFR sensors should be set to record the minimum fault currents. The DFR-monitored phase currents can be set at a typical overcurrent threshold of 150% of nominal currents. This setting should also be above the maximum emergency loading. The DFR-monitored neutral currents can be set at 20% above maximum loading. DFR-monitored voltage channels can also be set at an overvoltage threshold of 110% of nominal values.

1.8 DFR AND NUMERICAL RELAY SAMPLING RATE AND FREQUENCY RESPONSE

Obsolete oscillograph devices produce oscillograms having a frequency response of about 1000 to 1200 Hz. Newer DFR frequency responses can reach a much higher value than 1200 Hz and normally are a function of sampling rate and the device low-pass filter interface. To avoid aliasing of the analog signal sampled, a low-pass sampling filter is used in numerical relays. The filter blocks any frequency that is higher than half the sampling rate, and thus affects the relay oscillograph record when based on filtered samples. For example, a numerical relay with a sampling rate of 20 samples/cycle will have a frequency response of 600 Hz ($=0.5 \times 20 \times 60$), whereas a DFR with a sampling rate of 64 samples/cycle will have a frequency response of 1920 Hz ($=0.5 \times 64 \times 60$).

A DFR can be specified to have a typically higher sampling rate than that used in numerical relaying. Sampling rates in most DFRs are programmable and range from 64 samples/cycle to 320 samples/cycle, depending on the manufacturer. Sixty-four samples/cycle is a typical sampling rate, and it is more than enough to provide sufficient resolution to verify 60-Hz short-circuit study simulation models. The DFR record length is normally set at 1 s, with the number of pre-fault cycles being programmable and normally set at 5 cycles with a range of 1 to 10 cycles.

A DFR sampling rate of 5760 samples/s per channel can also be defined as 96 samples/cycle ($= 5760/60$) at 60 Hz. The sampling rate can also be defined in terms of electrical degrees as $360°/96 = 1$ sample every 3.75°. The sampling rate can also be written as 1 sample every 173.6 μs ($= 16.666$ ms/96).

1.9 OSCILLOGRAPHY FAULT RECORDS GENERATED BY NUMERICAL RELAYING

The multifunction numerical relays used for protection also provide valuable oscillograph fault records. Multifunction numerical relays used to protect generators and transformers have benefited power system disturbance monitoring functions by providing valuable oscillograph fault records for postmortem analysis. Transformer high-, low-, and tertiary-side currents as well as generator neutral- and system-side currents are normally not monitored by a substation DFR, due to the device's limited number of analog channels. Built-in numerical relaying capability can assist in the analysis of power system disturbances. The information contained in these records can be used to identify the type of system testing needed to identify the cause of the tripping. This will speed up the return of equipment to service. For example, it will provide the necessary data to keep a generating machine either off-line for testing and inspection, when necessary, after an electrical tripping event, or to return the unit to service with minimum delay. Oscillographic monitoring of generators provides invaluable information that will enhance a utility's decision-making process. In addition, numerical relay fault records provide fault type and fault location to assist in fast restoration of faulted transmission lines.

Numerical relays are designed with antialiasing filters, which are required to prevent frequency folding and aliasing of multiples of the sampling rate from appearing with the original samples, which represent the true signal. Antialiasing filters will fulfill the *Nyquist criterion*, which states that frequencies above one-half of the sampling rate must be removed to avoid aliasing error in the signal sampled. Sampling of digital channels has a normal scanning rate that permits at least 1 ms of resolution time.

1.10 INTEGRATION AND COORDINATION OF DATA COLLECTED FROM INTELLIGENT ELECTRONIC DEVICES

DFRs, revenue metering, and numerical relaying IEDs have the ability to transmit fault records and events to designated locations for further analysis. One of the challenges for power system analysis is to coordinate the overwhelming amount of data gathered by IEDs used for protection, fault recorders, meters, and SERs. There exist software and hardware systems that can collect and analyze the IEDs furnished by various manufacturers using different data formats to select the most relevant and vital data required for the analysis. However, the use of a standard format such as COMTRADE can help to promote remote collection of data using sophisticated software.

Many difficulties were encountered during analysis of a major system blackout in 2003, due to the fact that many fault records had time stamps that were difficult to correlate. To avoid similar problems in the future, it is now mandated that time stamps be synchronized. This can be accomplished by time synchronization of all IEDs in the bulk power system using an IRIG-B signal from GPS receivers.

1.11 DFR SOFTWARE ANALYSIS PACKAGES

Most DFRs are equipped with software packages that can provide various system calculations and manipulations to simplify the analysis functions. Some of the software analysis packages currently available are described below.

1.11.1 Phasor Analysis

Phasor analysis helps in assessing the quality of a three-phase system in terms of voltage or current magnitude and phasor relationships and in understanding power system phenomena. When one phase is selected as a reference, the remaining phases will then be plotted with their angles shown relative to the reference for a known phase sequence. Phasors can then be shown with their magnitudes and arguments (polar form), so that three-phase analysis can then be executed to calculate active power (MW), reactive power (MVAR), and current and voltage phasors.

Case Study 1.1: Use of DFR Phasor Analysis Software Figure 1.2 shows an independent power producer (IPP) connected to a bulk power system through the use of 115-kV line L1 as a plant startup ac source. The cogeneration facility at plant Y was on a reserve shutdown status, with the understanding that the 115-kV system is the sole source for the IPP station service facility. A phase B to-ground (B-g) fault occurred on the 230-kV bus 1 at substation X, with the DFR-monitored voltages and currents shown in Fig. 1.2. The DFR record shown in Fig. 1.3 at substation X reveals that the zero-sequence current contribution trace L1-In from the plant Y unit transformer to the fault was interrupted when 115-kV CB A1 opened. However, immediately following the CB A1 opening, three-phase balanced voltage traces L1-Va-n, L1-Vb-n, and L1-Vc-n appeared on 115-kV line L1 at substation X. The voltages stayed on the line for 3.5 cycles, followed by an opening transient. It should be noted that as according to the design of the IPP facility and the

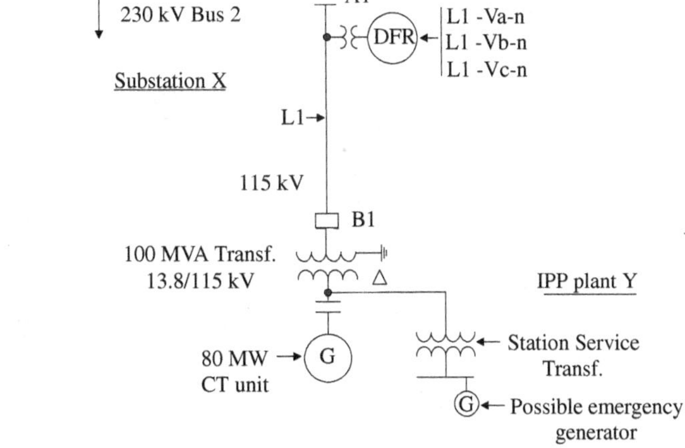

Fig. 1.2 System one-line diagram showing DFR-monitored voltages and currents.

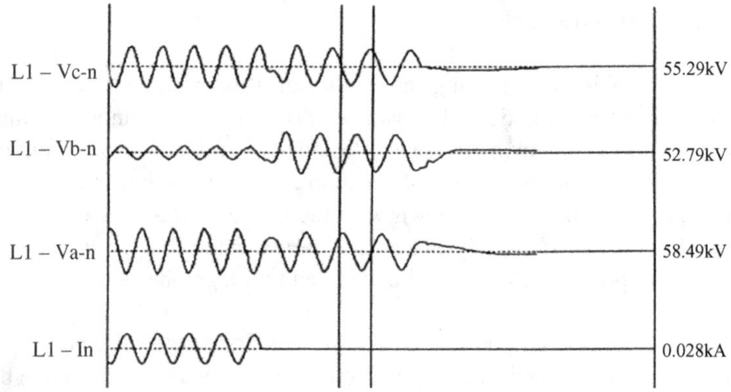

Fig. 1.3 DFR record for the 115-kV line L1 voltages and ground current during clearing of the 230-kV B-g fault.

operating agreement with the utility, this line voltage should be zero (a deenergized system). The removal of the line voltages coincided with the direct transfer trip signal, which was generated by the opening of CB A1 and sent over the fiber optic system to trip CB B1 at plant Y. This feature was incorporated in the design to guarantee the opening of the interconnection CB B1 following an accidental opening of CB A1 at substation X. Hence, it will facilitate resynchronization of the IPP units using circuit breakers at plant Y.

The DFR records were analyzed further to make sure that no distribution source at the IPP facility was placed in parallel with the transmission system. An off-line DFR software phasor analysis program was used to investigate the source of this line voltage. As shown in Fig. 1.3, the depressed B phase voltage was back to normal after the opening of CB A1, whereas phases A and C revealed a slight phase shift at the opening moment. Figure 1.4 illustrates the pre-fault voltages, which are balanced (120° between

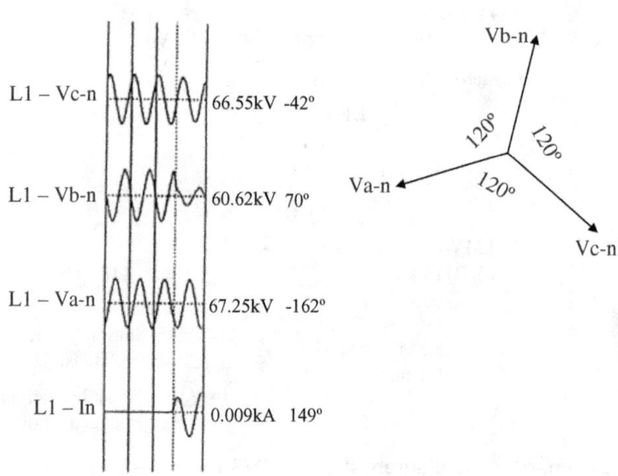

Fig. 1.4 115-kV voltage phasor diagram for the pre-fault condition.

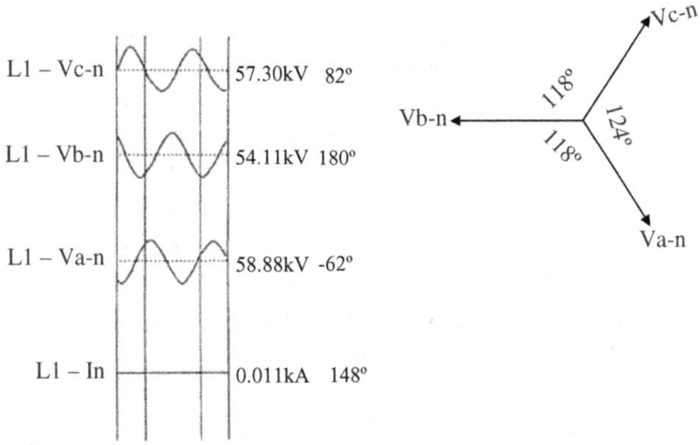

Fig. 1.5 115-kV voltage phasor diagram for the post-fault condition.

phases) with an approximate average phase-to-neutral magnitude value of 67.5 kV and a period indicating 60-Hz system frequency. As shown in Fig. 1.5, analysis of the line L1 voltages following the line trip at substation X revealed an unbalanced voltage with an approximate average magnitude value of 56.5 kV and a slightly greater period for this line voltage than the nominal 60 Hz. Therefore, the frequency of this voltage is lower than 60 Hz and is calculated to be about 51 Hz. The lower magnitude of voltage and frequency may imply that the source is a small emergency generator that was connected at the IPP plant auxiliary bus prior to the phase B-g line fault. The source appears to be very weak, with no significant positive-sequence contribution that has any effect on the contribution to the initial B-g fault on the 230-kV bus and the zero-sequence current flow on the 115-kV system during the fault.

1.11.2 RMS Calculation

The root-mean-square (RMS) value of continuous periodic current signals when sampled N times per cycle is defined as

$$I_{\text{RMS}} = \sqrt{\frac{1}{N}\sum_{k=1}^{N}(i_k)^2} \tag{1}$$

where I_{RMS} is the root-mean-square current value, N the number of samples per cycle, and i_k the data points sampled.

One of the uses for RMS values of monitored DFR signals is to confirm the power system model. This can be done by using fault type and fault location to run a software package to simulate faults, using short-circuit studies and comparing the results calculated with the measured (via calculations) DFR RMS values recorded. RMS calculations can be executed for all recorded signals through the entire record by positioning two cursors separated by a fixed 1-cycle length (16.666 ms). The RMS calculation algorithm grabs the digital sampled data within the DFR memory (between the two cursors) and executes an accurate RMS software calculation. This can be done by using one of the phasor estimation techniques, depending on the DFR manufacturer. One technique could be discrete Fourier transform (DFT) formula (1) with the algorithm window defined as the 60-Hz period, which is equal to the distance between the two cursors. The results are posted as a part of the record in primary values. Figure 1.3 illustrates a DFR record with the RMS currents calculated for the 1-cycle-width cursors shown at the right side of the record. For phase-to-ground faults, RMS values for ground currents can then be used to validate the short-circuit model by comparing the RMS calculated from the DFR record versus the RMS values obtained from short-circuit study simulations.

Example 1.1: RMS Calculation of Continuous Periodic Current Signals

When Sampled Eight Times per Cycle Let

$$i(t) = 10 \sin \omega t \tag{2}$$

Then the RMS value is

$$\frac{I_{\text{peak}}}{\sqrt{2}} = \frac{10}{\sqrt{2}} = 7.07 \text{ A} \tag{3}$$

Assume that the analog current was sampled eight times per cycle (Table 1.1). Using equation (2), the samples can first be deduced and digital RMS formula (1) can be applied:

$$I_{\text{RMS}} = \sqrt{\tfrac{1}{8}\sum_{0}^{8} (i_k)^2} = \sqrt{\frac{0+50+100+50+0+50+100+50}{8}} = \sqrt{\frac{400}{8}} = 7.07 \text{ A}$$

which is the same result as that obtained in equation (3).

TABLE 1.1 Data for Current Sampled Eight Times per Cycle

Sample	ϖt	Sample Value ($10 \sin \varpi t$)	$(i_k)^2$
1	0°	$10 \sin 0° = 0$	0
2	45°	$10 \sin 45° = 10/\sqrt{2}$	50
3	90°	$10 \sin 90° = 10$	100
4	135°	$10 \sin 135° = 10/\sqrt{2}$	50
5	180°	$10 \sin 180° = 0$	0
6	225°	$10 \sin 235° = 10/\sqrt{2}$	50
7	270°	$10 \sin 270° = -10$	100
8	315°	$10 \sin 315° = -10/\sqrt{2}$	50

TABLE 1.2 Data for Current Sampled Four Times per Cycle

Sample	(ϖt)	Sample Value ($10 \sin \varpi t$)	$(i_k)^2$
1	0°	$10 \sin 0° = 0$	0
2	90°	$10 \sin 90° = 10$	100
3	180°	$10 \sin 180° = 0$	0
4	270°	$10 \sin 270° = -10$	100

When Sampled Four Times Per Cycle Assume that the analog current was sampled four times per cycle (Table 1.2). Using equation (2), the samples can first be deduced and digital RMS formula (1) can then be applied. It should be noted that this sampling rate is above the Nyquist criterion of ≥ 2 of the frequency of the signal needed (60 Hz):

$$I_{\mathrm{RMS}} = \sqrt{\tfrac{1}{8}\sum_{0}^{4} (i_k)^2} = \sqrt{\frac{0 + 100 + 0 + 100}{4}} = \sqrt{50} = 7.07\ \mathrm{A}$$

which is the same result as that obtained above.

1.11.3 Calculation of Active and Reactive Powers

Calculation of the active and reactive power is very useful for a power plot during power system oscillation to determine the oscillation frequency. Generator or transmission-line active power flows can be calculated using the active power formula, $P = \sqrt{3}V \times I \cos\phi$, and then plotted as a function of time. Generator reactive power flows can be calculated using the formula $Q = \sqrt{3}V \times I \sin\phi$, and then plotted as a function of time. In addition, reactive power flow can be recorded for further analysis. Generator out-of-step or oscillation conditions can then be confirmed by analyzing active and reactive power plots. For faults near generator terminals, plotting reactive powers can illustrate the sudden change of power factor from load to fault condition where the phase angle for load is near 30° and greater than 60° for faults.

1.11.4 Data Display Manipulation

All DFR records can be displayed for analysis on computer screens for true RMS calculation, phasor analysis, harmonic analysis, and trace manipulation, including expanding, compressing, selection, and movement. Any user-selected group of channels up to the total traces involved in a given disturbance can be displayed. The DFR start sensors or triggers will start the data storage process of all inputs. Software packages will provide support for data display and the analysis required. This software will include conversion of data to a COMTRADE format, allowing the records to be used for testing, troubleshooting, and verification purposes. In addition to all software analysis functions noted earlier, the following functions are also normally available:

1. Immediate display of all abnormal channels, with automatic separation between traces
2. Selection and relocation of traces during record display
3. Overlay of certain traces for balanced display of a three-phase system
4. Manipulation of traces for compression, expansion, or change of amplitude
5. Grouping of certain channels to provide easy display (e.g., three-phase-to-neutral voltages and three-phase and neutral currents for each transmission line monitored).
6. Synchronization of fault records at one end of the line with records obtained at the remote end of the line (can be used to execute double-ended fault location algorithms, if required)

Case Study 1.2: Use of DFR Data Display Manipulation Figure 1.6 shows a one-line diagram for a 115-kV system connecting substations Y and Z and the DFR signals monitored at substation Y. An apparent lightning strike caused a three-phase fault on 115-kV line L1. Cursor X shown in Fig. 1.7 makes the fault appear to be a simultaneous three-phase fault that lasted for 5.5 cycles. Phase A clears at point a, followed by phase B clearing at point b after 120°, followed by phase C at point c after 120°, thus confirming the correct phase sequence as ABC. However, a lightning creation mechanism of a simultaneous three-phase fault requires further analysis.

DFR technology permits expantion of the recorded traces in the direction of both the time and magnitude axes. The expansion of the time axis of 115-kV line L1 three-phase and neutral current traces, shown in Fig. 1.8, reveals that the fault started as an apparent direct lightning hit on phases A and B, as indicated by cursor a, causing an A-B-g fault and then evolving into a three-phase-to-ground fault, as indicated by cursor b 0.7 ms after cursor a.

Therefore, the incident can be explained as a direct lightning hit on phases A and B where the fault current caused by the lightning stroke went to ground via the tower footing resistance. The voltage buildup across the tower footing resistance caused a back flashover from the ground to phase C about 0.7 ms later.

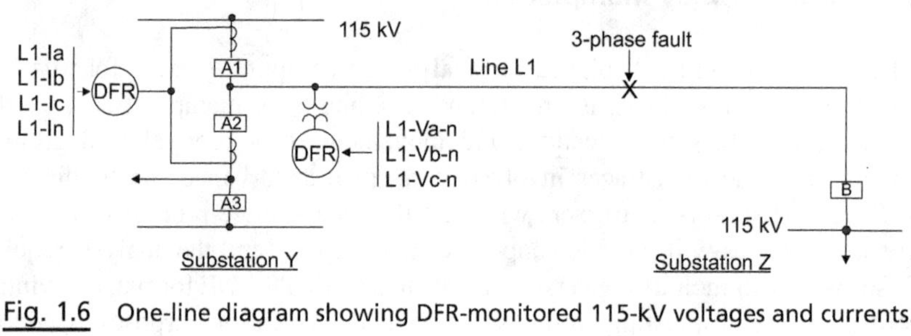

Fig. 1.6 One-line diagram showing DFR-monitored 115-kV voltages and currents.

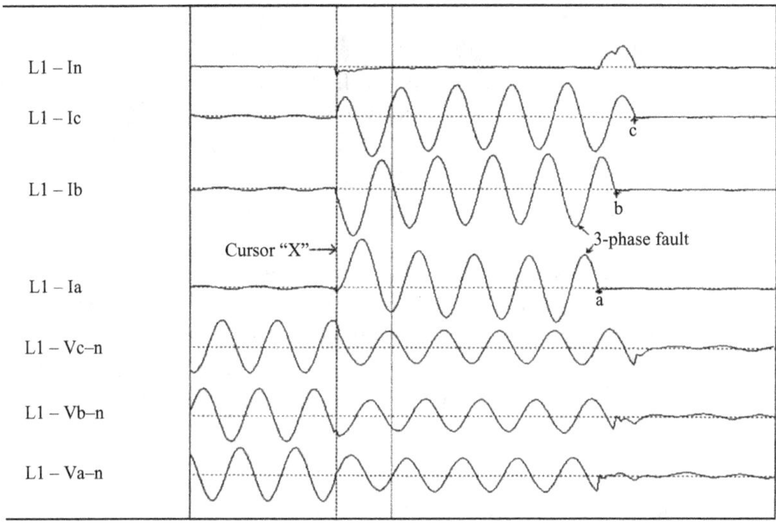

Fig. 1.7 Substation Y DFR record for voltages and currents revealing a simultaneous three-phase fault.

1.11.5 Fault Location

Fault Location Using a DFR Record and PC-Based Short-Circuit Simulation Software For bolted faults (fault resistance = 0) with symmetrical currents, the fault location can be determined by sliding the fault application point on the faulted line using short-circuit simulation studies until a result match is established between DFR-recorded versus calculated currents and voltages.

Fault Location Using a DFR Record and a Software Package When a transmission line's three voltages and four currents are monitored by a DFR data

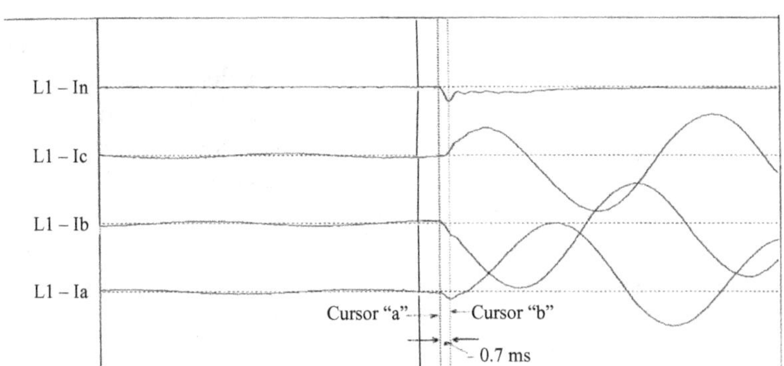

Fig. 1.8 Substation Y DFR time-expanded record for currents confirming a phase A-B-g fault, then evolving into a three-phase-to-ground fault.

acquisition unit, a fault location function can be performed using either a single- or a double-ended fault location algorithm residing within the DFR. A fault classification subroutine is needed to provide the correct faulted impedance loop. This is one of the six loops that cover all types of 10-line faults. A software subroutine is also needed to determine the faulted line so that the faulted-line impedance can be calculated automatically and sent to operating or control centers. Presently, this function is also provided by dedicated numerical distance line protection.

Fault Location Using Numerical Relaying Numerical relaying provides very accurate fault location results as a part of event record output from the relay. Numerical relaying has provided good results for fault location since its inception in the late 1980s. Some DFR and numerical relay ocillograph records provide means to synchronize data obtained from both ends of a faulted line to execute more accurate double-ended fault location algorithms.

1.11.6 Harmonic Analysis of a Power System

A DFR can be used as a power quality instrument by recording the harmonic and total harmonic distortion (THD) profiles of voltage and current waveforms. A DFR can perform harmonic analyses on recorded currents and voltages and use the information to measure trends over time or to compare harmonic distortion at different locations. Depending on the manufacturer and the type of interface filtering, most DFRs can calculate from dc up to the twenty-fourth harmonic of currents and voltages. Harmonic analysis and THD profiles can be used for power quality monitoring of various monitored power sources and to assess the harmonic sources for removal purposes.

1.11.7 Symmetrical Components Analysis

Symmetrical components analysis software can be used on DFR-recorded currents and voltages to obtain positive-, negative-, and zero-sequence components. During ground faults the zero-sequence currents can be used for the validation of short-circuit 60-Hz models. The phase currents are not suitable for use, due to the effect of the positive-sequence component of load flow. Current or voltage sequence components can trigger a DFR to capture certain fault records. In addition, the current and voltage sequence components can help in studying the response of special relaying elements to system faults. This type of analysis can also be used to confirm the occurrence of low-grade equipment faults. Case Study 5.7 is a postmortem analysis of sampled DFR records for the transformer fault sources. A symmetrical component transform operation was performed on 138-kV system contributions to a transformer ground fault, indicating the presence of positive-, negative-, and zero-sequence current components during the fault, thus confirming the occurrence of a winding ground fault. Pre- and post-fault analyses revealed only positive-sequence currents. This matches the off-line simulation for the strength of the 138-kV system behind the transformer.

1.12 VERIFICATION OF DFR ACCURACY IN MONITORING SUBSTATION GROUND CURRENTS

For substations where all ground currents from all transmission lines and transformers zero-sequence sources are monitored by a DFR, confidence can be established for the correct connection and scaling of all DFR currents. This can be accomplished by applying Kirchhoff's first law for a faulted line. This method is applicable to symmetrical fault currents that contain no dc offsets. This is normally the case when the fault incident point is at the voltage peak, indicating the slow fault-creation mechanism that normally accompanies insulation failure. In this case, the RMS current calculated by the DFR software is sufficiently accurate to use this method to verify DFR recorded signal accuracy. All zero-sequence sources monitored will be selected in one record, and the 1-cycle RMS calculation window will be selected for the entire record so that results are obtained at the same time. The following case study illustrates a simple procedure that can be implemented to verify DFR accuracy.

Case Study 1.3: Verification of DFR Accuracy Based on the one-line diagram for substation X in Fig. 1.9, and since all ground currents (zero-sequence) feeding the C-g fault from transmission lines and autotransformer tertiary currents are monitored by a DFR, the following check can be made to verify the accuracy of the DFR recording of the currents monitored.

Figure 1.9 shows the DFR-monitored ground currents and the transformer delta tertiary. Figure 1.10 shows the 230- and 115-kV systems around substation X where a phase-to-ground fault occurred on line L1. The DFR record in Fig. 1.11 reveals all

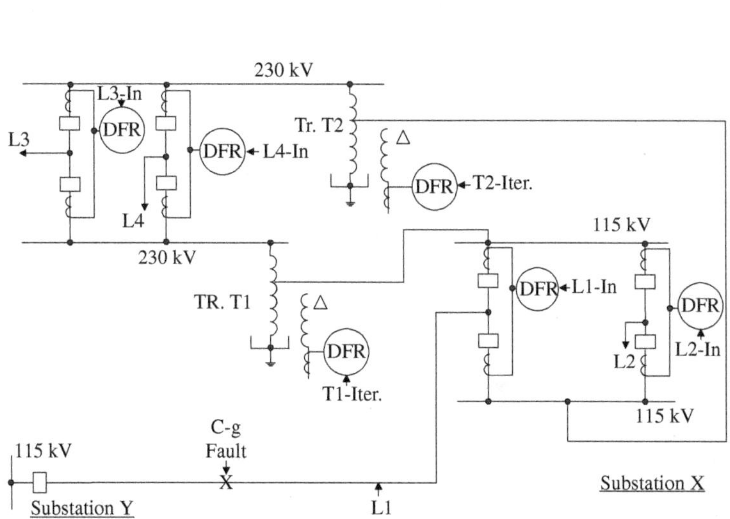

Fig. 1.9 Substation X one-line diagram showing DFR-monitored currents.

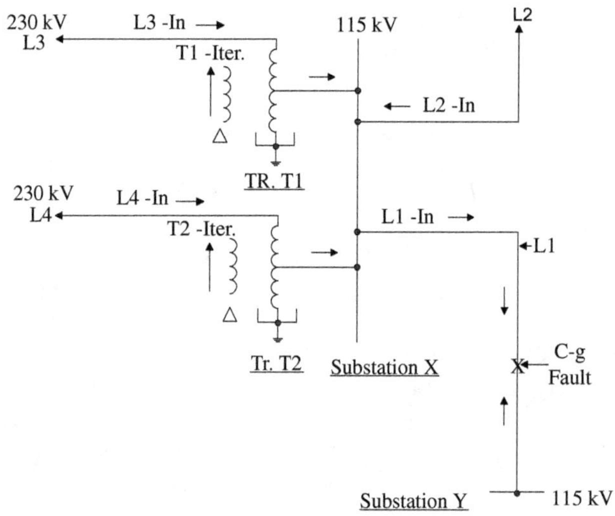

Fig. 1.10 One-line diagram showing ground current flows to the C-g fault.

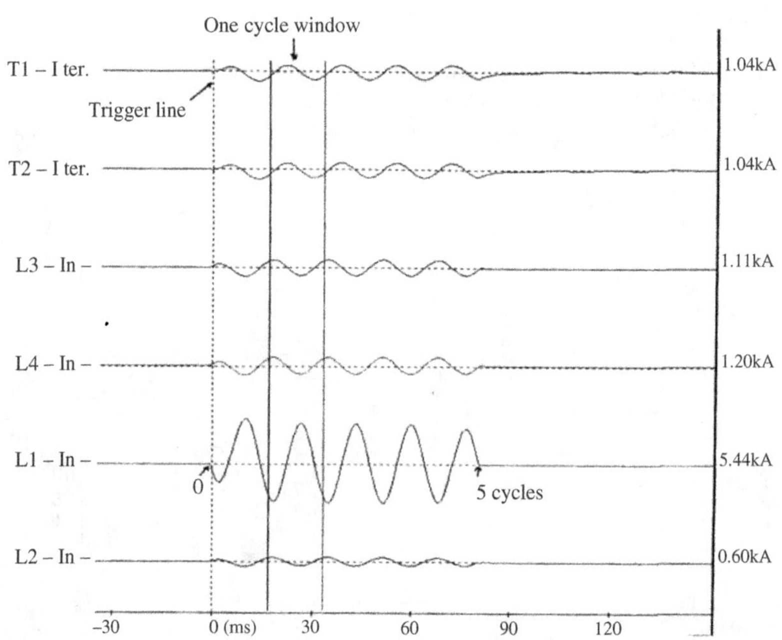

Fig. 1.11 Substation X DFR record for ground currents showing the one-cycle RMS calculation.

ground current sources feeding into the line L1 C-g fault. The sources from substation X are: 230-kV line L3, 230-kV line L4, 115-kV line L1, 115-kV line L2, and transformer T1 and T2 delta tertiary windings. The DFR shown in Fig. 1.11 also reveals symmetric fault currents, which results in accurate RMS current calculations by the DFR software package.

From symmetrical components analysis, the following equation is satisfied if all currents are in per-unit (pu) values on a common base, which in this case is 100 MVA:

$$I_n(\mathrm{L1}) = I_n(\mathrm{L2}) + I_n(\mathrm{L3}) + I_n(\mathrm{L4}) + I_{\mathrm{ter}}(\mathrm{T1}) + I_{\mathrm{ter}}(\mathrm{T2}) \tag{1}$$

At the beginning of the fault, while the line is still fed from substation X, DFR RMS calculation routine can be executed for a one cycle window which is positioned about one cycle from fault initiation. RMS currents are then calculated and the results become part of the DFR record. The currents shown at the right side of the DFR record in Fig. 1.11 are as follows:

$$\begin{aligned}
\text{L1-In} &= I_n(\mathrm{L1}) = 5440\ \mathrm{A}\\
\text{L2-In} &= I_n(\mathrm{L2}) = 600\ \mathrm{A}\\
\text{L3-In} &= I_n(\mathrm{L3}) = 1110\ \mathrm{A}\\
\text{L3-In} &= I_n(\mathrm{L4}) = 1200\ \mathrm{A}\\
\text{T1-Iter} &= I_{\mathrm{ter}}(\mathrm{T1}) = 1040\ \mathrm{A}\\
\text{T2-Iter} &= I_{\mathrm{ter}}(\mathrm{T2}) = 1040\ \mathrm{A}
\end{aligned}$$

To convert all these currents to pu values, we need to define the base current for 230- and 115-kV lines as well as the 13.8-kV delta tertiary currents:

$$\begin{aligned}
I_{\mathrm{base}}(115\ \mathrm{kV}) &= \frac{100 \times 10^6}{\sqrt{3} \times 115 \times 10^3} = 502\ \mathrm{A}\\
I_{\mathrm{base}}(230\ \mathrm{kV}) &= \frac{100 \times 10^6}{\sqrt{3} \times 230 \times 10^3} = 251\ \mathrm{A}\\
I_{\mathrm{base}}(13.8\ \mathrm{kV}) &= \frac{100 \times 10^6}{3 \times 13.8 \times 10^3} = 2415\ \mathrm{A}
\end{aligned}$$

The 13.8-kV base current is calculated using a dividing factor of 3 for phase current that flows inside the delta and is equal to $I_{\mathrm{line}}/\sqrt{3}$.

$$I_{\mathrm{pu}} = \frac{I}{I_{\mathrm{base}}}$$

where I is the DFR recorded current. Therefore, currents in pu are

$$I_{\mathrm{n}}(L1) = \frac{5440}{502} = 10.83\ \mathrm{pu}$$

$$I_{\mathrm{n}}(L2) = \frac{600}{502} = 1.2\ \mathrm{pu}$$

$$I_{\mathrm{n}}(L3) = \frac{1110}{251} = 4.42\ \mathrm{pu}$$

$$I_{\mathrm{n}}(L4) = \frac{1200}{251} = 4.78\ \mathrm{pu}$$

$$I_{\mathrm{ter}}(\mathrm{T1}) = \frac{1040}{2415.4} = 0.431\ \mathrm{pu}$$

$$I_{\mathrm{ter}}(\mathrm{T2}) = \frac{1040}{2415.4} = 0.4.31\ \mathrm{pu}$$

Substituting these values in equation (1) gives us

$$\text{right-hand side (RHS)} = 10.83\ \mathrm{pu}$$

$$\text{left-hand side (LHS)} = 1.2 + 4.42 + 4.78 + 0.431 + 0.431 = 11.26\ \mathrm{pu}$$

$$\%\ \text{error} = \frac{\text{current difference in pu}}{\text{the smaller of the two}} \times 100$$

$$= \frac{\text{LHS} - \text{RHS}}{\text{the smaller of the two}} \times 100$$

$$= \frac{11.26 - 10.83}{10.83} \times 100 = 4.2\% < 10\% \qquad \text{O.K.}$$

It can then be concluded that the DFR accuracy of recording is acceptable.

1.13 USING DFR RECORDS TO VALIDATE POWER SYSTEM SHORT-CIRCUIT STUDY MODELS

DFR and numerical relay event records can be used to validate 60-Hz power system models. During ground faults the ground (neutral) $3I_0$ currents and the faulted phase-to-neutral voltage can be used for validation of the short-circuit models. It is not recommended that phase currents be used, due to the effect of the positive-sequence component of the load flow. The DFR recorded will reflect the effect of load flow; however, quasi-steady-state short-circuit study will not reflect the same effect. Phase-to-phase and three-phase fault cases can be used when the positive-sequence measured current I_1 is calculated and the effect of load flow is removed. When the fault incident point is at the voltage peak, the fault currents will be symmetrical, containing no dc offset. The resulting currents from all the sources can then be used to validate the short-circuit model. This can be accomplished by comparing the RMS values calculated from the DFR record versus the RMS values obtained from the short-circuit study simulation.

For solid ground faults, the only unknown that must be simulated for model validation is fault location. The unknown fault location point can be solved by sliding the fault on the faulted line until a match is established between the simulated fault and the DFR analog values recorded. For high-resistance (tree) ground faults where the fault location is known, the fault can be simulated at the known location while varying the fault resistance values until a match is established between the simulated fault and the DFR analog values recorded.

Case Study 1.4: Bolted 115-kV Phase-to-Ground Bus Fault This case study is ideal for verification of the power system short-circuit model. This is due to the known fault location and the symmetrical nature of the fault current. The solid phase B-g fault was caused by the failure of a station post insulator, which supports a disconnect switch during heavy rain. Monitoring of all ground current sources and neutral currents at substation X will allow verification of the modeling of the power system. In this case study, autotransformer T9 and T10 ground currents on the 115-kV bus are not monitored by the DFR, thus allowing only a partial assessment to be made. Figure 1.12 is a symbolic substation X one-line diagram showing the DFR monitored and unmonitored neutral lines and transformer currents.

At the beginning of the disturbance the B-g fault was symmetrical, containing no dc offset. This portion of the fault duration can therefore be used to verify the accuracy of the modeling of the generators and their step-up transformers. In addition, partial verification for the other recorded sources of the ground fault can also be verified. This can be done by comparing the DFR RMS current values measured to the values calculated from a short-circuit simulation study. All ground current sources can be compared to the 115-kV bus fault, with the exception of the

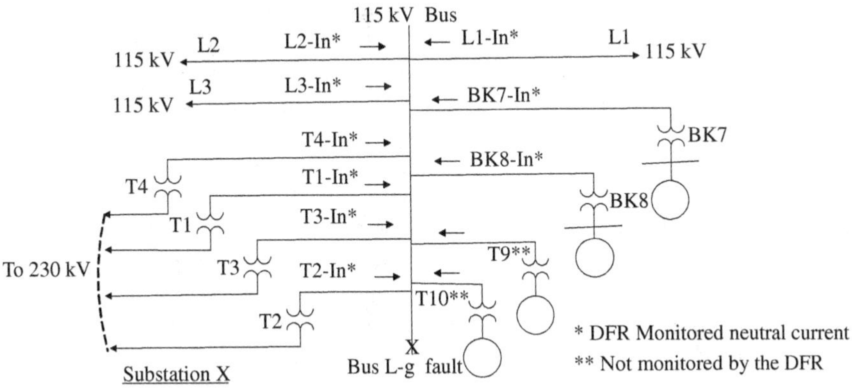

Fig. 1.12 Substation X symbolic one-line diagram showing DFR-monitored currents.

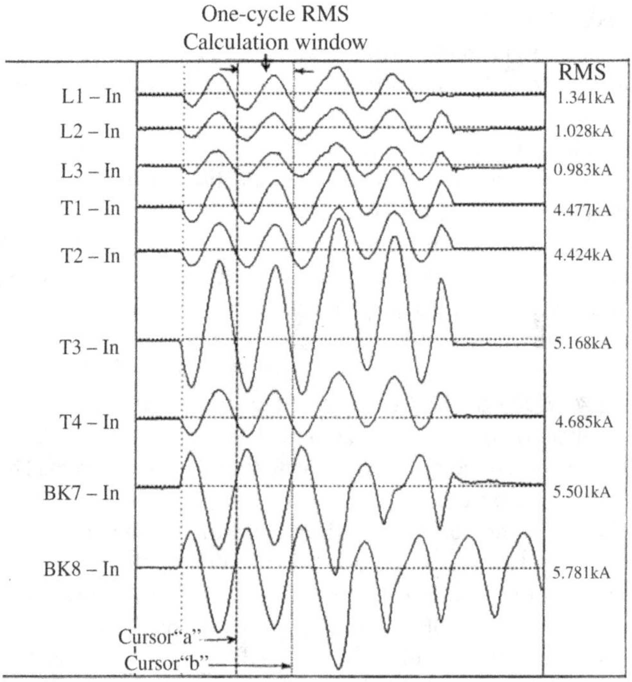

Fig. 1.13 DFR record for ground currents showing the one-cycle RMS calculation window.

T9 and T10 contributions, which are not monitored by the DFR, due to the lack of available DFR analog channels.

Figure 1.13 shows the DFR record for some of the neutral current contributions to the 115-kV bus fault. In addition, the figure reveals the RMS values of these fault currents as calculated by the DFR calculation software for a 1-cycle data window starting 1 cycle after initiation of the fault. Ground currents are not affected by load flow and can therefore be used to compare simulated versus actual fault current values.

A solid phase-to-ground bus fault was then simulated in the short-circuit study program to provide calculated system values which could then be compared with the DFR record. The study provides a total 115-kV bus ground fault of 44.59 kA, as shown in Fig. 1.14. Table 1.3 shows the RMS values for the measured and calculated current values and the comparison needed to ascertain the validity of the model. The analysis also defines an error formula to obtain the accuracy of the system model. The errors should not exceed ±10% (±15% is also acceptable for special cases). This threshold is considered reasonable to accommodate errors in the primary sensing equipment (CTs and PTs) and in the DFR equipment and the parameters used to calculate impedances for short-circuit study simulations. Based on these criteria, the neutral current of unit transformer BK8 trace BK8-In shown in Fig. 1.13 has an error of 2%, which confirms that the modeling in the short-circuit study of the generating units and their associated step-up transformer is correct. In addition, the maximum error for the

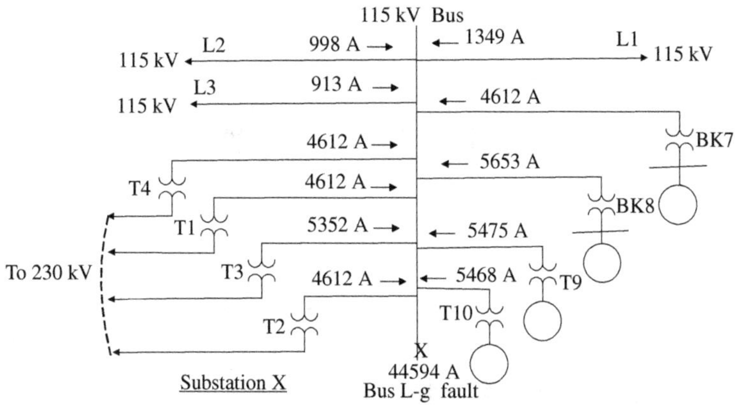

Fig. 1.14 Short-circuit study simulation producing zero-sequence currents ($3I_0$) for a solid 115-kV L-g bus fault.

feeders is 7.7% for the line L3 contribution to the fault, which is less than the error threshold and therefore is also acceptable.

Case Study 1.5: High-Resistance 230-kV Phase-to-Ground Line Fault This case study is not an ideal case for verification of the power system short-circuit model, due to the resistive nature of the fault and the unknown fault location. The high-resistance C-g fault occurred on 230-kV circuit line L1, as shown in Fig. 1.15. The fault was caused by a tree that was located where arcing was observed between the conductor and the tree. The fault was therefore (luckily) located at about 94% of line

TABLE 1.3 Comparison Between Measured and Calculated Fault Currents

Feeders Contributing to the Bus Fault	DFR RMS Values Measured	RMS Values Calculated from the Short-Circuit Study	Error Values[a]	Percent Error[b]
L1-In	1341 A	1349 A	−8	−0.6
L2-In	1028 A	998 A	30	3.0
L3-In	983 A	913 A	70	7.7
T1-In	4477 A	4612 A	−135	−3
T2-In	4424 A	4612 A	−188	−4.25
T3-In	5168 A	5352 A	−184	−3.6
T4-In	4685 A	4612 A	73	1.6
BK7-In	5501 A	5653 A	−152	−2.7
BK8-In	5781 A	5665 A	116	2.0
T9-In	Not monitored	5475 A		
T10-In	Not monitored	5468 A		

[a]Error = DFR RMS values measured −RMS values calculated from short-circuit study.
[b]% Error = [(DFR RMS values measured −RMS values calculated)/the smaller of the two] × 100.

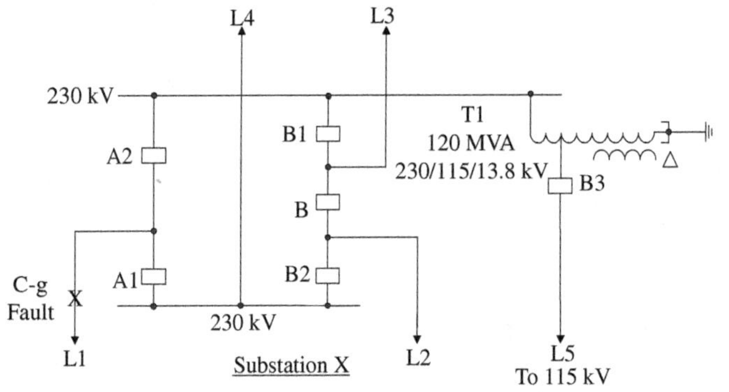

Fig. 1.15 Substation X one-line diagram.

length from substation X. The fault location information was then entered into the short-circuit study to try to estimate the remaining fault resistance unknown parameter. At this fault location the tree fault resistance was varied until a best match was achieved between the DFR-recorded line L1 currents and voltages and the simulation results. A tree fault resistance of 46 primary ohms provided the best fit between the two results. The sources from substation X are: 230-kV line L3, 230-kV line L4, 115-kV line L2, transformer T1 tertiary, and 115-kV line L5.

Since all ground currents for substation X to the fault are monitored and their RMS values are shown in Fig. 1.16 on the right side of the DFR record, the first step will be

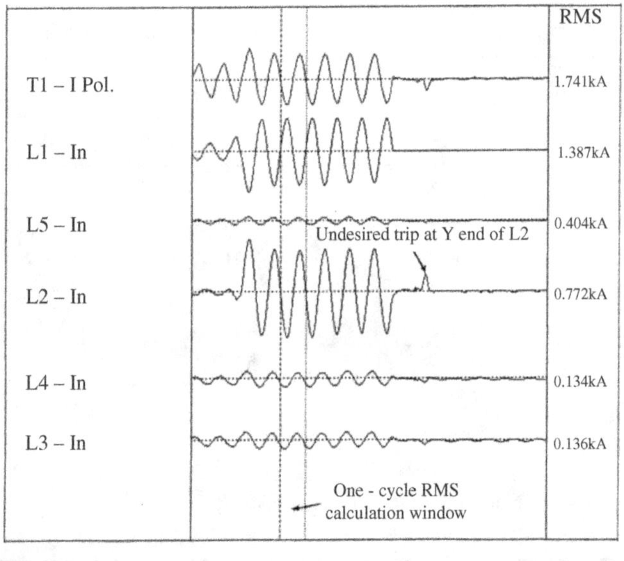

Fig. 1.16 Substation X DFR record for ground currents revealing the one-cycle RMS calculation.

to confirm the accuracy of the DFR in recording the analog signals. As defined in Section 1.10 (in pu):

$$
\begin{aligned}
I_{\mathrm{n}}(\mathrm{L1}) &= I_{\mathrm{n}}(\mathrm{L2}) + I_{\mathrm{n}}(\mathrm{L3}) + I_{\mathrm{n}}(\mathrm{L4}) + I_{\mathrm{n}}(\mathrm{L5}) + I_{\mathrm{ter}}(\mathrm{T1}) \\
I_{\mathrm{base}}(115\,\mathrm{kV}) &= \frac{100 \times 10^6}{\sqrt{3} \times 115 \times 10^3} = 502\ \mathrm{A} \\
I_{\mathrm{base}}(230\,\mathrm{kV}) &= \frac{100 \times 10^6}{\sqrt{3} \times 230 \times 10^3} = 251\ \mathrm{A} \\
I_{\mathrm{base}}(13.8\,\mathrm{kV}) &= \frac{100 \times 10^6}{3 \times 13.2 \times 10^3} = 2525.2\ \mathrm{A} \\
I_{\mathrm{pu}} &= \frac{I}{I_{\mathrm{base}}}
\end{aligned}
\tag{1}
$$

where I is the current recorded. Therefore, currents are

$$
\begin{aligned}
I_{\mathrm{n}}(\mathrm{L1}) &= \frac{1387}{251} = 5.53\ \mathrm{pu} \\
I_{\mathrm{n}}(\mathrm{L2}) &= \frac{772}{251} = 3.08\ \mathrm{pu} \\
I_{\mathrm{n}}(\mathrm{L3}) &= \frac{136}{251} = 0.54\ \mathrm{pu} \\
I_{\mathrm{n}}(\mathrm{L4}) &= \frac{134}{251} = 0.53\ \mathrm{pu} \\
I_{\mathrm{n}}(\mathrm{L5}) &= \frac{404}{502} = 0.80\ \mathrm{pu} \\
I_{\mathrm{ter}}(\mathrm{T2}) &= \frac{1741}{2525.2} = 0.67\ \mathrm{pu}
\end{aligned}
$$

Substituting these values in equation (1) gives us

$$
\begin{aligned}
\mathrm{RHS} &= 5.53\ \mathrm{pu} \\
\mathrm{LHS} &= 3.08 + 0.54 + 0.53 + 0.80 + 0.67 = 5.62\ \mathrm{pu} \\
\%\ \mathrm{error} &= \frac{\mathrm{LHS} - \mathrm{RHS}}{\text{the smaller of the two}} \times 100 \\
&= \frac{5.62 - 5.53}{5.53} \times 100 = 1.6\% < 10\% \qquad \text{O.K.}
\end{aligned}
$$

It can then be concluded that the DFR accuracy of analog signal recording is acceptable.

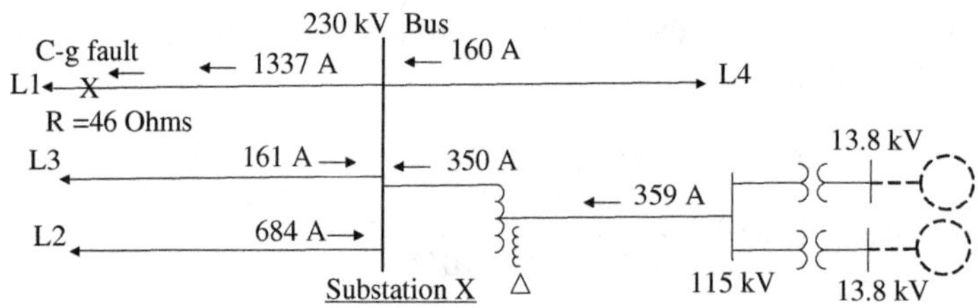

Fig. 1.17 Short-circuit simulation results for a high-resistance 230-kV L-g fault on the line L1 fault location point.

The model can now be verified similar to Case study 1.4 using the DFR measured values shown in Fig. 1.16 and the RMS values calculated from the short-circuit study simulation one-line output diagram shown in Fig. 1.17. Table 1.4 summarizes the analysis results.

It can be concluded that this is a difficult case to match. When contribution errors are calculated as defined in formula (2), the errors exceed the safe threshold of 10%. However, when the errors are normalized, as in formula (3), to the total calculated short-circuit value of 1337 A, the errors are below 10%. In addition, when the normalized errors of the feeders as defined in footnote b are added, 6.6 – 1.9 – 1.95 + 1.4 = 4.15% which is very close to the normalized error of the total current of 3.8%.

TABLE 1.4 Comparison Between Measured and Calculated Fault Currents

Feeders Contributing to the Bus Fault	DFR RMS Values Measured	RMS Values Calculated from the Short-Circuit Study	Error Values	Percent Error[a]	Percent Error[b]
L1-In	1387 A	1337 A	50 A	3.8	3.8
L2-In	772 A	684 A	88 A	12.9	6.6
L3-In	136 A	161 A	−25 A	−18.4	−1.9
L4-In	134 A	160 A	−26 A	−19.4	−1.95
L5-In	404 A	359 A	45 A	12.5	3.4
T1-Ipol	1741 A	Not given			
L5-In + T1-Ipol	Not monitored[c]; can be calculated as 369 A	350 A	19 A	5.4	1.4

[a]% Error = DFR RMS values measured – RMS values calculated)/the smaller of the two.
[b]% Error = DFR RMS values measured – RMS values calculated)/total current calculated.
[c]Can be calculated from the DFR-recorded traces of L5-In and T1-Ipol:

$$\text{L5-In} + \text{T1-Ipol} = \frac{404}{502} + \frac{1741}{2525.2} = 0.8 + 0.67 = 1.47 \text{ pu}$$

$$= 1.47 \times I_{\text{base}} = 1.47 \times 251 = 369 \text{ A}$$

This case study illustrates that it is very difficult to confirm power system short-circuit models for high-resistance ground faults, and that if verification is required, several definitions of normalized errors should be employed to get a feeling for maximum errors.

1.14 COMTRADE STANDARD

The IEEE Standard C37.111 defines a common format for transient data exchange (COMTRADE) for power systems, which was developed to provide a standard file format for sampled analog wave forms and event data collected by a variety of monitoring equipment. The COMTRADE standard format is not a communication protocol for transferring data files, but it enables the exchange of data files between varieties of incompatible devices. The standard can link incompatible devices for the purpose of interchanging fault records and relay testing data. Playing back fault events with the help of the COMTRADE standard can flag equipment problems, hidden failures, and relay setting errors. The availability of COMTRADE format and fault record playback capability is a valuable tool in troubleshooting and assessing undesired operations. The system restoration phase following undesired operations can be enhanced and performed faster by using these tools.

Computer-based test equipment in conjunction with the COMTRADE file format can help in recreating specific system disturbances for testing and developing new relay products. In addition, fault events that present a challenge to existing relaying systems can be applied, using the COMTRADE standard as a benchmark to select and approve new numerical relaying systems supplied by various manufacturers.

Case Study 8.1 describes an application for the COMTRADE format where an incident was triggered by a switching transient that was recorded by a DFR. The COMTRADE format of the record was transferred to computer-based test equipment. The DFR device and test equipment were manufactured by two different companies. The transient activated an undesired breaker failure operation, which resulted in the tripping of a 150-MW combined-cycle plant. Analysis of the DFR records did not reveal a fault, and therefore the plant was returned to service postulating incorrect operation of a 138-kV breaker failure relay. To analyze the incident further, the DFR record for the incident was first converted to COMTRADE format and then played back to test the breaker failure relay, after its isolation, using an automated test set. The playback confirmed that the breaker failure relay produced a consistent output and that the output was present only when the dc power was switched on, energizing the relay. The playback testing led to a review of the schematic and wiring drawings and the discovery of a welded contact in one of the breaker failure initiation auxiliary relays.

REFERENCES

Application and Evaluation of Automatic Fault Recording Devices. IEEE Power System Relaying Committee Report. *IEEE Transactions on Power Apparatus and Systems*, Vol. PAS-84, No. 12, December 1965, pp. 1187–1203.

Application of Fault and Disturbance Recording Devices for Protective System Analysis. IEEE Power System Relaying Committee, Special Publication 87 TH 0195-8-PWR, 1987.

Disturbance Monitoring and Reliability Requirements. NERC Reliability Standard PRC-002-02.

Elmore, W. A., Ed. *Pilot Protective Relaying*. New York: Marcel Dekker, 2000.

Fault and Disturbance Data Requirements for Automated Computer Analysis. IEEE Power System Relaying Committee Working Group 111, Special Publication 95 TP 107, 1995.

Ibrahim, M. A., and F. Stacom. Adirondack 230-kV substation outage of July 1, 1995. Presented at the Georgia Tech Protective Relaying Conference, May 1–3, 1996, Atlanta, GA, and at the 23rd Annual Western Protective Relay Conference, October 14–16, 1996, Spokane, WA.

IEEE Standard Common Format for Transient Data Exchange (COMTRADE) for Power Systems. IEEE Standard C37.111-1999, a revision of IEEE Standard C37.111-1991.

Phadke, A. G., T. Hlibka, and M. A. Ibrahim. A digital computer system for EHV substations: analysis and field tests. *IEEE Transactions on Power Apparatus and Systems*, Vol. PAS-95, No. 1, January–February 1976, pp. 291–301.

Phadke, A. G., and J. S., Thorp. *Computer Relaying for Power Systems*. New York: Wiley, 1988; Chichester, UK: Research Studies Press, 1988.

Summary of the special publication: Application of Fault and Disturbance Recording Devices for Protection System Analysis. *IEEE Transactions on Power Delivery*, Vol. 4, No. 3, July 1989.

Supplement to Application and Evaluation of Automatic Fault Recording Devices. IEEE Power System Relaying Committee Report. *IEEE Transactions on Power Apparatus and Systems*, Vol. PAS-90, No. 2, March–April 1971, pp. 751–755.

Understanding Microprocessor-Based Technology Applied to Relaying. IEEE Power System Relaying Committee, Report of Working Group 116 of the Relaying Practice Subcommittee, February 2004.

2

PHENOMENA RELATED TO SYSTEM FAULTS AND THE PROCESS OF CLEARING FAULTS FROM A POWER SYSTEM

Power system disturbances are classified primarily as shunt faults. In this chapter we describe and analyze phase and ground shunt faults and their causes. Arc-over at the voltage peak represents a slow fault-creation mechanism when insulation fails, whereas lighting hits are random and may produce fault incident points at zero voltage. Faults occurring in power systems can therefore be either symmetrical or asymmetrical, depending on the fault incident angle. In addition, analysis of the nature of fault currents and their different modes of clearing from a power system is also covered and supported by DFR records and oscillograms. Fault modes of clearing, which can be either high speed, sequential, delayed local backup (breaker failure), or remote delayed backup, are also included. Power system phenomena related to the fault clearing process are described based on actual DFR records and oscillograms.

2.1 SHUNT FAULT TYPES OCCURRING IN A POWER SYSTEM

As shown in Fig. 2.1, 10 different types of shunt faults can occur in a power system: three-phase faults, three phase-to-phase faults, three phase-to-ground faults, and three double-phase-to-ground faults. Protection against faults can be grouped as phase

Disturbance Analysis for Power Systems, First Edition. Mohamed A. Ibrahim.

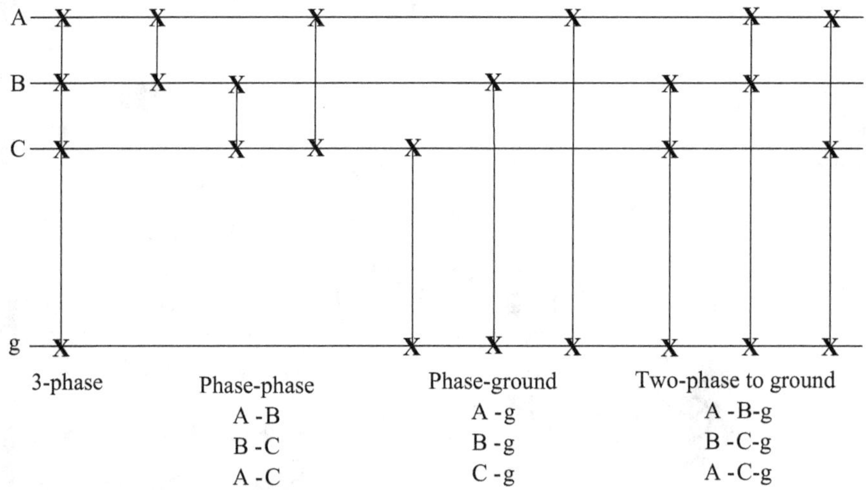

Fig. 2.1 Power system shunt faults.

protection and ground protection. Traditionally, six electromechanical distance relays, three for phase connections and three for ground connections, are required to provide complete line protection against all types of these 10 faults. With the advent of numerical relaying where the faulted loop becomes very important, fault classification becomes very important in calculating the correct faulted loop impedance. Failure to determine the correct faulted loop will result in failure of the relay to clear the fault. Other numerical line relays execute six impedance calculation loops simultaneously to cover all types of phase and ground faults.

2.2 CLASSIFICATION OF SHUNT FAULTS

All faults are described using Fig. 2.2, where the DFR monitors three line-to-neutral voltages (line side), three line currents, and the neutral current. The DFR receives the input from the transmission line through current transformers (CTs) and voltage transformers (PTs) or capacitive voltage transformers (CVTs). Current and voltage signal designations shown on the symbolic one-line ac diagram of Fig. 2.2 match the DFR labels in Figs. 2.3 and 2.4. Figure 2.2 is also used to describe phase-to-ground

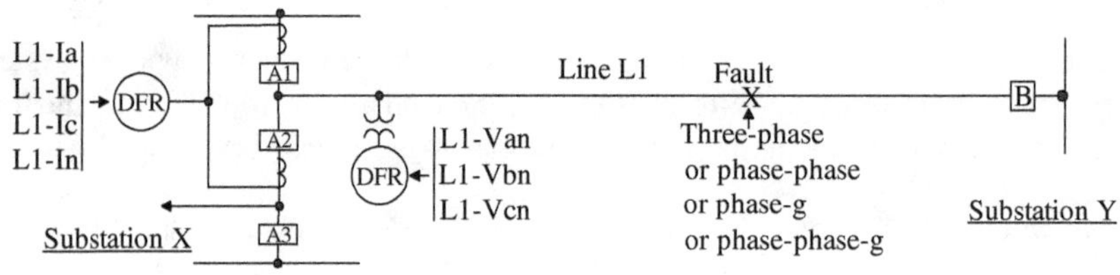

Fig. 2.2 Simple power system one-line diagram displaying DFR-monitored voltages and currents.

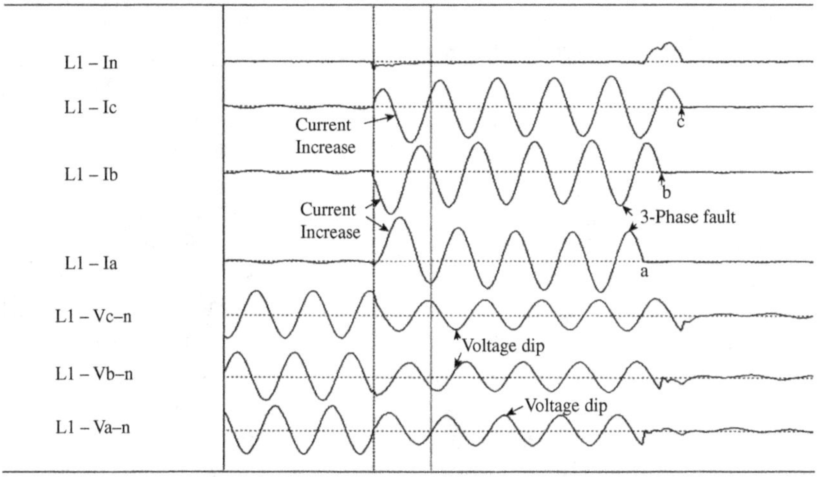

Fig. 2.3 Actual DFR record indicating a power system three-phase-to-ground fault.

and double-phase-to-ground shunt faults, with Van representing Va-n or Va. Similar representations are also valid for phases B and C.

2.2.1 Three-Phase Faults

Figure 2.3 shows a DFR record for a three-phase fault. A three-phase fault is normally accompanied by the following parameter variations:

1. An increase in current (i.e., over current) in phases A, B, and C
2. In the absence of dc offset or after the dc component dies out, balanced three-phase currents where phase A leads phase B by 120° and phase B leads phase C by 120°
3. A decrease in voltage (i.e., undervoltage) in phases A, B, and C (for bolted close-in three-phase faults, the voltages dip to zero value)

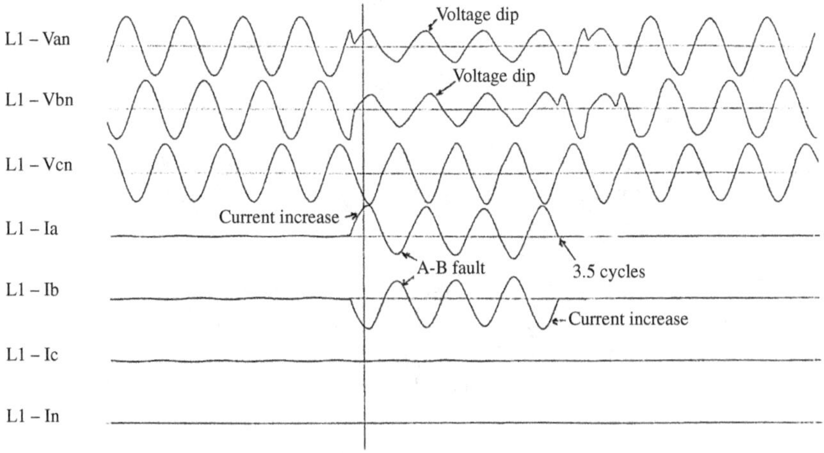

Fig. 2.4 DFR record showing a power system phase A-B fault.

4. Balanced three-phase voltage where phase A leads phase B by 120° and phase B leads phase C by 120° (for nonbolted faults).
5. No residual current unless asymmetrical currents exist at the beginning of the fault or CT saturates on some current phases due to heavy fault current, or during the process of current interruption when the first phase current interrupts at the zero-crossing point

The DFR record in Fig. 2.3 shows the three phase currents with phase sequence ABC, where I_a reaches peak before I_b, while I_b leads I_c by 120°. Therefore, the three-phase current inputs to the DFR must be wired in accordance with the phase sequence ABC. The dc offset component of three-phase currents start to die in accordance with the faulted loop time constant for this fault. In about two cycles from the fault incident point, dc offset starts to die out, thus transforming the three-phase currents into a balanced set. The three-phase-to-neutral voltages are dipped to a lower value from normal during the fault. The sequence of circuit breaker interruptions of currents at zero crossings in the order A, B, C leads to the appearance of a neutral current that is different from 60 Hz (i.e., non-60 Hz). A small decaying dc offset neutral current at the initiation of the fault is also shown in Fig. 2.3, with non-60-Hz neutral components at the interruption of the fault. Phase A starts to interrupt first, followed by phase B 120° later [$(120/360) \times 16.66 = 5.55$ ms]. Phase C interrupted at 120° from the clearing point of phase B. The three-phase fault lasted for 5.5 cycles, when phase C interrupted and cleared the fault.

2.2.2 Phase-to-Phase Faults

Three different phase-to-phase faults (A-B, B-C, and C-A) can occur in a power system, requiring three dedicated phase distance relays for the use of old electromechanical component relays or static relays, one selected fault loop, or three phase-to-phase loop equations calculated simultaneously by numerical relays. The phase fault is normally classified based on the following:

1. An increase (overcurrent) of current in the faulted phases
2. A phase shift of 180° between the faulted phase currents
3. A decrease in voltage (undervoltage) in the faulted phases
4. A voltage drop that is about equal for the two faulted phases
5. No neutral current, due to the absence of zero-sequence current (no connection to ground)

The DFR record in Fig. 2.4 shows phase A current equal and opposite to phase B current, while phase C carries no fault current. The voltage dip in phase A is similar to that in phase B. The fault duration is 3.5 cycles, and the fault is cleared simultaneously in phases A and B, due to interruption of the faulted loop by the clearing of any of the phases at current zero.

2.2.3 Phase-to-Ground Faults

Three different ground faults (A-g, B-g, and C-g) can occur in a power system, requiring three dedicated ground distance relays for the use of old electromechanical component relays or static relays, one selected fault loop, or three phase-to-ground loop equations calculated simultaneously by numerical relays. The ground fault is normally classified based on some or all of the following system parameter changes:

1. Dip (drop) in faulted phase voltage (undervoltage)
2. High faulted phase current (overcurrent)
3. Presence of zero-sequence current (residual current), $3I_0$, I_n, or I_g
   In addition, more system changes can accompany the ground fault:
4. Presence of negative-sequence current, I_2
5. High or low unfaulted voltages, depending on an X_0/X_1 equivalent behind the fault. If the ratio X_0/X_1 is greater than 1, higher voltage will occur on the unfaulted phases. As an example, for a ratio of 3, the overvoltage will reach about 126% of normal values. If the ratio X_0/X_1 is less than 1, undervoltage will be experienced on the unfaulted phases. If $X_0 = X_1$, the unfaulted-phase voltages will stay undisturbed, with their pre-fault voltage values.
6. High or low unfaulted phase current, depending on the value of I_0–I_1. For two-end feed (network), the current in the unfaulted phases will be a function of the positive- and zero-sequence distribution factors. If $I_0 = I_1$, as in the case with a stub fault (a fault fed from one end only), unfaulted phase current flows will be zero.

Analysis of the DFR record in Fig. 2.5 reveals that the A-g fault duration is 4.5 cycles (1 cycle = 16.66 ms). It can be observed that phase A exhibits overcurrent while the

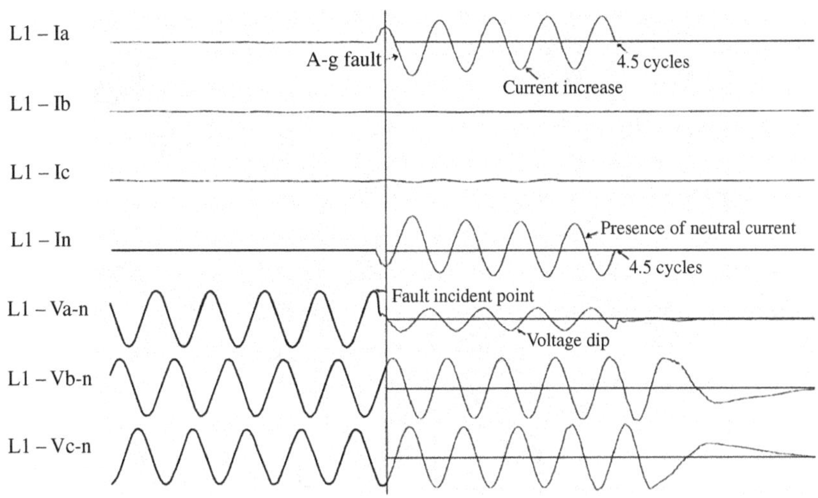

Fig. 2.5 DFR record showing a power system phase A-g fault.

neutral reflects a ground current flow of $3I_0$. The CT neutral current shunt (if required) is wired to the DFR with the neutral current trace set either in or out of phase with the faulted phase current for line internal faults (in this example it is out of phase). Phase A-to-neutral voltage shows a voltage dip during the fault. The DFR record reveals a very slight effect on unfaulted current and voltage phases, indicating that X_0 is very close to X_1 behind the fault.

2.2.4 Double-Phase-to-Ground Faults

Three different phase-to-phase-to-ground faults (A-B-g, B-C-g, and C-A-g) can occur in a power system. These faults are cleared by the operation of one or more of the three phase or three ground distance relays or numerical relay elements used to cover phase-to-phase and phase-to-ground faults, respectively. The phase-to phase-to-ground fault is normally classified based on:

1. An increase of current (over current) in the faulted phases
2. Faulted phase currents that are in phase (with no phase shift)
3. A decrease of voltage (undervoltage) in the faulted phases
4. About an equal voltage drop in the faulted phases
5. The appearance of neutral (zero sequence $= 3I_0$) current, which can flow either in or out of phase with the phase fault currents, depending on the system grounded sources behind the line terminal

The DFR record in Fig. 2.6 shows an increase in phase A and phase C current, while phase B carries a loading current. The two faulted phase currents are in phase. It also

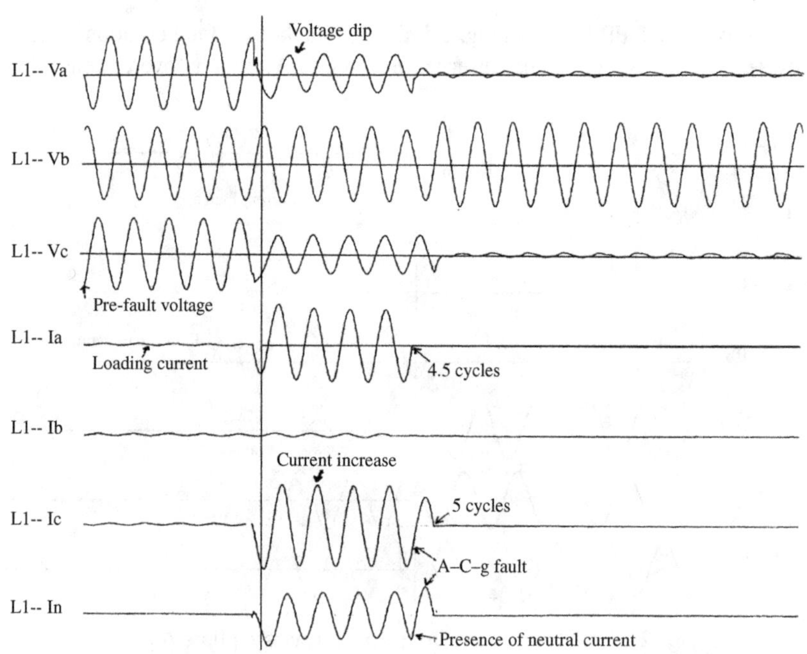

Fig. 2.6 DFR record showing a power system phase A-C-g fault.

shows the appearance of neutral current that can be in or out of phase with the phase currents (depending on the system ground sources). The voltage dip on phase A is similar to that on phase C. The fault duration was 4.5 cycles for phase A and 5 cycles for phase C.

2.3 TYPES OF SERIES UNBALANCE IN A POWER SYSTEM

Series unbalance can occur during an open-phase condition with sufficient load flow. Common causes of an open phase are a broken conductor, a disconnect switch blade that is not in a secure closed position, or a blown fuse. This may result in zero-sequence current flow provided that there are ground returns and ground sources at each side of the open conductor. If one side is ungrounded, the zero-sequence network will become open, resulting in no zero-sequence current flow. Negative-sequence current will also flow during an open phase condition (provided that there are generation sources at each side of the open phase), which may stress the generator rotor. As a result, the generator's negative-sequence $I_2^2t = k$ limit may be exceeded. Detection of an open-phase condition is normally carried out by ground time overcurrent relays employed in transformer neutrals and neutral connections of CTs applied to line relaying of the transmission system. In addition, an open-phase condition may be detected by generator negative-sequence relays when the unit continuous negative-sequence limit-setting thresholds are exceeded.

2.4 CAUSES OF DISTURBANCE IN A POWER SYSTEM

Faults are considered shunt unbalances that occur in a power system and are classified as either phase or ground faults. The most common fault is the phase-to-ground, and generally about 80 to 85% of faults are phase-to-ground and the rest are split between phase-to-phase and three-phase faults. Shunt faults occur in a power system when insulation fails. Insulation failures can be attributed to design defects, improper manufacturing, and aging of the material used. Faults can also be due to electrical causes initiated by lightning strokes (hits), switching surges, and overvoltage stress. In addition, faults can be due to mechanical failures caused by high wind, snow or ice, or insulation contamination. Faults can also be initiated by thermal stress caused by overcurrent. Faults can also be caused by galloping conductor phenomena where different line phases come close to each other and arc-over, causing phase-to-phase faults. This phenomenon is normally began by wind forces and can lead to cyclic faults. This can cause multiple faults to occur in a limited time span, leading to circuit breaker equipment failures if not detected and isolated by locking out the circuit breaker and preventing further automatic reclosing action. Control systems exist that can count the number of faults that can occur within a time span, and upon reaching a threshold setting, automatic reclosing is locked out.

Faults can be caused by insufficient insulation between phases or between phase and ground. The fault mechanism may be triggered by moisture or humidity, where minor arcing may start. For confined spaces, ionized gases may grow, and eventually, the air surrounding the live phases may be broken by arc-over, causing faults. This may force faults to evolve from one fault type to another type when the ionized cloud originated from the first fault grows to involve other phases. A phase-to-ground fault can evolve to either double-phase-to-ground or three-phase-to-ground. In a confined space, a phase-to-phase fault can evolve to a phase-to-phase-to-ground, three-phase, or three-phase-to-ground fault. This normally occurs due to spread of the ionized cloud.

Faults can be caused by hot gases originating from a fire or brush fire. This can ionize the air surrounding the transmission-line phase conductors. This will cause the air to conduct, causing a short circuit to occur between the live phases and ground. Faults can also be caused by lightning surges (induced voltage or direct strike), causing the high voltage to flash over the insulator. Ground faults on transmission lines usually occur due to flashover of the insulator during lightning hits (direct or by induction) or insulator failure due to pollution or during the melting of ice. Especially at lower voltages, wind can cause faults by forcing trees to contact a live conductor.

Other causes are ice, earthquakes, fires, explosions, falling trees, flying objects, physical contact by animals, and contamination. Faults also result from accidents such as vehicles hitting line poles or live equipment and digging that contacts underground cables. Faults that are caused by tree contact can be due either to high loading, forcing the line conductor to sag to the tree; high wind, moving the tree to the live conductor; or simply the tree growing to reach the live conductor. Tree faults can be of high resistance or intermittent, causing the fault current to be variable and sometimes difficult to detect. A high-impedance fault syndrome can be a challenge for any relay system to detect. This type of fault may be caused by a conductor falling and lying on ground consisting of dry soil, rocks, asphalt, or concrete.

A strain insulator used for line entrance to a substation may fail mechanically, causing the live conductor to fall to ground, causing L-g faults or falling on another live conductor, causing a phase-to-phase fault (depending on the system configuration). Faults can also occur due to vandalism caused by shooting at an insulator, throwing a piece of wire on the line, or loosening the steel pole bolts at the foot of a tower. Mayflies may nest and accumulate around substation insulators, causing faults through flashover to ground.

2.5 FAULT INCIDENT POINT

Faults can be caused by high-speed phenomena such as a lightning hit, and as a result, the fault incident point can become random. If a lighting hit occurs on or near the zero point on a faulted phase voltage, maximum dc offset will be generated, resulting in a total asymmetrical fault current. The dc offset current component is created because current cannot rise instantly in an inductive circuit. In addition, the circuit power factor will be honored by the presence of the dc component. On the contrary, if a lightning hit occurs near or on the peak point of the voltage, the fault current will become symmetric.

For a slow fault-creation mechanism, arc-over will always occur at the peak voltage, and hence fault currents will be symmetric with no dc offset present. From a statistical point of view, faults appear more often at voltages large enough to initiate insulation breakdown (i.e., close to the maximum point on the wave). Several examples are described in Chapters 3 and 4. The fault incident point can be found by performing a cycle-by-cycle comparison around the triggering time. When no match exists between the present and preceding (1 cycle ago) cycles, the point at which an over- or undercurrent or voltage can be designated as the fault incident point (time = 0 point or reference point). It is the occurrence of the fault that caused over- or undercurrent or voltage to occur.

2.6 SYMMETRIC AND ASYMMETRIC FAULT CURRENTS

The presence of dc offset depends on the fault incident point and is generated due to the fact that fault loop current cannot increase instantly through an inductance, and current must also lag the voltage by the natural power factor angle of the system. The magnitude of the initial dc offset is a function of the fault incident angle at which the fault begins. A fault occurring at voltage zero crossing will result in full dc offset. The dc offset decays with a time constant equal to *L*/*R*, where the inductance *L* is obtained from the *X*/*R* ratio of the system supplying the fault.

Figure 2.7 illustrates a system one-line diagram with the DFR-monitored cable current and voltage. The DFR record in Fig. 2.8 shows the fault incident point occurring at the phase A voltage peak. The A-g fault was caused by the insulation failure of the 345-kV cable, and therefore the fault mechanism is slow, causing flashover to occur at the voltage peak, as shown. Hence, the fault current is symmetric with no dc offset component.

Figure 2.9 shows a 138-kV line with DFR-monitored voltage and currents. Figure 2.10 shows the phase C line voltage as well as the phase and neutral currents. The C-g fault is caused by a lightning strike, and since a lightning hit is a random

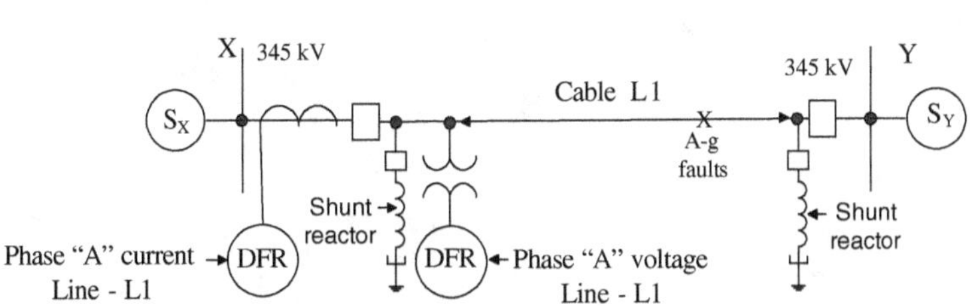

Fig. 2.7 Cable L1 DFR-monitored voltages and currents.

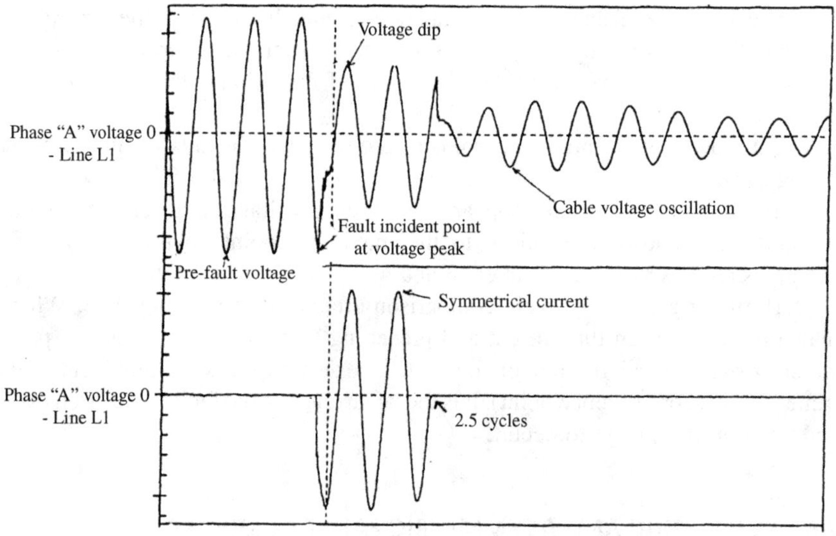

Fig. 2.8 Substation X DFR record showing the symmetrical fault current with the fault incident point occurring at the voltage peak.

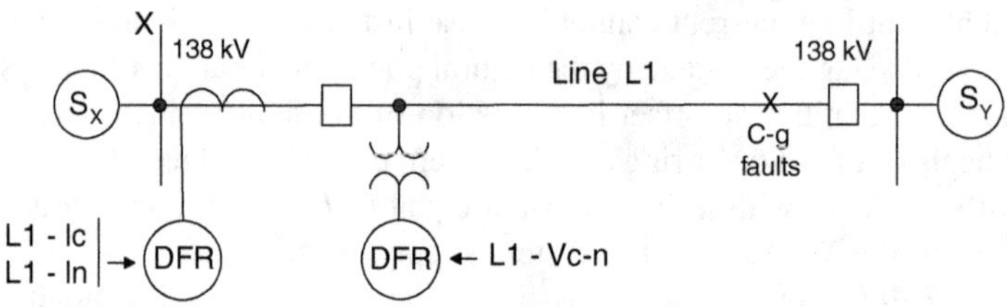

Fig. 2.9 Line L1 DFR-monitored voltages and currents during the C-g fault.

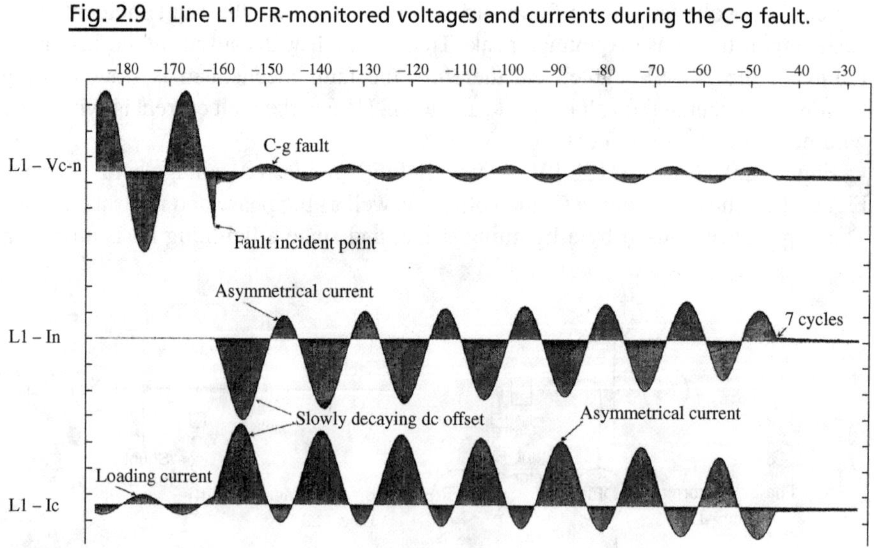

Fig. 2.10 Shaded asymmetrical line L1 faulted phase current and voltage, showing a slowly decaying dc offset with the fault incident point away from the voltage peak.

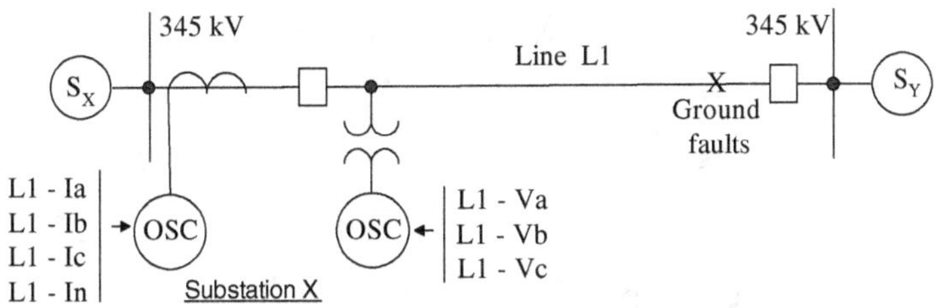

Fig. 2.11 Line L1 oscillograph-monitored voltages and currents.

phenomenon, the fault can occur at any incident point on the deriving voltage. In this case the fault is occurring at an incident angle other than at the voltage peak, and the phase C fault current is therefore asymmetric and contains a dc current causing current offset. The C-g fault lasted for 7 cycles.

The oscillograph-monitored currents and voltages for line L1 during both ground faults is shown in Fig. 2.11. The oscillogram of Fig. 2.12 reveals the line-side three-phase-to-neutral voltages and three-phase and neutral currents of a 345-kV transmission line during a ground fault. The A-g fault was caused by the presence of heavy wet snow. The insulation failure mechanism will be slow, and therefore the arc will be established at the voltage peak, as shown. The fault current is therefore symmetric with no dc offset. The fault was cleared by zone 1 from one end in 3.5 cycles and then followed by remote-end clearing after a 1.5-cycle delay. Figure 2.13 also reveals an oscillogram of the three phase voltages and three-phase and neutral currents of a 345-kV transmission line L1 during a ground fault. However, this time the C-g fault is caused by a lightning strike. Since a lightning hit is a random phenomenon, the fault can occur at any incident point on the driving voltage. In this case the fault is occurring at an incident angle other than at the voltage peak and the phase C fault current is therefore asymmetric and contains dc current, causing a current offset. The fault was cleared by zone 1 from one end in 3 cycles and then followed by remote-end clearing after a 1.5-cycle delay.

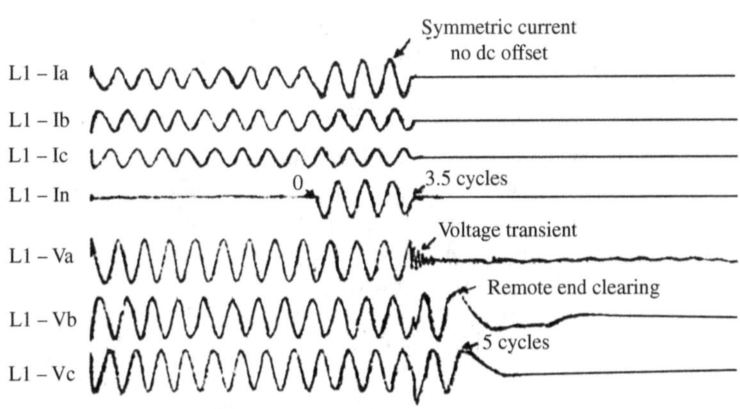

Fig. 2.12 Oscillogram for an A-g fault with symmetrical fault currents.

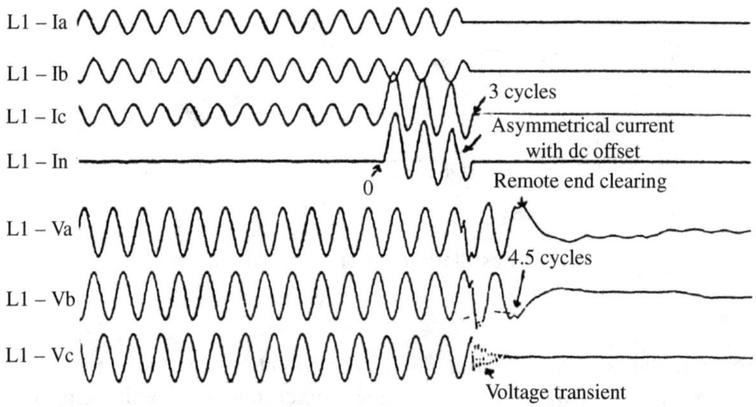

Fig. 2.13 Oscillogram for a C-g fault with asymmetrical fault currents.

2.7 ARC-OVER OR FLASHOVER AT THE VOLTAGE PEAK

For a slow fault-creation mechanism, the arc-over will always occur at the peak voltage when insulation fails, and hence fault currents will be symmetric with no presence of dc offset. At peak voltage the electric stress will be at its highest level to start the ionization process leading to arc-over, eventually causing faults to occur. This can happen by any of the following fault scenarios:

1. When a heavily loaded transmission line sags to a tree in the right-of-way
2. When a heavy wind moves a tree to contact the energized line
3. When a tree grows onto an energized line in the absence of an adequate tree trimming program
4. When a grounded object moves toward an energized line
5. When a live conductor falls to ground or on a grounded object
6. An insulation flashover failure due to pollution or ice
7. An insulation failure of a power element (cable, transformer, etc.)
8. An ionized cloud caused by minor arcing in a confined space
9. An ionized cloud caused by gas eruption or a brush fire

The fault-creation mechanism in scenarios 1 through 9 is due to insulation failures due to a process that can be described as slow movement, in a time frame of seconds. Power equipment failures can cause fires and generate ionized clouds that can also cause additional faults with symmetric fault currents. The phenomenon of arc-over at the voltage peak produces a symmetrical fault current with no dc offset present in the current wave. The absence of asymmetrical (dc offset) current makes the distance relay measure the correct impedance to the fault. For this reason, numerical distance relays are equipped with an algorithm to remove dc offset from an asymmetrical fault current to enhance the impedance and fault location algorithm.

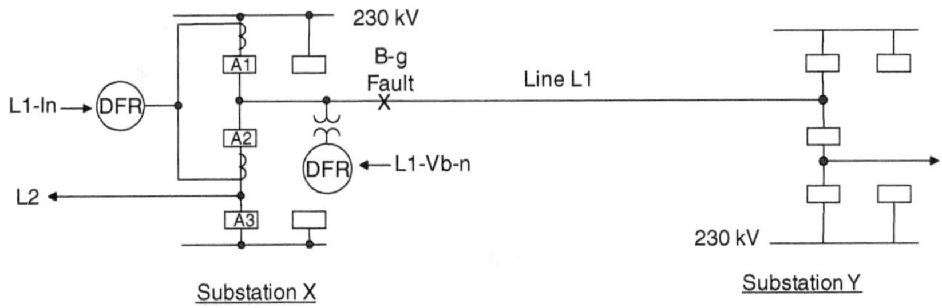

Fig. 2.14 Line L1 DFR-monitored analog signals.

The following four case studies document ground faults with their fault incident points occurring at the voltage peak.

Case Study 2.1: Live Conductor Falling to Ground The DFR-monitored currents and voltages for line L1 during a B-g fault are shown in Fig. 2.14, while the initial fault incident point, shown in Fig. 2.15, is at the voltage peak, which is typical for an arc-over of a live conductor moving to ground. The fault was solid to ground because the live conductor fell over the disconnect switch steel structure support, which is connected to the substation ground mat. As a result, the fault current is symmetrical with no dc offset component.

Case Study 2.2: Tree Fault Figure 2.16 shows the DFR-monitored currents and voltages for line L1 during an A-g fault. Figure 2.17 is a DFR record showing automatic reclosing onto a tree of line L1 at substation X, causing a phase A-g fault. The fault incident angle is at the phase A-n voltage peak point, resulting in a symmetric phase A and neutral currents that lasted for 5 cycles. The DFR record also shows that $I_A = 3I_0$, which is always the case for stub faults (faults fed from one end only).

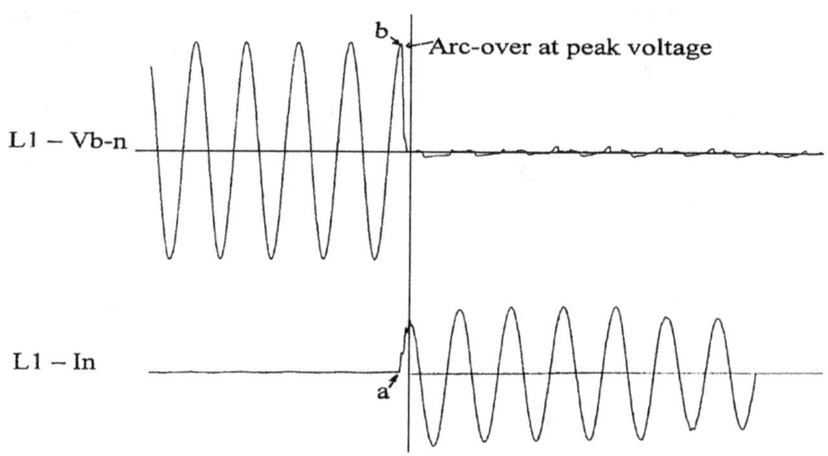

Fig. 2.15 Initial fault incident point shown at the voltage peak for a B-g fault caused by a live conductor fall.

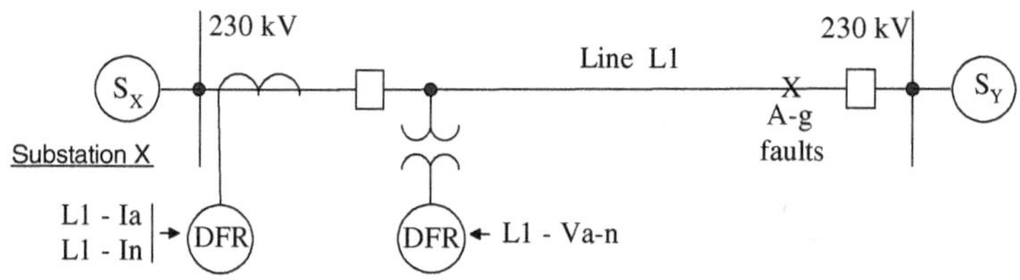

Fig. 2.16 Line L1 DFR-monitored analog signals.

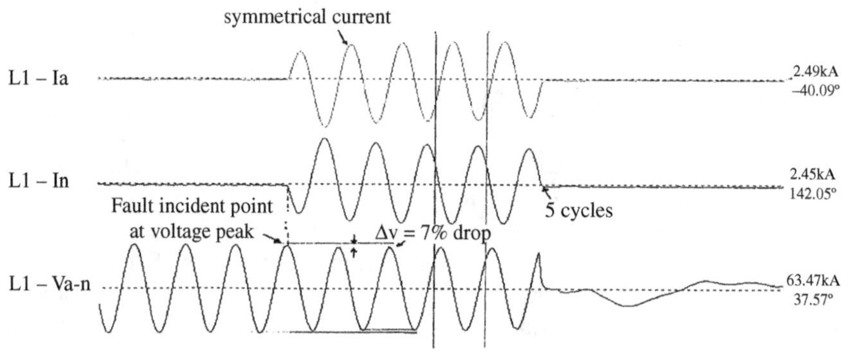

Fig. 2.17 DFR record documenting a reclosing onto the fault at the voltage peak with symmetrical fault current.

Case Study 2.3: 345-kV Submarine Cable Insulation Failures Figure 2.18 shows a simplified system one-line diagram with DFR-monitored cable voltages and currents. Figure 2.19 shows the fault incident point occurring at the phase A voltage peak. The A-g fault was caused by an insulation failure of the 345-kV cable, and therefore the fault mechanism is slow, causing flashover to occur at the voltage peak, as shown. Hence, the neutral fault current is symmetric with no dc offset component.

Case Study 2.4: 345-kV Pipe-Type Cable Pot Head Failure Figure 2.20 shows the 345-kV cable connections between generating unit G4 and the 345-kV substation as well as the phase A bus-monitored voltage. The faulted voltage is shown in Fig. 2.21, where an arc-over occurred at the voltage peak, due to the pot head failure

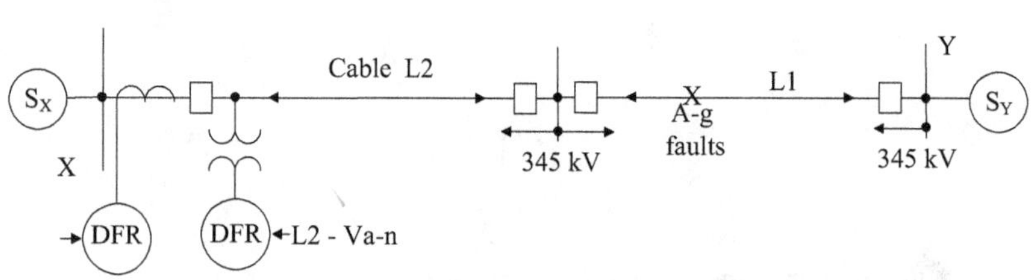

Fig. 2.18 Cable L2 DFR-monitored voltages and currents.

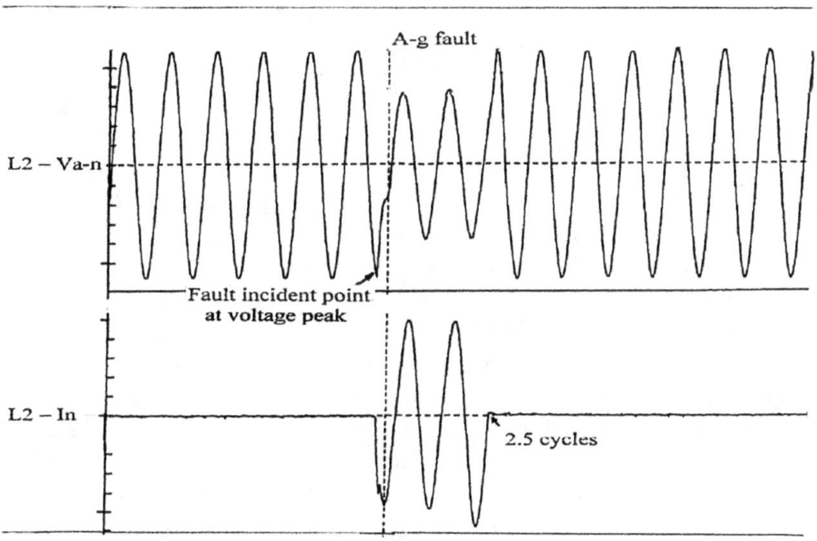

Fig. 2.19 Initial fault incident point shown at the voltage peak for an A-g fault caused by a cable insulation failure.

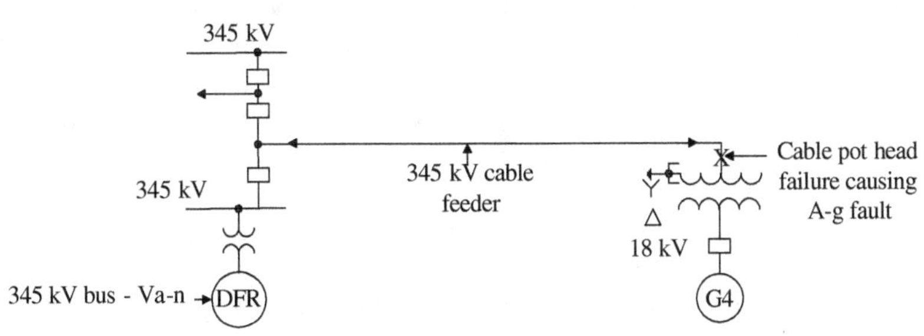

Fig. 2.20 One-line diagram showing the DFR-monitored voltage.

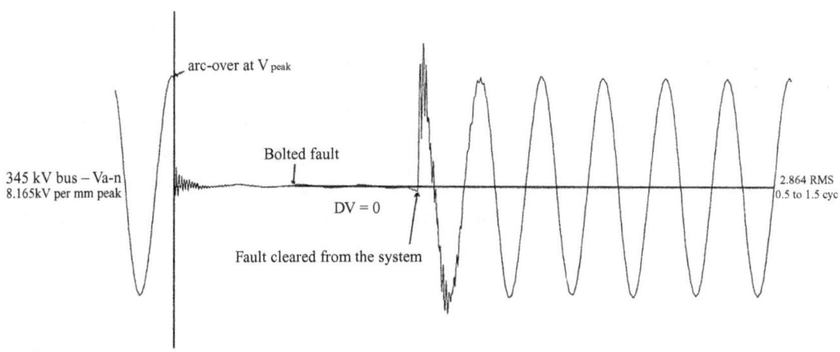

Fig. 2.21 Initial arc-over point shown at the voltage peak for an A-g fault caused by a cable pot head failure.

for the 345-kV cable connection. The voltage dipped to zero, indicating a close-in phase A-g fault.

2.8 EVOLVING FAULTS

For confined spaces, ionized gases may grow, and eventually the air surrounding the live phases may be broken by an arc-over, causing faults. This may force faults to evolve from one fault type to another type when the ionized cloud originated from the first fault grows to involve other phases. The evolving fault process can also take place in open space when light wind moves the ionized cloud from the faulted phase to an adjacent live phase. It can also occur in substations during power equipment failure when the ionized cloud and fire smoke rise to involve the substation live buses. A phase-to-ground fault can evolve to either double-phase-to-ground or three-phase-to-ground. A phase-to-phase fault in a confined space can also evolve to a phase-to-phase-to-ground or three-phase or three-phase-to-ground fault. This normally occurs due to the spread of an ionized cloud.

Evolving faults may be critical for numerical distance-based line protection. For some numerical relays, classification of the fault is crucial for correct determination of the faulted loop. The relay must select one of the six optional impedance calculation loops. The impedance calculations that cover all 10 types of faults consist of three phase (A-B, B-C, C-A) and three ground (A-g, B-g, C-g) loops. Numerical relays must be designed and tested thoroughly to cope with varying degrees of evolving faults to guarantee that one of the six fault impedance calculation loops will clear the fault without excessive delay. For this reason, some numerical line relays execute six continuous impedance calculations, and in this case the evolving faults may not be critical. Evolving faults are less critical for power system elements protected using differential relaying principles. The following case studies illustrate some evolving faults.

Case Study 2.5: Phase B-C Fault Evolving to a Three-Phase High-Side (13.8-kV) Fault Inside a Transformer Enclosure Confined Space Figure 2.22 shows a power system one-line diagram and numerical relay–monitored currents during the evolving faults. The presence of snow and moisture inside the transformer enclosure caused a phase-to-phase fault that evolved to a three-phase fault by spreading of the ionized cloud. The relay oscillography fault record shown in Fig. 2.23 for W2 current illustrates a phase B-C fault occurring at point a and lasting for 0.75 cycle, then evolving to a three-phase fault within the transformer-confined space at point b which lasted for an additional 4 cycles and cleared at point d by tripping the generator breaker.

Case Study 2.6: Phase B-g Fault Evolving to a B-A-g Fault in an Open Space for a 345-kV Line Figure 2.24 shows the system one-line diagram with the numerical relay–monitored voltages and currents at substation X as well as the

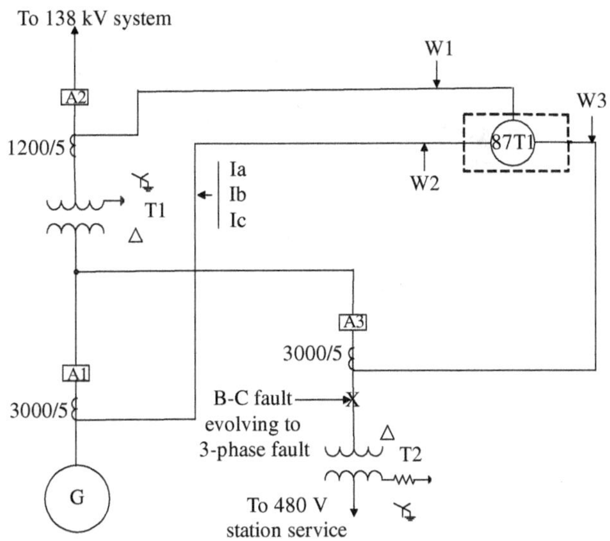

Fig. 2.22 Plant one-line diagram showing numerical relay–monitored currents.

DFR-monitored voltages at substation Y. A 345-kV phase B-g fault occurred on line L6 due to an insulation failure. The fault incident point occurred at peak voltage point a in Fig. 2.26. The ionized cloud began to spread in the open air to involve phase A after 1.5 cycles, as illustrated at point b in both Figs. 2.25 and 2.26. The initial phase B-g fault lasted for 1.5 cycles, then evolved to a phase A-B-g fault, which lasted for an additional 2 cycles and cleared at point c. Figure 2.25 also

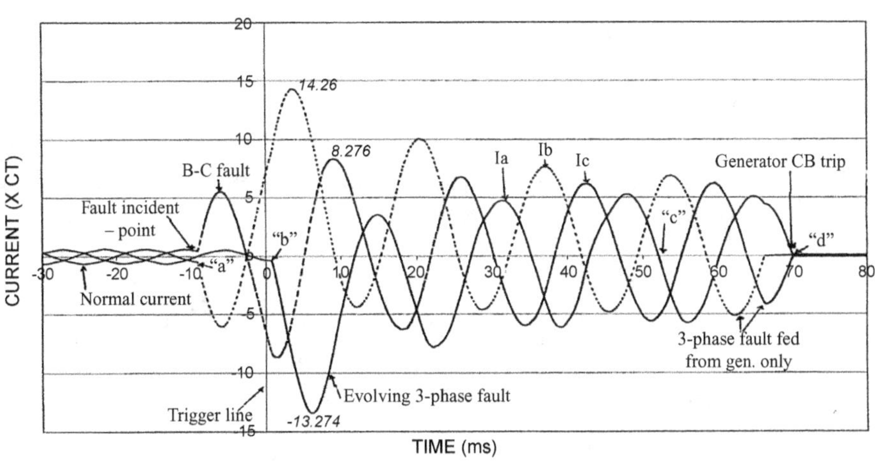

Fig. 2.23 Numerical relay W2 current record showing a phase B-C fault evolving to a three-phase fault.

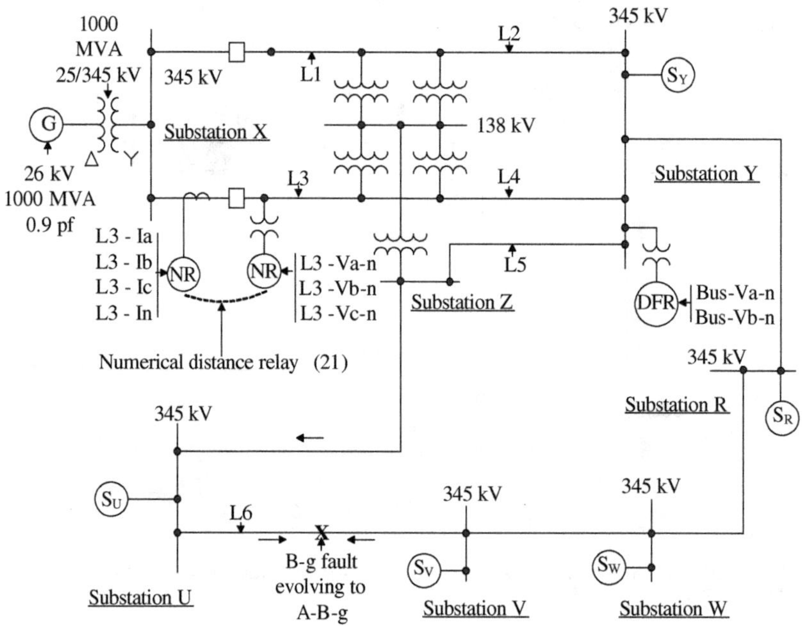

Fig. 2.24 One-line diagram showing the numerical relay–monitored voltages and currents at substation X and DFR-monitored voltages at substation Y.

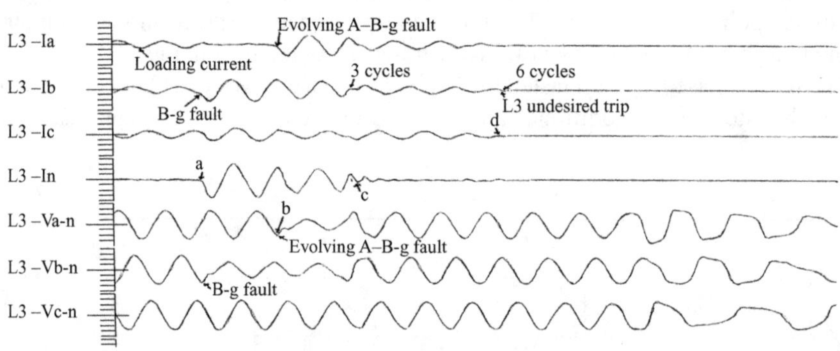

Fig. 2.25 Substation X numerical relay oscillography fault record confirming a B-g fault evolving to an A-B-g fault.

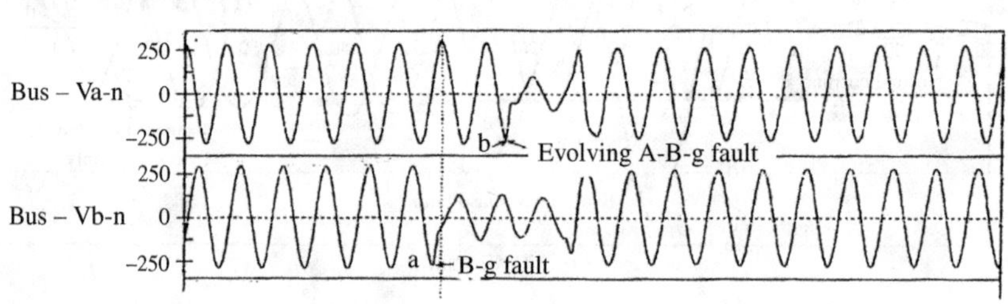

Fig. 2.26 Substation Y DFR record for the 345-kV bus voltages, displaying an evolving fault.

reveals that line L3 tripped undesirably after 6 cycles, due to the wrong setting for the quad element of a numerical relay.

2.9 SIMULTANEOUS FAULTS

Simultaneous faults occur primarily on parallel transmission circuits. The example described here documents a flashover to ground occurring on the two top insulators of a double-circuit 230-kV tower. The flashover probably occurred due to a lightning strike that created simultaneous B-g faults on the two 230-kV circuits L1 and L2, shown in Fig. 2.27. The fault was located at a tower where one of the static (ground) wires was missing. During an ice storm that occurred earlier, one of the static wires for several towers failed and the line had to be restored in an emergency without repairing the failed wire. Mechanical forces caused by high fault currents blew the L1 phase B insulator string, dropping the line on the lower arm, and the fault also damaged the insulator on circuit L2.

The DFR record in Fig. 2.28 shows the simultaneous nature of the faults, where the cursor line intersects the line L1 and L2 faulted phase voltage at the same fault incident points for both lines. In addition, the DFR record illustrates that the line L2 B-g fault was cleared at substation X by tripping CBs B and B1 in 3.5 cycles, while the faulted phase voltage confirms that substation Y cleared the fault by tripping CBs C and C1 in 4.5 cycles. Line L1 was cleared from substation Y by tripping CBs D and D1 in 3.5 cycles, forcing the current contribution from X to increase (stub fault at this time). The B-g fault on line L1 was finally cleared from substation X by tripping CBs A and A2 at 5 cycles.

The simultaneous fault criteria are used to assess the power system stability of a bulk power system. It becomes a part of the basic criteria for design and operation of interconnected power systems. In these power system regions, it is mandated to provide normal clearing of simultaneous phase-to-ground faults occurring in different phases of two adjacent transmission circuits on multiple-circuit tower configurations.

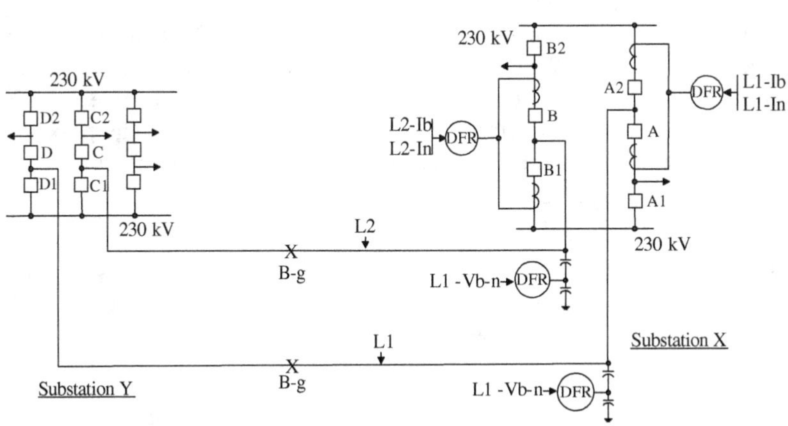

Fig. 2.27 One-line diagram showing DFR-monitored voltages and currents at substation X.

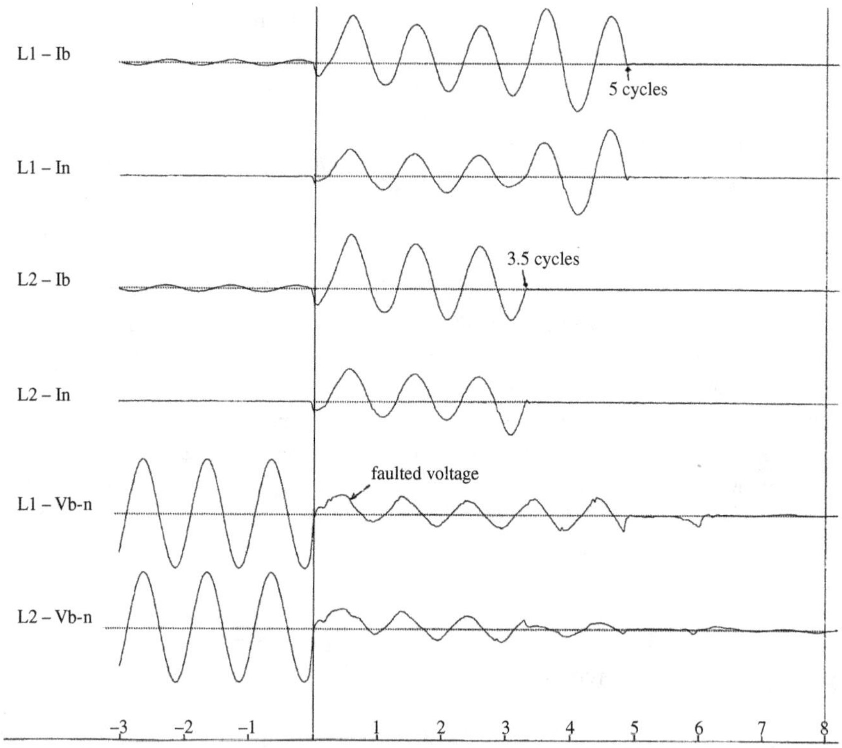

Fig. 2.28 Faulted voltages and currents for lines L1 and L2 at substation X showing simultaneous B-g faults.

2.10 SOLID OR BOLTED ($R_F = 0$) CLOSE-IN PHASE-TO-GROUND FAULTS

Faults can occur without any fault resistance. In this case the faulted phase voltage near the fault point will collapse to zero at high speed. Figure 2.29 shows a simple one-line diagram with line L2 DFR-monitored voltage and currents during a line L1 ground fault. The DFR record in Fig. 2.30 shows a close-in solid phase B-to-ground fault that lasted for several cycles. The fault incident point is at the peak voltage, implying a slow fault-creation mechanism of an insulation failure, with the voltage

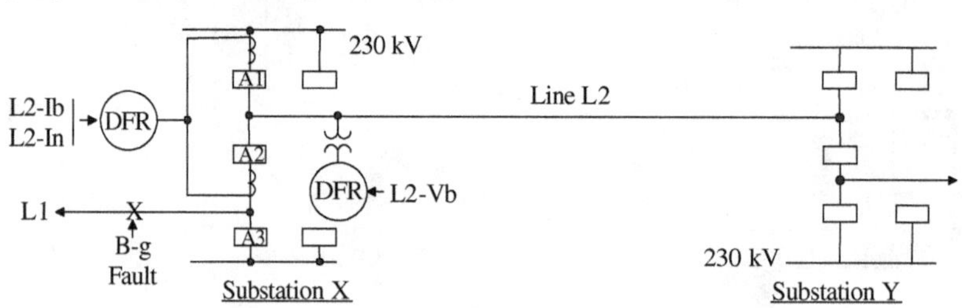

Fig. 2.29 Line L2 DFR-monitored analog signals.

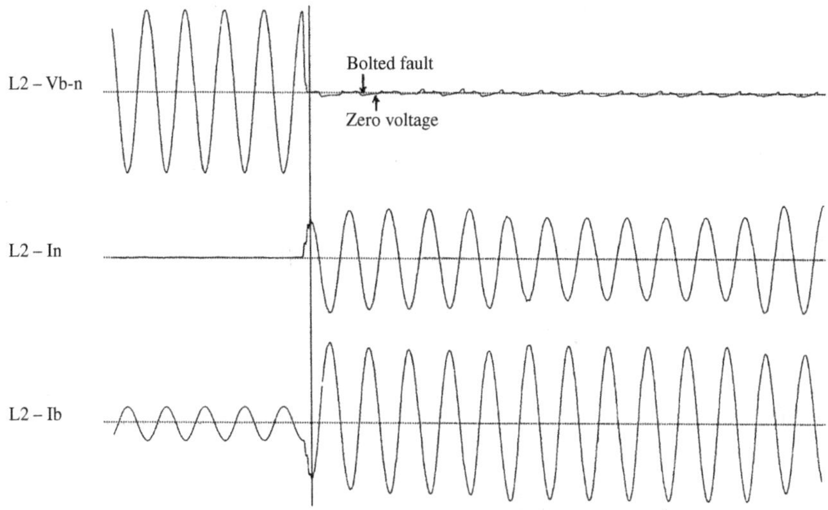

Fig. 2.30 Solid B-g close-in fault at substation X.

magnitude dipping to zero. Bolted L-g faults do not affect the directionality of ground distance relaying since relay polarization is derived from either sound (nonfaulted) phases or from memory or a combination of phase and memory stored quantities.

2.11 SEQUENTIAL CLEARING LEADING TO A STUB FAULT THAT SHOWS A SOLID ($R_F = 0$) REMOTE LINE-TO-GROUND FAULT

When line-side voltage is monitored by a DFR and one terminal of the line clears a ground fault prior to the remote end, a local recording of zero voltage upon clearing will confirm a solid fault. Figure 2.31 shows a simplified system one-line diagram with DFR-monitored voltages and currents used to analyze the A-g fault. The DFR record in Fig. 2.32 illustrates an A-g fault occurring on line L1 at point a, which was cleared from substation X in 3 cycles at point b. Now the fault is defined as a stub fault being fed only from substation Y for an additional 4 cycles to be cleared at point c. The phase A faulted phase voltage dropped to a value at the fault incident point during the 3-cycle clearing duration for substation X. At point b, phase A voltage trace L1-VA dropped further, to

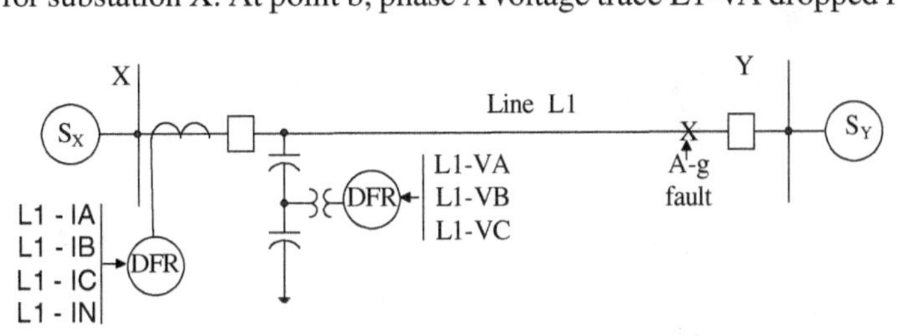

Fig. 2.31 Line L1 DFR-monitored voltages and currents.

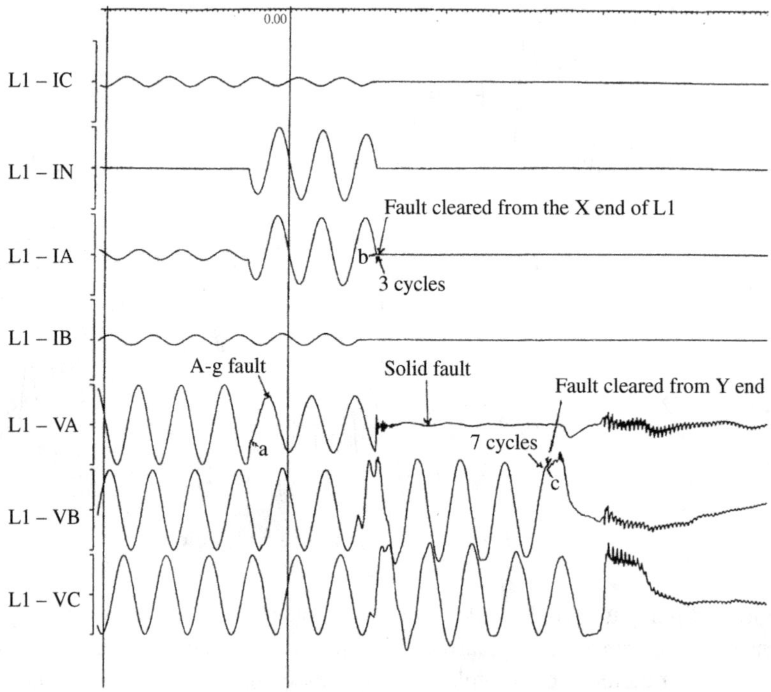

Fig. 2.32 DFR record for substation X showing a solid fault during sequential clearing.

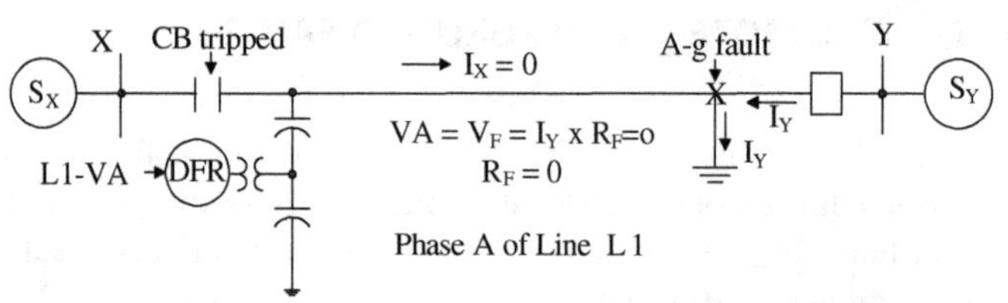

Fig. 2.33 Stub fault out of the Y end of the line, displaying the voltage calculated at X.

zero, indicating that the fault is solid ($R_F = 0$). Figure 2.33 shows the stub fault being fed for an additional 4 cycles, with the voltage being measured at X with its zero value.

2.12 SEQUENTIAL CLEARING LEADING TO A STUB FAULT THAT SHOWS A RESISTIVE REMOTE LINE-TO-GROUND FAULT

When line-side voltage is monitored by an oscillograph, and one terminal of the line clears a ground fault prior to the remote end, a local recording of a voltage upon clearing will confirm that this voltage is measured across a fault resistance (R_F) as a result of fault current flow from the remote end. Figure 2.34 shows the simplified

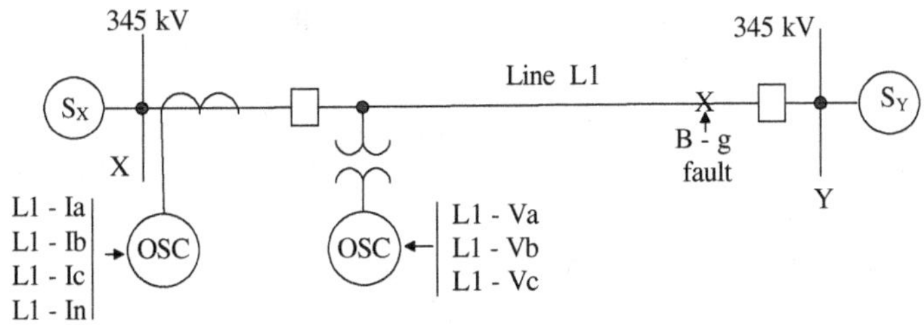

Fig. 2.34 Line L1 oscillograph-monitored voltages and currents.

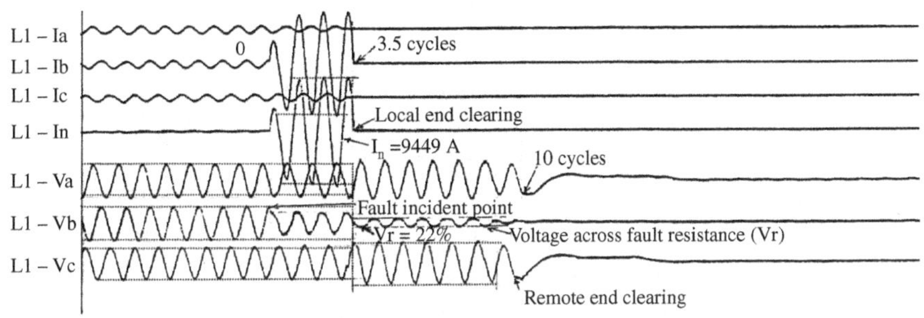

Fig. 2.35 Substation X oscillograph confirming fault resistance during sequential clearing.

system with oscillograph (OSC)-monitored voltages and currents. The oscillograph record in Fig. 2.35 shows a B-g fault occurring on line L1 which was cleared from substation X in 3.5 cycles. Now the fault is defined as a stub fault being fed only from substation Y for an additional 6.5 cycles. Phase B faulted phase voltage dropped to a value at the fault incident point during the 3.5-cycle clearing duration for substation X. Phase B voltage trace L1-Vb dropped further to a lower value of 22%, indicating the presence of fault resistance. Figure 2.36 shows the B-g fault as a stub fault being fed from the Y end only. The voltage measured at the line end at substation X is defined as the voltage drop across the fault resistance.

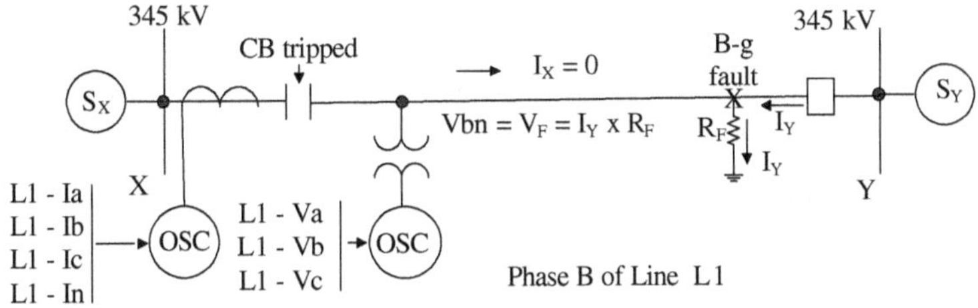

Fig. 2.36 Stub fault out of the Y end of the line with the voltage calculated at X.

2.13 HIGH-RESISTANCE TREE LINE-TO-GROUND FAULTS

Tree faults can be caused either by trees growing into a line, high wind moving tree branches near the line, or a heavily loading line sagging into trees in the right-of-way of the line. Tree faults can be of a high-resistance nature, depending on how dry the tree and the ground soil (dry or wet) are as a return for fault currents. High-resistance tree faults may plot outside the *R–X* diagram characteristics of the line ground distance relays. Therefore, clearing high-resistance ground faults can be either by a ground time overcurrent backup relay or a ground distance relay after the time delay required for the fault current to increase slowly after the current finds an easy path to

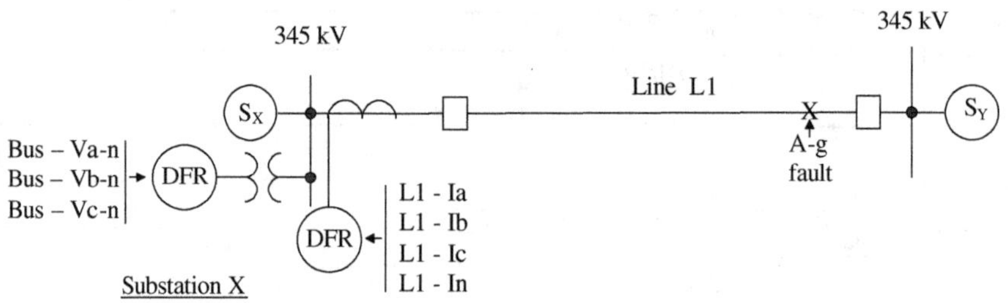

Fig. 2.37 Line L1 DFR-monitored voltages and currents.

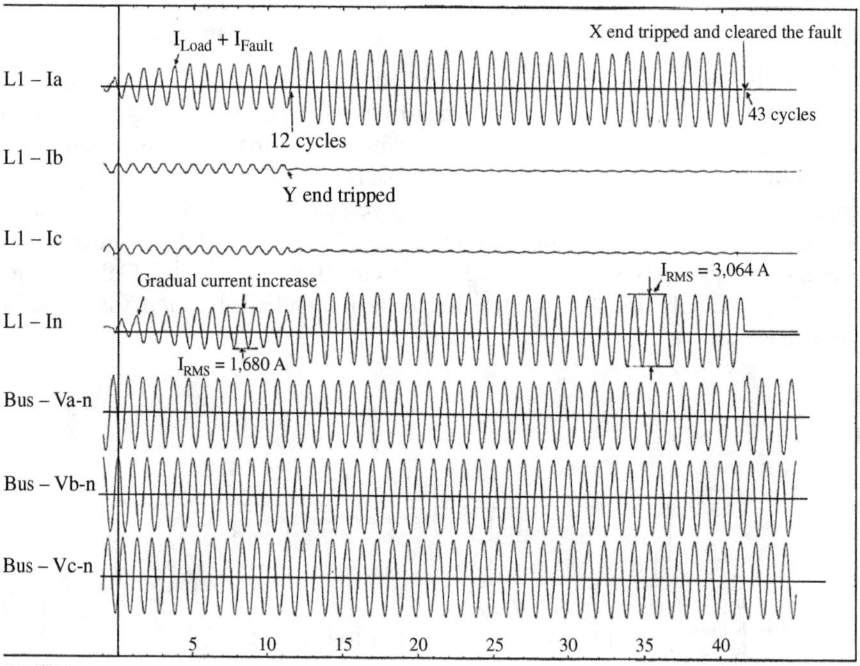

Fig. 2.38 DFR record for line L1 currents and bus voltage during a high-resistance ground fault.

return through true ground. Application of ground time overcurrent relays for line protection play an important role in providing protection for a power system against high-resistance (tree) faults and open-phase conditions. The following two case studies illustrate high-resistance ground faults.

Case Study 2.7: High-Resistance A-g Fault Figure 2.37 shows the 345-kV system simplified one-line diagram and DFR-monitored currents and voltages during a high-resistance fault that occurred when the line sagged on to a tree. Figure 2.38 shows a substation X DFR record for the high-resistance A-G fault occurring on line L1 with gradual increase in the ground fault current. The fault was cleared from the substation Y end of the line in 12 cycles. The stub nature of the fault forced the ground current contribution from substation X to increase from 1680 A to 3067 A. The fault was eventually cleared by a ground time overcurrent element in 43 cycles.

Case Study 2.8: High-Resistance C-g Fault Figure 2.39 shows the 345-kV system simplified one-line diagram and DFR-monitored currents and voltages during a high-resistance tree fault. The substation X DFR record shown in Fig. 2.40 reveals a

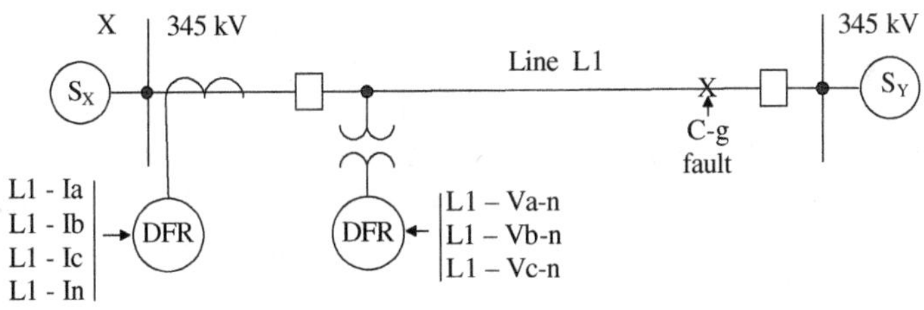

Fig. 2.39 Line L1 DFR-monitored voltages and currents.

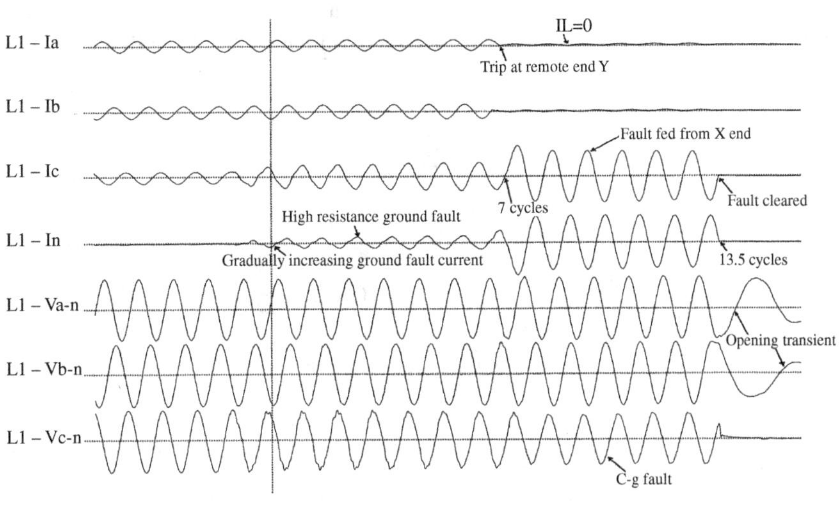

Fig. 2.40 Substation X DFR record indicating a high-resistance C-g fault.

high-resistance C-g tree fault occurring on line L1. The fault was cleared from the substation Y end of the line in 7 cycles. The substation X current contribution was initially insufficient to trip and clear the fault. The stub nature of the fault forced the ground current contribution from substation X to increase to twice its initial value. The fault was eventually cleared by the ground distance relay in 13.5 cycles when the relay-measured impedance traveled to a point inside the relay characteristic.

2.14 HIGH-RESISTANCE LINE-TO-GROUND FAULT CONFIRMING THE RESISTIVE NATURE OF THE FAULT IMPEDANCE WHEN FED FROM ONE SIDE ONLY (STUB)

Faults may occur either as a solid or involving varying amounts of resistance. The ground (static) wire circuit is tied to ground at each tower, and these connections to Earth will have a mostly resistive impedance. Normally referred to as a tower footing resistance, this should be kept as small as possible through good design. Faults involving tree contact can be of high resistance with variable values. The arc impedance is essentially resistive but can appear to protective relays to be nonresistive (impedance with an inductive reactance), due to an out-of-phase current contribution from the remote end. When the fault is fed from currents at both ends, the fault resistance appears to the distance relays at either end as a much larger value, and with load flowing over the line, as an impedance can tilt either upward or downward, and not as a resistance.

Figure 2.41 shows a 345-kV simplified system one-line diagram and DFR-monitored currents and voltages during a high-resistance fault. Figure 2.42 confirms that the L-g fault is going to ground with a resistive nature when fed from one side only. The A-g fault was cleared earlier from substation X, followed by clearing from substation Y. The clearing of the fault is at zero voltage, confirming that the voltage drop ($V = I \times R_F$) across the fault resistance is in phase with the phase A current at the interruption point.

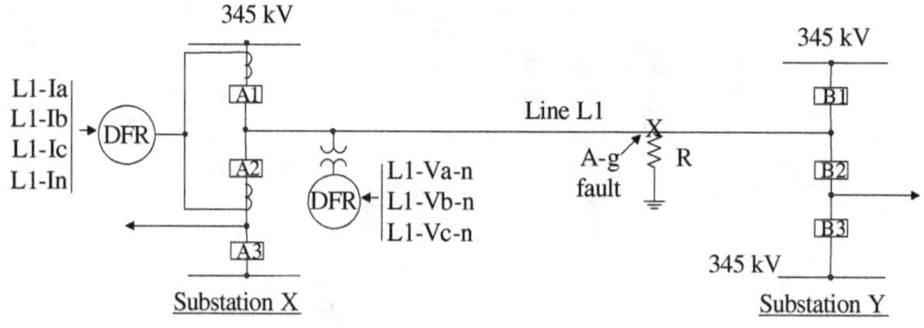

Fig. 2.41 Line L1 DFR-monitored voltages and currents.

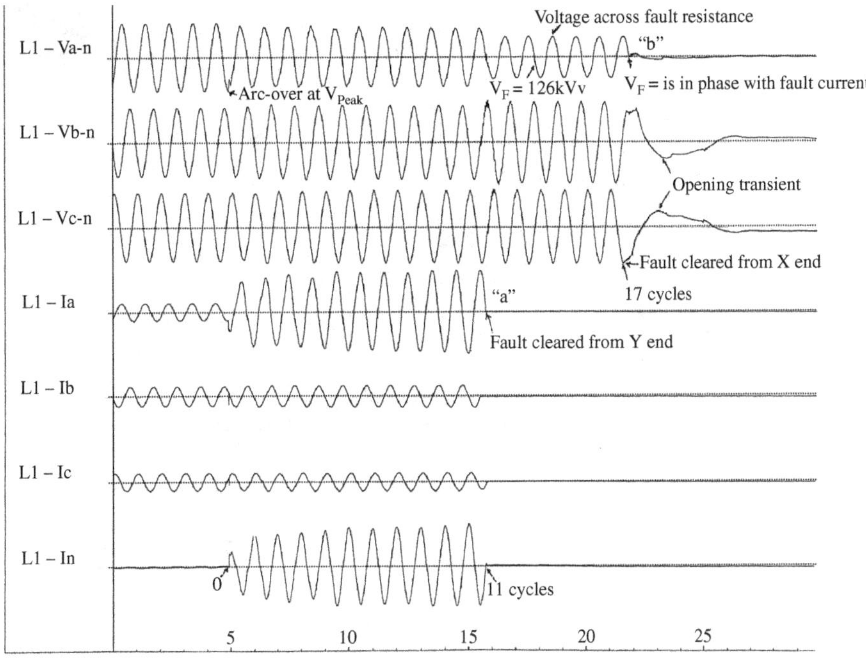

Fig. 2.42 DFR record for line L1 currents and voltages during a high-resistance A-g fault.

2.15 PHASE-TO-GROUND FAULTS ON AN UNGROUNDED SYSTEM

The currents for ungrounded system faults are very low, thus causing minimum equipment damage and continuation of service. Figure 2.43 shows a 13.8-kV ungrounded system being fed from a delta tertiary winding of an autotransformer and DFR-monitored voltages. As shown in Fig. 2.44, line A-to-ground faults on ungrounded systems shift the normal balanced voltage triangle, forcing phase A-n voltage to collapse to zero and the unfaulted phases to increase to the line-to-line voltage value. During the ground fault, the voltage difference between neutral and ground will be equal to the zero-sequence

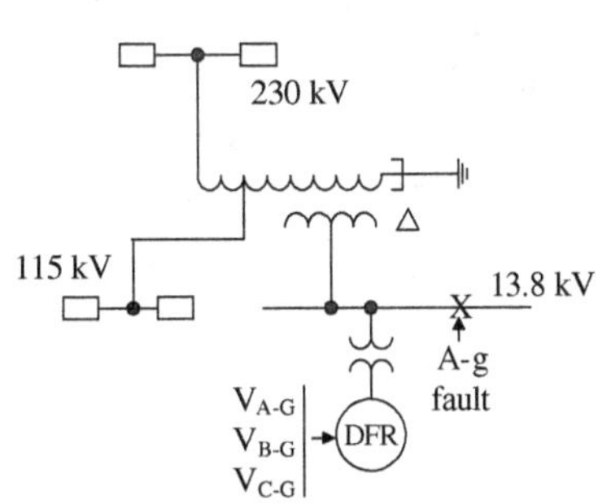

Fig. 2.43 System one-line diagram showing DFR-monitored analog signals.

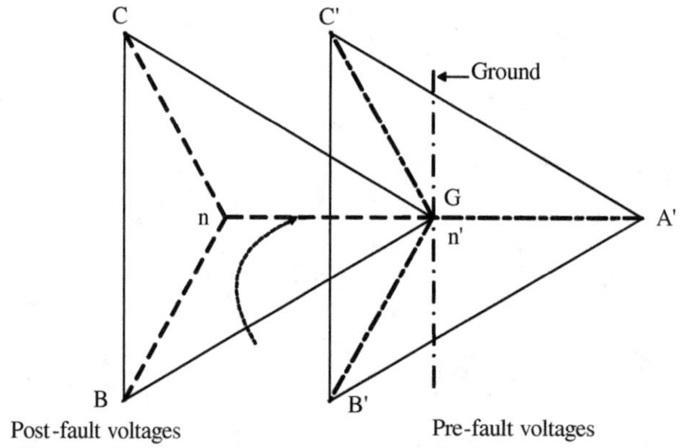

Fig. 2.44 Neutral shift during ground faults on ungrounded systems.

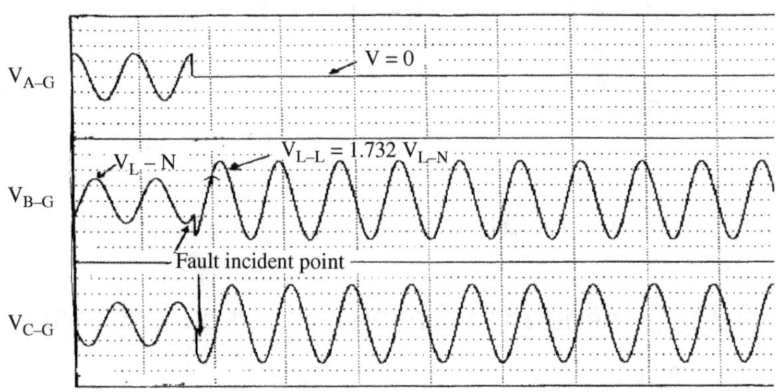

Fig. 2.45 Phase-to-neutral voltage record for a ground fault on an ungrounded system.

voltage $= V_0 = \frac{1}{3}(V_a + V_b + V_c) = V_{L-n}$. Therefore, ground fault detection is accomplished using three PTs connected as grounded wye/broken delta. During ground faults, the voltage relay applied across the broken delta will sense three times $V_{L\text{-}n}$. The DFR record in Fig. 2.45 confirms the finding in Fig. 2.44 by revealing that during the phase A-g fault, the faulted phase-to-neutral voltage $V_{A\text{-}G}$ collapses instantaneously to a zero value, and the voltage of the unfaulted phases $V_{B\text{-}G}$ and $V_{C\text{-}G}$ increase essentially by $\sqrt{3}$ to be equal to line-to-line voltage $V_{L-L}(= \sqrt{3}V_{L-n})$.

2.16 CURRENT IN UNFAULTED PHASES DURING LINE-TO-GROUND FAULTS

Current flow in the unfaulted phases will be driven mainly by the positive- and zero-sequence source ratios at each end of the feeders. As a result, currents can flow on unfaulted phases as long as the ground fault is fed from both ends of the line.

Assuming a phase A-g fault, the faulted phase current $I_a = I_0 + I_1 + I_2$, where I_0 is the zero-sequence current, I_1 the positive-sequence current, and I_2 the negative-sequence current.

As $a = 1$ at 120° and $a^2 = 1$ at 240°, the unfaulted phase current

$$I_b = I_0 + a^2 I_1 + a I_2 \tag{1}$$

$$I_c = I_0 + a I_1 + a^2 I_2 \tag{2}$$

$$I_1 = I_2 \qquad a^2 + a = -1 \qquad I_0 \text{ is not equal to } I_1 \tag{3}$$

Substituting (3) in (1) and (2) yields

$$I_b = I_c = I_0 - I_1$$

When the ground fault is fed from one end only (stub),

$$I_1 = I_2 = I_0 \quad \text{and} \quad I_b = I_0 + a^2 I_1 + a I_2$$

Since $1 + a + a^2 = 0$,

$$I_b = I_0(1 + a + a^2) = 0 = I_c$$

Therefore, as should be true, there are no fault current flows in the unfaulted phases for a ground fault fed from one end (stub) only.

The following two case studies illustrate the unfaulted current flows during ground faults.

Case Study 2.9: Phase A-g Fault Figure 2.46 shows the system one-line diagram and the numerical differential relay 87LP–monitored currents used to analyze an A-g fault. Symmetrical component current flows are shown in the three-line ac diagram shown in Fig. 2.47 for phases A, B, and C of the connected system. The numerical relay

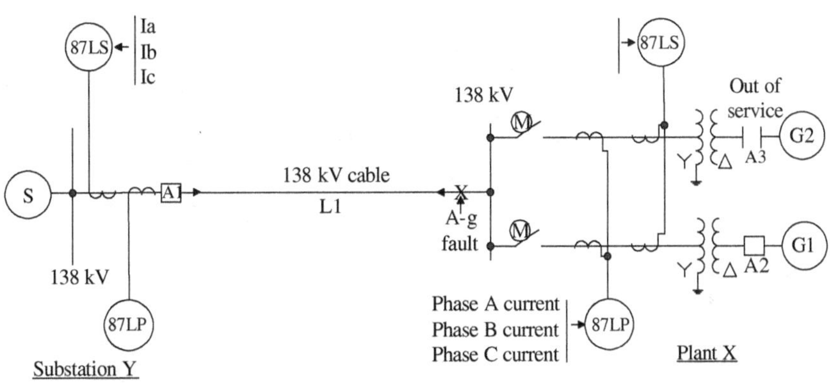

Fig. 2.46 One-line diagram showing numerical differential relay–monitored currents.

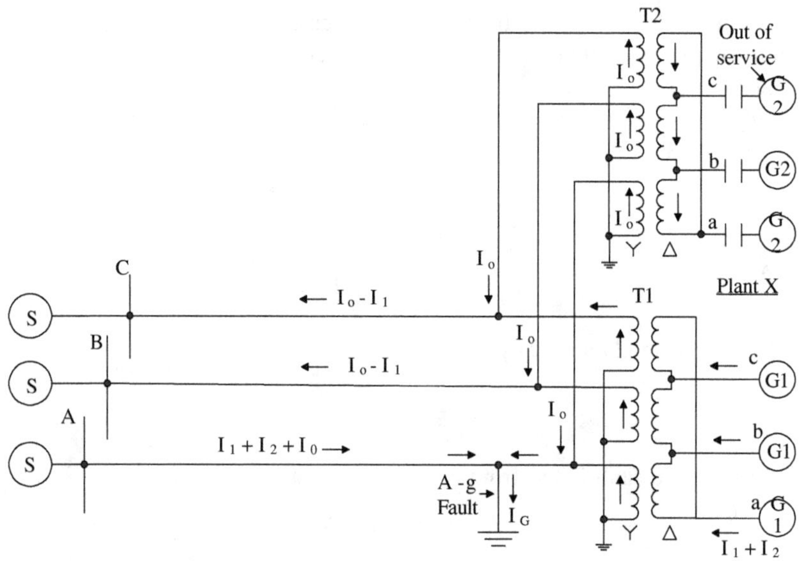

Fig. 2.47 Symmetrical components current flow on the system three-line ac diagram.

87LP oscillograph record in Fig. 2.48 illustrates the flow of currents on unfaulted phases B and C at substation X for 3.5 cycles, thus illustrating these phenomena. In addition, the flow began at point a and stopped at point b at exactly 3.5 cycles, when the remote end at Y clears first. From 3.5 to 5 cycles the current is fed from the G1 generator at substation X only (stub). The DFR record confirms that no fault current flows in the unfaulted phases for the ground fault when fed from generator G1 only, as it should.

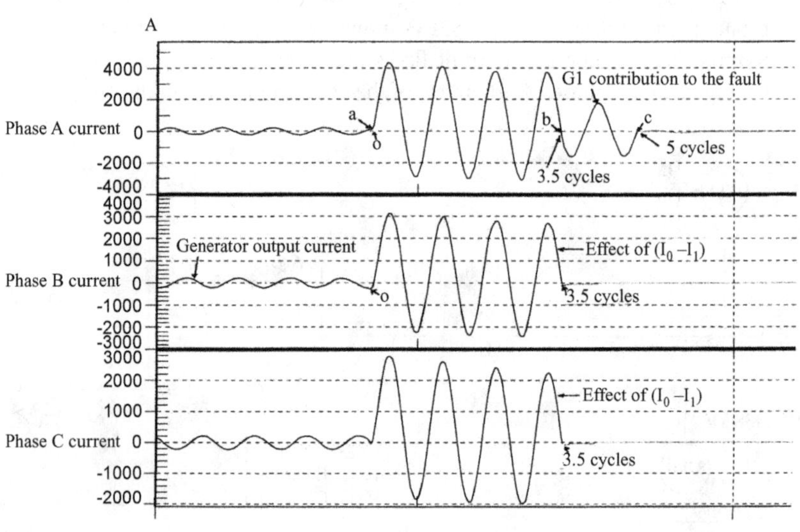

Fig. 2.48 Plant X numerical relay 87LP fault record showing unfaulted current flow.

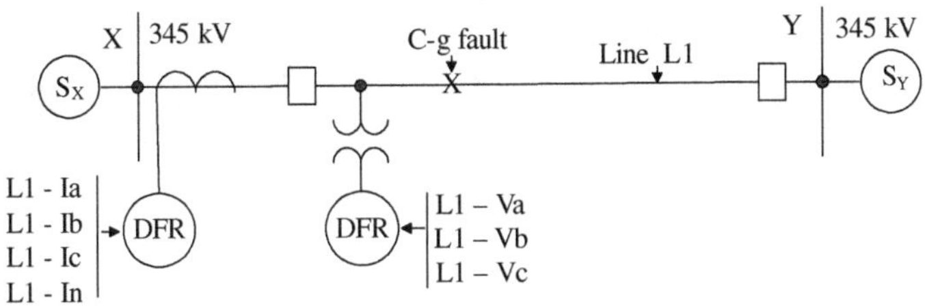

Fig. 2.49 Line L1 DFR-monitored voltages and currents.

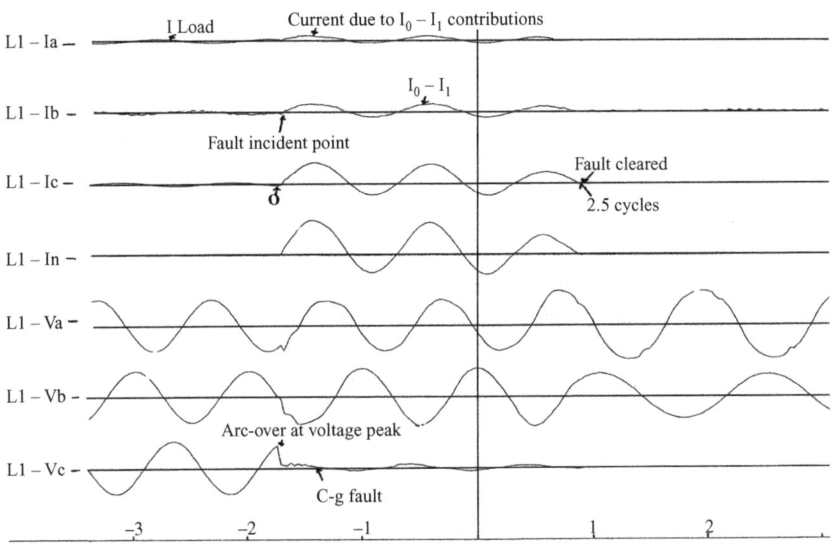

Fig. 2.50 Substation X DFR record for line L1, indicating unfaulted current flow.

Case Study 2.10: Phase C-g Fault Figure 2.49 shows the simplified 345-kV system one-line diagram and the DFR-monitored currents and voltages during a C-g fault. Figure 2.50 illustrates the effect of component $I_0 - I_1$ on unfaulted phases A and B during a C-g fault occurring on line L1, which lasted for 2.5 cycles. The pre-fault balanced-loading current flow on phases A, B, and C is also illustrated in the figure.

2.17 LINE-TO-GROUND FAULT ON THE GROUNDED-WYE (GY) SIDE OF A DELTA/GY TRANSFORMER

For a phase a-g fault on the grounded-wye side of the transformer shown in Fig. 2.51, the a-g fault will be seen as a phase A-to-phase C fault on the delta side and if we assume that the total fault current is 1.0 pu and that the winding phase-to neutral

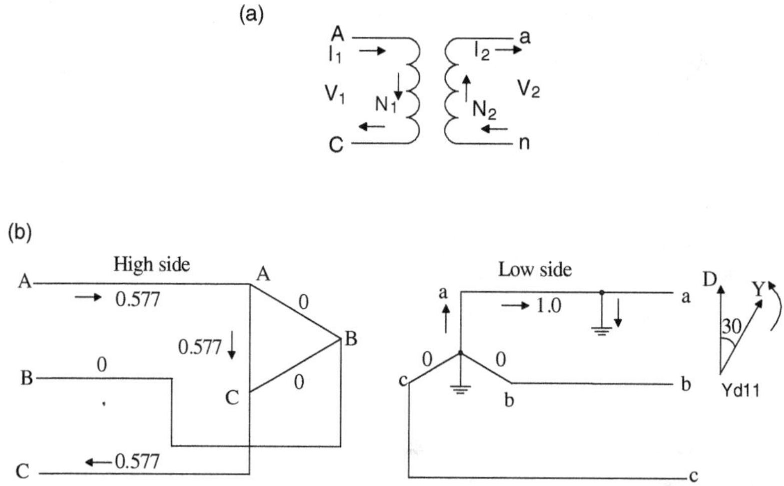

Fig. 2.51 Phase-to-ground fault on the grounded-wye transformer side.

voltage of the wye side at point a is assumed to be = 1 pu. The corresponding current can be deduced by applying the ampere-turns coupling principle. For the two-winding transformer shown in Fig. 2.51(a):

$$I_1 \times N_1 = I_2 \times N_2$$

or

$$I_1 \times V_1 = I_2 \times V_2$$

By applying the same principles as those for the delta/wye transformer shown in Fig. 2.51, we obtain

$$I_a \times 1 \text{ pu} = I_{AC} \times \sqrt{3} \text{ pu}$$

$$I_{AC} = \frac{1}{\sqrt{3}} = 0.577 \text{ pu}$$

The same result can be obtained when we apply symmetrical components to calculate the positive- and negative-sequence currents and transform the currents to the delta side. In this case the high-side (delta) positive-sequence current leads the low side (wye) by 30°, and the high-side negative-sequence current lags the low side by 30°. The use of time overcurrent relays on the delta side will see only 57.7% of the low side per unit current during low-side ground faults. Care should be taken when applying fuses on the delta side to make sure that low-side ground faults can be detected.

2.18 LINE-TO-LINE FAULT ON THE GROUNDED-WYE SIDE OF A DELTA/GY TRANSFORMER

For a phase b-c fault on the grounded-wye side of the transformer shown in Fig. 2.52, fault currents will flow on all the phases on the delta side. If we assume that the total fault current for three-phase operation is 1.0 pu, the phase current will be 0.866 pu for a phase-to-phase fault at the same location. If the winding phase-to neutral voltage at point a of the wye side is assumed to be = 1 pu, the delta voltage winding will be $\sqrt{3}$ pu. The corresponding current can be deduced as in Section 2.17, by applying the ampere-turns coupling principle for the transformer:

$$I_1 \times N_1 = I_2 \times N_2$$

or

$$I_1 \times V_1 = I_2 \times V_2$$
$$I_b \times 1\,\text{pu} = I_{AB} \times \sqrt{3}\,\text{pu}$$
$$I_{AB} = \frac{0.866}{\sqrt{3}} = 0.5\,\text{pu}$$

$$I_c \times 1\,\text{pu} = I_{BC} \times \sqrt{3}\,\text{pu}$$
$$I_{BC} = \frac{0.866}{\sqrt{3}} = 0.5\,\text{pu}$$

Applying Kirchhoff's law yields

$$I_B = I_{AB} + I_{BC} = 0.5 + 0.5 = 1.0\,\text{pu}$$

The same result can be obtained when we use symmetrical components to calculate the positive- and negative-sequence currents and transform the currents to the delta side, where positive-sequence current leads the low side by 30° and negative-sequence current lags the low side by 30°. The flow of currents will dictate the use

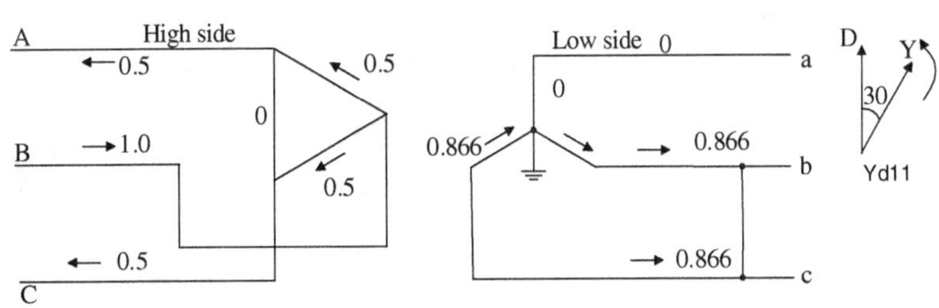

Fig. 2.52 Phase-to-phase fault on the grounded-wye transformer side.

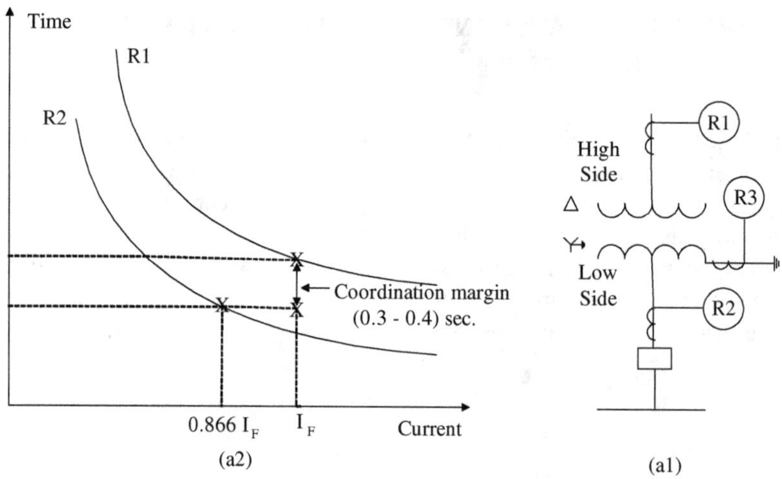

Fig. 2.53 Transformer overcurrent protection.

of three time overcurrent relays on all phases of the transformer high side to make it possible to coordinate with relaying downstream located at the low side. The time overcurrent transformer protection relays are shown in Fig. 2.53(a1). Coordination of relays R1 and R2 should be based on the low-side phase-to-phase faults as shown in Fig. 2.53(a2), where the low-side phase-to-phase curve is shifted upward to the three-phase curve. In other words, high-side phase relays seeing 1 pu of fault current should coordinate with low-side phase relays seeing only 0.866 pu of the fault current.

2.19 LINE-TO-LINE FAULT ON THE DELTA SIDE OF A DELTA/GY TRANSFORMER WITH NO SOURCE CONNECTED TO THE DELTA WINDING

Phase-to-phase faults occurring on the leads of the delta winding can be classified using currents from the grounded-wye side of the transformer. The following steps can be employed to deduce the type of delta lead faults using recorded current information on the wye side:

1. Obtain the transformer phasing diagram from its nameplate.
2. Express the currents obtained from the numerical relay oscillograph record in per-unit values.
3. Assign the known wye winding per-unit current oscillograph or DFR records expressed in per-unit values with their direction on the transformer grounded-wye winding.
4. Deduce the currents inside the transformer delta winding using the polarity rules and ampere-turns coupling for each phase between the two winding sides of the transformer.

5. Combine the delta per-unit currents using Kirchhoff's law to classify the type of phase-to-phase fault. The delta currents should confirm the current spilt in the delta winding to I and $2I$ in accordance with the winding impedance path ratio $2Z$ and Z, respectively.

These steps are used to classify the following delta lead phase-to-phase faults.

Phase B-to-Phase C Fault The transformer shown in Fig. 2.54 is a Yd1 IEC connection where the high-side wye leads the low-side delta by 30°. By knowing the current flow in the wye winding and following steps 1 through 5 above, current flow through the transformer shown in Fig. 2.54 can be determined to confirm that the low-side delta lead has a phase b-c fault.

Phase A-to-Phase C Fault The transformer shown in Fig. 2.55 is a Yd1 IEC connection where the high-side wye leads the low-side delta by 30°. By knowing the current flow in the high-side wye winding and following steps 1 through 5 above, current flow through the transformer shown in Fig. 2.55 can be determined to confirm that the low-side delta lead is a phase a-c fault.

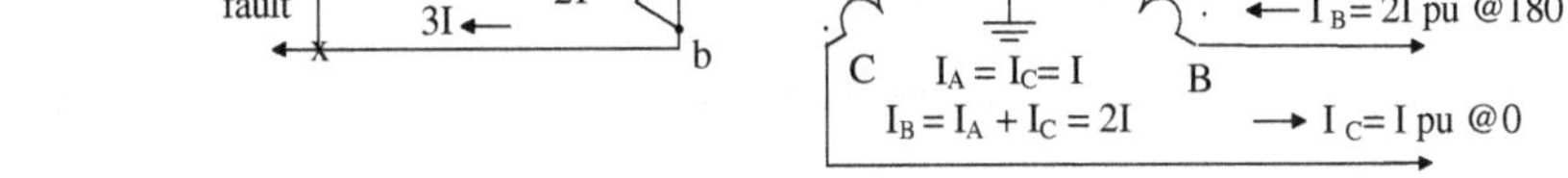

Fig. 2.54 Transformer phasing diagram with a phase b-c fault on the delta leads.

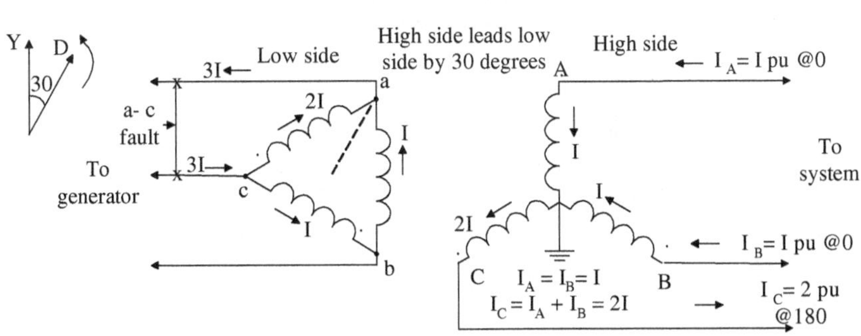

Fig. 2.55 Transformer phasing diagram with a phase a-c fault on the delta leads.

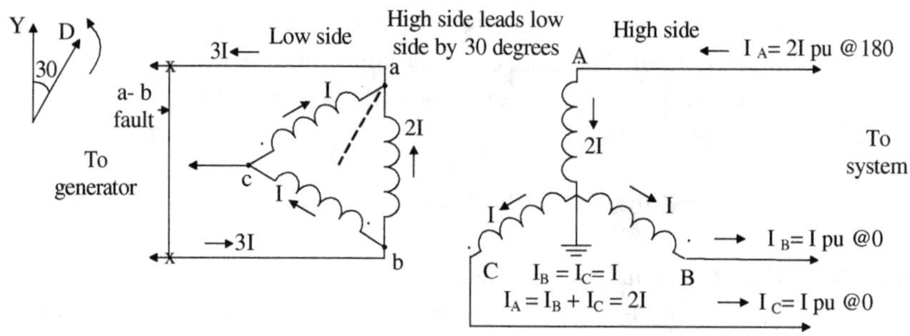

Fig. 2.56 Transformer phasing diagram with a phase a-b fault on the delta leads.

Phase A-to-Phase B Fault The transformer shown in Fig. 2.56 is a Yd1 IEC connection where the high-side wye leads the low-side delta by 30°. By knowing the current flow in the wye winding and following steps 1 through 5 above, current flow through the transformer shown in Fig. 2.56 can be determined to confirm that the low-side delta lead is a phase a-b fault.

2.20 SUBCYCLE RELAY OPERATING TIME DURING AN EHV DOUBLE-PHASE-TO-GROUND FAULT

The subcycle relay operating time during an EHV double-phase-to-ground fault has resulted in a total clearing time of 2 cycles for one of the phases and 2.5 cycles for the other phase of the fault. The total fault clearing time consists of the relay operating time plus the breaker interrupting time. The EHV circuit breaker A has a 2-cycle specified nominal interrupting time. The subcycle operating time to clear this A-C-g fault is carried out by the electromechanical induction cup relay. Electromechanical induction cup or cylinder relays can rotate in less than 8 ms in the presence of sufficient fault current. The performance of these relays in the presence of sufficient fault current is difficult to match with the present state-of-art numerical relaying technology.

Figure 2.57 shows the 765-kV system simplified one-line diagram and DFR-monitored currents and voltages used to illustrate subcycle clearing of the A-C-g fault. The DFR record shown in Fig. 2.58 illustrates the A-C-g fault and confirms a clearing

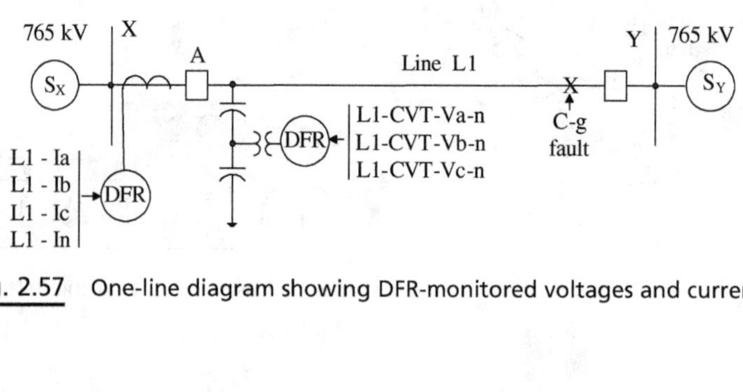

Fig. 2.57 One-line diagram showing DFR-monitored voltages and currents.

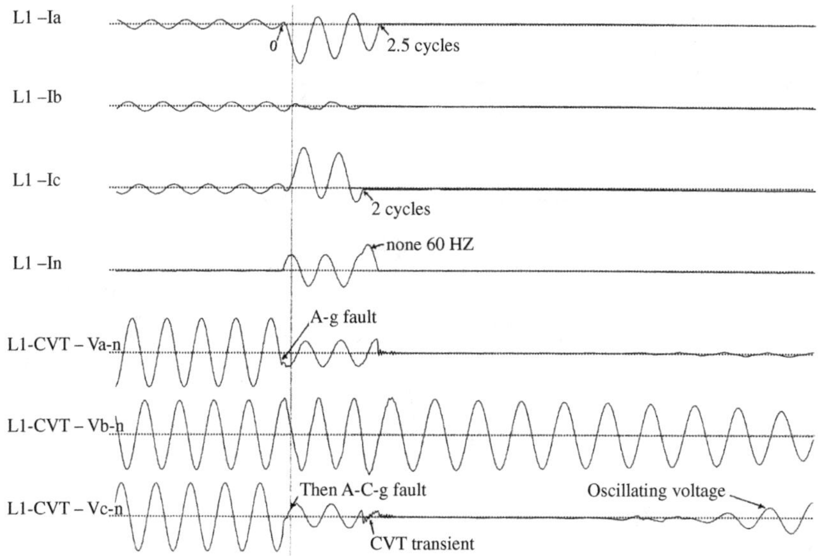

Fig. 2.58 Line L1 DFR record confirming subcycle clearing during the A-C-g fault.

time of 2 cycles for phase C and a clearing time of 2.5 cycles for phase B. Based on a nominal interrupting time of 2 cycles, it can be postulated that a total clearing time of 2 cycles is accomplished by a relay operating time of less than 1 cycle (subcycle).

2.21 SELF-CLEARING OF A C-g FAULT INSIDE AN OIL CIRCUIT BREAKER TANK

Figure 2.59 shows a 230-kV power system one-line diagram and DFR-monitored currents and voltages during a C-g fault. The DFR record of Fig. 2.60 illustrates that

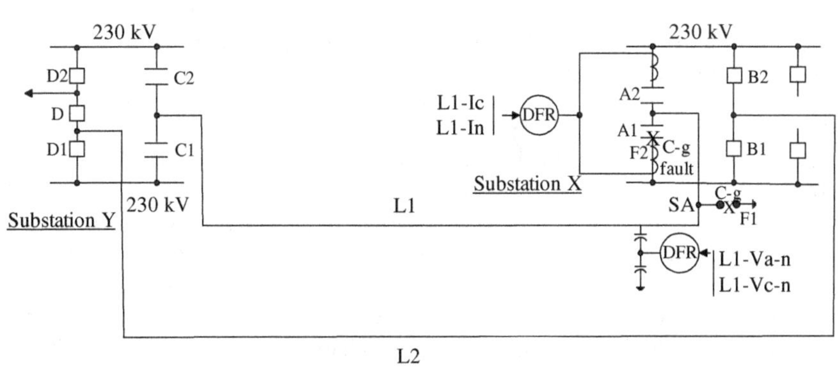

Fig. 2.59 System one-line diagram showing the DFR-monitored analog signals.

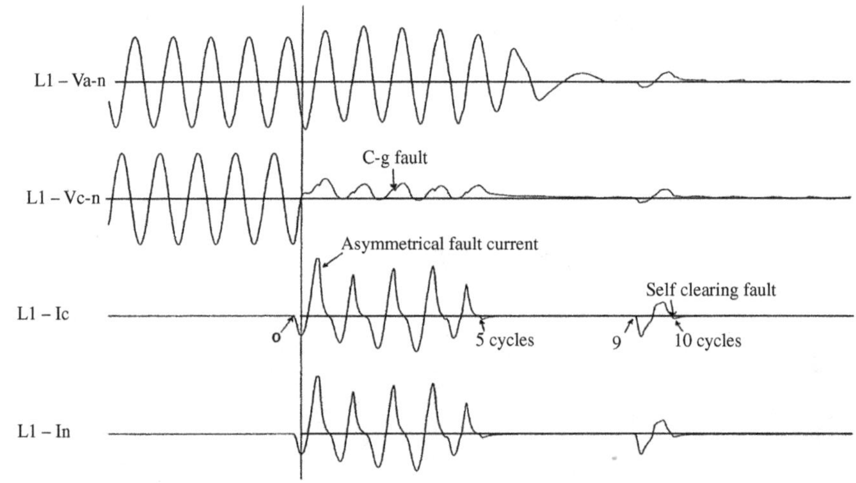

Fig. 2.60 DFR record for faulted phase voltage and currents, showing the self-cleared 1-cycle fault.

the first fault lasted for 5 cycles and the second fault lasted for 1 cycle. Line L1 faulted phase I_C and neutral current I_N illustrates the sequence of the two faults. The first fault was caused by the failure of a conventional line surge arrester, and the second fault occurred inside the line circuit breaker when the main interrupter failed, resulting in a flashover to the tank wall. The second fault was self-cleared by either the oil movement inside the interrupter or the heavy metal parts of the interrupter falling inside the breaker tank. Figure 2.60 also reveals that the 1-cycle fault duration is interrupted at zero current. However, the 230-kV high-impedance electromechanical bus differential relaying system operates in 6 ms and deenergizes the 230-kV bus by tripping all bus-associated circuit breakers. Apparently, the 1-cycle duration was long enough to cause high-speed operation of the bus relay.

2.22 SELF-CLEARING OF A B-g FAULT CAUSED BY A LINE INSULATOR FLASHOVER

Figure 2.61 shows a simplified system one-line diagram and L1 DFR-monitored currents and voltages during a B-g fault. The DFR record in Fig. 2.62 shows that the flashover occurred at peak voltage, indicating an insulator flashover at the fault incident point marked “arc-over at voltage peak”. The voltage dipped on phase B, indicating a phase B-to-ground fault. The transmission-line voltage is 115 kV and the zero-sequence contribution sources at substation X appear to be weak. The fault lasted only about 1.5 cycles and had a self-clearing mechanism process at point b without the tripping action of 115-kV CBs A1 or A2 for line L1 at substation X. The ground fault occurred during an ice storm. It can only be postulated that the mechanism of the

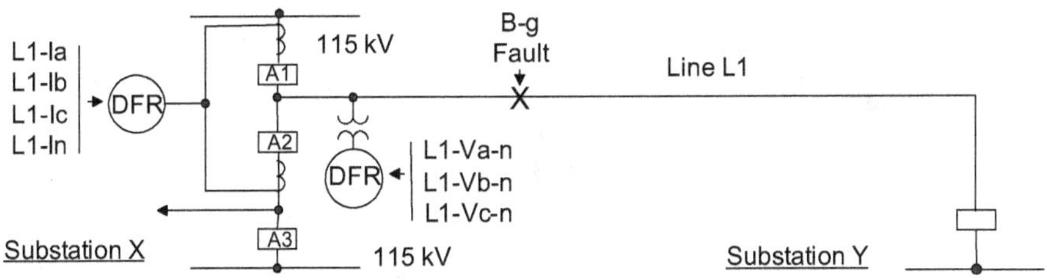

Fig. 2.61 Line L1 DFR-monitored voltages and currents.

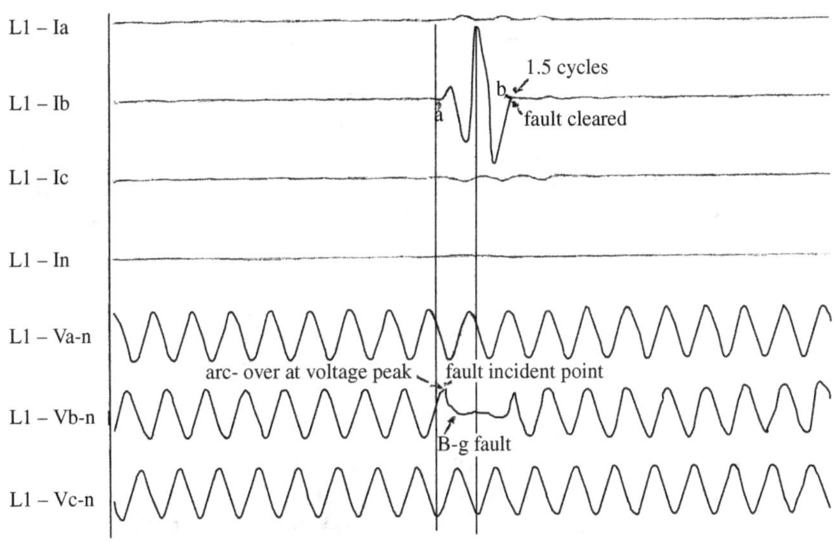

Fig. 2.62 DFR record for L1 showing the self-cleared B-g line fault.

self-clearing process was through the sudden breaking of the ice in the ionized path of the flashover. It can also be postulated that ice movement could be started by the mechanical force generated by the fault current. The DFR-monitored line voltages were restored to normal at point b following the disappearance of the fault.

2.23 DELAYED CLEARING OF A PILOT SCHEME DUE TO A DELAYED COMMUNICATION SIGNAL

Figure 2.63 shows a 345-kV simplified one-line diagram and oscillograph-monitored currents and voltages during an A-g fault. Figure 2.64 illustrates a 345-kV A-g fault occurring near substation Y at the end of line L1. The Y end cleared the fault in about 2 cycles by a zone 1 element, while the X end of the line continued to feed the fault

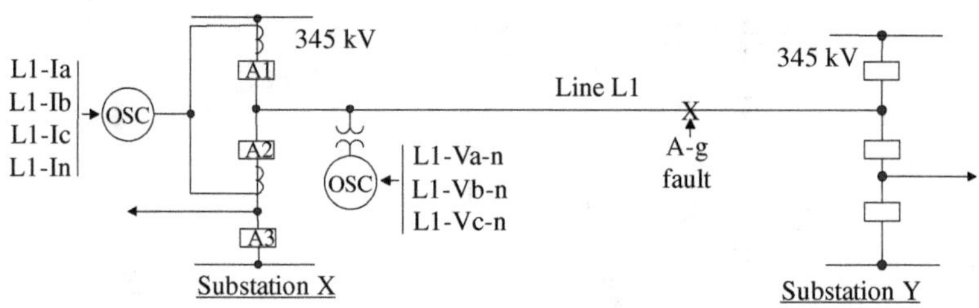

Fig. 2.63 Line L1 oscillograph-monitored voltages and currents.

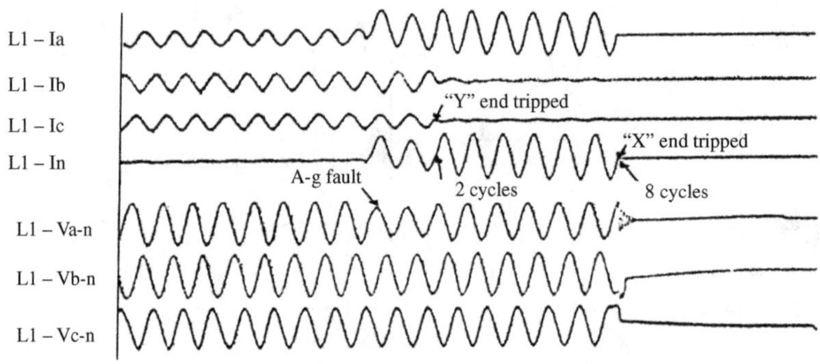

Fig. 2.64 Substation X oscillogram showing delayed clearing (8 cycles).

current for an additional 6 cycles. The slow clearing of the line end at substation X was caused by delay of the communication system at substation Y to send a permissive transfer trip signal to substation X for the pilot permissive overreach transfer trip (POTT) scheme to operate. This delayed clearing can be partially mitigated by allowing zone 1 at each end of the line to prolong keying of the permissive transfer trip signal.

2.24 SEQUENTIAL CLEARING OF A LINE-TO-GROUND FAULT

Sequential clearing occurs for systems with a weak contribution from one (local) end and a strong current contribution from the other (remote) end. Ground fault levels are affected by the positive- and zero-sequence sources. When the strong end clears first, the ground fault will be fed as a stub from the weak end, forcing the ground current to increase slightly and reach the relay pickup setting to operate and clear the fault sequentially.

Sequential clearing should be avoided for certain power system configurations, to avoid prolonging faults in the system. This can be accomplished by applying sensitive current differential pilot relay schemes or a distance-based pilot scheme with weak-in-feed logic. The following two case studies illustrate sequential clearing of line-to-ground faults.

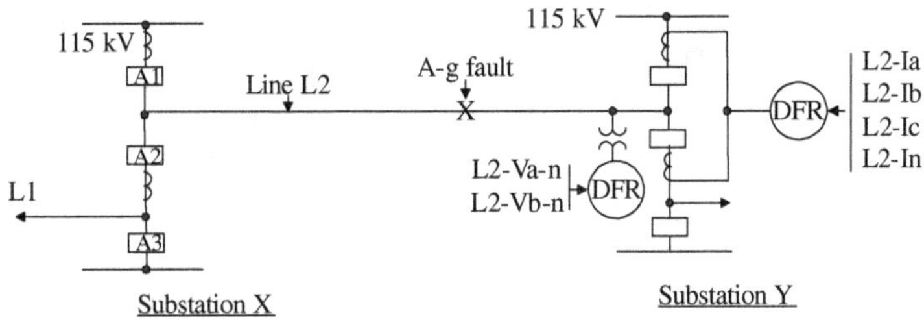

Fig. 2.65 Line L2 DFR-monitored voltages and currents.

Case Study 2.11: Phase A-to-Ground Fault Figure 2.65 shows the simplified 115-kV system one-line diagram and DFR-monitored currents and voltages during the A-g fault. Line L2 is protected by step-distance relays for phase and time overcurrent relays for ground. Figure 2.66 shows a line L1 DFR record where substation X clears the fault in 4 cycles by operation of the instantaneous element of the ground time overcurrent relay. The current contribution to the fault for substation Y goes up sequentially, allowing substation Y to clear the fault after an additional 13 cycles by the ground time overcurrent element, resulting in a total clearing time of 17 cycles (recording stopped at this point).

Case Study 2.12: Phase C-to-Ground Fault Figure 2.67 shows the simplified 345-kV system one-line diagram and DFR-monitored currents and voltages used to analyze a C-g fault. Figure 2.68 shows a line L1 DFR record where substation Y clears the fault in 7.5 cycles by a zone 1 (Z1) ground distance element. The current contribution to substation X rises, allowing substation X to clear the C-g fault

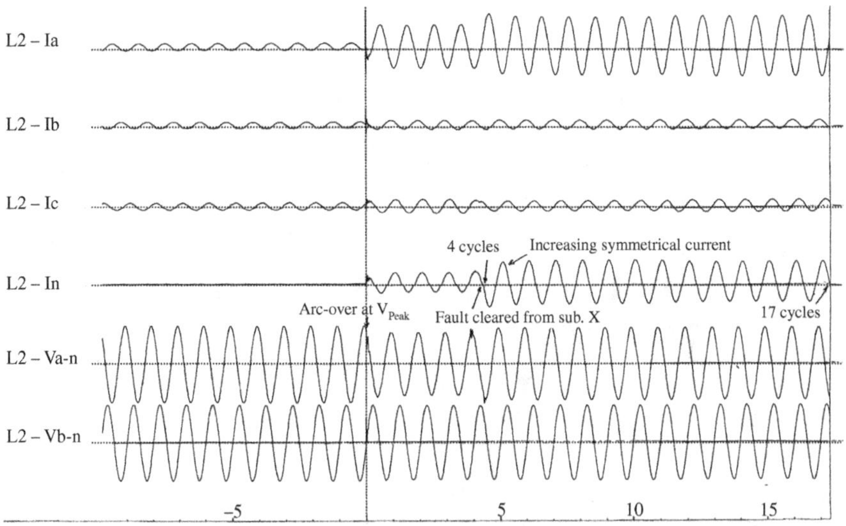

Fig. 2.66 Substation Y DFR record confirming sequential clearing of the A-g fault.

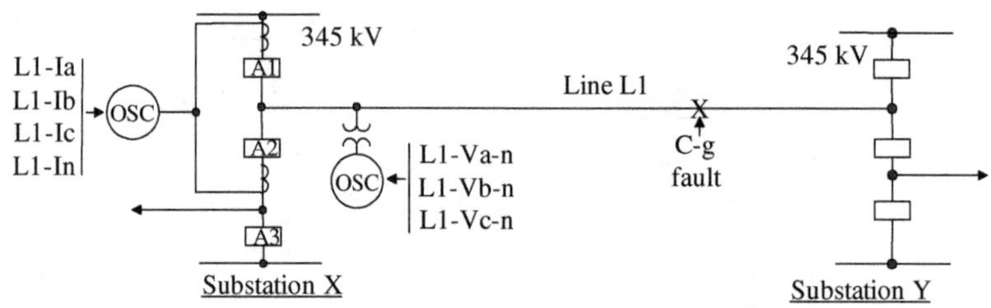

Fig. 2.67 Line L1 DFR-monitored voltages and currents.

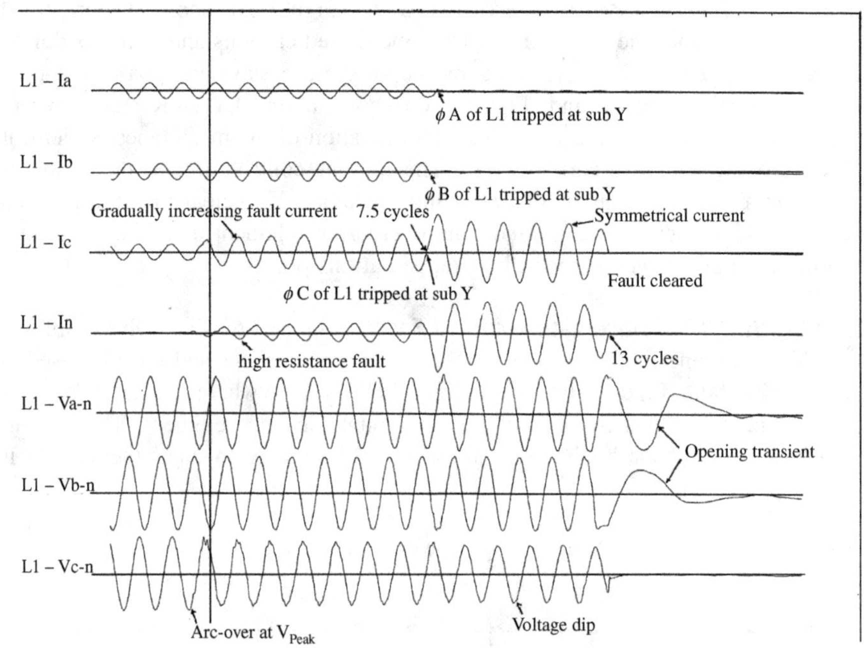

Fig. 2.68 Substation X DFR record showing sequential clearing during the C-g fault.

sequentially by the pilot ground distance element after an additional 5.5 cycles, with a total clearing time of 13 cycles.

2.25 STEP-DISTANCE CLEARING OF AN L-g FAULT

Step-distance relaying is employed due to the lack of communication channels between the two ends of the transmission line. Step-distance relaying will provide high-speed clearing of faults within the overlapping Z1 zones as shown in Fig. 2.69.

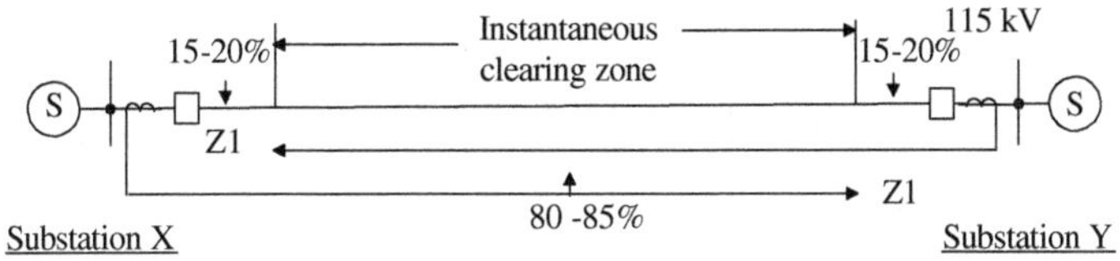

Fig. 2.69 Instantaneous Z1 coverage for a step-distance scheme.

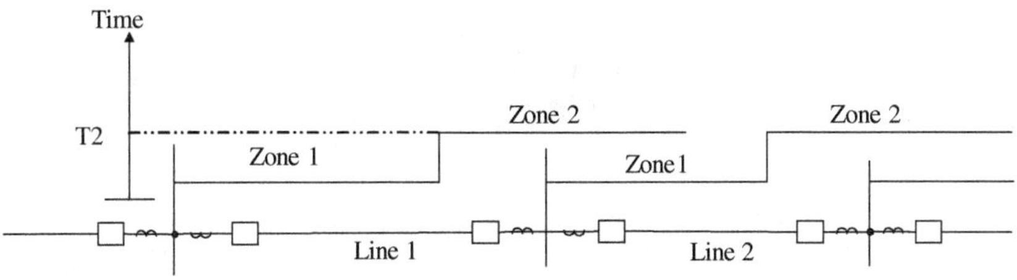

Fig. 2.70 Step-distance scheme using two protection zones.

For faults within 15 to 20% of any of the line terminals, Z1 will clear locally at high speed, and Z2 (step distance) will clear remotely by delayed T2 time at about 0.4 to 0.5 s, as shown in Fig. 2.70.

Figure 2.71 shows a simplified 115-kV system one-line diagram and DFR-monitored currents and voltages used to analyze an A-g fault. The DFR record in Fig. 2.72 reveals that substation Y, which is at the close end to the fault, clears by Z1 in 4.5 cycles at point a, where the ground current contribution from substation X begins to increase. The far end at substation X clears by Z2 ground distance relaying in 22.5 cycles at point b. A tap on the line is connected to a generator through a delta/delta transformer. The phase A-g fault after 22.5 cycles becomes a fault on an ungrounded system. As a result, for an additional 3 cycles of recording, DFR trace L1-Va-n displays a low voltage value ($V_{a\text{-}n} = 0$), whereas DFR traces L1-Vb-n and L1-Vc-n are equal to the line-to-line voltage, as should be true ($V_{b\text{-}n} = V_{c\text{-}n} = V_{L\text{-}L} = \sqrt{3} \times V_{L-n}$).

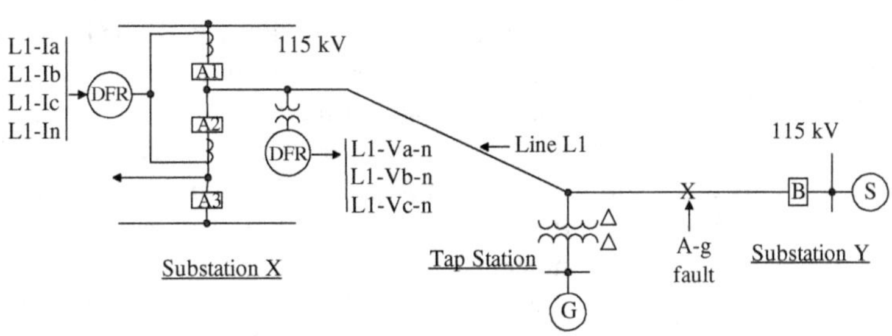

Fig. 2.71 One-line diagram showing DFR-monitored voltages and currents.

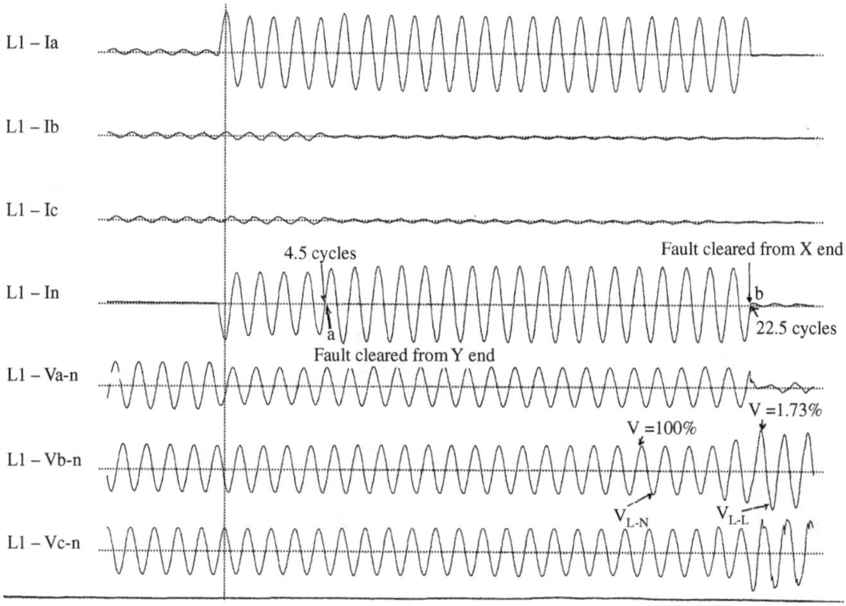

Fig. 2.72 Substation X DFR record for line L1 voltages and currents during step-distance clearing of an A-g fault.

2.26 GROUND FAULT CLEARING IN STEPS BY AN INSTANTANEOUS GROUND ELEMENT AT ONE END AND A GROUND TIME OVERCURRENT ELEMENT AT THE OTHER END

For older electromechanical relays, the step distance was normally employed for phase fault protection, while for ground fault protection a time overcurrent (TOC) relay with an instantaneous element (if it can be set) was employed. Ground faults in the overlapping 60% zone (Fig. 2.73) will be cleared by instantaneous protection at each end of the line. The ground relay instantaneous element is set to cover about 80% of the protected line length. The fault at F1 near the end of the line at substation Y will be cleared by the instantaneous element of relay R2 at substation Y and the time overcurrent element of relay R1 at substation X. Numerical protection includes a step distance phase and ground elements as well as ground time overcurrent elements. This will be illustrated here for a 115-kV line that is protected using this concept.

Figure 2.74 shows a simplified 115-kV system one-line diagram and DFR-monitored currents and voltages during an A-g fault. The DFR record in Fig. 2.75 shows a fault away from the overlapping zone. The fault is cleared in 3.5 cycles from the Y end of the line by the instantaneous element, and the fault ground current from the remote end begins to increase from a value of 685 A to 970 A, thus allowing the X end of the line to clear the A-g fault in 21.5 cycles by the ground time overcurrent element.

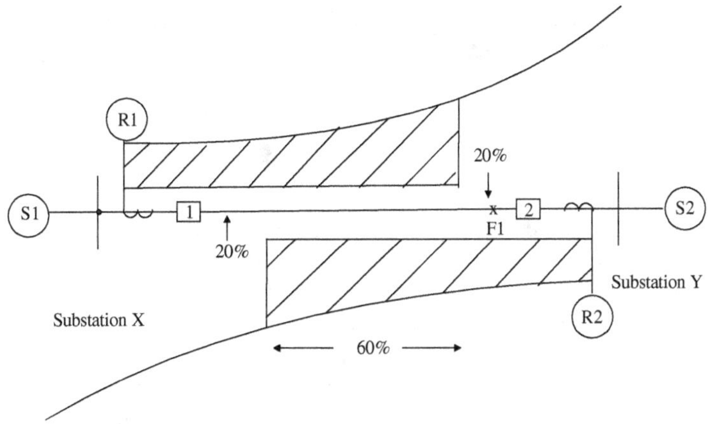

Fig. 2.73 Line protection using ground instantaneous and TOC relay elements.

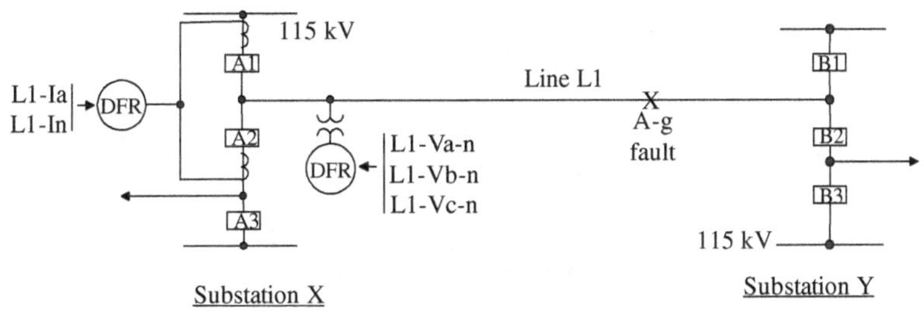

Fig. 2.74 Line L1 DFR-monitored voltages and currents.

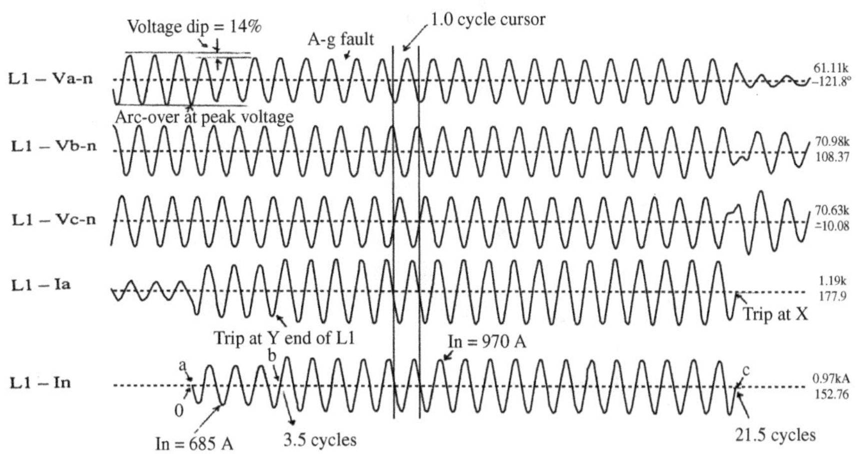

Fig. 2.75 Substation X DFR record showing the TOC relay clearing of an A-g fault in steps.

2.27 GROUND FAULT CLEARING BY REMOTE BACKUP FOLLOWING THE FAILURES OF BOTH PRIMARY AND LOCAL BACKUP (BREAKER FAILURE) PROTECTION SYSTEMS

Remote backup relaying is designed to clear system faults when primary and local backup (breaker failure) fail. This will occur either during extreme power system contingencies or when redundancy concepts are not used, leading to a single failure that can disable primary protection. This will be illustrated here for a 230-kV line fault coupled with modification of the associated 230-kV breaker trip coil assemblies to add a second trip coil, leading to magnetic flux cancellation.

Figure 2.76 shows a system one-line diagram and DFR-monitored currents and voltages used to analyze a line L1 B-g fault. The DFR record in Fig. 2.77 shows a phase B-g fault occurring on line L3. One cycle later, the primary and secondary numerical relays of line L1 protection are operated to energize the primary and

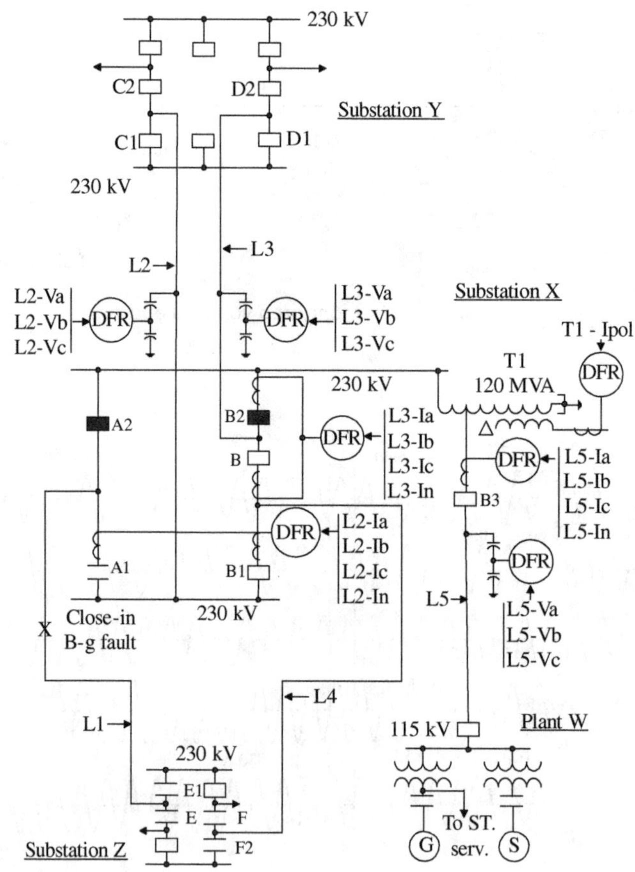

Fig. 2.76 One-line diagram showing DFR-monitored voltages and currents.

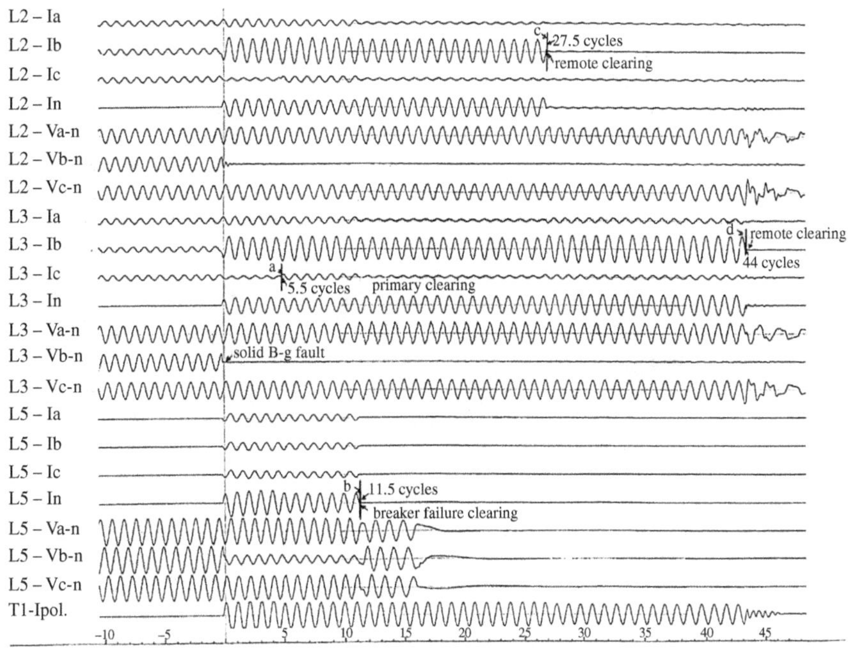

Fig. 2.77 Substation X DFR record showing voltages and currents during the ground fault with primary, local, and remote clearing times.

secondary trip coils of CBs A1 and A2. Three cycles later (4 cycles after $t = 0$), CB A1 at substation X opened. One and a half cycles later (5.5 cycles after $t = 0$), CBs E1 and E2 at substation Z opened. Eight and a half cycles after $t = 0$, the CB A2 breaker failure relaying system timed out to energize its associated breaker failure lockout relay. Three cycles later (11.5 cycles after $t = 0$), circuit breaker (115 kV) B3 at substation X opened. Remote CBs C1 and C2 of line L2 at substation Y tripped 27.5 cycles after $t = 0$. Forty-four cycles after $t = 0$, remote 230-kV CBs D1 and D2 of line L3 at substation Y opened and isolated substation X from the bulk power system. Remote backup relaying had to clear the initial B-g fault on line L1, as designed, following the failure of fault clearing by both the primary and breaker failure relaying systems. The failures were attributed to magnetic flux cancellation due to design errors for CBs A2 and B2 at substation X.

2.28 BREAKER FAILURE CLEARING OF A LINE-TO-GROUND FAULT

Figure 2.78 shows a system one-line diagram and DFR-monitored currents and voltages during a 345-kV bus 1 C-g fault. The DFR record in Fig. 2.79 shows a phase C-g fault that occurred inside the CB C1 capacitor bank breaker. The ground fault occurred when the 345-kV CB C1 flashed-over to ground. The high-impedance bus

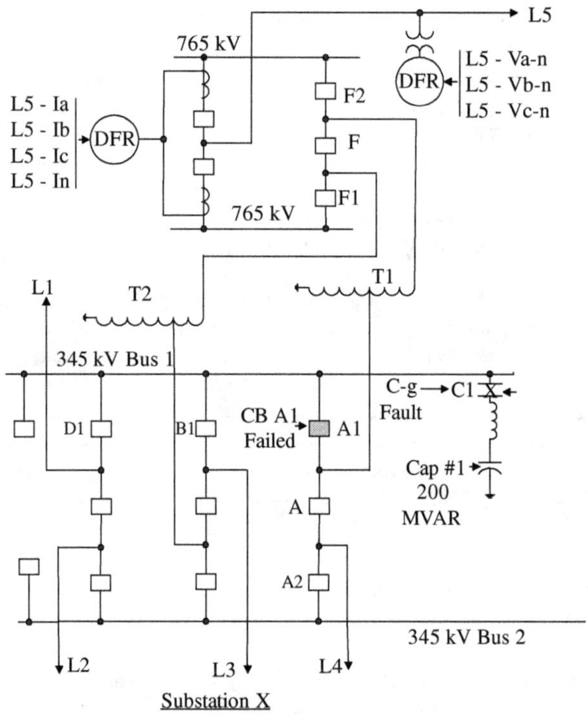

Fig. 2.78 One-line diagram showing line L5 DFR-monitored voltages and currents.

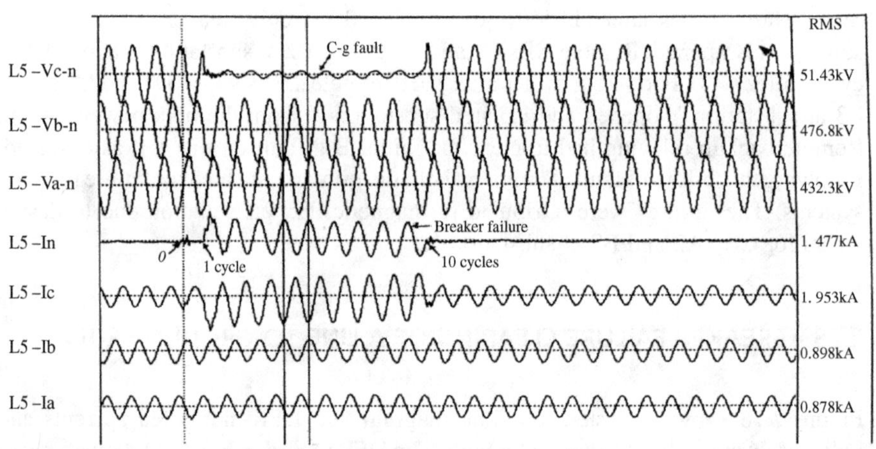

Fig. 2.79 Line L5 DFR record showing breaker A1 failure protection clearing of the C-g fault.

differential relay operated and tripped 345-kV CBs B1 and D1, which cleared the fault in 3 cycles. However, 345-kV CB A1 failed, and its breaker failure protection operated to trip 345-kV CB A and 765-kV CBs F and F2. The DFR record in Fig. 2.79 shows a total breaker failure clearing time of 9 cycles. This failure time consists of 0.5 cycle of bus differential relay operation to initiate breaker failure, 0.5 cycle of bus differential lockout relay, 5.5 cycles of breaker failure timer, 0.5 cycle of CB A1 breaker failure lockout relay, and 2 cycles of breaker interrupting time.

2.29 DETERMINATION OF THE FAULT INCIDENT POINT AND CLASSIFICATION OF FAULTS USING A COMPARISON METHOD

Determination of the fault incident point and classification of faults using a comparison method can be used for oscillograph records obtained from numerical differential relays having only two current inputs. Analysis of the fault record for the differential relay shown in Fig. 2.80 can be useful when done on a per-phase basis. Since the fault involves phases B and C only, the phase A current of the generator system side is equal and opposite to the generator neutral-side current. The sum of the system- and neutral-side currents represents the differential relay operating current. This component is zero for phase A current, indicating that phase A current is related primarily to unit output feeding the system load. The relay system is designed to have $I_S = -I_N = -I$, where I_S is the system-side connection to the numerical differential

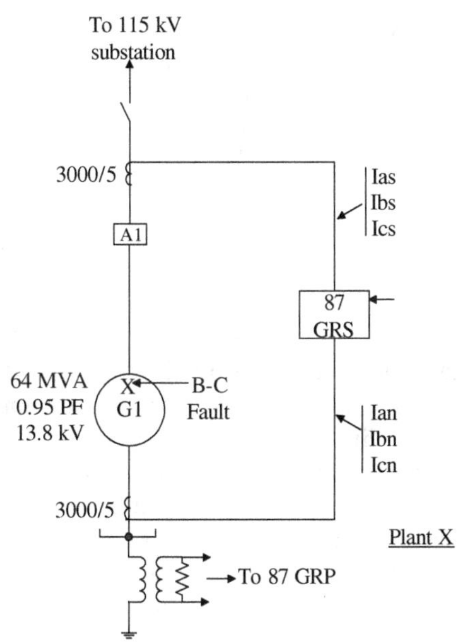

Fig. 2.80 One-line diagram with the generator G1 numerical relay–monitored currents shown.

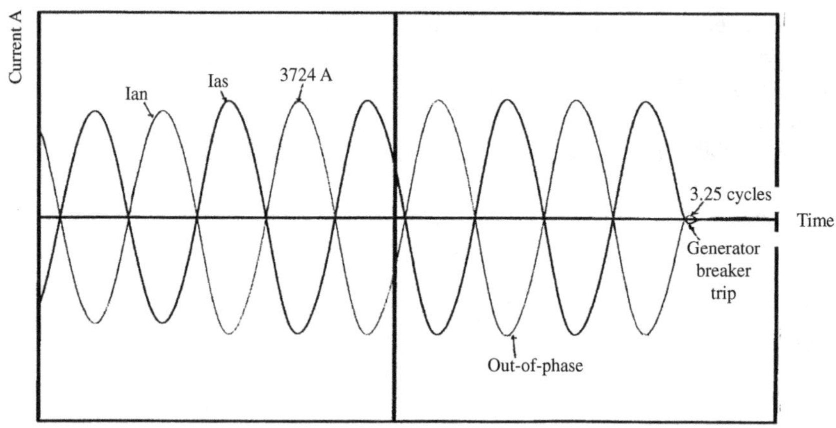

Fig. 2.81 Generator numerical relay oscillography fault record showing phase A currents for the generator neutral and system sides.

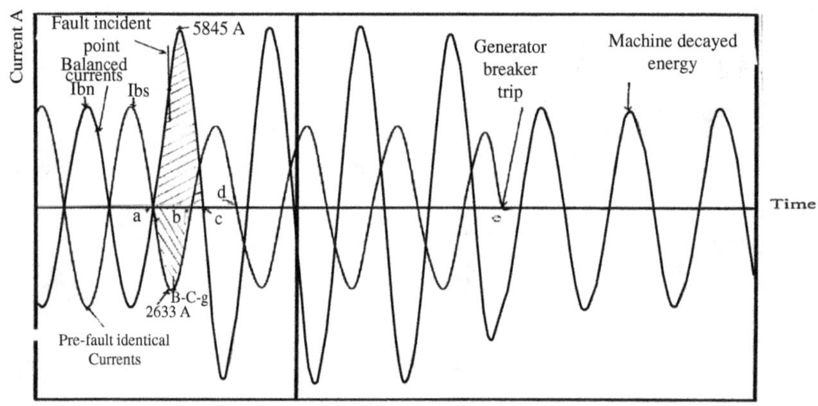

Fig. 2.82 Fault incident point shown on the fault record for phase B currents for the generator neutral and system sides.

relay and I_N is the neutral side to the relay, for through power flow and external fault conditions. Therefore, Fig. 2.81 shows two identical currents, with one of them reversed for phase A, providing $I_{diff} = I_S + I_N = 0$. Using this fact, analysis of this fault can best be accomplished by making plots for individual generator phases for each pair of neutral and system differential relay inputs, as shown in Figs. 2.81, 2.82, and 2.83. The three-phase currents from the generator neutral side will be compared with those from the generator system side to deduce the relay differential current and hence the correct sequence of events. Therefore, analysis can be simplified and there is no need to analyze the relay voltages to classify the fault. A fault incident point can therefore be detected by doing half-cycle comparison for currents around the beginning of the fault. A half-cycle can be declared as containing a fault incident

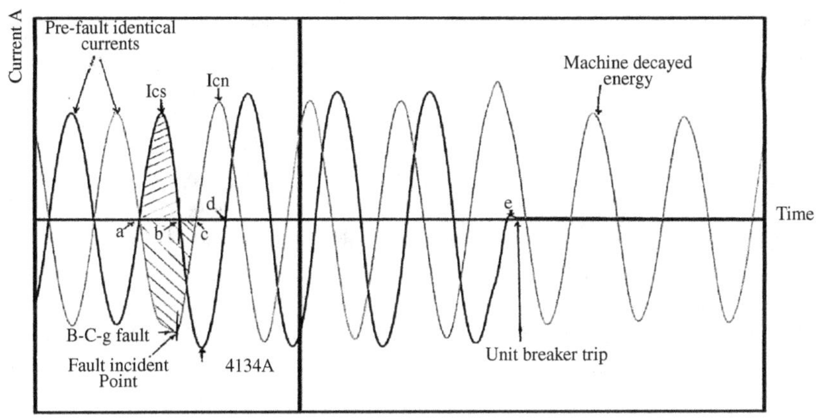

Fig. 2.83 Fault incident point shown on the fault record for phase C currents for generator neutral and system sides.

point if the time between its associated zero crossing is different from (smaller or greater than) that in the preceding half-cycle (i.e., a half-cycle ago).

In Fig. 2.82, the I_{bs} current hashed half-cycle (negative) has a zero crossing at points a and b. The width of the half-cycle time between a and b is smaller than the previous unhashed half-cycle. In addition, for current I_{bn} the hashed half-cycle area (positive) width of ac is larger than that in the preceding half-cycle. This will imply that a phase angle shift has occurred for currents I_{bn} and I_{bs} in the overlapping hashed areas. As a result, the fault incident point can be deduced and is shown in Fig. 2.82. Similarly, in Fig. 2.83, the I_{cs} current hashed half-cycle (positive) has a zero crossing at points a and b. The width of the half-cycle time between a and b is smaller than that of the preceding unhashed half-cycle (negative). In addition, for current I_{cn} the hashed half-cycle (negative) area width of ac is larger than that in the preceding half-cycle. This implies that a phase angle shift has also occurred for currents I_{cn} and I_{cs} in the overlapping hashed areas. As a result, the fault incident point can be deduced and is shown in Fig. 2.83. By comparing the time distance between the fault incident point and the trigger line in Figs. 2.82 and 2.83, it can be deduced that a simultaneous B-C fault has occurred (equal time distances).

REFERENCES

Blackburn, J. L. *Applied Protective Relaying*. Pittsburgh, PA: Westinghouse Electric Corporation, 1979.

Blackburn, J. L. *Protective Relaying Principles and Applications*. New York: Marcel Dekker, 1987.

Blackburn, J. L. *Symmetrical Components for Power Systems Engineering*. New York: Marcel Dekker, 1993.

Elmore, W. A., Ed. *Protective Relaying Theory and Applications.* New York: Marcel Dekker, 2000.

Fault and Disturbance Data Requirements for Automated Computer Analysis. IEEE Power System Relaying Committee Working Group 111 Report, Special Publication 95 TP 107, 1995.

Greenwood, A. *Electrical Transients in Power Systems.* New York: Wiley-Interscience, 1971.

Ibrahim, M. A. St. Lawrence FDR 230-kV substation disturbance of March 6, 1996. Presented at the Georgia Tech Protective Relaying Conference, April 30–May 2, 1998, Atlanta, GA, and at the 25th Annual Western Protective Relay Conference, October 26–28, 1998, Spokane, WA.

Ibrahim, M. A., and F. Stacom, Adirondack 230-kV substation outage of July 1, 1995. Presented at the Georgia Tech Protective Relaying Conference, May 1–3, 1996, Atlanta, GA, and at the 23rd Annual Western Protective Relay Conference, October 14–16, 1996, Spokane, WA.

Neuenswander, J. R. *Modern Power Systems.* Scranton, PA: International Textbook Company, 1971.

Phadke, A. G., T. Hlibka, and M. A. Ibrahim. Fundamental basis for distance relaying with symmetrical components. *IEEE Transactions on Power Apparatus and Systems,* Vol. PAS-96, No. 2, March–April 1977, pp. 635–646.

3

POWER SYSTEM PHENOMENA AND THEIR IMPACT ON RELAY SYSTEM PERFORMANCE

Power system phenomena are normally captured and defined during the analysis of power system disturbances. Power system phenomena provide useful and basic information and knowledge that can enhance the required background for sound relaying basics. In addition, it can enhance the system disturbance analysis process to arrive at the correct time line that makes it possible to isolate the faulted element and restore service to the balance of the power system. The effects on the relaying system are also highlighted, so that the relay system design and its intended function can be optimized.

In this section we document most of these phenomena as they appear on DFR and numerical relay fault records. The power system phenomena and related behaviors of the protection systems are supported by DFR and numerical oscillography records. Involved are generator rotor oscillations with the power system during out-of-step conditions and synchronization events; appearance of 120 Hz in the rotor surface during faults and unbalanced conditions; flow of the third harmonic and its variations in a generator neutral grounding system; negative-sequence current flow in generators; active (MW) and reactive (MVAR) generator outputs during machine terminal faults; inadvertent generator energization; generator loss of excitation; generator-trapped decayed energy; generator neutral zero-sequence voltage coupling

Disturbance Analysis for Power Systems, First Edition. Mohamed A. Ibrahim.

through step-up transformer interwinding capacitance; transformer inrush currents; energizing of a transformer with a high-side fault; zero-sequence current sources; zero-sequence mutual coupling and its effect on the performance of bus differential relaying; current transformer (CT) saturation; nonzero current crossing during asymmetrical faults; capacitive voltage transformer (CVT) and bushing potential device transients; circuit breaker re-strikes; circuit breaker pole disagreement; capacitor bank closing transient and capacitor bank outrush into faults; voltage oscillations due to shunt reactor application; Ferranti voltage rise; magnetic flux cancellation; breaker failure backfeeding; spillover of surge arrestors; overscaling of A/D converters; subsidence current; and wrong formation of the neutral for motor stator winding.

3.1 POWER SYSTEM OSCILLATIONS LEADING TO SIMULTANEOUS TRIPPING OF BOTH ENDS OF A TRANSMISSION LINE AND THE TRIPPING OF ONE END ONLY ON AN ADJACENT LINE

Electric power systems constantly experience small stable power oscillations. They occur as the generator rotor accelerates or slows down while rebalancing electric output power to mechanical input power for individual units to respond to changes in load, generation, or network condition. Oscillations that can grow in amplitude are called *unstable oscillations*. Oscillations are sometimes called *power swings*, and once initiated, they flow back and forth across the system.

Power system oscillations occur during any of the following conditions:

1. Sudden changes in load
2. Switching of transmission lines
3. Tripping or removal of generators from service
4. Occurrence of faults in a system

3.1.1 Generator Oscillation Scenario

A generator oscillation scenario can occur through the following steps:

1. A change takes place in the power system (e.g., line open, generator trip, addition or removal of a block of loads).
2. The rotors of the system begin to change speed (from synchronous) in response to the system change.
3. The speed change to the rotor leads to a change in the unit angle spread (rotor angle).
4. When the rotor angle changes, the power outputs of generators change.

5. When the power outputs of system generators change, the rotors change speed again, and as a result, oscillation has begun.
6. The generators will oscillate about an operating point until the system positive damping reduces the amplitude of the oscillation and achieves a stable system condition.

3.1.2 Description of an Out-of-Step Oscillation Incident

During a major power system blackout, oscillation took place between neighboring utility systems. The onrushing power swing caused voltage and current oscillations, leading to active power, reactive power, and apparent impedance oscillations. Zone 1 distance relay elements, applied with no intentional time delay, may react to the swinging apparent impedance and may operate during power swing conditions.

The electrical center of oscillation around substation X shown in Fig. 3.1 resulted in the nearly simultaneous trip of six 345-kV transmission lines when the swing sliced through them. Figure 3.1 also reveals the DFR-monitored currents and voltages for lines L1, L2, and L3 during the incident. The DFR record in Fig. 3.2 monitors the three-phase and neutral currents as well as the line-side phase-to-neutral voltages of 345-kV lines L1 and L2. The DFR record was triggered by the neutral current appearing as a result of interrupting three-phase loading currents of a balanced system with 120° between phases. The DFR record pre-trigger length is 5 cycles and the post-trigger length is 35 cycles. The DFR record reveals that line

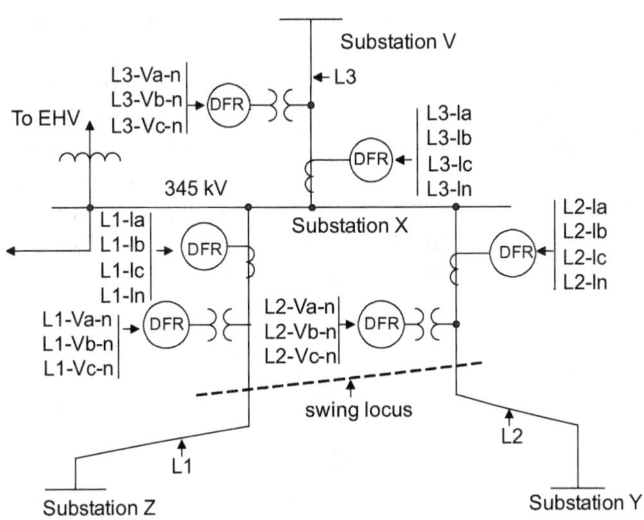

Fig. 3.1 System one-line diagram showing DFR-monitored voltages and currents.

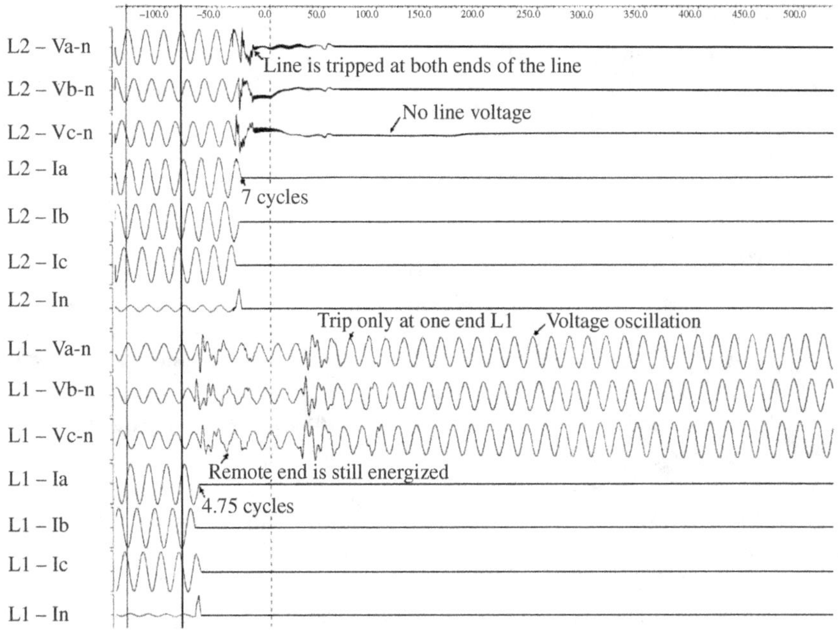

Fig. 3.2 Substation X DFR record showing 345-kV line L1 and L2 voltages and currents during power system oscillation conditions.

L1 tripped at substation X, as indicated by the interruption of line L1 phase currents at about 4.75 cycles. However, voltages still appeared at the line L1 end, indicating that the remote end of L1 at substation Z did not trip (observe the oscillating nature of L1 voltage).

Line L2 tripped 2.25 cycles after the tripping of line L1. Less than a cycle after the line L2 current interruption, the line-side voltage disappeared, confirming that line L2 tripped simultaneously at both ends by operation of the individual zone 1 element at each end of the line. This will confirm that the center of the out-of-step (swing) locus has passed through the overlapping protection zone area of Z1 characteristic in the *R–X* diagram at both ends of line L2, as illustrated in Fig. 3.7. This is analogous to placing a three-phase fault at the electrical center where the swing impedance locus travels to intersect line L2.

3.1.3 Analysis of Z1 Tripping for Line L1

The reported tripping of line L1 at substation X by a zone 1 element can be confirmed using DFR symmetrical component calculations to obtain the positive-sequence voltages and currents. The following secondary values are from the results of an analysis of symmetrical components as provided by the DFR software package:

$$\begin{aligned} V_1 &= 36\text{ V at } 98.8^\circ \\ I_1 &= 6.3\text{ A at } 54.5^\circ \\ V_2 &= 0.5\text{ V} \\ I_2 &= 0.2\text{ A} \\ V_0 &= 0.1\text{ V} \\ I_0 &= 0 \end{aligned}$$

The secondary positive-sequence impedance can then be calculated as

$$Z_1 = (V_1/I_1) = 5.7\,\Omega \text{ at } 44.3^\circ \; (= 36\text{ V at } 98.8^\circ/6.3\text{ A at } 54.5^\circ)$$

The primary impedance [$= 57\,\Omega (= 5.7 \times \text{PT/CT})$] is superimposed on the line L1 zone 1 (Z1) mho distance relay characteristic shown in Fig. 3.3. It confirms operation of the Z1 relay at substation X for line L1 during the power oscillation condition. Analysis was also carried out for line L3, which was not tripped during the oscillation incident, and symmetrical component analysis was performed at the same 1-cycle window where lines L1 and L2 were tripped. The DFR record in Fig. 3.4 shows the voltages and

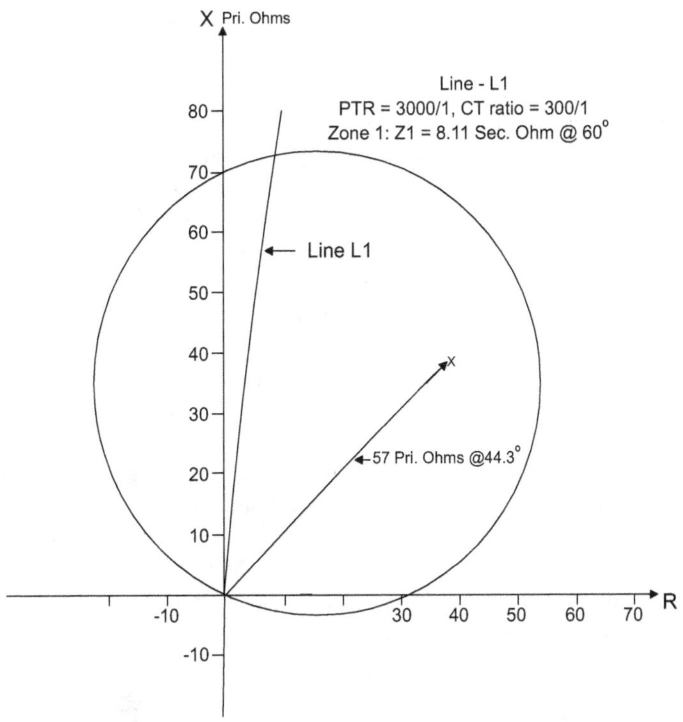

Fig. 3.3 Distance relay characteristics for line L1 at substation X showing the apparent impedance swing inside the Z1 relay operating region.

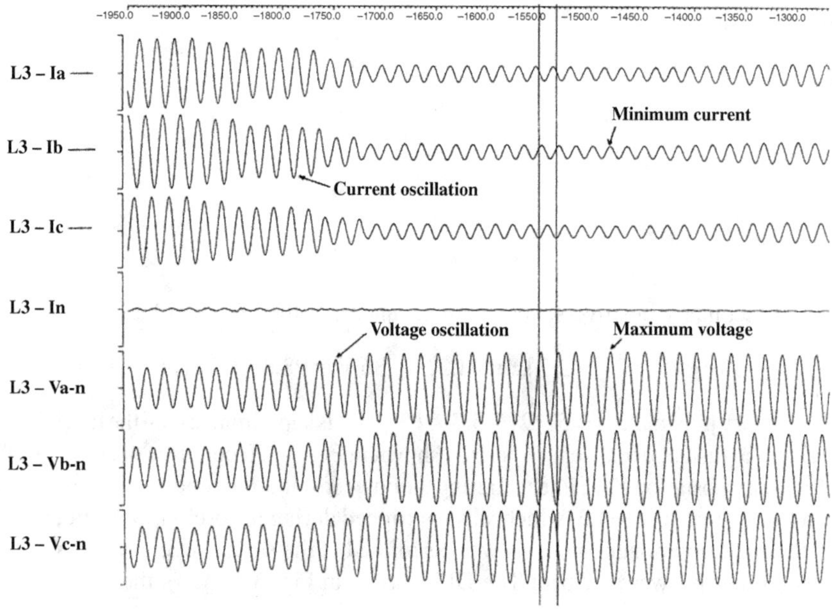

Fig. 3.4 Substation X DFR record showing 345-kV line L3 voltages and currents during a power oscillation condition.

currents for line L3 during the disturbance. Figure 3.5 illustrates the impedance calculated for line L3, which is 165 Ω at −122° and is shown to be outside the Z1 setting for line L3, and hence no tripping took place for the L3 terminal at substation X.

It can be concluded that power system oscillation will also lead to impedance oscillation. Depending on the location of the electrical center of oscillation, out-of-step tripping of the zone 1 direct trip element will depend on the path of the locus of the oscillating apparent impedance measured by Z1 at each end of the line. One end only of line L1 is tripped due to the apparent impedance locus path 2 going through only Z1, characteristic of one end of the line, as shown in Fig. 3.6. Apparent impedance path 3 traveling outside the operation area of Z1 happened with the nontripping of line L3 during the same power system oscillation. Apparent impedance path 1 traveling outside the tripping area of Z1 will not cause tripping of the associated line, similar to path 3.

Figure 3.7 shows apparent impedance path 4 traveling at the overlapping protection area, resulting in simultaneous tripping of both ends of L2 by the zone 1 relay element. The oscillation process is equivalent to placing a three-phase fault at the intersection of the swing locus with the line. It should be noted that all three elements (A-B, B-C, and C-A) associated with distance phase protection operated to trip the line at both ends during the swing condition.

3.1.4 Effect on Relaying Systems

Severe power system oscillations that will cause voltage and current oscillation will affect the operation of the Z1 direct (no intentional time delay) trip element. Setting the

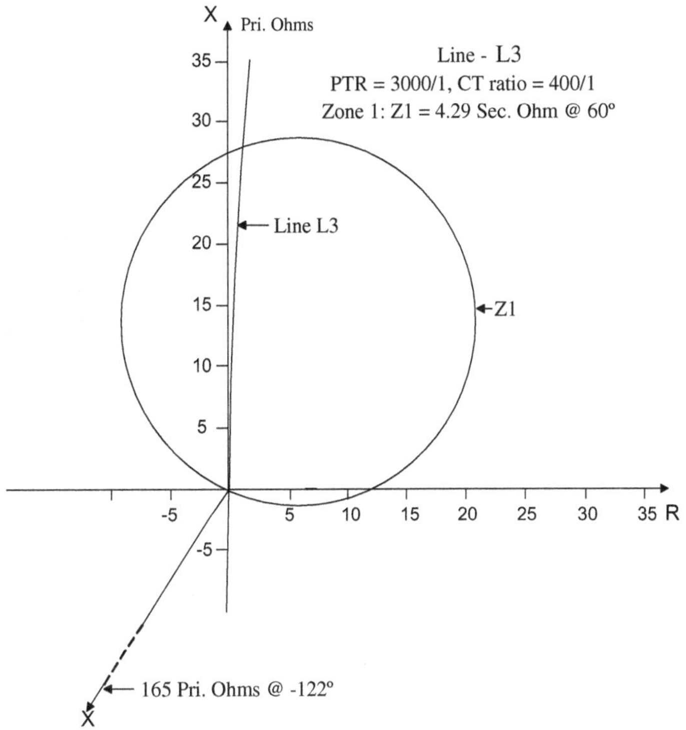

Fig. 3.5 Distance relay characteristics for line L3 at substation X showing the apparent impedance swing outside the Z1 relay operating region.

Z1 phase element at 80 to 85% and the Z1 ground element at 75 to 80% (due to ground loop impedance) is ideal and should not be altered due to the effects of power system oscillation phenomena on distance relaying. The Z1 element is only reacting to phenomena that should be eliminated or reduced in severity so that problems cannot be exported from one end of a system to its neighboring system. Out-of-step tripping or blocking of Z1 requires detailed system stability studies for proper islanding with balanced generation and load. It should be noted that when one mho element is blocked from tripping during power system oscillation, another element must be permitted to trip to remove the danger of unstable swing on the connected power system.

3.2 GENERATOR OSCILLATIONS TRIGGERED BY A COMBINATION OF L-g FAULT, LOSS OF GENERATION, AND UNDESIRED TRIPPING OF THREE 138-KV LINES

3.2.1 Description of the Incident

A combustion turbine unit GA of the 500-MW combined-cycle plant shown in Fig. 3.8 was tripped due to a phase C surge arrester failure associated with the 138-kV side of

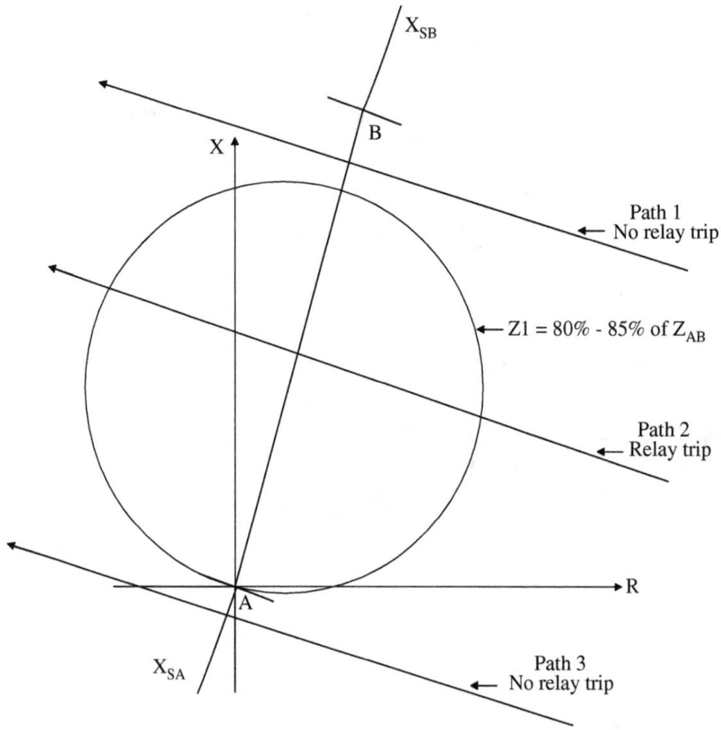

Fig. 3.6 *R–X* diagram showing the Z1 element in relation to apparent impedance oscillation.

GSU transformer T1. The ground fault was cleared from the system successfully in about 5 cycles by the tripping unit GA generator breaker and tripping of 138-kV remote CBs A1 and A2 at substation Y. Unit GA was subsequently shut down by activation of lockout relays associated with unit protection systems. About 63.5 cycles (1.054 s) after $t = 0$, CT unit GB was tripped mechanically by the trip signals originating from the turbine lube oil header pressure being low, while the unit is oscillating against the power system. No other relay targets were recorded for the unit GB trip.

3.2.2 Theory and Conclusions of the Analysis of a Unit GB Trip

Figure 3.8 shows a system one-line diagram and DFR-monitored currents for unit GB and active power calculated for units GA and GB during the disturbance. As illustrated in Fig. 3.9, the active power of unit GA began to decrease at the inception of the fault until it reaches zero at the clearing point of the fault. The cause of unit GB swing (being out of step) with the system is attributed to a combination of events that cause the unit to oscillate. The solid C-g fault close to the high side of unit GA, coupled with the simultaneous loss of three 138-kV feeders (L1, L2, and L4) within substation Y,

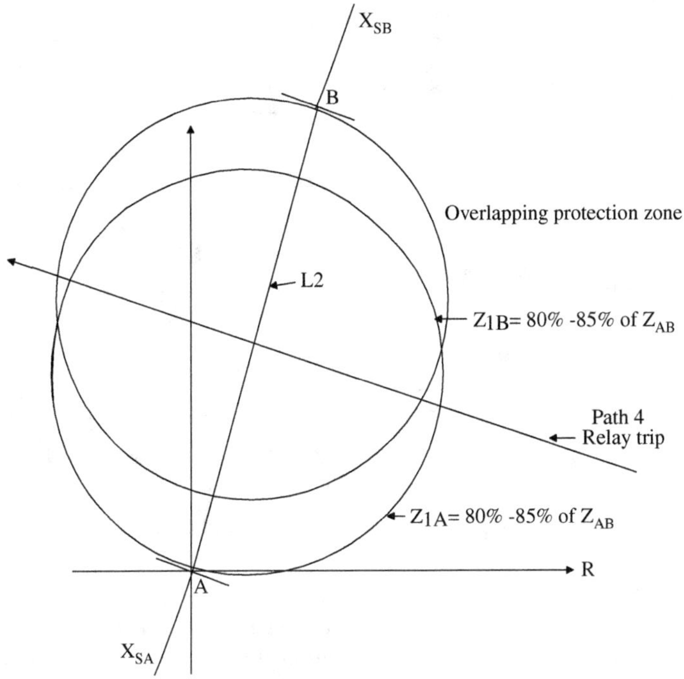

Fig. 3.7 *R–X* diagram indicating the Z1 overlapping characteristic in relation to apparent impedance oscillation.

affected the equivalent transfer impedance X between the unit and the system as reflected in the power transfer formula:

$$P = \frac{\{(V_1 \times V_2)\sin\theta\}}{X}$$

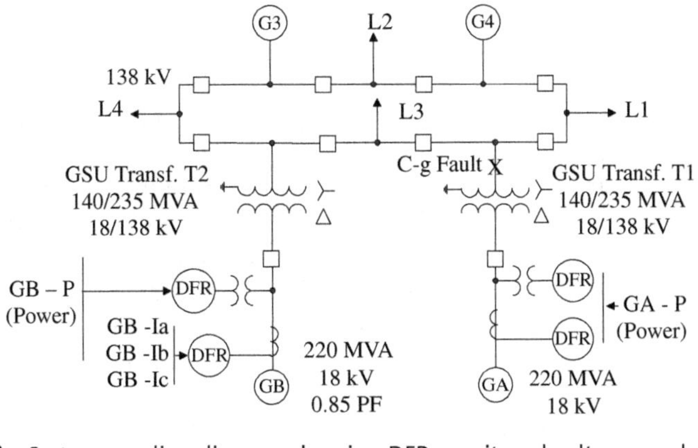

Fig. 3.8 System one-line diagram showing DFR-monitored voltages and currents.

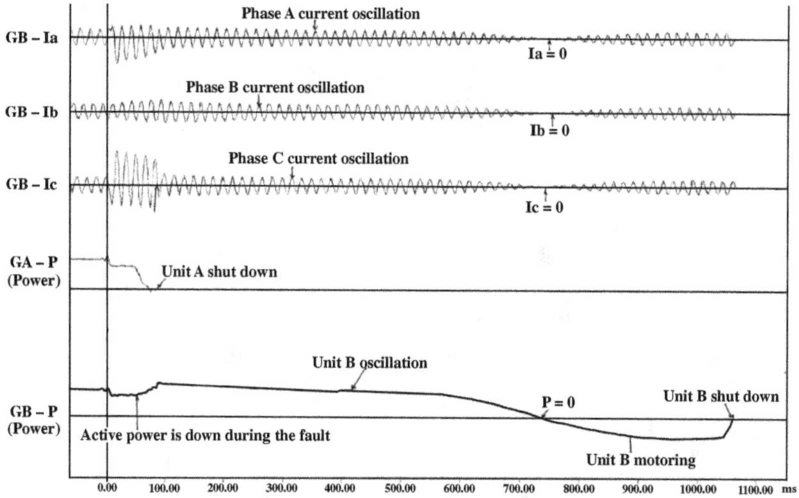

Fig. 3.9 Plant DFR record showing unit GA shutdown and GB oscillation with the system.

where θ is the unit rotor angle between the unit voltage V_1 and the system voltage V_2. In addition, the sudden deficiency in generation caused by the early loss of unit GA has further aggravated unit GB stable operations (unit rotor in step with the system) and may have contributed to its oscillation, as shown in Fig. 3.9. Generator three-phase currents also began to oscillate, reflecting unit GB oscillation with the power system.

Referring to trace GB-P in Fig. 3.9, the pre-disturbance unit GB output was affected by the fault and momentarily reduced its output level at the inception of the fault. The unit power output began to increase slowly during the last cycle of the fault duration, reaching a peak value upon fault clearance. The machine rotor angle with respect to the system also began to increase, so that the unit peak value output may be at a point where the unit rotor angle is near (or at) 90°, and at this point the unit may be at a point of no return. What also contributed to this out-of-step condition was an increase in the transfer impedance between the unit and the system by the simultaneous tripping of three 138-kV feeders at substation Y.

The unit power output level continues to decrease while the rotor angle continues to increase beyond 90°. The rotor angle reaches 180° and the unit output becomes zero ($P = 0$). The unit then enters the motoring region, and at this stage it absorbs MW from the system, which further accelerates the generator rotor. The unit remained in the motoring region for a total of 20 cycles before it was eventually tripped. It can be concluded that the unit GB rotor began to lose its magnetic bond with the system (i.e., out of step), which normally holds the rotor in-step with the stator, 5 to 6 cycles after $t = 0$, and began to increase its rotor angle beyond the stable point of 90°. Unit GB has lost control of its torque angle and entered an unstable condition. By definition the unit may have slipped poles. The oscillation of the unit with the system was stopped when unit GB was tripped mechanically by the turbine low oil header pressure system.

3.3 STABLE POWER SWING GENERATED DURING SUCCESSFUL SYNCHRONIZATION OF A 200-MW UNIT

When a generator is synchronized to the power system, its monitored voltage, current, active power, and reactive power will begin to oscillate. This generates an out-of-step condition that can lead to stable or unstable oscillation, depending on the synchronizing angle. The oscillatory frequency of the system during a power swing ranges from 0.25 to 3 Hz. The lower the frequency of oscillation, the more severe the impact on the stability of the power system will be.

Figure 3.10 shows a system one-line diagram and DFR-monitored currents, voltages, and calculated active and reactive powers during the synchronizing event. DFR trace G1-MW shown in Fig. 3.11 revealed that the hydro unit was connected to the power system at point a and starts to draw some power from the system. The unit begins to oscillate as shown at a frequency of oscillation of about 2.4 Hz. The oscillation can be declared as stable due to the fact that the peak current at point c is less than the peak current at point b, and the peak current at point d is less than peak current at point c. This unit synchronization is stable because the oscillation magnitude is reduced (does not grow) as time increases.

The stable power swing generated by successful synchronization of the 200-MW hydro unit therefore has no impact on relaying, since no CT stress is experienced when the synchronizing angle is not so excessive as to cause CT saturation by the presence of large asymmetrical currents. As a result, no excessive current differential CT mismatch is present that can generate undesired tripping of the generator-associated differential relay. Therefore, the stable damped power system oscillation during generator synchronization will have no further impact on power system differential protection elements.

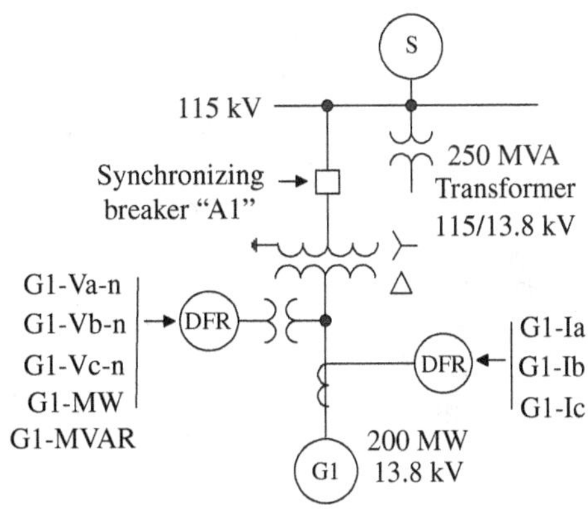

Fig. 3.10 System one-line diagram showing DFR-monitored voltages and currents.

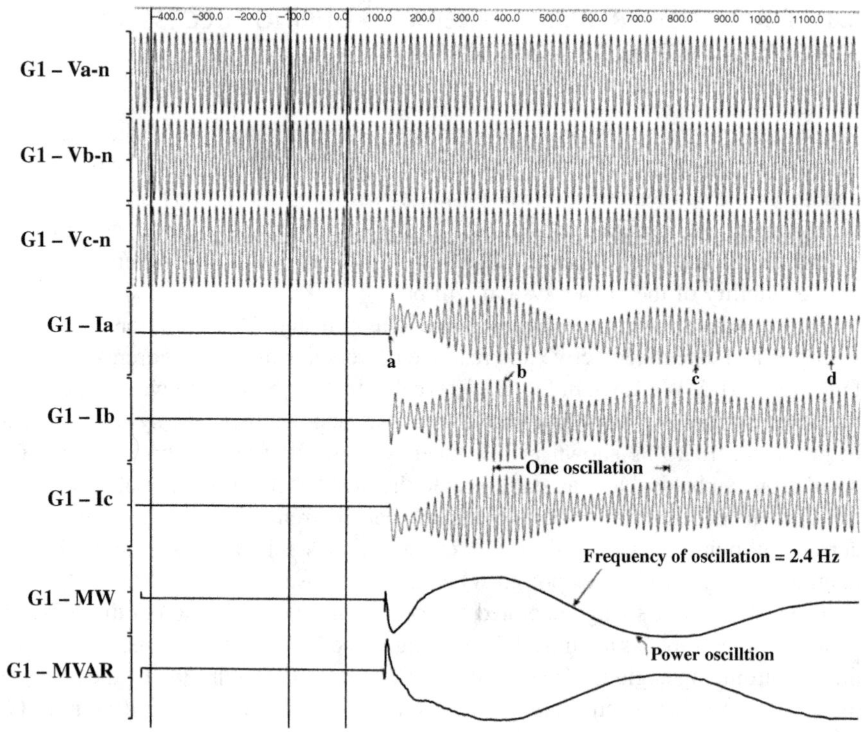

Fig. 3.11 Generator G1 DFR record showing unit stable oscillation with the system.

3.4 MAJOR SYSTEM DISTURBANCE LEADING TO DIFFERENT OSCILLATIONS FOR DIFFERENT TRANSMISSION LINES EMANATING FROM THE SAME SUBSTATION

During a major power system blackout, an inadvertent island was formed with hydro generation and loads. The total 2000-MW generation was connected to the power system shown in Fig. 3.12. The generation was a split connection to 115- and 230-kV systems. Figure 3.12 also shows the DFR-monitored currents and voltages for lines L1, L2, L3, and L4 and the active and reactive powers calculated for L4 during the disturbance. The DFR at substation X was triggered at the beginning of the oscillation process, with records generated for 230-kV lines L1, L2, L3, and L4.

Power system oscillation shape and frequency are a function of the system equivalents at each end of the line. The DFR figures for lines L1, L2, L3, and L4 reveal unique oscillations. The frequency of oscillation can be obtained from an analysis of DFR records as follows:

$$F_{\text{osc}} = \frac{1}{t} \qquad \text{hertz}$$

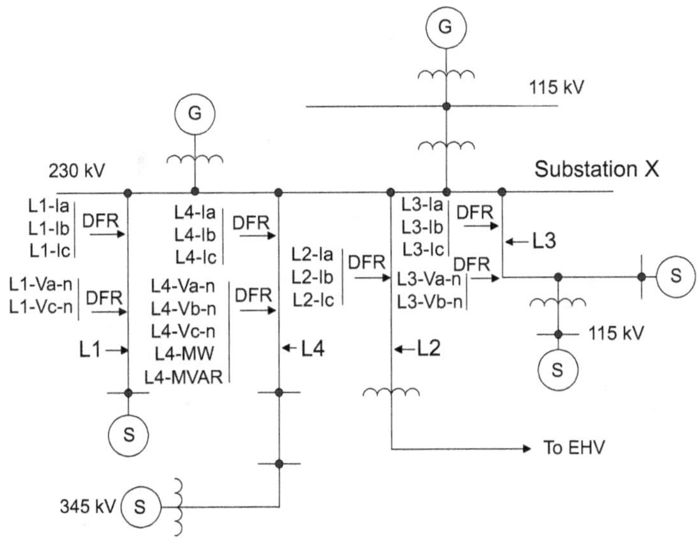

Fig. 3.12 System one-line diagram showing DFR-monitored voltages and currents.

where: F_{osc} is the swing oscillation frequency (hertz) and, t is (the number of cycles between current peak-to-peak or current minimum-to-minimum points) $\times$ 16.66 ms.

As shown in Fig. 3.13, line L1 exhibits an out-of-step condition with a frequency of oscillation of 1 Hz. The time for one oscillation is calculated as 2 $\times$ the time between the minimum and maximum points of the current trace L1-Ib. Figure 3.14 reveals the

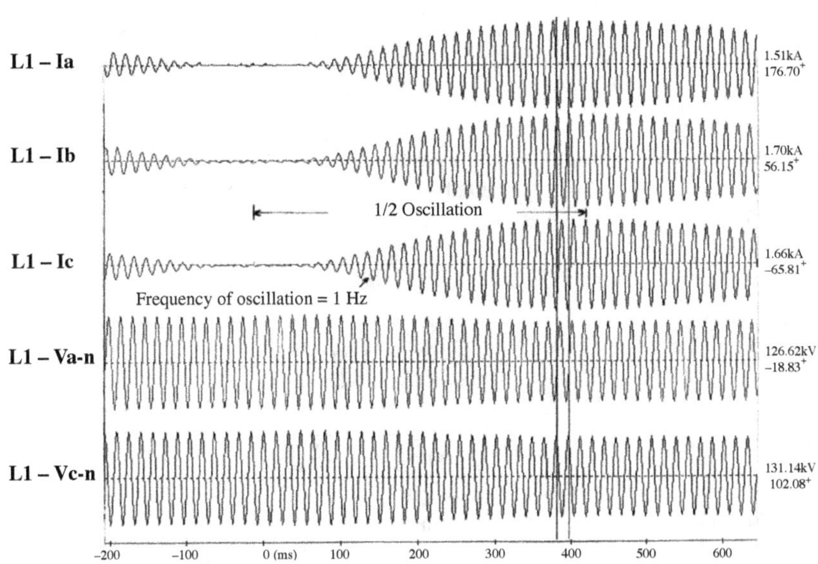

Fig. 3.13 Substation X DFR record showing line L1 voltage and current oscillations during power swing conditions.

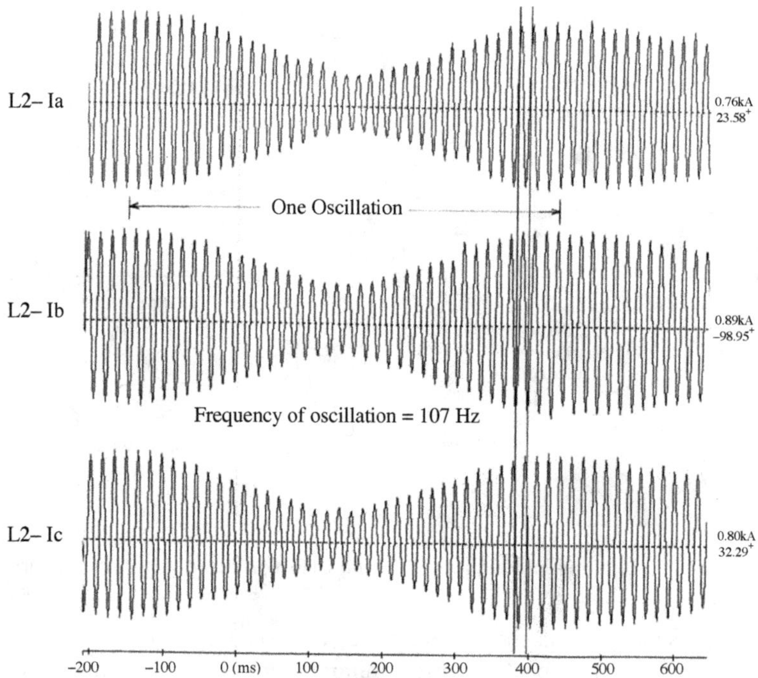

Fig. 3.14 Substation X DFR record showing line L2 voltage and current oscillations during power swing conditions.

fact that line L2 exhibits an out-of-step condition with a frequency of oscillation of 1.7 Hz, where the time for one oscillation is calculated as peak-to-peak for current trace L2-Ia. Figure 3.15 reveals the fact that line L3 exhibits an out-of-step condition with a frequency of oscillation of 2.7 Hz, where the time for one oscillation is calculated as minimum-to-minimum points for current trace L3-Ib. Figure 3.16 reveals the fact that line L4 exhibits an out-of-step condition with a frequency of oscillation of 0.75 Hz, where the time for one oscillation is calculated as peak-to-peak for the current trace L4-Ia. The same frequency of oscillation for L4 can be obtained from the MW or MVAR plots in Fig. 3.16.

3.5 APPEARANCE OF 120-Hz CURRENT AT A GENERATOR ROTOR DURING A HIGH-SIDE PHASE-TO-GROUND FAULT

A negative-sequence current is present during unbalanced system faults, open-phase conditions, and unbalanced loadings. Figure 3.17 shows the location of a ground fault that occurred on a 115-kV overhead structure that connects the unit transformer bushing and the 115-kV cable potheads. The unbalanced phase currents caused by the fault created negative-sequence current in the stators of generators G1 and G2. Negative-sequence current flow in the stator interacts with normal positive-sequence

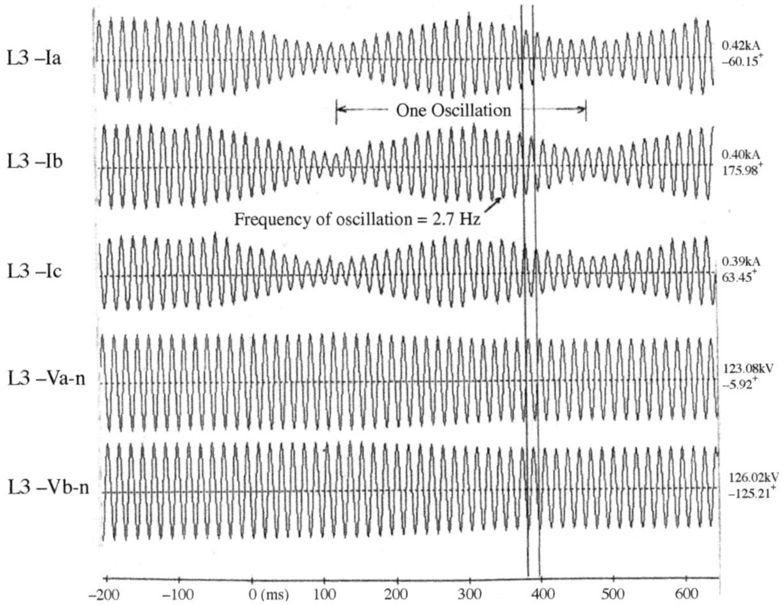

Fig. 3.15 Substation X DFR record showing line L3 voltage and current oscillations during power swing conditions.

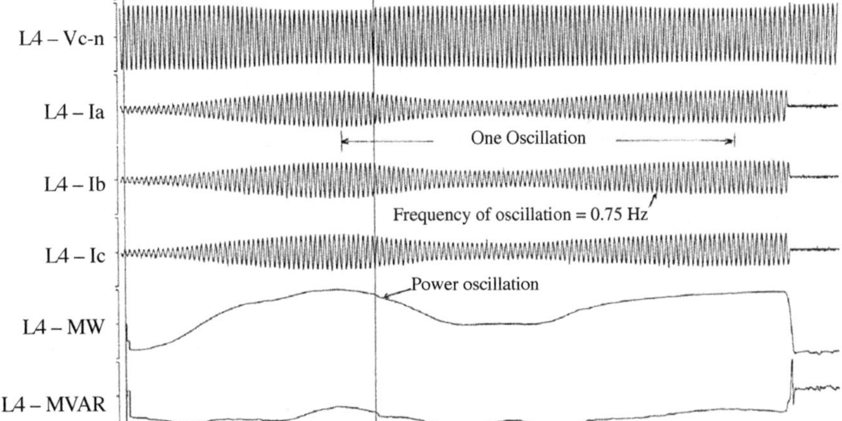

Fig. 3.16 Substation X DFR record showing line L4 voltages and currents and power oscillations during the power system out-of-step condition.

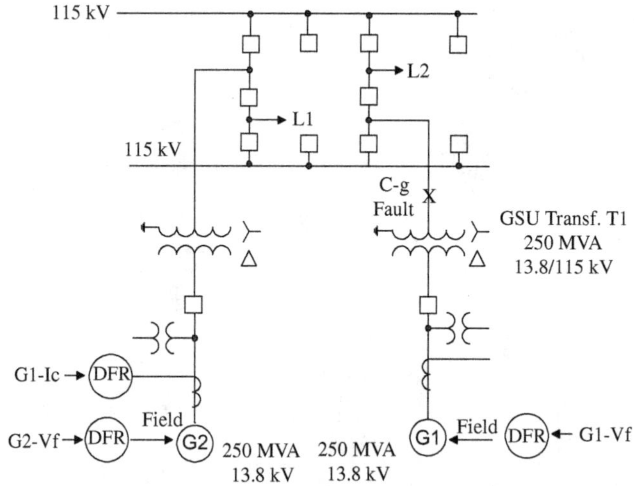

Fig. 3.17 System one-line diagram showing DFR-monitored voltages and currents.

current to induce a double-frequency current (120 Hz) in the rotor. Figure 3.17 shows the monitored DFR field voltages for G1 and G2 and phase C current for G2 during the 115-kV fault.

The generating unit G1 was shut down to clear the ground fault occurring on the 115-kV cable. This disturbance resulted in capturing an actual DFR record for the presence of double-frequency rotor current as a result of negative-sequence stator current flow. The C-g fault was cleared from the system in 3.5 cycles, and unit G1 decayed energy continued to feed the fault for several seconds.

By monitoring the generator rotor dc excitation voltage, the induced double frequency is captured and displayed. Figure 3.18 reveals the presence of double-frequency 12-Hz rotor-induced current as a result of the flow of the stator

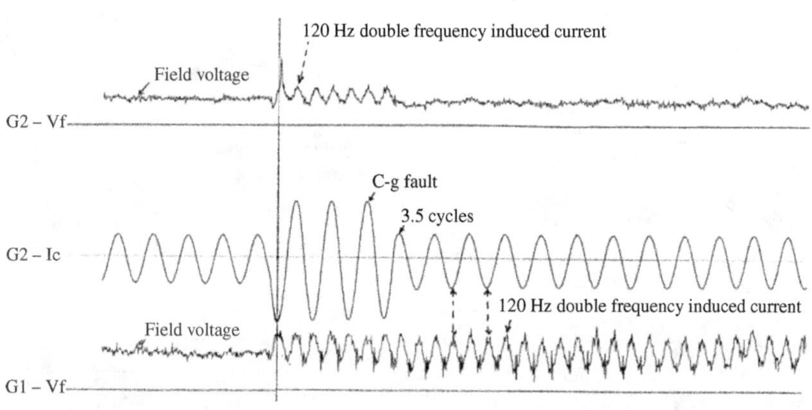

Fig. 3.18 Substation X DFR record showing generator field voltage with 120-Hz current induced in the rotor.

negative-sequence current that is normally present during system unbalanced faults. The 120 Hz is confirmed on the field voltage trace by the generation of 2 cycles of the rotor current for 1 cycle of 60-Hz stator current.

The negative-sequence current generated by system unbalance can cause current to flow in the retaining rings and rotor iron, which can generate harmful excessive heat if not detected. The 120-Hz current induced in the rotor will cause surface heating that can cause rotor damage if the negative-sequence current flow in the stator is not detected to trip the unit. Every generator is designed with a tolerance for negative-sequence current represented by a constant labeled as K equal to I_2^2t. Therefore, negative-sequence overcurrent element device 46 is employed for generator protection against unbalanced system faults and open-phase conditions. The elements are normally coordinated with the machine K constant.

3.6 GENERATOR NEGATIVE-SEQUENCE CURRENT FLOW DURING UNBALANCED FAULTS

During line-to-ground, phase-to-phase, and phase-to-phase-to-ground shunt unbalanced faults and open-phase series unbalanced conditions, generators contribute negative-sequence currents. Figure 3.19 shows the system one-line diagram and generator DFR-monitored currents and voltages during the B-g fault. Current trace G1-I2 in the DFR record in Fig. 3.20 illustrates the generator negative-sequence contribution to the 6.9-kV B-g fault. The DFR record was captured by the change in circuit breaker position, leading to missing the pre-fault duration for the DFR record. The initial system contribution shows the I_2 contribution for 7 cycles followed by 1 cycle for insertion of the 345-kV breaker opening resistor, and eventually, unit decayed energy continues to supply the fault. The negative-sequence current generated by system unbalance can cause current to flow on the rotor surface, which can cause harmful excessive heat if not detected.

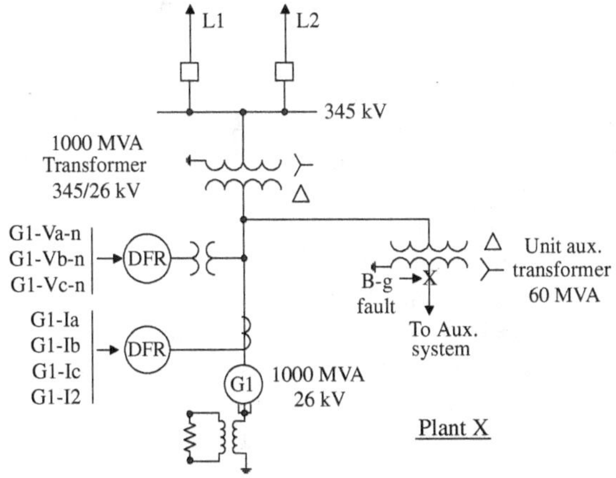

Fig. 3.19 System one-line diagram showing DFR-monitored voltages and currents.

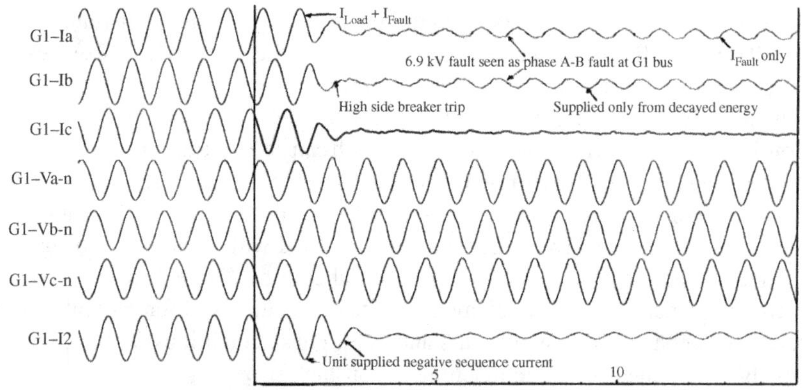

Fig. 3.20 Substation X DFR record showing G1 negative-sequence current during the fault.

3.7 INADVERTENT (ACCIDENTAL) ENERGIZATION OF A 170-MW HYDRO GENERATING UNIT

When a generator is inadvertently energized while off-line (at standstill), the generator stator current will cause high magnitudes of current flow in the machine rotor. This rotor current is initially at 60 Hz, but decreases in frequency as the rotor speed increases, due to induction motor action. Therefore, 60-Hz voltage induced in the rotor will confirm the following induction motor relationship:

$$\text{rotor frequency} = \text{slip} \times \text{stator frequency}$$

When the rotor is locked in place by hydraulic brakes, as is the case for hydro units, the rotor speed $= N = 0$. therefore, the

$$\text{slip} = \frac{(Ns - N)}{Ns} = \frac{Ns - 0}{Ns} = 1.0\,\text{pu} = 100\%$$

(where Ns is the synchronous speed), and hence 60-Hz voltage will be induced only in the rotor. Inadvertent energization of a generator connected to a strong system results in stator current in the range of three to four times the machine rating. The machine terminal voltage in this case can be about 50 to 70% of nominal system value.

The inadvertent energization phenomenon was displayed when a generator was accidentally energized by a three-phase 115-kV system while the rotor was locked in place using shaft brakes. In this situation, the generator behaves as an induction motor with a slip = 100% (rotor locked by applying brakes). During three-phase energization, with the machine at standstill, a rotating magnetic flux at a synchronous frequency (60 Hz) will be induced in the generator rotor. The resulting rotor current paths are similar to the negative-sequence rotor current paths during unbalanced system conditions. This will result in rapid rotor heating, which can quickly damage the rotor.

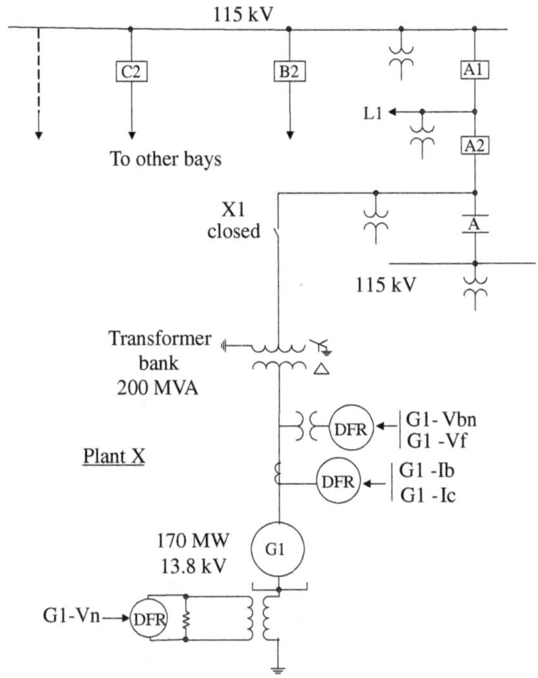

Fig. 3.21 One-line diagram showing unit DFR-monitored currents and voltages.

Figure 3.21 shows a system one-line diagram and the generator DFR-monitored currents and voltages during an incident. The inadvertent energization of unit G1 lasted for 11 cycles, as illustrated in Fig. 3.22 with high current magnitude for trace G1-Ib for phase B and trace G1-Ic for phase C and the presence of phase B voltage. The abnormal condition was detected by the unit loss of excitation protection. The unit field trace G1-Vf reveals induced 60-Hz voltage to the rotor at locked rotor

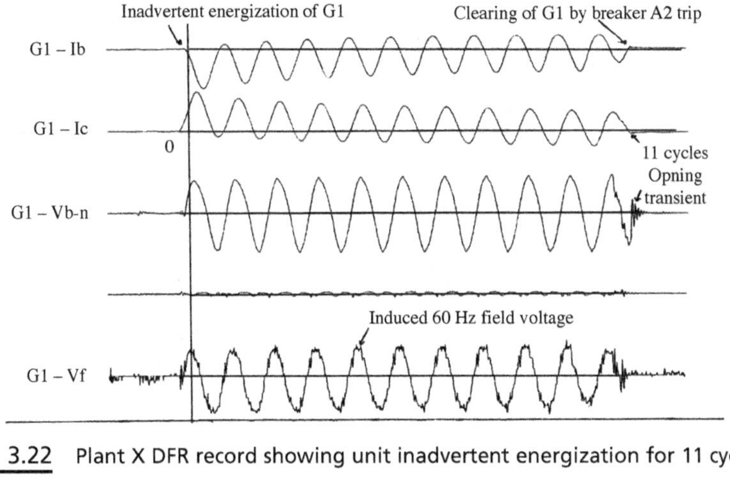

Fig. 3.22 Plant X DFR record showing unit inadvertent energization for 11 cycles.

conditions, forcing current flow in the rotor similar to induction motor action. The DFR record reveals an RMS value of 19.1 kA for phase B stator current and 18.2 kA for phase C stator current, which is about 2.7 times the unit rating. The generator phase B-n voltage was calculated by the DFR software at 5.02 kV, which is 62% of the machine-rated voltage. Dedicated generator protection elements may be used to protect the unit against inadvertent energization incidents.

3.8 APPEARANCE OF THIRD-HARMONIC VOLTAGE AT GENERATOR NEUTRAL

Presence of third-harmonic voltage at the generator neutral and its terminal can be considered as a sign of generator health, with no presence of stator ground fault near either the neutral or the terminal. The third-harmonic current will flow from the generator phases to ground via the capacitances of the stator, ISO-phase buses, surge capacitor, and transformer windings and back through the generator neutral. The third-harmonic voltage is produced by the nonlinearity within the generator. When the generator is healthy (no stator ground fault), the third-harmonic voltage will cause circulation of third-harmonic currents around the generator. This will result in third-harmonic voltage appearing across the generator neutral grounding resistor connected to the secondary winding of the distribution transformer between the neutral of the generator and ground. The value of the voltage will depend on many factors, as explained in Section 3.9.

Figure 3.23 shows a system one-line diagram and generator DFR-monitored currents and voltages. Figure 3.24 shows trace G1-Vn, representing the neutral of

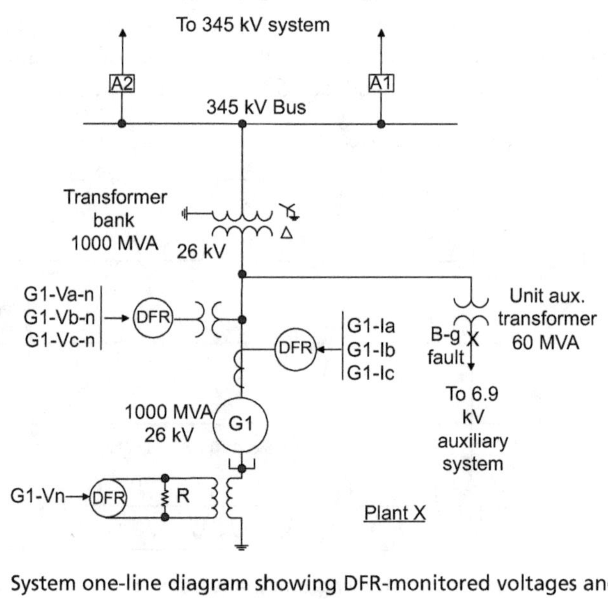

Fig. 3.23 System one-line diagram showing DFR-monitored voltages and currents.

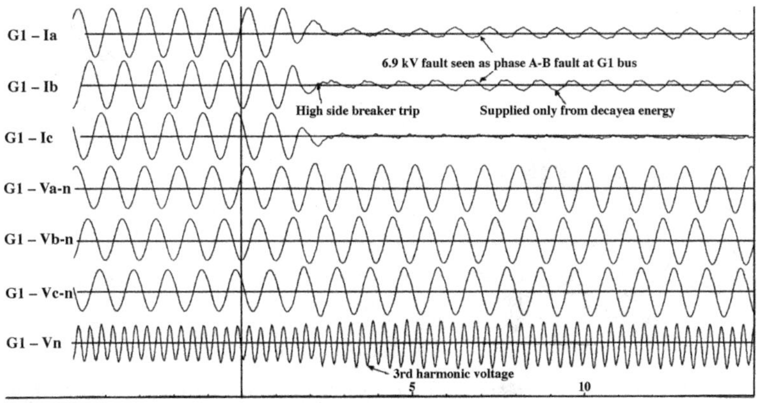

Fig. 3.24 Plant X DFR record showing the generator neutral third-harmonic voltage.

an 800-MW fossil unit during a phase-to-ground fault in the secondary winding of the unit auxiliary transformer. The DFR record also shows the third-harmonic voltage during the fault and after tripping of the high-side breakers. Generator trapped energy kept feeding the fault with an increase in the third-harmonic voltage level. Figure 3.25 shows a simplified one-line diagram for a steam unit with DFR-monitored voltages and currents. Figure 3.26 illustrates the generation of third-harmonic voltage at the neutral of the steam generating unit during the machine output loading condition. Points a and b represent 1 cycle for 60 Hz for phase C-n voltage and 3 full cycles for unit neutral voltage Vn, thus implying a third harmonic (180 Hz = 3 × 60 Hz).

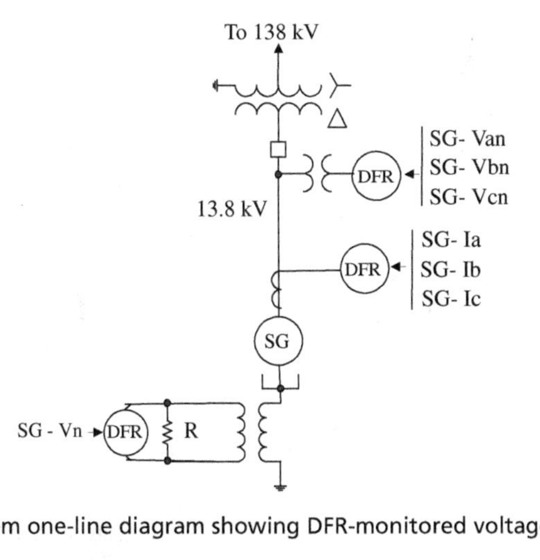

Fig. 3.25 System one-line diagram showing DFR-monitored voltages and currents.

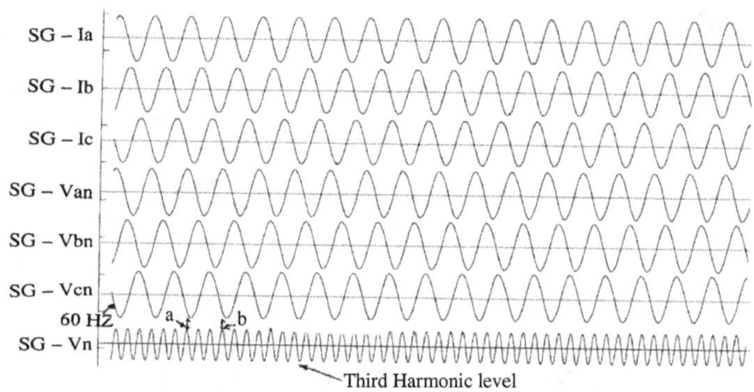

Fig. 3.26 Unit SG DFR record showing generator neutral third-harmonic voltage during normal operation.

3.9 VARIATIONS OF GENERATOR NEUTRAL THIRD-HARMONIC VOLTAGE MAGNITUDE DURING SYSTEM FAULTS

The level of third-harmonic voltage appearing at the neutral of a given generator depends on many factors. These factors include the generator design, the active (MW) and reactive (MVAR) generator power output, the relative values of the impedance to ground, and the capacitances of the stator winding, bus work, cabling, and transformer winding. Figure 3.27 shows a system one-line diagram and DFR-monitored generator neutral voltage and unit transformer neutral current during a fault. Figure 3.28 shows a plot of third-harmonic voltage during a high-side 138-kV ground fault that lasted for 4.5 cycles. The generator neutral third-harmonic voltage magnitude shown in Fig. 3.28 varies as a function of pre-fault rated machine loading during the line-to-ground high-side fault as a result of variations of unit active and

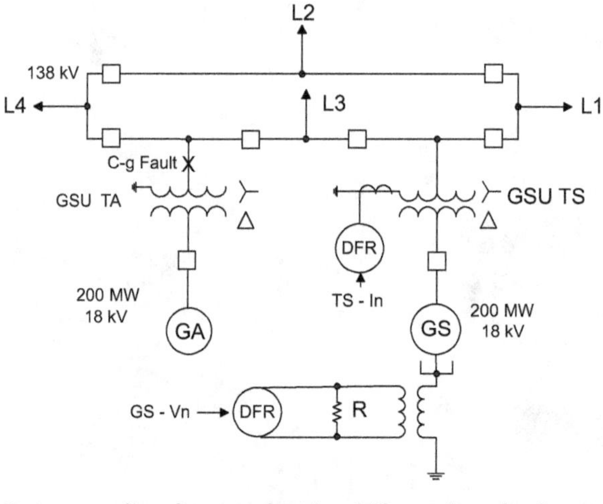

Fig. 3.27 System one-line diagram showing DFR-monitored voltage and current.

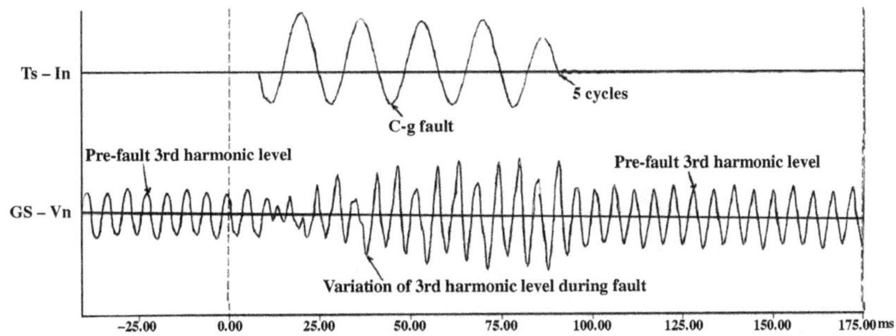

Fig. 3.28 Plant DFR record showing variation of the unit neutral third-harmonic voltage.

reactive power, and finally, post-fault, as no power is flowing following unit tripping (unit trapped energy).

Due to the varying nature of the generator third-harmonic voltage, an assessment must be performed for each machine to determine the minimum third-harmonic level generated as a function of the active and reactive power. This is required to safely apply 100% stator ground fault protection when employing the neutral third-harmonic undervoltage concept as an option.

3.10 GENERATOR ACTIVE AND REACTIVE POWER OUTPUTS DURING A GSU HIGH-SIDE L-g FAULT

The generator active power output will go down during system faults. The unit decreased power output during faults is confirmed by the system transfer formula, $P = [(V_1 \times V_2)\sin\theta]/X$, where θ is the angle between the two voltages V_1 and V_2 and X is the system transfer impedance. However, at the same time, machine reactive power output will increase due to the fact that the fault current will change its power factor angle from loading at about 30° to more than 60° during the fault and may lag the deriving voltage by an angle close to 90° for faults closer to the generating unit.

Figure 3.29 shows a simplified system one-line diagram and the generator DFR-monitored currents and active and reactive power calculated during the fault.

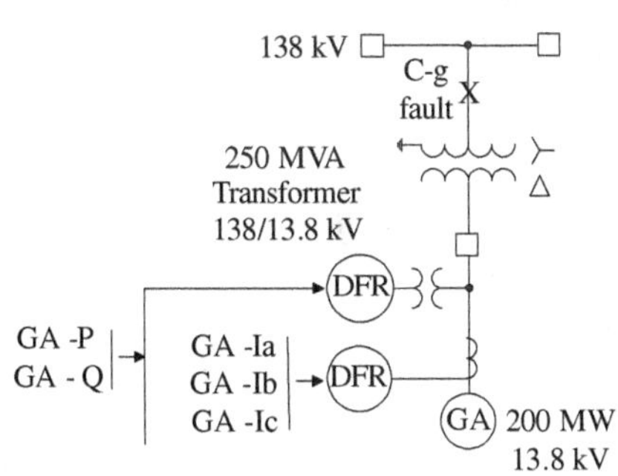

Fig. 3.29 System one-line diagram showing DFR-monitored voltages and generator power calculated.

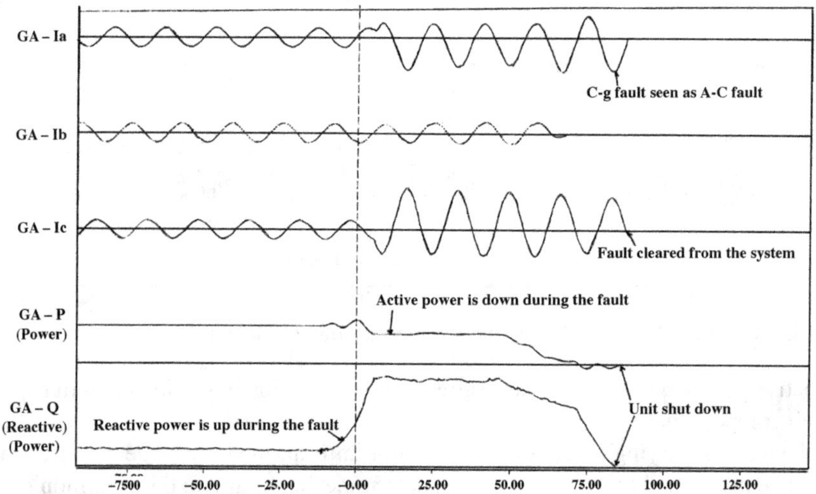

Fig. 3.30 Plant DFR record showing GA active and reactive powers and currents during the C-g fault.

Figure 3.30 shows generator active power output going down, due to decreased voltage during the fault, causing the output to go down instantly, and eventually, mechanical shutdown will remove the generator out of service as shown. Reactive power is increasing to supply the reactive power needed in the fault due to the current lagging the voltage by an angle close to 90°. The reactive power then eventually dies out when the fault is cleared as shown.

3.11 LOSS OF EXCITATION OF A 200-MW UNIT

Normally, generator excitation (field) is adjusted to provide the proper terminal voltage to supply reactive support (VAR) to a system. A loss of excitation of a generating unit is considered an abnormal system condition. During loss of excitation, the loss of a field relay will function to detect reactive power flow from the system into the generator. A loss of field will force the generator to run in the underexcited region and receive the needed reactive power from the system. This concept is illustrated by recording a loss of excitation incident and showing reactive power being absorbed by the generator at a value of −167 MVAR just prior to the trip of the unit.

Due to the loss of field, the underexcited condition normally increases the generator rotor temperature due to the induced eddy current in the field winding, rotor body, wedges, and retaining rings. The temperature increase can cause machine damage if not detected and mitigated. Therefore, this abnormal generator system operating condition must be detected to trip and isolate the unit.

The loss of field protection device 40 is normally applied using an offset mho relay, which is connected to the generator terminal and supplied with terminal voltages and currents. The relay measures the impedance as seen from the generator's

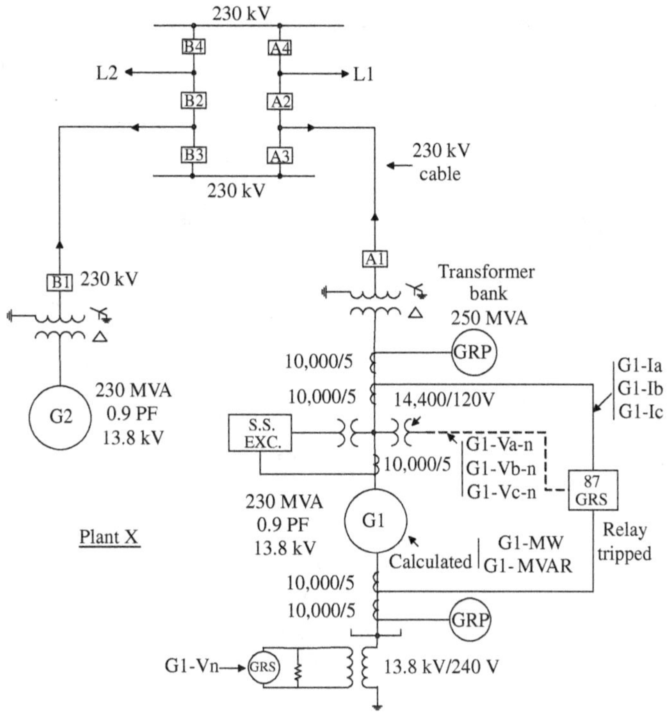

Fig. 3.31 One-line diagram showing generator numerical relay–monitored voltages and currents during the incident.

terminal and will operate to shut down the machine when impedance falls inside the characteristic of the relay mho circle.

A loss-of-excitation incident occurred when an operator reached and removed excitation from a generating unit while the machine was generating 190 MW. The loss of field began with the operator's action. The generator terminal voltage began to decay along with the real power level, and reactive power rose until the trip. The generator multifunction numerical relay 87GRS detected the loss of excitation in 0.5 s (as designed) and operated on field failure and immediately initiated the emergency lockout relay (86ES) to shut down the unit.

Figure 3.31 shows the system one-line diagram and generator numerical relay–monitored voltages and currents. Figure 3.32 illustrates the numerical relay–generated oscillography record during the loss of excitation incident. The relay oscillography record has less than 2 cycles prior to loss of the field event, where the excitation and the reactive machine power change sign, with 75 MVAR being absorbed instantly by the unit. The loss-of-excitation element operated to start a timer set for 30 cycles. Figure 3.32 shows a 30-cycle relay operating time to trip the unit, followed by a 3-cycle circuit breaker interrupting time. At the point of tripping, the unit was absorbing 167 MVAR from the connected power system.

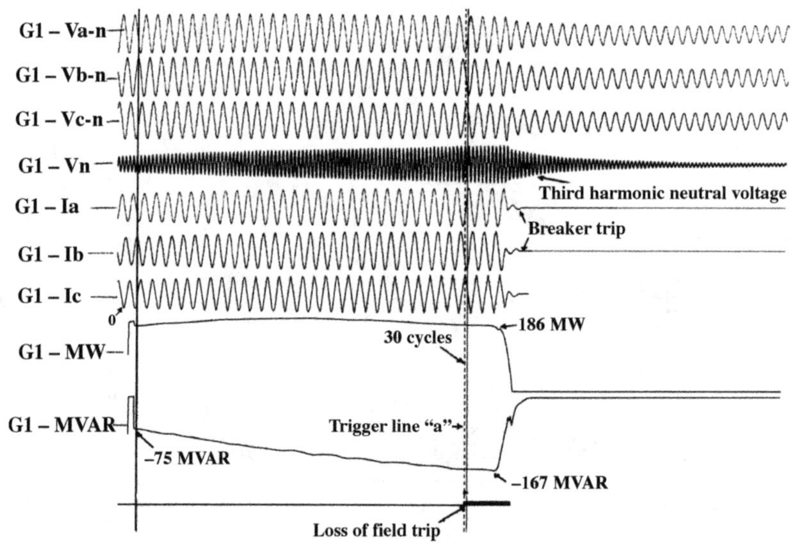

Fig. 3.32 Numerical relay oscillography record showing G1 currents and voltages and active and reactive powers calculated during the loss-of-excitation incident.

3.12 GENERATOR TRAPPED (DECAYED) ENERGY

Generator decayed energy is caused by the large stored kinetic energy of the machine following the removal of shaft power. The magnetic flux trapped in the field, coupled with the generator shaft mechanical inertia that is left, will cause the fault current to decay slowly (not abruptly) over time. This time could be several seconds after the generator is tripped. This decay of the generator energy could increase the duration of faults in the area between the interrupting device and the generator side, as shown in Fig. 3.34, and hence the ensuing damage will increase. Adding a generator breaker is therefore advantageous for removal of decayed energy fed to faults in this area. This will minimize the damage that can occur for faults between the generator breaker and the system, including a unit GSU transformer.

However, the machine continues to supply current into the fault until the generator magnetic flux has died out and voltage is no longer generated. Reduction of the generator magnetic flux is brought about by deenergization of the generator field, which is usually done in slowly through discharge resistances. It must be realized that, essentially, the rotor flux of the generator is maintained for seconds regardless of the reduction in the rotor field current because of the magnetizing action of rotor damper windings and other closed circuits. The generator decayed energy will have no impact on the generator relay system, since the relay system has been operating since early in the initial fault. However, this stored energy can result in more damage and ensuing fire, especially if the initial fault is between the generator and the system-interrupting device. This issue could be one justification for employing generator breakers.

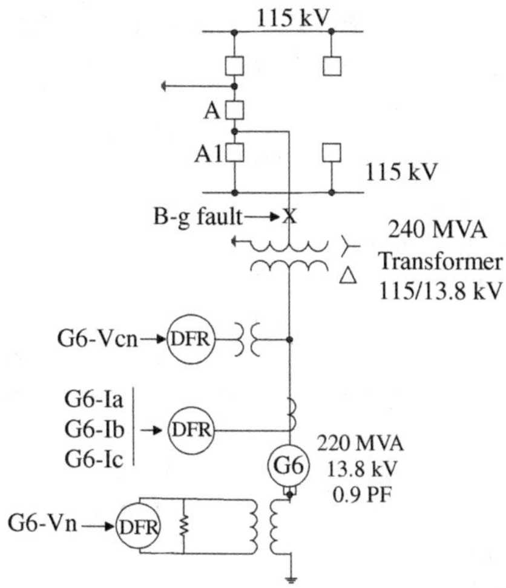

Fig. 3.33 System one-line diagram showing generator DFR-monitored currents and voltages.

Figure 3.33 shows a simplified system one-line diagram and the generator DFR-monitored currents and voltages during a B-g fault. Figure 3.34 shows a generator G6-Ia phase A current trace feeding the high-side B-g fault. Trace G6-Ib confirms generator separation from the system when the loading current is interrupted at point b. From the incident of fault initiation until its clearance from the system in 3.5 cycles, the generator phase A current contribution to the fault was lowered by the effect of load flow flowing opposite to the fault current component. As soon as the ground fault was removed from the system at point c, the generator decayed energy supported an increase in unit phase A current that lasted for several seconds. At this instant the generator phase A current is equal and opposite to phase C current. The B-g fault on the 115-kV high side of the GSU is seen as phase A-C on the delta 13.8-kV generator side.

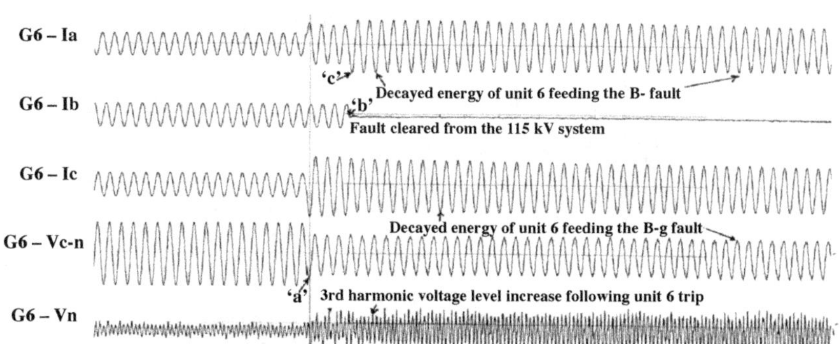

Fig. 3.34 Plant DFR record showing generator decayed energy feed to the initial B-g fault.

3.13 NONZERO CURRENT CROSSING DURING FAULTS AND MIS-SYNCHRONIZATION EVENTS

Nonzero crossing of current can be attributed to asymmetrical current containing dc offset. As a result, the circuit breaker has to wait for the current to reach the zero-crossing point for interruption to take place. This will delay the total clearing time until the current zero crossing is reached.

The steam unit of the combined-cycle plant X shown in Fig. 3.35 was in the process of being synchronized with the system following successful synchronization of the CT gas-fired unit. The automatic synchronization mode was selected, but due to a speed control problem the unit was not connected to the system. The unit was apparently oscillating and the synchronizer was unable to bring the unit within the allowable slip tolerance. The operator decided to place the unit in manual mode by changing the control switch position mode from "Auto" to "Off" to "Manual." During the mode change, the automatic synchronizer produced a false output that closed the unit breaker accidentally, thus energizing the generator at a wider angle with the system.

Figure 3.35 shows a combined-cycle plant one-line diagram and feeder L1 DFR-monitored currents during an incident. The DFR record in Fig. 3.36 reveals feeder L1 three-phase currents during the mis-synchronizing incident. The unit transformer differential relay reacted to CT mismatch caused by CT saturation due to the presence of severe dc offset. The phase C trace L1-Ic current was interrupted at zero crossing 8.5 cycles after incident initiation. Unlike phase C, phase A trace L1-Ia current and phase B trace L1-Ib current cleared in 11 and 14 cycles, respectively, due to the delay in reaching zero crossing. This non-zero-crossing phenomenon can delay the clearing process and can subject generating units to more stress, due to prolonging the duration of high current flow.

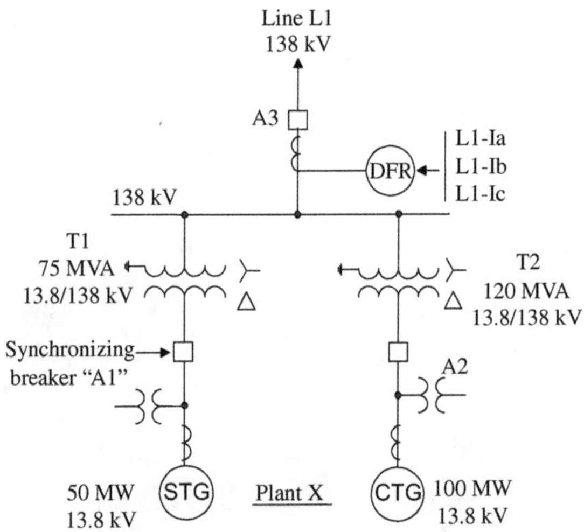

Fig. 3.35 System one-line diagram showing the feeder L1 DFR-monitored currents.

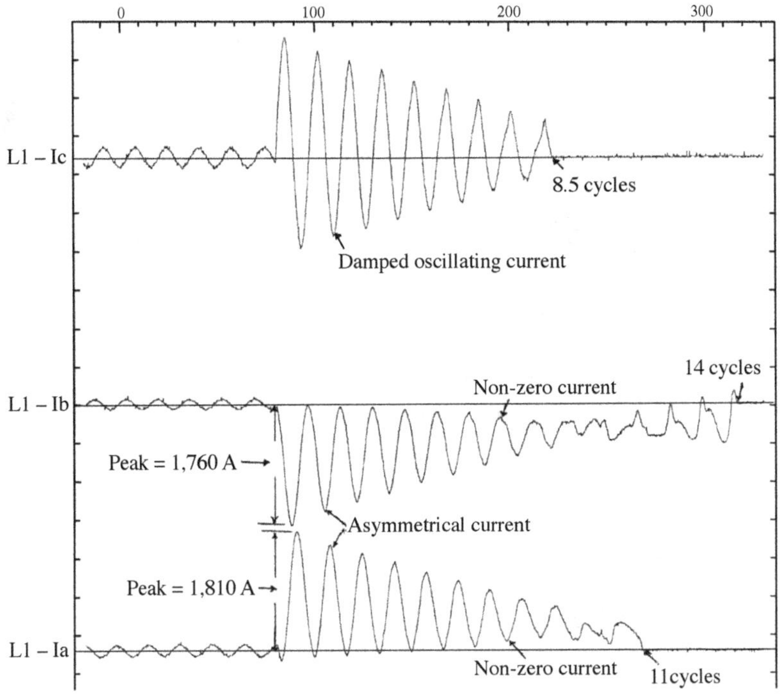

Fig. 3.36 DFR record showing asymmetrical and delayed zero crossing currents.

3.14 GENERATOR NEUTRAL ZERO-SEQUENCE VOLTAGE COUPLING THROUGH STEP-UP TRANSFORMER INTERWINDING CAPACITANCE DURING A HIGH-SIDE GROUND FAULT

Unit transformer capacitances between the primary and secondary windings permit coupling the high-side zero-sequence voltage, generated by ground faults, to the generator neutral. Figure 3.37 shows a one-line diagram for a unit generator connected to a 345-kV system with DFR-monitored voltages and currents. A phase B-g fault occurred on line L1, which is part of an EHV 345-kV system, generating the DFR record shown in Fig. 3.40, documenting the phenomena of coupling the zero sequence to the generator neutral.

The 345-kV B-g fault was cleared from the system in 3 cycles. Figure 3.38 shows a unit transformer connection with a lumped winding intercapacitance shown in only one phase. Figure 3.39 shows a single phase equivalent circuit that couples the high-side zero-sequence voltage to the generator neutral circuit. The high-side zero-sequence voltage is generated by the 345-kV phase B-g fault. The DFR record trace G1-Vn in Fig. 3.40 indicates that during the 3-cycle fault duration, 60 Hz (fundamental), which is coupled to the generator neutral by capacitive coupling through a transformer bank, is superimposed on the normal generator neutral third-harmonic voltage. After clearing

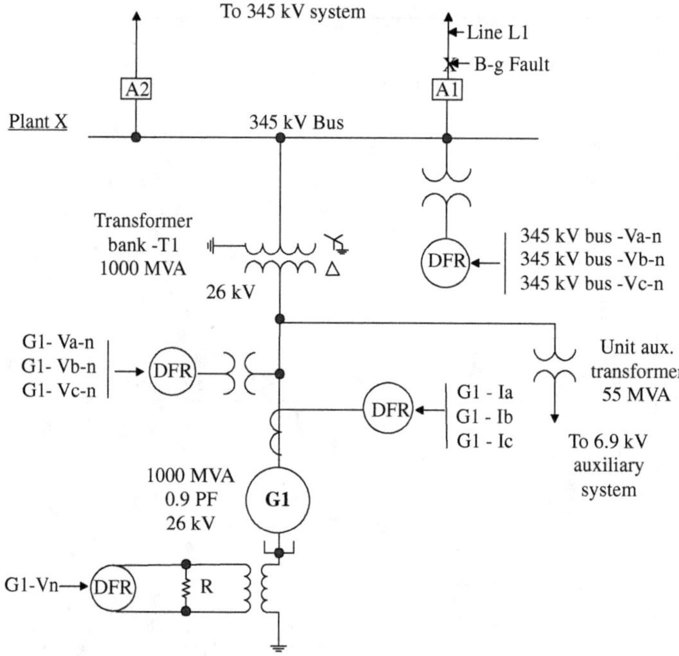

Fig. 3.37 System one-line diagram showing DFR-monitored voltages and currents.

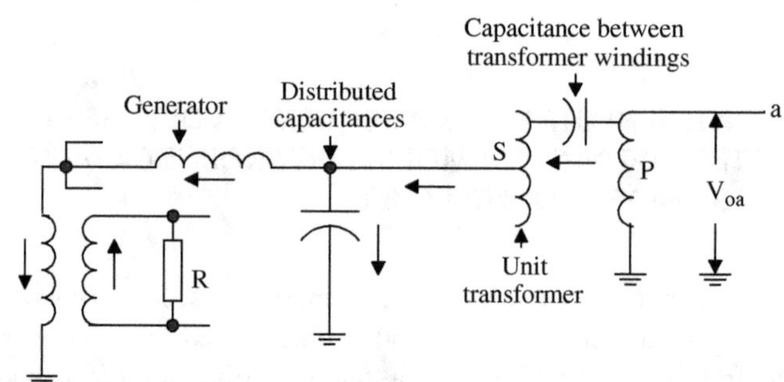

Fig. 3.38 Unit transformer connection with a lumped winding intercapacitance shown in only one phase.

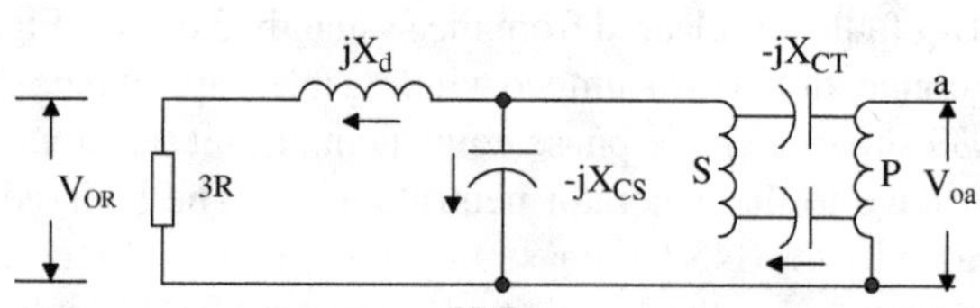

Fig. 3.39 Equivalent circuit representing unit transformer winding intercapacitances.

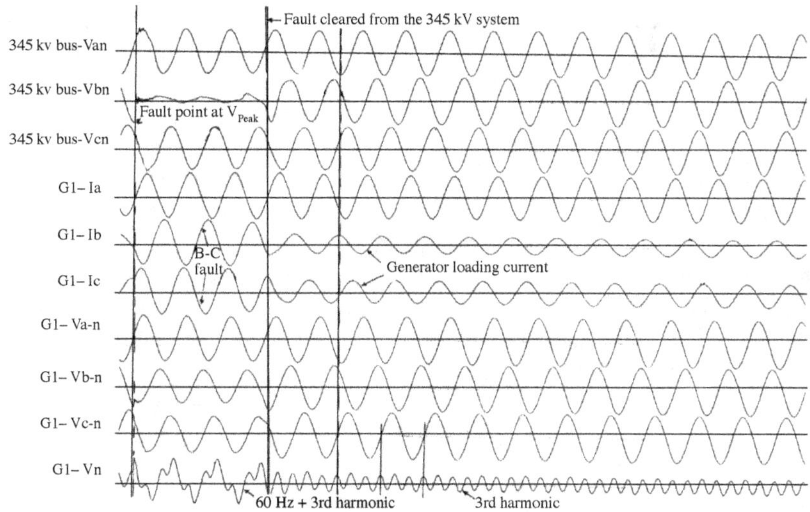

Fig. 3.40 DFR record showing zero-sequence voltage coupling via transformer winding capacitances.

of the fault, the same trace reveals the presence of third-harmonic voltage only (which is normally used as a sign of a healthy generator free of stator ground faults).

The 95% generator neutral tuned 60-Hz overvoltage relay 59G has to be set above this coupled voltage. Luckily, the transmission lines are normally protected using dual pilot protection systems, providing 100% line instantaneous coverage when breaker failure protection is provided. This results in fault clearance of system ground faults either in a primary time of 3 to 5 cycles or a breaker failure time of 10 to 16 cycles prior to operation of relay 59G, which is normally set at 0.5 s (30 cycles) or more, to coordinate with the fuses applied to the secondary side of the potential transformer connected to the generator terminal.

The 59N also must also be coordinated with the potential transformer (PT) secondary circuit fuses for ground faults on the low-voltage side for grounded-wye primary and secondary PT connection. If coordination cannot be accomplished, the PT secondary connection has to be changed to ground one of the secondary phases instead of grounding the neutral. This will convert ground faults to phase-to-phase faults and thus eliminate the generation of neutral zero-sequence voltage.

3.15 ENERGIZING A TRANSFORMER WITH A FAULT ON THE HIGH SIDE WITHIN THE DIFFERENTIAL ZONE

Energizing a faulty transformer or a transformer with a fault within its differential protection zone, where generation of harmonic restraint quantities during the inrush process may block operation of the differential element, has been an issue. This phenomenon was captured for a fault outside the transformer in the transformer

13.8-kV lead inside the differential zone. A 5-MVA station service transformer was energized into a phase-to-phase fault. Apparently, wind blown water intruded into an outdoor station service transformer enclosure and resulted initially into a phase-to-phase fault, which evolved to a three-phase fault. The transformer T2 differential relay instantaneous element and instantaneous backup element of the unit transformer T1 numerical differential relay, used to protect T2, operated to clear the fault. This incident documents an energization of a transformer into a fault within the differential relay protection zone. It reveals the generation of second-harmonic contents in the transformer inrush current which blocked tripping of the differential operating function of the numerical relay during the fault. It highlights the importance of an instantaneous (nonrestrained) differential function and time overcurrent independent backup element. It will also provide an analysis of the second-harmonic content of the transformer fault current. The fault probably occurred due to windblown water and moisture creating a humid environment, trapping humid air and moisture inside the 5-MVA 13.8/4.16-kV transformer enclosure. The failure occurred during heavy rain, with water trapped on a concrete slab. No water leaked from the top of the transformer enclosure.

Figure 3.41 shows a system one-line diagram and numerical transformer differential relay–monitored currents. The numerical relay–generated oscillography record is shown in Fig. 3.42, where a phase B-C fault occurred when 13.8-kV breaker A3 was closed to energize station service transformer T2. The fault evolved to a three-phase

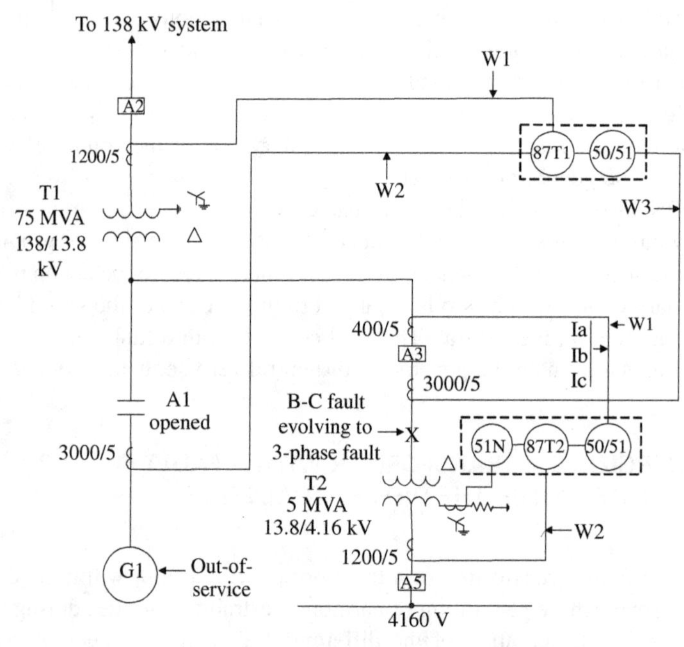

Fig. 3.41 One-line diagram showing numerical differential relay–monitored currents.

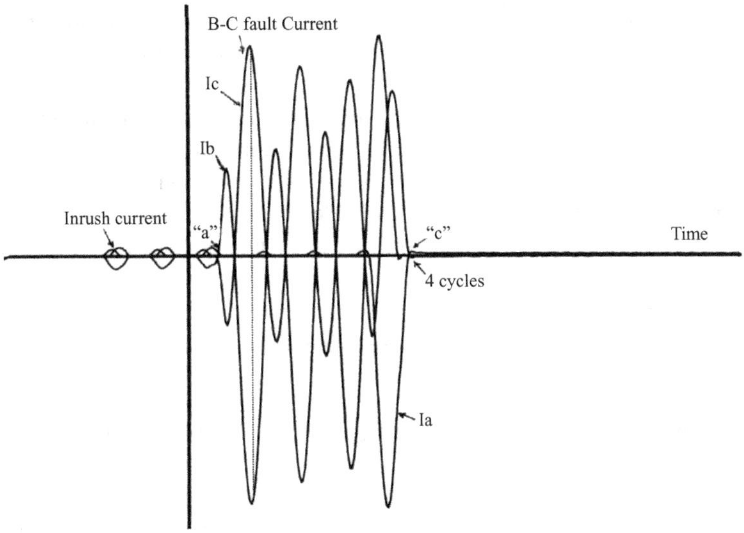

Fig. 3.42 T2 Numerical relay oscillograph record showing the initial inrush current followed by the B-C fault and the evolving three-phase fault.

fault when phase A got involved 3.25 cycles after initiation of the B-C fault. The T2 fault record reveals that the phase B current is about equal and opposite to phase C current. This will confirm the currents shown in the oscillograph record of Fig. 3.42, which is recorded by the relay for the interval between points a and c for the first three cycles of the B-C fault. The T1 fault record is generated by the instantaneous trip of the 13.8-kV input winding W3. The RMS value calculated for phase B of winding W3 is 6504 A, which exceeds the instantaneous setting threshold of 4440 A.

3.15.1 Analysis of the Harmonic Contents in the Fault Current

The T2 differential relay record reveals that at the beginning of the fault, phase A current has harmonic content. For a duration of 116 ms. the percentage second harmonic to fundamental (60 Hz) for phase A has a range of 90 to 43%. The percentage ratio range for phase B is 58 to 45% and for phase C is 91 to 31%. All values at the energization of T2 and during the fault are above the second-harmonic restraint toehold of 20%. As a result, the percent differential function of the T2 numerical relay was restrained from operation during the fault.

It is clear that energizing the transformer with a high-side fault has generated enough second harmonics to restrain the harmonic restraint percentage numerical differential relay element. However, energizing the transformer while moving the fault inside the transformer tank may generate a similar second harmonic/fundamental ratio that is either higher or lower than the high-side fault covered in this case. Therefore, energizing a transformer with an internal fault will remain

setting. This threshold is normally defined as a percentage ratio of the 60-Hz fundamental current.

3.16.3 Options Available for Second-Harmonic Detection Logic

Three options are available for the second-harmonic detection logic and their implementation depends on the specific relay manufacturer.

1. *Independent operation of each phase.* In most applications, similar to the use of individual per-phase electromechanical or static relays, inrush restraint logic operates independently for each phase. This option is also employed for numerical relay protection of EHV single-phase units or single-phase unit applications for GSU transformers, requiring optimization of the cost of spare units. In this case, four single-phase units (one spare) application will be preferred over the option of using two three-phase units (one spare three-phase unit). This option enhances the dependability aspect of the protection, allowing relay trips for a faulted phase during inrush conditions where the 60-Hz fault current will be much greater than that of the second harmonic.
2. *Cross blocking of tripping.* In this option one phase exceeding the threshold also issues a tripping block for the other two phases, enhancing the security of the relay.
3. *An averaging technique.* In this procedure a composite inrush detection threshold is calculated as $\frac{1}{3}$ (the sum of three individual thresholds). This single calculated composite threshold will be used to restrain all the phases, thus enhancing the security of the relay, especially for some of the new transformer designs where one phase may be very low on second harmonics during inrush energization. One of the advantages of numerical multifunction differential relaying is to provide a learning mode for the transformer inrush phenomenon prior to finalization of the relay design at the commissioning phase of the transformer.

3.17 INRUSH CURRENTS DURING ENERGIZATION OF THE GROUNDED-WYE SIDE OF A YG/DELTA TRANSFORMER

Figure 3.44 shows a system one-line diagram and DFR-monitored currents and voltages. The DFR record in Fig. 3.45 represents a transformer inrush current phenomenon during the energization of T1 from the high-side grounded-wye winding. The figure shows GSU transformer energization of the grounded-wye side, which is connected to the 115-kV system. The figure also shows neutral current trace T1-In, documenting current return to the system and unidirectional pulses for the currents shown in phases A and C.

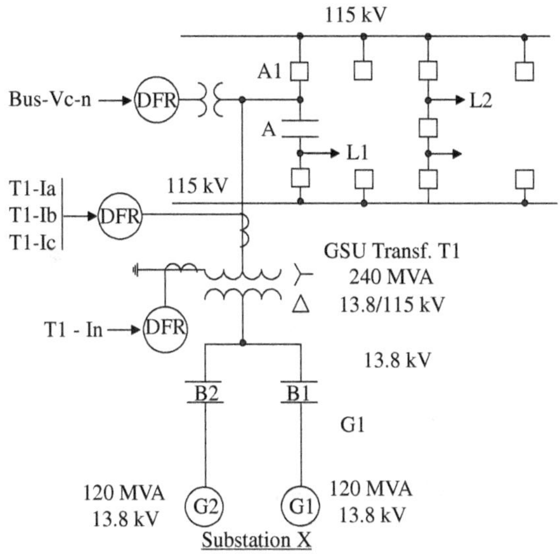

Fig. 3.44 One-line diagram showing DFR-monitored currents and voltages.

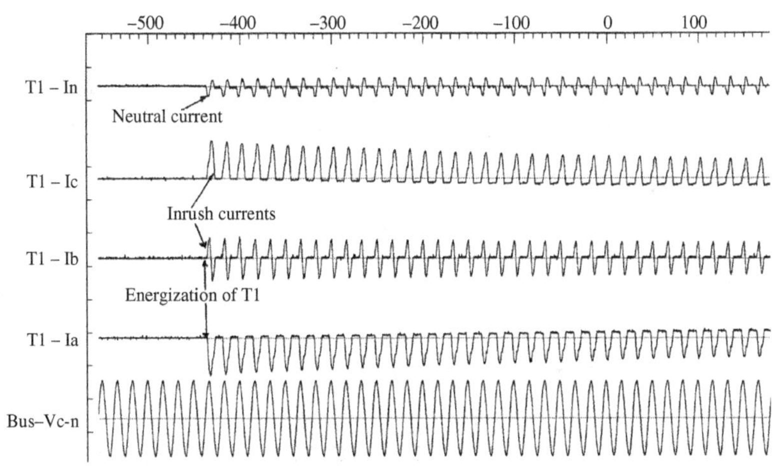

Fig. 3.45 Inrush currents at the grounded-side winding during energization of T1.

3.18 INRUSH CURRENTS DURING ENERGIZATION OF A TRANSFORMER DELTA SIDE

3.18.1 Energization of the Delta Side of a Delta/YG Transformer

In inrush for a delta/YG transformer, where the delta side has a ground relay connected to the CT neutral, the ground element will not sense any secondary current, because it

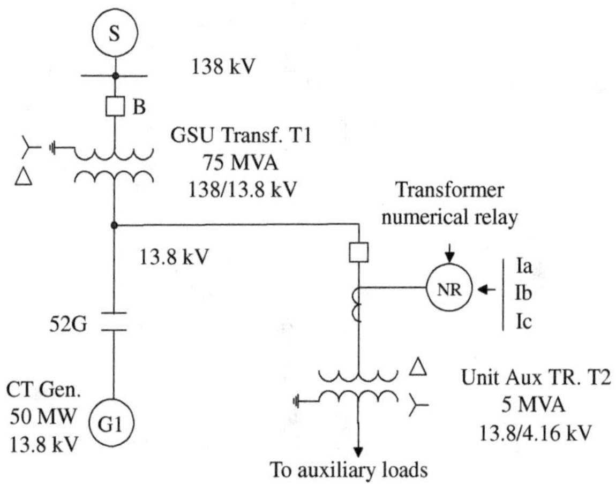

Fig. 3.46 One-line diagram showing numerical relay–monitored currents.

cannot support secondary current that flows in the high-side primary (delta-connected winding does not support a zero sequence; no connection to ground). Figure 3.46 shows a system one-line diagram and numerical relay–monitored currents during energization of T2. The DFR record in Fig. 3.47 represents a transformer inrush current phenomenon during energization of T2 from the high-side delta winding. The 4.16-kV currents display the inrush unidirection phase currents with no neutral current flow due to the delta winding connection with any return to ground, and hence the sum of the three-phase currents is equal to zero.

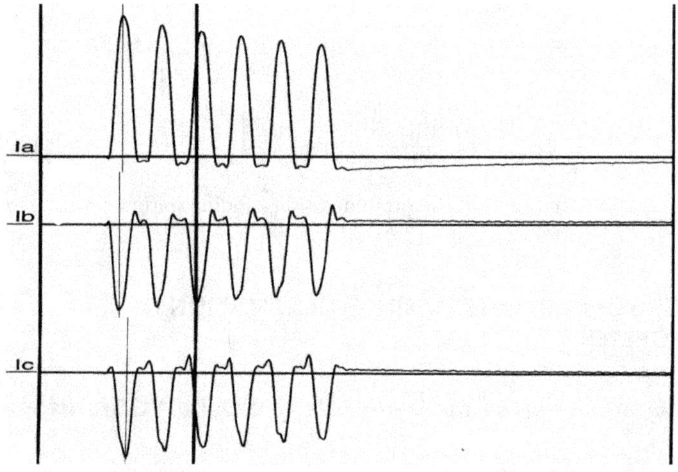

Fig. 3.47 Inrush currents at the delta-side winding during energization of T2.

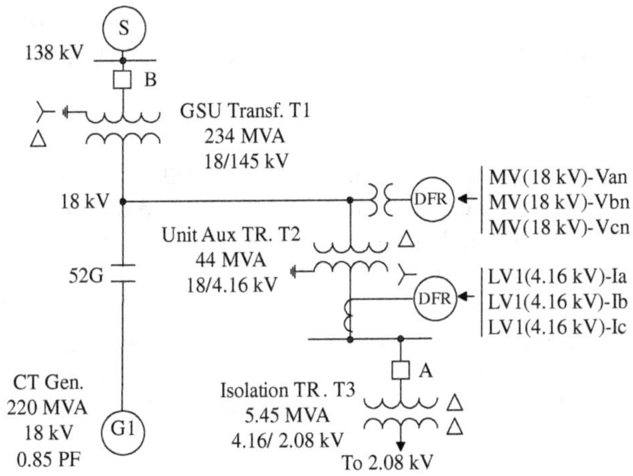

Fig. 3.48 One-line diagram displaying DFR-monitored currents and voltages.

3.18.2 Energization of the Delta Side of a Delta/Delta Transformer

Normally, this type of energization will not generate neutral current at the source delta side during the inrush period. Figure 3.48 illustrates a system one-line diagram and the T2 numerical relay–monitored 13.8-kV-side currents. The Energization of a T3 auxiliary transformer from the high-side 4.16-kV delta winding created inrush currents at the delta side with no neutral current present, similar to energization of the delta/YG transformer from the delta side. Figure 3.49 illustrates the unidirectional inrush current pulses for the 4.16-kV delta-side monitored currents. The sum of the three-phase currents is equal to zero, confirming no neutral current flow due to energization from the delta winding side.

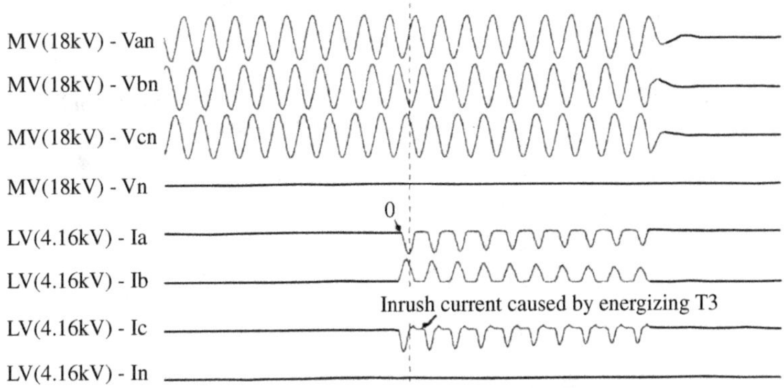

Fig. 3.49 Inrush currents at the delta-side winding during energization of T3.

3.19 TWO-PHASE ENERGIZATION OF AN AUTOTRANSFORMER WITH A DELTA WINDING TERTIARY DURING A SIMULTANEOUS L-g FAULT AND AN OPEN PHASE

Figure 3.50 illustrates a system one-line diagram as well as DFR-monitored voltages and currents. As shown, a close-in ground fault occurred on line L1 when a strain insulator failed, causing the B-phase conductor between the A-frame and the line disconnect switch to snap to ground. The fault was detected by both line relaying systems, which operated at the same speed to energize the CB A1 and A2 dual trip coil assemblies. Circuit breaker A1 opened successfully; however, CB A2 failed to open and activated its breaker failure relay to isolate the failed breaker by tripping 230-kV CB B1 and 115-kV CB A3. Circuit breaker B1 also failed to open and the initial B-g fault had to be cleared by remote backup relaying at substations Y and Z, thus isolating

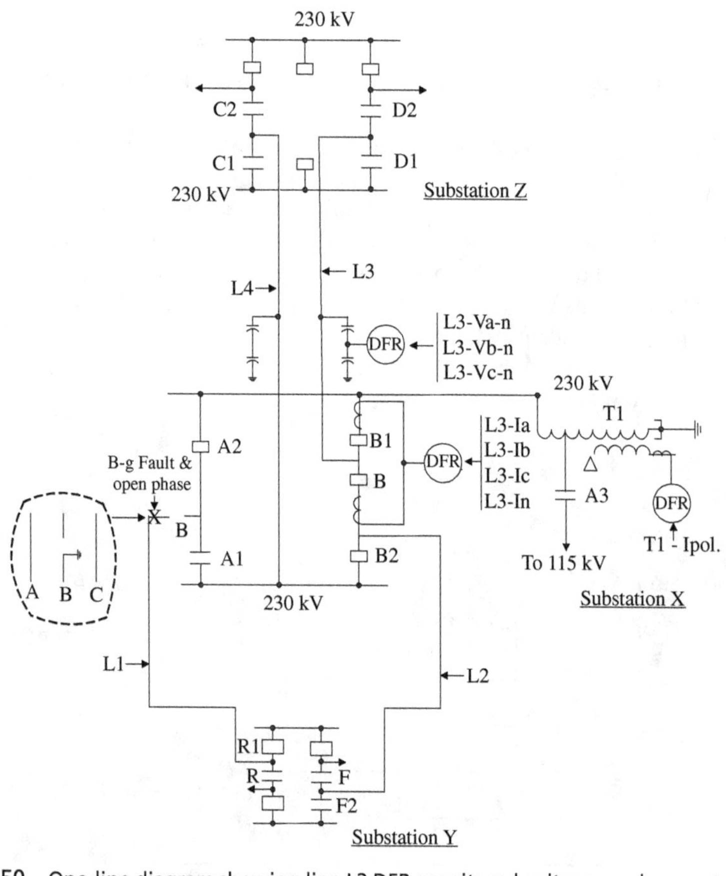

Fig. 3.50 One-line diagram showing line L3 DFR-monitored voltages and currents during automatic reclosing of line L1 from substation Y.

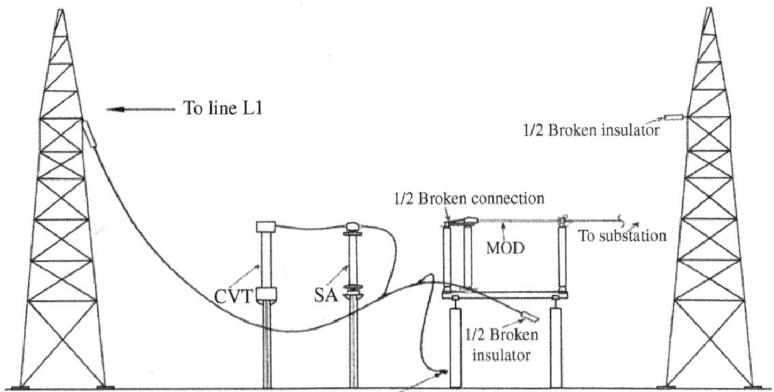

Fig. 3.51 Mechanism of simultaneous fault L-g toward line L1 and an open phase toward substation X.

substation X. CB R1 at Y reclosed automatically after 10 s into simultaneous faults, B-g toward the L1 line side and an open B phase toward the X substation. The automatic reclosure reenergized a permanent B-g fault toward substation Y and an open phase toward substation X. Figure 3.51 shows how the simultaneous fault mechanism was created. Apparently, the strain insulator failed, causing the conductor to snap to ground. The line conductor movement hit the connection to the line disconnect switch on its way down and broke it, creating the open phase toward substation X.

Phases A and C have energized autotransformer T1 at X and about 268 circuit miles of 230-kV lines between the Y and Z switchyards. Figure 3.52 shows the 230-kV line L3 voltages and currents as well as transformer T1 tertiary current during automatic reclosing by the line L1 end at substation Y. The voltages and currents on line L3 are much distorted with harmonic content. In addition, the reclosing of R1 has energized transformer T1 through phases A and C. The open-phase conditions on phase B forced zero-sequence current circulation between the transformer T1 neutral and the remainder of the connected system. Initially, the zero-sequence capacitance of the energized lines has helped to provide a path for the B phase current to flow through line and equipment capacitances to ground. As a result, sustained overvoltages occurred at X due to the energization of phase B line charging of three 230-kV circuits at X via only the delta winding of auto transformer T1. The severe overvoltages were restrained only by prolonged abnormal surge arrester operation at Y. The high voltage also caused saturation of the autotransformer, which was confirmed by alternative transient program (ATP) simulation. The high voltage on phase B at the open end of line L2 has apparently triggered phase B surge arrester operation at the substation Y end of the line.

The simultaneous faults were difficult to analyze, due to the unavailability of line L1 and L2 recorded voltages and currents. The analog quantities for the L1 and L2 circuits were inadvertently omitted in the design package. The missing L1 and L2 signals were obtained using ATP simulation of the simultaneous faults. Figure 3.53

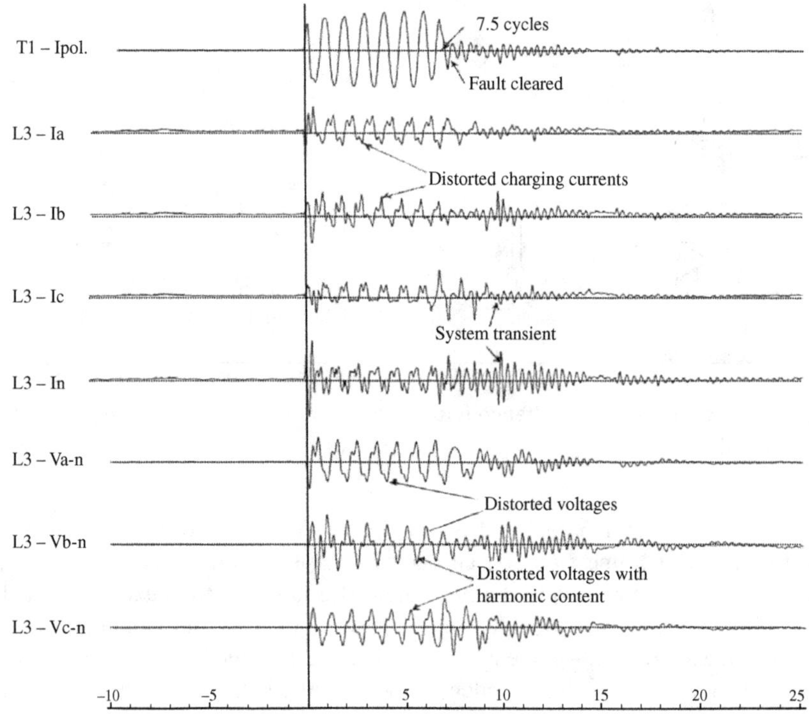

Fig. 3.52 DFR record for 230-kV line L3 during the simultaneous open phase and B-g fault.

shows ATP simulation output for line L2 phase B current. The simulation result confirmed that the energization of phase B line charging via only the autotransformer delta winding necessitated cyclic arrester discharge for the duration of the 7.5-cycle fault following the automatic reclosure. It also indicated that autotransformer residual fluxes caused cyclic saturation in phases A and B, contributing to further harmonic distortion of voltage and current waveforms.

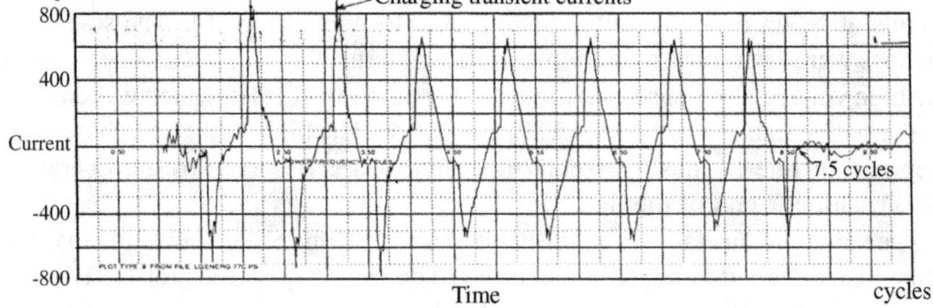

Fig. 3.53 Alternative transient program output for phase A current for 230-kV line L2.

3.20 PHASE SHIFT OF 30° ACROSS THE DELTA/WYE TRANSFORMER BANKS

Two winding power transformers are normally connected in accordance with NEMA standards, with the high-voltage side leading the low-voltage side by 30°. Figure 3.54 shows a system one-line diagram with the transformer high- and low-side DFR-monitored voltages. Figure 3.55 shows phase B-to-neutral voltage of the high side (138 kV) and phase B-to-neutral voltage on the low side (18 kV) of a GSU transformer. The time difference between the zero crossings of two waveforms can be

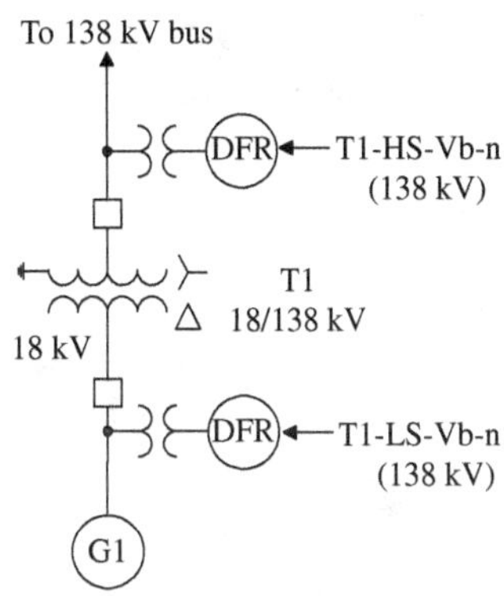

Fig. 3.54 DFR-monitored high- and low-side voltages.

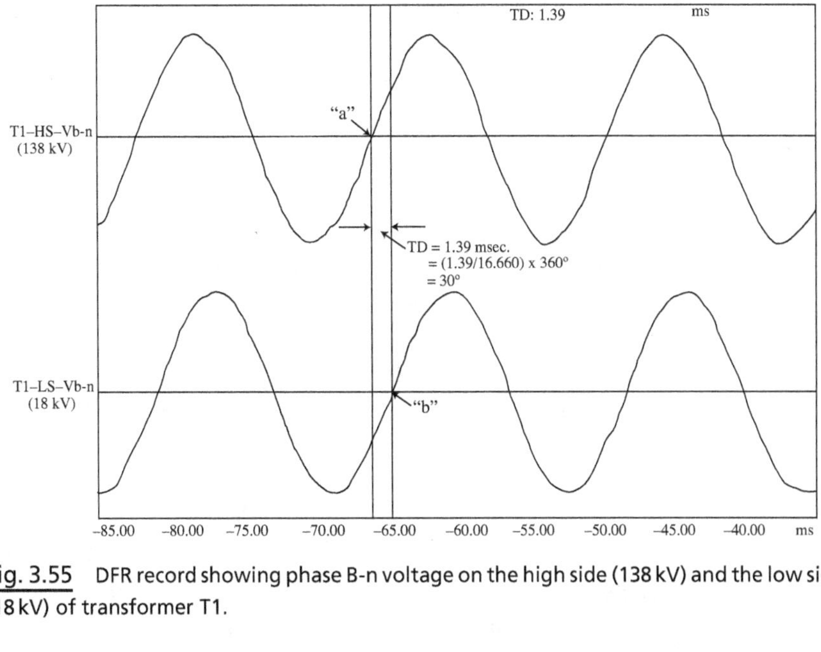

Fig. 3.55 DFR record showing phase B-n voltage on the high side (138 kV) and the low side (18 kV) of transformer T1.

converted to phase angle measurement if the waveforms are of a single frequency. Therefore, the time difference between the cursor a 138-kV-side voltage zero crossing and the cursor b generator-side voltage zero crossing of Fig. 3.54 is 1.39 ms (as stamped by the DFR software). This time can be translated into a phase angle in electrical degrees equal to $30° = (1.39/16.666) \times 360°$. Hence, the high side leads the low side by 30°, in accordance with the NEMA standards. According to an IEC standard, this transformer is defined as Yd1, where the high-voltage Y side leads the low-voltage delta side by 30° ($= 1 \times 360°/12$, where the face of the clock represents 360°).

3.21 ZERO-SEQUENCE CURRENT CONTRIBUTION FROM A REMOTE TWO-WINDING DELTA/YG TRANSFORMER

It is common to refer to the delta/grounded-wye or similar transformer banks as "ground sources" to ground faults on the system. In general, the ground fault is the source of the system unbalance. However, the transformer can be designated in this case as a ground source since ground ($3I_0$) current flows up the grounded neutral, through the system, and down the fault into ground. Figure 3.56 shows a system one-line diagram and line L1 DFR-monitored currents and voltages at substation X during the B-g fault on line L2. Figure 3.57 shows the contribution of zero-sequence currents from the delta/grounded-wye unit transformers at plant W. The DFR record shows that currents on phases A, B, and C are equal (about 658 A) and in-phase and thus represent zero-sequence components (I_0). The neutral current (I_n), which is the sum of the three

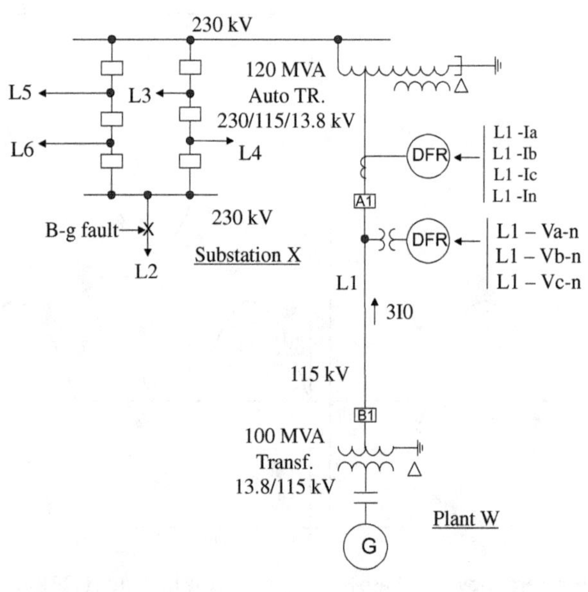

Fig. 3.56 One-line diagram showing the line L1 DFR-monitored currents and voltages.

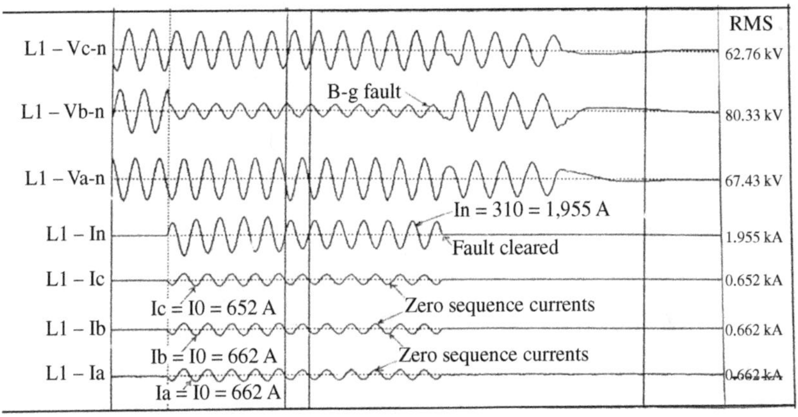

Fig. 3.57 DFR record showing the zero-sequence current flow during the ground fault.

phase currents, is shown to be about 1952 A and opposite in direction to the phase currents ($I_n = 3I_0$). This illustrates the flow of zero-sequence currents from ground sources, confirming the neutral point connection to be a zero-sequence current filter for ground fault detection.

3.22 CONVENTIONAL POWER-REGULATING TRANSFORMER CORE TYPE ACTING AS A ZERO-SEQUENCE SOURCE

Conventional phase angle–regulating transformers (PARs) have an exciting unit and a series unit. In a wye/wye three-phase three-legged core construction, the direction of flux induced by zero-sequence current is the same in all three legs. This results in a flux return path through air, resulting in a relatively low exciting impedance to zero-sequence current. The three-legged, three-phase core construction thus has the effect of providing a fictitious delta tertiary winding of relatively high impedance and allows the flow of zero-sequence current. Therefore, a conventional phase angle–regulating transformer with an exciting unit having a core design will act as a zero-sequence source for system phase-to-ground faults.

The zero-sequence equivalent circuit for a PAR is thus similar to a grounded-wye/delta/grounded-wye three-winding transformer. The one-line diagram in Fig. 3.58 shows the DFR-monitored 345-kV line L1 neutral current and zero-sequence voltage as well as the PAR exciting unit neutral current. The one-line diagram also shows the C-g fault occurring on the 345-kV cable feeder. Figure 3.59 shows the DFR record for line LI faulted phase voltage and neutral current as well as the PAR exciting primary neutral current. The DFR record reveals that the C-g fault lasted for 2.5 cycles. The trace PAR1-EXC-Inp confirms a zero-sequence ($3I_0$) current contribution for 2.5 cycles during the C-g fault. The flow of current on the neutral of the primary exciting unit during the C-g fault confirms that the core design of the exciting unit acts as a zero-sequence source during system ground faults on either the source or load side of the

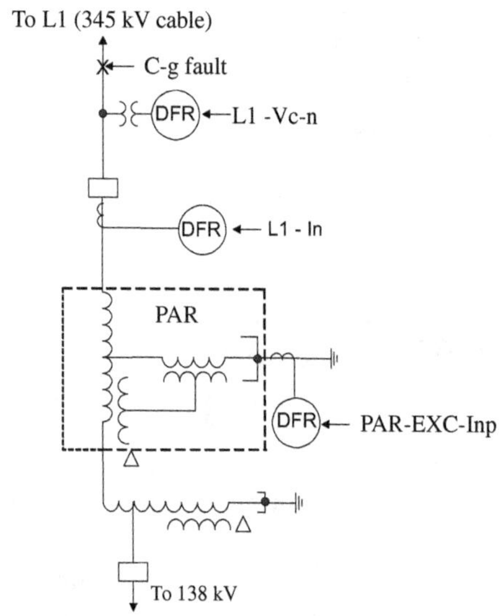

Fig. 3.58 One-line diagram showing PAR-monitored currents and voltages.

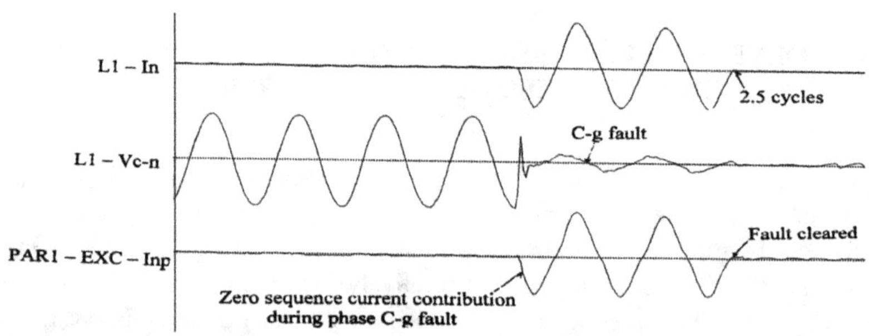

Fig. 3.59 Zero-sequence current flow in the PAR exciting unit neutral during the C-g fault.

PAR. The PAR zero-sequence impedance therefore needs to be modeled to calculate this current accurately when a neutral ground overcurrent is inserted in the neutral of the primary winding of the exciting unit to provide ground backup protection.

3.23 CIRCUIT BREAKER RE-STRIKES

A breaker re-strike phenomenon may occur when the recovery characteristic of the dielectric strength or the voltage-withstanding capability between the parting contacts might be lower than the rate of rising of voltage across the breaker. In other words, the rate of building the dielectric strength between the breaker parting contacts is slower

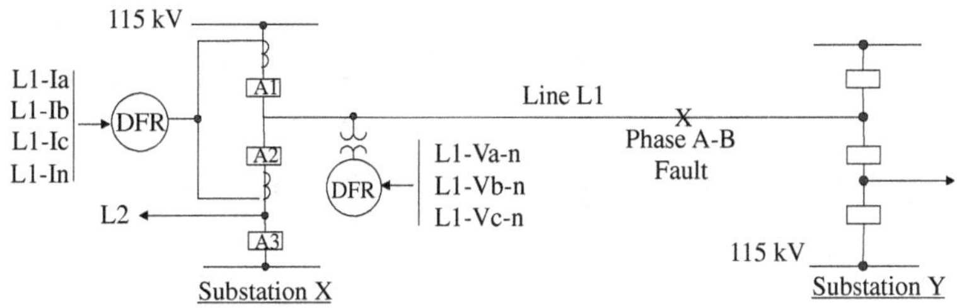

Fig. 3.60 Line L1 DFR-monitored voltages and currents.

than the buildup of the transient recovery voltage across the breaker. When a fault occurs on a system, a substantial fault current flows and the circuit breaker will attempt to interrupt at current zero. The parting of the circuit breaker contacts does not itself interrupt the current, because an arc will be established between the parting contacts, through which the current will continue to flow. Successful interruption depends on controlling and finally, on extinguishing the arc. Different types of circuit breakers use a variety of media to interrupt the fault current. If the rate of application of the recovery voltage should exceed the rate of buildup of dielectric strength in the medium between the contacts, the breaker will be unable to hold off the voltage and a reignition or re-strike will occur. *Re-strike* is defined as breaker conduction of a fault current half-cycle after successful interruption at current zero.

Figure 3.60 shows a simplified system one-line diagram and DFR-monitored voltages and currents for line L1. The DFR record shown in Fig. 3.61 reveals that the phase A-to-B fault was cleared after 3.5 cycles, as indicted by trace L1-Ia at point a and trace L1-Ib. Half a cycle later, at point b, the currents are reestablished for an additional half a cycle, and finally, the fault was cleared at point c. This is an

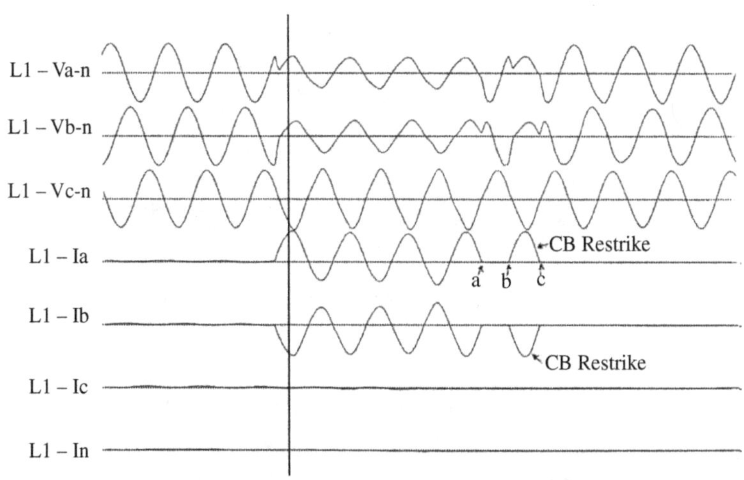

Fig. 3.61 DFR record showing breaker re-strike during clearing of the A-B fault.

indication of a breaker re-strike and requires evaluating the breaker for possible maintenance work.

3.24 CIRCUIT BREAKER POLE DISAGREEMENT DURING A CLOSING OPERATION

Figure 3.62 shows a simplified system one-line diagram and DFR-monitored voltages and currents for line L1. In addition, the one line of Fig. 3.62 represents automatic reclosing of CB A1 onto a phase-to-ground fault. The current traces in the DFR record in Fig. 3.63 show that the A-g fault was cleared in 4.5 cycles. The voltage traces in the DFR record in Fig. 3.63 indicate that phases B and C closed within 1 ms of each other. However, phase A voltage trace L1-Va-n shows that CB pole A closed 19 ms later, confirming the CB A1 pole disagreement condition. Such a condition requires

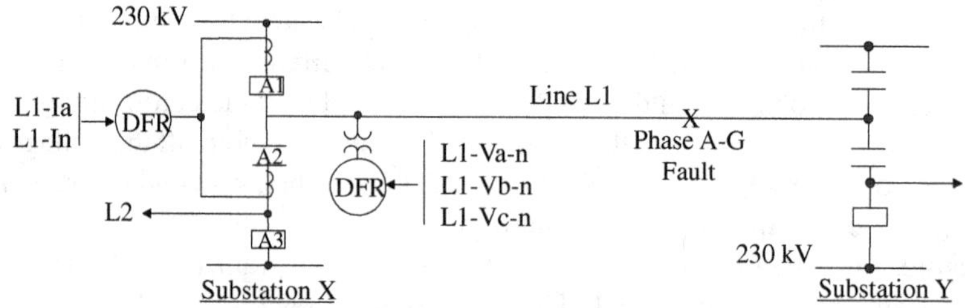

Fig. 3.62 Line L1 DFR-monitored voltages and currents during the phase A-g fault.

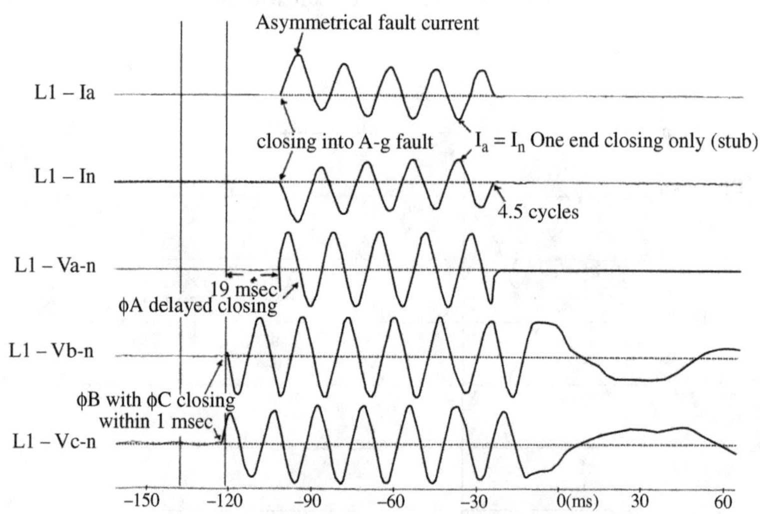

Fig. 3.63 DFR record showing CB pole disagreement during automatic reclosing onto the A-g fault.

investigation and, possibly, maintenance to correct a breaker A1 problem. The correct automatic three-phase closing is for phase A to close at exactly the same time, as exhibited by phases B and C.

3.25 CIRCUIT BREAKER OPENING RESISTORS

In certain types of circuit breakers the main contacts are shunted by resistors, as shown in Fig. 3.64. The resistor can help circuit breakers with multiple break contacts to distribute the transient recovery voltage more uniformly across the several breaking contacts. In addition, opening resistors can reduce the severity of the transient recovery voltage at the time of interruption by introducing resistor damping. Figure 3.64 shows a typical resistance switching circuit where resistor R is used to modify the recovery transient. During fault interruption, the breaker main contact A will open, interrupting the fault current and transferring a residual current to flow through the resistor. Subsequently, this current must be interrupted by opening auxiliary interrupter B.

Figure 3.65 shows a simplified system one-line diagram and DFR-monitored voltages and currents for line L1. The DFR record in Fig. 3.66 shows the result of inserting an opening resistor. It reveals a phase A-B fault evolving to A-B-g after 4.5 cycles, and the entire fault lasting for 6 cycles through the main interrupter path. Traces L1-Ia and L1-Ib indicate that the fault current was then transferred to the

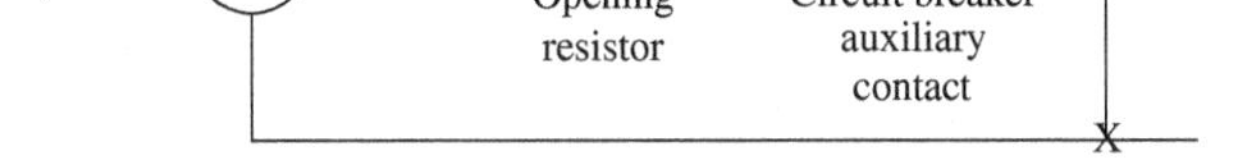

Fig. 3.64 Circuit breaker with an opening resistor during fault clearing.

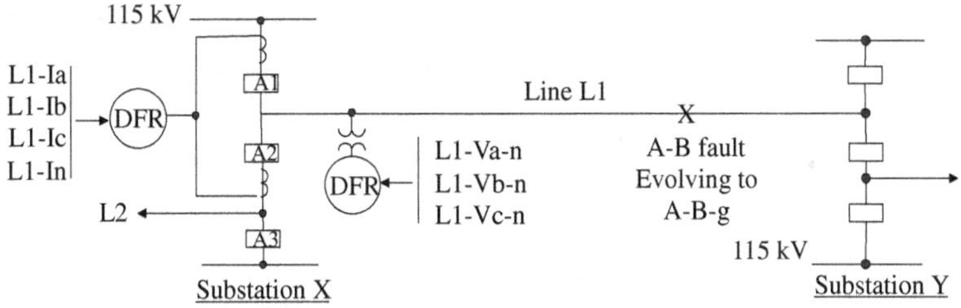

Fig. 3.65 Line L1 DFR-monitored voltages and currents.

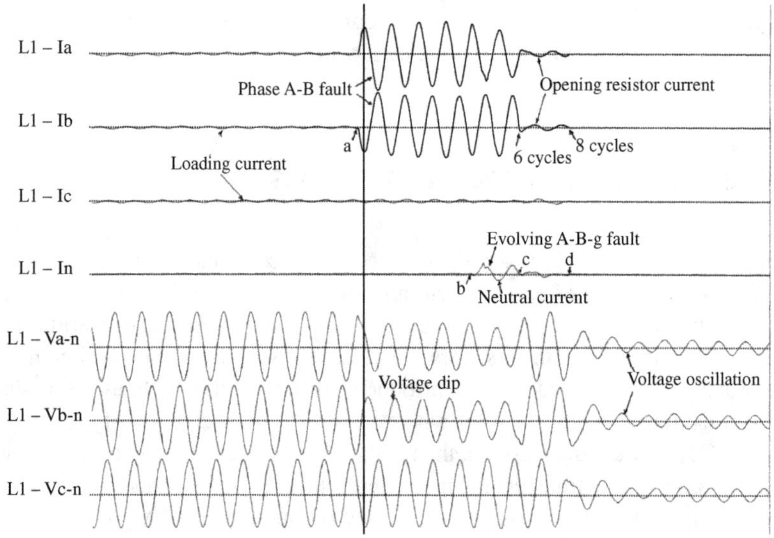

Fig. 3.66 Substation Y DFR showing the line L1 opening resistor currents.

opening resistor branch after 6 cycles at current zero. The current, reduced by the insertion of the opening resistor, was interrupted after an additional 2 cycles by opening of the auxiliary breaker contacts. The fault was therefore cleared from the system in 8 cycles.

3.26 SECONDARY CURRENT BACKFEEDING TO BREAKER FAILURE FAULT DETECTORS

The breaker failure backfeeding phenomenon can occur for systems having a ring or breaker-and-a-half bus configuration. The phenomenon can result in the undesired activation of breaker failure relaying without the flow of currents in the primary circuits. Breaker failure fault detectors are energized by secondary current flow due to voltage impressed across the CT associated with the opened breaker. Figure 3.67 shows line L1 DFR-monitored current and voltage during the C-g fault. The phenomenon is illustrated using an actual example.

As indicated in the DFR record shown in Fig. 3.68, the initial line L1 phase C-g fault was cleared from line L1 at substation X in 5 cycles by the opening of CBs B1 and B2. Four cycles later, CB B1 failed, causing a C-g bus fault as shown. No primary current was flowing through CB B2 when the CB B1 C phase failed to ground. However, CB B2 breaker failure relaying was activated by energization of CB failure fault detectors and initiation by the bus 2 differential lockout relay. Therefore, breaker failure relay operation for CB B2 was due to backfeeding to its breaker failure current detector, resulting in secondary current flow that exceeded the relay ground element pickup of 2 A.

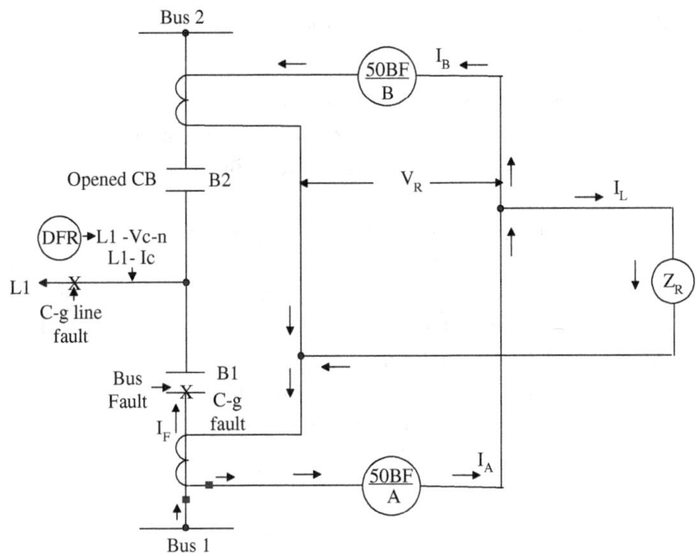

Fig. 3.67 One-line diagram illustrating the breaker failure backfeeding phenomenon.

Figure 3.67 shows a source of difficulty for breaker failure relaying in ring and breaker-and-a-half bus systems. CBs B1 and B2 have both opened successfully and cleared the initial fault. However, when CB B1 faulted to ground, causing a second fault, all fault current I_A was flowing from the breaker's bus-side CT to the line relaying. The flow of current I_L through line relay burden Z_R will apply voltage V_R across the CB B2 current transformer. The voltage V_R apparently had approached the ANSI relaying accuracy class voltage for the current transformer connected tap of 800: 5. In addition, the initial fault current contained a large dc offset, as shown in Fig. 3.68, and the secondary line current revealed that saturation occurred on the CT associated with either CB B2 or B1 or both. As a result, a remanent magnetic flux could have been left. This residual flux, which degrades the quality of the CT, could have contributed to the excessive current flow I_B through the fault detectors. This will force the CT associated with CB B2 to draw excessive magnetizing current flow I_B, enough to pick up the fault detectors and cause the breaker failure relaying scheme to operate.

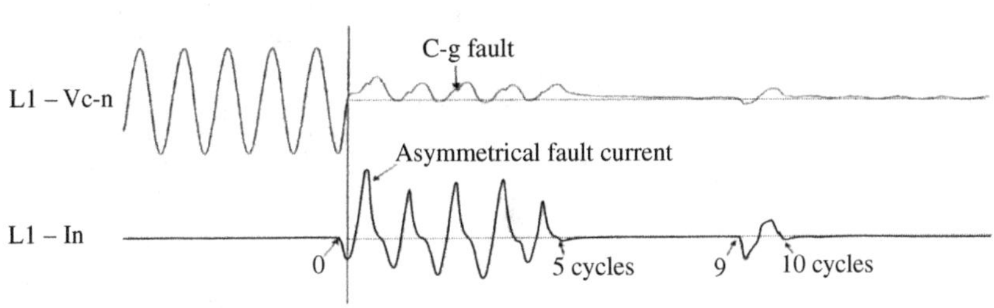

Fig. 3.68 DFR record for voltage and current during the initial and second faults.

3.27 MAGNETIC FLUX CANCELLATION

The magnetic flux cancellation phenomenon may appear during the modification process of some existing single-trip coil assemblies to add a second trip coil. Dual-trip coil assemblies may be required to adhere to relaying system redundancy criteria, especially when dual dc batteries and dual relaying systems fed from separate CTs and separate potential sources secondary windings are added. Figure 3.69 shows a modified trip coil assembly with dual identical trip coils added. Figure 3.70 indicates that when a coil is excited with a direct current, a magnetic field energy is introduced which will generate a force that will move the trip rod by overcoming the spring holding force. It is also obvious that swapping the plus (+) and minus (−) leads of the dc input source will change the orientation of the north and south poles without any effect on the force generated to move the trip rod. The force generated is normally

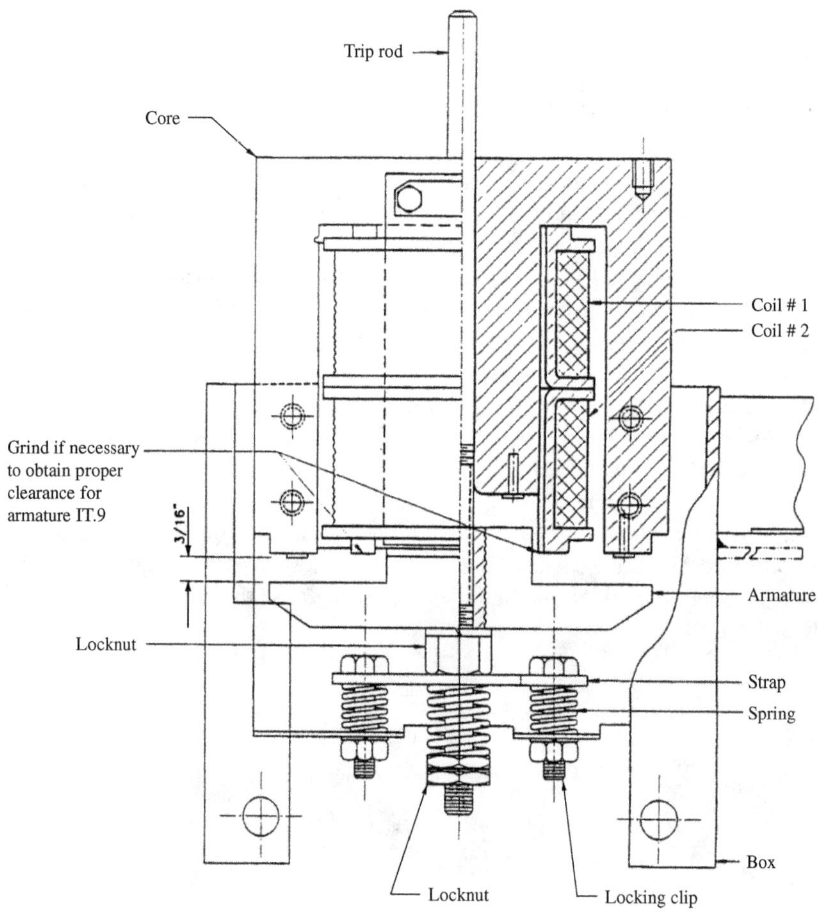

Fig. 3.69 Modified trip coil assembly with dual identical trip coils.

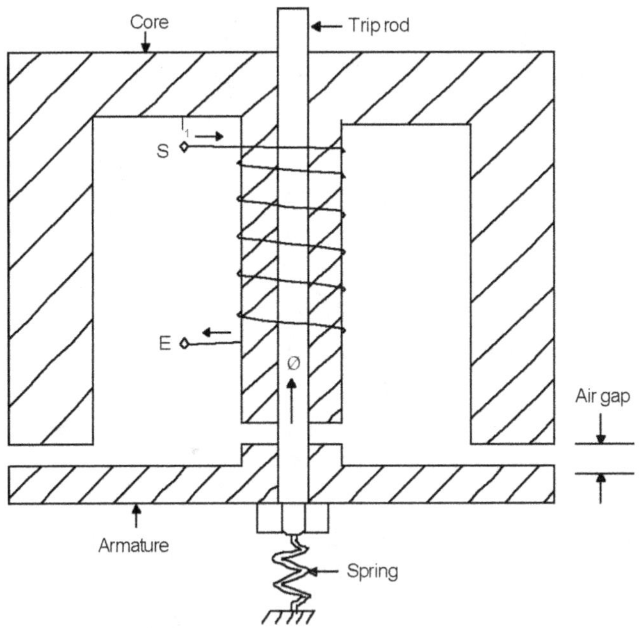

Fig. 3.70 Basic magnetic structure.

proportional to the square value of the coil voltage. Therefore, it can be deduced that this assembly is not polarized. The assembly can become polarized only when a second trip coil is added to the structure.

Figure 3.71(a) illustrates the correct way of assembling the dual trip coils. The windings should be mounted to produce added flux when the coils are energized simultaneously. Figure 3.71(b) illustrates that when the coils are assembled opposite to each other, a magnetic flux cancellation can take place when the coils are energized simultaneously with no movement of the armature. Figure 3.72 illustrates the dual trip coil structure showing incorrect assembly of the coils.

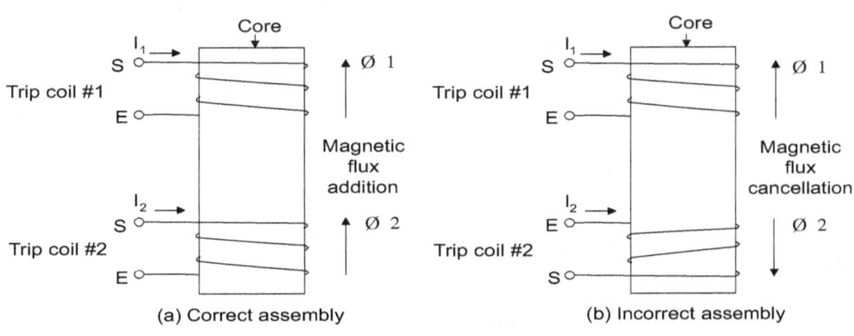

Fig. 3.71 Magnetic flux cancellation for a dual-coil assembly.

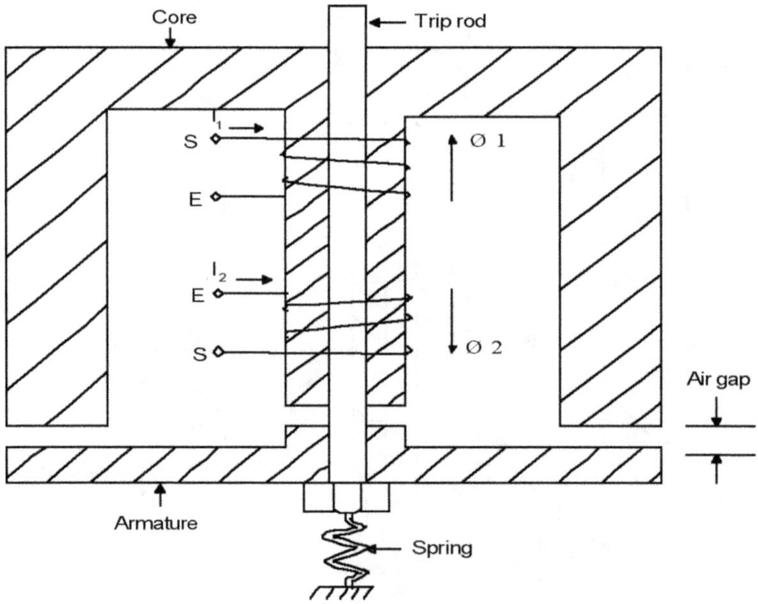

Fig. 3.72 Dual trip coil structure showing incorrect assembly of the coils.

Quite often, circuit breakers are placed in service with their trip coil assemblies wired in a flux cancellation mode. The assemblies may be put together by the original circuit breaker manufacturers and placed in service, with the field testing done by energizing individual trip coils. Therefore, the flux cancellation phenomenon will not be uncovered prior to placing the facility in service. The correct thing to do is to test the system in accordance with the testing criteria: "that the design of a breaker with two trip coils must be such that the breaker will operate if both trip coils are energized simultaneously, and verified by tests." The assembly testing can be done on the bench by paralleling the two coils (through the connection of positive and negative leads) and in service by simultaneous energization of both trip coils.

Magnetic flux cancellation phenomenon for trip coil assemblies will produce no trip rod movement (the breaker will not be told to move), and fault clearing must be done by either local and/or remote backup relaying systems. If design deficiency leading to magnetic flux cancellation is not uncovered prior to placing the facility in service, the occurrence of line faults leading to equal operating time of both numerical relaying systems, and hence energization of auxiliary output relays at the same time, will lead to a delayed fault clearing, which should be prevented.

3.28 CURRENT TRANSFORMER SATURATION

Asymmetrical current can influence the performance of a current transformer during power system faults. The presence of dc offset in fault currents coupled with a large

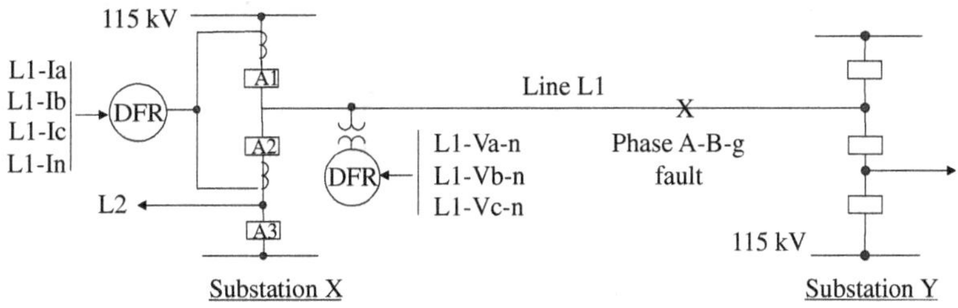

Fig. 3.73 One-line diagram showing line L1 DFR-monitored currents and voltages.

current magnitude may result in CT saturation. The dc component has far more influence than ac fault current components in producing severe saturation. The combination of high fault current, high burden, and a low connected CT ratio can stress the CT to be near its nonlinear region and enter into saturation.

CT saturation time is normally a function of fault current, core flux density, CT parameters, CT connected burden, and dc time constant. CT saturation normally occurs after a time delay from the fault inception point, thus allowing high-speed relaying to detect faults prior to CT collapse. CT saturation during a heavy current for faults outside closed differential zones can cause a current mismatch that may result in false differential relay operations. Sizing the CT to have an adequate knee-point voltage with suitable allowance for possible dc components and remanence can reduce the effect of saturation.

Figure 3.73 shows a simplified system one-line diagram and line L1 DFR-monitored currents and voltages during the fault. As shown in the DFR record in Fig. 3.74, a phase A-B-g fault occurred on line L1, which lasted for 4 cycles. The phase B current trace L1-Ib begins to saturate 1.5 cycles after the fault incident point, due to the nature of

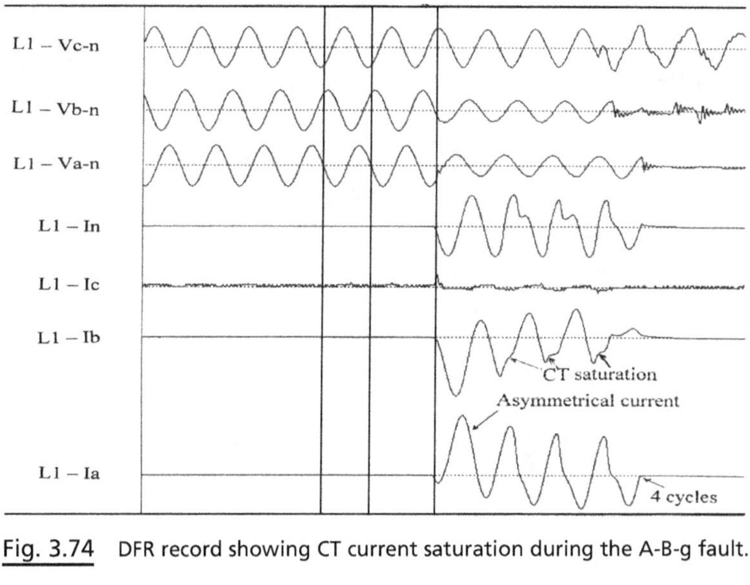

Fig. 3.74 DFR record showing CT current saturation during the A-B-g fault.

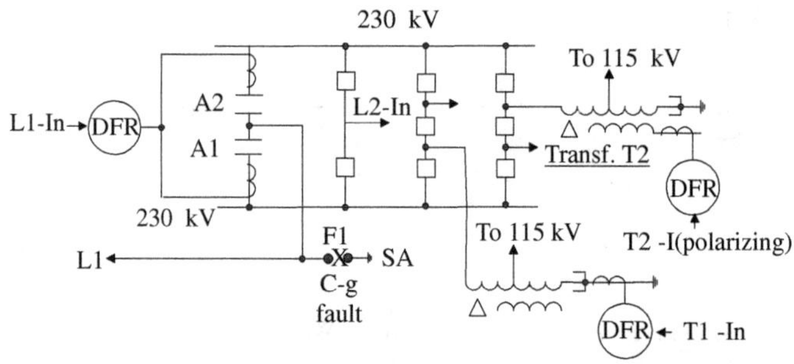

Fig. 3.75 One-line diagram showing DFR-monitored line and transformer currents.

the asymmetrical fault current. The phase A current trace L1-Ia begins to saturate 2 cycles after the fault incident point, also due to the asymmetrical fault current.

Figure 3.75 is another system one-line diagram with the DFR-monitored currents for lines L1 and L2 and transformer T1 neutral and T2 polarizing. This is another CT saturation case during which a surge arrester failure caused a close in a C-g fault on line L1 at F1 which was cleared in 5 cycles. Figure 3.76 reveals that the time to CT saturations for the T1 neutral current is about 2 cycles, while the time to saturate T2 polarizing current during the C-g fault is about 1 cycle.

The line L1 trace L1-In shown in Fig. 3.77 reveals that one line L1 CT has a period of about 1.5 cycles to reach the saturation flux density. The secondary fault currents for the line L1 circuit are not a faithful replica of the primary fault currents. The initial phase C-g fault has a large dc component with a slow decay rate, due to the large *X*/*R* ratio for the generating station. The accuracy class C800 for the 2000/5 CTs associated with CBs A1 and A2 was downgraded to C200 for the 800 : 5 connected tap. The L1 electromechanical line relaying, CT cable leads, CT winding, and CT lead resistances have imposed a high burden across the CT and with a flow of asymmetrical high fault current impressed voltage across the CT which exceeded the knee point for the 800: 5 CT tap. The line L2 trace L2-In shown in Fig. 3.77 reveals no CT saturation.

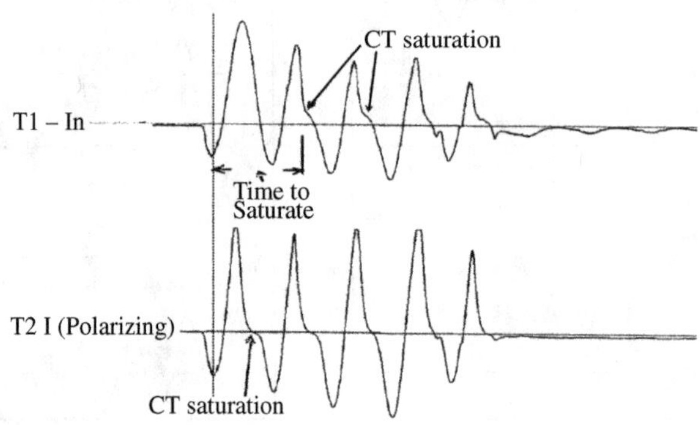

Fig. 3.76 DFR record showing saturation of T1 neutral and T2 polarizing currents.

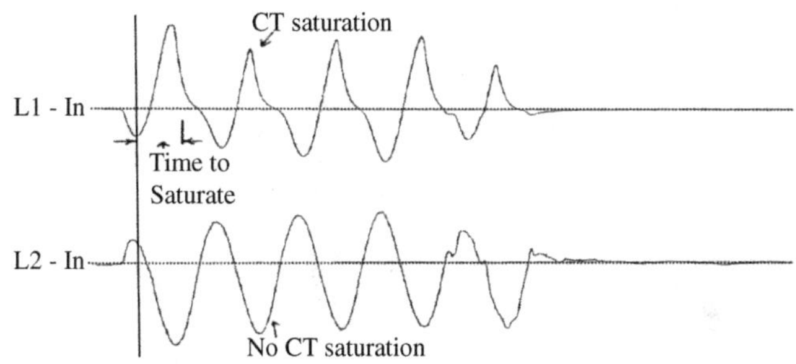

Fig. 3.77 DFR record showing line L1 current saturation.

3.29 CURRENT TRANSFORMER SATURATION DURING AN OUT-OF-STEP SYSTEM CONDITION INITIATED BY MIS-SYNCHRONIZATION OF A GENERATOR BREAKER

Normally, differential relaying schemes are immune to power swing phenomena. This is due to the fact that the current, being the only input when the swing begins, is nearly equal to the current output (provided that the center of oscillation is outside the differential protection zone). Figure 3.78 shows a system one-line diagram and DFR-monitored STG generator voltages and currents. The DFR record in Fig. 3.79 shows high current levels in phases A, B, and C when the STG unit began to oscillate against the system and the CTG unit at the mis-synchronization instant. In addition, due to the asymmetrical nature of the currents, trace STG-Ib reveals CT saturation after 1 cycle, while trace STG-Ia reveals CT saturation after 3 cycles.

The original swing between the STG unit and the system has also caused high asymmetrical currents which entered the differential relay characteristics of three differential relays: transformer 87T, generator 87G, and overall relay 87OA associated with the STG unit protection systems. Only the 87OA overall differential relay

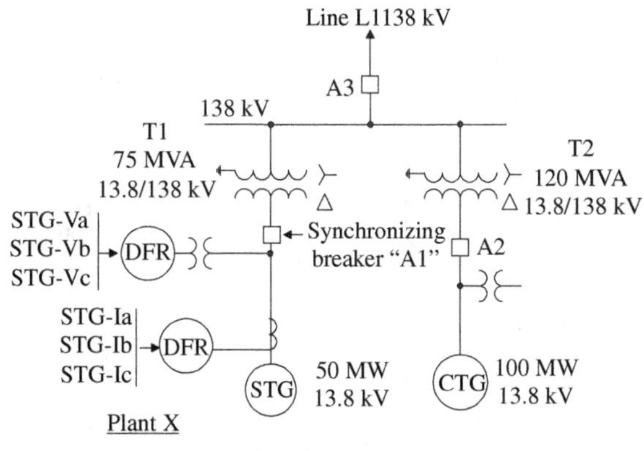

Fig. 3.78 One-line diagram showing DFR-monitored generator voltages and currents.

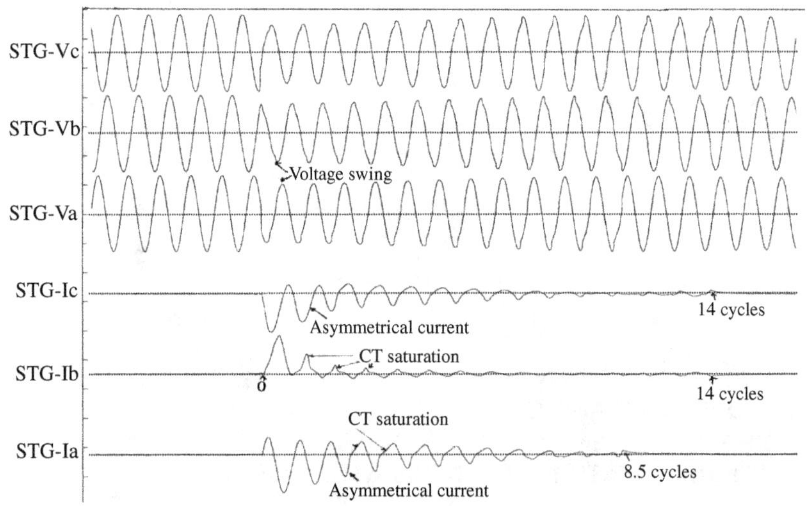

Fig. 3.79 DFR record showing CT current saturation during mis-synchronization of the STG unit.

operated undesirably and shut down the plant. As shown in Fig. 3.80, the overall differential relay 87OA receives its 13.8-kV input from the generator neutral wye–connected CTs. The high-side input to the 87OA differential relay is from the delta-connected transformer bushing CTs. The neutral CTs were supplied by the generator manufacturer with a single ratio of 3000:5 and IEC accuracy stated as B250-200. The high-side transformer bushing CTs have a multiratio 1200:5 and ANSI standard accuracy of C800. There is obviously no match between the performance of

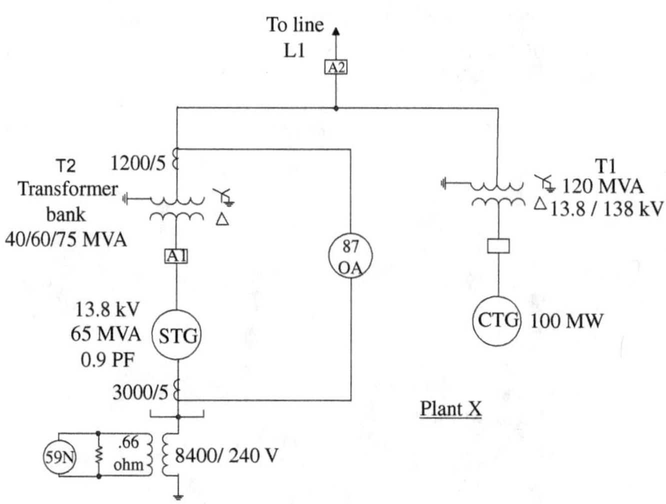

Fig. 3.80 One-line diagram showing the unit STG differential relay 87OA connection.

the high- and low-side CTs in the presence of asymmetrical current that contains dc offset, causing some of the CTs to saturate, hence causing undesired unit transformer differential relay operation during breaker closing at a wide angle.

3.30 CAPACITIVE VOLTAGE TRANSFORMER TRANSIENT

A capacitive voltage transformer (CVT) transient can occur with faults that cause very depressed voltage at the location of the device. The higher the system impedance ratio, the worse the CVT transient will be. The CVT transient occurs due to the internal CVT energy storage elements (capacitances and inductances). CVT transients reduce the fundamental component of the voltage during the fault. The decrease in the fundamental component in the voltage will result in a decrease in the calculated fault impedance ($Z = V/I$). This may cause elements such as zone 1 to overreach near the boundary reach of the element. CVT transient is severe when voltage drops from normal to near zero (5%). A voltage dip to 20% will not cause a severe CVT transient. A strong source and long line will not result in a severe CVT transient. A weak source and short line will produce a severe CVT transient.

The phenomenon can be mitigated by adding filters for the voltage input to the zone 1 (Z1) distance element. This solution will certainly delay the operating time for Z1. Figure 3.81 shows a system one-line diagram with DFR-monitored line voltages. The DFR record shown in Fig. 3.82 reveals the transient response of CVT voltage outputs from devices associated with the L1 and L2 circuits. The L2 CVT device

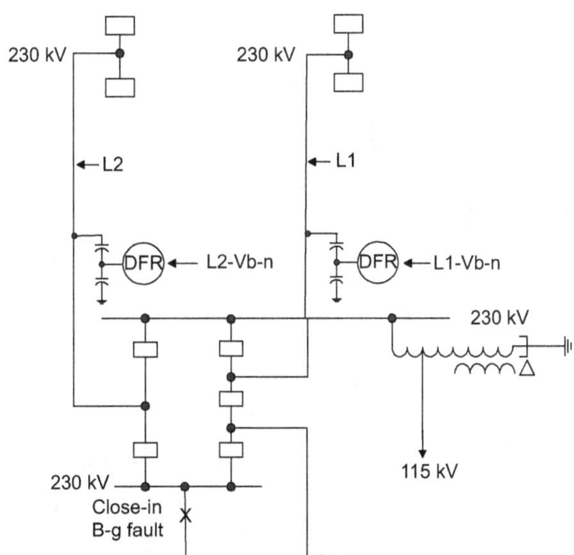

Fig. 3.81 One-line diagram with line L1 and L2 DFR-monitored voltages.

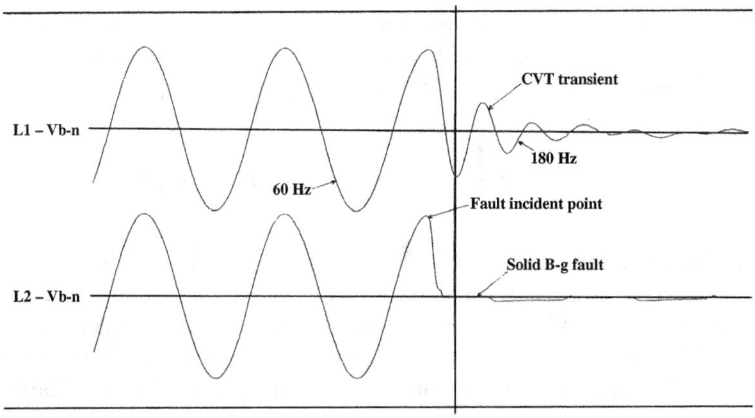

Fig. 3.82 DFR record showing the line L1 CVT transient voltages during the bolted B-g fault.

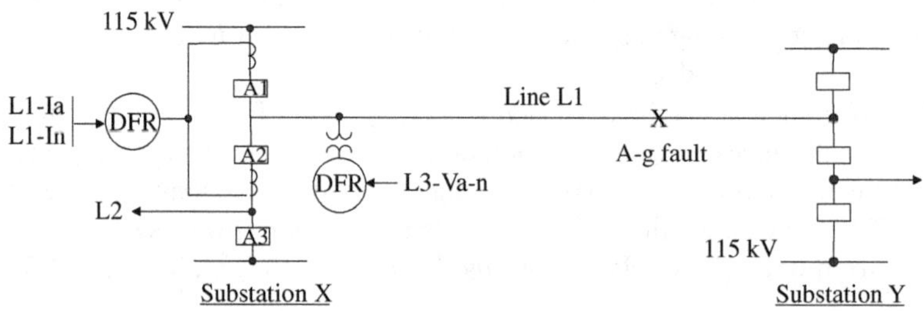

Fig. 3.83 Line L1 DFR-monitored voltages and currents.

tracks the primary voltage more accurately than the L1 CVT device. More oscillation of the secondary voltage is exhibited by the L1 device. As shown in Fig. 3.82, a voltage oscillation at a frequency of about 180 Hz was generated and lasted for more than 20 ms. These transient responses of the L1 and L2 devices are different despite identical manufacturing methods with similar connected burdens.

Figure 3.83 shows another symbolic system one-line diagram with line L1 DFR-monitored voltage and currents. The DFR record in Fig. 3.84 shows the voltage output from the CVT connected to line L1 during an A-g fault that was cleared after 2.5 cycles. Trace L1-Va-n of Fig. 3.84 reveals that the CVT voltage transient is occurring at the instant of clearing the phase A-g fault.

3.31 BUSHING POTENTIAL DEVICE TRANSIENT DURING DEENERGIZATION OF AN EHV LINE

Bushing potential devices derived from a calibrated capacitance tap of a circuit breaker bushing or CT column have a limited burden. They exhibit transients at the

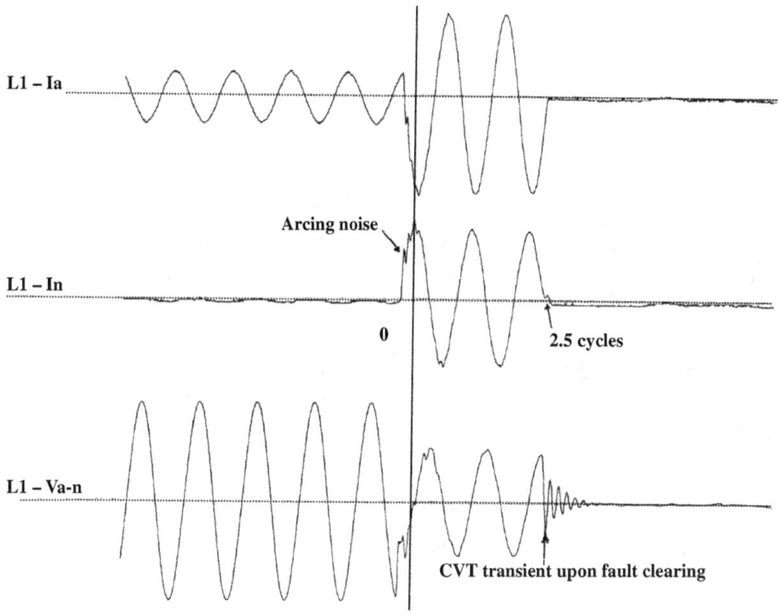

Fig. 3.84 DFR record showing the line L1 CVT transient at the clearing of the A-g fault.

inception of system faults as well as ringing upon the clearance of the fault. The ringing is caused by the exchange of energy stored in the capacitance of the device with inductances from connected transformers and the system.

A simplified 765-kV system one-line diagram as well as the DFR-monitored voltages and currents are shown in Fig. 3.85. The voltage fed to the relay system is derived from the bushing potential device (BPD) derived from a circuit breaker support column insulator. Figure 3.86 shows the C-g fault, which resulted in a phase C voltage dip and the flow of neutral (ground) current that lasted for 2.5 cycles. Upon the clearing of the C-g fault, a voltage transient appeared as an output of the BPD, which lasted for about 3.5 cycles. The BPD transient started immediately upon clearing of the C-g fault, and the frequency of oscillation is about 300 Hz.

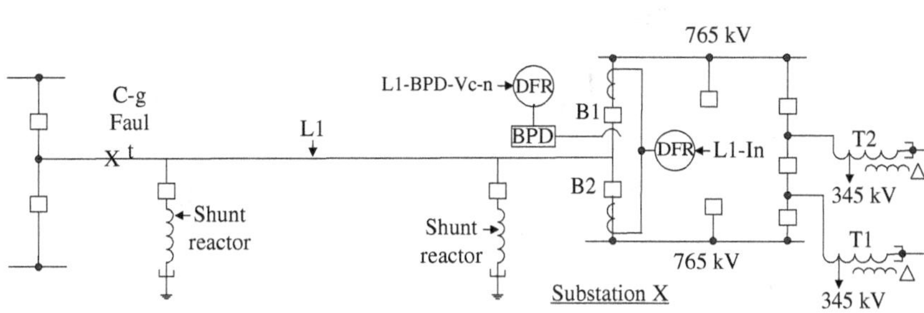

Fig. 3.85 One-line diagram showing DFR-monitored voltages and currents.

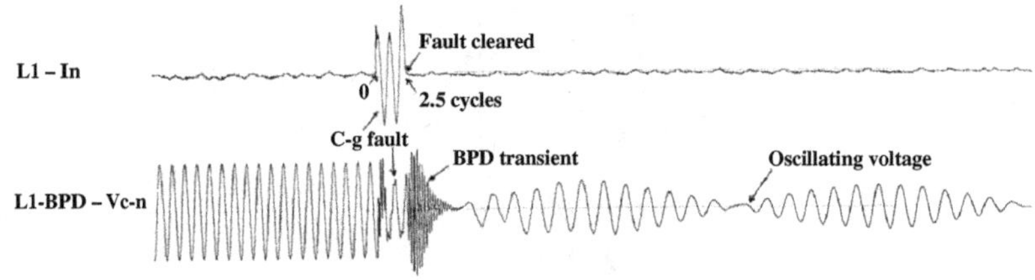

Fig. 3.86 DFR record showing the BPD transient and line voltage oscillation after fault clearing.

3.32 CAPACITOR BANK BREAKER RE-STRIKE FOLLOWING INTERRUPTION OF A CAPACITOR NORMAL CURRENT

The opening operations of a capacitor bank circuit breaker may introduce transitory phenomena on the system. Figure 3.87 shows a substation one-line diagram and the DFR-monitored 345-kV bus voltages and shunt capacitor currents. The DFR record shown in Fig. 3.88 reveals that the phase C current trace Cap. BK - Ic interrupts the capacitor bank nominal current at current zero at point a when the voltage trace Bus-Vc-n is at peak value. Then when the voltage is changing polarity going to the other half-cycle, phase C restruck, allowing high-frequency current to flow between points b and c for about 4 ms. The same transient is reflected in the phase C voltage. The frequency of oscillation is about 360 Hz, as shown in Fig. 3.89. During this relatively short time span of less than half a cycle, the recovery characteristic of the dielectric strength or the voltage withstand capability between the parting breakers contacts might be lower than the rising voltage across the breaker at a particular time instant, leading to a breaker re-strike. The capacitor bank circuit breaker dielectric strength eventually recovered and interrupted the high-frequency current transient at current zero.

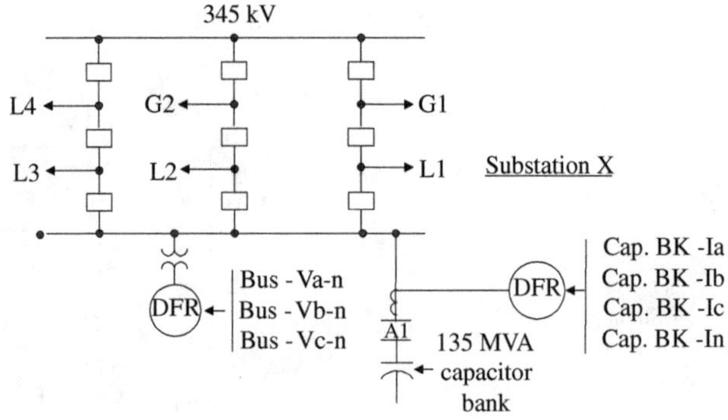

Fig. 3.87 One-line diagram showing DFR-monitored currents and voltages.

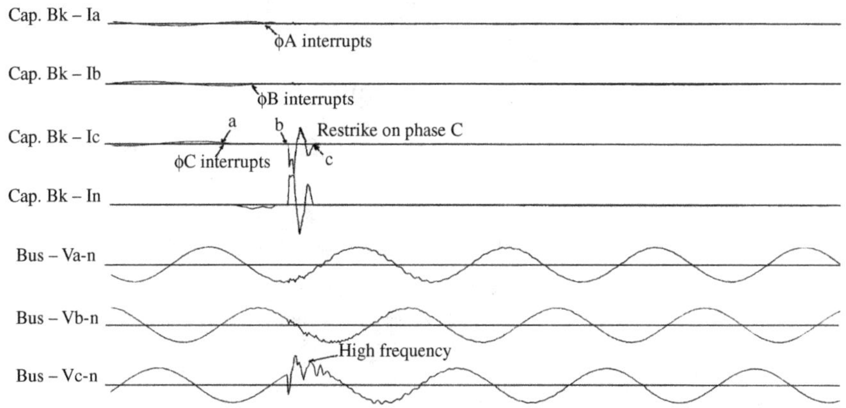

Fig. 3.88 Substation X DFR record showing re-strike on capacitor bank phase C current during opening of the capacitor circuit breaker.

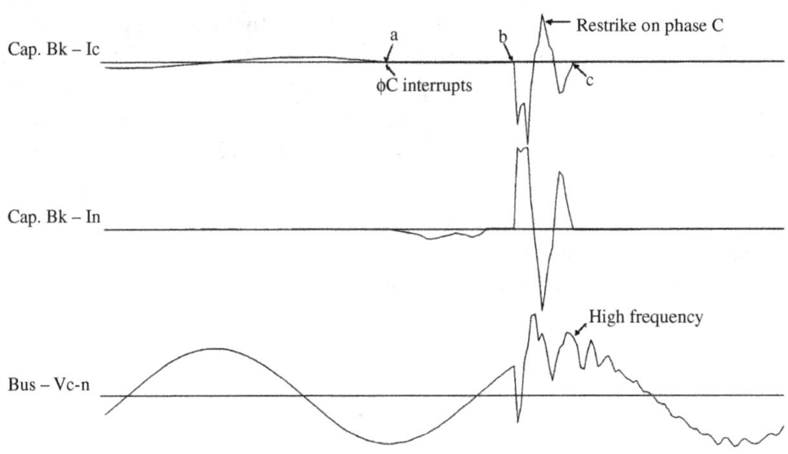

Fig. 3.89 Substation X DFR record showing re-strike on capacitor bank phase C current during opening of the capacitor breaker.

3.33 CAPACITOR BANK CLOSING TRANSIENT

During the switching of shunt capacitor banks, high-magnitude and high-frequency transients may occur. When a capacitor bank is energized, an energization transient will occur with a surge of current of high magnitude and with a frequency as high as several hundred hertz. Energizing a capacitor bank with an energy storage element will generate a current transient and induce a high-frequency voltage transient. The frequency of oscillation (f_S) will be a function of the capacitance (C) and the system equivalent inductance (L) and is defined as $f_s = 1/2\pi\sqrt{LC}$. Figure 3.90 illustrates the substation one-line diagram and the DFR-monitored 345-kV bus voltages and shunt capacitor currents during the closing of capacitor breaker A1. The DFR record in

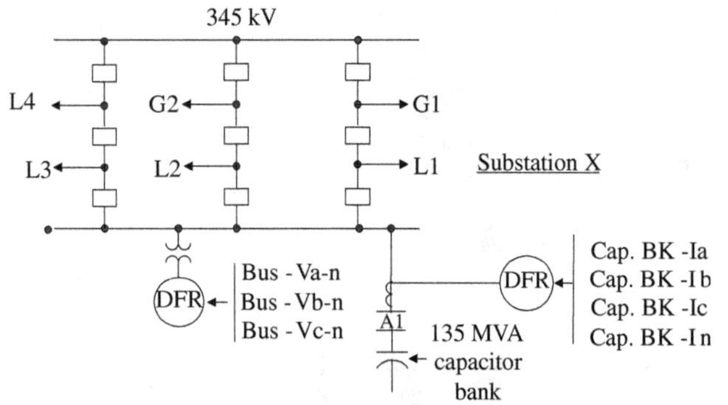

Fig. 3.90 One-line diagram showing DFR-monitored currents and voltages.

Fig. 3.91 shows the high-frequency transient in phases A, B, and C current traces and is also reflected in the neutral current trace. In addition, high-frequency transient noise is induced in the individual phases of the monitored bus voltage. The frequency of oscillation is about 540 Hz. The closing transient lasts for about 40 ms (2.5 cycles of 60 Hz). The capacitance value is about 30 μF, and using a system equivalent impedance of 3 mH (including the capacitor series inductor), the frequency of oscillation for the closing transient will be $f_s = 1/2\pi\sqrt{LC} \approx$ about 530 Hz. The maximum capacitor inrush current recorded is five times the normal capacitor 60-Hz current. The instantaneous 50-element set to protect the bank should always be set above the inrush current during the closing transient.

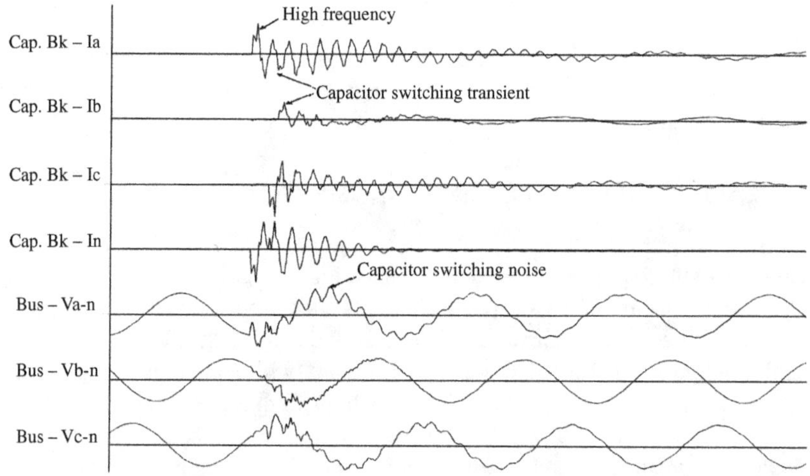

Fig. 3.91 Substation X DFR record showing capacitor bank transient currents during energization of the capacitor bank.

3.34 SHUNT CAPACITOR BANK OUTRUSH INTO CLOSE-IN SYSTEM FAULTS

3.34.1 Definition of the Outrush Phenomenon

The capacitor bank outrush phenomenon is defined as dumping of high-frequency (f_S) oscillatory capacitor current onto the fault with a sinusoidal damped current wave having a duration of a few milliseconds with the initial magnitude limited only by the surge impedance ($= \sqrt{LC}$) of the circuit. A capacitor bank feed to a close-in fault from its stored energy ($\frac{1}{2}CV^2$) is known as an outrush phenomenon. The capacitor bank outrush to close-in faults at the natural frequency of the LC equivalent circuit lasts for only a few milliseconds, as will be illustrated. Utilities often use in-line current-limiting inductors for capacitor bank installations to limit the severity of outrush currents from the bank(s) into close-in faults and capacitor inrush currents as well as back-to-back switching. A capacitor discharge into a fault results in a damped oscillation with a frequency defined as $f_s = 1/2\pi\sqrt{LC}$, where L and C are the equivalent inductance and capacitance of the fault respectively. To reduce this transient outrush, a current-limiting reactor is selected and added in series with the capacitor bank. This series reactance will limit the transient outrush current to a manageable level that should not stress the current transformer secondary circuits.

The outrush phenomenon can best be illustrated using DFR records. Figure 3.92 shows a substation X one-line diagram and DFR-monitored 345-kV bus voltages and shunt capacitor currents. The DFR record of Fig. 3.93 shows a close-in A-g fault that occurred on feeder F1 and cleared in 4 cycles. The DFR record also reveals that the capacitor outrush phase A current trace Cap. BK-Ia lasted for about 5.7 ms with a high-frequency oscillation of about 1140 Hz.

Figure 3.94 shows a substation X one-line diagram with a 135-MVAR capacitor bank and DFR-monitored 345-kV bus voltage and shunt capacitor currents. The DFR

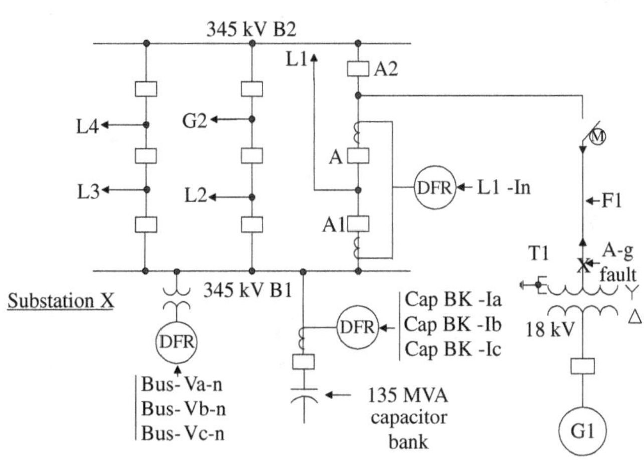

Fig. 3.92 One-line diagram showing DFR-monitored voltages and currents.

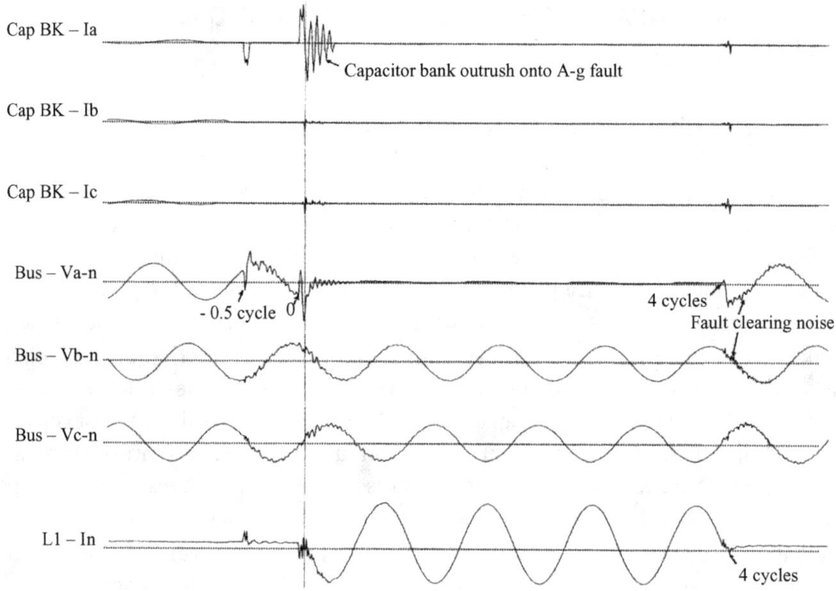

Fig. 3.93 DFR record showing the capacitor bank outrush current contribution to the A-g fault.

record in Fig. 3.95 shows that the monitored phase A-n voltage has a voltage dip lasting for 3 cycles. Capacitor bank phases A, B, and C and neutral currents show the high-frequency current contributions of the capacitor to the system being triggered by the A-g fault on feeder F1. The high-frequency current contributed by the capacitor bank is superimposed on the capacitor nominal phase current. The capacitor bank neutral shows the discharge current at a natural frequency of 540 Hz. The high-frequency current lasted for only 4 ms. The DFR record also reveals a transient

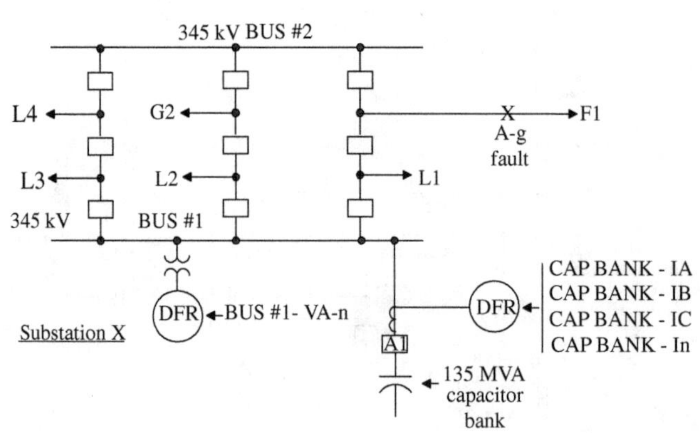

Fig. 3.94 One-line diagram showing DFR-monitored currents and voltages.

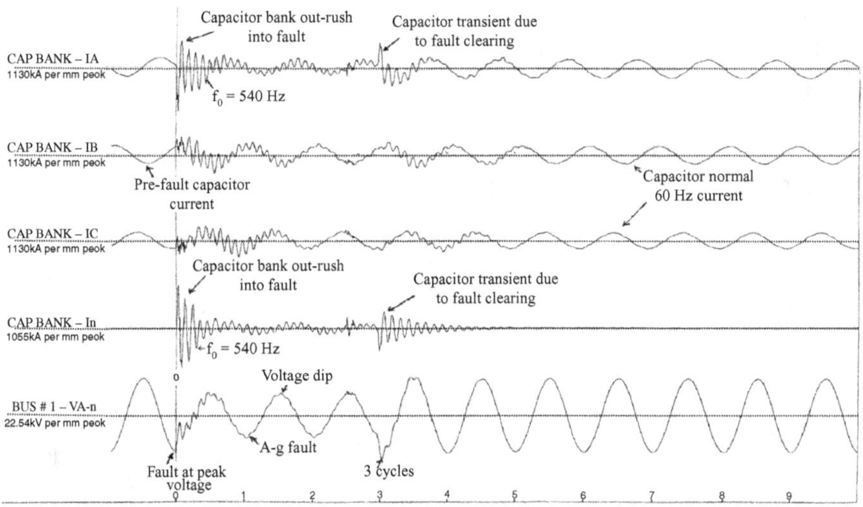

Fig. 3.95 DFR record showing capacitor bank currents and bus voltage during the capacitor bank outrush feeding the A-g fault.

triggered by the fault removal, where an oscillatory capacitor transient current is generated.

3.34.2 Capacitor Outrush Simulation Studies

Figure 3.96 shows a simulation study one-line diagram to determine capacitor bank outrush currents into a close-in fault. The simulation study equivalent circuit is shown in Fig. 3.97. Simulation studies are required to study the impact on current

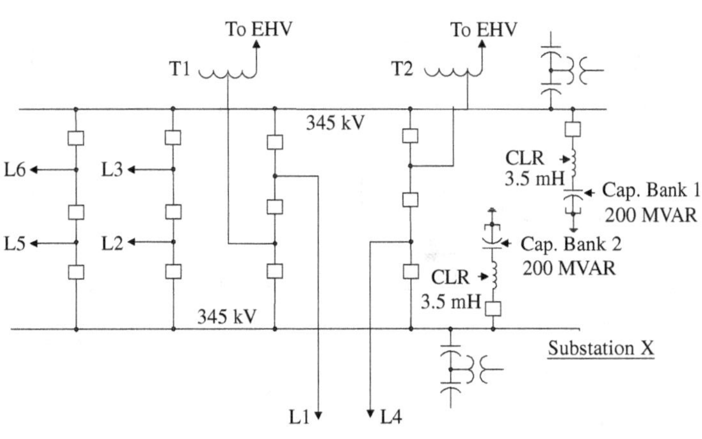

Fig. 3.96 Simulation study one-line diagram.

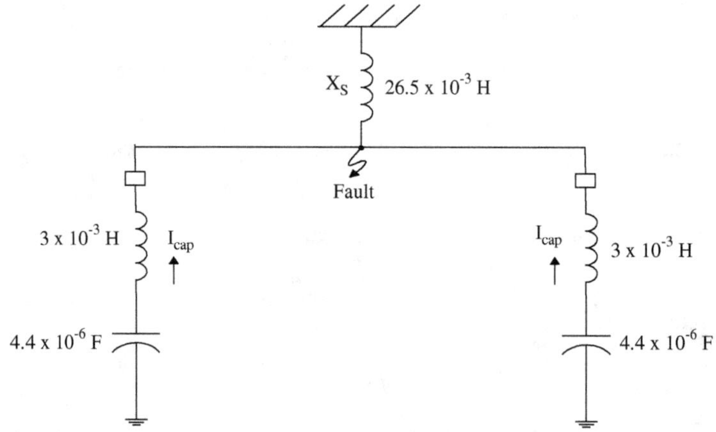

Fig. 3.97 Simulation system equivalent circuit.

transformers in terms of their overvoltage capability. The capacitor discharge current normally has a higher natural frequency (above 1000 Hz), thus causing a high CT secondary burden. The capacitor bank current natural frequency is calculated as

$$f_s = \frac{1}{2\pi\sqrt{LC}} = \frac{1}{(2 \times 3.14\sqrt{3} \times 4.4 \times 10^{-9})} = 1386\,\text{Hz}$$

The CT inductive connected burden will be high at this natural frequency, and when it is multiplied by the capacitor current will impose a high voltage across the CT and may cause the CT to fail during the capacitor bank outrush phenomena. Figure 3.98 reveals the capacitor bank discharge current as a function of time during outrush phenomena into close-in faults.

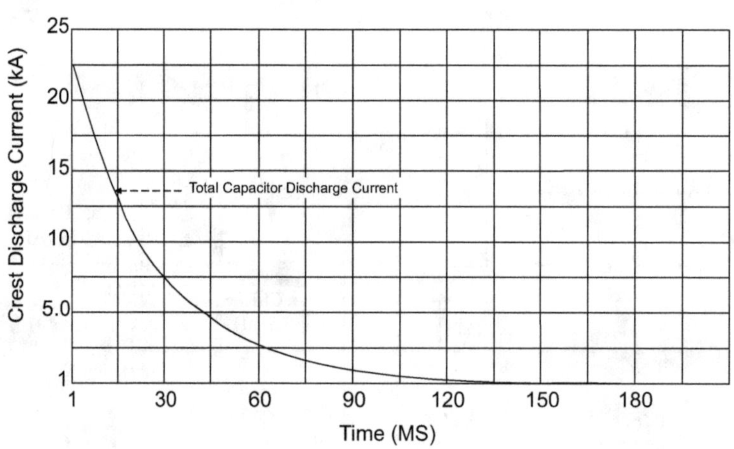

Fig. 3.98 Capacitor bank outrush phenomenon in close-in faults.

3.35 SCADA CLOSING INTO A THREE-PHASE FAULT

In some transmission systems, automatic reclosing is used to test the faulted line once or twice or more from each end of the line. If automatic reclosing of the line has failed and the line status will indicate an open position at the energy control center, operators may make another attempt to close after 5 minutes for one of the line breakers from the strong end of the line via SCADA. The strong end is defined as the end that will supply sufficient fault currents to clear the fault, assuming that the line is tested by reclosing into a permanent fault. Figure 3.99 shows a simplified system one-line diagram where CB A1 was closed into a three-phase fault via SCADA. In addition, the DFR-monitored voltages and currents are shown in Fig. 3.99. The DFR record in Fig. 3.100 illustrates the three-phase currents during the fault, with a total clearing time of 6.5 cycles. The phase B voltage reveals that the CB phase B pole closed 2 cycles ahead of the phase A and C poles (unequal pole closing), which implies the need for inspection and possible maintenance of the breaker.

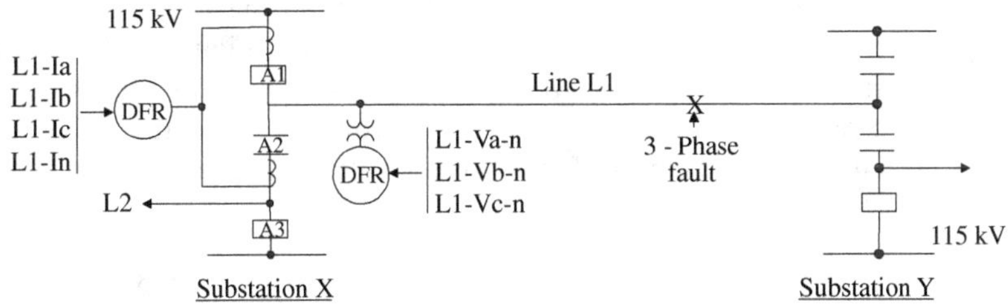

Fig. 3.99 Line L1 DFR-monitored voltages and currents.

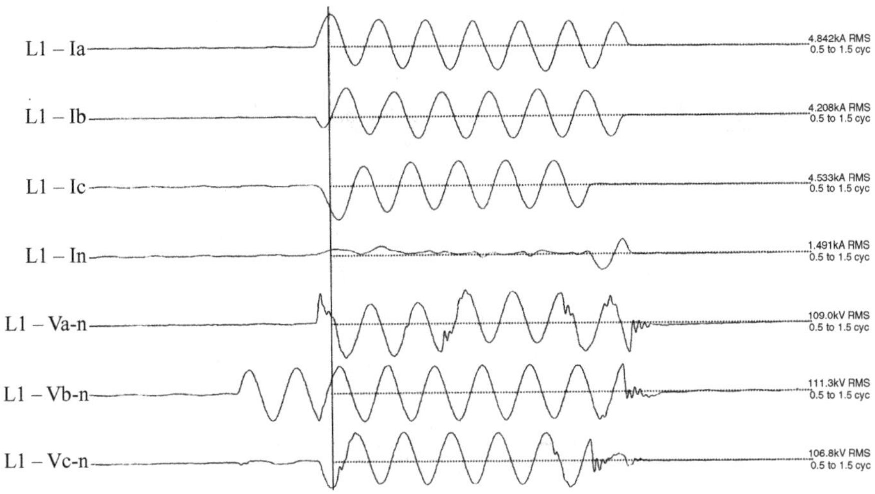

Fig. 3.100 Substation Y DFR record showing reclosing into a three-phase fault.

3.36 AUTOMATIC RECLOSING INTO A PERMANENT LINE-TO-GROUND FAULT

Automatic reclosing is implemented to be able to restore a system following temporary line faults. Radial circuits benefit most by automatic reclosing because there is only one source of power. Multiterminal lines or interconnections in a network require a considerable amount of study before a reclosing sequence can be prescribed. Multishot reclosing is frequently used on low-voltage lines (33 kV and below). For 765- and 345-kV systems with a strong transmission network, high-speed reclosing may be applied at both ends of the line and, in addition, two delayed (5 to 35 s) staggered automatic reclosings, either one from each end or both applied from the strong end, before going to lockout. Automatic reclosing does not apply to cables because the breakdown on insulation is usually permanent. Circuit breakers can be programmed via reclosing relays to close for one or more of the following modes: hot bus–dead line, hot line–dead bus, and hot line and hot bus with sync check. Load flow studies are normally used to set the angle of the sync check relay. A rapid restoration strategy for some critical systems may bypass the reclosing mode, allowing the system operator to close breakers directly via SCADA. Figure 3.101 shows a simplified system one-line diagram where CB A1 was closed into a line-to-ground fault via auto reclosing and the DFR-monitored voltages and currents. The DFR record shown in Fig. 3.102 reveals that as soon as voltages appeared on the phase A

Fig. 3.101 One-line diagram showing DFR-monitored voltages and currents.

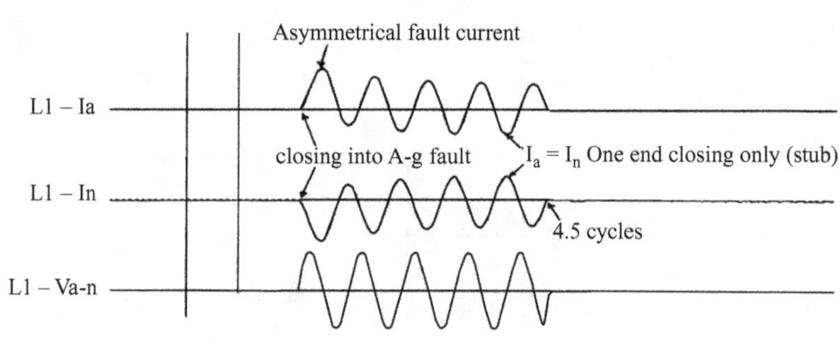

Fig. 3.102 Substation Y DFR record showing reclosing into an A-g fault.

voltage trace L1-Va-n, phase A-g was established with fault currents appearing on line L1 phase A and the neutral for a duration of 4.5 cycles.

3.37 SUCCESSFUL HIGH-SPEED RECLOSING FOLLOWING A LINE-TO-GROUND FAULT

A large majority of faults on overhead lines are transient and temporary and may be caused by lightning strikes. The fault may disappear as soon as the line clears the first fault successfully from the system by tripping and opening all CBs at all terminals of the line. This permits immediate resumption of service by use of automatic high-speed reclosing of the circuit breakers at both ends of the line. On overhead lines, the insulating value of the air is restored as soon as the fault current stops and the arc-over process disappears. This restoration is not immediate and depends on the voltage level of the line, fault duration, conductor spacing, and weather condition. Automatic high-speed reclosing must therefore be delayed until air deionization is under way. Typical high-speed reclosing dead time is 0.5 s (30 cycles) for 765-kV systems, 0.4 s (24 cycles) for 345-kV systems, and 0.25 s (15 cycles) for 138-kV systems. Probabilities have been calculated regarding the minimum dead time as well as laboratory testing before the circuit could be energized via high-speed reclosing. Figure 3.103 shows a simplified system one-line diagram and DFR-monitored voltages and currents for line L1.

The DFR record in Fig. 3.104 illustrates a phase B-to-ground fault that lasted 5 cycles, followed by high-speed reclosing in 28 cycles. The reclosing was implemented only at the X end of the line, as illustrated by the low currents on phases B and C following the successful reclosing, as compared with the line L1 pre-fault loading currents. High-speed reclosing is used primarily to preserve the stability of the system and must be employed at both ends of the line provided that 100% of the line is protected by pilot relaying schemes. Therefore, high speed reclosing should not be applied at one end only.

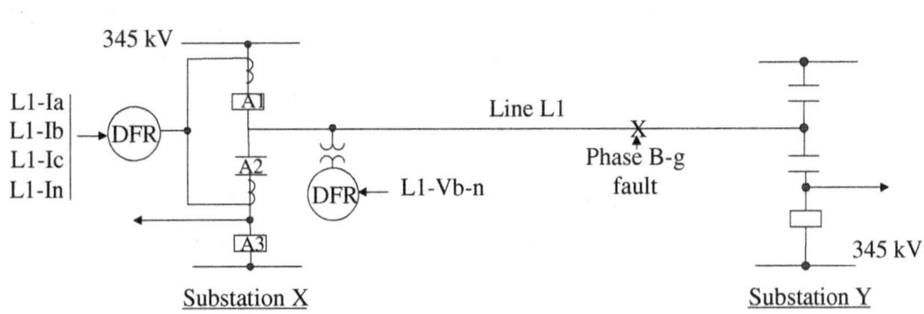

Fig. 3.103 Line L1 DFR-monitored voltages and currents.

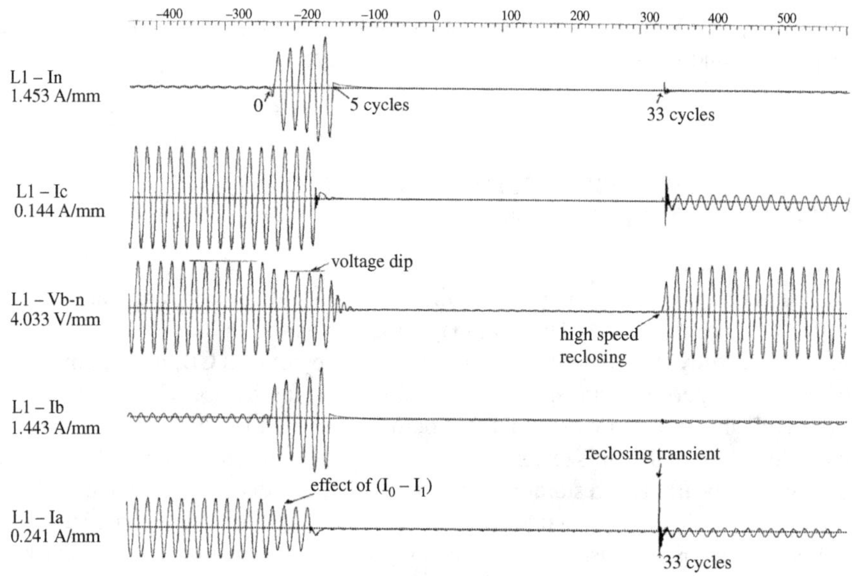

Fig. 3.104 Substation Y DFR record showing the initial fault and subsequent successful high speed reclosing from one end only.

3.38 ZERO-SEQUENCE MUTUAL COUPLING–INDUCED VOLTAGE

It is very well known that current flow in a transmission line produces magnetic fluxes in its proximity. Transmission lines are normally constructed in a common corridor, which creates parallel transmission lines to the line being protected. Ground fault currents are always composed of positive-, negative-, and zero-sequence components. The positive- and negative-sequence currents are balanced sets, and therefore the sum of magnetic fluxes produced by them are nearly equal to zero. As a result, positive- and negative-sequence mutual coupling between transmission lines is insignificant. However, the zero-sequence set of currents consists of three equal and in-phase phasors. This produces an equivalent flux, which will link adjacent lines and produce zero-sequence voltage. The zero-sequence voltages and currents are related by the zero-sequence mutual impedance Z_{0M}. The magnitude of the zero-sequence mutual impedance of two parallel lines (having the same voltage rating) could be 40 to 70% of zero-sequence self-impedance of the individual lines.

Zero-sequence mutual coupling phenomena were captured during the analysis of one of the 230-kV disturbances. Figure 3.105 shows a system one-line diagram and line L1and L2 DFR-monitored currents and voltages. The one-line diagram shows line L1 in an open position after it was tripped and locked out to clear an initial C-g fault, which was caused by phase C surge arrester failure. During the subsequent 1-cycle self-clearing C-g fault inside circuit breaker A1, ground current ($3I_0$) is flowing on feeder L2 as shown in trace L2-In in Fig. 3.106. The 230-kV circuits L1 and L2 are mutually coupled due to sharing the same transmission towers. As a result,

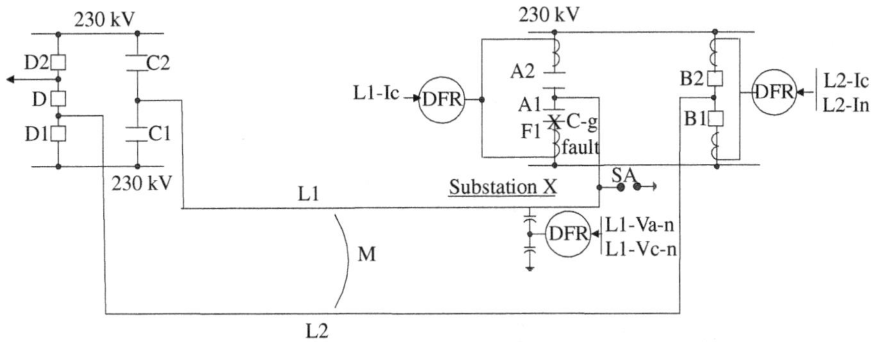

Fig. 3.105 One-line diagram showing line L1and L2 DFR-monitored currents and voltages.

ground current ($3I_0$) flows in the L2 circuit during the second fault, inducing a zero-sequence voltage in the parallel L1 line. CVTs at 230-kV substation X for phases A and C (phase B is not monitored) with line L1 opened at both ends revealed equal and in-phase voltage components. Traces L1-Va-n and L1-Vc-n in Fig. 3.107 show the zero-sequence voltages induced on phases A and C between the ninth and tenth cycles and confirm the existence of zero-sequence mutual coupling phenomena.

As shown in Fig. 3.108, when the fault currents in the two parallel lines are flowing in opposite directions, the distance relay R will measure the apparent impedance, which is less than the actual impedance of the line. The relay in this case will overreach, and if not set properly can react to external system faults. As shown in Fig. 3.109, when the fault currents in the two parallel lines are flowing in the same direction, the distance relay R1 will measure the apparent impedance, which is greater than the actual impedance of the line. The relay in this case will underreach, and if not set properly will not cover the part of the relay protection zone near the reach balance

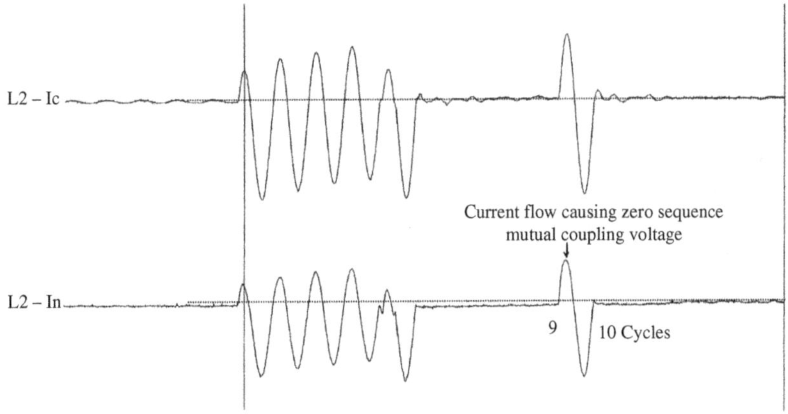

Fig. 3.106 DFR record showing line L2 neutral current that can cause mutual zero-sequence mutual coupling during the C-g fault.

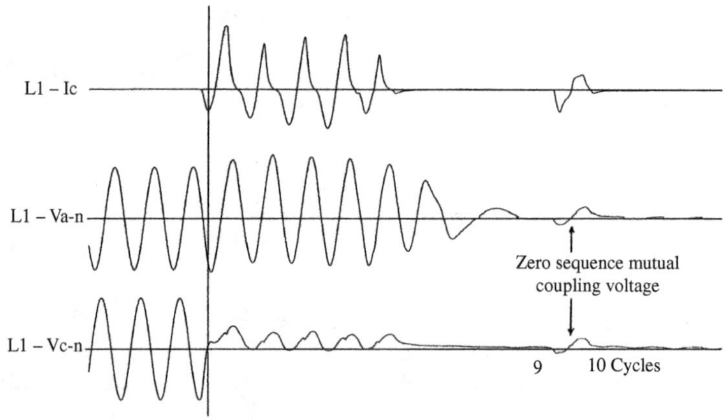

Fig. 3.107 DFR record showing zero-sequence mutual coupling voltage during the C-g fault.

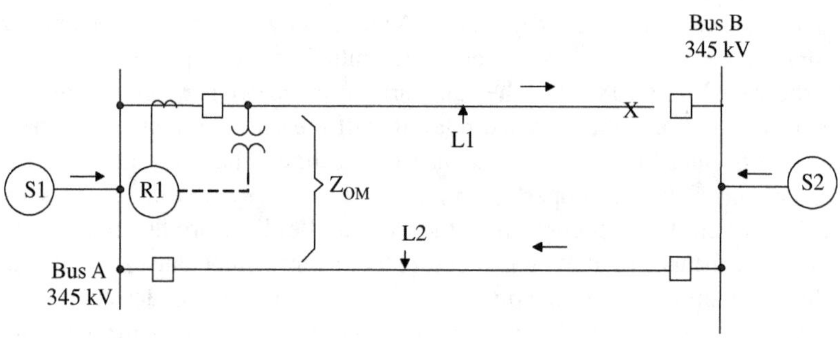

Fig. 3.108 Case of overreach when two zero-sequence currents flow in opposite directions.

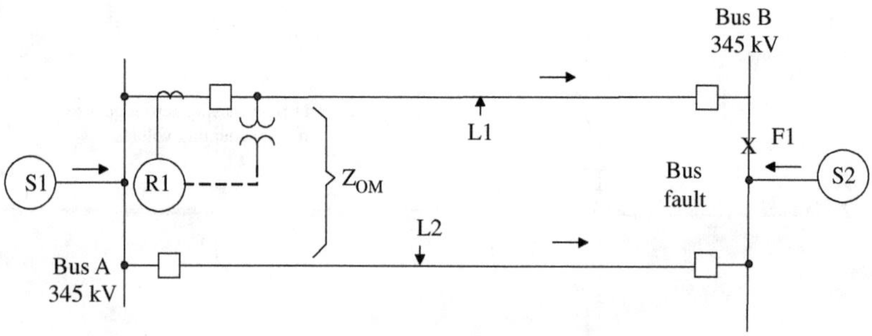

Fig. 3.109 Case of underreach when two zero-sequence currents flow in the same direction.

point. Case Study 6.17 provides more details about the overreach and underreach aspects of this topic.

3.39 MUTUAL COUPLING PHENOMENON CAUSING FALSE TRIPPING OF A HIGH-IMPEDANCE BUS DIFFERENTIAL RELAY DURING A LINE PHASE-TO-GROUND FAULT

Undesired operation of all the phases of a high-impedance bus differential relay occurred during a 345-kV phase-to-ground line fault. The line fault was cleared in three cycles from substation X. However, phases A, B, and C of the primary and secondary 345-kV high-impedance bus differential relays operated undesirably (falsely) within the 3-cycle fault. The undesired trip was due to zero-sequence mutual coupling phenomena, as will be shown.

Figure 3.110 shows a substation one-line diagram for the CB C1 maintenance condition, line L1–monitored currents, and bus 1–monitored voltage. The DFR record in Fig. 3.111 reveals that a line L1 phase A insulator failure to ground occurred at the voltage peak, causing an A-g fault. The fault was attributed to an insulator failure, causing it to flash over to ground. The voltage trace Bus No.1 A-n voltage indicates a voltage dip, which lasted for 3 cycles. The current trace line L1-Ia shows an increase in the magnitude of the current, and the current trace line L1-In confirms the presence of zero-sequence current flow. The DFR also indicates the disappearance of bus 1

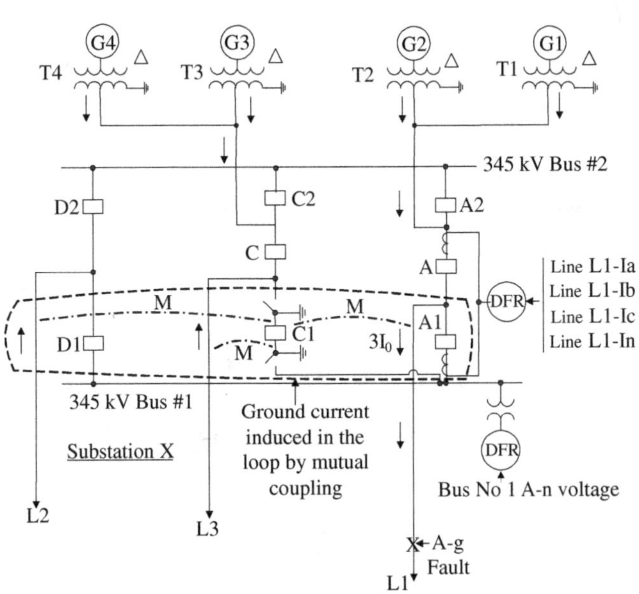

Fig. 3.110 System one-line diagram showing the CB C1 maintenance condition.

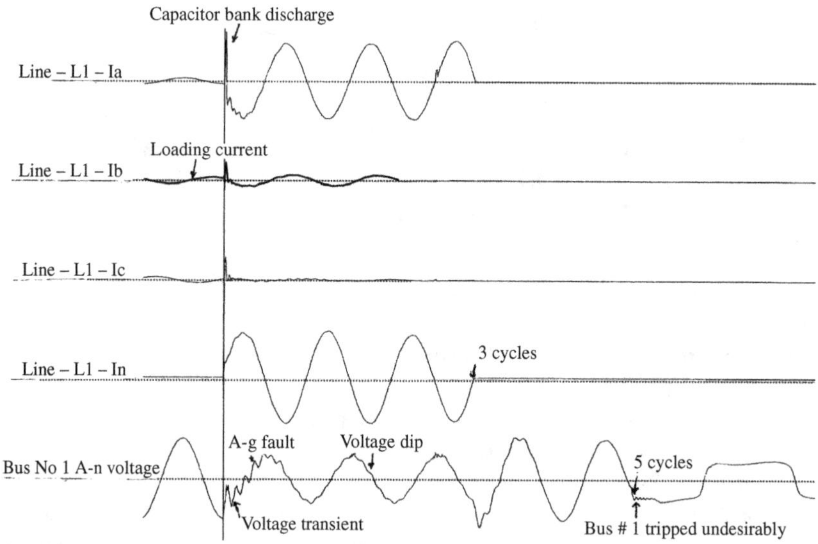

Fig. 3.111 DFR record showing the undesired trip of bus 1 during a line L1 ground fault.

voltages at 5 cycles, thus confirming the undesired operation of the bus differential relays.

Circuit breaker C1 was taken out of service for scheduled maintenance prior to the occurrence of the fault. This outage is normally accomplished by opening CB C1, followed by opening associated breaker disconnect switches and then applying personnel safety grounds around the breaker on both breaker disconnect switches, as shown in Fig. 3.112. Maintenance personnel then closed the breaker locally for testing purposes, forming three separate (phase A, B, and C) loops, including a ground mat. Phase A, B, and C loops include CT columns with secondary CT inputs to the primary and secondary bus 1 differential relays, as shown in Fig. 3.112. Any primary current flow in the denergized bus between the personnel safety grounds will be reflected in the secondary CT circuit, which feeds the bus differential relays.

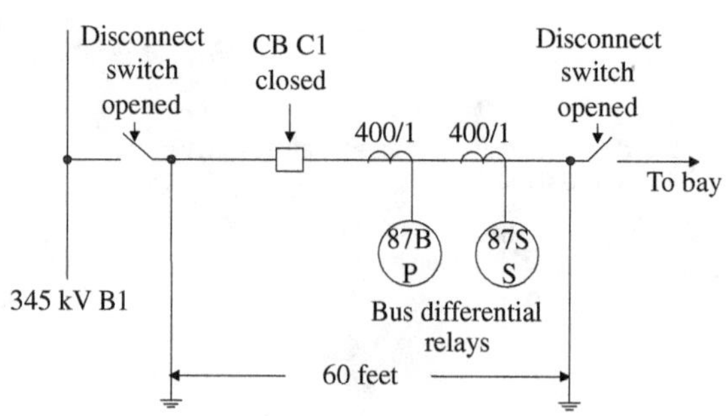

Fig. 3.112 CB C1 loop with the breaker closed, ground chains applied, and the disconnect switches opened.

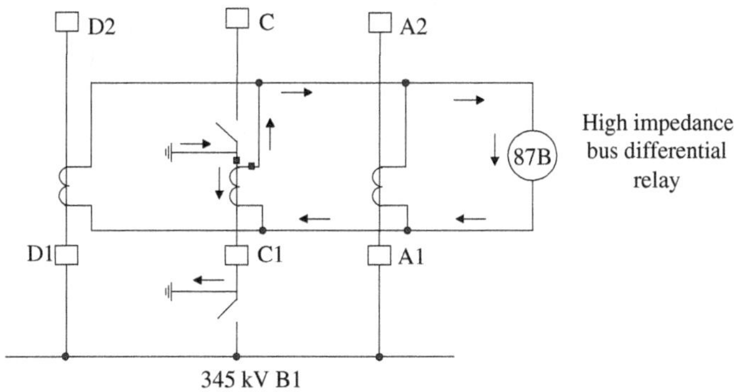

Fig. 3.113 One-line ac diagram for the CT connections to bus differential relay 87B.

As shown in Fig. 3.110, the current flow was due to zero-sequence current mutual coupling between the breaker loop and the 345-kV circuits running above the isolated breaker. When the A-g fault occurred on L1, zero-sequence voltage was induced in the breaker closed loop, and with ground chains applied at both ends, zero-sequence current was forced to flow on all the phases. The induced zero-sequence voltage was created by the flow of zero-sequence currents feeding the A-g fault.

Any current flow in the 345-kV isolated loop will be reflected in the CT secondary circuit feeding bus relays and will appear as an operating quantity that will flow in the high-impedance differential relay operating coil circuit. This secondary current flow may be small; however, it may generate sufficient voltage when it flows through the 2600-Ω resistance of the coil of the overvoltage relay. When the voltage generated exceeds the setting of the relay, it will cause undesired operation of the bus differential relays.

Figure 3.113 illustrates induced current flow from the breaker isolated loop to the relay. The zero-sequence current flow in the primary circuit will be transformed to the bus differential relay circuit through the CT secondary circuit. This current is considered a differential operating current that is not balanced by any other current and will therefore flow to the relay high-impedance operating coil of 2600 Ω, as shown in Fig. 3.113. Therefore, a voltage was created across the high-impedance coil that apparently exceeded the relay setting and caused the false operation. This type of phenomenon can be simulated in a short-circuit study to confirm the undesired operation caused by mutual coupling phenomena. A short-circuit study zero-sequence mutual coupling simulation produced a 74-A primary, and with a CT ratio of 400 : 1, the relay current is 0.185 A, yielding 481 V across the relay and thus exceeding the high-impedance relay setting of 150 V. The currents calculated are shown in Fig. 3.114.

The design implementation of a high-impedance bus relay is based on terminating all bus CT wiring at a terminal cabinet in the yard in a spot that produces equal CT cable runs to individual CT input to the scheme. A four-conductor cable will connect the yard termination cabinet to the high-impedance relay located inside the relay building. Therefore, there is no way to isolate individual CTs via test switches as a corrective action to prevent future undesired operation of the relay.

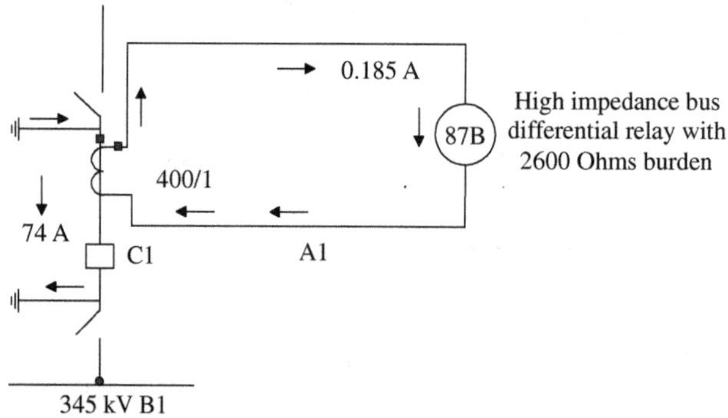

Fig. 3.114 Secondary CT current flow through the high-impedance bus differential relay.

3.40 APPEARANCE OF NONSINUSOIDAL NEUTRAL CURRENT DURING THE CLEARING OF THREE-PHASE FAULTS

Faults normally occur in a power system when insulation fails. The insulation failure mechanism can be slow and an arc will be established at the voltage peak. Fault currents are therefore symmetric and contain no dc offsets. The presence of dc offset depends on the fault incident point and is generated due to the fact that current cannot increased instantly through an inductor, and the circuit phase angle must be preserved. Although the majority (85%) of faults occurring in a power system are phase-to-ground faults, three-phase faults can also occur. Three-phase faults can occur when restoring a system following maintenance work with safety ground chains left on by mistake. Three-phase faults can also evolve from phase-to-ground and phase-to-phase faults during lightning hits followed by transmission tower voltage buildup and within a confined switchgear enclosure during the spread of ionized clouds. Asymmetrical currents appear with a different degree of asymmetry during three-phase faults, resulting in unequal asymmetrical current magnitude. The decay of dc components will eventually bring the three currents close in magnitude as they become symmetric.

Figure 3.115 shows line L1 DFR-monitored currents and voltages during a three-phase fault caused by a lightning hit. Figure 3.116 shows the DFR record for line L1

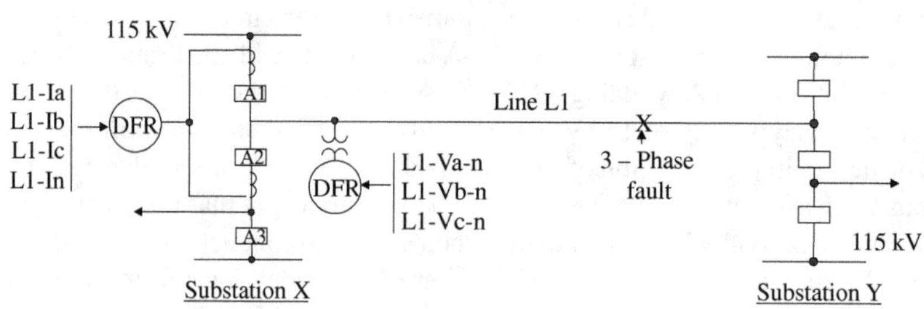

Fig. 3.115 Line L1 DFR-monitored voltages and currents.

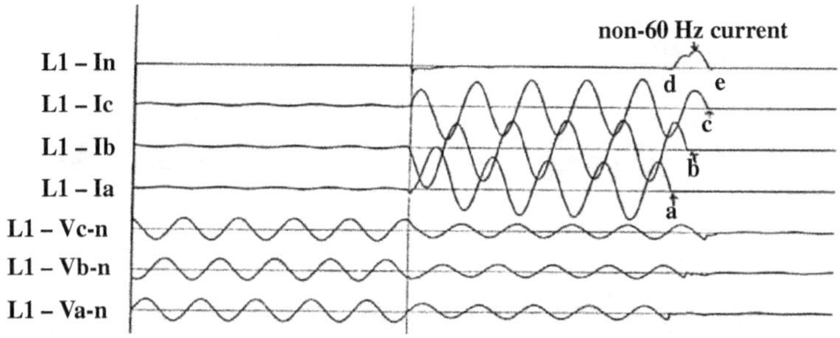

Fig. 3.116 DFR record showing line L1 non-60-Hz neutral current during interruption of a three-phase fault.

voltages and currents. The fault occurred first as a double-phase-to ground fault caused by a direct lightning strike, followed by a buildup of voltage through the tower footing resistance, raising the potential to flash over and evolving the fault to a three-phase fault. The fault lasted 4.5 cycles on phase A and cleared at point a, as shown in Fig. 3.116. Phase B cleared at point b after an additional 5.5 ms (120°). Phase C cleared at point c after an additional 5.5 ms (120°). As a result, the neutral current trace L1-In, which is the sum of Ia + Ib + Ic, is shown as a nonfundamental (60-Hz) component marked by points d and e in Fig. 3.116. Figure 3.117 documents phase and neutral currents during a three-phase fault. The neutral current trace L1-In illustrates a non-60-Hz component at the initiation of the fault, due to varying dc offset currents for the phases and at the interruption of the fault first on phase C at point a, followed by phase B at point b and phase A at point c. Generation of neutral currents during

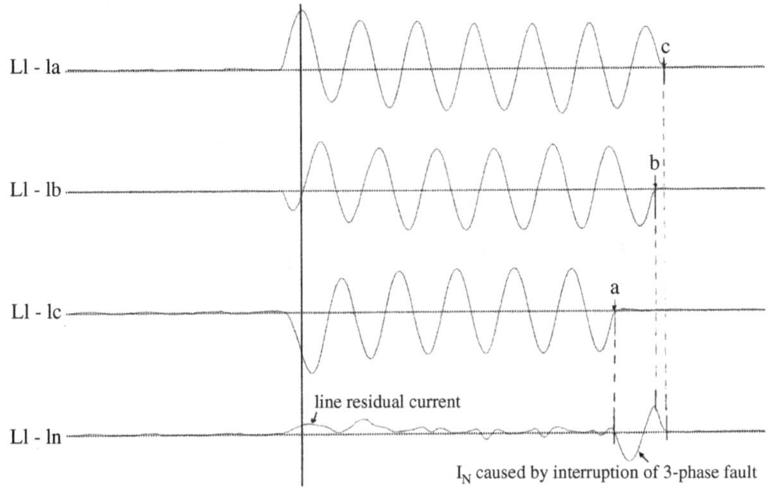

Fig. 3.117 DFR record showing line L1 phase and non-60-Hz neutral currents during inception and interruption of the three-phase fault.

interruption of a three-phase fault may trigger some ground instantaneous fault detector elements with the generation of targets and event records.

3.41 CURRENT REVERSAL ON PARALLEL LINES DURING FAULTS

On parallel lines, clearing external faults for one of the lines may result in a sudden reversal of fault current on that line. Figure 3.118 shows a system one-line diagram and line L2 DFR-monitored currents and voltages that can be used to illustrate the current reversal phenomenon. Since the C-g fault on line L1 was close to substation Y, the ground fault current first flowed on line L2 from substation X toward substation Y, as shown in Fig. 3.118. The DFR record shown in Fig. 3.119 documents the voltages and currents for line L2 at substation X for the C-g fault occurring on line L1. Therefore, a carrier blocking signal was generated on the line L2 pilot relay to prevent the X end from tripping falsely for the external C-g fault on line L1. As shown in Fig. 3.119, the Y end of line L1 cleared the fault first in 3 cycles by Z1 operation tripping CBs C1 and C2. However, the C-g fault is still in the system and is being fed from the remote end of L1 at substation X. This forces instantaneous reversal of the L2 ground current, shown in trace L1-In in Fig. 3.119, to flow from substation Y toward substation X to feed the C-g fault. The reversal of ground current on line L2 is also shown in the one-line diagram of Fig. 3.120 following the tripping of CBs C1 and C2

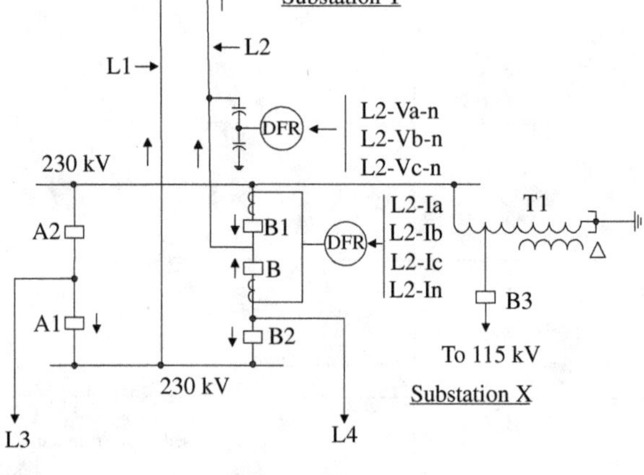

Fig. 3.118 One-line diagram showing the two parallel lines L1 and L2 and line L2 DFR-monitored analog signals at substation X.

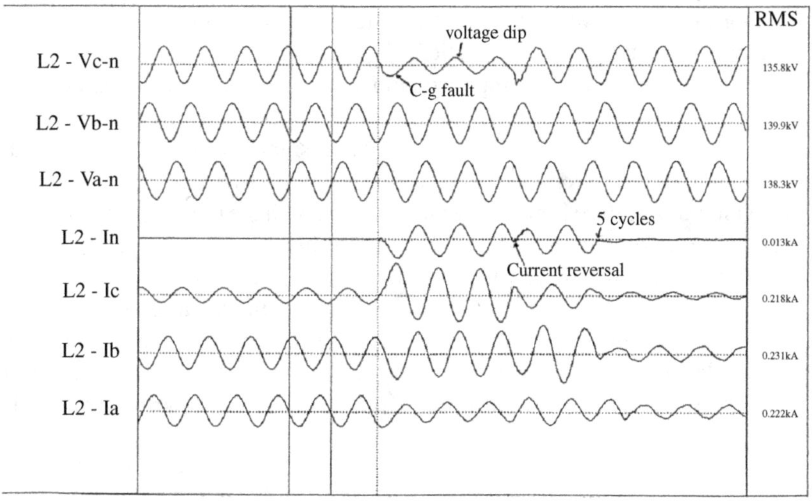

Fig. 3.119 Substation X DFR record showing current reversal on line L2 after tripping of the line L1 end at substation Y.

for the faulted line L1 at substation Y. The fault was eventually cleared in an additional 2 cycles by the line L1 pilot relay scheme to reflect a total clearing time of 5 cycles clearing for the X end of L1 by tripping CBs A1 and B2. The transient blocking feature for the line L2 terminal at substation Y kept the L2 relay terminal at Y honest by

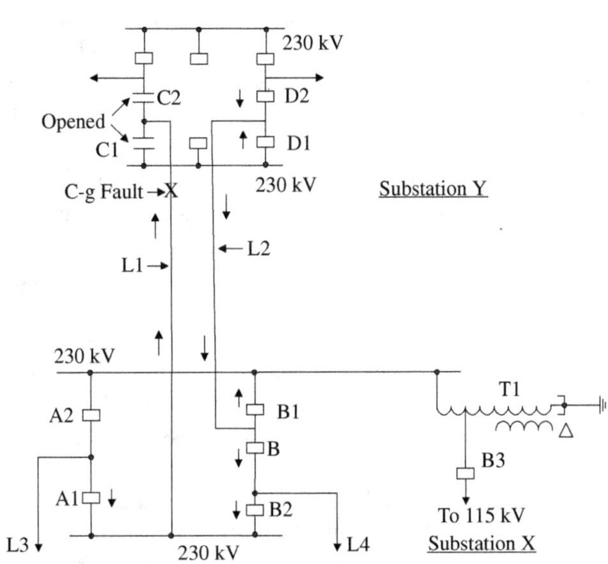

Fig. 3.120 One-line diagram showing current reversal on line L2 after the tripping of CBs C1 and C2 of the line L1 end at substation Y.

prolonging the carrier blocking signal for the directional blocking scheme despite current reversal in the tripping direction.

Therefore, for directional comparison blocking systems, the tripping elements at the end, being blocked, must reset before the blocking channel signal is removed. The opposite end of the line must transmit a blocking signal before the remote end trip elements operate. For overreaching directional comparison transfer trip systems, a transient blocking feature is normally incorporated with the pilot scheme to cope with any sudden reversal of fault current. A transient blocking feature provides coordination time for the pilot relay communication system. This is normally included to ensure that the channel trip system is removed before the local relay operates upon current reversal. Permissive underreaching transfer schemes offer an advantage in the use of parallel lines during current-reversal situations. This is due to the use of Z1 underreaching elements, which will not reach faults on the opposite parallel line.

3.42 FERRANTI VOLTAGE RISE

The Ferranti effect occurs primarily on high-voltage transmission lines and underground cables. The *Ferranti effect* is a rise in voltage occurring at the receiving end of a long transmission line relative to the voltage at the sending end, which occurs when the line is open at one end. This effect is due to the voltage drop across the line inductance (due to charging current) being in phase with the sending end voltages, resulting in a voltage rise at the receiving end. Therefore, both capacitance and inductance are responsible for producing this phenomenon. The Ferranti effect is more pronounced the longer the line and the higher the voltage applied. The relative voltage rise is proportional to the square of the line length. Due to the high capacitance, the Ferranti effect is much more pronounced in underground cables, even short cable lengths. High voltage can also occur during either light loading or when the load is disconnected.

The Ferranti rise effect is due to the absorption of a capacitance charging current when a line is energized but open-ended. The Ferranti rise effect is an overvoltage condition associated with high-voltage lines whose receiving ends are open. The Ferranti voltage phenomenon is a rise in voltage from the closed end to the open end. The magnitude of the overvoltage depends on the length of the open-ended line and the strength of the system tied to the closed end of the line. Overvoltages greater than 10% above normal can easily occur. The current flow into an open-ended line is the charging current for the natural capacitance of the line. The leading reactive current passes through the line's inductive reactance and causes a voltage rise. The highest voltage occurs at the open end.

Figure 3.121 illustrates a II model that approximates the impedance of an open-ended transmission line, with the natural capacitance of the line represented by placing shunt capacitance lumped at both ends of the line. In reality the resistance, inductive reactance, and capacitive reactance values of a line are distributed along the entire length of the line. The current flow is the charging current of the line-to-neutral

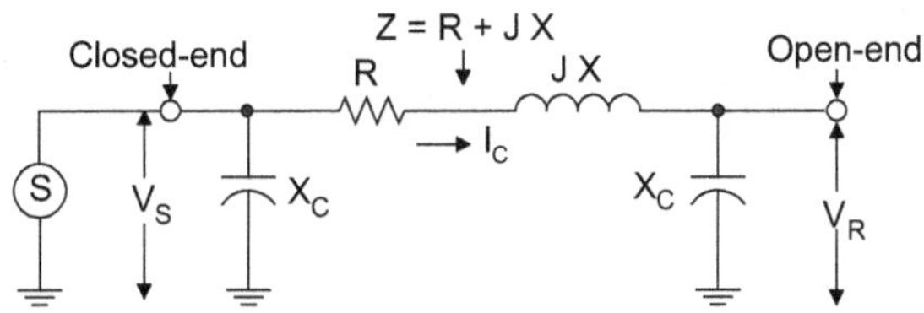

Fig. 3.121 Transmission-line equivalent circuit model.

capacitance, which is spread over the entire length of the line (no active power flow). Charging current is leading reactive current flow. When leading reactive current passes through the line's inductive reactance it causes a voltage rise from the closed end to the open end. Figure 3.122 shows a vector diagram representing line current, impedance, and voltage, while Fig. 3.123 reveals the voltage magnitude profile for an open-ended line.

For extra high voltage (EHV) transmission systems, shunt reactors are used to clamp the line open-ended voltage to a safer value. On EHV transmission systems, shunt reactors are considered to be an integral part of the lines and it is assumed that an outage of shunt reactors would necessitate removing some lines from service. In addition, overvoltage relays are applied at the EHV line ends with a level setting for voltage and time. The relays are coordinated with the equipment voltage capability and staggered with time coordination to avoid tripping many lines at the same time.

Figure 3.124 shows a simplified high-voltage system one-line diagram and the DFR-monitored voltages and currents for line L1 to demonstrate the Ferranti voltage rise phenomenon. The DFR record in Fig. 3.125 shows a C-g fault being cleared from substation X in 5 cycles. It also reveals that line voltage traces L1-Vb-n and L1-Vc-n have an overvoltage Ferranti rise effect value 22% above the nominal voltage as soon

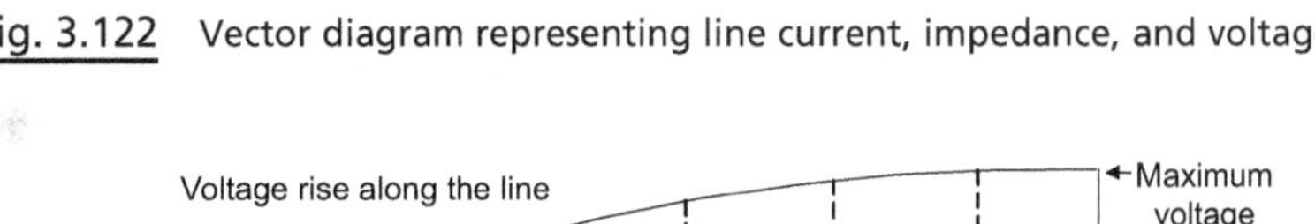

Fig. 3.122 Vector diagram representing line current, impedance, and voltage.

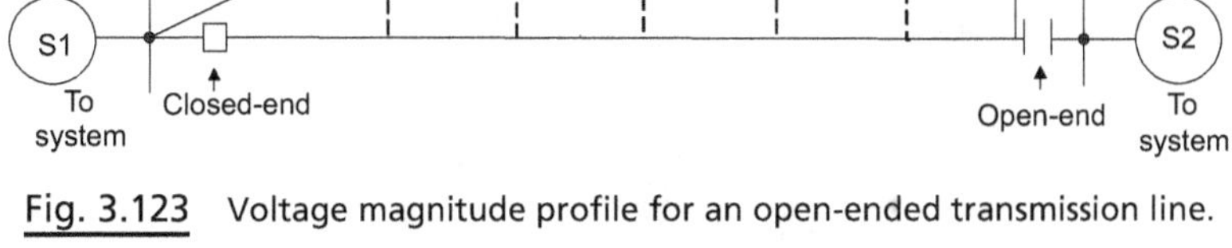

Fig. 3.123 Voltage magnitude profile for an open-ended transmission line.

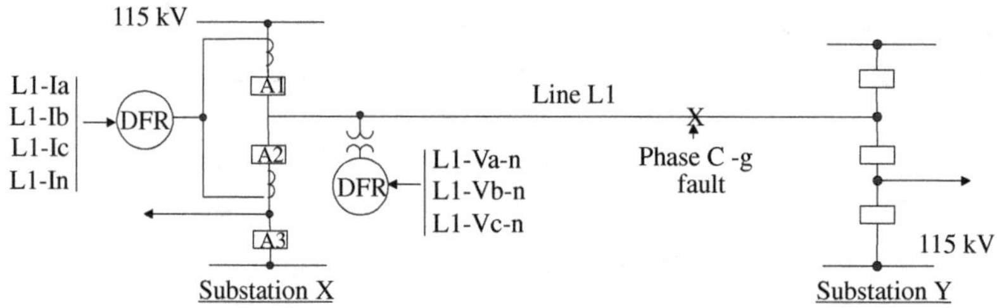

Fig. 3.124 One-line diagram showing line L1 DFR-monitored voltages and currents.

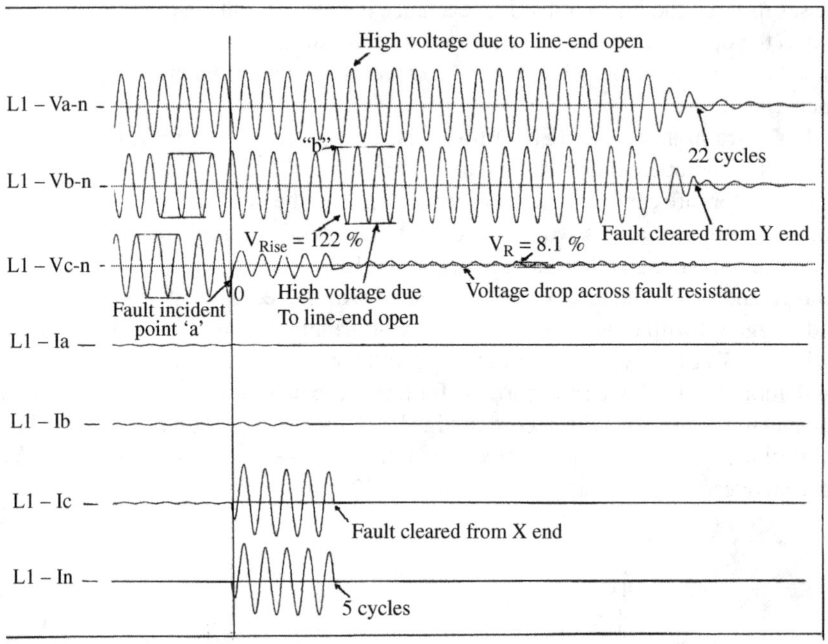

Fig. 3.125 DFR record showing high voltage due to an open line end during clearing of the C-g fault.

as the fault is cleared from the substation X end of the line at point b, thus simulating an open end that lasted for 14 cycles, until the remote end of the line at substation Y cleared the fault.

3.43 VOLTAGE OSCILLATION ON EHV LINES HAVING SHUNT REACTORS AT THEIR ENDS

Long EHV transmission lines are generally compensated by means of shunt reactor banks. Shunt reactors are added to the system to compensate for the capacitive charging current of EHV transmission lines. Shunt reactors are normally connected on the line side of the circuit breakers so that they can be dropped whenever the line is tripped or during heavy line-loading conditions. The reactive power (MVAR) load of a shunt reactor is proportional to the square of the phase-to-ground voltage and is independent of line loading.

Oscillation of energy between the line or cable charging system and the shunt reactors used for feeder compensation will take place following isolation of the feeder. Shunt reactor banks are used to absorb excessive reactive power from a power system to help reduce and control system voltages. Reactive power provided to the system from a transmission line's natural capacitance can be stated mathematically as

$$\text{MVAR supplied} = \frac{V^2}{X_C}$$

The reactive power used by a line can be stated mathematically as

$$\text{MVAR consumed} = I^2 \times X_L$$

where V is the system voltage, X_C the line equivalent capacitive reactance, I the line loading current, and X_L the line equivalent inductive reactance. Shunt reactor protection should be designed to cope with oscillation phenomena. Electrical pole disagreement detection, when used with shunt reactor breakers, should be set above the magnitude of the oscillating current.

3.43.1 Voltage Oscillation During an EHV A-C-g Fault

Figure 3.126 shows a simplified EHV system one-line diagram and DFR-monitored voltages and currents for line L1 to demonstrate voltage oscillation phenomena. The DFR record in Fig. 3.127 reveals an A-C-g fault occurring on line L1 and cleared successfully in less than 3 cycles from both ends of the line. Immediately following isolation of the fault, voltage oscillation caused by the exchange of line capacitive energy ($\frac{1}{2}CV^2$) with shunt reactor inductive energy ($\frac{1}{2}LI^2$). As a result, current flow in the loop formed by shunt reactors at both ends of the line is shown in traces

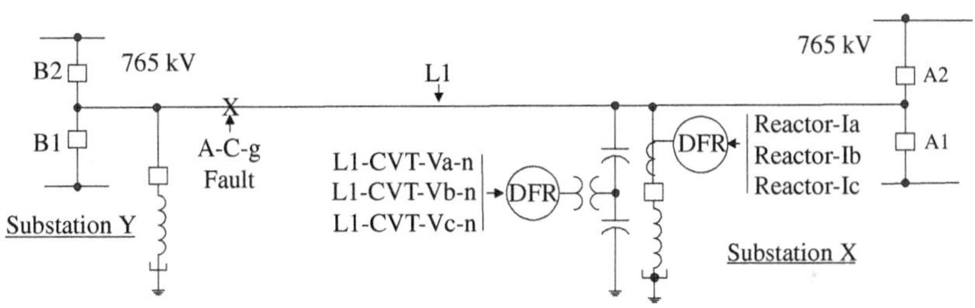

Fig. 3.126 One-line diagram showing line L1 DFR-monitored voltages and currents.

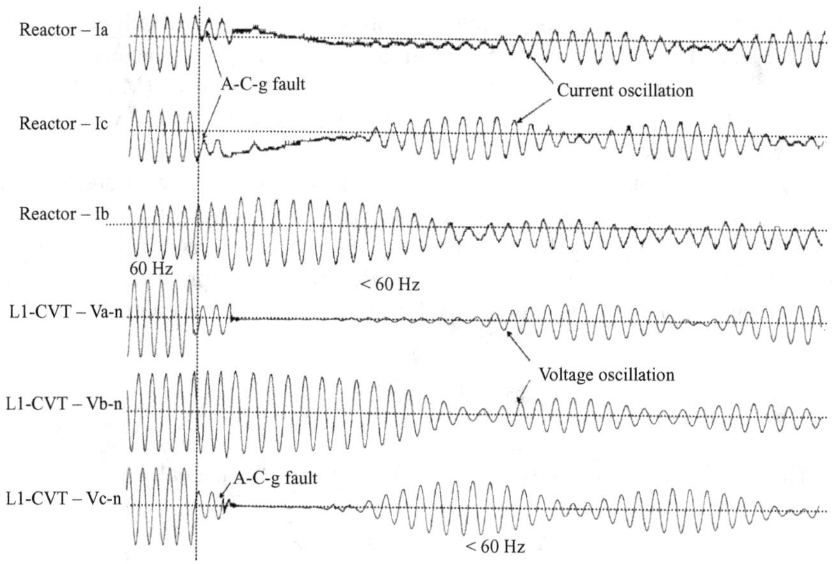

Fig. 3.127 DFR record showing line L1 voltages and shunt reactor current oscillations following clearing of the A-C-g fault.

reactor-Ia, reactor-Ib, and reactor-Ic. The frequency of the voltage oscillation is a function of the degree of shunt reactive compensation and is shown as less than 60 Hz with shunt reactors connected at both ends of the line.

3.43.2 Voltage Oscillation During an EHV A-g Fault

Figure 3.128 shows a simplified EHV system one-line diagram and DFR-monitored voltages and currents for line L2 to demonstrate voltage oscillation phenomena. The DFR record in Fig. 3.129 reveals an A-g fault occurring on line L1 and cleared successfully 2.5 cycles from both ends of the line. Immediately following the isolation

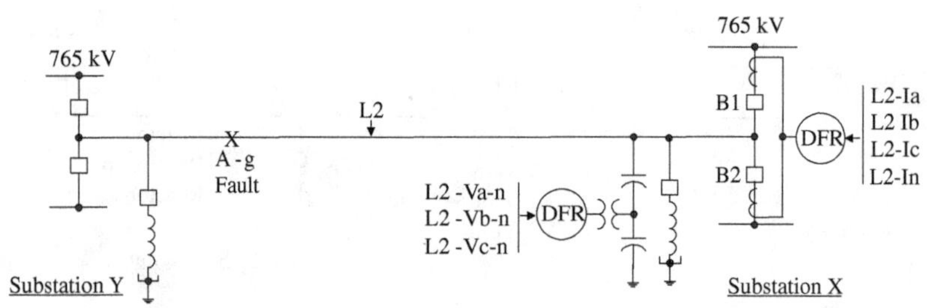

Fig. 3.128 Line L2 DFR-monitored voltages and currents.

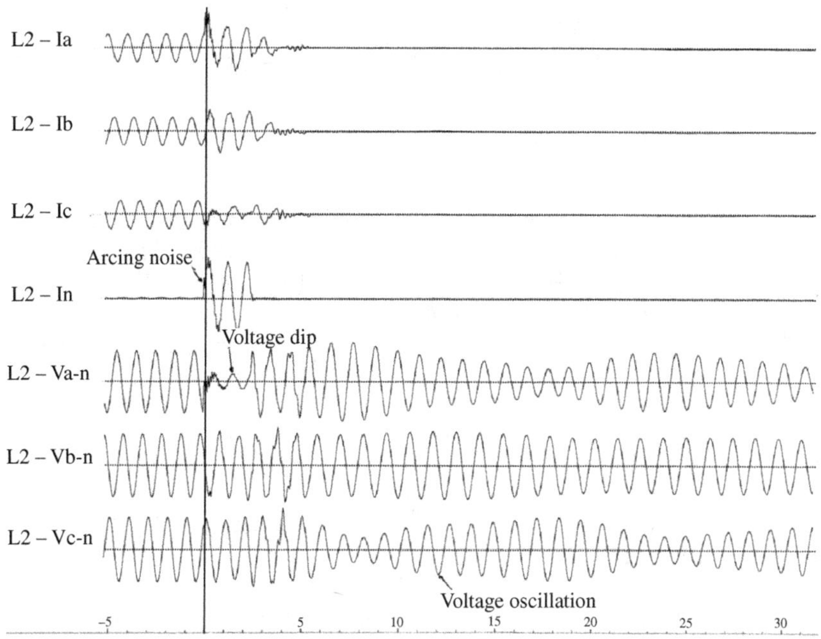

Fig. 3.129 DFR record showing line L1 voltage oscillations after clearing of the A-g fault.

of the fault, the DFR voltage traces for line L2 started to oscillate with a frequency lower than 60 Hz. The voltage oscillation is caused by the exchange of line capacitive energy with shunt reactor inductive energy.

3.43.3 Voltage Oscillation During an EHV Switching Operation

Figure 3.130 shows a simplified 765-kV one-line diagram and DFR-monitored voltages for line L1 and shunt reactor currents at substation X to demonstrate voltage oscillation phenomena.

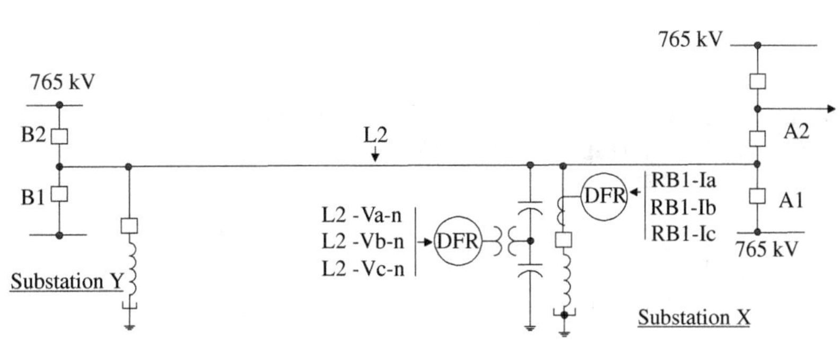

Fig. 3.130 One-line diagram showing line L2 DFR-monitored voltages and currents during the switching operation.

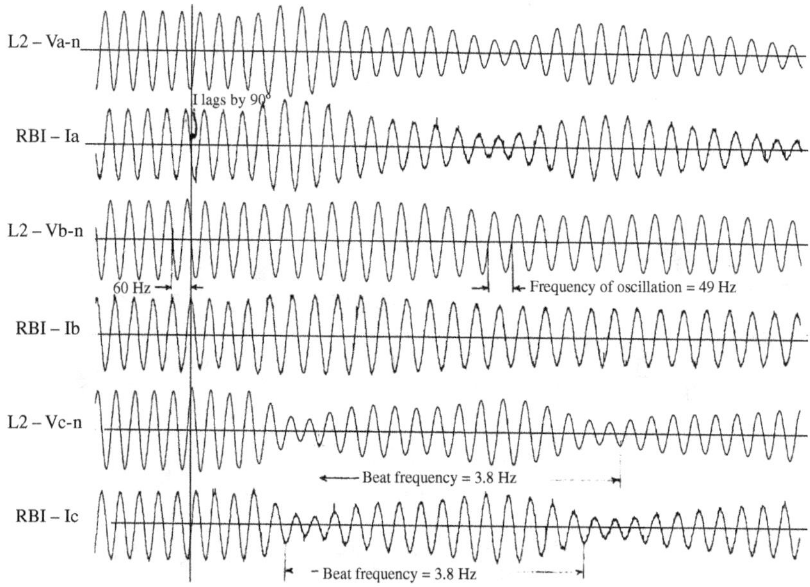

Fig. 3.131 DFR record showing line L2 voltage and shunt reactor current oscillations following a switching operation on the line.

The DFR record in Fig. 3.131 reveals a normal line switching operation that was executed by tripping breakers at both ends of the line. Immediately following isolation of the line, voltage oscillation caused by the exchange of line capacitive energy with shunt reactor inductive energy took place. As a result, current flow in the loop formed by shunt reactors at both ends of the line is shown in traces RB1-Ia, RB1-Ib, and RB1-Ic. The frequency of the voltage oscillation is a function of the degree of shunt reactive compensation and is shown to be 49 Hz. The beat frequency of the voltage oscillation is calculated and shown to be 3.8 Hz. The DFR record trace RB1-Ia confirms that pre-switching reactor current lags the deriving 60-Hz voltage by 90° as well as during the oscillation period, with frequency calculated as 49 Hz. The DFR record also confirms that on a per-phase basis, the line voltage oscillation is identical to the reactor current oscillation.

3.44 LIGHTNING STRIKE ON AN ADJACENT LINE FOLLOWED BY A C-g FAULT CAUSED BY A SEPARATE LIGHTNING STRIKE ON THE MONITORED LINE

Figure 3.132 shows a simplified system one-line diagram and the oscillograph-monitored voltages and currents for line L1. The oscillogram shown in Fig. 3.133 reveals a lightning strike on an adjacent line. The oscillograph was started by line L1 neutral current generated by a lightning strike on an adjacent line. Fourteen cycles later, line L1 was hit by another lightning strike, causing a phase C-g fault which was cleared locally in 3 cycles and remotely in 4.5 cycles. The fault current shown is asymmetrical,

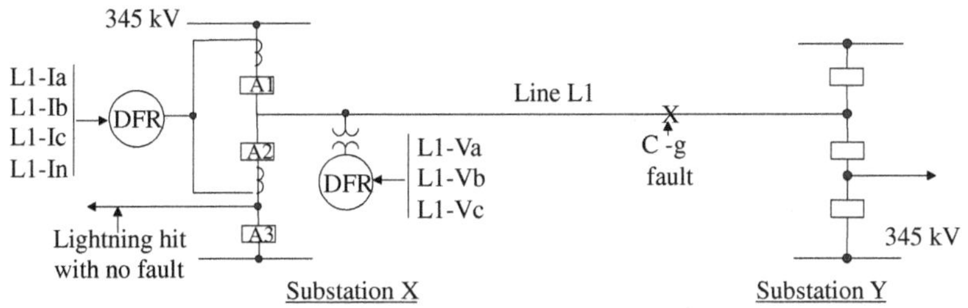

Fig. 3.132 One-line diagram showing line L1 DFR-monitored voltages and currents.

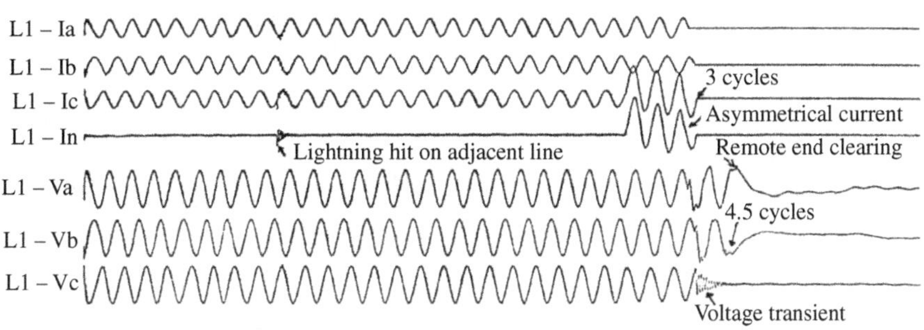

Fig. 3.133 Substation Y DFR showing L1 voltages and currents prior to and during the C-g fault.

which is normally dependent on the fault incident point, related to the random nature of the lighting strike and the lightning hit point on the faulted phase voltage.

3.45 SPILL OVER OF A 345-kV SURGE ARRESTER USED TO PROTECT A CABLE CONNECTION, PRIOR TO ITS FAILURE

A surge arrester employed to protect a high-pressure fluid-filled 345-kV cable failed, causing a phase-A-g fault. The fault was cleared from the system successfully. Half a cycle earlier the surge arrester was triggered from the trigger line at the voltage peak with high-frequency current flow in a process defined as surge arrester spillover. An arrester total failure then occurred, causing a solid phase-A-g fault. A surge arrester can begin a spillover process in which current will be conducted for a short duration. Conductions can take place and the arrester may not be able to reseal. As a result, the surge arrester may fail, causing a phase-to-ground fault. Figure 3.134 shows a system one-line diagram where a surge arrester is used to protect a 345-kV pipe-type cable connecting the unit transformer to the substation. In addition, Fig. 3.134 shows the DFR-monitored voltages and currents used to define the arrester spillover phenomenon. The DFR record of Fig. 3.135 shows the cable DFR-L1 monitored neutral current where the surge arrester began a spillover process that eventually led to its failure. The spillover process began 0.5 cycle prior to the arrester failure. The fault was a bolted (solid) phase A-g as revealed by the bus-Va-n voltage trace and lasted for 4 cycles.

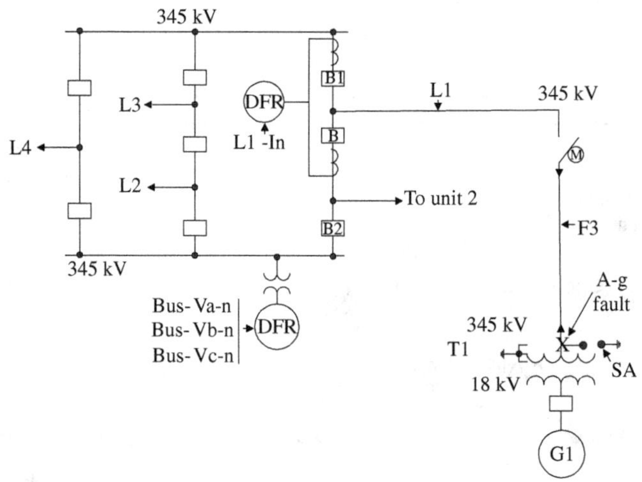

Fig. 3.134 System one-line diagram with DFR-monitored voltages and currents.

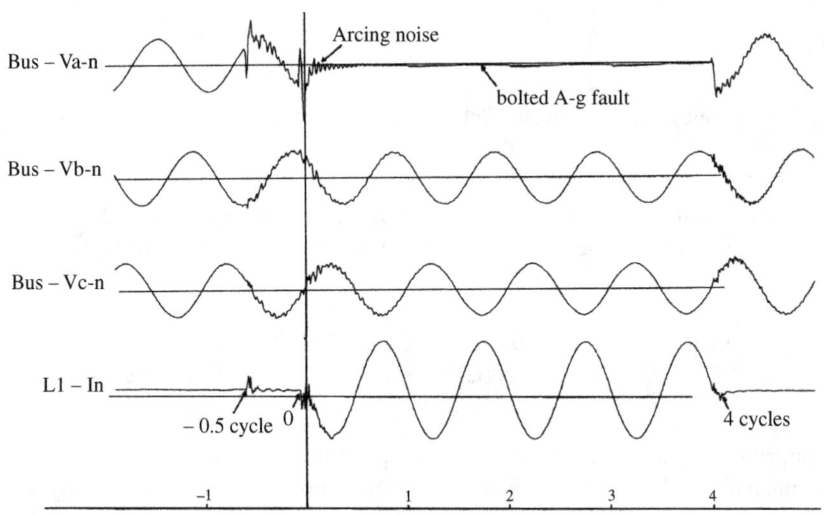

Fig. 3.135 Substation X DFR record showing surge arrester SA spillover prior to the A-g fault.

3.46 SCALE SATURATION OF AN A/D CONVERTER CAUSED BY A CALIBRATION SETTING ERROR

DFR analog signals are scaled using a software setting that represents either a voltage or a current. The scaling of currents should accommodate loading flow as well as maximum asymmetrical fault currents. This dynamic ratio between these two currents should be accommodated by the A/D converter range and should not exceed the full

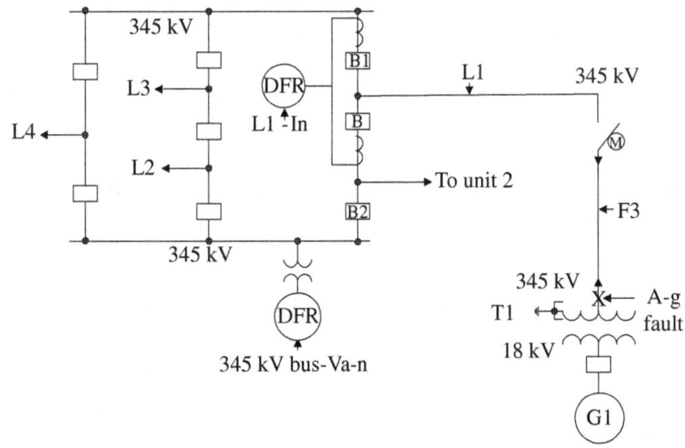

Fig. 3.136 System one-line diagram showing DFR-monitored voltages and currents.

range. The signals sampled will be truncated to fit in the A/D maximum range value only when the magnitude of the fault current instantaneous samples (represented as voltages) exceeds the maximum A/D range allocated. As a result, the instantaneous current value will be represented at a value that is less than the actual value.

Figure 3.136 shows a system one-line diagram where a phase A-g fault has occurred. In addition, Fig. 3.136 illustrates the DFR-monitored bus voltage and the L1 neutral current used to define overscaling of the A/D phenomenon. The DFR record in Fig. 3.137 shows the cable L1 monitored neutral current where the sine wave representing the current was clipped for both the positive and negative half-cycles. The fault was a solid A-g fault occurring at the voltage peak and resulting in symmetrical current that lasted for 4 cycles. This overscaling problem should be avoided to make

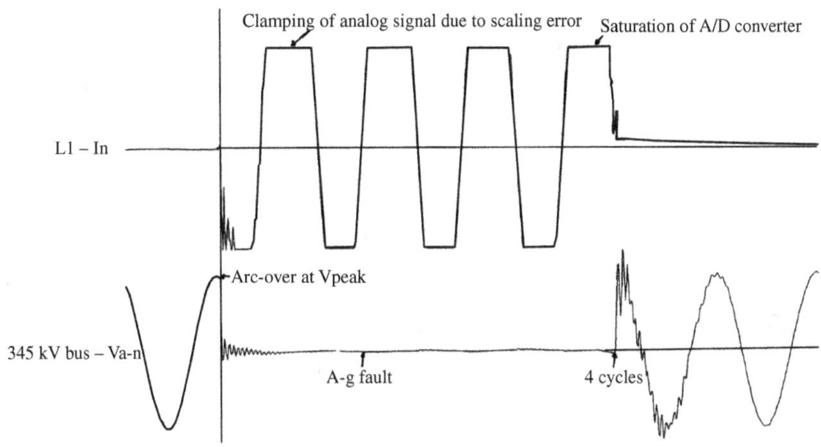

Fig. 3.137 Substation X DFR record showing clamping of the L1 neutral current during the A-g fault.

possible the faithful reproduction of the current sampled and an accurate RMS current value that is normally calculated using the samples. This will allow the distance elements to measure accurate fault impedance without element under or over-reaching.

3.47 APPEARANCE OF SUBSIDENCE CURRENT AT THE INSTANT OF FAULT INTERRUPTION

Figure 3.138 shows a system one-line diagram and the DFR-monitored currents. Secondary currents may not be interrupted immediately in many current transformers. Magnetic flux in the current transformer core may reach a high value, then begin to decay over time. The decaying current is characterized by nonzero crossings and eventually subsides to zero over an extended period of time. This decaying current in the transformer secondary circuit is referred to as *subsidence current*. Figure 3.139 illustrates ground currents and tertiary current during a phase B-g fault on line L1. The ground fault lasted for 3.75 cycles and was interrupted at current zero on line L1 and the remaining lines. However, the tertiary current exhibits a subsidence current that decayed over time to a zero value.

Subsidence current can cause the overcurrent element of the protective relay to remain picked up (indicating erroneously that primary current is still flowing) long after the circuit breaker has, in fact, actually interrupted the primary current. This extended pickup of overcurrent elements is called *delayed dropout*. If an overcurrent element is used to indicate breaker status and has a very long dropout delay, the control circuits of the power system might make a serious system error, indicating that the primary current has not been interrupted, and then initiating remedial and undesirable action, such as tripping additional breakers.

The subsidence current in the secondary of the current transformer creates the same problems regardless of the characteristics of the protective relay: that is, whether the relay elements are electromechanical in nature or in a microprocessor implementation in which successive samples of secondary current are processed using

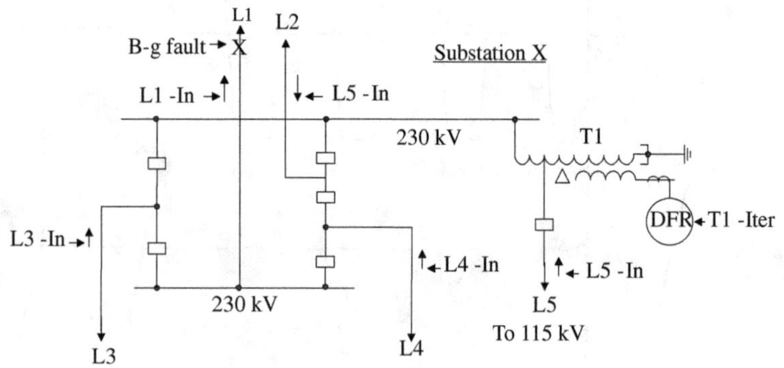

Fig. 3.138 Substation X one-line diagram showing DFR-monitored currents.

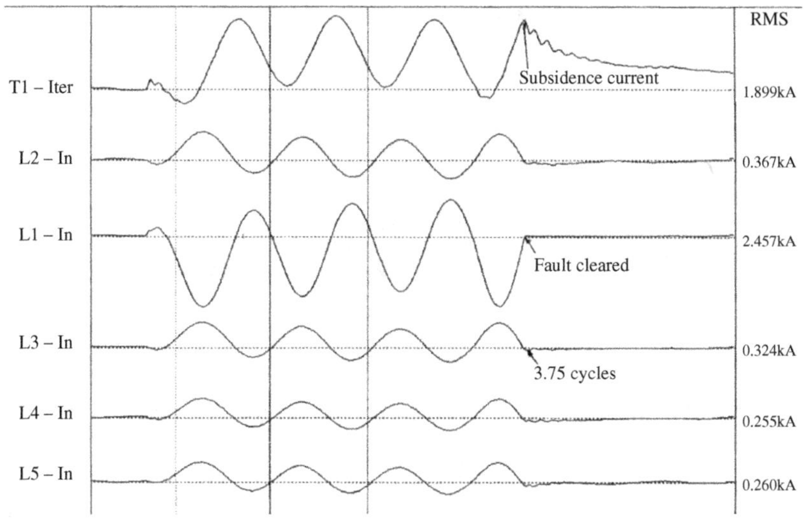

Fig. 3.139 Neutral current fault contributions from the lines as well as the subsidence current of the T1 tertiary.

computer algorithms. Subsidence current is neither dc nor a linear ramp signal but, instead, is a decaying exponential. Accordingly, the filtering methods for protective relays cannot fully reject the decaying exponential subsidence current, which in turn retains the overcurrent elements set to a low threshold. Hence, the problem with subsidence current remains, even with modern microprocessor relays.

3.48 ENERGIZING OF A MEDIUM VOLTAGE MOTOR THAT HAS AN INCORRECT FORMATION OF THE STATOR WINING NEUTRAL

A 3000-hp gas compressor for a combined-cycle plant arrived from the factory with the stator winding neutral unconnected. This gives the user the flexibility to determine the needed motor rotation direction based on the mechanical load connected. A mistake was committed in wiring and connecting the motor to the 4.16-kV supply. Upon energization of the motor during the commissioning phase, the motor-associated breaker was tripped by the motor numerical multifunction relay. The tripping of motor-associated 4.16-kV breaker A was initiated by the instantaneous overcurrent element of the motor numerical relay.

Figure 3.140 shows a system one-line diagram and the numerical relay–monitored currents. The numerical relay oscillography record is shown in Fig. 3.141, where the three phase currents recorded by the relay are not balanced. Analyzing the record, it can be concluded that the phase C current is equal to the negative sum of the other two currents, A and B ($I_c = -I_a - I_b$). Figure 3.142 shows the as-found condition, where the neutral is formed incorrectly using terminals T3, T4, and T5 instead of the correct

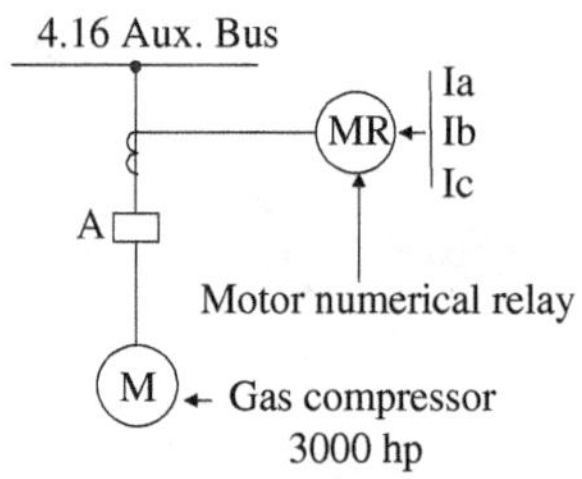

Fig. 3.140 One-line diagram showing a motor connection.

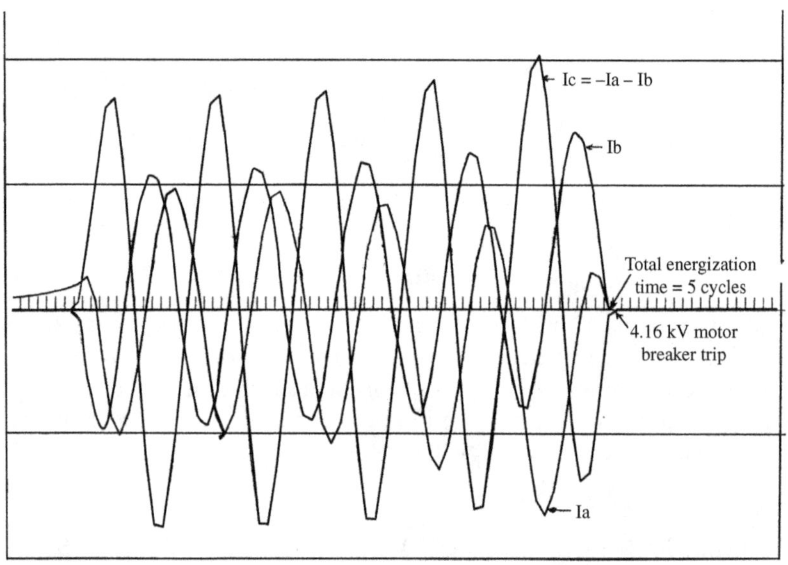

Fig. 3.141 Motor numerical relay oscillography record that generated a trip output.

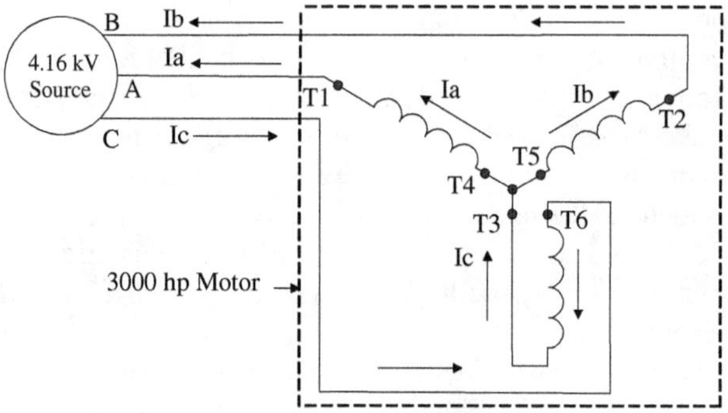

Fig. 3.142 Motor phase windings with the incorrect terminals to form the neutral.

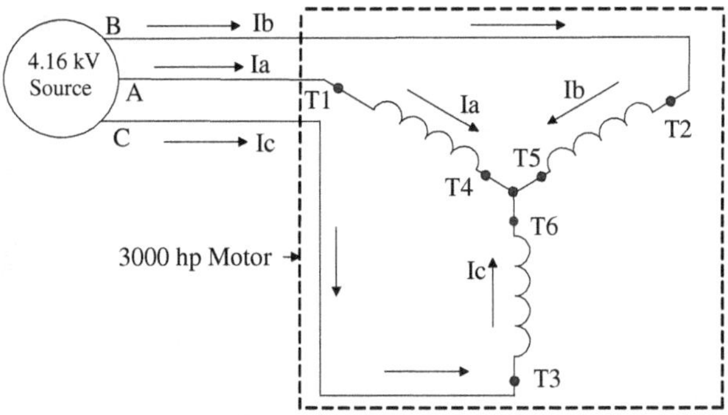

Fig. 3.143 Motor three-phase windings with the correct terminals to form the neutral.

arrangement: T4, T5, and T6. Figure 3.143 illustrates correct formation of the neutral for this particular application.

3.49 PHASE ANGLE CHANGE FROM LOADING CONDITION TO FAULT CONDITION

During line loading conditions the phase angle between the voltage and current is roughly between $-30°$ and $+30°$ (near the real axis), depending on current direction and VAR dispatch. When a fault occurs on a transmission line, an abrupt change of phase angle between voltage and current from a normal load flow to a fault condition will take place. The current will lag the voltage by an angle greater than 60° during line fault conditions, depending on the line impedance, which varies with voltage level. The phasor diagram shown in Fig. 3.144(a) documents the current phasor change, which during loading will lag the deriving voltage by an angle φ_{load}. The loading current will change abruptly to a fault current that lags by an angle φ_{fault} during a fault condition. The angle φ_{fault} will be equal to the line impedance angle if we assume that the fault resistance $R_F = 0$. Figure 3.144(b) shows the swinging of the impedance from Z_{load} during loading condition to Z_{fault} in high-speed time during faults.

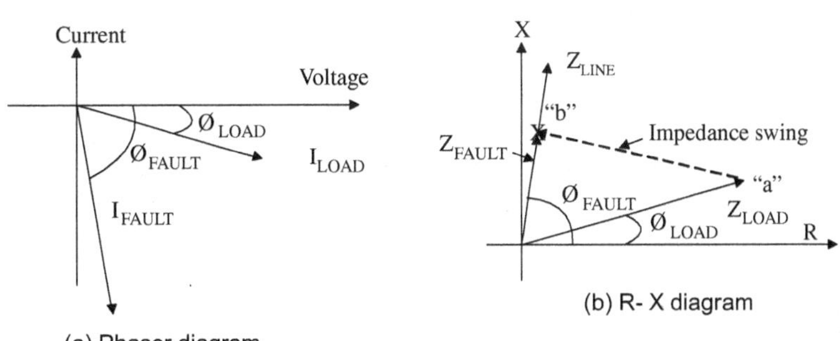

Fig. 3.144 Phasor and *R–X* diagrams showing the change from load to fault.

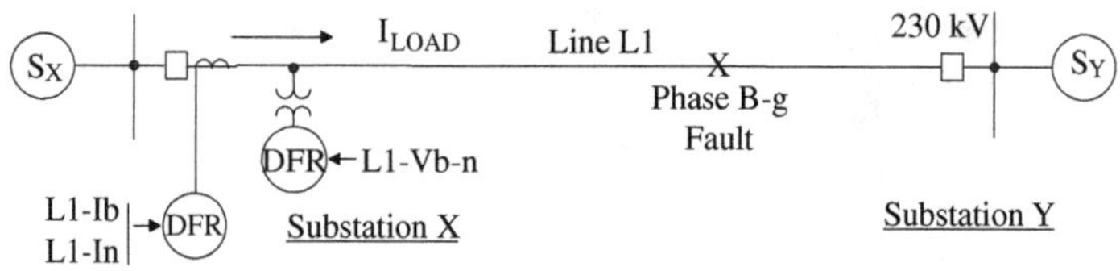

Fig. 3.145 System simplified one-line diagram with DFR-monitored currents and voltages.

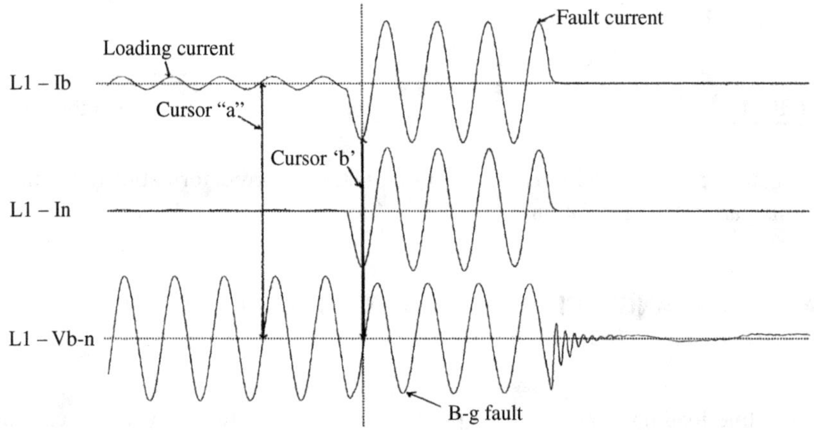

Fig. 3.146 DFR record showing the change from loading to fault conditions.

Figure 3.145 shows a simplified system one-line diagram with DFR-monitored currents and voltages at substation X during the phase B-g fault. Figure 3.146 confirms the load flow condition using cursor a, where the phase B current trace L1-Ia is nearly in phase with phase B-to-neutral voltage trace L1-Vb-n. Figure 3.146 also confirms the fault condition using cursor b, where the phase B current trace L1-Ia lags the phase B-to-neutral voltage trace L1-Va-n by nearly 90°. Therefore, DFR analysis confirms the concepts shown in Fig. 3.144 for the phasor and *R–X* diagrams during the abrupt change from load to fault for line L1.

REFERENCES

Blackburn, J. L. *Protective Relaying Principles and Applications*. New York: Marcel Dekker, 1987.

Blackburn, J. L. *Symmetrical Components for Power Systems Engineering*. New York: Marcel Dekker, 1993.

Elmore, W. A., Ed. *Protective Relaying Theory and Applications*. New York: Marcel Dekker, 2000.

Greenwood, A. *Electrical Transients in Power Systems*. New York: Wiley-Interscience, 1971.

Ibrahim, M. A. Mitigation of mutual coupling effects on the reach of ground distance relays protecting a 345-kV System. Presented at the Georgia Tech Protective Relaying Conference, May 3–5, 2000, Atlanta, GA.

Ibrahim, M. A. St. Lawrence FDR 230-kV substation disturbance of March 6, 1996. Presented at the Georgia Tech Protective Relaying Conference, April 30–May 2, 1998, Atlanta, GA, and at the 25th Annual Western Protective Relay Conference, October 26–28, 1998, Spokane, WA.

Ibrahim, M. A., and F. Stacom. Adirondack 230-kV substation outage of July 1, 1995. Presented at the Georgia Tech Protective Relaying Conference, May 1–3, 1996, Atlanta, GA, and at the 23rd Annual Western Protective Relay Conference, October 14–16, 1996, Spokane, WA.

IEEE Guide for AC Generator Protection. IEEE Power System Relaying Committee Publication, 1986, and follow-up revisions.

Khalifa. M. *High-Voltage Engineering*. New York: Marcel Dekker, 1990.

Neuenswander, J. R. *Modern Power Systems.* Scranton, PA: International Textbook Company, 1971.

Skilling, H. H. *Transient Electric Currents.* New York: McGraw-Hill, 1952.

Sonnemann, W. K., C. L. Wagner and G. D. Rockefeller. Magnetizing inrush phenomena in transformer banks. *Transactions of the IEEE*, Vol. 77, No 3, 1958, pp. 884–892.

4

CASE STUDIES RELATED TO GENERATOR SYSTEM DISTURBANCES

In this chapter we present case studies related to generator disturbances and abnormal system conditions. Case Study 4.1 presents an actual DFR record that reveals the presence of a double-frequency rotor current as a result of negative-sequence stator current flow, which is normally present during unbalanced system faults. Case Study 4.2 covers inadvertent energizing of a generating unit with induced 60-Hz current in the rotor circuit that can stress the rotor thermally and destroy it. Due to the increase in the frequency of inadvertent energizing occurrences, dedicated protective relay schemes developed and recommended by generator manufacturers are described. Case Study 4.3 documents a generator trip due to the loss of excitation caused by human error and its impact on the system and the required unit loss of field protection. Case Study 4.4 illustrates a unit trip due to the loss of its excitation system and the impact of the unit excitation loss on the surrounding system voltages. Case Study 4.5 highlights a missynchronization incident of a unit caused by a false automatic synchronizer output during the mode change which closed the unit breaker accidentally, thus energizing the generator at a wider angle with the system. Case Studies 4.6 and 4.7 describe missynchronization of units, resulting in power plant trips. The incidents apparently generated an out-of-step condition (power swing), causing the generating unit to start oscillating with the power system. This will result in high asymmetrical currents that

Disturbance Analysis for Power Systems, First Edition. Mohamed A. Ibrahim.

affect the performance of the CTs connected to the generator differential relay. Asymmetrical current flow can stress CTs connected to unit differential numerical relays, causing uneven responses that result in false relay trip output.

Case Study 4.8 documents a combined-cycle plant trip during an out-of-step power system condition leading to unit oscillations with the system triggered by a high-side phase-to-ground fault. Case Study 4.9 documents plant tripping during the synchronization of a unit at an excessive angle, which caused an out-of-step condition on the system that forced generators to an oscillating state. Case Study 4.10 illustrates the analysis of a external fault cleared successfully, leading to an awareness of many power system phenomena. Case Study 4.11 documents a unit trip during the failure of one of the generator surge capacitor leads. The absence of one of the phase capacitors resulted in a neutral shift, thus generating 60-Hz generator neutral unbalance with enough 60-Hz neutral voltage generated to cause operation of the overvoltage element of the 95% stator ground fault protection system. Case Study 4.12 describes the occurrence of a generator stator ground fault which was cleared by operation of the 95% stator ground fault protection system. Case Study 4.13 describes the occurrence of a three-phase fault at the terminal of a generating unit caused by failure of the hydrogen cooler water path, allowing the water to fill the potential transformer compartment, thus causing the fault.

Case Study 4.14 describes a three-phase fault at the terminal of a 50-MW generator due to a cable connection failure. Case Study 4.15 documents a generator stator phase-to-phase fault caused by the failure of a rotor fan blade. Case Study 4.16 describes an undesired tripping of a pump storage plant during a close-in phase B-g 345-kV line fault. Case Study 4.17 describes the tripping of an 800-MW plant and the associated 345-kV lines during a bus fault. Case Studies 4.18, 4.19, and 4.20 illustrate generator tripping during external system disturbances and faults. Case Study 4.21 describes the undesired tripping of an induction generator by a differential relay during energization of a capacitor bank connected within its protection zone. Case Study 4.22 illustrates three undesired trips of a steam turbine unit of a combined-cycle plant which were initiated by the generator differential elements of numerical multifunction relays during attempts of the unit to synchronize with the system. Case Study 4.23 covers sequential shutdown of a steam turbine unit as part of a combined-cycle plant design. Case Study 4.24 covers a commissioning unit trip and its analysis using a numerical relay fault record. Case Study 4.25 covers the phasing procedure of a new generator with the system prior to commissioning. Case Studies 4.26 and 4.27 illustrate the setting procedure for a 100% stator ground fault protection system using the third-harmonic undervoltage concept and the philosophy of setting generator relaying elements to provide system backup protection.

4.1 GENERATOR PROTECTION BASICS

Generator protection systems are applied to protect against faults as well as abnormal system operating conditions. Generator faults include stator phase and ground faults

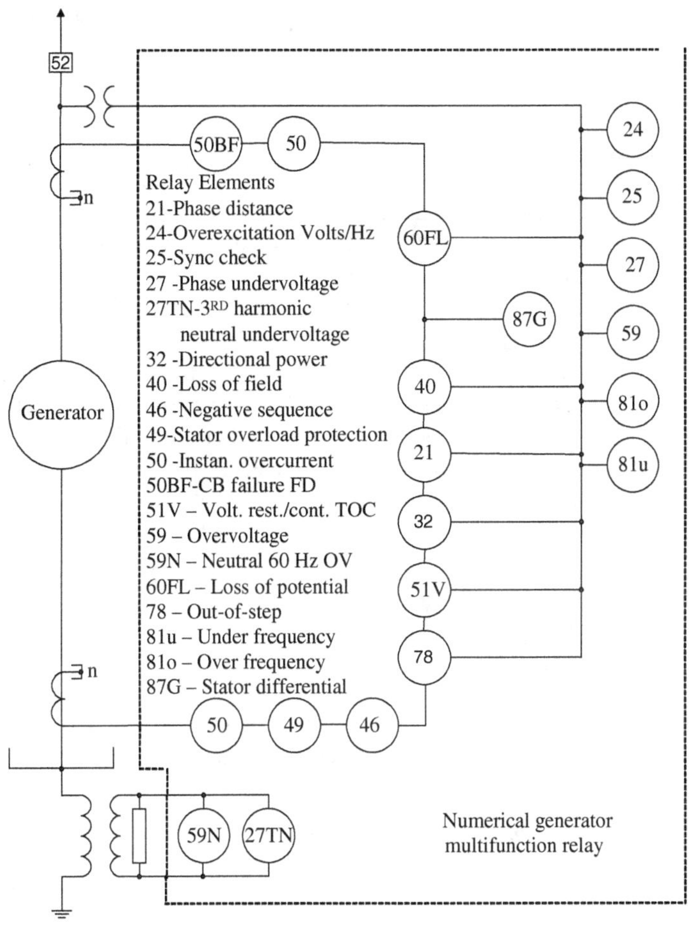

Fig. 4.1 Numerical generator multifunction relay.

as well as rotor ground faults. Abnormal system operating conditions include uncleared power system phase and ground faults, inadvertent generator energization, reverse power (motoring of a generator), mis-synchronizing of unit breakers, loss of excitation, unit oscillation during out-of-step events, overexcitation of GSU transformers, and system voltage collapse. Traditionally, generators were protected using discrete electromechanical and static component relays. Presently, protection of new units as well as replacements of obsolete component relays is accomplished using multifunction numerical generator devices. Figure 4.1 illustrates a functional diagram of one numerical generator multifunction relay with numerous protection elements shown. Numerical technology has benefited the generator protection area by the possibility of applying redundant numerical protection packages, due to the low cost. In addition, numerical technology has enhanced postmortem generator disturbance analysis by producing fault and oscillography records. This elaborate generator monitoring has resulted in faster restoration of units following their forced outages.

CASE STUDIES

Case Study 4.1: Appearance of Double-Frequency (120-Hz) Current in a Hydrogenerator Rotor Due to Stator Negative-Sequence Current Flow During a 115-kV Phase-to-Ground Fault

Abstract A generating unit rated at 170 MW was shut down due to a ground fault occurring on the 115-kV overhead structure that connects the generator step-up transformer bushing and the 115-kV cable pot heads. The fault was initiated when a worker apparently dropped an energized electrical cord onto the 115-kV overhead structure. In this case study we present an actual DFR record that reveals the presence of a double-frequency rotor current as a result of negative-sequence stator current flow, which is normally present during unbalanced system faults and unbalanced loading conditions. In addition, we describe the power system involved in the disturbance, protection of the high-voltage cable, the sequence of events, the ground fault mechanism, and power system phenomena.

Description of the Power System Figure 4.2 shows a unit G6 connection to a 115-kV system through a 200-MVA GSU transformer and a 1-mile 115-kV high-pressure fluid-filled (HPFF) cable that connects the powerhouse to the switchyard. Other generating units are also shown to illustrate their contribution and behavior to the unit G6 high-voltage 115-kV lead fault.

Description of Protection Systems for the High-Voltage 115-kV Cable The 115-kV high-voltage cable shown in Fig. 4.3 is protected by dual relaying systems. One system is an electromechanical pilot wire (current differential) relay 87 applied on a private hardwire. This system was later replaced by a numerical current differential relay applied on a fiber optic communication link. The second system consists of phase and ground electromechanical instantaneous overrcurent 50P/50N relays applied at the switchyard end of the 115-kV cable. The instantaneous element is set to cover the cable and 50% of the GSU transformer. The relays are made directional by setting them above the unit maximum short-circuit contribution to external system faults. This incident highlights the excellent performance of electromechanical instantaneous induction cup relays in providing high-speed clearing of 0.5 cycle in the presence of sufficient fault currents. Because some of these electromechanical relays were obsolete, they were replaced by numerical relays.

Description of the Incident A contractor was hired by a utility to repair the deteriorating grout between the tops of the joints of a generating facility's transformer bay concrete slab. In the process of erecting fall-protection guardrails, the contractor

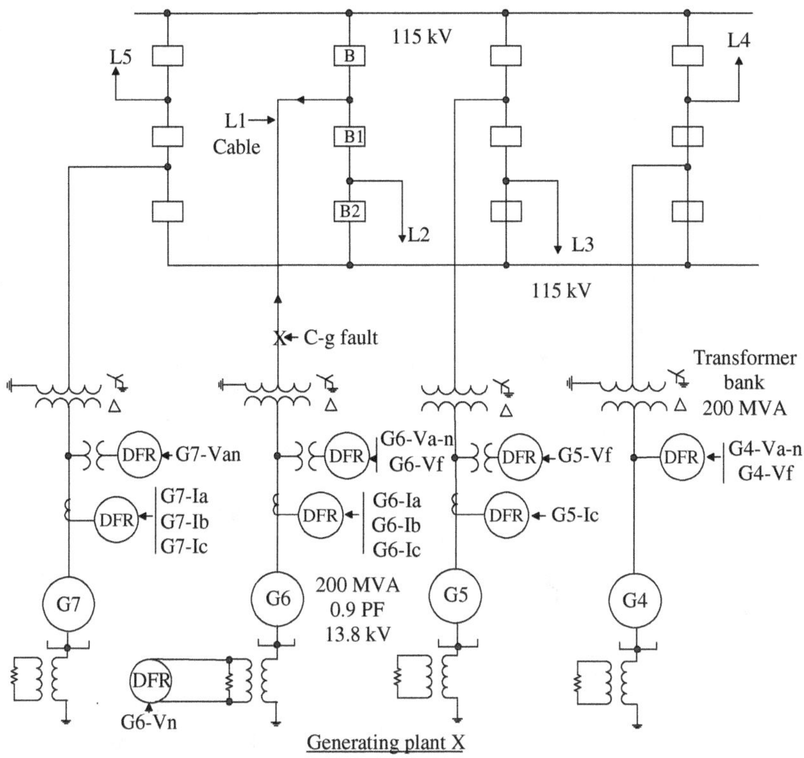

Fig. 4.2 One-line diagram showing DFR-monitored currents and voltages.

pulled an excessive amount of extension cord for his portable drill and inadvertently allowed the excess cord to fall beyond his fenced work area. The excess extension cord fell onto the energized 115-kV high-voltage overhead structure that connects the GSU transformer bushing to the 115-kV cable pot heads. A high-voltage flashover to ground occurred, also causing the low-voltage cable to fail to ground. Ground currents returned to ground via the concrete steel bars and the ground of the low-voltage receptacle. As a result, a phase C-to-ground fault occurred in the 115-kV overhead structures. Both the 115-kV and low-voltage 240-V faults were cleared from the system successfully. However, unit G6 decayed energy, which follows unit shutdown, kept feeding the initial fault for several additional seconds.

Analysis of the Incident The DFR device for the 115-kV system was being installed, and therefore no fault record was available for the 115-kV voltages and currents during this incident. For this reason, the analysis is based on DFR records obtained from the power plant for the 13.8-kV generator voltages and currents. Based on the unit transformer phasing diagram shown in Fig. 4.4, the high-voltage (115-kV) wye side leads the low-voltage (13.8-kV) delta side by 30° (in accordance with the NEMA standard). The high-voltage C winding of the transformer is magnetically

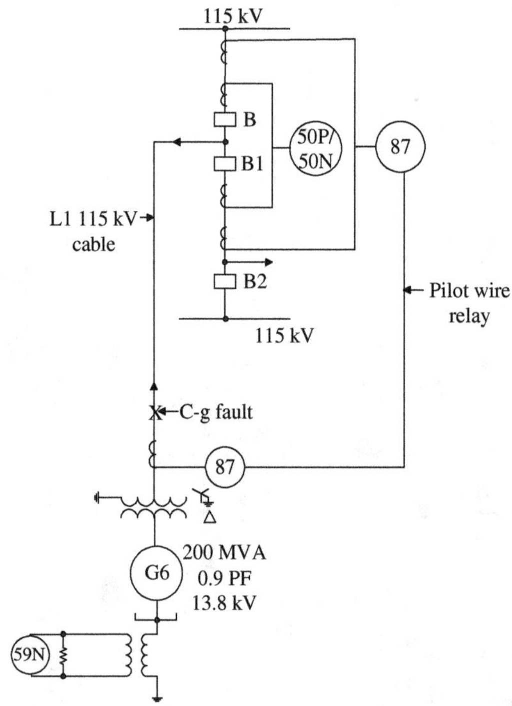

Fig. 4.3 One-line AC showing 115-kV cable L1 protection systems.

coupled to the low-voltage c-a winding. As a result, a phase C-g fault on the high side will appear as a phase c-a fault on the generator side. The DFR record in Fig. 4.5 confirms the fault by showing a voltage dip on phase a of trace G4-Va-n and phase c of trace G6-Vc-n. Phase c of generator G4 reveals that the voltage dip lasted for only 3.5 cycles, thus implying that the fault was cleared from the 115-kV system at this time.

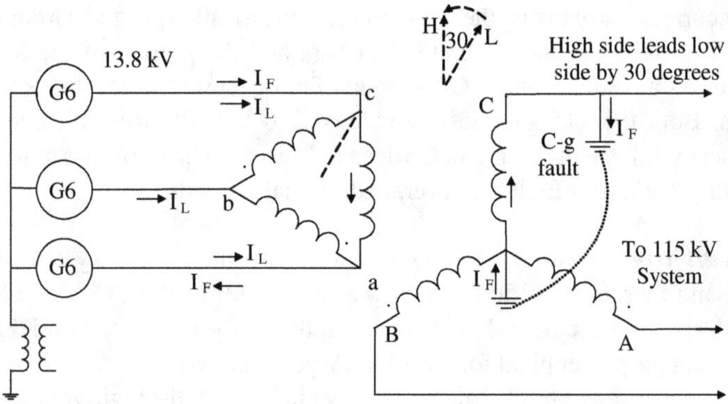

Fig. 4.4 Transformer phasing diagram showing the winding currents before tripping of the 115-kV system side breakers (0 to 3.5 cycles).

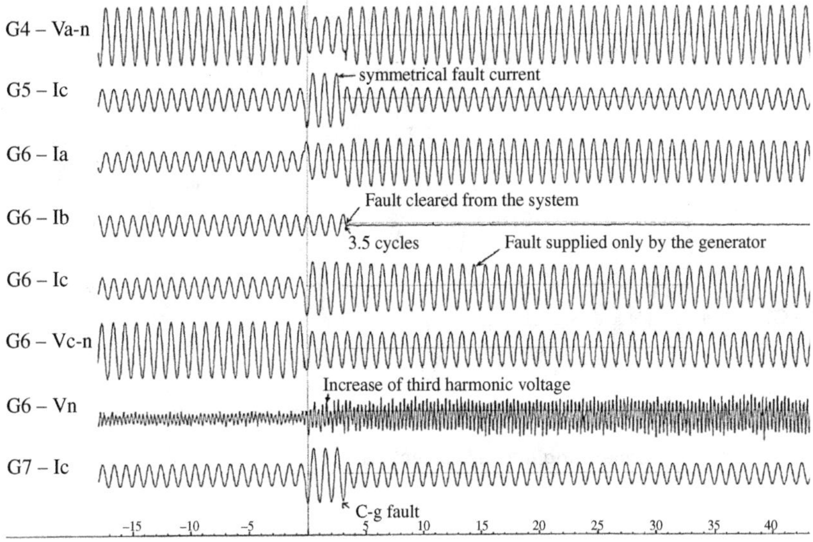

Fig. 4.5 Generator currents and voltages during the C-g fault.

This was also confirmed by interruption of the phase b loading current of unit G6, as confirmed by trace G6-Ib. Traces G6-Ia, G6-Ib, G6-Ic, G6-Vc-n, and G6-Vn shown in Fig. 4.5 indicate a continuation of the decayed energy feeding the initial fault after isolation of the fault from the system. The additional unit G6 decayed energy generally lasts for about 15 s.

Ground Fault Mechanism The low-voltage extension cord fell on the 115-kV phase C conductor. When the cord came in contact with the high-voltage conductor, arc-over to the low-voltage cable was established (Fig. 4.6). Ground current flow was

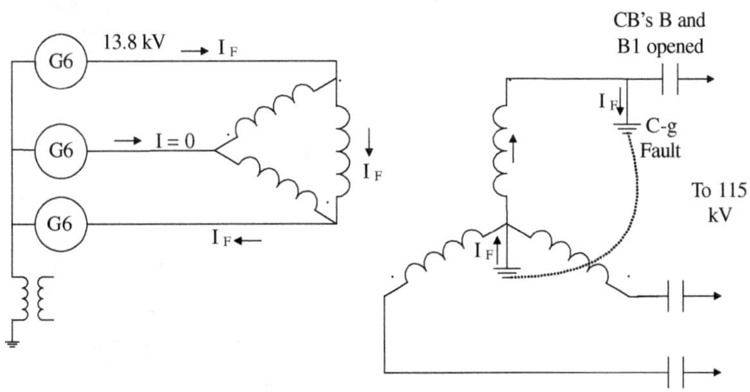

Fig. 4.6 Winding currents for the fault being fed from the generator only following tripping of the 115-kV breakers ($t > 3.5$ cycles).

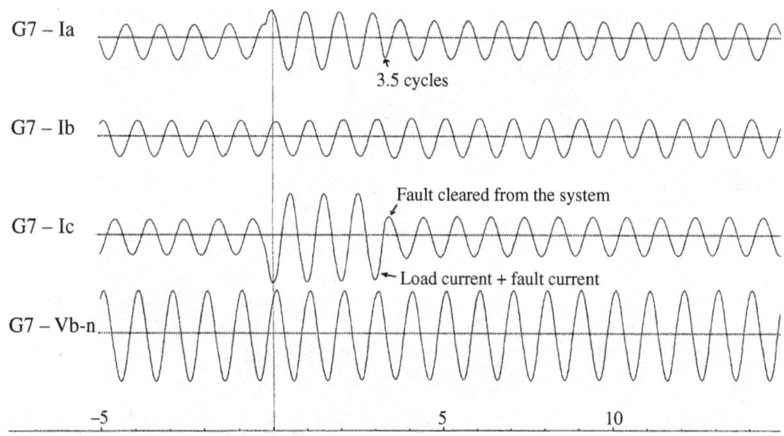

Fig. 4.7 Generating unit G7 currents and voltages during the C-g fault.

through the steel bars inside the concrete wall, where the cord is touching, and the low-voltage extension cord to the ground via the 240-V wall receptacles.

Sequence of Events The sequence of events for this incident as constructed from an analysis of the plant DFR records shown in Figs. 4.5, 4.7, and 4.8 is as follows. At $t = 0$, a phase C-g fault occurred on the 115-kV overhead structure. Figure 4.7 shows the unit G7 contribution to the external C-g fault on 115-kV cable L1, which was cleared from the system in 3.5 cycles. The ground fault was cleared from the 115-kV system by tripping the 115-kV CBs B1 and B2. Despite the initiation of unit G6 shutdown, decayed energy from unit G6 continues to feed the C-g fault for an

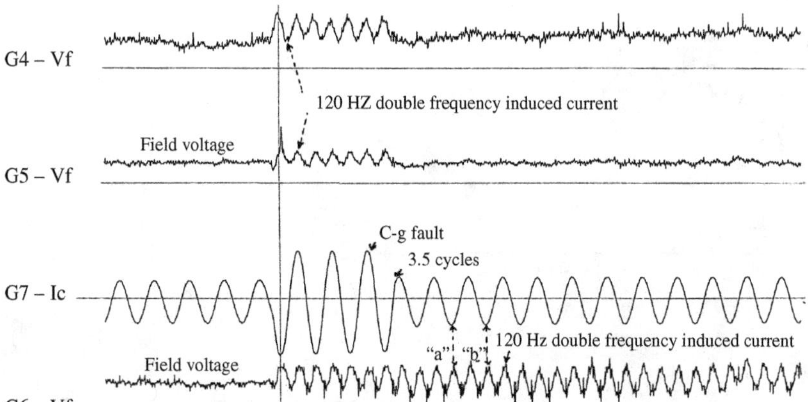

Fig. 4.8 Generating units G4, G5, and G6 field voltages showing 120-Hz induced rotor currents.

additional 15 s. Figure 4.5 shows unit 6 decayed energy to the C-g fault for an additional 41 cycles after clearing of the fault from the 115-kV system.

Analysis of the Appearance of Double-Frequency (120-Hz) Current in the Generator Rotor There are a number of system abnormal conditions and unbalanced faults that can produce negative-phase-sequence current components. In addition, when a generator is supplying an unbalanced load, the phase currents and terminal voltages deviate from the ideal balanced relationship, and a negative phase sequence armature current (I_2) is imposed on the generator. Therefore, unbalanced phase currents create negative-sequence current in the generator stator, which can be calculated using symmetrical components as follows:

$$I_2 = \frac{1}{3}(I_a + a^2 I_b + aI_c)$$

where I_2 is the negative-sequence current, I_a, I_b, and I_c are phase currents, $a^2 = 1$ at 240°, and $a = 1$ at 120°.

The negative-sequence current in the stator winding creates a magnetic flux wave in the airgap which rotates in opposition to the rotor at synchronous speed. The flux produced by the negative-sequence current as seen by the rotor has a frequency of 120 Hz, which is twice the synchronous speed as a result of reverse rotation of the negative-sequence current combined with positive rotation of the rotor. This can be summarized as follows: A stator negative-sequence current flow will interact with normal positive-sequence current to induce a double-frequency current (120 Hz) in the rotor. This flux induces currents in the rotor body iron, wedges, and retaining rings at twice the system frequency. The 120-Hz rotor current flows into the surface elements of the rotor, causing excessive rotor heating that can cause unit damage if prolonged. Heating will occur in these rotor areas, and the resulting temperature will depend on the level and duration of the unbalanced currents.

The 120-Hz frequency generation at the rotor surface is confirmed in Fig. 4.8 when trace G6-Vf is compared with the 60-Hz current of trace G7-Ic for the time duration between points a and b. As shown in Fig. 4.8, traces G4-Vf and G5-V5 also reveal the generation of a frequency of 120 Hz superimposed on the field dc voltages V_f during the phase C-g fault. This 120 Hz appears as a result of generators G4 and G5 supplying negative-sequence currents to feed the L1 ground fault. Trace G6-Vf in Fig. 4.7 shows a continuation of the generation of 120 Hz from unit G6 trapped energy after the unit is tripped that lasts for an additional 25 cycles after clearing of the fault from the 115-kV system.

Generator Negative-Sequence Protection Unbalanced phase currents create negative-sequence current in a generator stator, which stresses the generator rotor. Negative-sequence current generated by system unbalance can cause current to flow in the retaining rings and rotor iron, which if not detected can generate harmful excessive heat. The generator has an established short-time negative-sequence rating defined as $I_2^2 = K$, where K is the manufacturer factor (the larger the generator, the smaller the K value). The negative-sequence relay needs to be set to plot under the generator negative-sequence capability curve. The negative-sequence relay settings are normally calculated slightly below (70%) the machine's $I_2^2 t$ capability. The relay settings are calculated to operate when the machine is operating at its full negative-sequence current capability to ensure tripping before damage to the rotor occurs due to overheating. The old electromechanical relays had a minimum sensitivity of 60% (0.6 pu), and the relay normally crossed the generator capability curve $I_2^2 t$. The relay was designed primarily to detect unbalance created by uncleared system faults. The numerical negative-sequence relay element, which is based on symmetrical components, can sense negative-sequence current as low as 2% (0.02 pu) of the machine rating. The

$$\text{relay operating time} = \frac{\text{machine } I_2^2 t = K}{(I_2/I_{\text{nom}})^2}$$

where I_2 is the relay calculated negative-sequence current and I_{nom} is the generator rated current.

System Phenomena

1. Appearance of negative-sequence current in the generator rotor as a result of system unbalanced faults, as indicated in trace G6-Vf in Fig. 4.8.
2. Generation of third-harmonic voltage during normal generator operation, with its level increasing following unit trip, as indicated by trace G6-Vn in Fig. 4.5.
3. Appearance of high-side L-g fault as an L-L fault on the low side. Based on the unit transformer phasing shown in Fig. 4.4, the 115-kV C-g fault is seen as a phase a-c fault at the 13.8-kV transformer low side.
4. Unit load flow can affect fault current contributions from the generator during power system faults. The DFR record shown in Fig. 4.7 confirms that the phase c current of unit G7 has a larger current contribution during the fault, due to load current being added to the fault current. Phase b shows current lower than that of phase c, due to load current being subtracted from the fault current. Trace G7-Ia in Fig. 4.7 shows the current on phase a to be equal prior to the fault and following the isolation of unit G6 from the system.
5. Generator decayed energy feeds the generator lead faults for several seconds after unit shutdown, as indicated in Fig. 4.5 for unit G6 phase currents.

Corrective Actions Construction work cannot be permitted unless fall-protection fencing is erected for each work zone. In addition, nonconductive netting is required in each transformer bay. These actions will prevent the occurrence of similar incidents while performing maintenance work.

Case Study 4.2: Inadvertent (Accidental) Energization of a 170-MW Hydro Unit

Abstract A generator in a three-phase 115-kV system was accidentally energized while the rotor was locked in place using shaft brakes. In this situation the generator behaves as an induction motor with a slip = 100% (locked rotor by applying brakes). During three-phase energization, while the machine is at standstill, a rotating magnetic flux at synchronous frequency (60 Hz) will be induced in the generator rotor. The resulting rotor current path is similar to the negative-sequence rotor current paths during unbalanced system conditions. This will result in rapid rotor heating, which can quickly damage the rotor. Generators can become energized accidentally while off-line, by one or a combination of the following: operating errors, breaker head flashovers, control circuit malfunctions, or a combination of these causes. Due to the increase in the frequency of occurrences, dedicated protective relay schemes have been developed and are recommended by generator manufacturers. In this case study we describe the unit protection systems, the inadvertent energization incident, and the breaker reclosing modes. In addition, it provides an analysis of the DFR records for the incident, confirmation of the operation of the loss of an excitation relay during the incident, corrective actions, and lessons learned.

Description of the System and Associated Protection The generator protection was designed in the early 1960s and was based primarily on electromechanical relays. It includes generator differential relay device 87 protecting the stator against phase faults, generator tuned neutral overvoltage relay device 59G providing 95% for stator ground faults, loss of excitation relay device 40, and mho phase distance system backup device 21. The unit was scheduled for a protection upgrade based on numerical generator multifunction technology. Since the hydro unit is used frequently, none of the machine protection is taken out of service automatically while the unit is not running. Figure 4.9 shows the system one-line diagram and unit DFR-monitored voltages and currents that are used to analyze the incident.

Description of the Incident An outage was taken to remove 115-kV line L1 and generator G1 from service to install a new DFR. As shown in Fig. 4.10, the outage was accomplished by opening 115-kV breakers A1, A2, and A3. Generator G1 disconnect switch X1 was left in a closed position (as an operating error), thus leaving only one gap between the generator and the 115-kV system. Restoration of the system was initiated following completion of the DFR work. The first step in the sequence of restoration was accomplished by closing breaker A1, which energized line L1, as shown in Fig. 4.11.

The operator then proceeded to commit an operating error by closing breaker A2 manually, as shown in Fig. 4.12, thus inadvertently energizing generator G1. Disconnect switch X1 was apparently left in a closed position without proper acknowledgment. The generator G1 rotor was in a standstill condition, with hydraulic

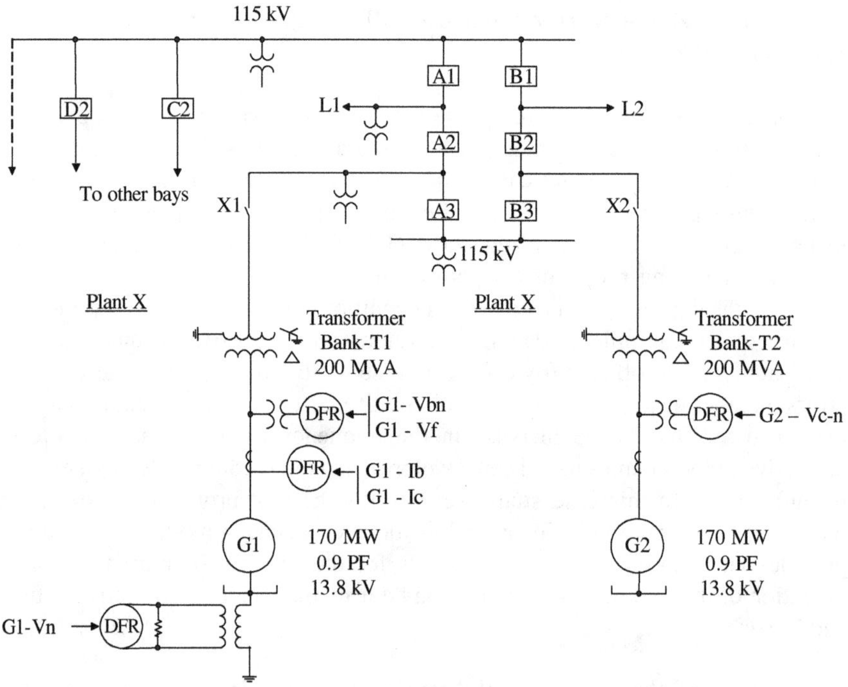

Fig. 4.9 One-line diagram showing G1 and G2 DFR-monitored currents and voltages.

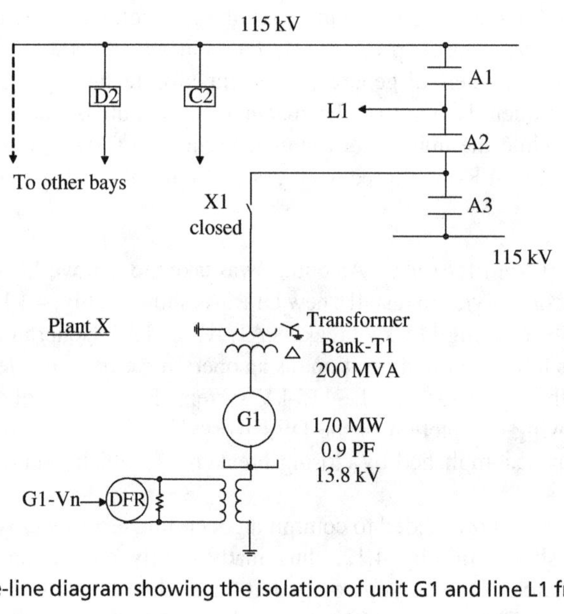

Fig. 4.10 One-line diagram showing the isolation of unit G1 and line L1 from the system.

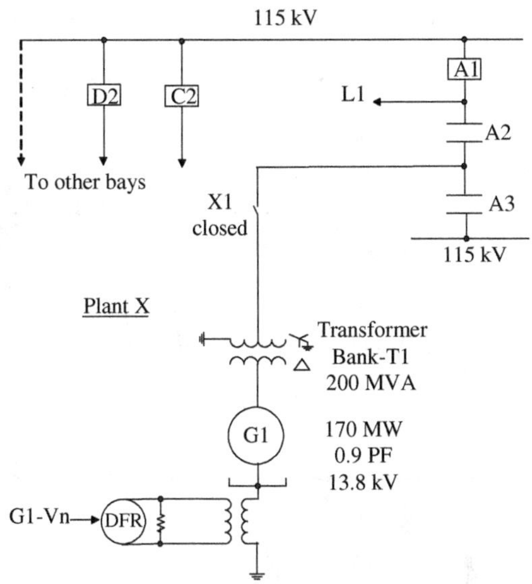

Fig. 4.11 Restoration of line L1 by closing CB A1.

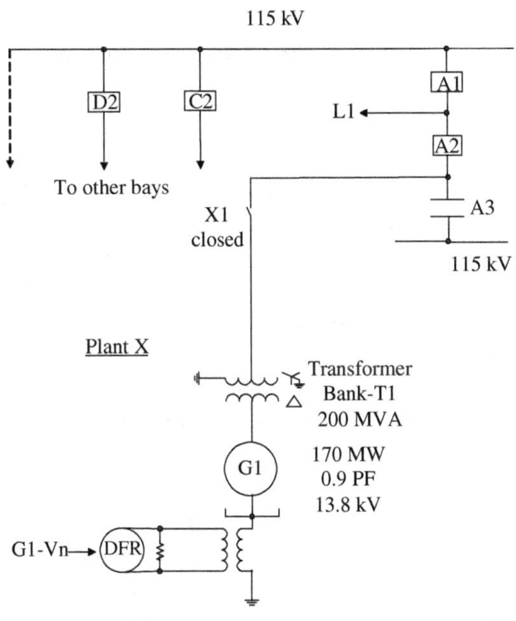

Fig. 4.12 One-line diagram showing inadvertent energization of unit G1 by mistakenly closing CB A2.

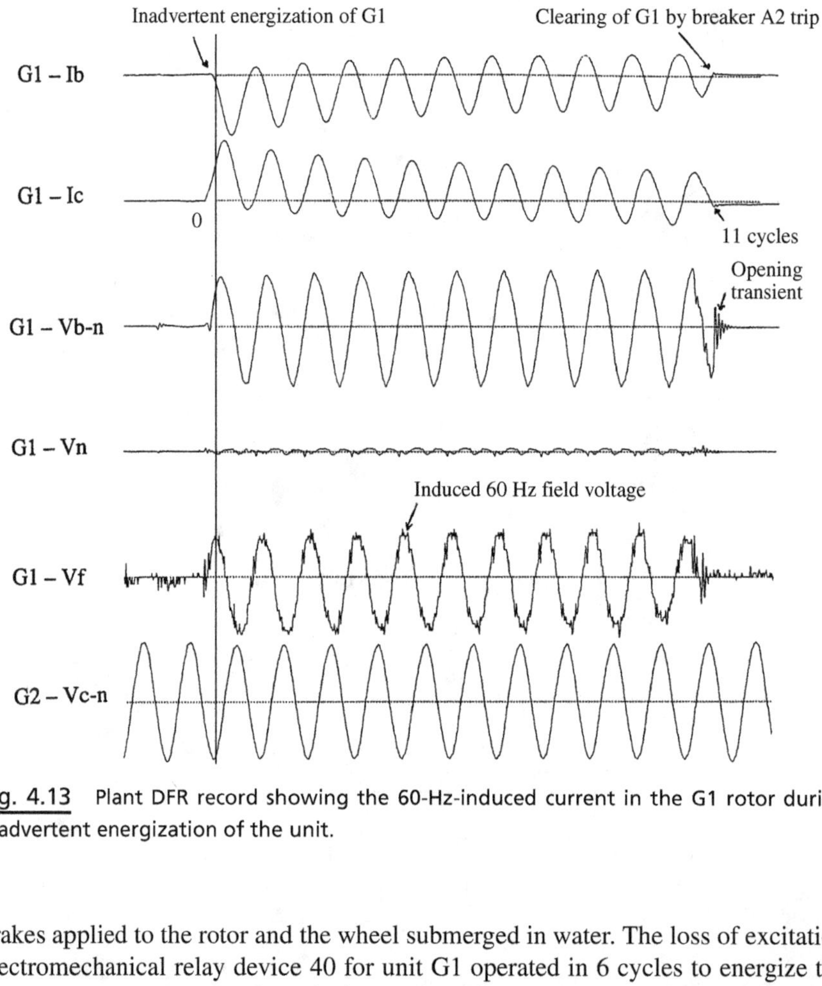

Fig. 4.13 Plant DFR record showing the 60-Hz-induced current in the G1 rotor during inadvertent energization of the unit.

brakes applied to the rotor and the wheel submerged in water. The loss of excitation electromechanical relay device 40 for unit G1 operated in 6 cycles to energize the generator emergency shutdown lockout relay. The generator was isolated from the system by tripping breaker A2 after a total incident duration of 11 cycles. Figure 4.13 shows the DFR record for the incident, with unit G1 three-phase currents and induced 60-Hz rotor current trace G1-Vf. The generator initial high stator current coupled with a low voltage across the machine resulted in secondary measured impedance by the loss of excitation relay, which fell in the relay impedance characteristic operating area.

Description of the Circuit Breaker A2 Reclosing Modes As shown in the CB A2 closing schematic in Fig. 4.14, the breaker can be closed via either of two modes, automatic or manual. The control switch 43 position determines the selection of the reclosing mode. The automatic mode is used to synchronize the unit and is supervised by sync check relay device 25. The manual mode is supervised by a closing logic and was followed in this incident to restore the three-breaker bay since the generator is out

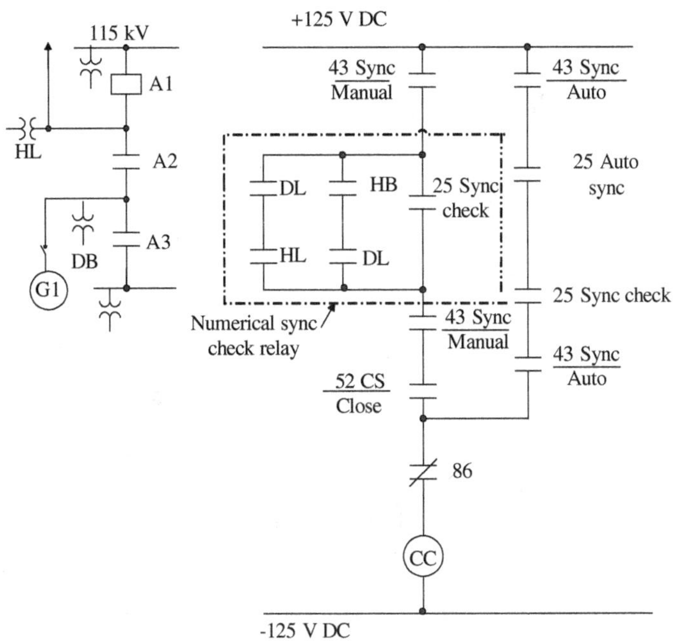

Fig. 4.14 DC schematic for the closing control circuit of CBs A2 and A3.

of service. In this mode the numerical sync check relay can provide one of three modes. The modes are driven by the availability of the bus and line potential sources. Circuit breaker A2 can close by sync check if both sources are hot. It can also close a hot bus dead line or hot line dead bus. In this incident the generator was out of service; therefore, the operator set the 43 mode switch to "manual." The operator then proceeded to close CB A2 by breaker control switch 52CS. No harm would have been done had the generator disconnect switch X1 been opened immediately when the generator was removed early from service (as it should be).

Analysis of the DFR-Monitored Currents and Voltages Figure 4.9 shows the system one-line diagram and the DFR-monitored currents and voltages for hydro units G1and G2 that are used to analyze the incident. The inadvertent energization of unit G1 lasted for 11 cycles, as shown in Fig. 4.13, with high current magnitude for trace G1-Ib for phase B and trace G1-Ic for phase C and the presence of unit phase B voltage. The unit field trace G1-Vf reveals 60-Hz voltage induced to the rotor, which forces current flow in the rotor similar to induction motor action. The DFR record in Fig. 4.15 shows RMS values of 19.1 kA for phase B current and 18.2 kA for phase C current. The RMS is obtained by the DFR software algorithm using discrete Fourier transform time-domain analysis for a 1-cycle window. The calculation was done at the ninth cycle after significant reduction of the initial dc offset component. The recorded generator phase B voltage was calculated by the DFR software at 5.02 kV, as shown in Fig. 4.15. The generator voltage was

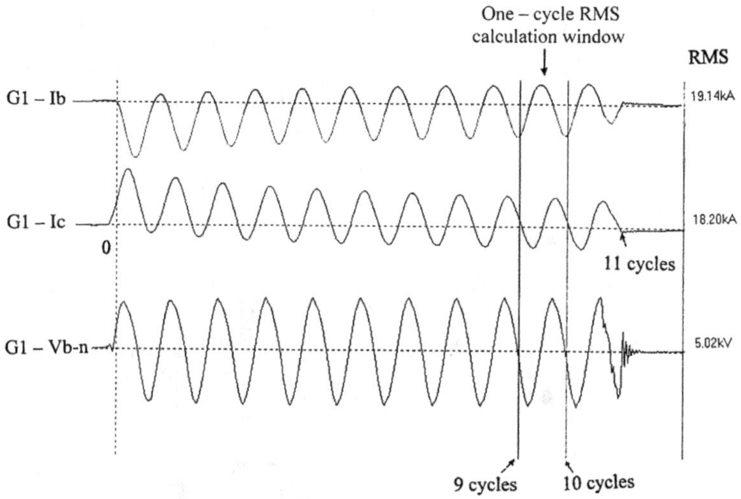

Fig. 4.15 Plant DFR record showing the RMS calculations for G1 currents and voltages near the end of the incident.

calculated using a circuit analysis of 5.2 kV and a current of 20.9 kA. The DFR recorded voltage and current are less than the values calculated by 4 and 11%, respectively, which could be attributed to pre-event voltage magnitude and system inaccuracies.

Calculation of the Stator Current During the Incident When a generator is connected inadvertently to a three-phase system voltage source while at standstill, it behaves like an induction motor. During this energization, a rotating magnetic flux at synchronous frequency (rotor at standstill) is induced in the generator rotor. The machine impedance at this 100% slip is equivalent to its negative-sequence impedance Z_{2g}. Its reactive part is equal to $\frac{1}{2}(X''_d + X''_q)$, where X''_d is the direct axis machine reactance and X''_q is the quadrature axis reactance. The machine terminal current and voltage during this energization period will be a function of the loop impedance formed by the generator, unit transformer, and system equivalent impedances. Since the plant is connected to a strong system, the stator current magnitudes are expected to be between three and four times rating, and the machine terminal voltage to be between 50 and 70% of the rated value:

$$I_{\text{rated}} = \frac{167 \times 10^6}{\sqrt{3} \times 13.8 \times 10^3} = 7022\ \text{A}$$

From a short-circuit study, for a three-phase fault at the plant 115-kV bus, $I_{\text{three-phase}} = 76.28$ pu at $-86°$. The system equivalent impedance $= X_S =$ 1/71.3 = 0.014 pu at $-86°$ on 100 MVA. By ignoring the resistive part and doing

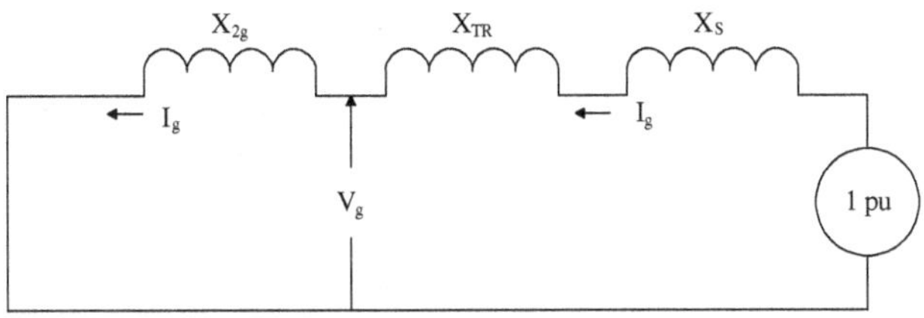

Fig. 4.16 Equivalent circuit for inadvertent energization of the G1 unit at standstill.

a simple current calculation for the circuit shown in Fig. 4.16 using reactance only, we have

$$X_{tr} = \text{j}0.058 \text{ pu on } 100\,\text{MVA}$$

$$X_{2g} = \text{j}0.1317 \text{ pu on } 100\,\text{MVA}$$

$$I_{gen} = \frac{(j1 \text{ pu})}{j0.014 + j0.058 + j0.1317} = 5 \text{ pu}$$

$$I_{base} = \frac{100 \times 10^6}{\sqrt{3} \times 13.8 \times 10^3} = 4183 \text{ A}$$

$$I_{gen} = I_{pu} \times I_{base} = 5 \times 4183 = 20.9 \text{ kA}$$

$$\text{voltage across the generator} = V_g = I_g \times X_g = 5 \times 0.1317 = 0.659 \text{ pu}$$

$$V_g = 0.659 \times V_{base} = 0.659 \times \frac{13.8}{\sqrt{3}} = 5.2 \text{ kV}$$

From the calculation, the generator current is 20.9 kA, which is close to three times its rated current, while the machine terminal voltage is 5.2 kV, which is 62% of the rated voltage.

Confirmation of the Operation of the Loss-of-Excitation Relay During the Event Machine loss-of-excitation protection is provided by an electromechanical single-phase single-zone-offset mho distance relay. The relay setting is shown in the *R–X* diagram in Fig. 4.17. The relay is connected to the system via a CT ratio of 2000 : 1 and a PT ratio of 120 : 1. The relay is set with a 2.5-Ω offset and a circle diameter of 16.8 Ω. The relay employs a fixed time-delay auxiliary unit set at 83 ms (5 cycles) to prevent undesired tripping due to shock or vibration. The current and voltage inputs for the loss-of-excitation relay is connected line to line between phases A and B.

At the time of the incident, only limited generator parameters were monitored by the plant DFR. As shown in Fig. 4.13, the monitored analog signals are generator phases B and C currents, generator phase B-to-neutral voltage, generator neutral voltage, and generator field voltage. Therefore, an attempt will be made to examine an impedance

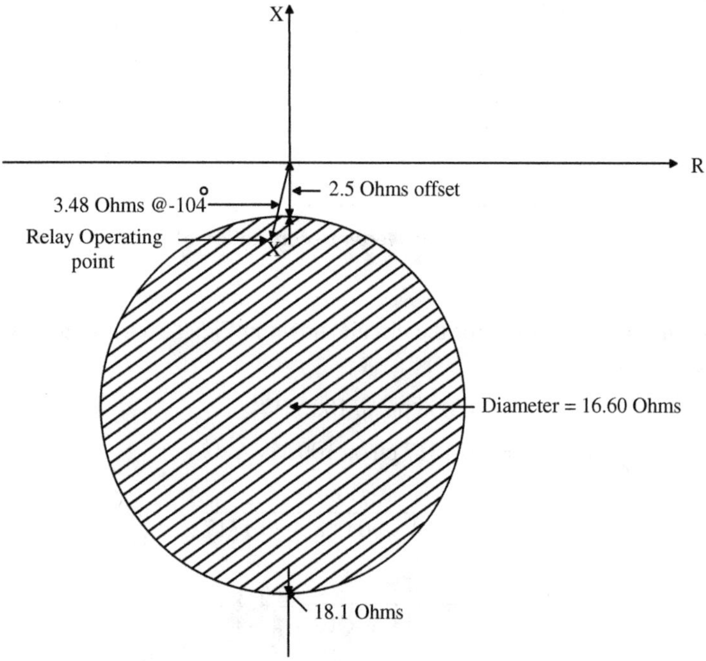

Fig. 4.17 *R–X* diagram illustrating the loss-of-excitation relay characteristic and its setting.

calculation using V_b and I_b, the recorded analog signals, rather then doing calculations based on line-to-line values. To confirm the relay response, an impedance, $Z_b = V_b/I_b$, will be calculated using RMS values. The DFR-stored samples will be examined to execute an accurate RMS calculation. Figure 4.18 shows the samples selected for traces G1-Ib and G1-Vb-n that are used for the RMS calculation for voltage and current to confirm operation of the loss-of-excitation relay during the event.

RMS Calculations for the Current and Voltage

RMS Value for the Phase B Current For the first energization cycle, the trace G1-Ib current shown in Fig. 4.18 has a minimum value at point (a) for sample 1325, where the instantaneous current is −56.97 kA. The maximum value is at point (b) for sample 1377, where the instantaneous current is 7.45 kA. The

$$\begin{aligned}\text{peak current value} &= \tfrac{1}{2}(\text{maximum value} + \text{minimum value})\\ &= \tfrac{1}{2}(56.97 + 7.45) = 32.21\ \text{kA}\end{aligned}$$

The RMS value = peak value/$\sqrt{2}$ = 32.21/$\sqrt{2}$ = 22.8 kA.

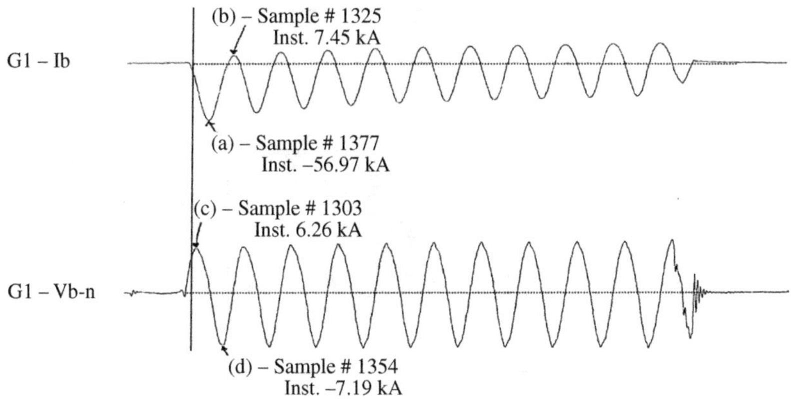

Fig. 4.18 RMS calculation for the G1 current and voltage to confirm operation of the loss-of-excitation relay during the incident.

RMS Value for the Phase B Voltage For the first energiztation cycle, the trace G1-Vb-n voltage shown in Fig. 4.18 has a maximum value at point (c) for sample 1303, where the instantaneous voltage is 6.26 kV. The minimum value is at point (d) for sample 1354, where the instantaneous voltage is −7.19 kV. The

$$\text{peak voltage value} = \tfrac{1}{2}(\text{maximum value} + \text{minimum value})$$
$$= \tfrac{1}{2}(6.26 + 7.19) = 6.73\ \text{kA}$$

The RMS value = peak value/$\sqrt{2}$ = 6.73/$\sqrt{2}$ = 4.76 kV.

Phase Angle Calculation The DFR has a sampling rate of 6000 samples/s or 100 samples/cycle. One complete cycle is equivalent to 360°. Therefore, each sample can represent a delay of 3.6° (= 360°/100). From Fig. 4.18, the minimum current is occurring at sample 1325 prior to the minimum voltage sample 1354. Therefore, the phase B current leads the phase B voltage by an angle equivalent to (1354 – 1325) × 3.6° or 104°.

Impedance Calculation Using Phase B Voltage and Current

$$Z_{pri} = \frac{V_{b-n}}{I_b} = 4.76\ \text{at}\ 0^\circ\ \text{kV}/22.8\ \text{kA at}\ 104^\circ = 0.209\ \Omega\ \text{primary at}\ -104^\circ$$

From the relay setting data, CT = 2000 : 1 and PT ratio = 120 : 1.

$$Z_{sec} = Z_{pri}(\text{CT}/\text{PT}) = 0.209 \times (2000/120) = 3.48\ \Omega\ \text{secondary at}\ -104^\circ$$

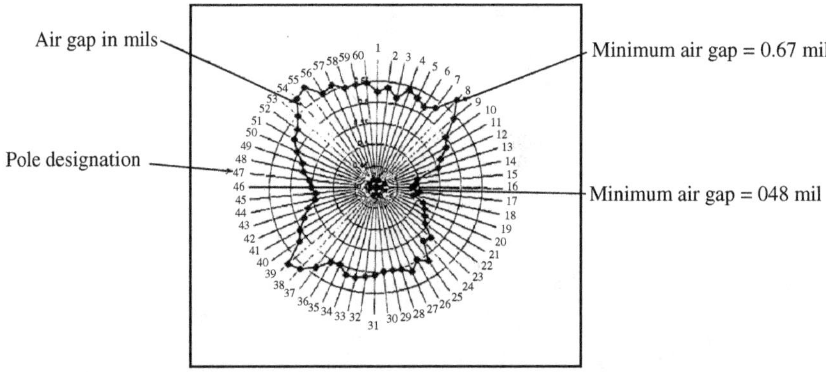

Fig. 4.19 Airgap as a function of past inadvertent energization of the machine poles.

This measured impedance and the operating point by the relay are plotted in the loss-of-excitation relay characteristics shown in Fig. 4.17.

Assessment of the Machine Airgap After the Incident The G1 unit is designed with an airgap of 600 to 650 mils with ± 5% tolerance. Measuring the machine airgap after the incident reveals that the shape of the stator–rotor system became oval rather than circular, as shown in Fig. 4.19. The maximum measured airgap is 0.678 mil, and the minimum airgap is 0.484 mil.

Dedicated Protection Schemes for Detection of Inadvertent Energizing Dedicated protection schemes to detect inadvertent energizing incidents have been developed and installed. Conventional generator relaying can have some limitations in the detection of inadvertent energization. This is true despite the fact, as demonstrated in this incident, that G1 loss-of-excitation relay device 40 may have detected and cleared the inadvertent energizing incident. Caution should be exercised when applying dedicated protection so that protection systems have dc power and so that relay input quantities are not removed when the generator is off-line. In this case, the protection system should trip the generator high-voltage and field circuit breakers and the unit auxiliary breakers. Many methods are available to design a dedicated scheme, but only two protection schemes are described here.

1. *Using a voltage-supervised overcurrent scheme.* As shown in Fig. 4.20, both voltage and current elements are monitoring the machine terminal. The undervoltage element device 27, shown in the logic diagram of Fig. 4.21, has adjustable pickup and dropout time delays and provides supervision of instantaneous overcurrent relay device 50. The undervoltage elements will arm relay 50 whenever the generator is taken off-line (no generator terminal voltage). When the machine is returned to service, relay element 50 will be disabled automatically, during the presence of voltage, by the undervoltage relay after the dropout time.

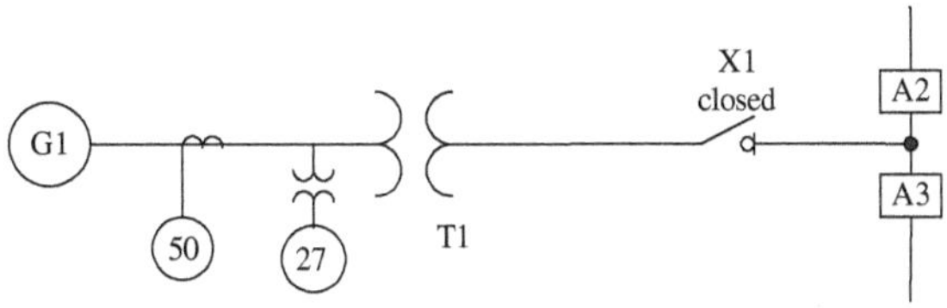

Fig. 4.20 Protection against inadvertent energization using voltage-supervised overcurrent relays.

2. *Applying a dedicated impedance relay located at the GSU high-voltage side.* Element 21 is polarized to look into the machine. The relay is set to detect the total impedance of the GSU transformer and the machine negative-sequence reactance.

System Phenomena

1. Current of 60 Hz induced in the rotor, caused by inadvertent unit energization, as shown in trace G1-Vf in Fig. 4.21.
2. Inadvertent energization of a generator connected to a strong system results in stator current in the range of three to four times rating. Figure 4.19 shows the equivalent circuit of inadvertent energization of G1. Accurate RMS current calculation using the DFR record in Fig. 4.15 reveals a value of 22.8 kA, which is 3.8 times the unit rated current of 6 kA.

Corrective Actions

1. Tear down the unit to reshrink the rotor and reround the stator.
2. Enhance operator training on the power system topology and emphasize the basics of system operations with safe rules.

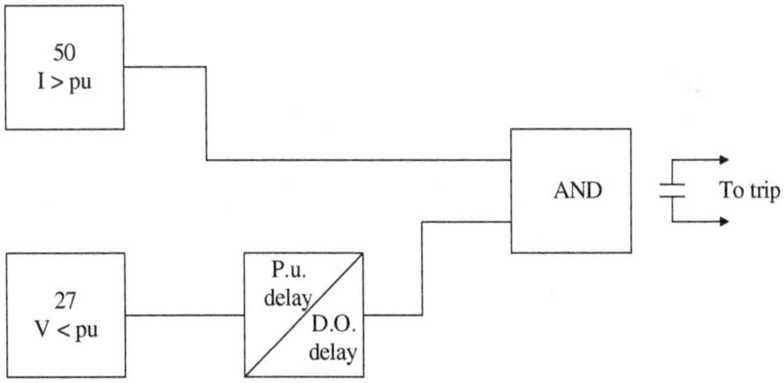

Fig. 4.21 Inadvertent energization protection function logic diagram.

3. Upgrade the unit by adding a 115-kV breaker, which will eliminate the vulnerability of the unit to future operating errors.

Lessons Learned

1. Generators should be separated by more than one gap when they are taken out of service.
2. Generator disconnect switches should be placed in an open position immediately following the removal of a generator from service.
3. Generator protective relaying should always remain in service regardless of the state of the generator. It is clear in this incident that if protection is taken out of service with the generator, more serious damage to the generating plant can certainly occur.
4. Inadvertent energizing of generator can be cleared using the generator loss-of-excitation protection relay. Apparently, the generator is behaving as an induction motor with excessive reactive power (VAR) initially being drawn, similar to the poor power factor of 15 to 20% upon starting induction motors.
5. Dedicated generator breakers can reduce the probability of the occurrence of inadvertent energizing of units.
6. Conventional generator protection elements can sense and isolate a unit following an inadvertent energizing incident. In this case study, loss-of-excitation device 40 detected the inadvertent phenomenon. This successful detection highlights and enforces the need to provide protection to keep a generator in service at all times.

Case Study 4.3: Loss of Excitation for a 200-MW Generating Unit Caused by Human Error

Abstract A generating unit rated at 200 MW was tripped out of service due to an operator error. The operator was in the process of completing the shutdown of unit G2 when he got interrupted by a phone call. Upon his return, he arrived at the unit G1 control board and inadvertently activated the removal of unit G1 by first shutting off the unit G1 excitation system. One of the generator multifunction numerical relays detected the condition and tripped the unit while the other relay failed to detect the loss-of-field condition. In this case study we describe the incident and analyze the failure of the generator primary relay to detect the loss of field incident. In addition, we provide a brief description of the loss of field protection, analysis of relay fault record, corrective actions, and lessons learned.

Description of the Unit Protection As shown in Fig. 4.22, the generator relaying zone is protected by dual numerical multifunction relays 87GRP and 87GRS. Redundant CT connections, lockout relays, and separate dc sources have been applied. Redundant protection is therefore offered for protection against generator

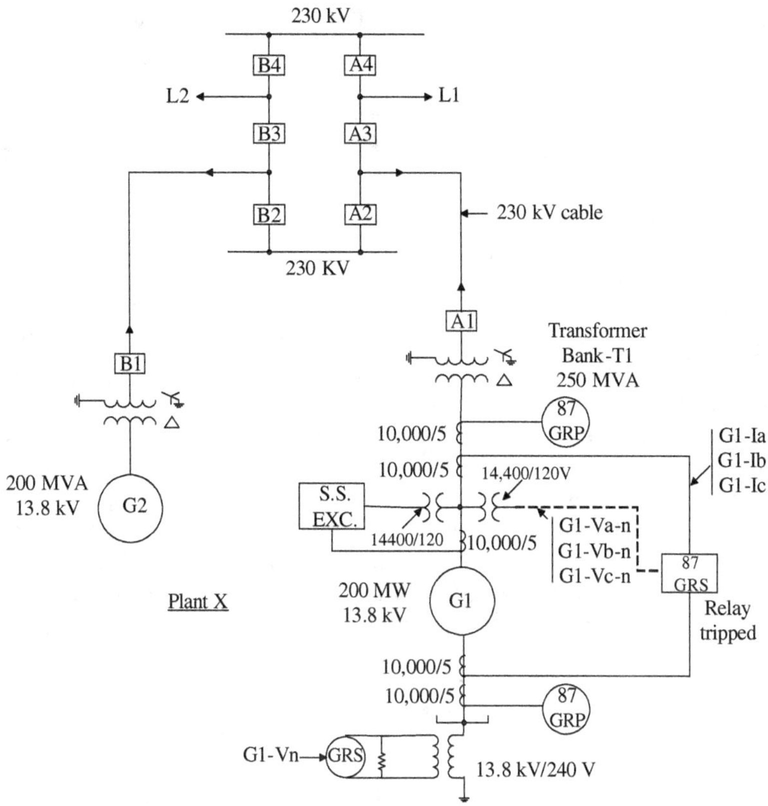

Fig. 4.22 One-line diagram with numerical relay–monitored generator voltages and currents.

faults as well as abnormal system conditions. One of these abnormal system conditions is loss of excitation, which is covered in this case study.

Description of the Incident An operator reached and removed excitation from a generator while the machine was generating 190 MW. The loss of field had begun by the operator's action, and as a result the generator terminal voltage began to decay along with the real power level, while the reactive power began to rise until the unit tripped. The generator multifunction secondary relay 87GRS detected the loss of excitation in 0.5 s (as designed) and operated on field failure and initiated the emergency unit shutdown immediately by energizing lockout relay 86ES, which tripped the unit. The primary relay 87 GRP failed to detect the condition and did not operate or generate an event record.

Performance of the Multifunction Numerical Generator Relays The loss of field was detected only by the secondary numerical relay 87GRS. The primary numerical relay 87GRP did not initiate any tripping, and hence no targets or event records were found. The failure of the 87GRP to detect the field loss was analyzed and will be explained in coming sections.

Analysis of the Secondary Relay 87GRS Oscillography Record Figure 4.22 shows the system one-line diagram with the 87GRS relay zone of protection and relay-monitored currents and voltages. The relay oscillography record shown in Fig. 4.23 reveals generator phase voltages and currents as well as the generator neutral voltage. The generator neutral voltage consists of a third harmonic with a changing magnitude due to the change in the reactive power absorbed by the generator upon the loss of field. The presence of the third harmonic at the generator neutral indicates a healthy generator. The current on all the phases reveals an increase in magnitude followed by an increase in the amount of VAR absorbed by the unit. As shown in Fig. 4.24, the reactive power calculated by the numerical relay indicates a level of −167 MVR. In a span of 28 cycles the VAR level increased from a value of −75 MVAR (2 cycles after the loss-of-excitation detection by the mho relay element) to a value of −167 MVR just prior to the unit trip. The relay record was generated by the relay trip output of the field failure element, as indicated by the solid line at trigger line a. The loss-of-field starting instant was lost due to insufficient pre-fault (event) data setting. The pre-fault setting was 0.5 s, which is less than the loss-of-field mho relay element operating time plus the timer setting of 0.5 s.

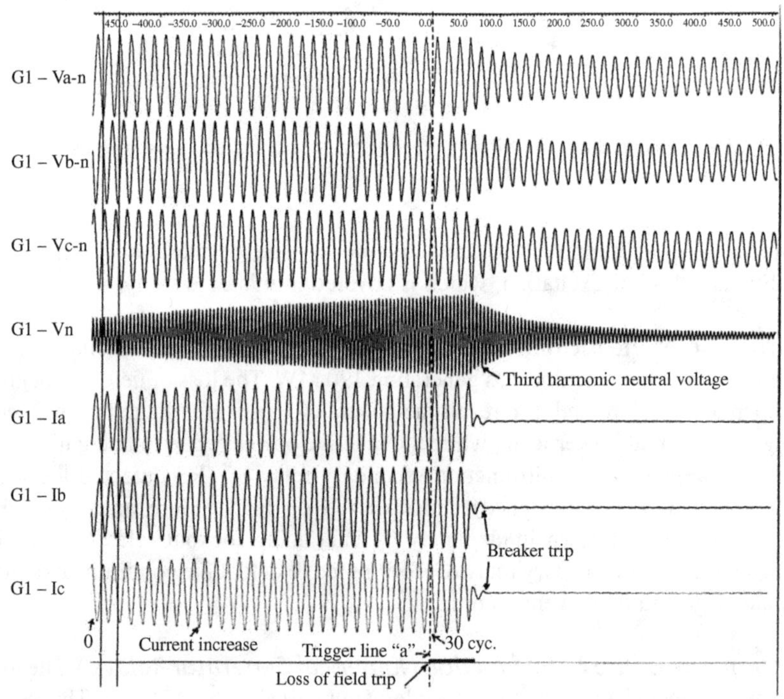

Fig. 4.23 Generator numerical relay record showing the loss-of-field event.

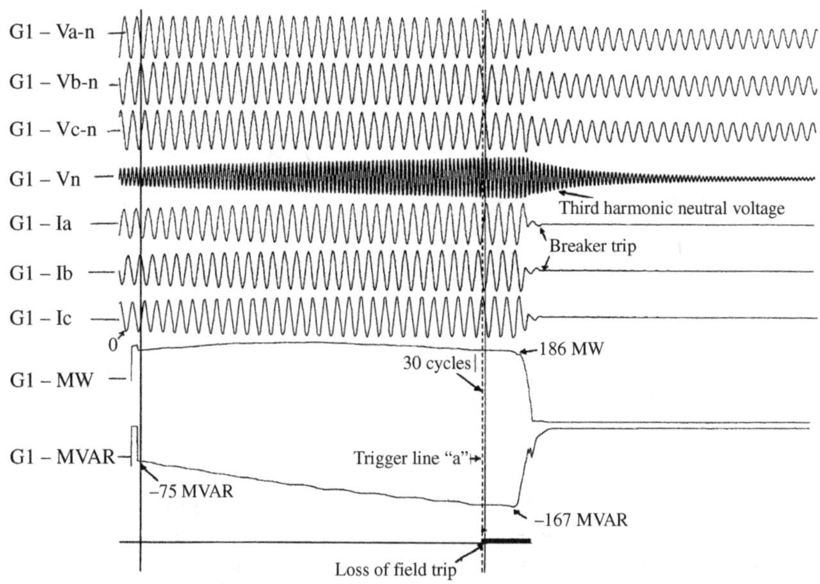

Fig. 4.24 Generator numerical relay record showing the loss-of-field event and generator active and reactive powers.

Explanation for the Primary Relay 87GRP Failure to Detect a Loss-of-Field Condition The generator loss-of-field condition was not detected by the 87GRP, due to the numerical relay loss-of-field condition being inhibited by the voltage supervision function. The enabling of the undervoltage element will allow a loss-of-field trip only if the machine terminal voltage is below the element voltage threshold setting. The numerical relay was commissioned with the 87GRP software default setting for the voltage supervision function enabled with an undervoltage setting of 70% of rated voltage. However, at the time of the loss-of-field incident, the terminal voltage recorded by the secondary numerical relay 87GRS was approximately 90% of rated voltage, confirming the fact that relay 87GRP was inhibited from operating.

Loss-of-Excitation (Field) Protection Normally, generator excitation (field) is adjusted to provide the proper terminal voltage to supply reactive support (VAR) to the system. Generator loss of field can be caused by either a field open circuit, a field short circuit, the accidental tripping of a field breaker, regulator control failure, or loss of the main exciter. Partial or total loss of field for a synchronous generator is detrimental to both the generator and the power system to which it is connected. The condition must be detected quickly and the generator isolated from the system to avoid generator damage. The generator drives excitation from the system, and during the loss-of-field event, the synchronous generator becomes an induction machine.

A loss of excitation of a generating unit is considered an abnormal system condition. A loss of field will force the generator to run in the underexcited region and receive the needed reactive power (VAR) from the system. This case study illustrates this concept

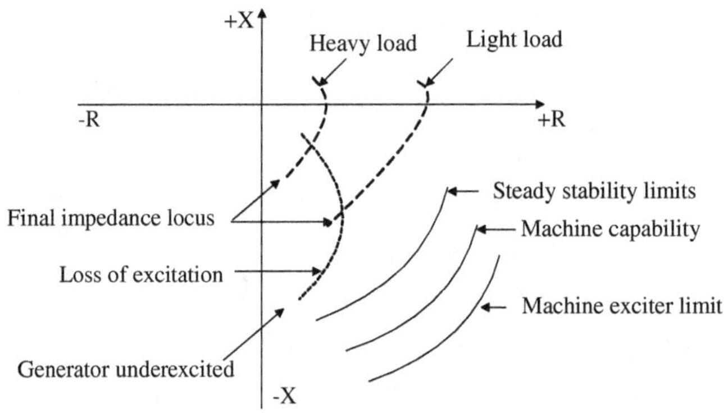

Fig. 4.25 Generator loss-of-field locus characteristics.

by recording a loss-of-excitation incident and showing reactive power being absorbed by the generator at a value of -167 MVAR just prior to the unit trip.

Due to the loss of field, the underexcited condition normally increases the generator rotor temperature due to the induced eddy current in the field winding, rotor, body, wedges, and retaining rings. The slip-induced eddy currents will heat the rotor surface, and the temperature increase can cause machine damage if not detected and isolated. The high reactive current drawn by the generator overloads the stator. It can also have an effect on the power system by decreasing the power system reactive resources, which can trigger system or area voltage collapse. Therefore, this abnormal generator system condition must be detected.

Figure 4.25 shows the loss-of-excitation locus during heavy- and light-load system conditions. During the loss of field, the impedance measured looking into the generator will vary. The loss-of-excitation relay characteristics have to be coordinated with the power system steady-state stability, the machine capability, and the machine minimum excitation limiter characteristics, as shown. The loss-of-field condition is normally detected by a mho relay element device 40, which is connected and supplied with generator terminal secondary voltages and currents. The relay measures the impedance as seen from the generator's terminal and will operate to shut the machine down when impedance falls inside the mho circle, as shown in Fig. 4.26.

Loss-of-Excitation Protection Options Loss-of-excitation protection of generators can be implemented using either a one- or two-element mho distance relay. The single element diameter shown in Fig. 4.27 is set equal to the machine synchronous impedance X_d with an offset of $\frac{1}{2}X'_d$ from the origin and a time delay of 0.5 to 1 s. The dual-element protection scheme is shown in Fig. 4.28; the larger mho circle is exactly similar to the single element described earlier, and the smaller circle will be instantaneous and is set at 1 per unit of machine impedance.

Loss-of-Excitation Relay Setting The loss-of-excitation condition can be detected using either one or two impedance elements. A trip will occur when the

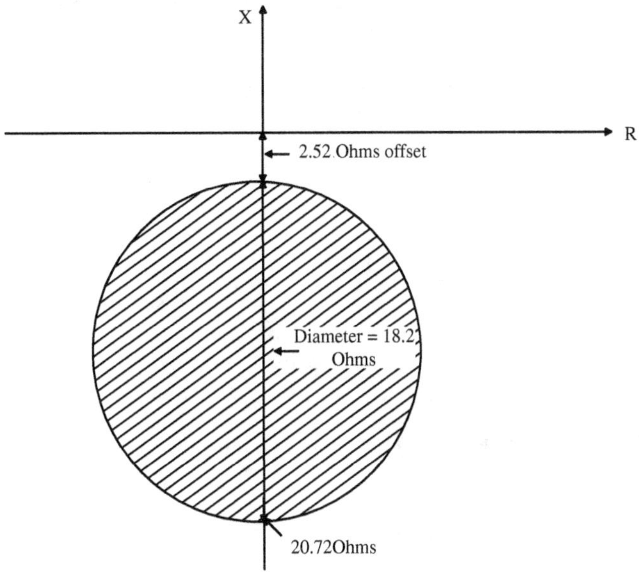

Fig. 4.26 *R–X* diagram showing the loss-of-field relay characteristic and setting.

impedance falls within the impedance circle for the delay time specified. As shown in Fig. 4.27, the relay circle diameter should be set equal to the machine synchronous reactance X_d, and the circle offset is equal to half the generator transient reactance X'_d. The associated timer could be set at 0.5 to 1 s. The diameter of the smaller circle shown in Fig. 4.28 (when used) should be set equal to the machine base impedance X_b (about 70% of X_d). As shown in Fig. 4.26, only one distance relay element was used in this case study and it is set as noted below:

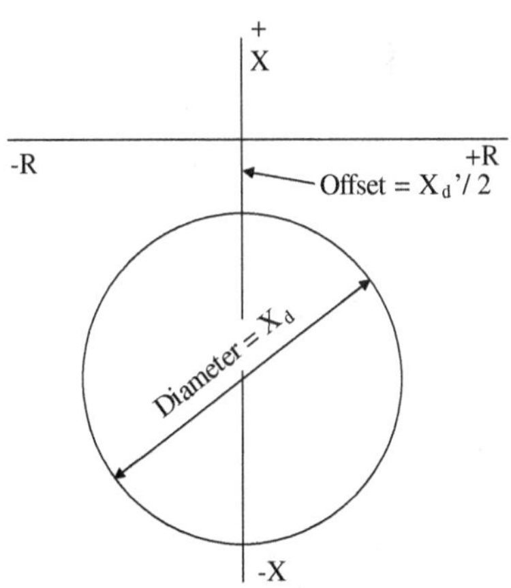

Fig. 4.27 Single-zone offset mho relay characteristics.

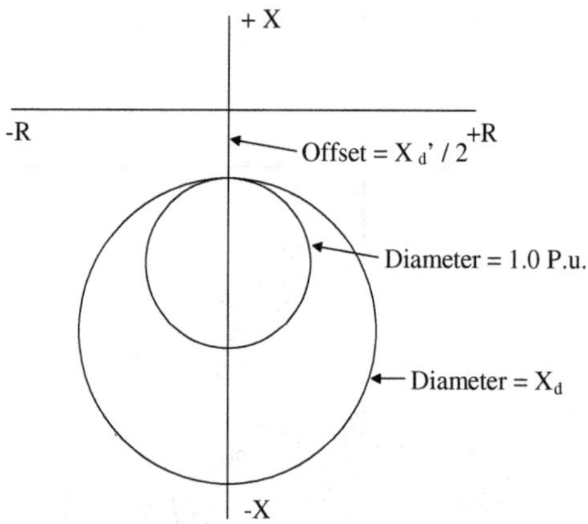

Fig. 4.28 Two-zone offset mho relay characteristics.

Generator data: 167 MVA, 13.8 kV, 0.9 PF.

$$X_d = 0.957\,\text{pu}\,(13.8\,\text{kV},\ 167\,\text{MVA})$$
$$X'_d = 0.265\,\text{pu}\,(13.8\,\text{kV},\ 167\,\text{MVA})$$

CT ratio: 10, 000 : 5

PT ratio: 14, 400 : 120

$$\text{offset} = \frac{X'_d}{2}$$
$$= \frac{0.265}{2} \times Z_b \times \frac{\text{CT}}{\text{PT}}$$

$$Z_b = \frac{\text{kV}^2}{\text{MVA}} = \frac{13.8^2}{167\,\text{MVA}} = 1.14\,\Omega$$

$$\text{offset} = \frac{0.265}{2} \times 1.14 \times \frac{2000}{120} = 2.52\,\Omega$$

$$\text{diameter of the mho relay} = X_d(\text{in pu}) \times X_b \times \frac{\text{PT}}{\text{CT}}$$
$$= 0.957 \times 1.14 \times \frac{2000}{120} = 18.2\,\Omega$$

The timer is set at 0.5 s. The distance element should be coordinated in the *R–X* diagram with the machine excitation limiter, the machine capability curve, and the steady-state stability limit. Figure 4.25 illustrates how the mho distance element of the multifunction relay is coordinated.

Power System Phenomena

1. Appearance of third-harmonic voltage at the generator neutral as a sign of its health, as shown in trace G1-Vn in Fig. 4.24
2. Increase in generator apparent current following loss of the generator excitation, as shown in the G1 phase current traces in Fig. 4.24.
3. Absorption of VAR from the power system by the generator upon the loss of its excitation system, as shown in trace G1-MVAR in Fig. 4.24

Corrective Actions

1. The generator control board should be reviewed thoroughly, including its operation philosophy and associated labels for switches and protective devices, with the operator who committed the error, as well as other operators.
2. The loss-of-excitation logic for generator primary relay 87GRP needs to be revised to eliminate the undervoltage requirement for a loss-of-excitation trip. The voltage supervision function of relay 87GRP should therefore be disabled.
3. The pre-fault data length should be increased from 0.5 s to .65 s to be able to show the instant of the occurrence of the excitation loss on the numerical relay oscillography record.

Lessons Learned

1. Use of the relay manufacturer's recommendation for undervoltage sensing for loss of excitation needs to be studied and justified prior to implementation.
2. Undervoltage condition sensing for loss of excitation for hydro units is not recommended because hydro units normally hold their voltage level when they lose excitation.
3. Protection redundancy for generators against abnormal system conditions, as offered by the new multifunction generator numerical relays, is an important concept. Failure to detect the loss-of-field condition could have subjected the unit to a lot of stress that could cause the generator to fail.
4. Use of generator numerical multifunction relaying concepts has provided an added level of protection against abnormal system conditions. The integration of all elements associated with generator protection against faults and abnormal system operating conditions has resulted in a reduction in the cost associated with the hardware and installations. As a result, redundant systems are affordable, as compared with the single application of discrete obsolete electromechanical and static generator protection relays.
5. The relay record pre-event length setting should be longer than the loss-of-excitation relay delay timer setting, to be able to capture the instant prior to loss of excitation. In this case study the loss-of-excitation incident point was missed because of insufficient pre-event record length, due to the relay

loss-of-field element delay time being set at 0.5 s, the same as the pre-event fault record length.

6. The impedance locus for the loss-of-excitation trajectory cannot be plotted, due to the missing pre-event loss-of-excitation data. The impedance locus plot needs to be defined from the start of the event to its end.
7. The use of voltage supervision for mho relay device 40 is intended to allow the generator unit to be run in an underexcited mode while possibly operating within the loss-of-field characteristics, but having nearly normal terminal voltage, around 90 to 95%. The undervoltage relay in this case will prevent unnecessary operations by the distance relaying scheme when the machine's absorbed reactive power level is approaching unit rating. This may be applicable to hydro machines when operated on occasions as synchronous condensers.

Case Study 4.4: Loss-of-Excitation Trip in an 1100-MW Unit

Abstract An 1100-MW 25-kV generating unit that was delivering full output to the system tripped due to operation of the loss-of-excitation relay, device type 40. In this case study we explain the cause of the field failure, display a chart for a voltage recorded two substations away from the generating plant, and provide lessons learned.

Description of the Machine Field Failure The generator tripped due to the field failure associated with the collector rings and brush rigging for the exciter generator. The ring surface deteriorated for unknown reasons at the time of analysis and resulted in brush arcing. This surface was contaminated with carbon dust and oil vapor and eventually flashed to the brush box, destroying the brush rigging and the exciter rotor collector rings. The electromechanical loss of excitation relay detected the condition and tripped the unit.

Analysis of the Voltage Chart The substation Y 345-kV bus voltage, shown in Fig. 4.29, dropped to 309 kV while the generator bus voltage dropped to 10.6 kV. The chart shown in Fig. 4.30 is for substation X, which is two stations away from the affected plant. It reveals a voltage decline from a value of 356.5 kV to a value of 342 kV in approximately 4.4 s, attributed to loss of excitation of unit G1 at plant Y. At 4.4 s from the start of the event, the locus of the impedance measured by device 40 landed in the circle of the relay on the *R–X* diagram, and the unit was tripped at this moment. When unit G1 stopped drawing its needed reactive power (VAR) from the system, the voltage started to rise and oscillate, due to the out-of-step condition caused by the tripping of the unit. The oscillations appear to be damped as their step gets reduced, as shown in the Fig. 4.30.

System Phenomena The power system voltage oscillation phenomena following the loss of the 1100-MW generating unit are shown in Fig. 4.30. The oscillation appears to be damped, leading to a stable system outcome.

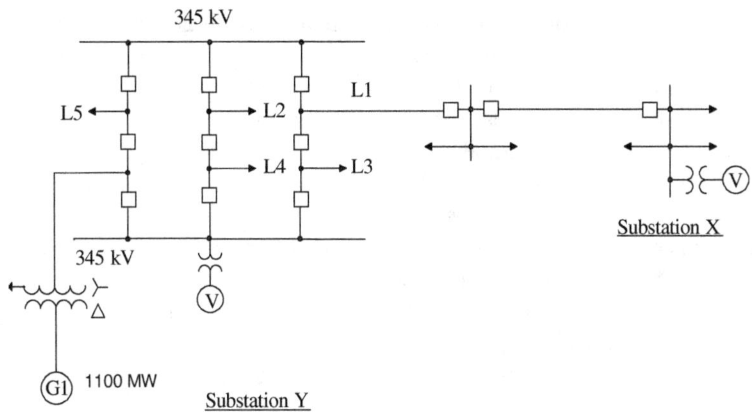

Fig. 4.29 Simplified 345-kV one-line diagram around substation Y.

Lessons Learned The severe depression in the system voltage that occurred before the trip of the unit, as a result of the generator absorbing reactive power upon the loss of the unit field excitation, can have a serious impact on a power system if not corrected. This illustrates the importance of the loss-of-field relay in preventing degraded system voltage problems by removing the generator from the system in a short period of time. It is very clear from this event that if the loss-of-field device 40 relay failed to detect the loss of excitation, a serious reactive power supply requirement may challenge operation of the system, due to the lack of protection redundancy.

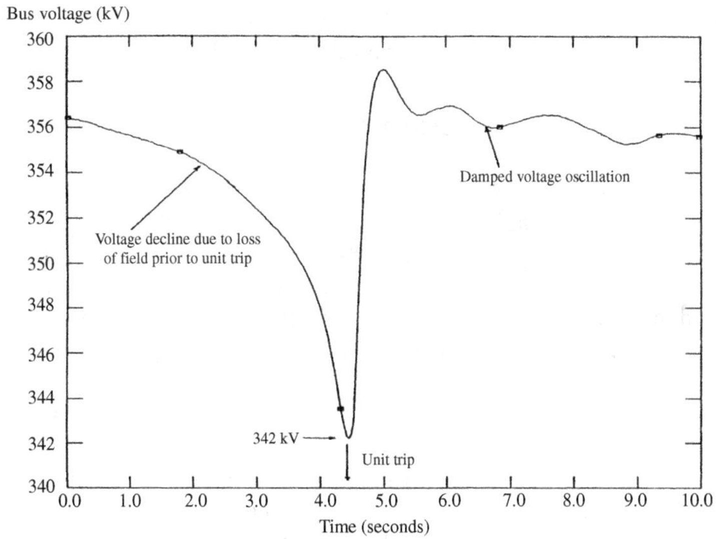

Fig. 4.30 Substation X 345-kV bus voltage for loss of unit G1 excitation at substation Y.

The loss-of-excitation event is normally treated as an abnormal power system operating condition. As such, in the past, most generating units were protected against abnormal system conditions, including loss of excitation by only one relay. Electromechanical or static relay use for generator protection was based on only a single relay to protect against abnormal operating conditions, including loss of excitation. Therefore, no redundancy against abnormal power system conditions was implemented with this protection philosophy.

New numerical multifunction generator relays offer reasonable cost and enhancements in the area of generator protection. The enhancement is accomplished by allowing the implementation of two redundant numerical relays to protect against system faults and abnormal power system operating conditions, in addition to the benefits offered by relay event and oscillography records.

Case Study 4.5: Mis-synchronization of a 50-MW Steam Unit for a Combined-Cycle Plant

Abstract The steam unit of a combined-cycle plant was in the process of being synchronized with the system following successful synchronization of the combustion turbine unit. The automatic synchronization mode was selected, but due to a speed control problem the unit was not connected to the system. The unit was apparently oscillating, and the synchronizer was unable to bring the unit within the allowable slip. The operator decided to switch the unit mode from "auto" to "off" to "manual." During the mode change, the automatic synchronizer produced a false output that closed the unit breaker accidentally, thus energizing the generator at a wider angle with the system. In this case study we describe the incident, oscillation of the unit against the system, analysis of the synchronizing relay failure, power system phenomena, corrective actions, and lessons learned. We also analyze the false operation of a differential relay that tripped the plant during the incident.

Description of the Power System and Associated Protection The combined-cycle plant, shown in Fig. 4.31, consists of an 100-MW combustion turbine generator (CTG) and a 50-MW steam generator (STG). Each generator is connected to the system via a delta/YG 13.8/138-kV transformer. Synchronization of the generators is done using dedicated 13.8-kV generator breakers. Therefore, the plant uses the 138-kV system as a starting station service source. The plant is connected to the system via a short gas-filled 138-kV cable. The plant protection is designed to fulfill the basic dual-element fault detection criteria, which will simply guarantee having more than one element to protect any given relay protection zone. Therefore, as shown in Fig. 4.32, a differential relay 87GS is dedicated to the STG. A harmonic restraint differential relay 87TS is dedicated to each unit GSU transformer. An overall harmonic restraint percentage restraint differential relay 87US is protecting the generator, ISO-phase bus, and unit GSU transformer. The 138-kV cable is protected by dual current differential relays applied over two dedicated fiber optic cables.

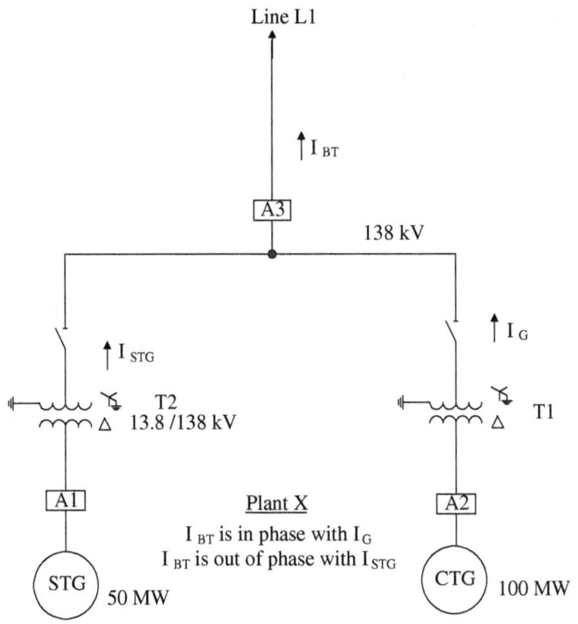

Fig. 4.31 Plant X main one-line diagram.

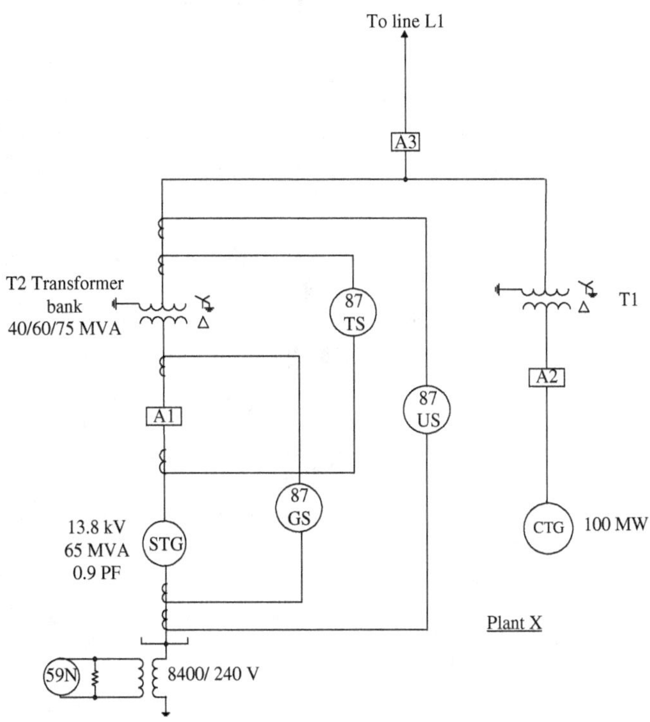

Fig. 4.32 One-line diagram showing the STG generator unit zones of protection.

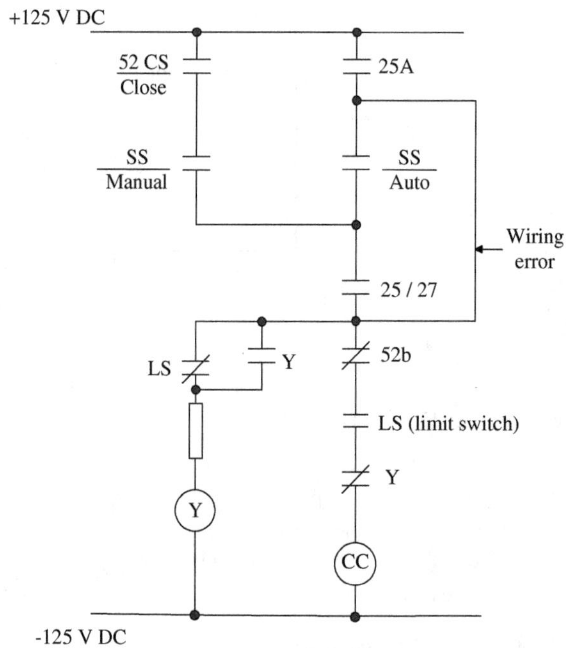

Fig. 4.33 DC schematic of the generator breaker closing circuit with the wiring error bypassing the auto sync mode.

Description of the Unit Synchronizing Scheme As shown in Fig. 4.33, a wiring error resulted in bypassing the sync check relay contact in the closing control circuit. The sync check relay normally provides added security to protect against any automatic synchronizing relay failure or human error during manual synchronization. The relay is normally set to provide a contact closure of around 10 to 20° difference between the generator and the system. This will prevent any closing of the breaker at angles wider than the setting of the sync check relay.

Figure 4.34 shows an automatic synchronizing relay device 25A connection to the system (bus) and generator voltages. Normally, a sync switch device SS with auto and manual positions is added to connect the bus and generator line-to-line voltages to device 25A, with the switch in the auto position. Due to a speed control problem, with the unit hunting around the normal speed with no success to synchronize with the system, the operator moved the SS switch from "auto" to "manual." The auto/manual panel-mounted switch was wired correctly to remove both the bus and generator sensing voltages when the switch is moved to the manual position (Fig. 4.35). During the transition of the mode change activation from automatic to manual, the 25A device issued a synchronization pulse with the generator and the system out of phase with an angle of 109°.

Description of the Incident The CTG unit of the combined-cycle plant was first synchronized to the system with the generator output at 22 MW. The STG unit was in

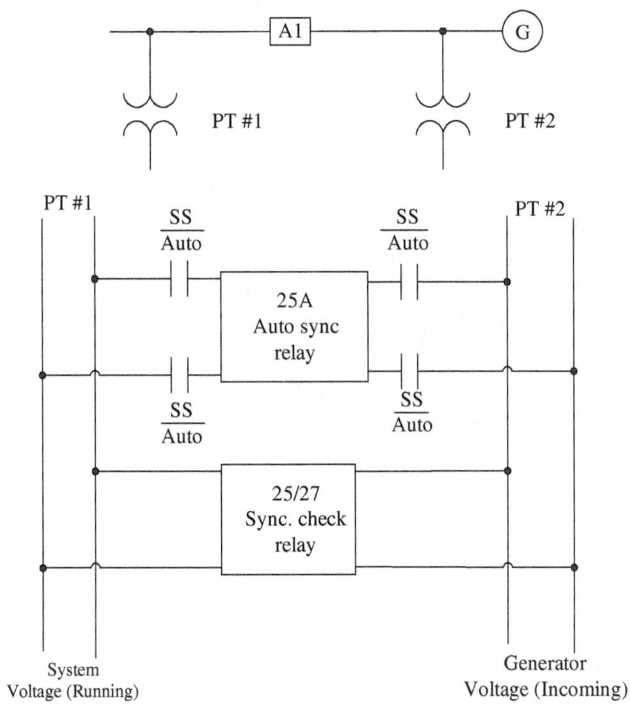

Fig. 4.34 Autosynchronizing circuit with SS switch interface.

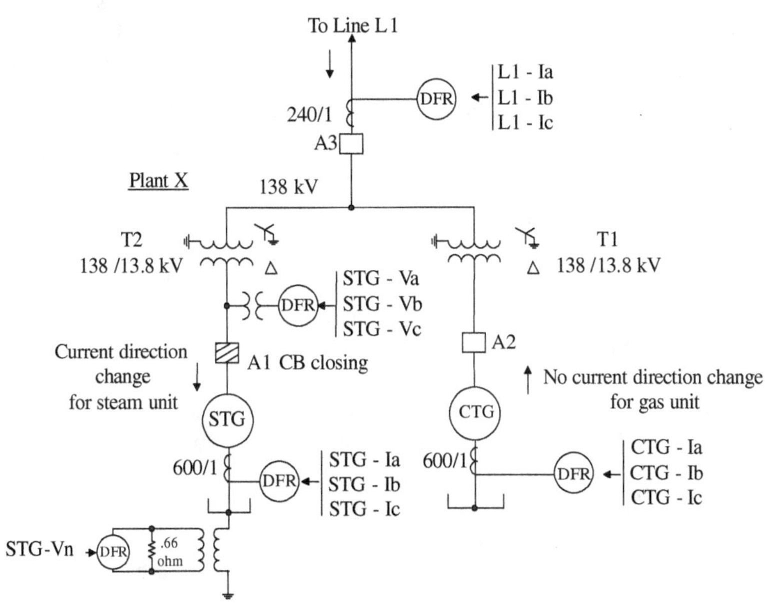

Fig. 4.35 Plant X main one-line diagram with DFR-monitored currents and voltages.

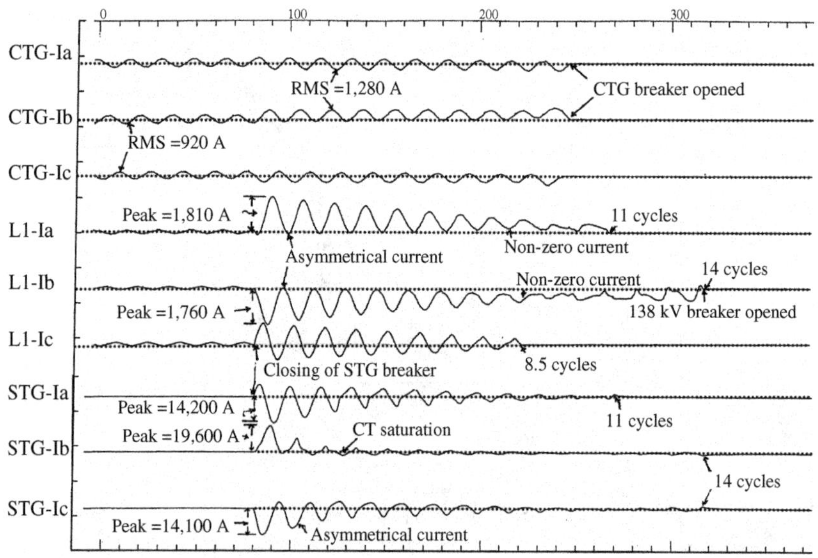

Fig. 4.36 Phase currents during the out-of-phase closing of the STG 13.8-kV breaker.

the process of being synchronized to the 138-kV system via the automatic synchronizing mode. Due to poor governor speed control, the synchronizer could not achieve the proper slip frequency, and hence the STG remained unsynchronized to the system with an idle speed near the synchronous speed of 3600 RPM. The STG oscillated slowly between speeds of 3590 and 3610 RPM. This condition lasted for 5 minutes, with no success in synchronizing the unit to the system. The operator then decided to synchronize the STG manually, moving the sync-selector switch from "automatic" to "manual." The unit breaker A1 was closed accidentally while the unit was out of step with the system. The unit started to oscillate with the system as shown in the DFR record in Fig. 4.36.

Analysis of the DFR Records The DFR record in Fig. 4.36 illustrates feeder L1 currents with normal pre-event CTG unit output for 5 cycles followed by accidental STG unit step current increase. The asymmetrical STG currents on phases A, B, and C started to oscillate until the plant was tripped by the undesired operation of the STG differential relay. The accidental closing of CB A1 at an excessive angle with the system caused the STG unit to oscillate with the power system. Feeder L1 CB A3 phases A, B, and C cleared in 11, 14, and 8.5 cycles, respectively, from the instant of accidental closing of CB A1. The DFR record shown in Fig. 4.36 reveals that the CTG unit output was stopped 10 cycles from the closing of CB A1.

Figure 4.37 reveals STG unit neutral voltage, phase-to-neutral voltages, and phase currents during the accidental unit breaker closing incident. Figure 4.38 shows a time expansion for the feeder L1 phase currents, where phase A current has a nonzero current crossing due to dc offset component. The asymmetrical phase A current lasted

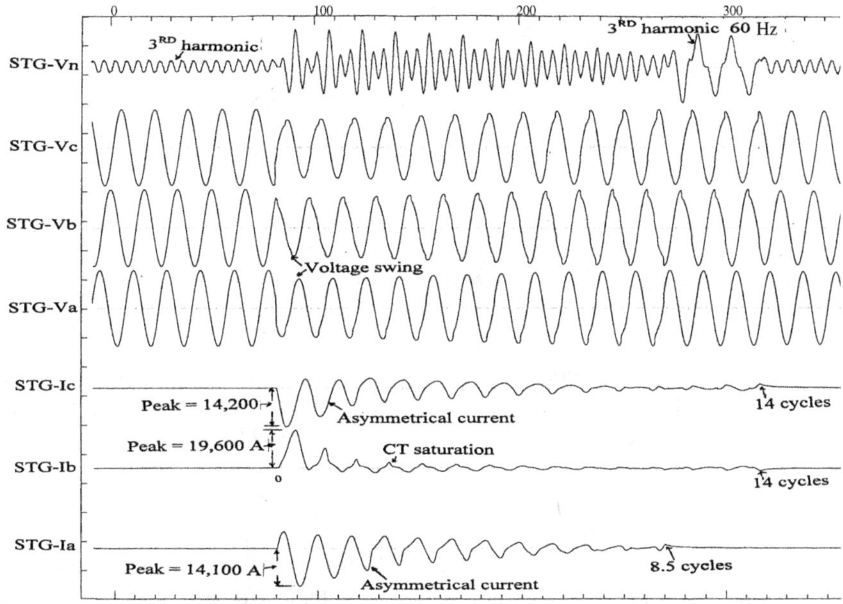

Fig. 4.37 STG voltages and currents during the out-of-phase closing incident.

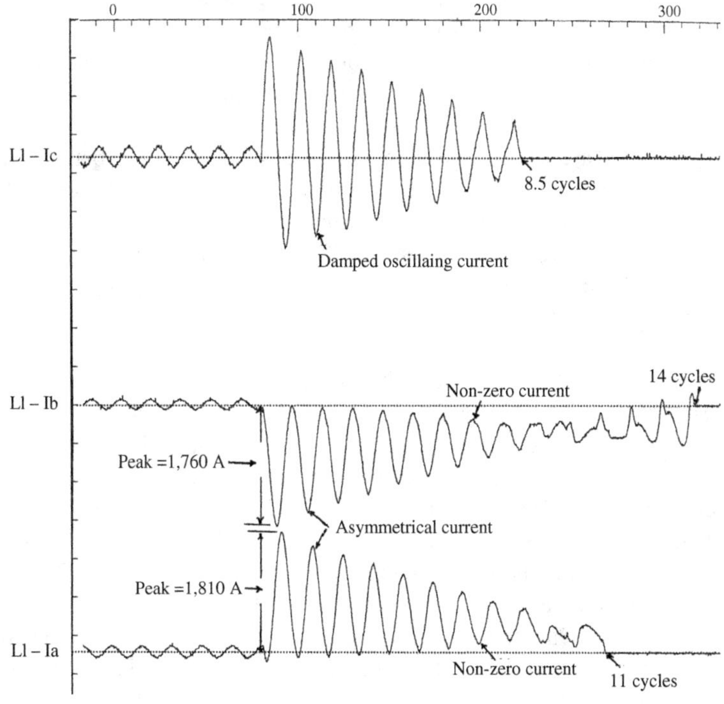

Fig. 4.38 Feeder L1 phase currents during the out-of-phase closing incident.

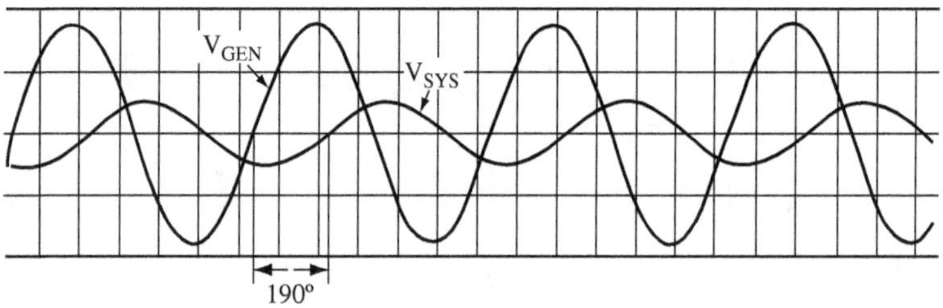

Fig. 4.39 Phase angle difference between the system and generator phase A-n voltages at the instant of closing.

for 11 cycles, with circuit breaker interruption at zero crossing. The phase B current lasted for 14 cycles, with behavior similar to that of the phase A current. The phase C current had a damped oscillatory current that lasted for 8.5 cycles.

Calculation of the Angle Between the Generator and the System As shown in Fig. 4.39, the system phase-to-neutral voltage is superimposed on the steam generator phase A-n voltage at the instant of closing. The difference between the two zero crossings (going positive) is a little more than a quarter-cycle. The exact measurement of the closing angle is 109°.

Power Swing During the Incident Closing a generator breaker with an excessive phase angle across the synchronizing breaker just prior to the closing action tends to cause an-out-of-step (power swing) condition. This power swing can sharply "bump" the generator and thus cause stress to the machine and its bearings. As illustrated in the DFR records of Figs. 4.36, 4.37, and 4.38, when the unit breaker was accidentally closed at an angle of 109°, an oscillation was generated between the STG rotor and the system and the CTG unit. Prolonging the presence of this oscillation can cause a mechanical shock to the machine, which may excite one of the torsion modes of the combined generator–turbine–shaft. The undesired operation of the overall differential relay saved the day, eliminating the likelihood of possible extensive machine damage.

Analysis of the Plant Shutdown by Operation of the Unit Transformer Differential Relay Normally, differential relaying schemes are immune to power swing phenomena, due to the through-flow nature of the swing, where the input matches the output to the relay. According to the DFR record in Fig. 4.36, when the swing occurred, the CTG unit contributed to the swing, as shown by the high current on phases A, B, and C. The original swing between the STG unit and the system caused high asymmetrical currents, which stressed the current transformer (CT) characteristics associated with the three differential relays 87T, 87G, and 87U associated with STG unit protection. Only the 87U overall differential operated undesirably and shut down the plant. As shown in Fig. 4.32, the overall differential relay 87U receives its 13.8-kV input from the generator neutral wye-connected CTs. The high-side input to the 87U relay is from the delta-connected transformer bushing CTs.

The neutral CTs were supplied by the generator manufacturer with a single ratio of 3000 : 5 and the IEC accuracy stated as B250-200. The high-side multiple ratio of 2000 : 5 transformer bushing CTs was specified as the ANSI standard accuracy of C800 and connected with a ratio of 1200 : 5. This resulted in no match between the performance of the high- and low-side CTs in the presence of an asymmetrical current that contains dc offset components. This type of undesirable trip is welcomed when system problems occur (mis-synchronization).

If the relay did not trip falsely due to CT saturation, two scenarios can be postulated:

1. Damage to the CTG generator if the oscillation continues to grow (unstable) at slip frequency without proper power system damping
2. A continuation of the swing, which can be assumed to be stable, until the unit recovers and can be pulled and synchronized to the system

Analysis of the Synchronizing Relay Failure The automatic synchronizer was declared inoperable until bench testing and manufacturer corrective actions could be taken. The 25A relay device was returned to the manufacturer for testing and failure analysis. A test setup was devised to re-create the condition that caused the problem. The relay manufacturer was able to emulate the condition described in the incident and was able to replicate the relay problem. This test circuit was used to simulate the auto/manual sync switch being cycled open and closed to recreate the problem and provide solutions.

Power System Phenomena

1. Power system oscillation (swing), which was prevented from progressing by the undesired differential relay tripping action, as shown in Fig. 4.36 for L1 current traces.
2. Current transformer saturation, as shown in trace STG-Ib in Fig. 4.37.
3. Nonzero crossing of current due to asymmetrical current caused by the presence of dc offset components, as shown in traces L1-Ia and L1-Ib in Fig. 4.38. As a result, the circuit breaker has to wait for the current to reach the zero-crossing point for current interruption to take place.

Corrective Actions

1. Modify the synchronizing circuit wiring by removing the jumper shown in Fig. 4.33 around the sync check relay output contact.
2. Megger the STG unit to confirm that no machine damage has occurred as a result of "bumping" the machine.
3. Test the STG unit generator circuit breaker.
4. Return the 25A device to the manufacturer for testing and duplication of the incident for problem definition and solutions.

Corrective Actions by the Relay Manufacturer Hardware and software changes were made to the relay to fix the problem. For hardware changes, the input filtering was increased for the bus and generator voltage inputs. This was accomplished by increasing the resistive and capacitive elements of the filter. Apparently, the hardware fix alone did not provide a complete fix for the problem. Therefore, a software solution was added to the phase angle calculation algorithm. Advance calculation and phase angle magnitude comparison were added to prevent operation if the angle algorithm detects a glitch, which could be caused by a sudden shift in the advance angle or in the phase angle between the two voltages. The new algorithm was tested successfully using abrupt angle changes, similar to those that occurred during the incident.

Lessons Learned

1. Undesired relay trips can benefit a system and reduce the stress exerted on equipment, especially when the system is subjected to serious problems. Undesired operation of the unit overall differential relay removes a condition stressful to the machine.
2. CT accuracy should be matched to differential relays among all inputs to avoid the occurrence of undesired trips.

Case Study 4.6: Mis-synchronization of a 200-MW Hydro Unit

Abstract An emergency generator unit trip occurred when an operator was trying to synchronize a 200-MW hydro unit to a system using the manual synchronizing mode. This apparently caused an out-of-step condition (power swing), forcing the G1 unit to start oscillating with the power system. This resulted in high asymmetrical currents, which affected the performance of the CTs connected to the generator differential relay. Asymmetrical current flow stressed the CTs connected to the unit differential numerical relay, causing uneven responses that resulted in a relay trip output. In this case study we describe the machine dynamic interaction between the unit and the system during the incident, and estimation of the unit closing angle. In addition, we explain the undesired differential relay trip, power system phenomena, and lessons learned.

Description of the Protection Systems The plant protection was designed to fulfill the basic dual-element fault detection criteria, which simply guarantee having more than one element to protect against faults within any given relay protection zone. Therefore, as shown in Fig. 4.40, the generator is protected by dual numerical multifunction differential relays 87GRS and 87GRP. The GSU transformer is also protected by dual numerical transformer multifunction relays.

Definition of a Power Swing (Out-of-Step) Phenomenon A power swing (or out-of-step) is a phenomenon that describes conditions in a power system when

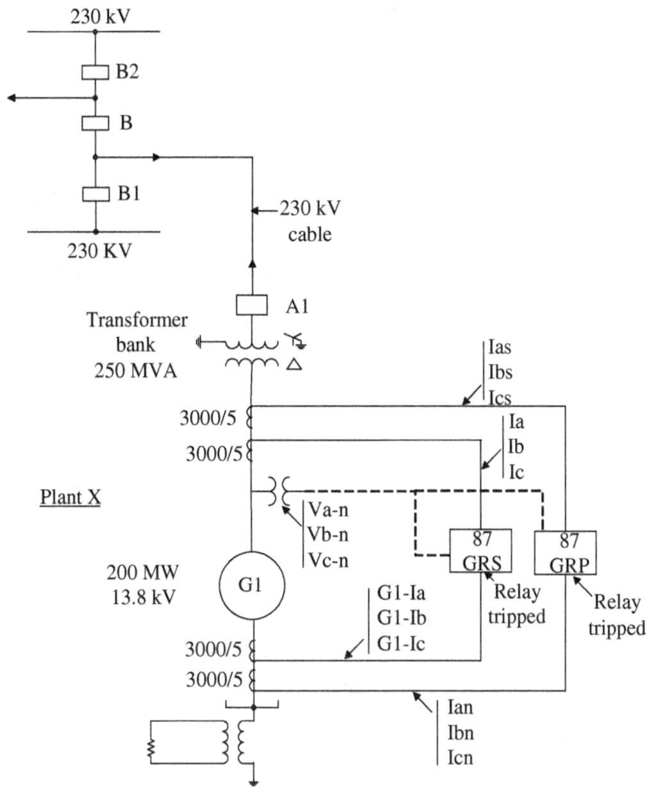

Fig. 4.40 One-line diagram showing numerical relay–monitored generator currents and voltages.

blocks of generation within the system are undergoing large changes of phase angle with respect to each other, leading to large changes in current, voltage, and power flow across the system as well as the apparent measured impedance.

Analysis of the Incident, Including the Sequence of Events Based on an analysis of the generator numerical relay fault records of Fig. 4.41, the 230-kV unit CB A1 was closed by the operator to synchronize the unit with the system. At the instant of closing it appears that the unit was at subsynchronous speed with its rotor angle lagging the system reference bus. The magnitude of the slip frequency at the time of synchronizing is unknown. Figure 4.41 also reveals the relay power calculated when the machine began to draw power from the system, reaching a peak value of 175 MW in 12 ms after the closing of CB A1. This may be considered a "bump," since a step change of 1 pu torque was applied to the machine. The electric motoring power drawn from the system by the machine produced an accelerating torque on the generator, working to increase the generator's frequency.

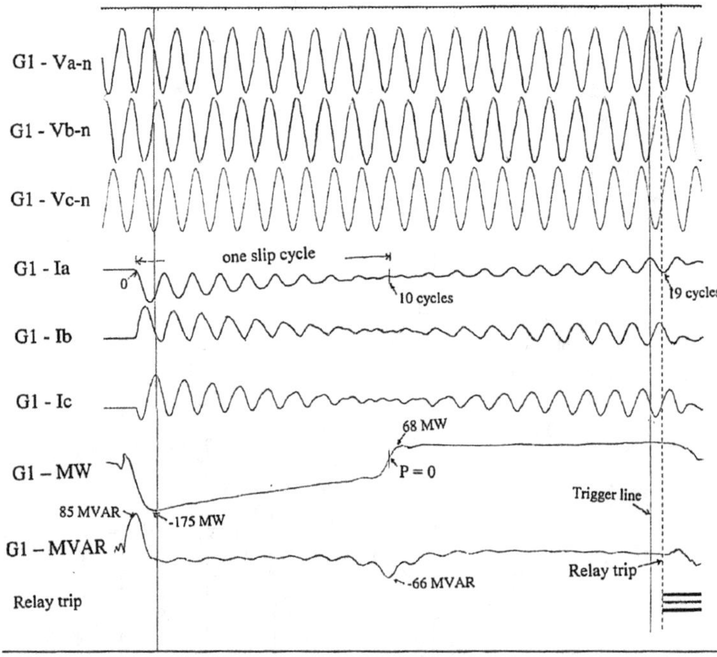

Fig. 4.41 Numerical relay 87GRS oscillography record showing generator oscillation with the system due to unit mis-synchronization.

As a result, the rotor's initial increasing lagging angle slowed down, stopped (i.e., $f_{gen} = f_{sys}$), reversed direction (i.e., $f_{gen} > f_{sys}$ and $P_{electric\text{-}motoring} > P_{mechanical\text{-}speed,\ no\text{-}load}$), and then begin to increase in the leading direction, shrinking the displacement between the unit and the system reference vectors. The synchronous equilibrium point is defined as the state in which the unit rotor angle matches the system angle and no power exchange takes place (i.e., $P = 0$, except for the slight motoring power needed to maintain speed no-load conditions). However, since the angular momentum of the unit could not be changed instantaneously at the equilibrium point to let $f_{gen} = f_{sys}$ (i.e., synchronous speed), the machine overshot the equilibrium point and would have continued to hunt the equilibrium point in a damped oscillatory manner had the unit remained connected to the system.

During the first part of this dynamic adjustment period (0 to 10 cycles), the machine currents decayed from their full offset values to zero and were 180° out of phase with their respective terminal voltages, confirming the motoring condition of the generator. Figure 4.42 illustrates the generator phase A current changes and voltage during the swing conditions, Fig. 4.43 shows the phase B current and voltage, and Fig. 4.44 shows the generator phase C current and voltage. It should be noted that when a machine is "bumped" and begins immediately to deliver power to the system, a deceleration torque will be generated to slow the machine rotor and hence generator frequency. In addition, when a new machine stable point is

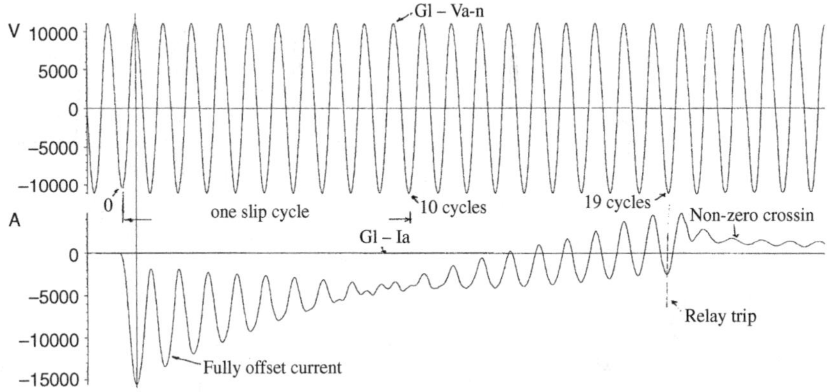

Fig. 4.42 Relay 87GRS record of unit phase A oscillating current and voltage.

reached, the machine will overshoot and oscillate back and fourth (hunting) with the system.

At 10 cycles from the closing of the unit breaker, the unit overshot the equilibrium state and the rotor angle continued to advance relative to the reference bus. As a result, the swing condition began to grow with the unit now delivering (i.e., generating) power to the system. Within this time frame, the machine current was in phase with the generator terminal voltage.

At 19 cycles from the closing of the unit breaker, the phase A differential relay for the primary and secondary generator multifunction relays 87GRP and 87GRS operated to initiate unit shutdown. The swing slip frequency can be calculated from the duration of each cycle. Since the period of the swing is 10 cycles, which is equivalent to a period of 0.166 s ($= 10 \times 0.0166$ s), the swing frequency will be equal $1/0.166 = 6$ Hz.

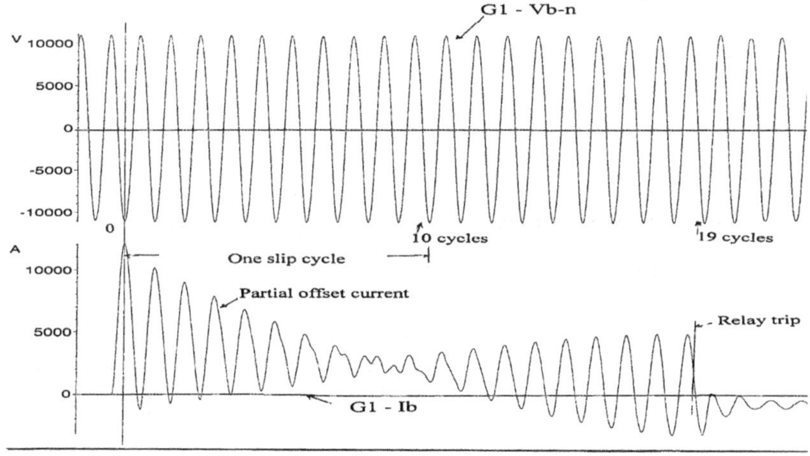

Fig. 4.43 Relay 87GRS record of unit phase B oscillating current and voltage.

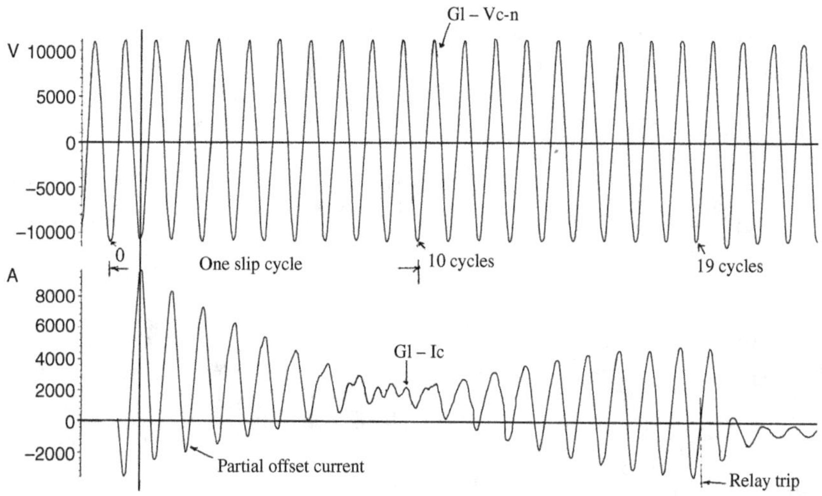

Fig. 4.44 Relay 87GRS record for unit phase C oscillating current and voltage.

Estimation of the Closing Angle Normally, the actual closing angle at the time of synchronization can be calculated accurately from an analysis of the recorded system voltage versus the machine voltage. Unfortunately, no DFR record was obtained for this event, so no system voltage was recorded. A rough electric circuit analysis was used to simulate a similar RMS synchronizing current. From this analysis, the closing angle was estimated to be between 15 and 20°.

Analysis of the Operation of the Differential Relay The operating characteristic of the 87GRS differential relay is shown in Fig. 4.45. The relay is set as follows:

$$\text{CT ratio} = 10{,}000 : 5$$

$$\text{relay pickup threshold} = I_{S1} = 0.25\ \text{A(secondary)}$$

$$\text{initial bias slope} = K_1 = 0$$

$$\text{restraint threshold level} = I_{S2} = 1.1 \times I_n = 5.5\ \text{A s}\ (I_n = 5\ \text{A})$$

$$\text{second bias slope} = K_2 = 40\%$$

I_{bias} = relay bias (restraint) current for phase A, defined as the average magnitude of $I_{an} + I_{as}$ where I_{an} = neutral-side current contribution and I_{as} = system-side current contribution

I_{diff} = differential current for phase A, defined as the vectrorial difference between I_{as} and I_{an}

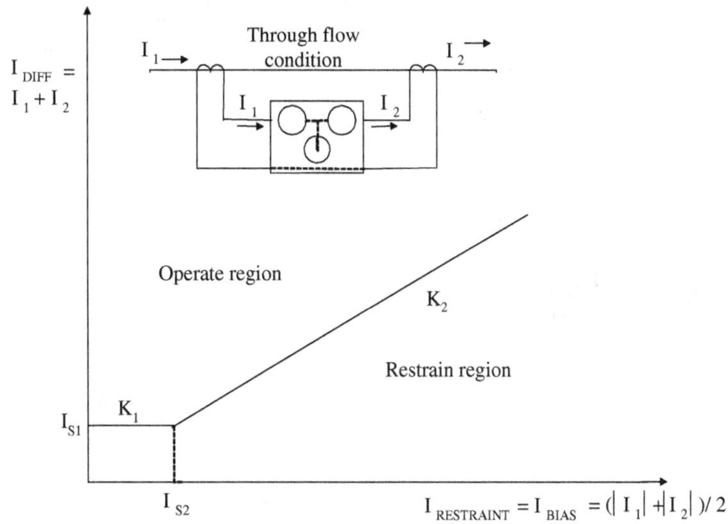

Fig. 4.45 Generator numerical relay 87GRS operating characteristics.

The relay has the following tripping criteria:

$$I_{diff} = \begin{cases} K_1 I_{bias} + I_{S1} & \text{for } I_{bias} < I_{S2} \quad (1) \\ K_2 I_{bias} - (K_2 - K_1) I_{S2} + I_{S1} & \text{for } I_{bias} > I_{S2} \quad (2) \end{cases}$$

From the 87GRS fault record:

$$I_{an} = 0.946 \text{ A s}$$
$$I_{as} = 1.19 \text{ A s}$$
$$I_{diff} = 0.462 \text{ A s}$$

I_{bias} can now be calculated as $= (0.946 + 1.19)/2 = 1.07$ A s. Since $I_{bias} = 1.07$ is less than the I_{bias} setting of 5.5 A, equation (1) can be used to check the relay operation:

$$I_{diff} = K_1 I_{bias} + I_{S1}$$

Since $I_{diff} = 0.462 \text{ A} > (K_1 I_{bias} + I_{S1}) > 0 \times 1.07 + 0.25$, a relay trip was issued.

Explanation of the Undesired Differential Relay Trips As previously explained, the out-of-step power swing between unit G1 and the system generated fully offset current on phase A and partial dc offset currents on phases B and C. The full dc component on phase A apparently stressed the phase A CT differential circuit. This caused different (dissimilar) CT transient responses for the generator neutral CTs and the generator terminal-side CTs. The uneven CT transient responses on phase A

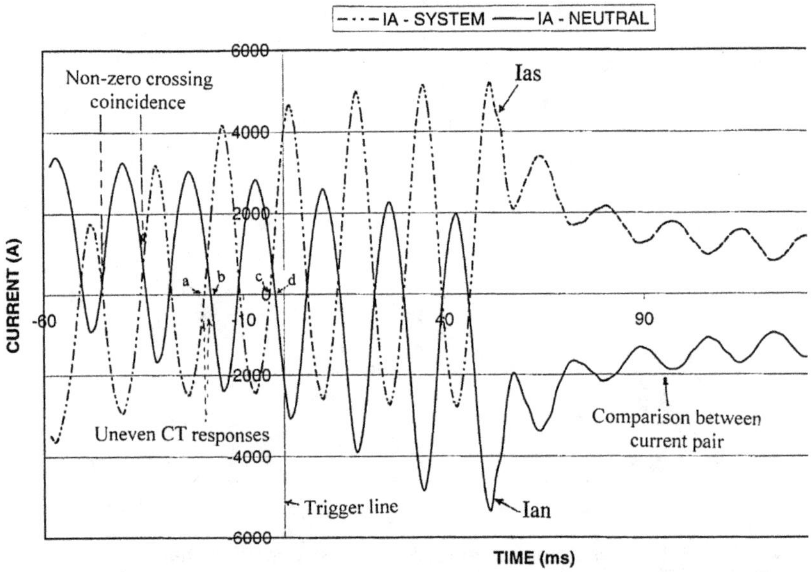

Fig. 4.46 Relay 87GRP record showing the 13.8-kV phase A generator neutral and system currents.

for both 87GRP and 87GRS generated a differential error current that exceeded the differential relay setting, causing a unit trip.

Figure 4.46 compares the two currents of the generator neutral end Ian and the generator system end Ias. Figure 4.47 shows a time expansion of the two current traces Ias and Ian prior to the relay trip output. The power swing phenomenon is considered as a through current flow condition for the generator phase differential relay. However,

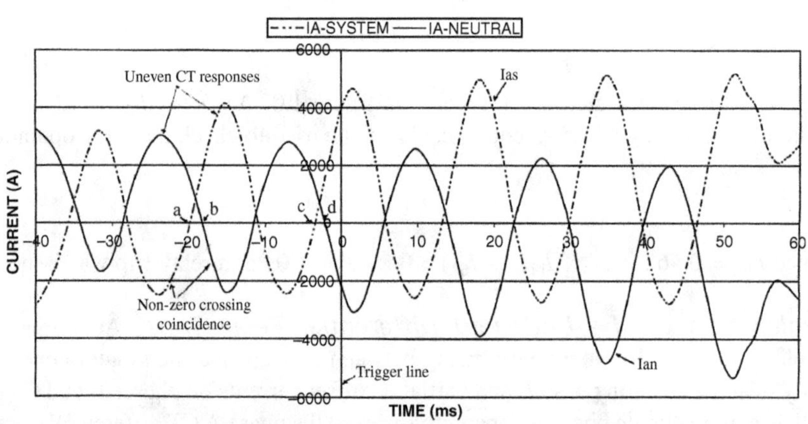

Fig. 4.47 Relay 87GRP record showing the time expansion of the 13.8-kV phase A generator neutral and system currents.

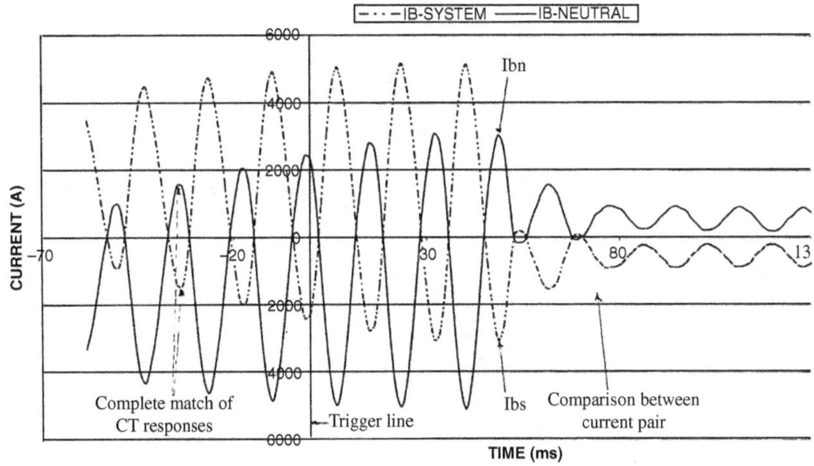

Fig. 4.48 Relay 87GRP record showing the 13.8-kV phase B generator neutral and system currents.

Figs. 4.46 and 4.47 reveal that the phase A neutral-side Ian reached its zero crossing at point a, while the system-side Ias crossed the zero point after a delay at point b. The same pattern is repeated a cycle later for Ian to cross at point c and Ias to cross the timing axis after a delay at point d. This generated a differential error current to fall in the operating region of the 87GRS relay operating characteristics shown in Fig. 4.45, and hence produced a differential relay trip.

Figure 4.48 compares two identical currents for phase B input to the relay, and Figs. 4.49 and 4.50 compare two identical currents for phase C. Therefore, since phases B and C exhibited partial dc offsets, their identical CT responses provide no differential (error) component and hence no relay trips. For phases B and C the swing phenomenon was a flow-through condition, as it should be.

Power System Phenomena

1. Non-zero-crossing current, caused by a full dc offset at the beginning, followed by a decay rate which keeps the current away from the zero crossing at the moment of breaker interruption, as shown in trace G1-Ia in Fig. 4.42.
2. Power swing as explained in the case study
3. Different secondary current responses for identical CTs used for differential protection to the same stimuli. This is due to different behavior of the dc transient and different levels of CT magnetic flux remanence.

Lessons Learned

1. This type of undesired differential relay trip provides benefits in stopping machine oscillations during out-of-step events related to the human–machine

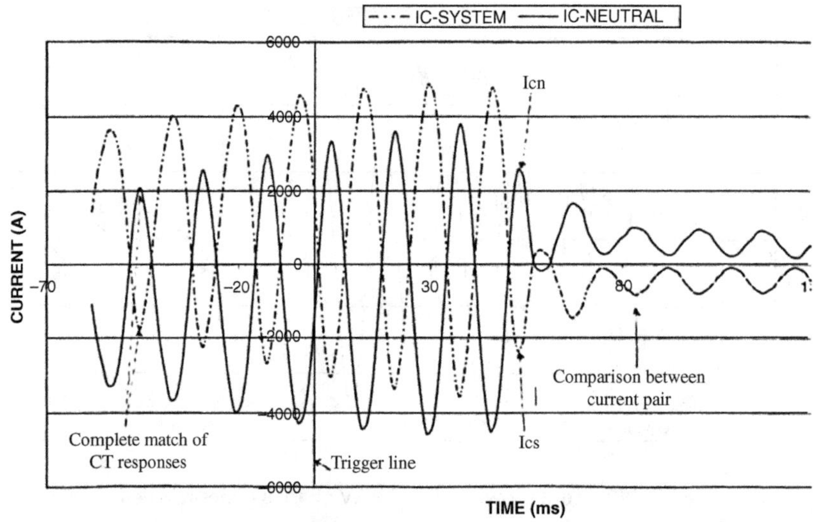

Fig. 4.49 Relay 87GRP record showing the 13.8-kV phase C generator neutral and system currents.

interface. The relay event records provide critical information to help develop lessons learned in order to optimize the operation of the system. It should be noted that the differential element for the numerical technology is designed to be sensitive (correctly so), to adequately protect the stator for any low-grade (high-resistance) type of fault. It is also known that the out-of-step conditions occurring by mis-synchronizations generate heavily offset currents which upset

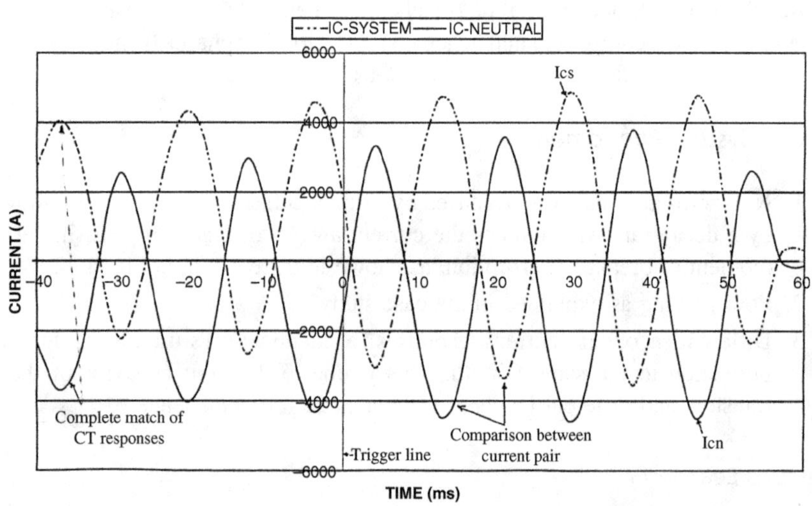

Fig. 4.50 Relay 87GRP record showing the time expansion of 13.8-kV phase C generator neutral and system currents.

CT response. This is a case of an undesired trip that may be acceptable, because depending on the closing angle, if it continues to grow, the swing may harm the generating unit involved.

2. The use of numerical sync check relays as a replacement for the existing static devices can safeguard closing at either an excessive angle or slip frequency. This can permit the operator to synchronize manually when both the angle and slip are within safe tolerances.
3. Many automatic synchronization events, amounting to 425 attempts, were carried out successfully for two of the generating units, with the new numerical generator protection applied for two units at substation X. No problem was reported during all these synchronizing attempts. Therefore, it can be concluded that the new sensitive numerical relay technology is performing exceptionally well while adequately protecting the associated units.
4. It is very well known that excessive phase angle across a sync breaker just prior to closing tends to sharply "bump" the machine. It has also been widely reported that closing at a static angle as low as 15 to 25° could cause as large a power swing as would closing at 0° with an excessively high slip frequency of 0.5 Hz. Since the automatic synchronizer safeguards both the angle and the slip, it is recommended that the automatic synchronization mode should always be the norm for the units.
5. For heavy through fault currents or full offset low through currents, it is unlikely that the CT outputs at each zone end will be identical, due to the effects of CT saturation. In this case a differential current can be produced. Normally, biasing will increase the relay setting such that the differential spill current is insufficient to operate the relay.
6. Stator generator phase differential relay is designed to fulfill two basic criteria: sensitivity for low-grade faults and stability for through-condition heavy external faults.
7. Superimposing the two current inputs for the generator differential relay on a per-phase basis can provide a clear indication of the causes of the undesired relay trips. The comparison method of overlaying a pair of generator neutral- and system-side currents provides a quick way of determining the cause of the differential undesired relay trips.
8. The initial immediate analysis of the event was declared as a power swing and a stator phase fault was ruled out, due to the near normality of all three phase-to-neutral voltages (no significant voltage dip). The initial magnitudes of the three phase currents also did not support a fault theory scenario.

Case Study 4.7: Undesired Tripping of a Numerical Differential Relay During Manual Synchronization of a Hydro Unit

Abstract An undesired trip of a generating unit occurred when an operator was attempting to synchronize the unit to the system. This case study is similar to

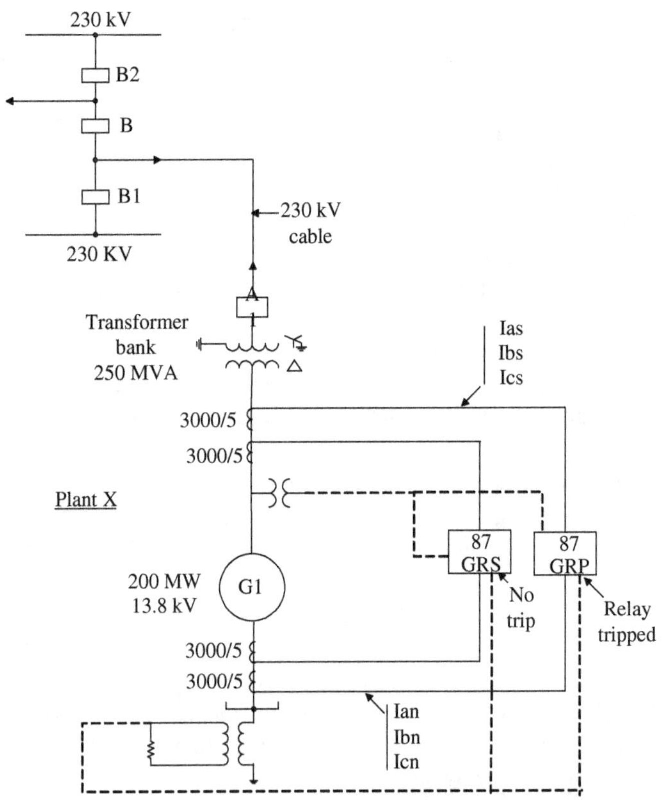

Fig. 4.51 One-line diagram showing 87GRP relay-monitored currents.

Case Study 4.6, which was described in complete detail. However, this case study illustrates a quick procedure to determine immediately the cause of the generator differential relay trip. Differential relay trips for through current flow conditions may sometimes be difficult to understand. However, the cause of such a trip is attributed primarily to the phenomena of uneven CT transient responses during through current flow conditions.

Description of the Protection System The plant protection is designed to fulfill the basic dual-element fault detection criteria. As shown on the one-line diagram in Fig. 4.51, the 200-MW hydro generating unit is protected by the dual numerical multifunction generator relays 87GRP and 87GRS.

Description of the Incident The operator apparently closed the generator breaker at an excessive angle (limited by the sync relay setting of 20°), which caused an out-of-step condition, forcing the unit to start to oscillate against the power system. This resulted in high asymmetrical currents, which affected the performance of the CTs connected to the generator differential relays. The uneven response of the neutral and system CTs caused a differential current component. The differential current reached the threshold trip of the numerical relay and caused an undesired trip in the relay, forcing a shutdown of the generating unit.

Numerical Relay Fault Record At the trip point, the numerical relay 87GRP indicated that the differential current for phase A is 400 A, for phase B is 600 A, and for phase C is zero. The relay trip target was phase B, which reached a secondary differential current of I_p/CT = 600/2000 = 1.2 A. No relay target was obtained for the secondary 87GRS differential relay system. The 87GRS is fed from separate CTs and its nonoperation will enforce the concept that CT transient response during saturation or partial saturation depends on many factors and that identical CTs may react differently to the same currents that contain dc offsets.

Analysis of the Relay Oscillography Record to Confirm the Cause of the Trip The cause of the trip can be found by superimposing the secondary currents of the generator neutral- and system-side CTs for each phase. Figure 4.52 compares the two currents for phase A. By first marking the intersection points of the two currents and then plotting the points in reference to the symmetry axis as a function of time, it is clear that the two currents have different shapes. This is attributed to the different transient responses of the current transformers, as they are subjected to the same through primary current conditions.

The phase B plots in Fig. 4.53 for the intersection points of the neutral and system sides of the differential relay show greater deviation from the axis of symmetry than

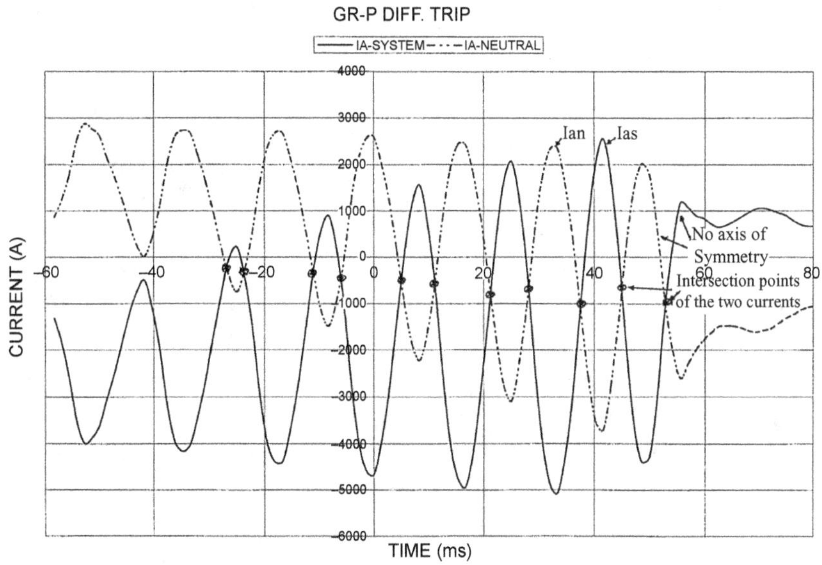

Fig. 4.52 Relay 87GRP record showing the phase A secondary currents for the generator neutral and system sides.

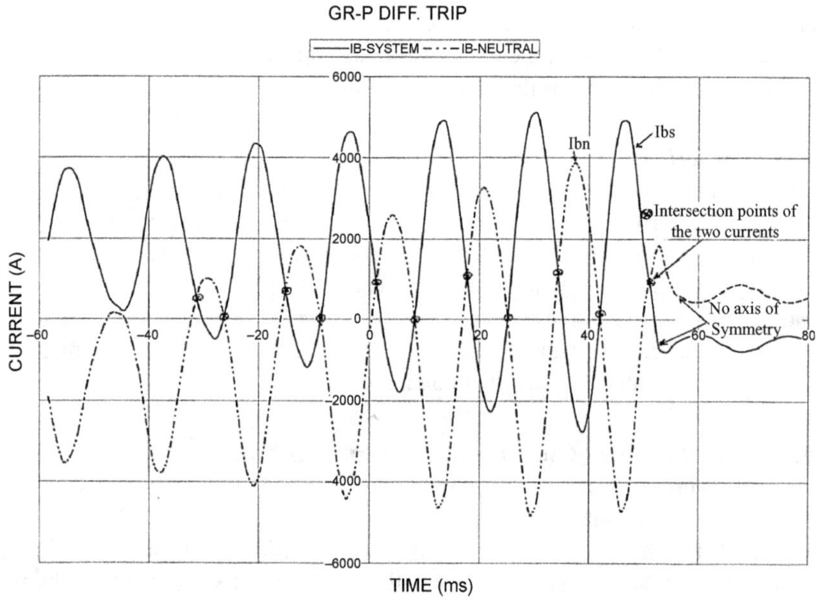

Fig. 4.53 Relay 87GRP record showing the phase B secondary currents for the generator neutral and system sides.

are shown for phase A. This explains the higher primary differential current of 600 A recorded by phase B and hence the relay target for the trip by phase B.

By contrast, as shown in Fig. 4.54, for phase C the intersection of the two superimposed currents is on the axis of symmetry for both the Icn and Ics currents.

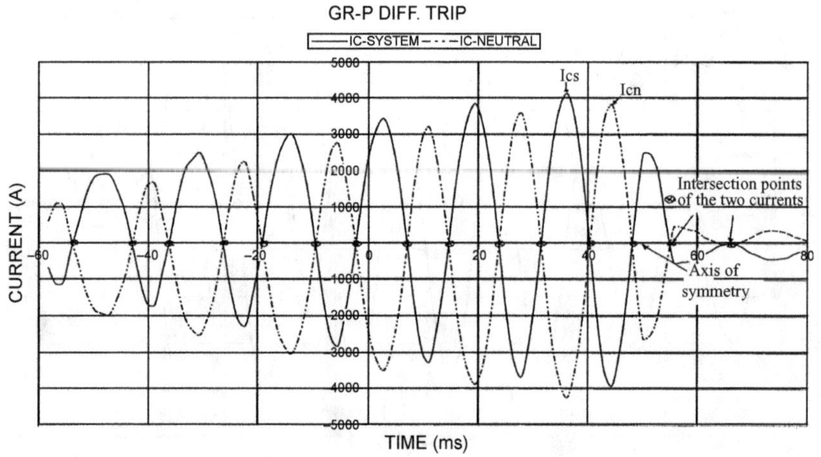

Fig. 4.54 Relay 87GRP record showing the phase C secondary currents for the generator neutral and system sides.

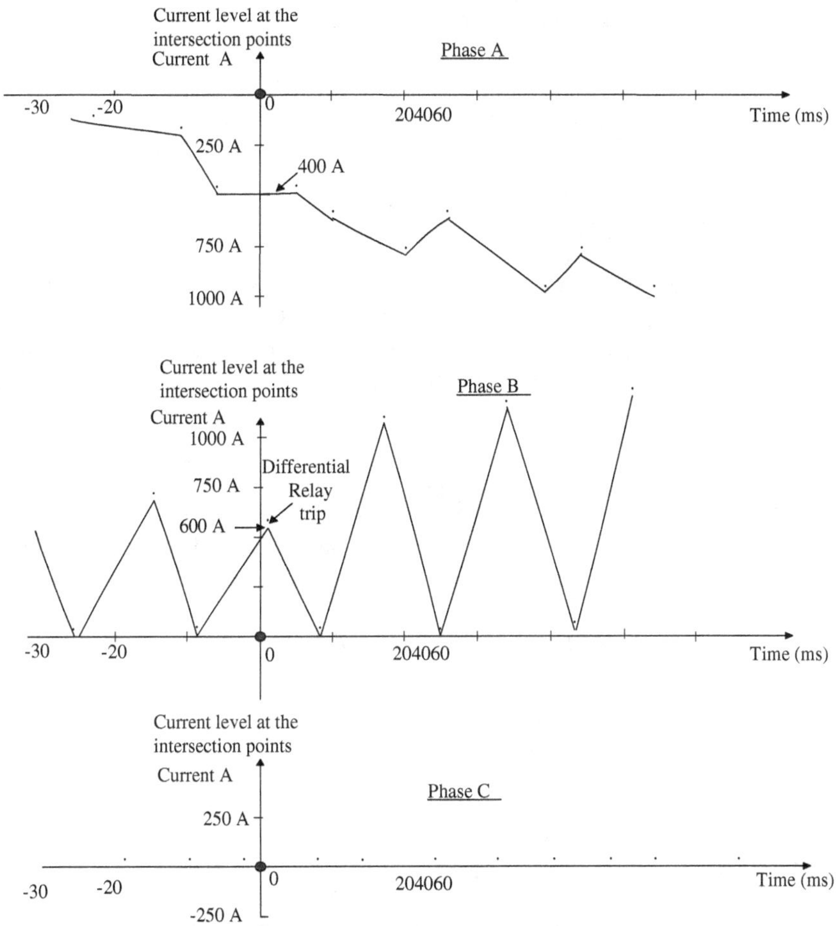

Fig. 4.55 Interpolation of overlapping of the generator 87GRP relay neutral- and system-side CT input currents of phases A, B, and C of the generating unit.

Therefore, the vectorial sum of the two currents is near zero, producing no differential current. By analyzing the locus of the points of intersection of the two input currents to the differential relay, as shown in Fig. 4.55, phase C points plot on a straight line that coincides with the symmetry axis of the two overlapping currents, whereas phase A reveals a varying magnitude of the differential current, as shown in Fig. 4.55. This explains the recorded primary differential current of 400 A at the time of tripping. The phase B points are plotted in a cyclic pattern when connected. The points plot far away from the axis of symmetry, which explains the differential current generated from the error of responses and the 600-A differential current that exceeded the trip threshold. Figure 4.55 shows the relay trip point for phase B at a current level of 600 A, while phase A recorded a differential current of 400 A, which was below the trip point threshold.

Lessons Learned Generator differential relay undesired trips for through current flow conditions can be explained and confirmed using the following steps:

1. Superimpose the generator neutral current and system current for each phase on top of one another. The sum of the two currents will produce the numerical differential relay current.
2. Find the locus of the intersection points between the two currents.
3. If the plotting of the points form a straight line that falls on the symmetry axis, no CTs transient response differential current will be falsely generated and no undesired trip is generated (similar to the phase C response). This can also be stated by saying that the axis of symmetry represents a mirror where the image above the axis is identical to the image below the axis. In other words, the currents above the axis can be folded and will eventually fall onto the top of the lower part.
4. When the points plot in a pattern other than the phase C plot shown in Fig. 4.55, violating the symmetry feature, a differential current will be generated, leading to undesired relay operation. This, in turn, can provide a quick explanation of the undesired trip of the differential relay for a through current flow condition. The cause of the trip can then be stated as uneven transient response of two identical CTs when subjected to an asymmetrical current containing dc offsets.

Case Study 4.8: Tripping of a 500-MW Combined-Cycle Plant Triggered by a High-Side 138-kV Phase-to-Ground Fault

Abstract A combustion turbine unit G1 of a 500-MW plant was tripped due to a phase C surge arrester failure associated with the 138-kV side of the G1 GSU transformer T1. The ground fault was cleared from the system successfully in about 5 cycles by tripping the unit G1 generator breaker A1 and the 138-kV remote breakers D1 and D2 at substation Y. Unit G1 was subsequently shut down by activation of lockout relays associated with unit protection systems. The steam unit GS was shut down 23.5 cycles (0.39 s) after initiation of the failure ($t = 0$). The shutdown of the GS unit was initiated by an excitation system trip. About 63.5 cycles (1.054 s) from $t = 0$, the CT unit G2 was tripped mechanically by trip signals originating from low turbine lube oil header pressure. No other relay targets were recorded for unit G2. This case study covers the analysis of an individual unit trip and behavior during and after a C phase-to-ground fault caused by surge arrester failure.

Analysis of the Plant Trip Incident The incident was analyzed using information from DFR records, the plant computer sequence of events (SOE), and relay targets. Preliminary analysis was performed to resynchronize CT unit G2 and ST unit GS back to the system. This was guided by basic information based only on relay targets, plant computer SOE, and familiarity with the plant relaying systems and

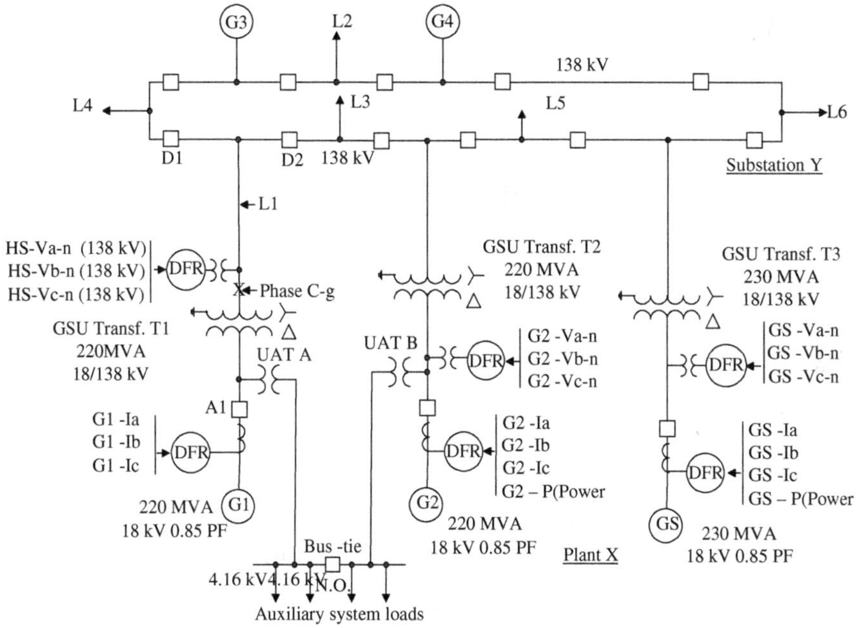

Fig. 4.56 Plant X one-line diagram showing all the monitored voltages, currents, and derived MW outputs.

drawings. Later, preliminary findings were confirmed using DFR record analysis. Figure 4.56 shows the plant X one-line diagram with the DFR-monitored voltages and currents and the MW generator outputs calculated.

Analysis of the Unit G1 Trip (0.079 ms after $t = 0$) A phase C 138-kV surge arrester associated with unit G1 GSU transformer T1 failed, causing a solid phase C-g fault within the 138-kV feeder protection zone, as shown in Fig. 4.56. The fault was cleared by the operation of the numerical primary 87-1 and secondary 87-2 current differential relays. The ground fault was cleared from the plant side of feeder L1 about 4.75 cycles (0.079 ms) after $t = 0$, via tripping of generator breaker A1, and from the 138-kV system 5 cycles after $t = 0$ by tripping 138-kV CBs D1 and D2 at substation Y. Feeder L1 lockout relays 86-P and 86-S at plant X activated additional generator protection lockout relays to shut down unit G1. Figure 4.57 shows that the fault occurred at the voltage peak, collapsing the voltage to zero value at high speed and producing symmetrical fault currents. This phenomenon also indicates that the arrester insulation failed internally to ground at the highest voltage stress point on the 60-Hz voltage sine wave.

Analysis of the Unit GS Trip (0.390 ms from $t = 0$) Unit GS was tripped by the energization of lockout relays 86-S1 and 86-S2 without any reported targets for any of the initiating protective relays. The only common element for the two lockout relays

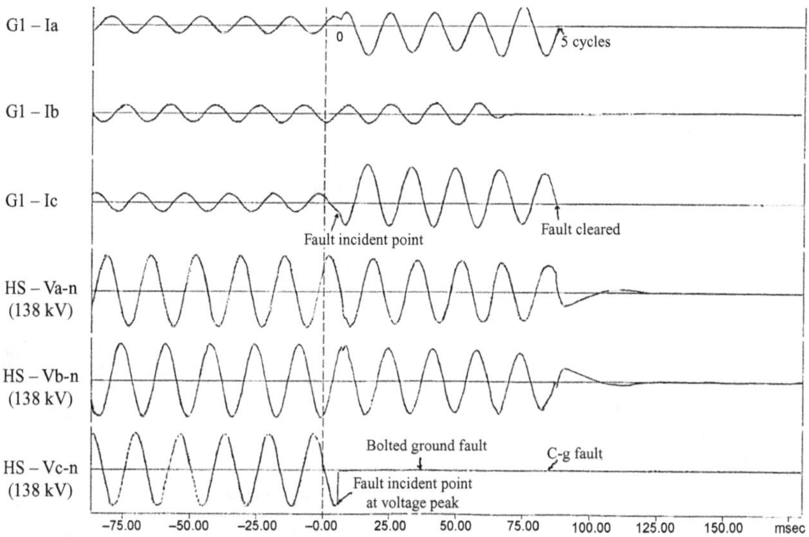

Fig. 4.57 CT unit G1 currents and 138-kV high-side voltages during the C-g fault.

that produced no obvious target is the excitation system trip signal. As shown in Fig. 4.58, the steam unit is monitored by the DFR unit, which was set incorrectly, producing a record length of only 19 cycles. Therefore, the unit was able to capture the first fault and only an additional 11 cycles of post-fault data.

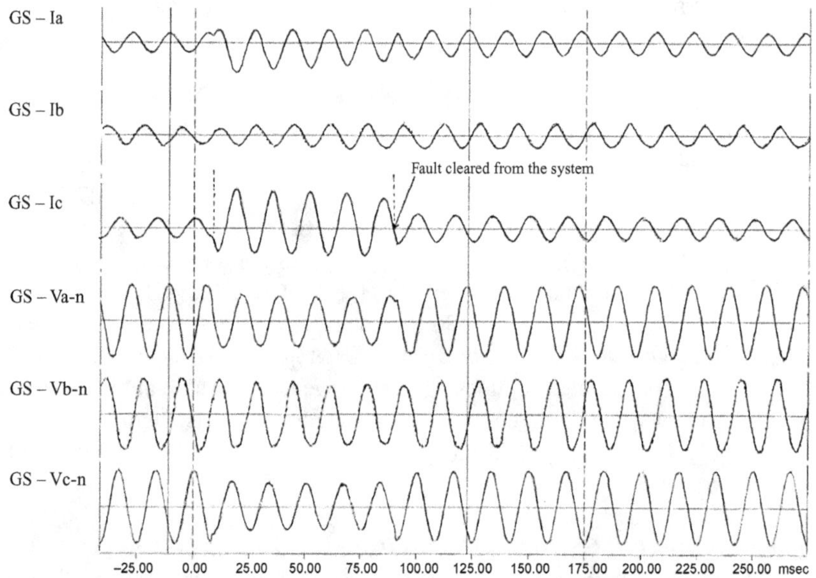

Fig. 4.58 Steam unit GS voltages and currents during the C-g fault.

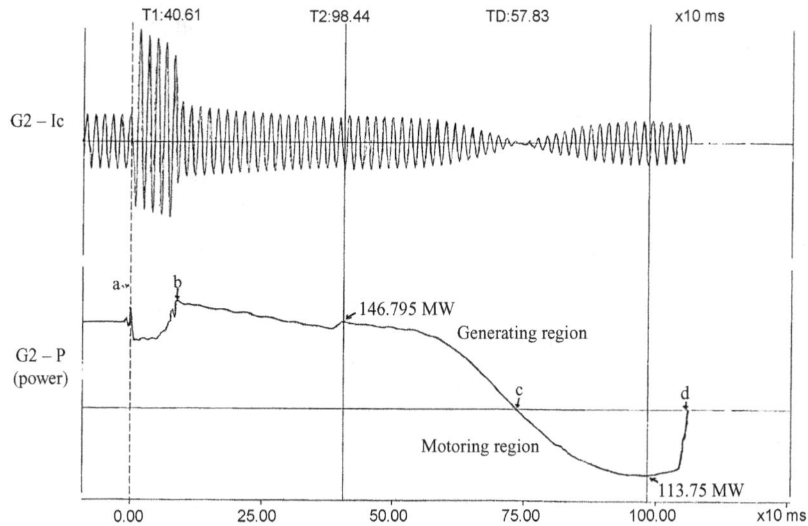

Fig. 4.59 CT unit G2 DFR record for phase C current and derived machine generating and motoring power output confirming unit oscillation.

The tripping of the steam unit was missed and was confirmed only by analyzing the plant computer event log. Through analysis of the relay operating time, the $t = 0$ event can be calculated. The tripping of the lockout relay 86-S1 and 86-S2 event time stamp with an added generator breaker time of about 0.058 ms (3.5 cycles) will produce the correct time stamp for the unit GS generator breaker open position. The difference from the $t = 0$ event is 0.390 ms (about 23.5 cycles), which is the same duration as that obtained by analysis of the unit G2 power plot of Fig. 4.59 where the power output value is 146.795 MW.

Figure 4.60 shows the DFR record during the first fault, documenting the unit GS phase A current and unit total power as calculated by the DFR software analysis tools.

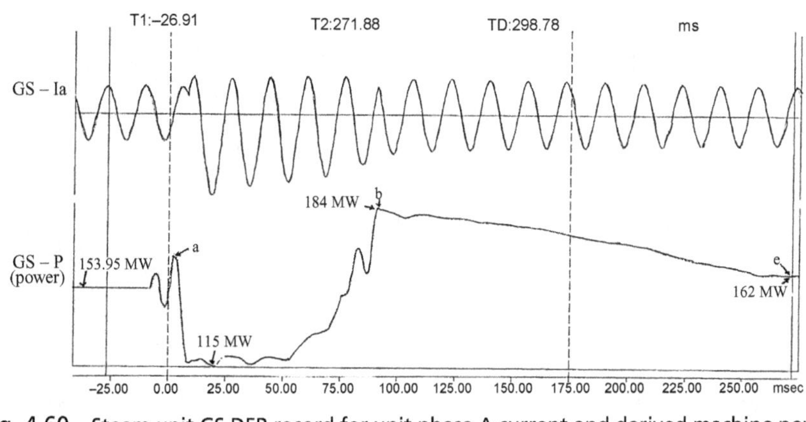

Fig. 4.60 Steam unit GS DFR record for unit phase A current and derived machine power output during and after the unit G1 high-side C-g fault.

According to the unit GS total power plot, the predisturbance unit output was about 153.95 MW, which was affected by the fault and reduced momentarily to about 115 MW at the inception of the fault. The unit power output slowly started to increase during the last cycle of the fault duration. Upon clearing of the fault, the unit ramped up its output to a post-fault maximum level of 184 MW. The unit power output is then ramped down to a value of 162 MW 11 cycles after fault clearing. This was the only recorded duration of unit GS power output due to an incorrect setting of a DFR record length of 19 cycles instead of 60 cycles or more.

Analysis of the Unit CT G2 Trip (1.054 s from $t = 0$) Based on an analysis of the fault record shown in Fig. 4.61, unit CT G2 was tripped from the system by opening its associated generator breaker 63.5 cycles (1.054 s) after $t = 0$. The plant computer even log generated "turbine lube oil header press low trip" as the only tripping event recorded in the system. The computer event time stamp is correlated as 18.5 cycles (0.308 ms) after $t = 0$.

Figure 4.62 also shows the oscillation of unit G2 with the system and its tripping in the second slip cycle when it was absorbing power (being motored) from the system. The figure also confirms the predisturbance unit G2 output to be 146.3 MW. The unit power was affected by the fault and was reduced momentarily to a level of about 130.8 MW at the inception of the fault. The unit power output began to increase slowly during the last cycle of the fault duration to reach a peak value of 176.8 MW upon fault clearing.

Figure 4.59 is a DFR record made during the first fault, documenting unit G2 phase C current as well as unit total power as calculated by the DFR software analysis tools.

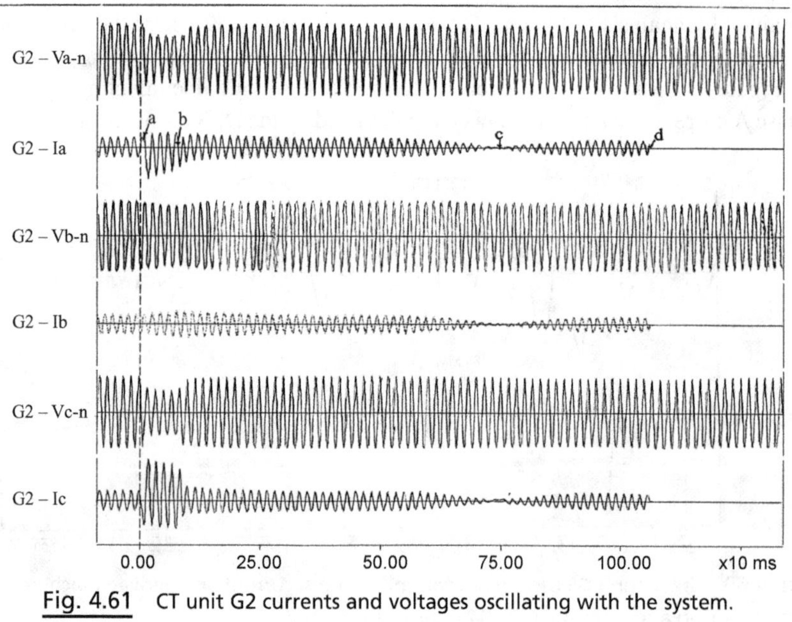

Fig. 4.61 CT unit G2 currents and voltages oscillating with the system.

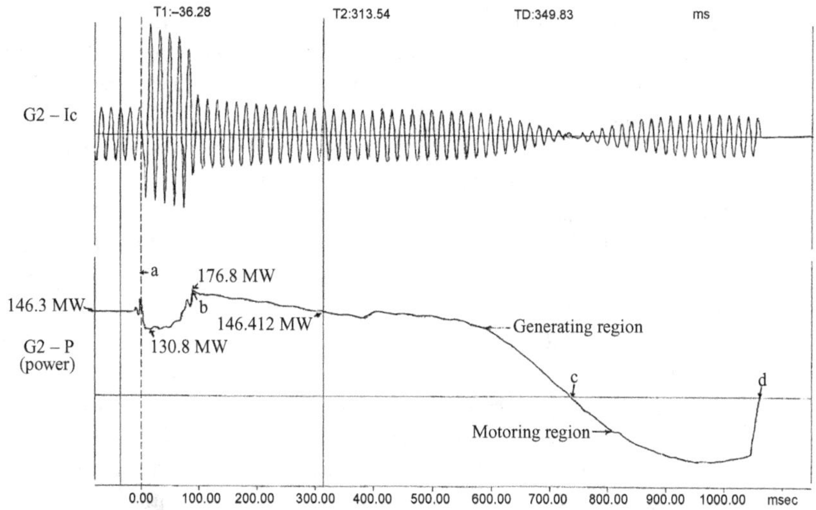

Fig. 4.62 CT unit G2 DFR record for phase C current and derived machine generating and motoring power output confirming unit oscillation.

Figure 4.59 also shows a small pump of unit G2 power output (146.795 MW) at the instant of unit GS trip and removal from the system. The unit power output continues to go down, finally reaching zero value. The unit was absorbing power from the system in the next slip cycle, with a level calculated as 113.75 MW 59 cycles (0.98 s) after $t = 0$.

Theory and Conclusions of Analysis of the Unit G2 and GS Trips The cause of the unit G2 swing (being out-of-step with the system) is attributed to a combination of events that normally cause the unit to oscillate. The solid C-g fault close to the high side of unit G2, coupled with the simultaneous loss of three 138-kV feeders in substation Y, affected the equivalent transfer impedance (X) between the unit and the system as reflected in the power transfer formula:

$$P = \frac{(V_1 \times V_2)\sin\theta}{X}$$

In addition, the sudden deficiency in generation caused by the early loss of unit G1 has further aggravated units GS and G2 stable operations (the unit rotor in step with the system) and may have contributed to their oscillations, as shown in Figs. 4.60 and 4.62, respectively. Referring to Fig. 4.62, the predisturbance unit B output of 146.3 MW was affected by the fault and reduced momentarily to a level of about 130.8 MW at the inception of the fault at point a. The unit power output began to increase slowly during the last cycle of the fault duration, reaching a peak value of 176.8 MW upon fault clearing. The machine rotor angle with respect to the system also began to increase. The unit output of 176.8 MW at point b may be at a point where the unit rotor angle is near (or at) 90°, and at this point the unit may be in a nonreturning state. What also contributed to this out-of-step condition was the increase in the transfer impedance between the unit and the system by the simultaneous tripping of three 138-kV feeders in substation Y.

The unit power output level continued to decrease while the rotor angle continued to increase beyond 90°. The rotor angle reached 180° and the unit output became zero at point c. The unit then entered the motoring region, and at this stage it absorbed active power from the system, which further accelerated the generator rotor. The unit remained in the motoring region for a total of 20 cycles before it was eventually tripped at point d, as shown.

It can be concluded that the unit G2 rotor began to lose its magnetic bond with the system, which normally holds the rotor in step with the stator (out-of-step) 5 to 6 cycles from $t=0$ and began to increase its rotor angle beyond the stable point of 90°. Unit G2 lost control of its torque angle and entered an unstable condition. By definition, the unit may have slipped poles. Unit G2 tripped mechanically and stopped its oscillation with the system. The frequency of oscillation of unit G2 is estimated to be 2 Hz.

Figure 4.60 shows that unit GS behaved exactly as did unit G2. But again, luckily, the unit tripped earlier, 23.5 cycles after $t=0$, and stopped the oscillation process of the unit with the system. The tripping of unit G2 by the turbine low oil pressure and unit GS by the excitation trip can be explained as symptoms of the early C-g fault at the GSU T1 high-side 138 kV.

Analysis of the Unit GS System Excitation Trip Postulation that the unit GS lockout relay trips by the excitation system needed to be confirmed by examining the unit excitation system computer to see if the excitation system auxiliary relays have issued the tripping of 86-S1 and 86-S2. The alarm history and time stamps were examined using the excitation system HMI. The first alarm was related to the trip, which means that the exciter controller lost power. This was determined because the only way to clear exciter history is to reboot the controller. The 125-Vdc circuit breaker feeding the exciter was found in the off position. The normal feed for exciter power is via an MCC from a load center that is fed from the 4160-V bus associated with unit G1. The loss of the unit G1 auxiliary 4160-V bus activated the automatic throw-over scheme to switch to the unit G2 auxiliary bus. The transfer was delayed for several seconds instead of the design time of a few cycles, resulting in the excitation trip.

Analysis of the Unit G2 Mechanical Trip The cause of the low lube oil turbine pressure trip was fully confirmed to be an extensive loss of auxiliary power for several seconds. Therefore, the issue needed to be examined fully to arrive at the cause. When the surge arrestor failed and unit G1 and the associated 138-kV feeder were tripped, power was lost to the load center feeding the lube oil pumps. The automatic throw-over scheme malfunctioned, resulting in an extensive delay in transferring the tie breaker (several seconds instead of few cycles) to closed. During this time delay the lube oil pressure continued to decay, reaching the trip level. Automatic start of the backup lube oil pump is set for 83 psi, and the CT lube oil trip has occurred at 8 psi. It is expected that the bus tie will normally pick up the load after feeding the lube oil pump immediately after clearing the C-g fault, while the lube oil pressure is still high enough to permit the backup lube oil pump to start.

Performance of the 500-MW Plant Protection Systems The protection systems used in the 500-MW plant and associated 138-kV feeders performed successfully during the initial C-g fault and the subsequent SCADA closing of the cable feeder L1 at substation Y. No undesired operations or any failure to operate were recorded for any of the plant relaying systems. However, three 138-kV feeders tripped at substation Y by the false operation of their associated pilot wire current differential relaying systems.

Corrective Actions

1. The protection and control scheme for the 4160-V tie breaker and associated automatic throw-over scheme need to be reviewed, and corrective actions need to be taken.
2. The unit G2 oscillation needs to be studied further, simulating the contingencies that contributed to the unit out-of-step condition with the system. For example, what if the unit was not tripped mechanically? Will the unit oscillation end up in a stable unit swing? If not, will the units' out-of-step protection element (device 78) detect the unstable swing and trip the associated unit?
3. The steam unit DFR total record length is set to about 19 cycles (0.315 s), with 3 cycles pre-fault and 16 cycles post-fault. This is the reason for the loss of valuable information at the unit trip point. The record length should be set at 1.5 s (90 cycles) with a pre-fault setting of 5 cycles.
4. The undesired simultaneous trips of three 138-kV feeders at substation Y during the C-g fault need to be analyzed for their causes and their impact on the oscillations that occurred to units G2 and GS.

Issues Raised by the Incident Analysis

1. If CT unit G2 did not trip mechanically, will the oscillation be detected before the unit becomes unstable and start to slip poles? The units are equipped with an out-of-step element (device 78) as part of a unit numerical multifunction protection package. Can this element detect this type of swing and eventually trip and protect the unit? No answers were received to be reported in this case study.
2. The oscillation may be started with the combination of a solid phase-to-ground fault, coupled with the loss of CT unit G1. Therefore, was the impact of this event alone sufficient to cause the instability of unit G2 as simulated by the system transient stability study? Was the impact of this event on the stability of steam unit GS also studied by running a system transient stability study? No answers were received to be reported in this case study.

Case Study 4.9: Tripping of a 110-MW Combustion Turbine Unit in a Combined-Cycle Plant During a Power Swing

Abstract Synchronization of a 120-MW unit at an excessive angle caused an out-of-step condition on the system that forces generators to an oscillating state. Oscillation of a combustion turbine (CT) unit of a combined-cycle plant resulted in an increase in the unit output reaching the trip limit for the unit. In this case study we analyze a DFR record for a portion of the event oscillation prior to the trip, describe the incident, and provide some corrective actions.

Description of the Incident Plant Y shown in Fig. 4.63 consists of a combined cycle with a CT unit nominal rating of 110 MW and a steam unit rated at 58 MW. A CT unit trip occurred when a generator rated at 120 MW at plant V shown in Fig. 4.64 was synchronized approximately 30° out of phase.

Analysis of the DFR Record Figure 4.63 shows the system one-line diagram during the power swing incident and the DFR-monitored voltages and currents at

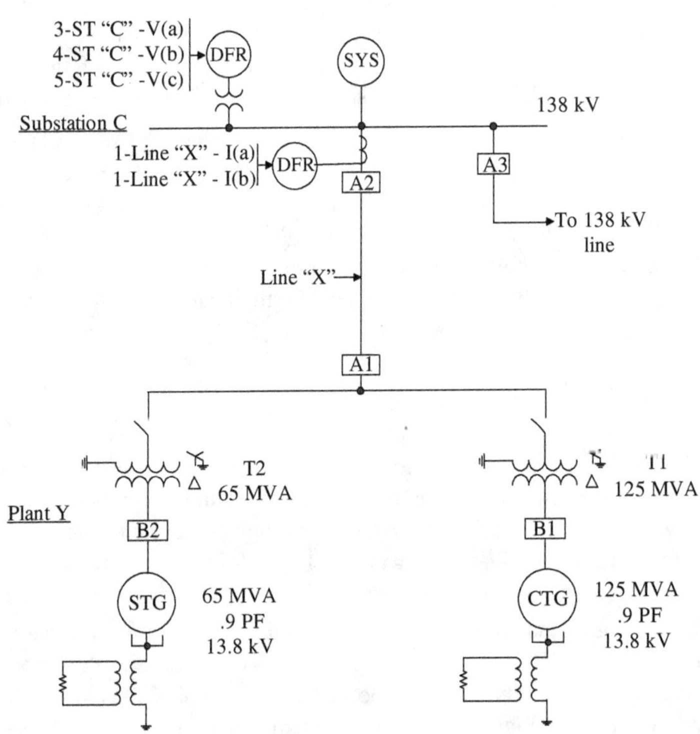

Fig. 4.63 One-line diagram showing the system involved in the power swing and DFR-monitored voltages and currents.

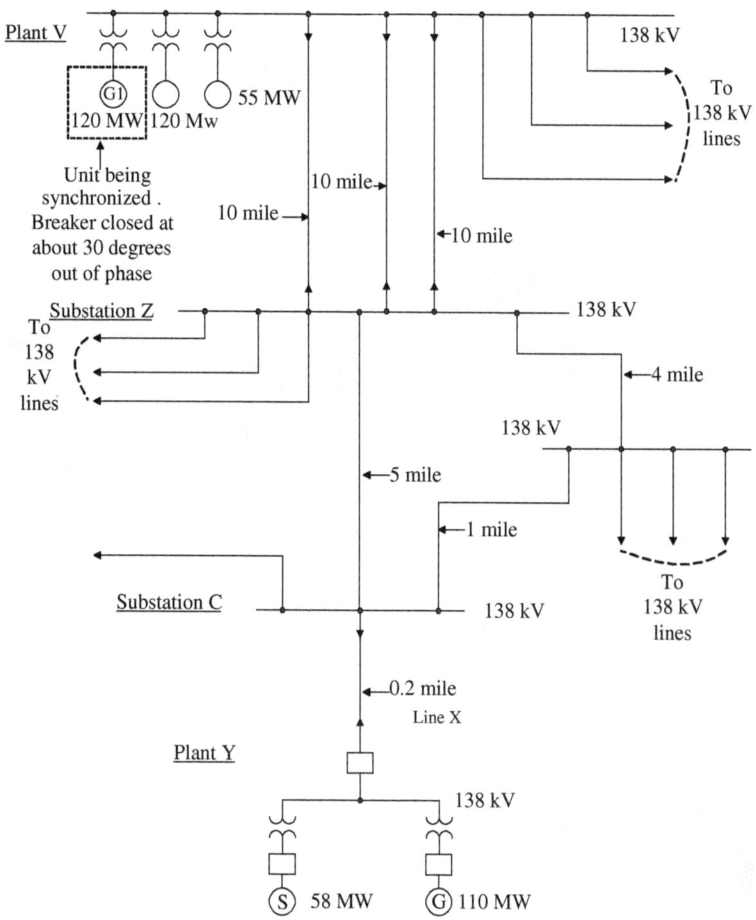

Fig. 4.64 One-line diagram of the power system involved during the power swing between units at plant Y and units at plant V.

substation C. The DFR record of Fig. 4.65 shows the only recording of this event for a duration of about 0.5 s at substation C. Traces 1 and 2 in Fig. 4.65 represent line X phase A and B currents, and traces 3, 4, and 5 represent 138-kV bus phase C, B, and A voltages, respectively. The recorded three-phase current values are the combined power output of the combustion and steam turbine units.

From an analysis of Fig. 4.65, the following conclusions can be reached. At point (f), predisturbance:

$$I = 648\text{ A}$$
$$V = 139.7\text{ kV}$$
$$S = \sqrt{3} \times V \times I = 156.9\text{ MVA}$$

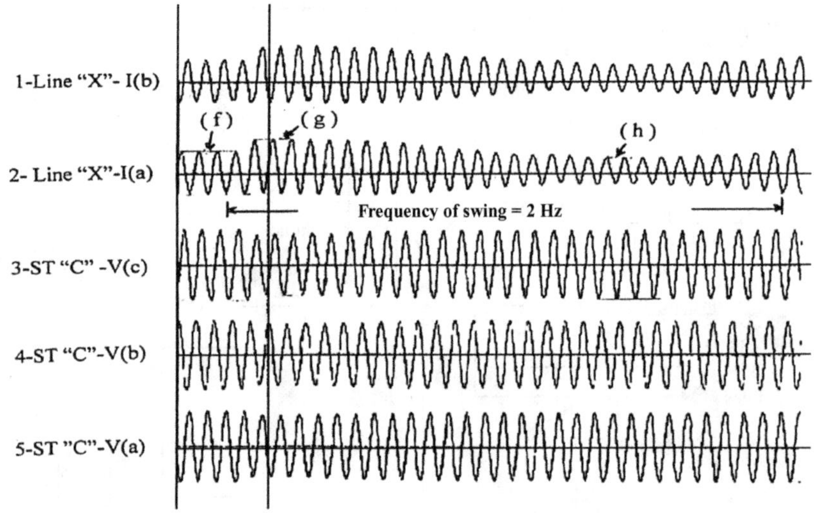

Fig. 4.65 Substation C DFR record showing the power oscillation between the units.

At point (g), the beginning of the power swing:

$$I = 938\ \text{A}$$
$$V = 123.1\ \text{kV}$$
$$S = \sqrt{3} \times V \times I = 200\ \text{MVA}$$

At point (h), the lowest generator power output during the swing:

$$I = 435\ \text{A}$$
$$V = 134.5\ \text{kV}$$
$$S = \sqrt{3} \times V \times I = 101.3\ \text{MVA}$$

The duration of the swing is 0.5 s. The frequency of swing oscillation = 1/0.5 = 2 Hz.

Analysis of the Plant Y Trip Figure 4.63 shows the one-line diagram for plant Y and substation C with the DFR-monitored voltages and currents that are used to analyze this disturbance. The closing of the breaker associated with the plant V 120-MW unit at about 30° during the synchronization process caused an out-of-step condition on the grid, causing a power swing that lasted for about 15 s, forcing unit oscillations on the system. The turbine trip at plant Y occurred about 5 to 6 s after the beginning of the swing. The generator output current value before the occurrence of the event was approximately 650 A. During the 0.5-s recording interval, the plant Y output current varied between a maximum value of 941 A and a minimum value of 426 A. The occurrence of an out-of-step condition caused the power sensor to feed an unstable load signal to the gas unit controller, which at this time had a load set point of 105 MW. The predisturbance plant output was approximately 157 MW. Based on a combustion turbine load set point of 105 MW, the CT was responsible for about 66% of the total plant output. Once the power swing occurred, the total power output then jumped up to approximately 200 MVA. Based on 66% sharing, the CT output jumped suddenly to approximately 132 MVA. Once the actual MW output from the combustion turbine started to change, the controller compensated for this by repositioning the valve to try to maintain the 105-MW load controller set point. This action caused the load output from the combustion turbine to decrease accordingly. The combined power output of the station then dropped to approximately 101 MVA, and with 66% sharing the CT output was about 67 MVA. This caused the controller to reposition the control valve once again to try to maintain the 105-MW load controller set point. Because the actual load signal was fluctuating due to the power swing on the grid, there was no possible way for the controller to be able to compensate for the load swings because of the severe oscillation in the load feedback signal. The power sensor eventually output a load signal greater than 125 MW, and the turbine control prediction circuit initiated a unit trip when the load signal rose above the set value.

Corrective Actions

1. No action was required at plant Y because increasing the sensitivity of the load controller in order to respond to this type of transient would also cause the controller to be too sensitive during normal operation. The control system responded correctly to the unit oscillation that was encountered and protected the machine, as it was designed to trip above 125 MW.
2. It is necessary to add synchronizing check relays to supervise operator action at plant V during the synchronization process to avoid a similar occurrence of closing generator breakers during synchronization of the unit at wide angles.

System Phenomena The power swing shown in Fig. 4.65, which appears clearer in generator currents than in voltages. The frequency of oscillation = 1/duration of the swing = 1/0.5 s = 2 Hz.

Case Study 4.10: Analysis of an 800-MW Generating Plant DFR Record for a Normally Cleared 345-kV Phase-to-Ground Fault

Abstract A 345-kV phase-to-ground fault occurred on line L1 shown in Fig. 4.66, near substation Y. A DFR record documenting the ground fault was captured at plant X. The fault was cleared successfully from both ends of the line. Analysis of the plant X DFR record for the L-g fault revealed valuable educational power system phenomena and basic relaying concepts. In this case study we describe the fault creation mechanism, analyze the fault, provide relevant power system phenomena, and describe how a ground fault on the grounded side of the transformer is seen as a phase-to-phase fault on the delta (generator) side.

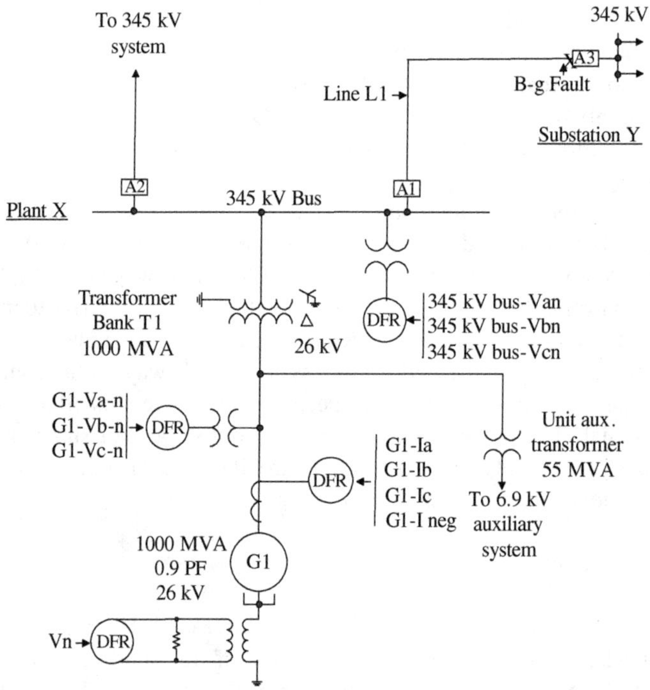

Fig. 4.66 One-line diagram showing DFR-monitored currents and voltages.

Description of the Fault Figure 4.66 shows the system one-line diagram with the DFR-monitored currents and voltages for unit G1 at plant X. The DFR record in Fig. 4.67 reveals a phase B-to-ground fault occurring on 345-kV line L1. The fault occurred due to a flashover of one of the 345-kV bus post insulators connecting line L1 at substation Y during ice-melting conditions (Fig. 4.68). Only 0.5 cycle of pre-fault data (instead of the normal 5 cycles) was captured, due to the triggering of the DFR by circuit breaker auxiliary contacts upon the opening of CB A1 to clear the fault. Analog triggering levels were adjusted to provide proper recording of pre-fault data. The fault was cleared from the 345-kV system in 3 cycles by tripping of CBs A1 and A3 by line relaying systems. The DFR trace G1-Ineg reveals a generator negative-sequence current contribution during the ground fault. This operation displays a normally cleared fault and relaying systems at plant X restrained correctly from operation during the external fault. However, valuable relaying and system information can be gathered from analyzing this fault, as demonstrated below.

System Phenomena Analyzing this normally cleared line-to-ground fault on the EHV system reveals the following valuable power system phenomena:

1. Arc-over at the voltage peak as demonstrated by the voltage trace 345 kV bus-Van shown in the DFR record in Fig. 4.67. This confirms a flashover of the

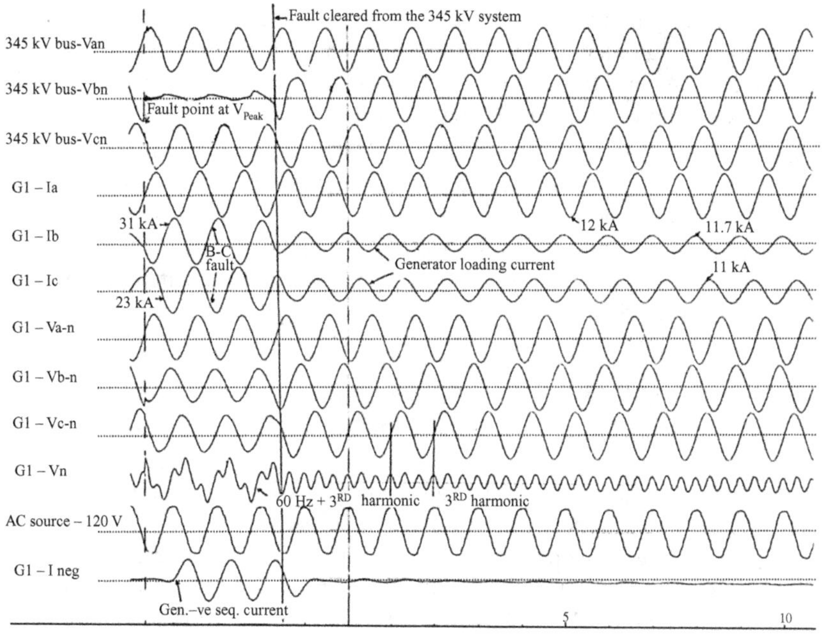

Fig. 4.67 Generator DFR record for the 345-kV line L1 B-g fault.

345-kV substation post insulator occurring at the voltage peak and thus indicating a slow fault-creation mechanism.

2. High-side (wye) phase-to-ground faults seen as low-side (Δ) phase-to-phase faults as shown on Fig. 4.67. Due to the wye/delta transformation, the B-g high-side fault will be seen as a phase b-to-c fault on the low side (generator). The DFR record in Fig. 4.67 reveals that phase B current trace G1-Ib is out of phase (approximately a 180° phase shift) with phase C current trace G1-Ic, indicating

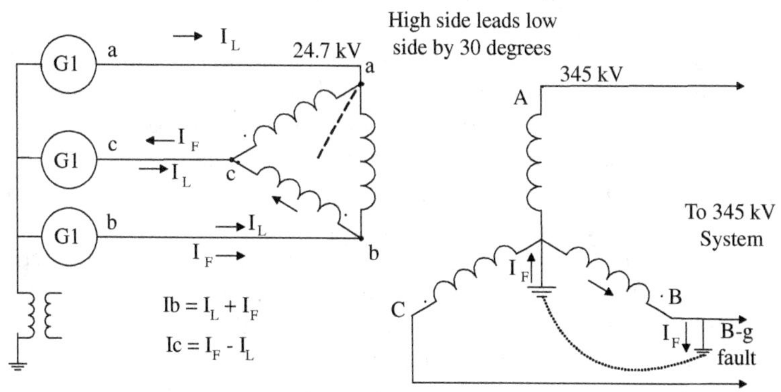

Fig. 4.68 Phasing diagram of generator step-up unit transformer T1.

a phase-to-phase fault, during the 3-cycle fault. In addition, the phase B current at the generator terminal is about 30.8 kA and is larger than the phase C current of 22.8 kA, due to the combined effects of the generator loading and fault currents shown in the transformer phasing diagram.

3. Coupling of high-side zero-sequence voltage generated during system ground faults to the generator neutral. The phenomenon is displayed by the trace G1-Vn, where the 60-Hz zero-sequence voltage is superimposed on the neutral third-harmonic generator voltage during the high-side B-g fault. This coupling phenomenon is accomplished via unit transformer capacitance between the transformer primary and secondary windings.

Case Study 4.11: Tripping of a 150-MW Combined-Cycle Plant Due to a Failed Lead of One Generator Terminal Surge Capacitor

Abstract In this case study we describe an incident at a combined-cycle plant where one lead of a generator terminal surge capacitor fails in an opened state. The absence of one of the phase capacitors resulted in unbalanced loading, forcing current to flow in the neutral and generating 60 Hz of generator neutral unbalance. Enough 60-Hz neutral voltage was generated to cause operation of the overvoltage tuned 60-Hz element of the stator 95% ground fault protection system which tripped the generating plant. Relay operation was confirmed by calculations and actual measurements. We also describe DFR records and system phenomena during the incident. In addition, we cover troubleshooting of the incident, lessons learned, and corrective actions.

Description of the Incident A combustion turbine generating unit (CTG) of the combined-cycle plant shown in Fig. 4.69 was tripped via an emergency lockout relay device 86E. The lockout relay was energized by the 59N stator ground fault overvoltage element (95%), which is tuned to 60 Hz. Several attempts were made within 8 hours to bring the plant back online. These many unsuccessful repeated attempts were made due to the pressure of trying to bring back to service a very efficient and low-cost plant.

As mentioned earlier, all these attempts to bring the unit up to speed in a no-load state were unsuccessful and generated similar relaying trips and similar relaying targets. Finally, at the eleventh trial, the field personnel connected a voltmeter to measure 60-Hz voltage across the secondary of the generator neutral grounding distribution transformer prior to starting the CTG unit. The voltmeter recorded a 60-Hz voltage having an RMS value of 13.8 V, which may be indicative of a stator ground fault. Since the maximum generator neutral voltage for a fault at its terminal is calculated as 138 V, it can be postulated that the 13.8-V reading indicates a fault located about 10% from the neutral. This assumes that the neutral voltage is linearly proportional to the stator winding size and therefore that the fault location is 10% from neutral (= 13.8 V/136 V). Based on this information, it was finally decided to examine the health of the stator by applying a meggering test. The test results indicated that

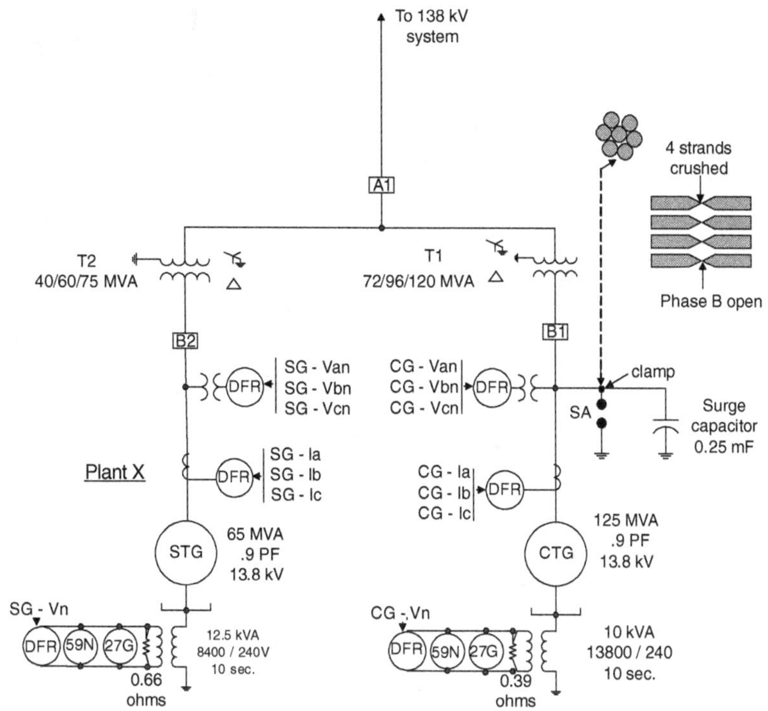

Fig. 4.69 Plant one-line diagram showing the CTG terminal surge capacitor and DFR-monitored voltages and currents.

the generator stator is healthy with no ground faults on any of the phases. While examining the generator terminal equipment it was observed that the phase B lead of one of the surge capacitor was broken. An analysis was then performed to see if the generator trip could be co-related to the capacitor lead failure. Subsequent analysis and calculations confirmed the cause of the trip to be the lead failure, and the plant was placed online after the capacitor lead was repaired.

Sequence of Events and Analysis of the Incident Figure 4.69 illustrates the system one-line diagram and DFR-monitored voltages and currents for the CTG and STG units. The incident began when a DFR operation alarm was received. Twelve minutes later the plant was tripped off-line by the operation of CTG unit stator ground fault protection overvoltage element 59N via emergency shutdown of lockout relay 86E. The DFR record in Fig. 4.70 shows phase voltages and currents as well as neutral voltages for the CTG at the beginning of the first CTG unit trip. The CTG neutral voltage trace CG-Vn shows a small fundamental 60-Hz component flowing in conjunction with the third-harmonic voltage. The fundamental 60-Hz component began to increase until it reached the 59N trip level. Figure 4.71 shows the STG phase voltages and currents as well as neutral voltage recordings where trace SG-Vn shows only third-harmonic voltage as a sign of STG unit health.

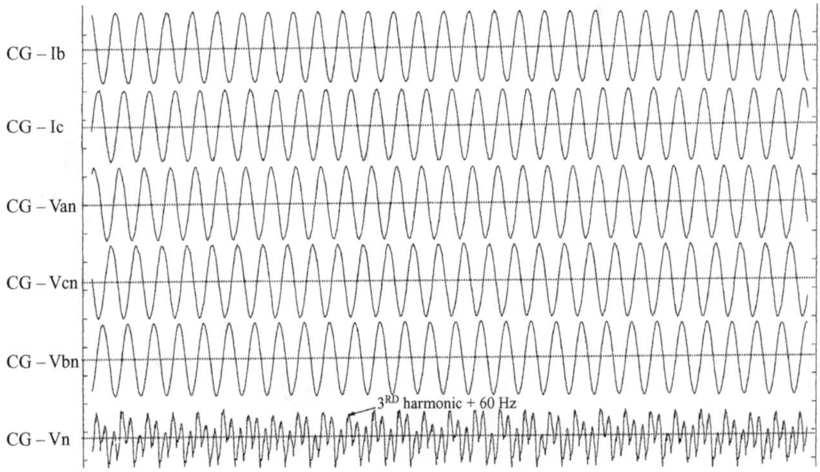

Fig. 4.70 Voltages and currents for the CTG unit at the beginning of the failure.

No investigation was undertaken following the early unit trip to find the real cause of the CTG unit first trip. Instead, the plant was restarted 2 hours later. The plant was tripped again by the same 59N relay system 10 minutes after energization. An additional seven unsuccessful attempts were tried in a span of 5 hours. During each attempt the plant is tripped by the same relaying system, similar to the early trips.

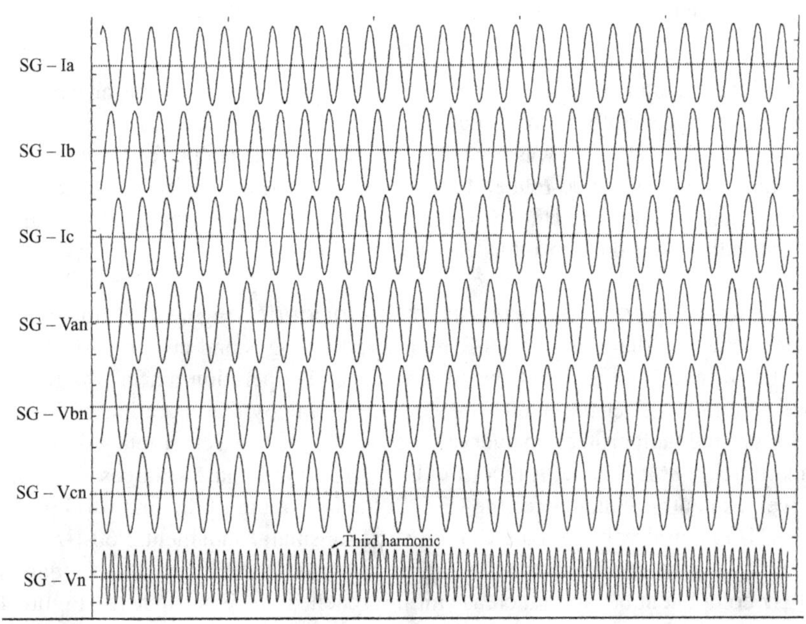

Fig. 4.71 Voltages and currents for the STG unit at the beginning of the failure.

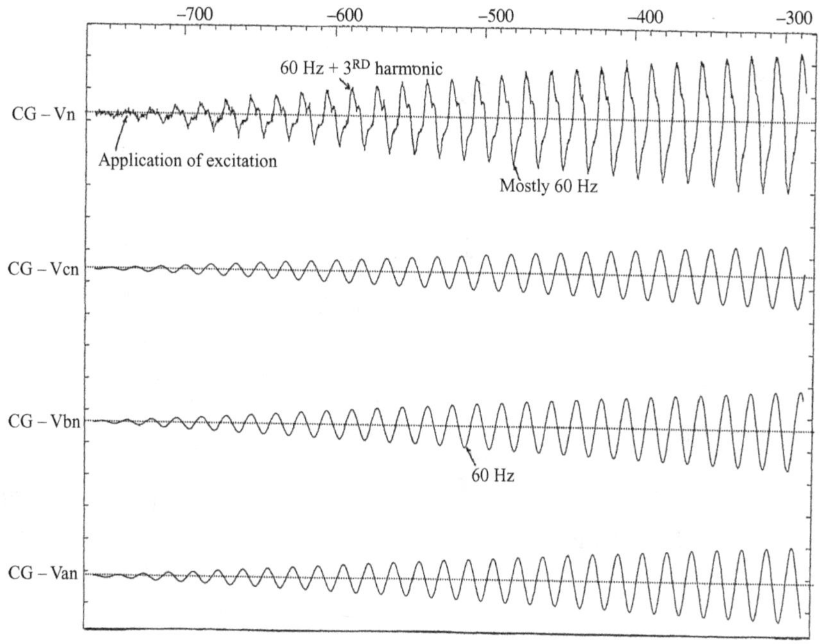

Fig. 4.72 Voltages and currents for the CTG unit during restart, showing the gradual application of excitation.

Figure 4.72 indicates gradual buildup of gas unit excitation followed by a unit trip by the 60-Hz tuned overvoltage relay element of the 59N relay. The operator apparently assumed mistakenly that there was a problem with the machine speed sensors that prevented synchronization of the unit. In all attempts, once the generator reached a certain speed, the field was slowly applied. The machine was then tripped by stator ground relaying system 59N when the neutral voltage approached the relay trip set point.

Troubleshooting the Incident Actual troubleshooting of the incident took place 7 hours after the first plant trip. The delay was caused by an operator's incorrect assumption that the main problem resided with the machine speed sensors, which prevented 10 attempts to synchronize the CTG unit generator with the system. After the tenth try, with the help of engineering analysis, a recommendation was issued to the field that the following procedure be followed to troubleshoot the problem.

1. Connect a spectrum analyzer to the unit neutral across the 0.39-Ω neutral resistor shown in Fig. 4.69 to obtain a 60-Hz voltage reading during the tenth try. A voltmeter connected to obtain an RMS voltage reading across the 59N relay registered a reading of 13.8 V at the time of the trip. With this reading in mind, it was postulated that a stator phase-to-ground fault may have occurred

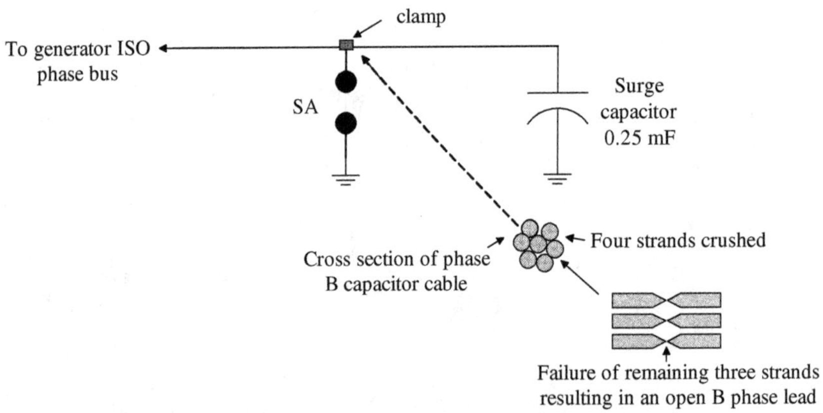

Fig. 4.73 Failure of phase B of the surge capacitor lead.

near the generator neutral at a location 13.8 V/138.6 V = 10% from the neutral point. Based on the 59N setting, this neutral voltage will cause relay operation after a 30-s delay.

2. Megger the generator to confirm if a ground fault has occurred in the stator. The testing results were good, indicating no presence of any faults inside the generator stator.
3. Inspect the generator isolated-phase (ISO) bus and the neutral grounding system. The inspection revealed that a phase B surge capacitor lead was found open near the connection clamp, as shown in Fig. 4.73.
4. Perform an analysis, supported by calculations, to confirm that finding the broken lead can contribute to the presence of the RMS 60-Hz voltage of 13.8 V measured (Fig. 4.74). Calculations indicated that a 60-Hz neutral component of 11.9 V will appear at the neutral. By adding a few RMS volts for normal third-harmonic presence, a match can be established between measured and calculated values.
5. With this explanation the capacitor lead was repaired and the unit was synchronized successfully back to the system.

Description of the 100% Stator Ground Fault Protection The stator ground fault protection consists of two elements: overvoltage relay 59N tuned to 60 Hz and protecting 95% of the stator winding starting from the generator terminal, and undervoltage relay 27 tuned to the third harmonic to sense open neutral and ground faults near the neutral, thus providing 100% stator ground fault protection. The undervoltage element is connected to an alarm, whereas the overvoltage element is wired to trip. The alarm generated by the 27G element will be followed by an operating procedure to confirm a stator ground fault near the neutral and to plan for unit shutdown if needed.

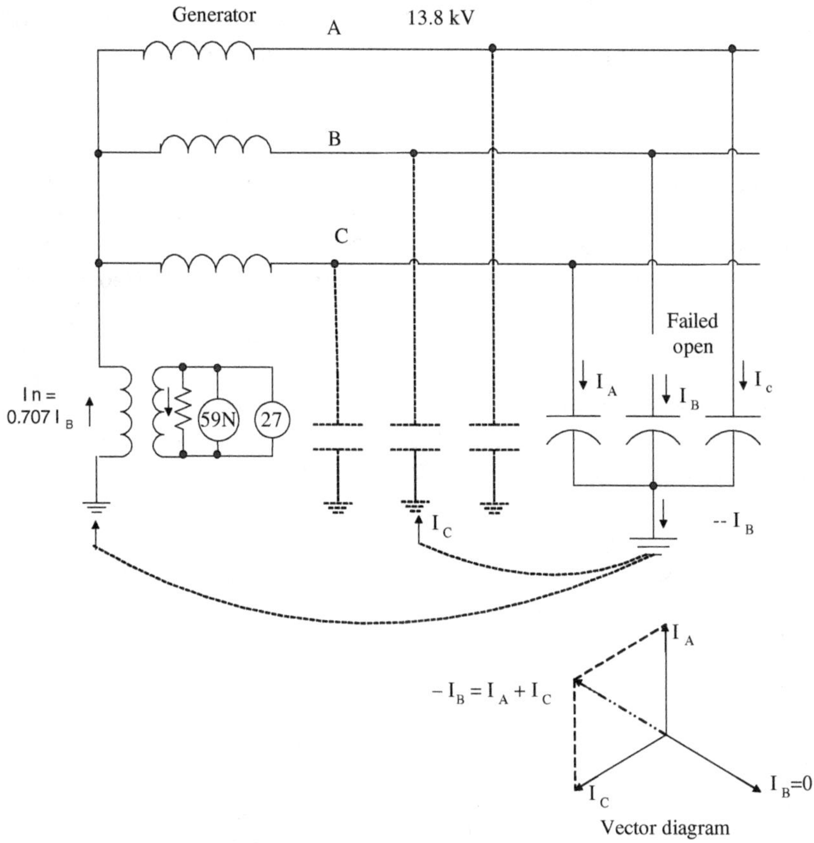

Fig. 4.74 Current flow around the generator due to opening of the phase B surge capacitor lead.

Establishing the Basis for 95% Ground Fault Protection As shown in Fig. 4.69, the generator is grounded through high resistance using a distribution transformer and a secondary resistor. The criterion normally used to size the resistor is the matching of the charging reactive power around the generator to the power loss through the secondary neutral resistor.

Generator rating: 125 MVA; 13.8 kV

Distribution transformer rating: 7.5 MVA, 13.8 kV/240 V

Generator capacitance (three-phase): 1.14 μF

Generator surge capacitor: 0.25 μF

Generator lead capacitance: 0.06 μF

Total capacitance: $\frac{1}{3}(1.14) + 0.25 + 0.06 = 0.69\,\mu F$

Capacitive reactance: $10^6/2 \times 3.14 \times 60 \times 0.69 = 3846\,\Omega$

To minimize the transient overvoltage, $3R_n$ must be $= 3846 = X_C$, so $R_n = 3846/3 = 1282\,\Omega$. Using a distribution transformer with a ratio of 13.8:0.24 yields

$$R_{\text{n-sec}} = 1282 \times n^2 = 1282 \times (0.24^2/13.8^2) = 0.388 = 0.39\,\Omega$$

Therefore, a secondary resistor of $0.39\,\Omega$ is used correctly.

For a phase-to-ground fault at the terminal of the generator, the neutral will be shifted by an amount $=$ L-g $= 13.8/\sqrt{3} = 7.97$ kV. The neutral relay will see a voltage $= 7.97$ kV/n, where n, the transformer turns ratio $= 13{,}800/240 = 57.5$.

$$V_{\text{n-sec}} = 7967/57.5 = 138.5\text{ V} \tag{1}$$

The neutral voltage can also be calculated using fault currents.

For a fault at the generator terminal, Neglecting positive- and negative-sequence impedances (very small compared to zero sequence), we have

$$\begin{aligned} I_0 &= \frac{V_{\text{l-n}}}{R} = \frac{13,800}{\sqrt{3} \times 3 \times 0.39 \times n^2} \\ &= \frac{7967}{3846} = 2.06\text{ A} \\ I_{\text{g-sec}} &= 3 \times I_0 \\ V_{\text{n-sec}} &= I_{\text{g-sec}} \times R_{\text{sec}} = 3 \times 2.06 \times \frac{13,800}{240} \times 0.39 = 138.6\text{ V} \end{aligned}$$

which is essentially the same value as that obtained in equation (1).

The overvoltage element of 59N is set to see 95% of the stator winding:

$$V(\text{set}) = 0.05 \times 138.6 = 6.93\text{ V}$$

when the overvoltage element is set at 5 V:

$$\text{relay coverage} = \frac{138.6 - 5}{138.6} = 96.4\% \text{ from the generator terminal}$$

The overvoltage element time setting curve should always be coordinated with the secondary potential transformer fuses.

Setting the Undervoltage Third-Harmonic Element of 100% Stator Ground Fault Protection The element is normally set based on a complete profile of the generator third-harmonic voltage at the machine neutral as a function of the active (MW) and reactive (MVAR) power generated. The objective is to find the minimum third-harmonic voltage generated by the machine so that the undervoltage (UV) third-harmonic element can be set at 50% of this level. This will guarantee that the UV element of the stator ground relay will not provide a false alarm for any normally generated low third-harmonic level. When the system is first installed, a preliminary setting can be selected based on an assumed third-harmonic generated level of 2 to 3% of the machine neutral voltage for a L-g fault at the terminal of the generator.

Setting the UV initially at 3% of 138.6 V = 4.16 V, at 50% the UV can be set at 2 V.

Since the UV relay 27 operation will alarm only for faults near the neutral of the generator, it is not critical to set the relaying element operation time delay. The UV element timer can be set at 3 s to avoid nuisance alarms. Therefore, relay 27 can be set at 2 V with a time delay of 3 s. The UV element alarm for the CTG unit was mistakenly left unconnected to the event log of the plant computer system. The setting of the UV relay 27 element can then be optimized by gathering data for the third-harmonic neutral voltage generated as a function of active and reactive machine power outputs, as explained before and as outlined in Case Study 4.26.

Setting the Undervoltage Inhibiting Element To avoid undesired alarms for stator ground fault protection during startup, an inhibit signal must be generated during generator startup operation. The inhibit threshold can be set at 85% of generator rated voltage. Using a PT ratio of 14,400 : 120 V, the stator ground protection package can be enabled at the following level:

$$V\,(\text{inhibit}) = 0.85 \times 13,800 \times \frac{120}{14,400} = 98\ \text{V}$$

Analysis of the Failure of the Capacitor Lead The capacitor lead, which consists of seven strands, with each strand capable of carrying 10 A, was found broken at the clamp connection, as shown in Fig. 4.73. No burn marks were found at the clamp location, which rules out any arcing phenomena that can generate heat, which could cause the lead to fail open. However, it appears that the original installation of the clamp was done improperly, causing excessive stress and resulting in breakage of the majority of the strands. It is postulated that a resistance was introduced in series with the phase B capacitor, causing a slight current unbalance in the generator neutral at the beginning of the incident, as shown in Figs. 4.69 and 4.71. This is evident at the first generator trip, where the 60-Hz neutral voltage was low compared to that in the later trips. In the several unsuccessful generator energization attempts, the phase B capacitor lead broke open, causing the 60-Hz neutral voltage to increase and dominate the third harmonics shown in Fig. 4.71. The broken lead caused the portion of the missing phase B 60-Hz current component to flow in the neutral, causing predominant 60-Hz neutral voltage.

Calculation of Generator Neutral Voltage During Failure of the Capacitor Lead Prior to the failure of the phase B capacitor lead, the sum of the capacitor phases A, B, and C currents was equal to zero. Therefore, the balanced nature of these currents has resulted in no 60-Hz current flows in the generator neutral, as shown in Fig. 4.70 for the STG unit. The capacitance of each phase is 0.25 μF;

$$\text{capacitive reactance } X_C = \frac{1}{2 \times \pi \times 60 \times 0.25 \times 10^{-6}} = 10{,}616\,\Omega$$

$$I_C = \frac{V_{\text{L-n}}}{X_C} = \frac{13{,}800}{\sqrt{3} \times 10{,}616} = 0.75\text{ A}$$

Since the phase B capacitor lead opened, the sum of the three phase currents drawn by the surge capacitor has no longer been balanced. As a result, the missing phase B current will flow to ground and back to the generator neutral. As shown in Fig. 4.74, $I_A + I_B + I_C = 0$ for a balanced three-phase system:

$$I_A + I_C = -I_B = I_n = 0.75\text{ A}$$

This 60-Hz current will flow to ground and return via the various capacitances of the generator and associated equipment and the generator neutral. As shown in Fig. 4.75, the distribution factor for the neutral component is as follows: Since $3R$ is in parallel with the equivalent capacitive reactance X_{co}, the

$$\text{distribution factor (d.f.)} = \frac{X_{co}}{3R - jX_{co}}$$

This will imply that the distribution factor for the generator neutral current I_n will be

$$\text{d.f.} = \frac{1}{\sqrt{2}} = 0.707$$

$$I_n = 0.75 \times 0.707 = 0.53\text{ A}$$

This 60-Hz current will flow through the secondary grounding resistor and develop neutral voltage, which will cause operation of the overvoltage relay element (if above the relay pickup):

$$\text{neutral voltage} = V_{\text{g-n}} = I_n \times R_{\text{sec}} = 0.53 \times \frac{13{,}800}{240} \times 0.39 = 11.9\text{ V}$$

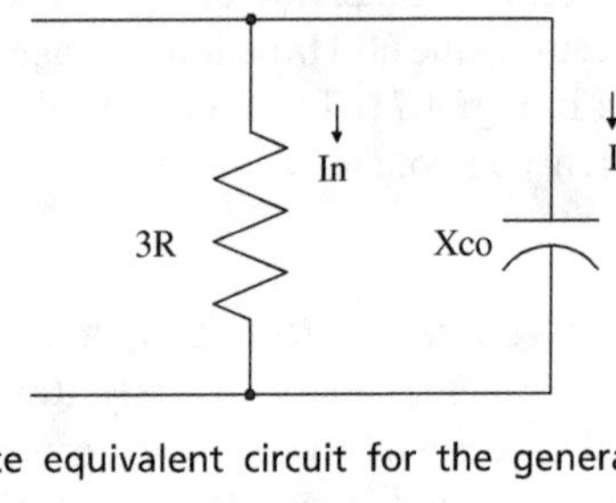

Fig. 4.75 Zero-sequence equivalent circuit for the generator high-resistance neutral grounding system.

which is above the relay setting of 5 V. This explains the operation of the stator ground fault protection overvoltage 59N element, tuned to 60 Hz, when the phase B capacitor lead failed open. Adding the third-harmonic voltage to this calculated 60-Hz value of 11.9 V will closely match the actual voltmeter recorded RMS reading of 13.8 V during the last trip.

System Phenomena

1. Appearance of third-harmonic voltage as a sign of generator health, as shown in trace SG-Vn in Fig. 4.71 for the steam generator STG unit.
2. Startup of generator units by gradual increase of the unit excitation near the machine synchronous speed, as shown in the voltage traces of Fig. 4.72

Corrective Actions

1. Connect the undervoltage element of the 100% stator ground fault protection system to alarm. The connection of the UV element to the plant computer system during this incident could help in faster troubleshooting of the problem associated with the trip of relay 59N. Apparently, no alarms were activated during the incident, which rules out stator faults and problems near the generator neutral. The initiation of an alarm by the third-harmonic undervoltage element should be followed by an assessment of the generator neutral voltage using a spectrum analyzer. Disappearance of the neutral third-harmonic voltage will confirm either the presence of a stator ground fault near the neutral or the fact that neutral integrity and its connection to ground no longer exist.
2. Repair the surge capacitor lead connection to the generator terminal surge arrestor using proper installation methods.

Lessons Learned

1. Following a forced outage, generating units should be restarted only after a through investigation of the causes of the trip and complete troubleshooting and diagnosis of the problems.
2. Understanding relaying targets in conjunction with generator neutral voltage measurements should be used to assess the cause of the operation of the stator ground fault protection system.
3. A spectrum analyzer should be used to assess the generator neutral content. The device readings of the fundamental 60 Hz and third harmonic should be used to confirm the presence of stator ground faults and their locations prior to energizing the generator unit following forced outages.
4. The setting of the overvoltage element of relay 59N is normally based on detecting stator ground faults in most of the stator winding (95 to 96%). Therefore, the setting of 5 V for the element was adequate in detecting the

problem that occurred during this incident. Raising the setting to a value above the 60 Hz generated during this incident to avoid a future trip in a similar incident is not prudent and should be avoided. The relay system main function is to provide as much protection coverage to the stator winding as possible.

Case Study 4.12: Generator Stator Ground Fault in an 800-MW Fossil Unit

Abstract In this case study we describe the occurrence of a generator stator ground fault that was cleared by the operation of 95% stator ground fault protection. We also describe the fault mechanism and provide calculations and an explanation of how to locate a ground fault. In addition, we analyze DFR records and corrective actions.

Description of the Fault A phase A-g stator ground fault occurred inside the 800-MW unit shown in Fig. 4.76. The unit stator conductors, which are embedded in the stator slots, are cooled using ionized water. The stator end conductors are cooled using hydrogen. The stator ground fault was caused by leaky water from a cracked hose in the stator conductor cooling system. Unit vibration was apparently the major cause of the cracking of the water hose, and as a result, the end winding began to crack and arc over. Arcing began to melt the copper conductors, which mixed with the

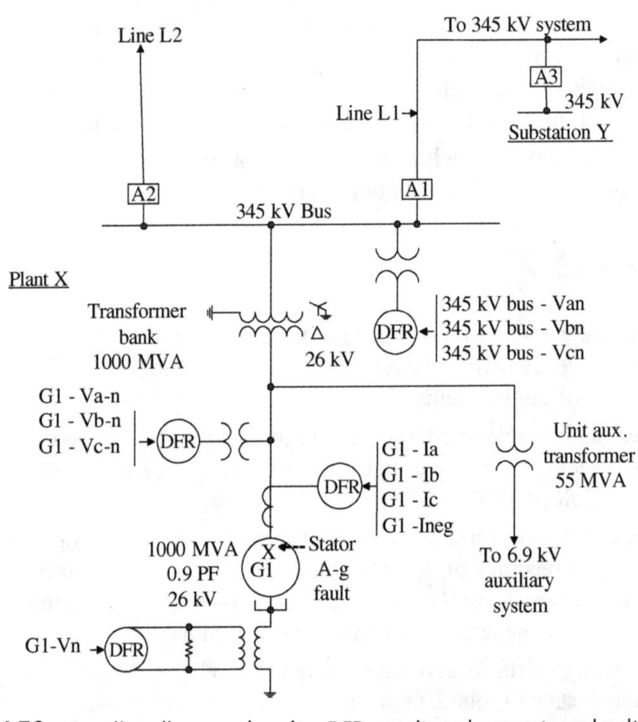

Fig. 4.76 One-line diagram showing DFR-monitored currents and voltages.

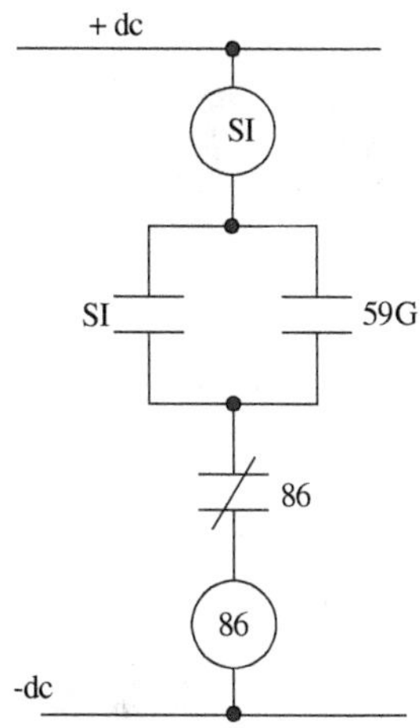

Fig. 4.77 Stator ground fault protection dc tripping circuit.

ionized water and formed a conductive path to ground. The generator was tripped and isolated from the 345-kV system by a lockout relay which was energized by the stator ground fault protection (Fig. 4.77). The relay is an electromechanical overvoltage relay 59N tuned to the 60-Hz frequency and covering 95% of the stator for ground faults. Figure 4.76 documents the dc schematic circuit for relay 59N.

Analysis of the DFR Record Figure 4.76 shows the system one-line diagram and the DFR-monitored voltages and currents for generator G1. The DFR record in Fig. 4.78

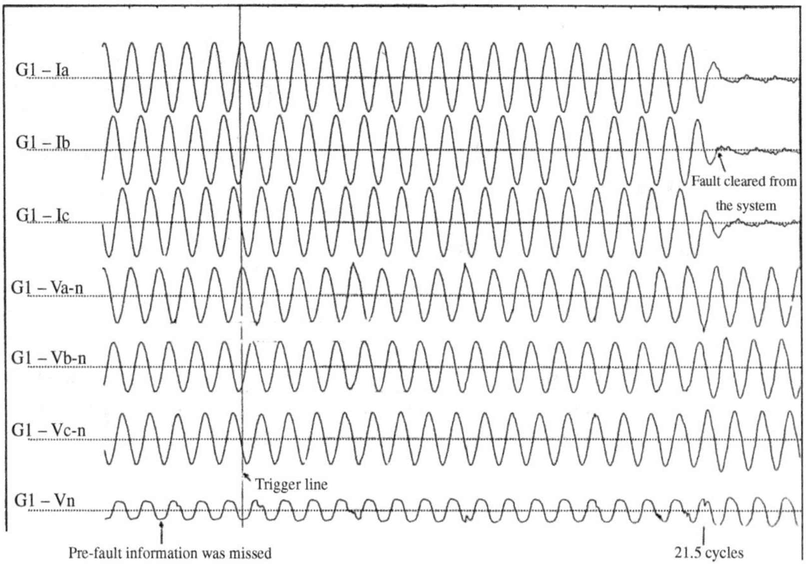

Fig. 4.78 DFR record for the G1 unit, showing the generator stator ground fault.

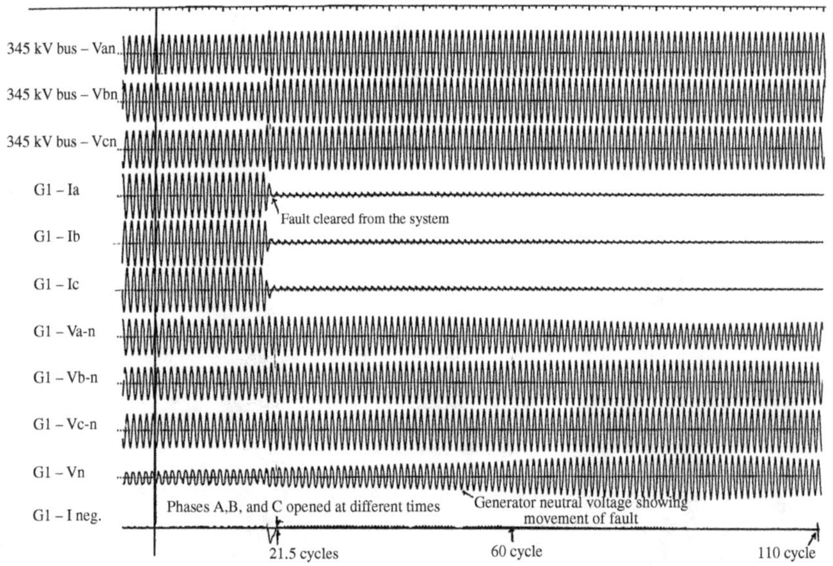

Fig. 4.79 Compressed DFR record showing unit decaying energy feeding the stator ground fault.

indicates a recording of 0.36 s (21.5 cycles) of a stator ground fault current. The total clearing time from the 345-kV system is about 30 cycles (a 27-cycle relay time + 1-cycle lockout relay + 2-cycle breaker interrupting time). The overvoltage relay 59N calculated operating time is about 0.48 s (29 cycles), which implies that 7.5 cycles (0.125 s) of faulted data as well as 5 cycles of pre-fault data were lost. Loss of these data was attributed to the triggering level setting of the generator neutral voltage. The DFR outputs in Figs. 4.79 and 4.80 show the feeding of the stator ground fault by machine decayed trapped energy, which lasted for 7.6 s. Also indicated is the movement of the stator ground fault location to different portions of the faulted phase. Voltage trace G1-Va-n of Fig. 4.79 reveals a lower voltage magnitude around a time frame of 65 cycles after the tripping of the generator in the 345 kV system. This will confirm the occurrence of the ground fault in the generator stator phase A winding.

Location of the Stator Phase A-to-Ground Fault The ground fault location can be approximated by performing voltage calculation and analysis of the DFR output. Based on information shown on Fig. 4.81 for the generator neutral grounding, the machine is grounded using a distribution transformer with a ratio of 26, 000/240 V with a resistor of 0.35 Ω connected across the secondary winding:

$$\begin{aligned} R_{\text{sec}} &= 0.35\ \Omega \\ R_{\text{pri}} &= 0.35 \times n^2 \\ &= 0.35 \times (26{,}000/240)^2 = 4107.6\ \Omega \end{aligned}$$

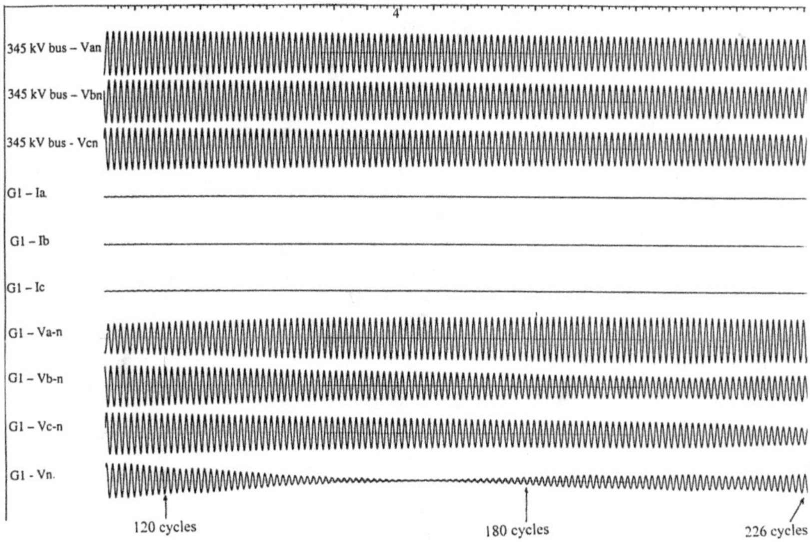

Fig. 4.80 Continuation of feeding of the initial fault by unit decayed energy.

The high-resistance generator neutral grounding is optimized by selecting the grounding resistor to be related to the capacitive reactance of the generator bus as follows:

$$3R = X_{co} = 3 \times 4107.6 = 12{,}322.9\,\Omega$$

The equivalent impedance can be calculated in the zero-sequence network shown in Fig. 4.82. The

$$\begin{aligned} \text{equivalent } Z_0 &= \frac{(3R)(-jX_{co})}{3R - jX_{co}} \\ &= \frac{12{,}322.9}{\sqrt{2}} \\ &= 8715\,\Omega \end{aligned}$$

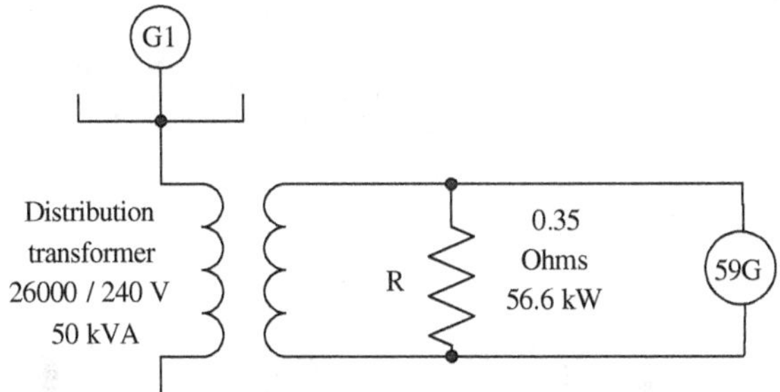

Fig. 4.81 Generator neutral high-resistance grounding system.

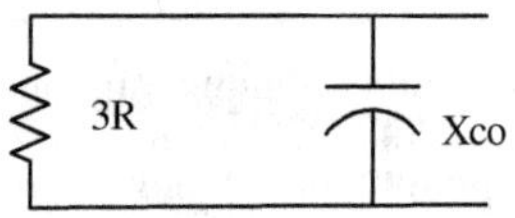

Fig. 4.82 Zero-sequence network equivalent circuit.

For an L-g fault at the generator terminal,

$$I_F = \frac{V_{\text{p-n}}}{Z_0} = \frac{26 \times 10^3}{\sqrt{3} \times 8715} = 1.722\ \text{A}$$

$$I_{\text{or}} = 1.722 \times \frac{1}{\sqrt{2}} = 1.218\ \text{A}$$

The neutral current $= I_{\text{n}} = 3 \times I_{\text{or}} = 3 \times 1.218 = 3.654$ A or

$$I_{\text{n}} = \frac{V_{\text{p-n}}}{R_{\text{pri}}} = \frac{26 \times 10^3}{\sqrt{3} \times 4107.6} = 3.654\text{A (same answer)}$$

$$V_{\text{sec}} = V_R = I_{\text{n}} \times n \times R_{\text{sec}}$$

$$= 3.654 \times \frac{26{,}000}{240} \times 0.35$$

$$V_R = 138.55\ \text{V}$$

Trace G1-Vn of the DFR record in Fig. 4.81 reveals a voltage value $V_{(\text{F})} = 29.6$ V.

The unit neutral voltage is linearly proportional to the fault position in the stator winding. Therefore, the percent location of fault from the neutral is

$$\frac{V_{(\text{F})}}{V_R} = \frac{29.6}{138.56 \times 100} = 21\%$$

The overvoltage relay is set at 5.4 V and curve 1. The relay pickup = 29.6/5.4 = 5.5 V. From curve 1 the time of relay operation = 0.46 s. Using the published ANSI curve formula for an inverse relay, we have

$$t = \left\{ \left[\frac{8.9341}{(V/V_p)^{2.0938} - 1} \right] + 0.17966 \right\} \times D$$

where t is the trip time; V, the fault voltage = 29.6 V; V_p the pickup setting = 5.4 V; and D, the time dial setting = 1. Substituting these values in the formula gives us

$$t = \left\{ \left[\frac{8.9341}{\{(29.6/5.4)^{2.0938} - 1\}} \right] + 0.17966 \right\} \times 1 = 0.475\ \text{s}$$

The percent location of the stator ground fault is 21% from the generator neutral. The numerical result obtained from the formula for the stator ground fault is more accurate and is very close to the value obtained from the relay curve.

Corrective Actions

1. The trigger level setting of the generator neutral voltage was lowered to enable recording all fault data.
2. A redundant 100% stator ground fault protection was added to the existing 95% electromechanical overvoltage relay. It is always a safe policy to have two independent relaying systems covering for protection against any faults in their zone of protection.
3. As a result of the early detection of the fault, the unit was slightly damaged and unit repair was considered a feasible option. However, the stator was rewound completely to prolong the unit life. The rewinding of the unit stator was done on site and the job was completed in one month with the generator back in service.

Case Study 4.13: Three-Phase Fault at the Terminal of an 800-MW Generator Unit

Abstract The 800-MW fossil generator shown in Fig. 4.83 was subjected to a three-phase fault while generating 200 MW of power. The fault was cleared from the 345-kV system successfully, and the unit was shut down. In this case study we describe the fault mechanism, sequence of events, fault analysis, corrective actions, and lessons learned.

Description of the Power System and Associated Protection The 800-MW fossil plant generator step-up transformer (GSU) is connected to the system via a single bus connecting the 345-kV lines. As shown in Fig. 4.83, the generator and the unit transformer are individually protected by differential relays. An overall harmonic restraint differential relay covers the generator and the GSU, including the single 345-kV bus. The 26-kV ISO-phase bus is protected by the overall differential relay on an instantaneous level and by the generator backup protection devices 21 and 46 on the time-delay level. The stator is protected against ground faults by an individual 95% relay applied to the generator neutral via a distribution transformer. The generator has a radio-frequency transducer that is installed in the generator neutral for incipient corona detection research during arcing on a trial basis.

Sequence of Events The sequence of events as reconstructed from an analysis of plant X and substation Y oscillograms and the plant computer log is as follows. Figure 4.84 shows the system one-line diagram and oscillograph-monitored voltages and currents for generator G1 and substation Y. At the beginning of the incident, the plant oscillograms revealed 60-Hz voltages at the generator neutral at varying magnitudes between 8 and 23 V and varying durations between 3 and 23 cycles.

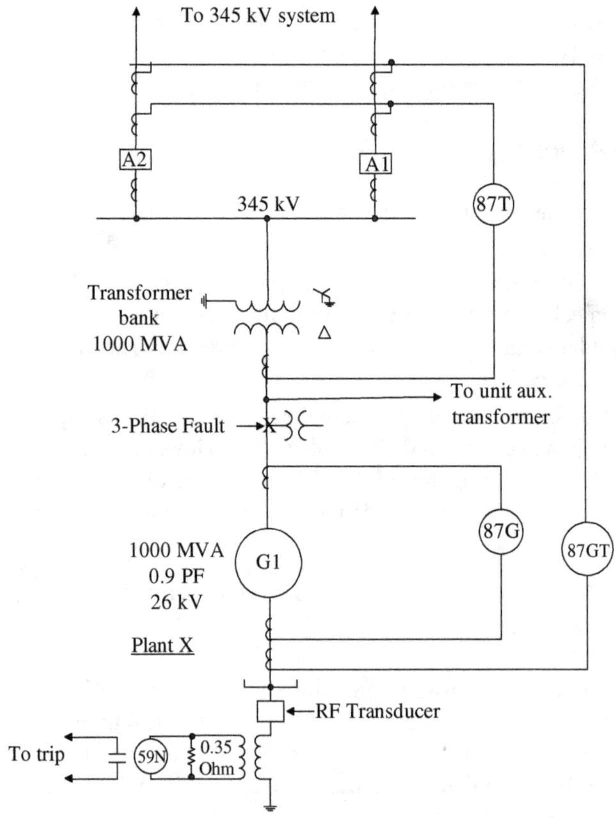

Fig. 4.83 One-line A-C diagram for generator and GSU transformer protection.

The plant computer indicated a generation level of about 200 MW for the plant. Twenty-three seconds later, a simultaneous three-phase fault occurred on the generator 26-kV ISO-phase bus. Substation Y oscillograms, shown in Fig. 4.85 reveal the presence of a three-phase fault. The plant oscillograph jammed and the fault record was lost; only a time stamp was generated. The fault was cleared at the plant by operation of the plant overall differential relay 87GT. Thirty-one seconds from $t = 0$ the plant computer indicated that the phase A and B currents decayed to zero and the phase C current is 2450 A. The presence of generator decaying current on phase C after only 9 s from the initial three-phase fault indicates bypassing of the high-impedance neutral grounding system.

Analysis of the Plant Oscillogram During the three-phase fault, the plant oscillograph jammed and no information about the neutral integrity was available during the fault. Prior to the fault the oscillograph revealed the presence of third-harmonic voltage, which confirms the integrity of the generator high-resistance neutral grounding system. Figure 4.86 shows the beginning of the intermittent

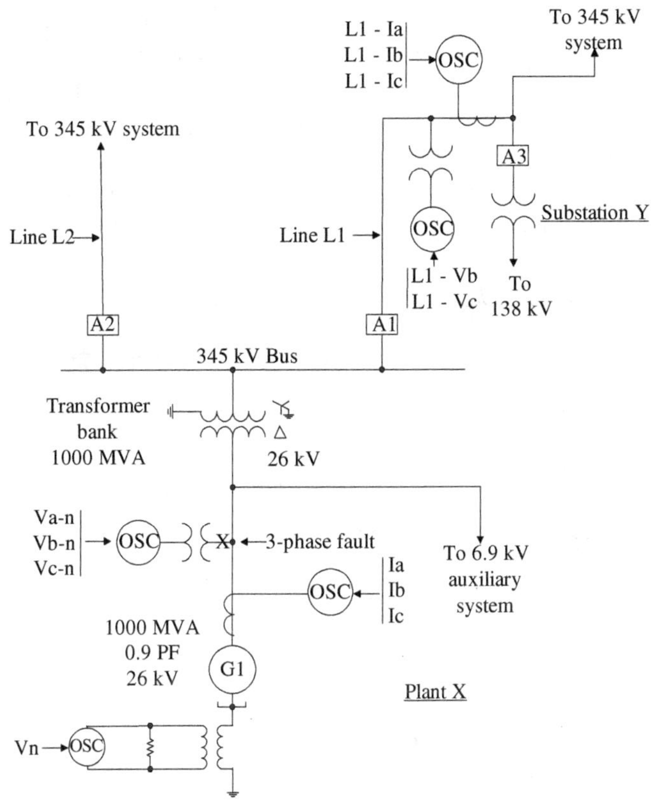

Fig. 4.84 One-line diagram showing oscillograph-monitored currents and voltages.

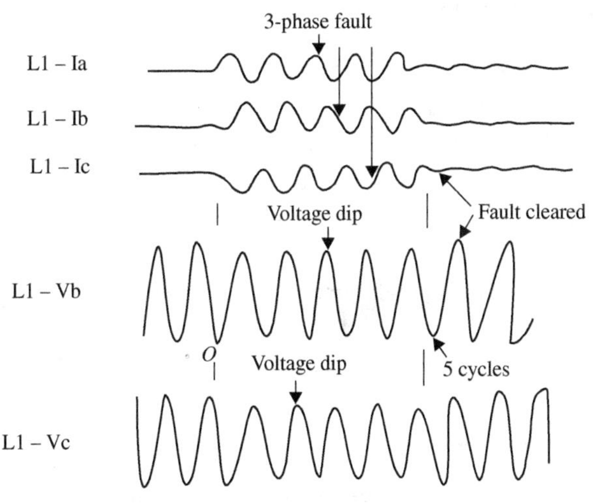

Fig. 4.85 Substation Y oscillograph record for line L1, illustrating the three-phase fault.

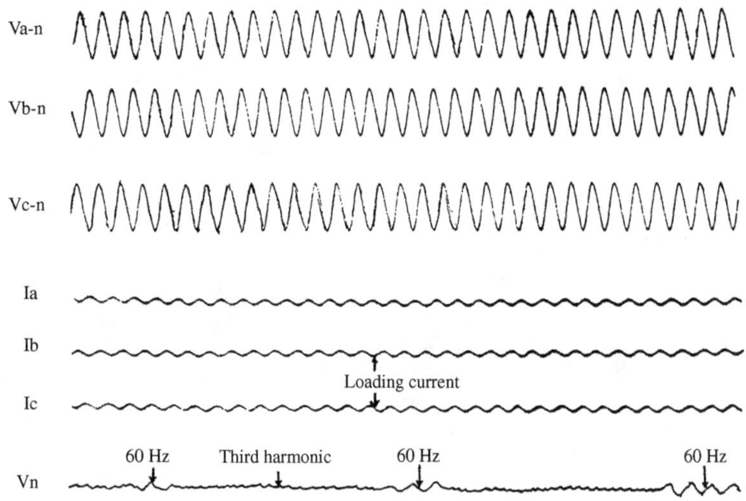

Fig. 4.86 Generator voltages and currents during the initial intermittent stator ground faults.

ground fault and generation of the 60-Hz generator neutral voltage. The oscillogram also shows the machine terminal voltages and loading currents. Figure 4.87 shows seven different oscillograms for the generator neutral voltage only after their extraction from seven individual oscillograms with the amount of time separation between them for a span of 16 s. It reveals a variation in 60-Hz neutral voltage magnitude and duration.

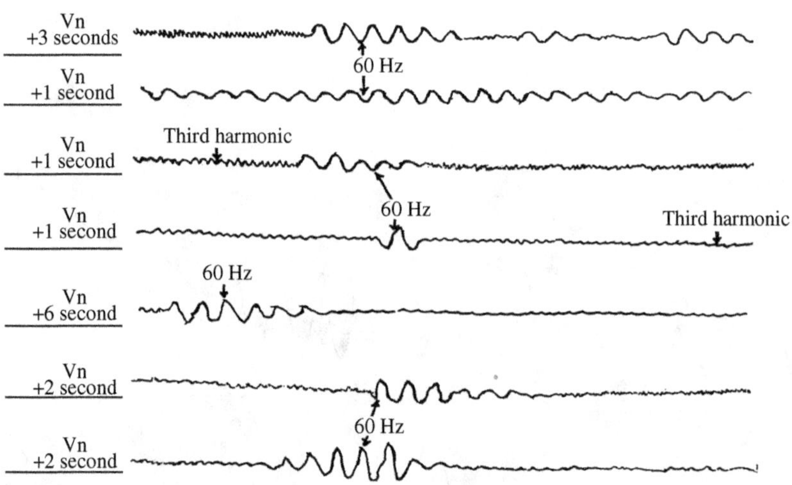

Fig. 4.87 Generator neutral voltages for several oscillograms as recorded individually during the intermittent ground faults for an interval of 16 s.

Visual inspection of the neutral compartment, the current transducer, and the coaxial cable lead of the radio-frequency (RF) equipment shown in the neutral of the generator clearly indicated a phase-to-ground fault. Therefore, it can be concluded that bypassing of the neutral grounding was caused by the failure of the RF monitoring system to ground. The RF equipment was added on a trial basis to assess the integrity of the machine insulation system by continuously trending the presence of RF signal through a spectrum analyzer.

The purpose of the generator neutral grounding through an impedance is to mitigate the damage at the point of the fault (in this case the PT bus compartment), by limiting the neutral current to a value below 10 A (4 A for this unit), limiting transient overvoltages, and providing a means of detection for ground faults. The former function was eliminated by the RF equipment failure; that is, ground current was not limited, contributing to the severity of the intermittent ground faults caused by leaking water into the PT compartment at the generating unit terminal.

Analysis of the Substation Oscillogram The oscillogram record for substation Y shown in Fig. 4.85 reveals a phase A-to-phase C fault, evolving in few milliseconds to a three-phase fault for a duration of 5 cycles. The substation Y oscillogram reveals that the three-phase fault was cleared from the 345-kV side by tripping of the plant's 345-kV breakers A and A1. The oscillogram also reveals overcurrent on phases A, B, and C and voltage dip on phases B and C for line L1 leading to three-phase fault classification (phase A voltage was not monitored).

Machine Terminal Fault Mechanism The machine terminal three-phase fault was caused by failure of the hydrogen cooler-water path. The water started to flow on the plant floor, with collection at the potential transformer compartment connection to the ISO-phase bus, and thus caused intermittent L-g faults at the terminal of the generator. The fault evolved to a phase A-to-phase C fault and then evolving in few milliseconds to a simultaneous three-phase fault at the generator 26-kV ISO-phase bus. Machine decayed energy eventually resulted in failure of the neutral compartment. The failure was attributed to the installation of RF monitoring equipment in the neutral of the generator. The RF equipment (current transformer and cables) failed due to voltage stress and faulted to ground, bypassing the neutral high-resistance grounding system. The plant oscillograph jammed and failed to record the machine decayed energy supply to the fault after isolation from the system.

Analysis of the Performance of Stator Ground Fault Relay 59N The generator neutral grounding shown in Fig. 4.83 is classified as a high-resistance grounding system. For a solid L-g fault at the terminal of the generator, the secondary neutral resistor of 0.35 Ω can be referred to the primary side as

$$R_{\text{pri}} = R_{\text{sec}} \times n^2 = 0.35 \times \left(\frac{26}{0.24}\right)^2 = 4107.6\,\Omega$$

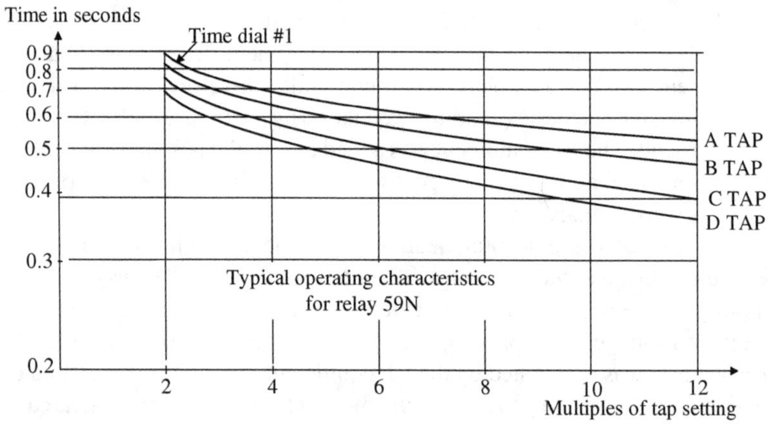

Fig. 4.88 Time voltage curves for 59N generator stator 95% ground fault protection.

The primary current for an L-g fault is

$$\frac{V_{\text{L-n}}}{R_{\text{pri}}} = \frac{26,000}{\sqrt{3} \times 4107.6} = 3.65\ \text{A}$$

The neutral voltage across the secondary resistor is

$$I_{\text{sec}} \times R_{\text{sec}} = 3.65 \times \frac{26}{0.24} \times 0.35 = 138.4\ \text{V}$$

From the calculation of 138 V across the neutral resistor for a solid-phase ground fault at the generator terminal, we can deduce that the neutral voltages corresponded to very high resistance intermittent ground faults on the generator phase connection.

The electromechanical 60-Hz tuned overvoltage relay device 59N is set at 5.4 V (a tap) and at time dial 1. Based on the relay operating characteristics shown in Fig. 4.88, the relay has a minimum operating time of about 0.55 s for a setting multiple of 12 × (12 × 5.4 = 65 V). The recorded oscillogram values for V_n vary between 8 V (8/5.4 = 1.5×) and 24 V (24/5.4 = 4.3×) for a duration between 3 cycles (3/60 = 0.05 s) and 23 cycles (23/60 = 0.38 s). Therefore, the 60-Hz voltage generated during the intermittent ground faults, which were caused by the dripping of water inside the PT's compartment, did not reach the relay set point and hence never caused operation of the 59N relay. The stator ground fault relay was unable to trip the unit because it did not detect the early intermittent terminal ground faults before evolving to a three-phase fault.

Corrective Actions Remove the RF monitoring equipment from the generator neutral compartment.

Lessons Learned

1. Three-phase fault can occur at the ISO-phase bus at the generator terminal; therefore, generators should be designed to withstand three-phase terminal faults.
2. Equipment should be specified with an adequate voltage-withstanding rating prior to their installation on the power system. Apparently in this case study, the RF current transformer with its interface cable did not have enough insulation to withstand the generator neutral potential rise during intermittent ground faults.

Case Study 4.14: Three-Phase Fault at the Terminal of a 50-MW Generator Due to a Cable Connection Failure

Abstract A three-phase fault occurred at the terminal of a 50-MW combustion turbine (CT) generator. The fault was detected by the generator differential relays. The relays separated the generator from the system by tripping the generator breaker, field breaker, and prime mover. The system contribution to the fault was stopped in three cycles. The generator current contribution to the three-phase fault cannot be stopped instantly, due to the energy stored in the rotating machine shaft. With time the machine current will slowly decrease in an exponential decaying function. This residual (trapped) energy continued to flow for several additional seconds, causing the initial phase-to-ground fault to evolve to a three-phase fault on the generator breaker system side. In this case study we describe the failure mechanism, sequence of events, relay DFR fault record, and lessons learned.

Description of the Power System and Associated Protection The 50-MW CT generating unit, shown in Fig. 4.89, is connected to the 138-kV system via a solid dielectric cable. The 13.8-kV generator circuit breaker is housed in an outdoor switchgear enclosure. The closing of the high side 138-kV breaker on an energized system, while the plant units are not generating, will provide startup station service power source to feed the generator auxiliary loads. The combustion turbine unit is protected by dual multi-function generator numerical relays fed from the same neutral- and system-side CTs (due to CT limitation). The transformer GSU and the 13.8-kV generator bus are protected by two separate transformer numerical differential relays. The unit and its associated GSU protection devices are shown in Fig. 4.90.

Sequence of Events The following events are based on an analysis of the generator numerical relay 87GP oscillography record (Fig. 4.91) and the transformer differential relay event record. At zero time, 60-Hz generator neutral voltage appears on trace Vn and continues for an additional 5 cycles. Based on the generator terminal voltages during the ground fault (no significant voltage dip), and due to the use of open delta-connected PTs grounded at phase B for the secondary winding as shown in Fig. 4.89, the ground fault occurred on phase B.

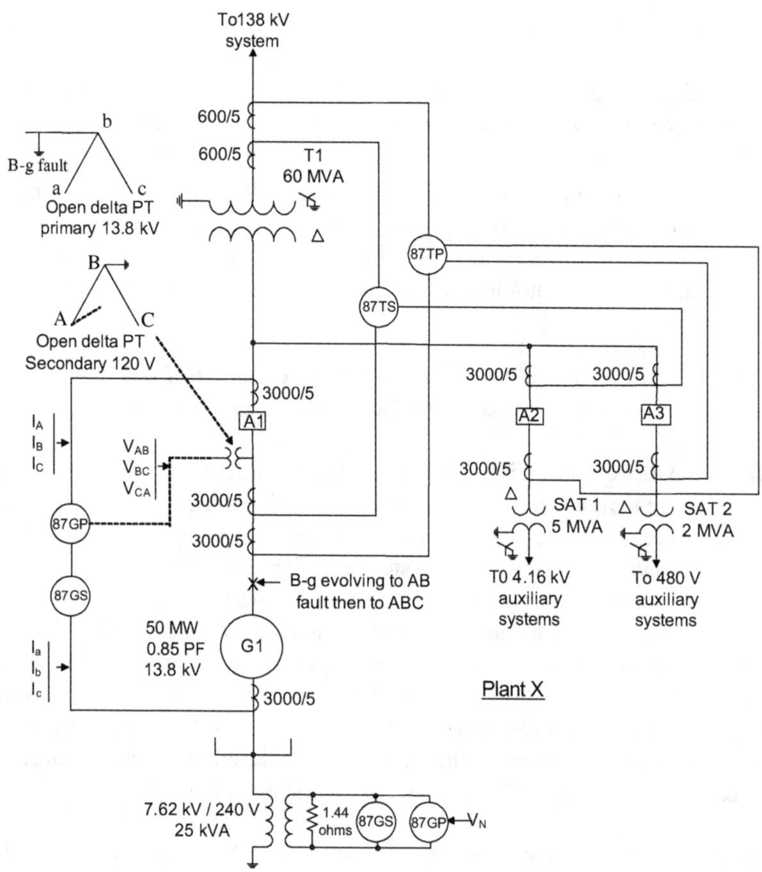

Fig. 4.89 Plant one-line A-C showing the unit and transformer protection and the location of the terminal fault.

At 5 cycles, the fault evolved to phase A-B for 4 ms, then to a solid three-phase fault at 8 cycles from $t=0$. The system-side contribution to the fault was stopped by tripping the generator breaker 52G. Several cycles later (not known, due to the lack of synchronized time), the initial ionized cloud started to grow and the initial three-phase fault is now being fed again from the system. This was due to the arc-over of the generator-side breaker studs to the system-side breaker connection, behind the opened 13.8-kV generator breaker 52G. The system feed to the fault location was stopped by tripping of the 138-kV circuit breaker. Several seconds from time zero (about 8 s), unit G1 trapped energy kept feeding the fault and eventually stopped when the machine shaft stopped.

Analysis of the Fault Record The numerical relay 87GP oscillography record shown in Fig. 4.91 displays a three-phase fault occurring inside the

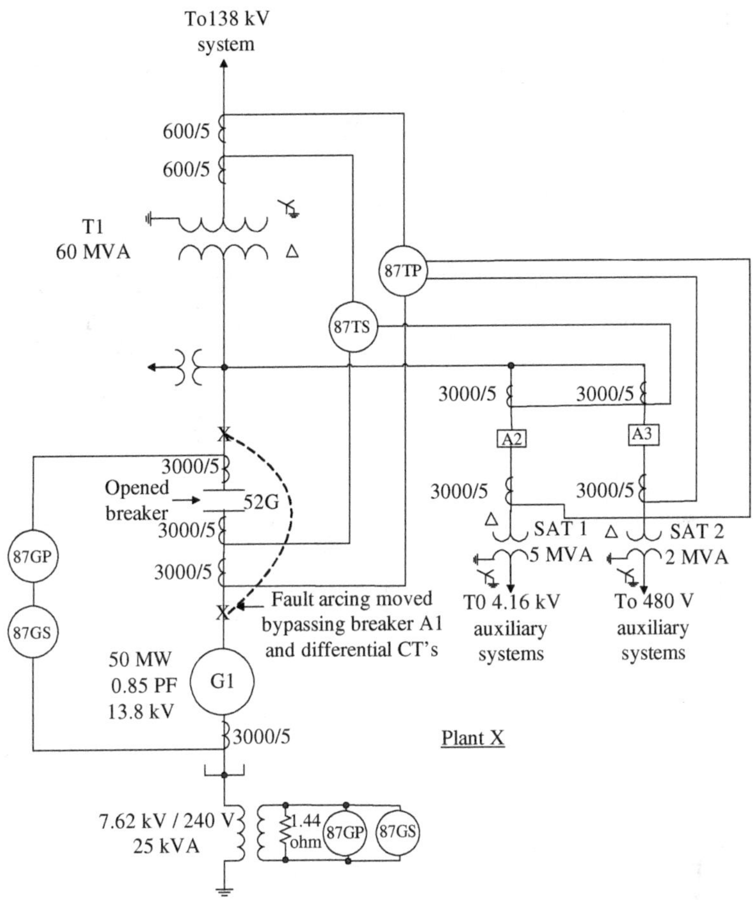

Fig. 4.90 Plant one-line diagram showing fault movement from the generator side to the system side around the 52G breaker.

13.8-kV switchgear. The fault was caused by the failure of a bus connector. The fault started as a line-to-ground fault on phase B, but after 5 cycles it evolved into phase A-B for 4 ms, then to a solid three-phase fault. The system currents I_A, I_B, and I_C were interrupted when the generator breaker was opened by the operation of generator differential relays in about 3 cycles. The generator-side currents I_a, I_b, and I_c continued to flow after the unit was shut down. The relay was programmed to provide a fault record only for an additional 6 cycles after the tripping. This fault records length setting prevented the display of additional fault records during the total length of generator decayed energy feeding the fault current flow, which is about 8s after tripping.

The movement of the fault to the system side of the 13.8-kV breaker is in the transformer differential zone. The DFR records of the 87TP and 87TS were lost when an operator panicked when smoke and an ionized cloud emerged from the 13.8-kV

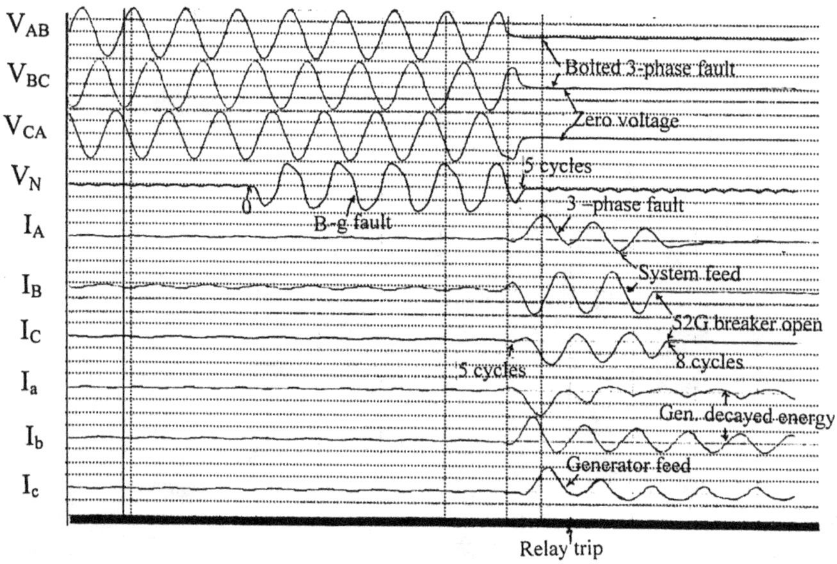

Fig. 4.91 Generator numerical relay 87GP oscillography record covering evolving of the B-g fault to the three-phase fault.

switchgear due to the feed from the generator, and removed the dc power to all relays. Some first-generation numerical relays can lose their oscillography fault records when the relay dc power is removed.

Analysis of the Fault Mechanism The numerical relay oscillography record shown in Fig. 4.91 displays a three-phase fault that occurred on the terminal of the generator. The three-phase fault began as a B-g fault, which was apparently caused by the initial arcing and tracking that caused an ionized cloud. The arcing began due to loose connections in the 13.8-kV generator cable terminations at the generator breaker switchgear enclosure. The connection of the breaker studs to the 13.8-kV switchgear bus is secured using several bolts. The bolts were not secured with the proper torque force, leaving a gap between the two connecting plates while carrying a 3000-A current. As a result, arcing and arcing by-products began to ionize the air inside the enclosed 13.8-kV switchgear, which caused phase B initially to arc to ground, leaving burn marks inside the switchgear enclosure and causing the ionized cloud to grow and the initial fault to evolve to a phase B-C fault for a few milliseconds and, finally, into a three-phase fault.

Transformer Differential Relay Fault Record (87TP) The following fault record was obtained from the numerical transformer differential relay. The high-side (138-kV) primary currents $I_a = 1513$ A, $I_b = 1498$ A, and $I_c = 1505$ A. The low-side input currents were all zero. This was due to shorting of the generator breaker terminals in the back by arc phenomena, thus bypassing the CT input to the

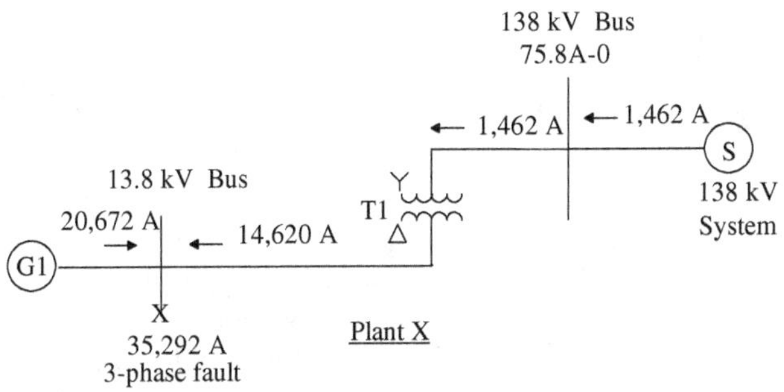

Fig. 4.92 Short-circuit study simulation of the three-phase fault at the generator G1 terminal.

transformer differential relay. This prevented the unit decayed energy from going to the 87TP relay circuit.

Confirmation of the Magnitude of the Current During the Fault From the numerical relay fault record, the high-side 138-kV fault currents for the second three-phase fault were $I_a = 1513$ A, $I_b = 1498$ A, and $I_c = 1505$ A. Taking an average value, we have

$$I_{av} = \tfrac{1}{3}(1513 + 1498 + 1505) = 1505 \text{ A}$$

From Fig. 4.92, short-circuit simulation for a three-phase fault at the 13.8-kV bus, the 138-kV system current contributions is 1462 A. The percent error between the relay fault record and calculated values can be defined as

$$\%\ \text{error} = \frac{(\text{fault record value} - \text{value measured}) \times 100}{\text{the smaller of the two}}$$

$$\text{current error } (\%) = \frac{1505 - 1462}{1462} \times 100 = 3\% \text{ less than } 10\% \qquad \text{O.K.}$$

Power System Phenomena Appearance of normal secondary line-to-line voltages on all phases during phase-to-ground faults occurring on systems utilizing open delta PT connections. This is due to the use of open delta-connected PTs grounded at the faulted phase as shown in Fig. 4.89, when the line-to-line voltages are monitored by the relay oscillography record. The ground fault occurred initially at phase B on the primary 13.8-kV side as shown, while the open delta-connected PTs are grounded at phase B of the secondary 120-V side. This resulted in line-to-line normal voltage reading in Fig. 4.91, due to no shift in the PT secondary phasor diagram.

Corrective Actions The failed switchgear was replaced and all 13.8-kV bus connections were examined for proper tightness of all bolts, with the proper torque force applied for each bolt.

Lessons Learned

1. Generating units are designed to withstand three-phase faults occurring on their terminals. This case study provided verification of this capability.
2. Extended flow of fault current is the reason that internal multiphase faults can possibly damage equipment around the unit to the point where it cannot be repaired economically. There is no means of stopping the flow of the generator current in this case. The fault current flow from decayed generator energy, accompanied by a long decay time constant, results in the vast majority of equipment damage occurring after tripping and shutdown of the units.
3. A concerted effort should be made in the generator and generator terminal design to avoid the occurrence of multiphase faults. The design should only permit ground faults to occur. The generator is normally designed to reduce ground fault current substantially to minimize equipment damage.
4. Use of a generator breaker will reduce the likelihood of catastrophic failures of the generator unit step-up (GSU) transformer by eliminating the generator-decayed energy feed for faults in the GSU transformer.
5. The dc power for numerical relaying should not be switched off. This may affect the retention of some oscillography fault records and event capturing for some first-generation numerical relays.

Case Study 4.15: Generator Stator Phase-to-Phase-to-Ground Fault Caused by Failure of the Rotor Fan Blade

Abstract One of the blades of the rotor fan of a 64-MVA generating unit became loose and started to hit the stator coil, causing a phase B-C-g fault. The fault was detected and cleared by the operation of both generator multifunction numerical relays. In this case study we describe the unit protection upgrade, the fault mechanism, power system phenomena, lessons learned, and the selection of fault records that can provide a quick explanation for a preliminary sequence of events.

Description of the Protection System As shown in Fig. 4.93, the unit is protected by dual generator multifunction numerical relays, labeled 87GRP and 87GRS, which provide protection against generator phase and ground short-circuit faults as well as abnormal power system conditions. The redundant relays are fed from separate CTs and powered from separate dc batteries. Unit shutdown is accomplished via separate emergency shutdown lockout relays. For generator stator phase or ground faults the relays initiate the release of CO_2. The relay CT ratio is normally designed to allow for a maximum secondary generator full output current of 5 A and to accommodate a unit full-load current of 2678 A ($= 64\ \text{MVA} \times 10^3/\sqrt{3} \times 13.8\ \text{kV}$). Therefore, the relay-connected CT ratio is 3000 : 5. There is no dedicated DFR monitoring of the unit. Full reliance is placed on the numerical relay oscillography and event records to analyze unit disturbances.

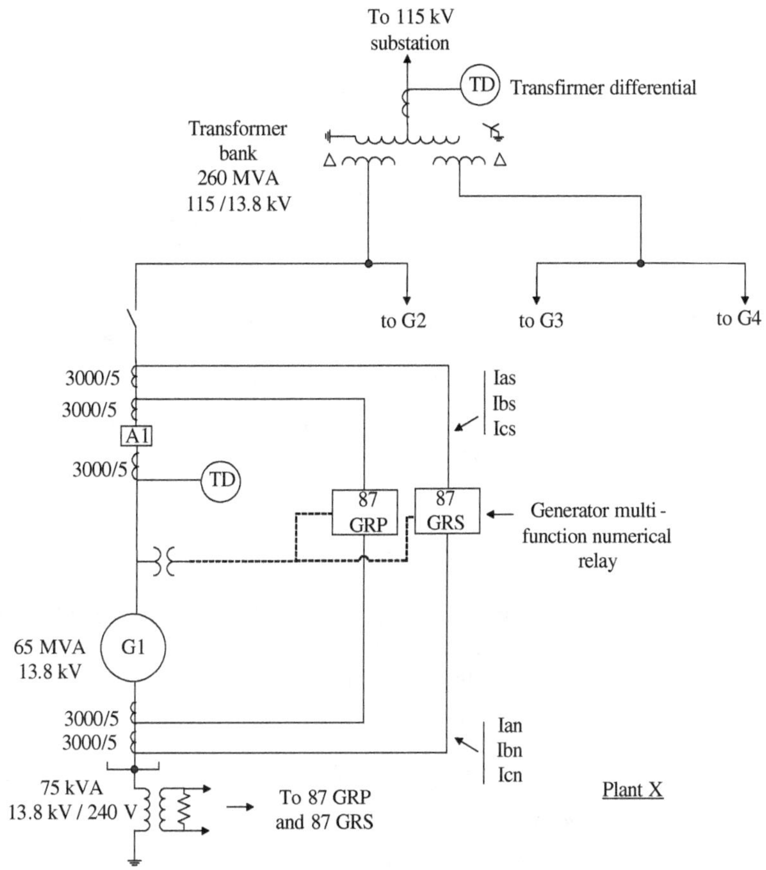

Fig. 4.93 One-line diagram showing the generator G1 numerical relays with monitored currents and voltages.

Generator Stator Phase Fault Primary Protection System Phase differential protection is normally carried out using elements that have either one-slope or dual-slope characteristics. The low slope setting will enhance the relay sensitivity, especially for low fault current values. The high slope setting will provide better relay stability and security for high fault currents by making the relay less sensitive during external phase faults.

Biased Differential Operating Characteristics of the 87GRP Relay The numerical relay 87GRP operating characteristics are shown in Fig. 4.94. The differential element is set as follows:

CT ratio: 3000 : 5

Differential trip minimum pickup: $0.1 \times \mathrm{CT} = 0.5$ A; CT = 5 A

Differential trip slope 1 : 10%

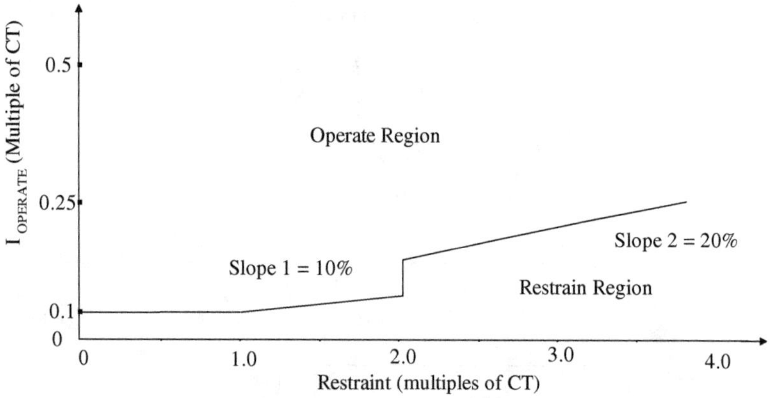

Fig. 4.94 Operating characteristic for the 87GRP differential relay.

Differential trip slope 2 : 20%
Differential trip delay: 0 cycles

Biased Differential Operating Characteristics of the 87GRS Protection System The characteristics of the differential element of secondary numerical relay 87GRS are shown in Fig. 4.95. The secondary differential element is set as follows:

CT ratio: 3000 : 5
Generator differential I_{S1}: 0.25 A
Generator differential K_1: 0%
Generator differential I_{S2}: 5 A
Generator differential K_2: 120%

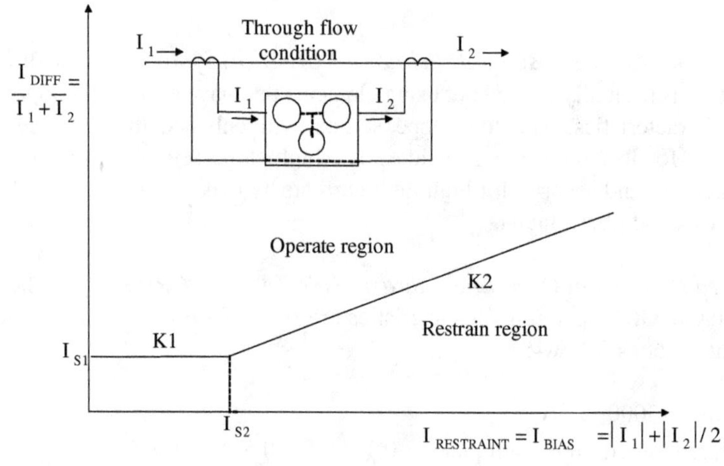

Fig. 4.95 Operating characteristic for the 87GRS differential relay.

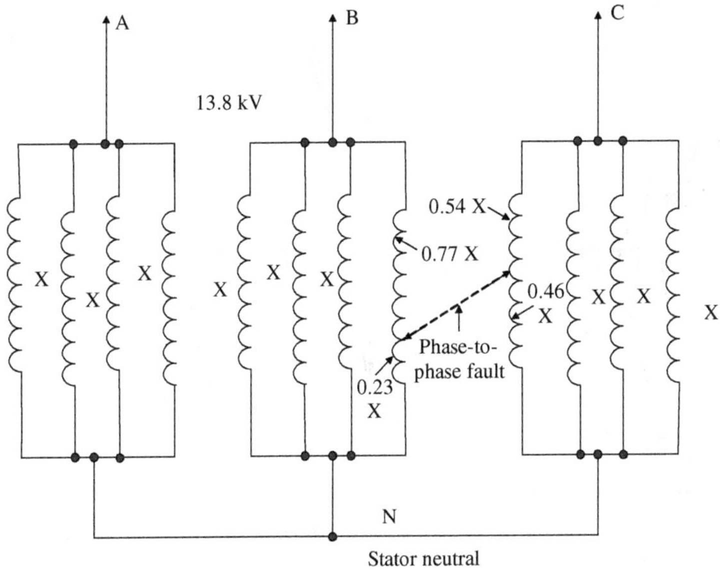

Fig. 4.96 Unit stator winding showing the generator phase A-B fault location.

Description of the Incident A fan blade mounted at the bottom of the unit rotor became detached and hit the stator winding, causing a generator stator phase-to-phase-to-ground fault. The primary and secondary differential relays detected the fault and tripped the associated breakers and the emergency lockout relays. An emergency shutdown was initiated and unit-associated CO_2 was released to prevent further stator winding damage.

Description of the Stator Windings As shown in Fig. 4.96, each stator phase consists of four parallel paths. Each path consists of 35 coils. Each coil has three turns and each turn has 11 layers. Due to the rotor fan failure, phase B was hit at 23% from the neutral side and phase C was hit at 46% from the neutral side. A phase-to-phase fault occurred when the rotor fan blade connected the two coils with the stator iron. The fan blade cut five layers from the winding turn. Short-circuit study simulation of this type of fault is not easy, and analysis of a combination of machine output and fault current requires plotting of individual currents.

Sequence of Events

1. At zero time, a phase B-C-g fault occurred when the fan blade hit the windings and connected the two coils.
2. At 3.25 cycles, phase A of CB A1 interrupted the flow of current between the generator and the system, as shown in Figs. 4.97 to 4.99.
3. At 3.5 cycles, phases B and C of CB A1 interrupted and cleared the system connection to the machine, as shown in Figs. 4.97, 4.100, and 4.101.

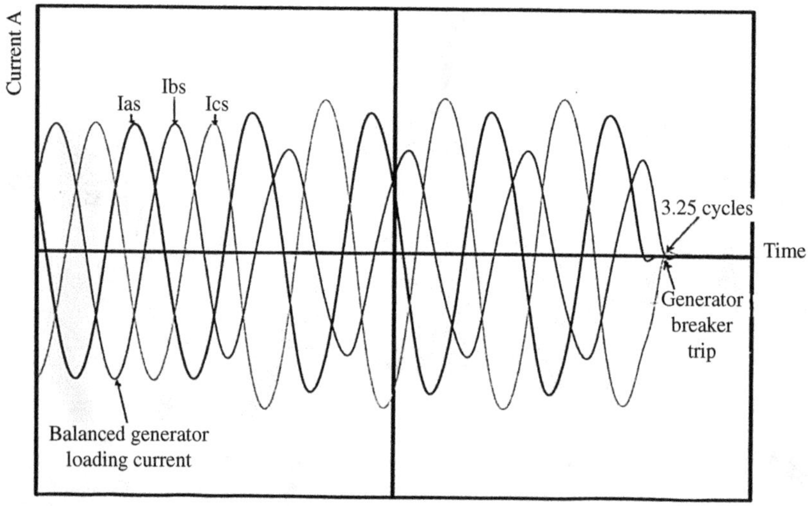

Fig. 4.97 Numerical relay 87GRS fault record showing the three-phase currents for the generator system side during the fault.

4. From 3.5 cycles to 8 s the unit decayed energy kept feeding the initial B-C fault for an additional 8 s until the energy was fully dissipated in supplying the fault and other machine coast-down losses.

Analysis of the Secondary Relay Oscillography Fault Records Analysis of the fault record for a differential relay can be useful when done on a per-phase basis.

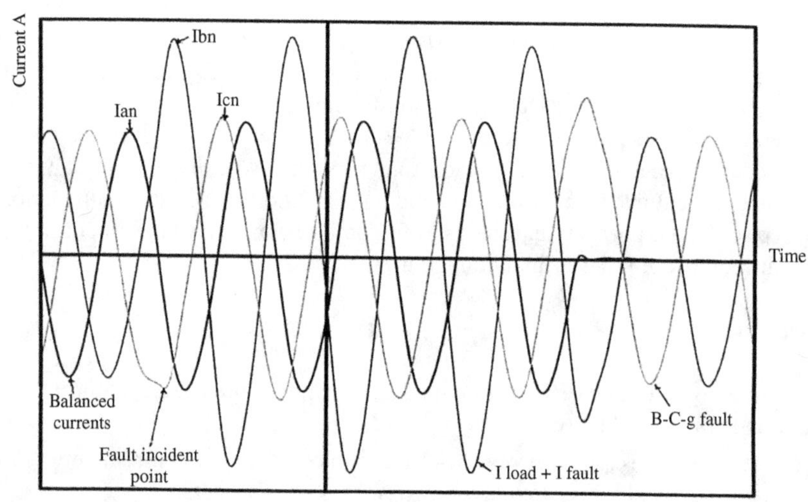

Fig. 4.98 Numerical relay 87GRS fault record showing the three-phase currents for the generator neutral side.

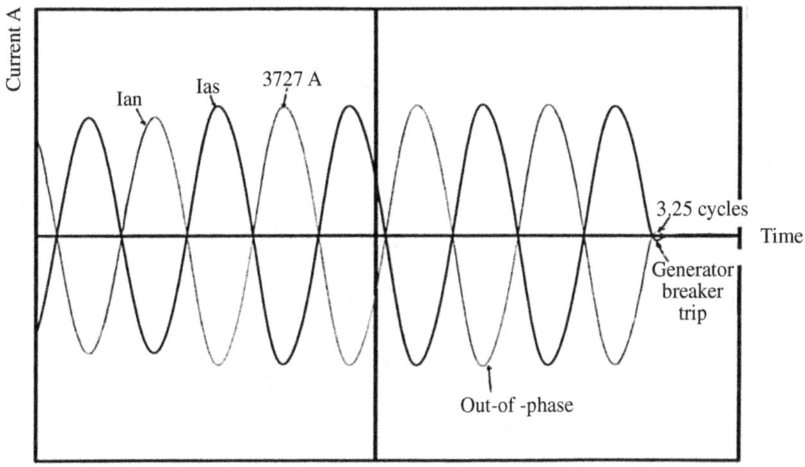

Fig. 4.99 Relay 87GRS fault record with phase A secondary currents for the generator neutral and system sides.

Since the fault involves phases B and C only, phase A of the generator system side is equal and opposite to the generator neutral side, as shown in Fig. 4.99. The sum of these two currents represents the operating differential relay current.

This component is zero for phase A current, indicating that phase A current is related primarily to unit output feeding the system load. The relay system is designed to have $I_2 = -I_1 = -I$ for through power flow and external fault conditions. Therefore, Fig. 4.99 shows two identical currents, with one of them reversed for phase A, providing $I_{\text{diff}} = I_1 + I_2 = 0$. Using this fact, the fault can best be analyzed by making plots for individual generator phases for each pair of neutral and system differential relay inputs, as shown in Figs. 4.99, 4.100, and 4.101. The three-phase currents from the generator's neutral side will be compared with those from the system side to deduce the relay differential current and hence the correct sequence of events. Therefore, the analysis can be simplified because there is no need to analyze the relay voltages for fault classification. Operation of the generator neutral 60-Hz overvoltage element of the 87GRP confirms that the B-C fault involved the stator grounded frame. Operation of both generator neutral overvoltage relays is analyzed in detail in the relay system response section.

Analysis of the Fault Incident Point The fault incident point can be determined through a half-cycle comparison for currents. A half-cycle can be declared as containing a fault incident point if the time between its associated zero crossing is different from (less than or greater than) that of the preceding half-cycle. In Fig. 4.100 the Ibs current hashed half-cycle (negative) has a zero crossing at points a and b. The width of the half-cycle time between a and b is smaller than that of the preceding unhashed half-cycle (positive). In addition, for current Ibn the hashed half-cycle area (positive) width of ac is larger than that of the preceding cycle

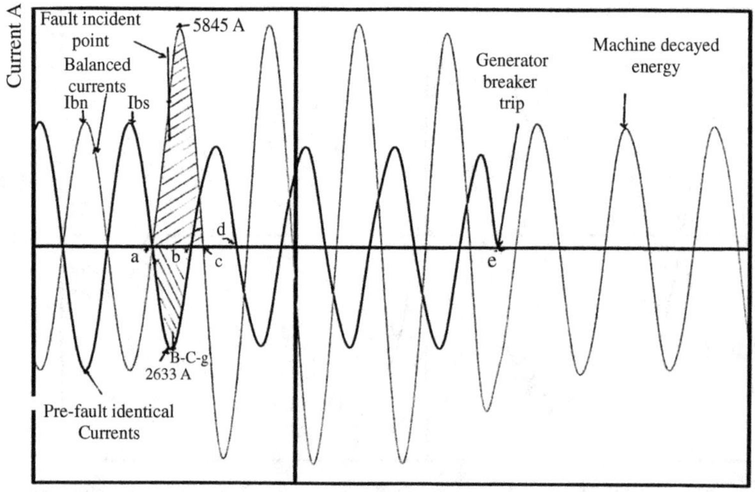

Fig. 4.100 Relay 87GRS fault record with phase B secondary currents for the generator neutral and system sides.

(negative). This implies that a phase angle shift has occurred for currents Ibn and Ibs in the overlapping hashed areas. As a result, the fault incident point can be deduced and is shown in Fig. 4.100.

Similarly, in Fig. 4.101, the Ics current hashed half-cycle (positive) has a zero crossing at points a and b. The width of the half-cycle time between a and b is smaller

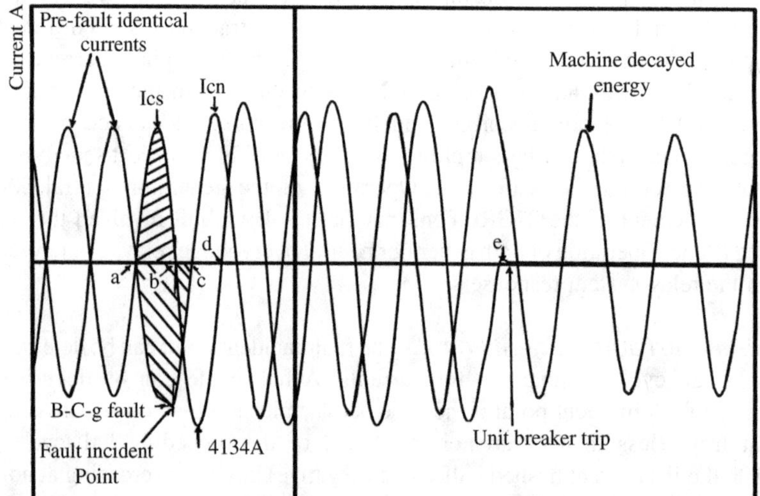

Fig. 4.101 Numerical relay 87GRS oscillography fault record with phase C currents for the generator neutral and system sides.

than that of the preceding unhashed half-cycle (negative). In addition, for the current Icn the hashed half-cycle (negative) area width of ac is larger than that of the preceding cycle (positive). This implies that a phase angle shift has occurred for currents Icn and Ics in the overlapping hashed areas. As a result, the fault incident point can be deduced and is shown in Fig. 4.101. By comparing the time distance between the fault incident point and the trigger line in Figs. 4.100 and 4.101, it can be deduced that a simultaneous B-C-g fault has occurred (equal time distances).

Analysis of the Numerical Relay Event Fault Records Analysis of the incident will also be based on the fault records of the generator primary and secondary numerical relays. The fault records reflect the resulting currents of the machine output power and phase-to-phase-to-ground fault. Figures 4.97 and 4.98 illustrate the three phase currents for the system and generator neutral sides, respectively.

Fault Record for the 87GRS Protection System The following information was recorded by the generator secondary numerical relay as part of the relay event record.

Neutral-side current contribution:

$$I_{\mathrm{an}} = 2634\ \mathrm{A}; I_{\mathrm{bn}} = 4134\ \mathrm{A};\ I_{\mathrm{cn}} = 2645\ \mathrm{A}$$

System-side current contribution:

$$I_{\mathrm{as}} = 2627\ \mathrm{A};\ I_{\mathrm{bs}} = 1884\ \mathrm{A};\ I_{\mathrm{cs}} = 2924\ \mathrm{A}$$

Differential currents:

$$I_{\mathrm{a}}\ \text{differential} = 26\ \mathrm{A};\ I_{\mathrm{b}}\ \text{differential} = 2999\ \mathrm{A};\ I_{\mathrm{c}}\ \text{differential} = 2994\ \mathrm{A}$$

Terminal voltages:

$$V_{\mathrm{an}} = 9027\ \mathrm{V};\ V_{\mathrm{bn}} = 7691\ \mathrm{V};\ V_{\mathrm{cn}} = 6473\ \mathrm{V};\ V_{\mathrm{n}}\ \text{measured} = 1154\ \mathrm{V}$$

Output power:

$$\text{three-phase watts} = 66.2\ \mathrm{MW};\quad \text{three-phase VAR} = 20.7\ \mathrm{MVAR};$$
$$\text{three-phase PF} = 0.954$$

Relay output:

Unit emergency shut down: on
General trip: on
Release of CO_2: on
Differential trip: on

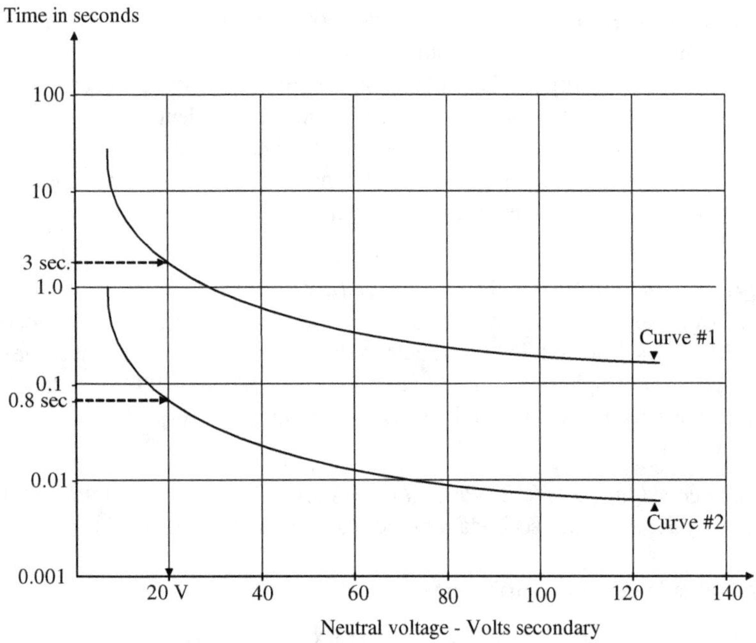

Fig. 4.102 Setting curves for the primary and secondary 59N elements.

Analysis of the Secondary Relay System 87GRS Response During the Stator Phase-to-Phase-to-Ground Fault The secondary numerical relay issued trip output only through the differential elements on phases B and C. No generator neutral overvoltage element on the secondary was recorded. This was due to a longer time-delay setting for the element, as shown in curve 1 of Fig. 4.102. The operation of the secondary differential numerical element can be examined based on the fault record described above and shown in the figure:

Examination of phase B differential operation:

Neutral current contribution, I_{bn}: 4134 A primary
System current contribution, I_{bs}: 1884 A primary
Differential current: 2999 A primary

Convert to secondary values by dividing by the CT ratio of 600:

Neutral current contribution, I_{bn}: 6.89 A secondary
System current contribution, I_{bs}: 3.14 A secondary

Differential current: 5 A secondary

$$I_{bias} = I_{restraint} = \frac{1}{2}(I_{bn} + I_{bs}) = \frac{6.89 + 3.14}{2} = 5.01 \text{ A}$$

Based on the relay characteristics and setting shown above, since $I_{\text{bias}} = 5.01$ A is greater than $I_{S2} = 5$ A, the relay will operate if I_{diff} is equal or greater than $(k_2 \times I_{\text{bias}}) - (K_2 - K_1)I_{S2} - I_{S1}$. By substituting the setting parameters provided above and the fault record currents, the

value of $(k_2 \times I_{\text{bias}}) - (K_2 - K_1)I_{S2} - I_{S1} = (1.2 \times 5.01) - (1.2 - 0)5 - 0.25 = 0.13$ A

The differential current of 5 A is far greater than the calculated expression of 0.13 A. This implies a relay operation.

Analysis of the Primary Relay System 87GRP Response During the Stator Phase-to-Phase-to-Ground Fault The primary numerical relay issued trips by the differential trip on phases B and C as well as by the generator neutral overvoltage element. The primary system neutral overvoltage relay was set at a faster time delay, shown as curve 2 in Fig. 4.102. The neutral overvoltage element responds to the fundamental frequency (60-Hz) voltage at the generator neutral side and normally provides ground fault protection for approximately 95% of the stator winding. The primary relay fault record indicated a primary neutral voltage of 1154 V. The secondary neutral 60-Hz voltage $= V_{\text{pri}}$/PT ratio $= 1154 \times (240/13{,}800) = 20$ V secondary. From the neutral overvoltage element operating curves of Fig. 4.102, the primary relay system operated in 0.8 s, while the secondary system required 3 s.

Power System Phenomena

1. Unit decayed energy feeding the initial fault and lasting for more than 8 s
2. Slow creation of a fault (fan blade movement), leading to arc-over at the voltage peak and resulting in symmetrical fault currents containing no dc offset, as shown in fault records

Corrective Actions

1. Based on the failure analysis, it was concluded that the rotor fan blade failed in high cycle fatigue (approximated as millions of cycles). A high vibration level due to centrifugal force occurring for many cycles, and cracking under the bolt, may have caused fatigue in the fan blade, forcing it to separate. Therefore, all existing fan blades (more than 150 blades) were replaced. The old blades need to be analyzed further to learn more about the failure mode.
2. The winding damage was repaired, thus avoiding the cost of rewinding the generator stator, and a high potential test was administered to the unit.
3. Based on the fast operation of the stator ground primary system and the limitation of stator ground fault current to 15 A by the use of high-resistance grounding, the time setting of the secondary system was altered to be identical

to that of the primary system. This will provide better coordination with the secondary circuit fuses of the PTs.

Lessons Learned

1. Analysis of complicated faults for generators, during their delivery of power, can be simplified by creating plots of neutral and system generator current inputs to the stator differential protection on a per-phase basis.
2. The use of multifunction generator numerical protection technology can provide a wealth of information in their fault and oscillography records, which simplifies the analysis.
3. The analysis of generator winding faults can become complicated in the presence of generator output power.

Case Study 4.16: Undesired Tripping of a Pump Storage Plant During a Close-in Phase-to-Ground 345-kV Line Fault

Abstract A line-to-ground fault occurred on one of the 345-kV lines close to a generating plant. The fault was caused by a flashover of one of the transmission-line insulators. The fault was cleared successfully at both ends of the line. A few cycles later the generating plant's four units tripped. In this case study we analyze the cause of the undesired trips of the plant units and analyze the oscillograms and fault calculations to confirm the analysis. In addition, we describe system phenomena, corrective actions, and lessons learned. We show how to combine the loading current with the fault current to examine the relay response to the resulting current.

Description of the Power System and Associated Protection Figure 4.103 shows the system one-line diagram and the oscillograph-monitored voltages and currents at substations X and Y. The hydro plant consists of four units, each rated at 280 MVA and 16.2 kV. The plant is normally run in the generation mode during the day, to reduce the load curve cycle. At night, with available surplus network power, the plant is put in the pumping mode. In this mode, the unit rotation is reversed by interchanging phases B and C using a motorized reversing switch. Each unit is connected to the 345-kV system via a dedicated unit step-up transformer and a 345-kV cable. The combination of two unit transformers is connected at a single point to a conventional breaker-and-a-half 345-kV substation. The plant and its associated substation are connected to the system via four 345-kV overhead lines: L1, L2, L3, and L5.

The plant is protected by dual relaying systems. However, apparently, when the plant was designed in the early 1970s, an instantaneous time overcurrent relay designated as 51M was added for the protection of each generator and activated only during the pumping mode, thus treating the unit as a motor and ignoring the fact that this element is in addition to the existing redundant conventional generator protection systems.

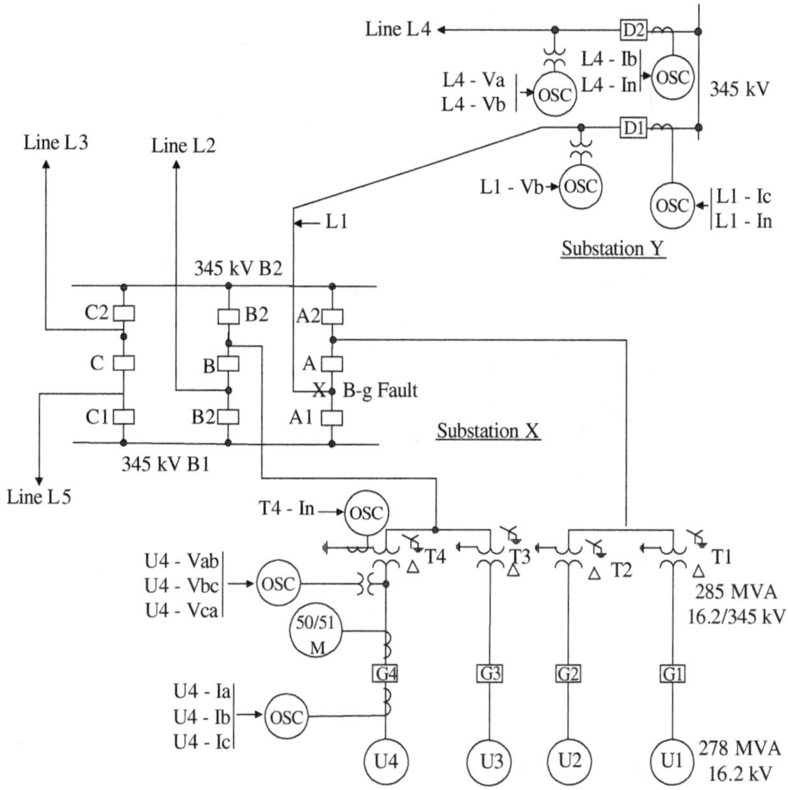

Fig. 4.103 One-line diagram showing oscillograph-monitored voltages and currents.

Analog Quantities Monitored by the Plant Oscillograph The plant oscillograph monitors each unit's three-phase currents and line-to-line voltages. In addition, it monitors the neutral current of each step-up transformer. Figure 4.103 shows the monitoring of U4, which is typical for the remaining three units, as well as substation Y oscillograph monitoring for line L1 phase C and neutral currents and phase-B-to-neutral voltage. In addition, the substation oscillograph monitors line L4 phase B and neutral currents and phases A and B to neutral voltages. Other lines are not shown, for reasons of simplicity.

Analysis of the Recorded Oscillograms The sequence of events as reconstructed from an analysis of the hydro plant and substation Y oscillograms shown in Figs. 4.104 and 4.105, respectively, is as follows. A phase B-g fault occurred on 345-kV line L1 close to the switchyard X end of the circuit. The fault occurred at point a on the neutral current trace T4-In in Fig. 4.104. The B-g fault appears as a phase B-C fault at the 16.2-kV generator bus. Two and a half cycles later, at point b in Fig. 4.104, 345-kV CBs A and A1 at substation X were tripped by the primary and secondary relaying systems. Three cycles later (from $t = 0$), at point c in Fig. 4.105, 345-kV CB

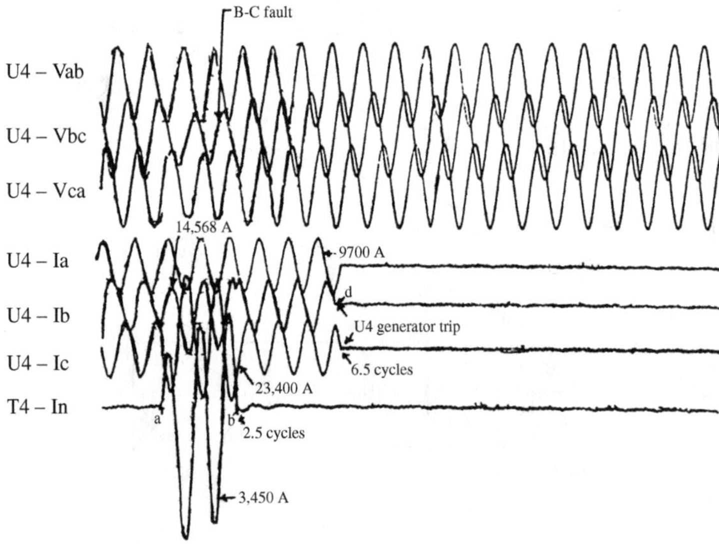

Fig. 4.104 Plant oscillogram fault record showing the B-g 345-kV fault.

D1 at substation Y tripped and cleared the fault. Six and one-half cycles later (from $t = 0$), at point d in Fig. 4.104, 16.2-kV plant breakers G1, G2, G3, and G4 were tripped falsely by operation of the instantaneous units of time overcurrent relay device 51M.

Analysis of the Causes of the Plant Shutdown The false tripping of the four units is attributed to misapplication of the instantaneous element of the 50/51M relay applied on the 16.2-kV unit leads. The B-g fault on the 345-kV switchyard system is seen as a phase B-C fault on the 16.2-kV transformer side, due to the wye/delta

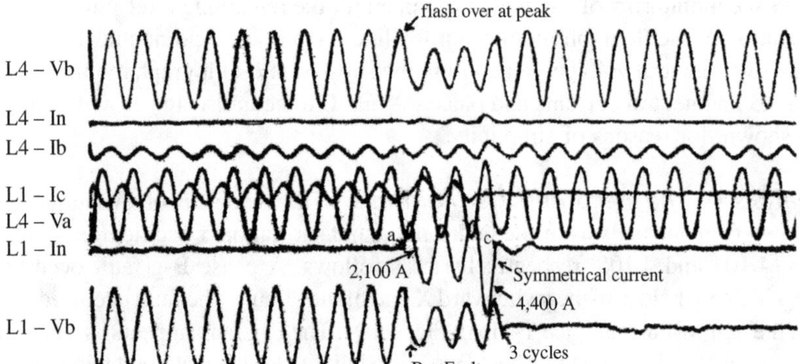

Fig. 4.105 Substation Y oscillogram fault record for the B-g 345-kV fault.

transformation. The fault current in phase C is added vectorially to the load component of the unit (pumping mode), resulting in sufficient current to operate the phase C instantaneous element of the 51M relay. Confirmation of the operation of phase C only is explained in the next section.

Fault Analysis Calculation to Confirm the Recorded Unit Currents and Hence Relay Operation The predisturbance condition at the plant was simulated in the short-circuit study for a close-in phase-to-ground fault at the 345-kV switchyard bus. The study indicates ground current ($3I_0$) contributions of 3600 A from the main transformer neutral and 2059 A from substation Y. The transformer neutral current measured by the plant unit oscillogram is about 3450 A, and the ground current measured for the station Y contribution is 2100 A. The errors between the calculated and measured values are less than 10%. This will confirm that the initial phase B-g fault was close to the plant end of line L1.

Since the fault currents recorded match the currents calculated for a fault located at substation X, analysis to verify operation of the unit 50/51M instantaneous element can be performed using the short-circuit study results. The instantaneous element of overcurrent relays 50/51M on phase C have operated on all four units, which have equal current contributions to the 345-kV fault. Therefore, only one unit can be analyzed. The unit selected for analysis is U4. Since the relays 50/51M are located at the unit terminals, the fault current contribution from the 16.2-kV unit transformer side needs to be calculated. In addition, the loading current effect must be considered when analyzing relay operation.

First we need to calculate the unit full loading (I_{fl}) current in the pumping mode:

$$I_{\mathrm{fl}} = \frac{S}{\sqrt{3} \times V_{\mathrm{L\text{-}L}} \times \mathrm{PF}}$$

$$S = 278\ \mathrm{MVA}$$

$$V_{L\text{-}L} = 16.2\ \mathrm{kV}$$

$$\text{power factor (PF)} = 0.9\ \text{lagging}$$

Therefore,

$$I_{\mathrm{fl}} = \frac{278 \times 10^6}{\sqrt{3} \times 16.2 \times 10^3 \times 0.9} = 10{,}900\ \mathrm{A}$$

Try to use the actual recorded unit current prior to the occurrence of the 345-kV B-g fault.

From the substation oscillogram, the pre-fault load (I_{load}) is $I_{\mathrm{load}} = 9700$ A, and assume a power factor of 0.9 lag: $I_{\mathrm{load}} = 9700$ A at $-25.8°$.

Next, we need to calculate the line-to-ground fault. Since the fault is very close to substation X, a close-in fault (I_{F}) was simulated at X: $I_{\mathrm{F}} = 119.9$ pu (on a

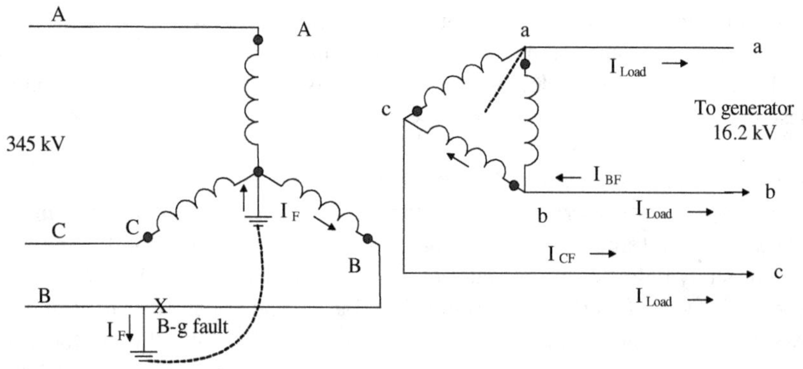

Fig. 4.106 Fault and load current flowing through the generator unit transformer.

100-MVA base), since at the point of the fault, $I_1 = I_2 = I_0 = 111.9/3 = 37.3$ pu. For unit U4, the

$$\text{high-side positive-sequence current}(I_{1\text{HS}}) = I_1 \times (\text{current distribution factor})$$
$$= (37.3) \times \frac{8.14}{110.9} = 2.736 \text{ pu}$$

It can be assumed that $I_{1\text{HS}} = I_{2\text{HS}} = 2.736$ pu.

Since the unit transformer shown in Fig. 4.106 is connected according to NEMA standards, this will imply that for the positive-sequence current, the

$$\text{low-side current}(I_{1\text{LS}}) = \text{high-side current}(I_{1\text{HS}}) \text{ at } -30^\circ \text{ pu}$$

For the negative-sequence current, the

$$\text{low-side current}(I_{2\text{LS}}) = \text{high-side current}(I_{2\text{HS}}) \text{ at } 30^\circ \text{ pu}$$

Since $I_{1\text{HS}} = I_{2\text{HS}} = I_1$, the current in the low-side lead

$$I_{\text{delta}} = I_{1\text{LS}} + I_{2\text{LS}} = I_1 \text{at} -30^\circ + I_1 \text{ at } 30^\circ$$
$$= I_1(I \text{at} -30^\circ + I \text{ at } 30^\circ)$$
$$I_{\text{delta}} = I_1 \times \sqrt{3} = \sqrt{3} \times 2.736 = 4.739 \text{ pu}$$
$$I_{\text{base}} = \frac{100 \text{ MVA}}{\sqrt{3} \times 16.2 \text{ kV}} = 3564 \text{ A}$$
$$I_{\text{delta}} = 4.739 \times 3564 = 16{,}803 \text{ A}$$

For correct phase angle referencing, we start from the fault point. The high-side phase B fault current will lag the faulted phase voltage by 90° (to simplify the

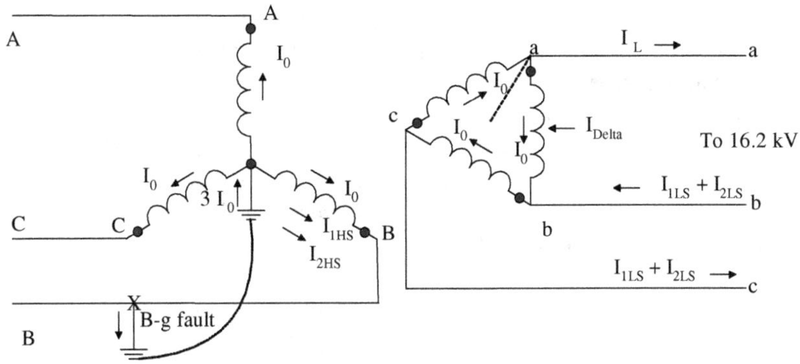

Fig. 4.107 Symmetrical components sequence current flow through the unit transformer during the fault.

calculations). This current is coupled to the low-side delta winding at the B-C leg of the delta. Applying the polarity rule for the system shown in Fig. 4.107: The phase B high-side current flowing out of polarity is substantially in phase with the low-side phase B line current flowing in polarity. Therefore, the low-side phase B current will lead by 90° and phase C will lag by 90°.

$$I_{delta} = I_{pu} \times I_{base} = 4.739 \times 3564 = 16{,}803 \text{ A}$$

$$\text{phase B current in the 16.2-kV lead} = I_{BF} = I_{delta} = 4.739 \times 3564 = 16{,}803 \text{ A at } 90^\circ$$

$$\text{phase C current in the 16.2-kV lead} = I_{CF} = -I_{delta} = 4.739 \times 3563.9 = 16{,}803 \text{ A at } -90^\circ$$

$$\text{total current} = \text{vectorial sum of (fault current + loading current)}$$

For phase B:

$$\text{total phase B current in the 16.2-kV lead} = I_B = I_{BF} + I_{load} = 16{,}803 \text{ A at } 90^\circ + 9700 \text{ A at } -25.8^\circ = 8730 + j12{,}581 = 15{,}313 \text{ A}$$

For phase C:

$$\text{total phase C current in the 16.2-kV lead} = I_C = I_{CF} + I_{load} = 16{,}803 \text{ A at } -90^\circ + 9700 \text{ A at } -25.8^\circ = 8730 + j21{,}025 = 22{,}765 \text{ A}$$

The relay is set at 8 A of secondary current. With a CT ratio of 2400 : 1, the setting is 19,200 A primary. From the analysis above, the phase C current of 22,765 A has exceeded the 51M setting of 19,200 A. The undesired trip of phase C of only the 50/51M relays on all the units is confirmed by the calculation shown, which is typical for the remaining three units.

Comparison of the Calculated and Measured Phase B and C Currents

From the plant oscillogram shown in Fig. 4.104, the phase B and C currents recorded are 14,568 A and 23,400 A, respectively. The currents calculated are 15,313 A for phase B and 22,765 A for phase C. The percentage error is defined as

$$\%\ \text{error} = \frac{100 \times (\text{current measured} - \text{current calculated})}{\text{the smaller of the two}}$$

$$\%\ \text{error for phase B} = \frac{100 \times (14{,}568 - 15{,}313)}{14{,}568} = -5.1\%$$

$$\%\ \text{error for phase C} = \frac{100 \times (23{,}500 - 22{,}765)}{22{,}765} = 3.2\%$$

Since the error is within the tolerance of $\pm 10\%$, the values calculated are O.K.

Corrective Actions Remove the instantaneous element of relay 51M from service on all four units according to the schematic shown in Fig. 4.108. The units are protected by redundant differential relaying as well as being protected against abnormal operating conditions.

Lessons Learned

1. Instantaneous relays can be used to protect synchronous machines in their motoring mode only, if they can be set above motor backfeeding into system faults. This is a synchronous machine that is protected like any generator of

unit transformer during the fault

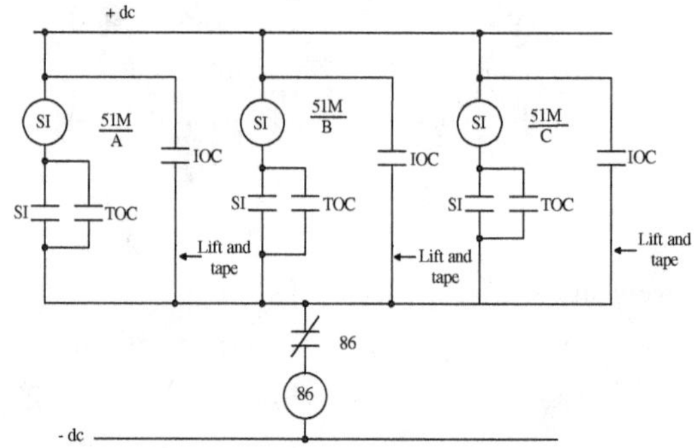

Fig. 4.108 DC schematic for time overcurrent protection for the unit during the pumping mode.

conventional size, with little or no benefit added by this instantaneous TOC relay element.

2. The effect of loading currents on fault analysis and examination of relay behavior cannot always be ignored. The combination of fault and loading currents can lead to an undesired relay outcome, as illustrated in this case study.

Case Study 4.17: Tripping of an 800-MW Plant and the Associated EHV Lines During a 345-kV Bus Fault

Abstract A phase-to-ground fault occurred on the 345-kV bus of plant X. Apparently, the ground fault was caused by flashover to ground during adverse weather conditions. The 800-MW unit decayed energy lasted for several seconds and caused the phase-to-ground fault to evolve into a phase-to-phase-to-ground fault. The flashover location was later determined to be the B and C phases of the 345-kV bus potential transformers. In this case study we provide fault analysis, power system phenomena, and illustrate the disadvantages of employing a single-bus configuration for generator connection.

Description of the Power System and Associated Protection A 800-MW fossil plant generator step-up transformer (GSU) is connected to the system via a single bus connecting the two 345-kV lines (Fig. 4.109). The generator and the unit transformer are protected individually by differential relays. An overall harmonic-restraint differential relay covers the generator, the GSU transformer, and the single 345-kV bus, as shown in Fig. 4.110.

Sequence of Events Based on Analysis of DFR and Oscillograph Records The sequence of events as constructed from an analysis of the plant DFR of Figs. 4.111 and 4.113 and the adjacent substation Y oscillogram of Fig. 4.112 is as follows. At time $t = 0$; a phase B-g fault occurred on the 345-kV bus. Figure 4.113 reveals that the fault incidence point is at the voltage peak, indicating a slow mechanism insulation failure, leading to flashover to ground during the storm. Three cycles later, the 345-kV CBs A1 and A2 at the plant were tripped by transformer differential and unit overall differential relaying via respective lockout relays. At this time the plant was isolated from the 345-kV system. Three to twenty-one cycles later, the unit continues to feed the initial B-g fault via decayed unit energy. The B-g fault on the 345-kV side is seen as a B-C fault on the 26-kV generator side. Twenty-one cycles from $t = 0$, the sustained unit decayed energy through the ground fault has created an ionized cloud that caused the initial B-g fault to evolve into a B-C-g fault, as shown in Fig. 4.111. The unit trapped energy continues to feed the B-C-g fault for 21 to 340 cycles. At 340 cycles, the unit current contributions to the fault ceased.

Analysis of the Performance of the Relay Systems All relay systems operated properly to clear the initial ground fault. Tripping the 345-kV breakers A1 and A2 in 3 cycles by harmonic-restraint differential relays is a fast recorded total clearing time

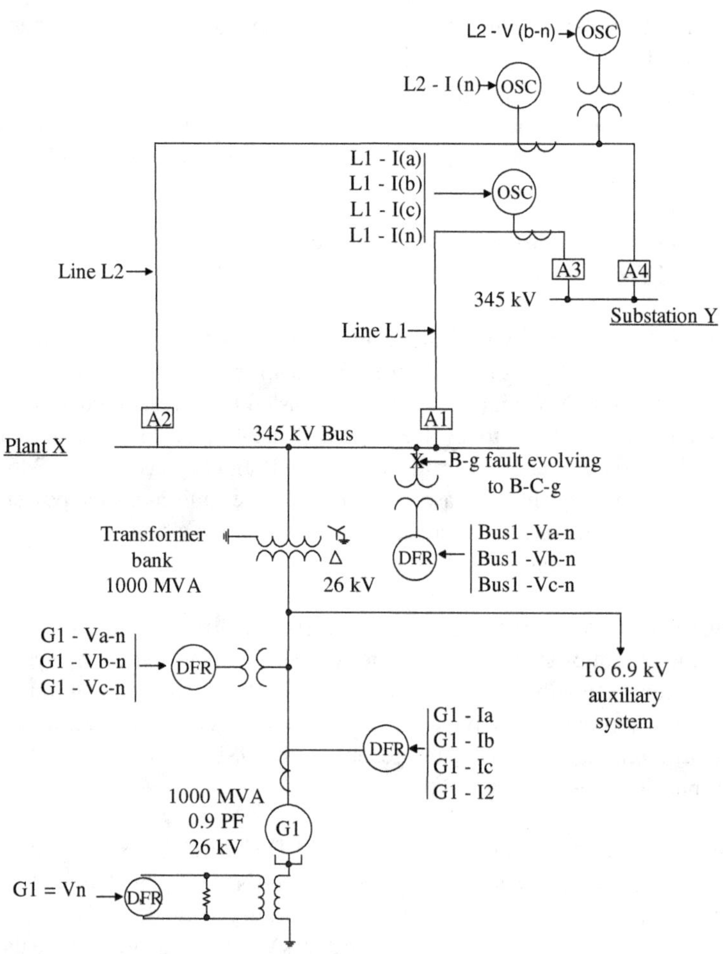

Fig. 4.109 One-line diagram with DFR- and OSC-monitored currents and voltages.

for the harmonic-restraint electromechanical transformer differential relay. The recorded total clearing time consists of the differential relay operating time + lock-out relay pickup time + circuit breaker interrupting time. The generator neutral trace G1-Vn reveals normal third-harmonic voltage during the pre-fault period. The trace then shows the 60-Hz component superimposed on the third harmonic due to unit transformer interwinding capacitance coupling to the high-side 345-kV ground fault.

Power System Phenomena

1. A fault incident point occurring at the voltage peak, indicating a slow mechanism insulation failure leading to flashover to ground as shown in trace Bus1-Vb-n in Fig. 4.113.

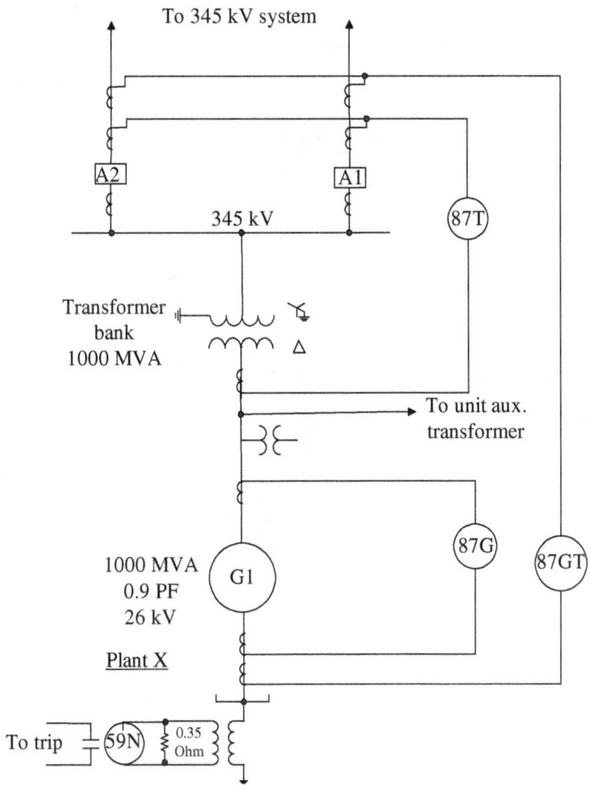

Fig. 4.110 One-line diagram for plant protection with discrete relays.

2. Unit decayed energy feeding the fault after shutdown and separation from the 345-kV system, which lasted for several seconds, as shown in generator current traces in Fig. 4.114.
3. High-side phase-to-ground fault seen as a low-side phase-to-phase fault.
4. Evolving fault to B-C-g occurring due to the spread of the ionized cloud caused by the initial B-g fault as shown in traces Bus1-Vb-n and Bus1-Vc-n in Fig. 4.111.
5. Coupling of high-side zero-sequence voltage to generator neutral. As shown in trace G1-Vn in Fig. 4.113, the pre-fault neutral voltage consists only of third-harmonic voltage, and during the fault the fundamental (60-Hz) zero-sequence voltage is superimposed on the third-harmonic voltage.

Lesson Learned Connecting the generating unit to the system via single bus configurations affects the availability of the unit during single contingency system faults.

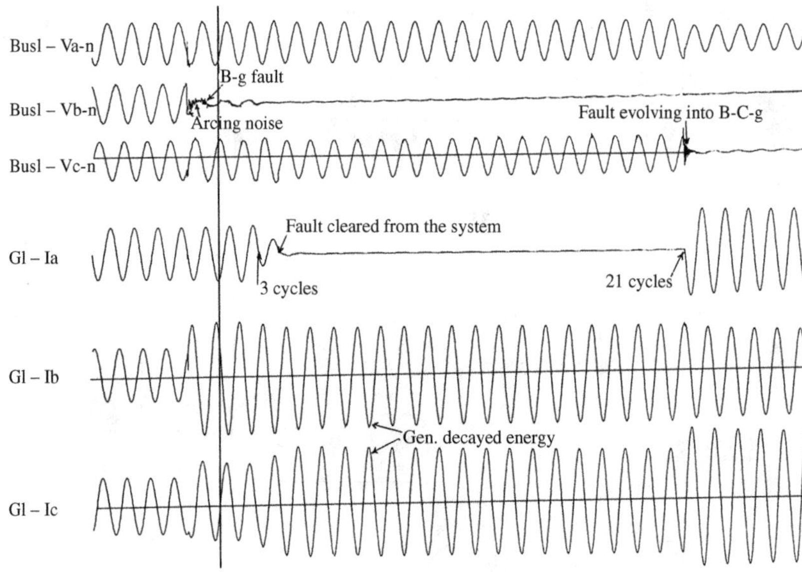

Fig. 4.111 Plant X DFR record showing the 345-kV B-g fault evolving to B-C-g.

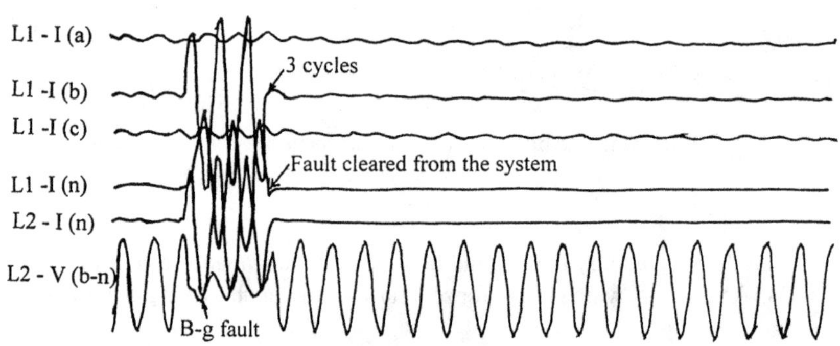

Fig. 4.112 Substation Y oscillogram confirming the 345-kV bus B-g fault.

Case Study 4.18: Tripping of a 150-MW Combined-Cycle Plant During an External 138-kV Three-Phase Fault

Abstract A combined-cycle plant rated at 150 MW was tripped during a three-phase system fault. The trip was caused by an undervoltage relay which operated instantly (without time delay) and initiated the plant shutdown. An undervoltage relay was installed by the gas turbine manufacturer at the 480-V station service supply to the combustion turbine generator (CTG) MCC cabinet. In this case study we explain the incident, misapplication of the undervoltage relay by the manufacturer, system phenomena, corrective actions, and lessons learned.

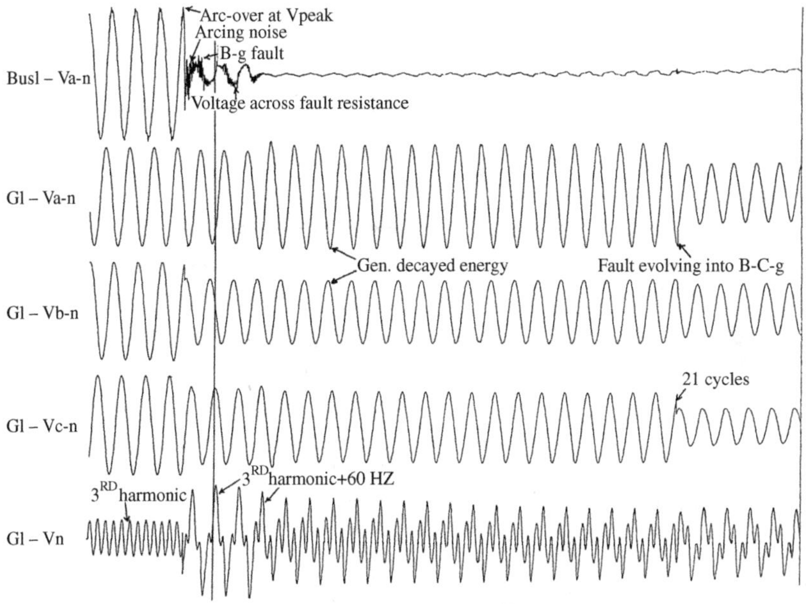

Fig. 4.113 Plant X DFR record showing the 345-kV faulted phase voltage and generator G1 voltages.

Description of the Undervoltage Protection for 480-V System Auxiliaries

Figure 4.115 shows the plant one-line diagram, where the 138-kV system is used as a startup station service source. The 4160- and 480-V station service loads are supplied from the 13.8-kV generator lead through station service transformers. The CTG critical

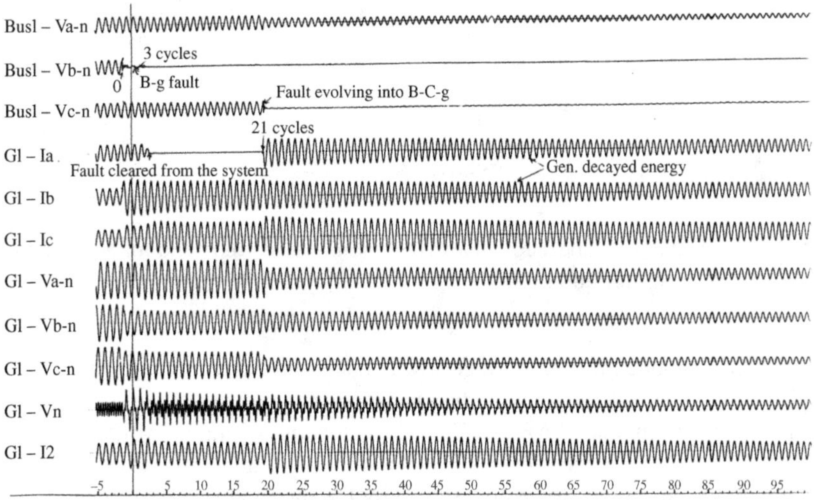

Fig. 4.114 Compressed DFR record showing the unit decayed energy feeding the 345-kV bus B-g fault.

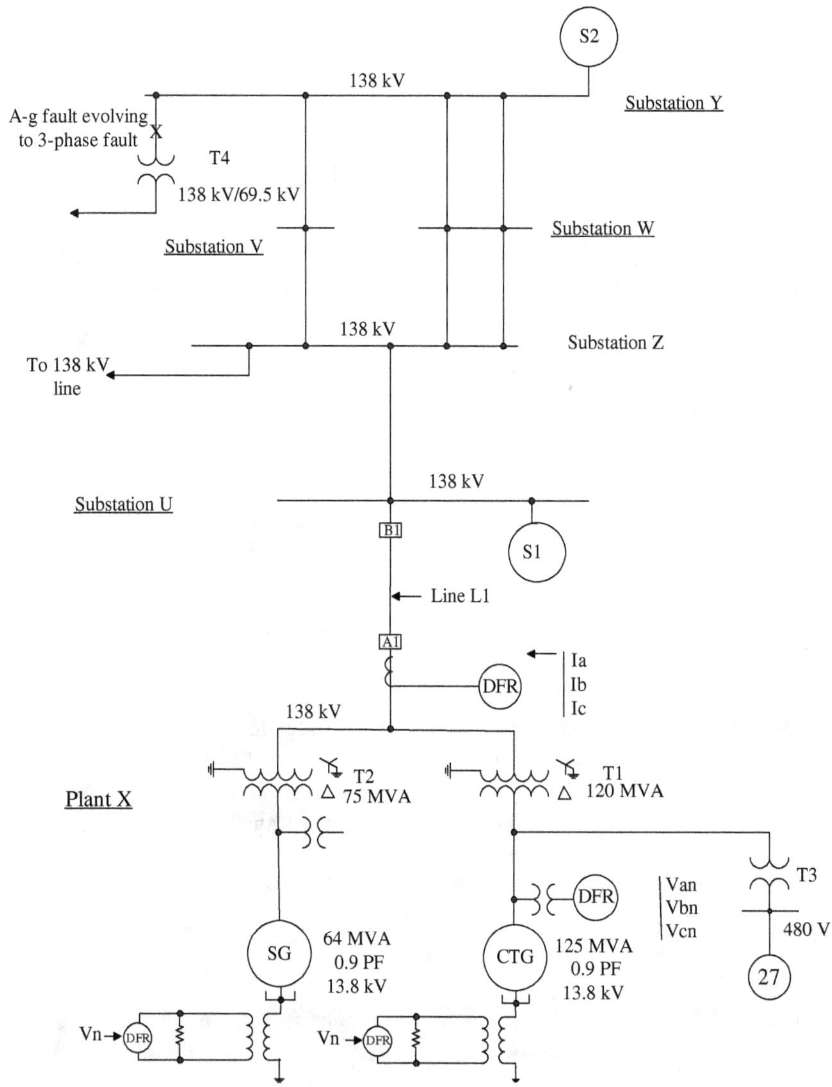

Fig. 4.115 System one-line diagram with DFR-monitored currents and voltages for plant X during the fault at substation Y.

MCC bus is supplied by transformer 3, which steps down the voltage to 480 V. The critical 480-V bus is monitored by two instantaneous undervoltage relay devices 27 connected between phases AB and BC. The detection of an undervoltage condition, when the 480-V bus drops below 80%, will initiate plant shutdown (as designed). This plant trip was recommended and designed by the turbine manufacturer without any delay. The one-line diagram in Fig. 4.115 also shows the DFR-monitored voltages and currents for the 138-kV feeder L1, the CTG, and the steam turbine generator (STG).

Description of the Tripping Incident The 138/69.5-kV transformer bank T4 at substation Y shown in Fig. 4.115 failed, causing an initial A-g fault that lasted for 1 cycle, then evolved into a three-phase-to-ground fault. During the fault, the system voltage at plant X was depressed to a value of 73% of normal. The fault also caused the 480-V auxiliary bus at plant X to decrease to a value lower than the 80% setting value of the undervoltage relays. This activated the undervoltage auxiliary bus trip scheme and resulted in the initiation of plant shutdown.

Analysis of DFR Records to Obtain the Sequence of Events As shown in Fig. 4.116, an initial A-g fault occurred due to a failure of transformer T4 at substation Y. The ground fault lasted for one cycle and is classified based on voltage trace CTG-Va-n going down and current trace L1-Ia going up. A few milliseconds later, the transformer bushing exploded, causing the initial L-g fault to evolve into a three-phase-to-ground fault. The three-phase voltages dipped while the three-phase currents increased during the three-phase fault. The fault lasted for 6 cycles and was cleared by the transformer T4 differential relays at substation Y. The three-phase fault resulted in lower voltage at plant X, which activated an undervoltage relay (27) at the CTG MCC 480-V bus. Relay device 27 operated within a few cycles and activated shutdown of the combined-cycle plant via the plant computer.

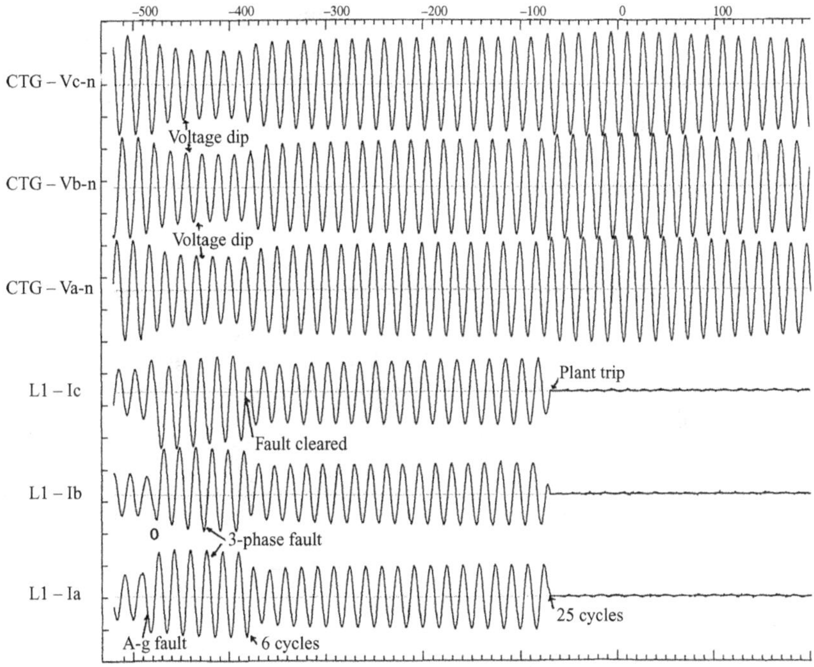

Fig. 4.116 138-kV line L1 currents and CTG generator voltages confirming the plant trip following the external three-phase fault.

The plant 138-kV L1 feeder breaker A1 tripped after an additional 19 cycles (25 cycles total). The plant computer added the extra time delay after initiation of the plant shutdown by the undervoltage tripping scheme.

Confirmation of the Magnitude of the Current and Voltage During the Fault

Measured DFR Currents and Voltages Figure 4.117 shows the time expansion of the faulted voltages and currents. Actual DFR-calculated recordings of voltages and currents are:

$$\text{CTG voltages: } V_A = 6.22\ \text{kV};\ V_B = 6.12\ \text{kV};\ V_C = 6.16\ \text{kV}$$

$$\text{Feeder L1 currents: } I_A = 1063\ \text{A};\ I_B = 1040\ \text{A};\ I_C = 1146\ \text{A}$$

$$\text{average DFR measured CTG voltage} = \tfrac{1}{3}(6.22 + 6.12 + 6.16) = 6.16\,\text{kV}$$

$$\text{average DFR measured L1 current} = \tfrac{1}{3}(1063 + 1040 + 1146) = 1083\ \text{A}$$

This current consists of the sum of the machine load current and three-phase fault current contribution.

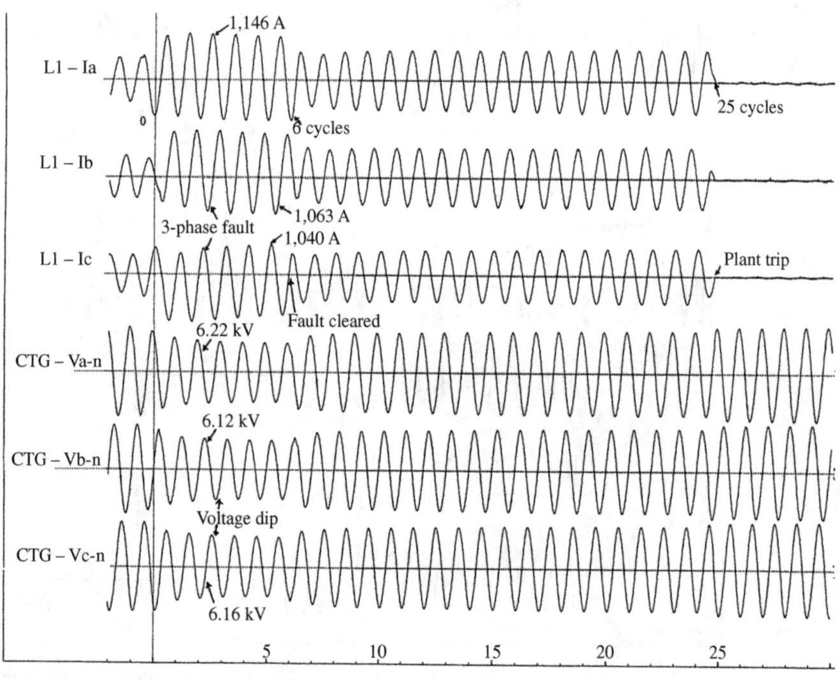

Fig. 4.117 Time expansion of 138-kV line L1 currents and CTG generator voltages.

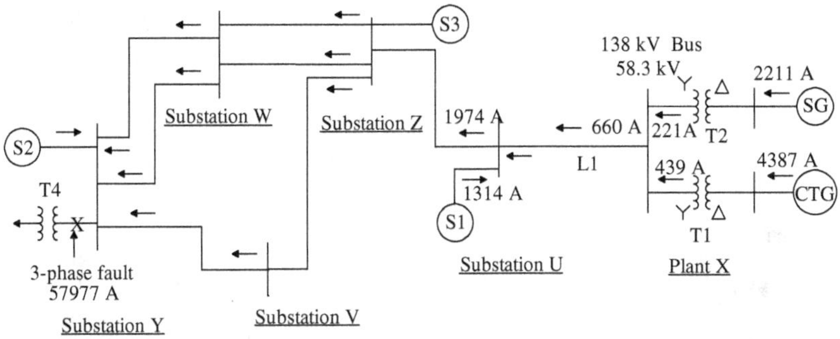

Fig. 4.118 Short-circuit study simulation for a three-phase fault at the Y bus.

Calculated Currents and Voltages From the simulated short-circuit study output shown in Fig. 4.118 for a three-phase fault at substation Y:

$$I_{\text{fault}} = 660\,\text{A}\ at\ -87^\circ = 660(0.0523 - j0.9986) = 34.5 - j659 \qquad (1)$$

For the loading current, the plant was producing 149.4 MW at 0.9 power factor (PF):

$$\text{plant apparent power(MVA)} = \frac{\text{power (MW)}}{\text{power factor}} = \frac{P}{\text{PF}}$$

$$= \frac{149.4}{0.9} = 166\,\text{MVA}$$

The loading current flowing in the 138-kV L1 feeder is

$$I_{\text{load}} = \frac{166 \times 10^6}{\sqrt{3} \times 138 \times 10^3} = 695\,\text{A at } -25.8^\circ \text{ (0.9 PF lagging)}$$

$$= 695(0.9 - j0.435) = 626 - j302 \qquad (2)$$

Using the superposition principle yields

$$I_{\text{total}} = I_{\text{fault}} + I_{\text{load}} = \text{eq. (1)} + \text{eq. (2)}$$

$$= 626 + 34.5 - j(302 + 659) = 660.5 - j961$$

The magnitude of feeder L1 current calculated = 1166 A as compared to the 1083 A current measured.

The CTG voltage calculated is 5.83 kV, compared to the 6.16 kV voltage measured.

The percent error between the measured and calculated values can be defined as

$$\% \text{ error} = \frac{(\text{value calculated} - \text{value measured}) \times 100}{\text{The smaller of the two}}$$

$$\text{current error } (\%) = \frac{1166 - 1083}{(1083)} \times 100 = 7.6\% \text{ less than } 10\% \qquad \text{O.K.}$$

$$\text{voltage error } (\%) = \frac{6.16 - 5.83}{5.83} \times 100 = 5.7\% \text{ less than } 10\% \qquad \text{O.K.}$$

System Phenomena

1. Third-harmonic-generated machine neutral voltage at varying magnitude as shown in traces CTG-Vn and STG-Vn in Fig. 4.119. The third-harmonic voltage profile goes up during the three-phase fault and at the trip of the units. The third-harmonic voltage profile for several seconds is shown in Fig. 4.120, confirming the level variation as a function of unit active and reactive power outputs.
2. Asymmetrical fault current containing dc offset, which normally occurs when the fault incident point is at an angle other than at voltage peak, as shown in traces L1-Ib and L1-Ic in Fig. 4.116.

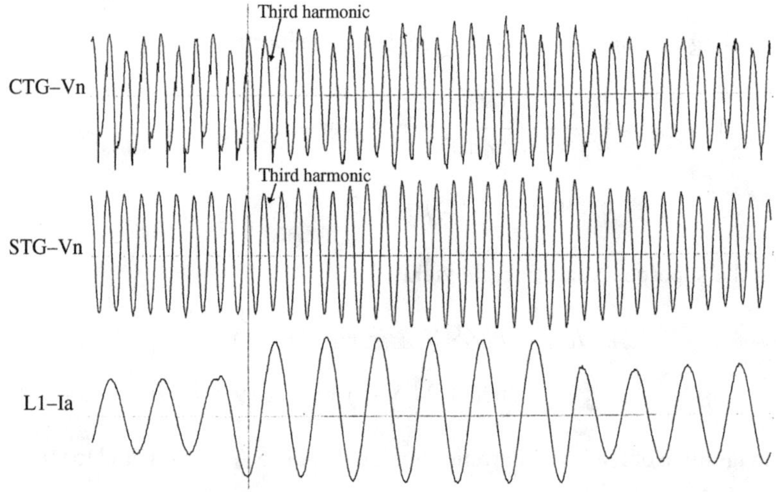

Fig. 4.119 Time expansion of third-harmonic unit neutral-generated voltages.

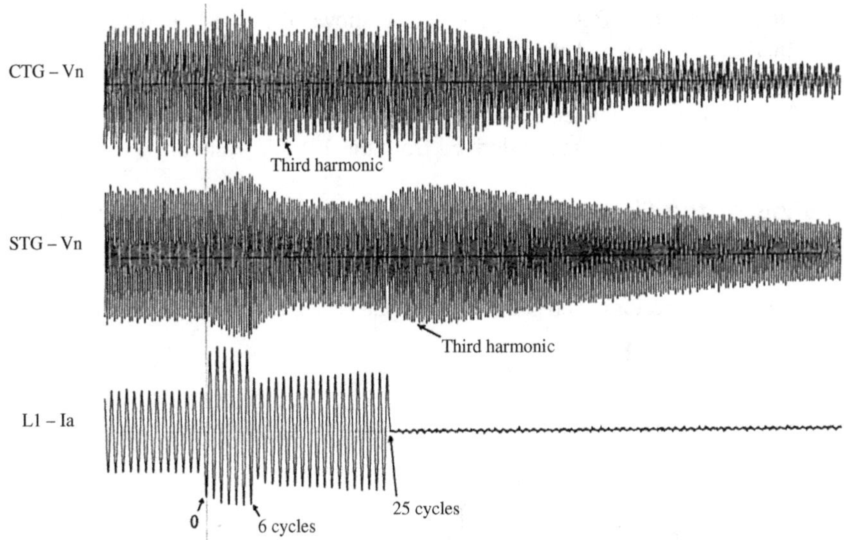

Fig. 4.120 Profile of third-harmonic neutral voltages for the CTG and STG units.

Corrective Actions The output of the undervoltage element should be delayed by 0.5 s to provide coordination with the normally cleared faults and delayed cleared breaker failure faults. This was accomplished by wiring the undervoltage relays output contacts to a separate timer and using the timer output contact to energize the plane shutdown lockout relays.

Lessons Learned The setting of timers associated with the 480-V bus undervoltage condition should be based on how long the critical 480-V motors can tolerate the low-voltage conditions. A margin of 0.3 s should be added to eliminate the reaction to normally cleared system faults. This will require a time-delay total of about 0.5 s.

Case Study 4.19: Tripping of a 150-MW Combined-Cycle Plant During a Disturbance in the 138-kV Transmission System

Abstract A line-to-ground fault occurred in the 138-kV system and evolved into a three-phase fault. The fault was cleared from the system successfully in 5.5 cycles. A combined-cycle plant tripped out of service by reacting to the disturbance. The stable power swing caused by the fault forced the CT gas-fired unit control to react unfavorably and trip the plant upon detection of the fast acceleration. In this case study we provide a sequence of events, power system phenomena, corrective actions, and lessons learned. We also explain the generator active (MW) and reactive (MVAR) power outputs during the three-phase fault and post the clearing of the fault.

Description of the Plant Figure 4.121 shows the one-line diagram of the combined-cycle plant and the surrounding 138-kV transmission system. The CT gas-fired unit (G) is rated at about 100 MW, while the steam turbine unit (S) is rated at 50 MW. The plant is connected to the system via a short 138-kV cable. The plant station service startup source is derived from the 138-kV system.

Analysis of the Plant DFR Record Figure 4.121 shows the system one-line diagram and DFR-monitored voltages and currents. The DFR record in Fig. 4.122 reveals a phase A-g occurring in the system due to the failure of a transformer-mounted surge arrester at substation Y. The phase A-g fault at location F1 lasted for

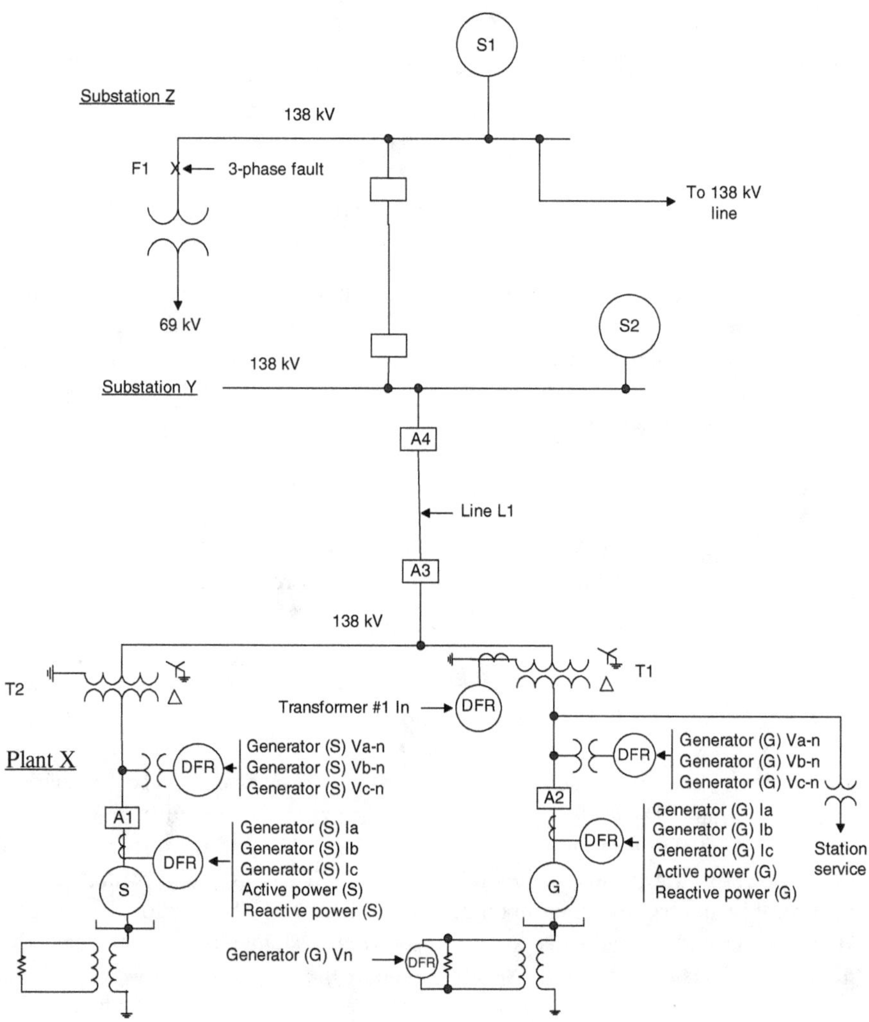

Fig. 4.121 One-line diagram with DFR-monitored currents and voltages.

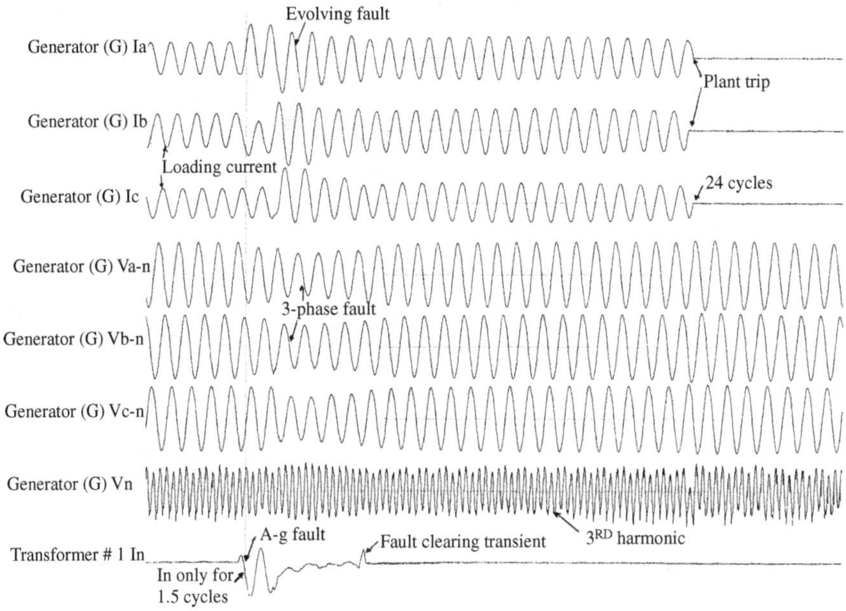

Fig. 4.122 DFR record showing CTG unit voltages and currents during the fault and subsequent unit trip.

1.5 cycles, as shown in the transformer 1 In neutral current trace. The A-g is confirmed by an increase in the phase A current, the presence of neutral current in the transformer neutral trace, and a decrease (dip) in the phase A voltage.

The ionized cloud and initial arcing by-product grew, forcing L-g to evolve into a three-phase fault that lasted for an additional 4 cycles. The three-phase fault is confirmed by the increase in currents and the decrease in voltage (dip) in all the phases. The generator G voltage traces shown in Fig. 4.123 reveal that the plant voltages during the three-phase fault in the 138-kV system dipped to 65% of the normal value. As shown in Fig. 4.124, the gas unit output decreased from 105 MW to 44 MW in 65 ms. Upon clearing of the fault, the machine output went from 44 MW to 133.5 MW in about 120 ms. The CT unit tripped when the rate of change of the power output exceeded the set limit. This rate of change of power (MW/s) exceeded the gas unit control setting and the machine was tripped. At 21 cycles the plant control system initiated shutdown. The 13.8-kV CT unit breaker A1 was opened in 24 cycles, while the steam unit breaker A2 opened in 25 cycles.

Behavior of Generating Units and Their Outputs During System Shunt Faults Figure 4.123 shows that during the three-phase fault, the CT gas unit and steam unit active output (MW) decreased while the reactive power (MVAR) increased to supply the fault current. The CT unit output decreased from its output of 105 MW to a value of 44 MW within 65 ms of the initiation of the fault. While

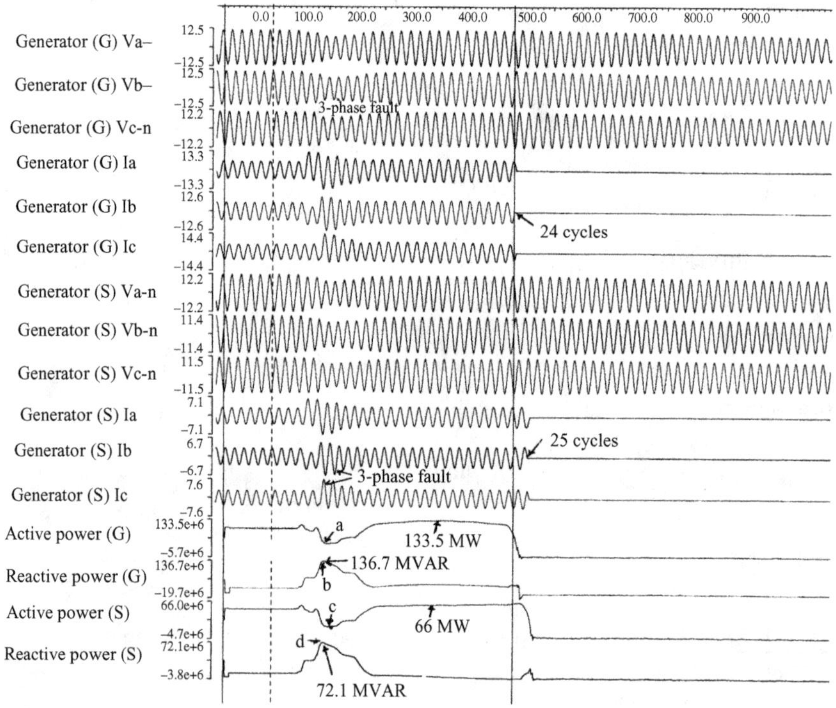

Fig. 4.123 CTG and STG unit voltages, currents, power, and reactive power during the fault and subsequent unit trip.

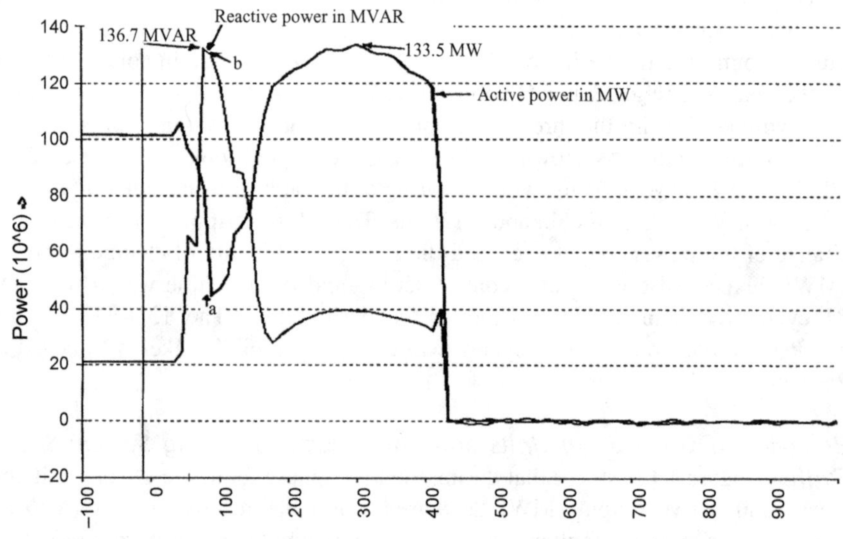

Fig. 4.124 DFR plot for the CTG active and reactive powers.

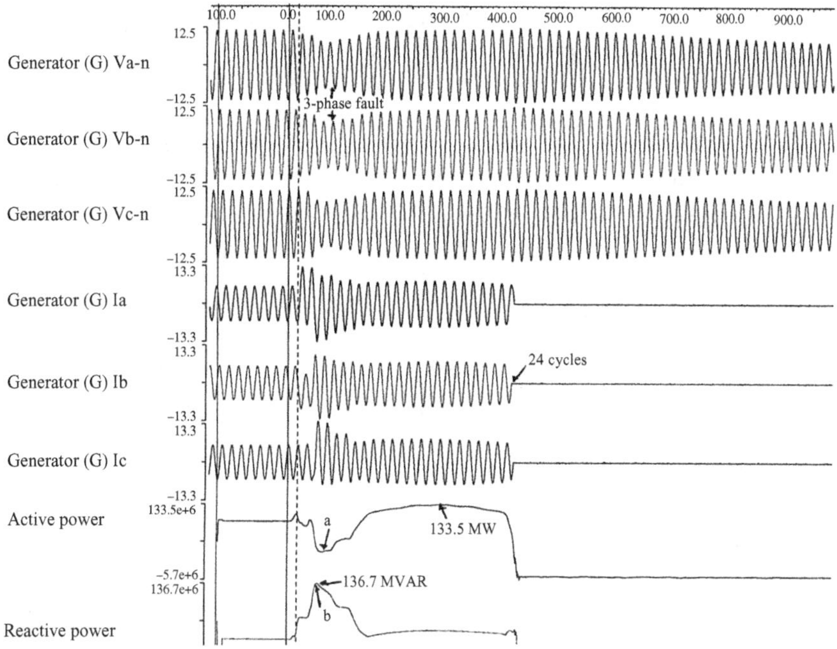

Fig. 4.125 DFR record confirming the three-phase fault and subsequent unit trip.

the MW output decreased at point a of Fig. 4.124, at the same instant the reactive machine output reached a maximum value of 136.7 MVAR at point b, to feed the three-phase fault. Upon clearing the fault, the reactive output decreased and the MW increased to 133.5 MW in about 120 ms, going to zero when the unit tripped. Figure 4.125 shows that during the three-phase fault, the CT unit active (MW) output goes down while the reactive power (MVAR) goes up to supply the fault current.

Power System Phenomena

1. Confirmation through the power flow formula that power flow is proportional to (voltage)^2. When the voltage decreased to a value of 0.65 pu, the power was proportional to the square of the voltage. Therefore, $P = (0.65)^2 \times 100\% = 42\%$ of the unit predisturbance output. Since the unit output prior to the fault was 105 MW, the power output during the fault $= 0.42 \times 105 = 44.1$ MW, which is very close to the DFR calculated value of 44 MW.
2. Variations of third-harmonic neutral voltage as a function of active (MW) and reactive (MVAR) power change in an unpredictable way. As shown in Fig. 4.122, the trace generator (G) Vn reveals higher third-harmonic voltage during lower machine power output.

Corrective Action The manufacturer of the CT unit was approached to analyze the undesired machine trip. The manufacturer presented options to delay the detection algorithm to override the time during successful clearance of the three-phase fault. However, no modification of the control system for the unit upon detection of unit acceleration was recommended. Concern was voiced due to overspeed and turbine flame stability that could occur while a gas unit controller was trying to mediate and reduce the unit MW output versus the unit trip, as occurred in this incident.

Lessons Learned

1. Utility users should be aware of the generating unit control feature set by the manufacturer. In addition, the impact on gas-fired unit controls by system conditions during faults needs to be recognized and understood.
2. Generator controls should be designed to withstand normally cleared system faults.
3. Manufacturers' recommendations for generating units should always be followed, even if this results in overtripping. Apparently, detection of a 20- or 50-MW instantaneous increase in unit output may lead to high acceleration, resulting in overspeed and concern regarding flame stability. This prompted the manufacturer in this case not to change its logic and set to trip the unit for successful high-speed clearing of three-phase faults. According to the unit manufacturer, this acceleration level needs to be eliminated by tripping the unit. No options other than tripping existed.

Case Study 4.20: Undesired Tripping of a 150-MW Combined-Cycle Plant Following Successful Clearing of a 138-kV Double-Phase-to-Ground Fault

Abstract A double-phase-to-ground 138-kV bus fault occurred at substation Y due to falling debris from an adjacent building hitting the high side of one of the yard transformers. The fault was cleared from the system successfully. A few cycles later a 150-MW combined-cycle plant was tripped undesirably by a gas turbine speed sensor that detects machine acceleration. The device activated a lockout relay leading to a "drop load" trip signal. In this case study we describe the transient stability phenomena following the clearing of system faults. We also highlight the fast action of gas turbine controls to successful high-speed clearing of a system fault. In addition, the digital fault record for the double-phase-to-ground fault is shown and analyzed and corrective action is suggested.

Sequence of Events The fault at substation Y started as an A-g bus fault occurring at the voltage peak. The fault evolved into an A-B-g fault a half-cycle later. The bus fault was cleared from the 138-kV system in 4.5 cycles. The combined-cycle plant at

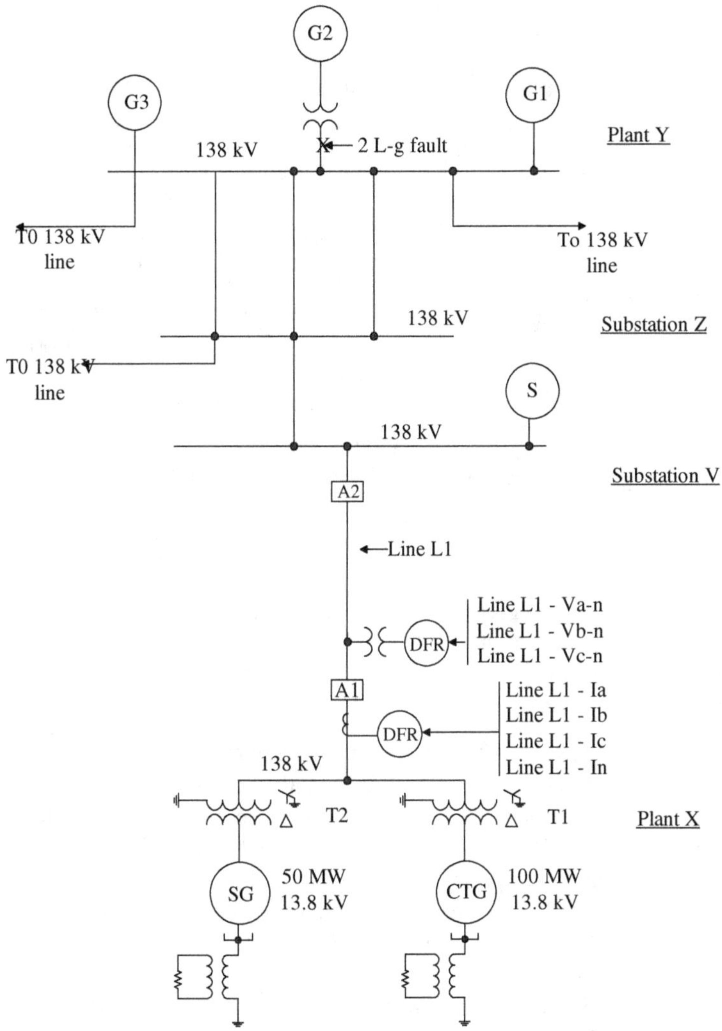

Fig. 4.126 Plant X DFR-monitored currents and voltages.

X tripped undesirably at 24 cycles from initiation of the fault by activation of the machine speed sensors.

Analysis of the DFR Record Figure 4.126 shows the system one-line diagram and line L1 DFR-monitored currents and voltages at plant X. The DFR record in Fig. 4.127 shows that the trace line L1-Va-n indicates a drop in the phase A voltage at point a coupled with an increase in the phase A current in the trace line L1-Ia and the neutral current in the trace line L1-In. This will lead to the classification of an initial phase A-g fault. The fault incident point is at the

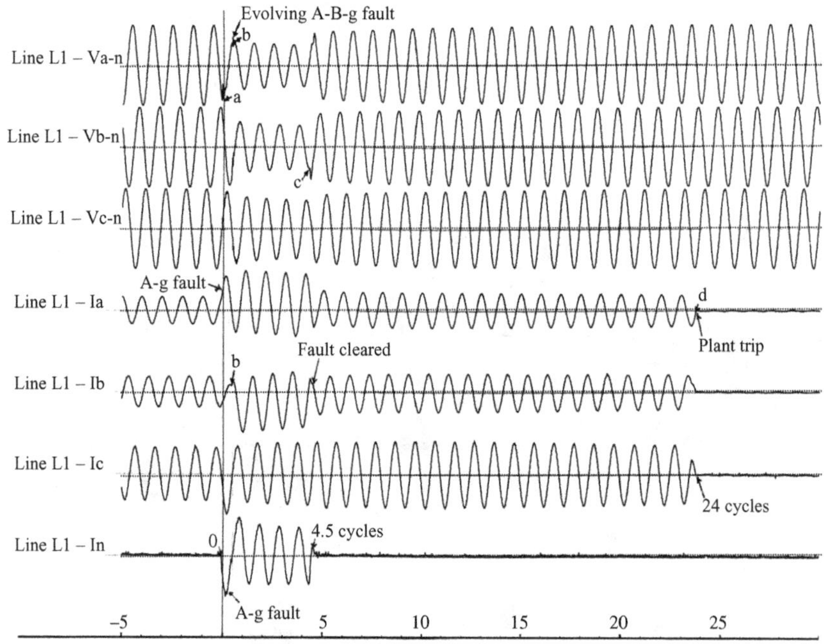

Fig. 4.127 Line L1 voltages and currents during the A-B-g fault and plant trip.

voltage peak, due to the slow fault-creation mechanism, establishing an arc-over at the peak voltage. The trace line L1-Ib reveals a current increase in phase B at point b, and the trace line L1-Vb-n reveals a voltage dip, confirming that the fault has evolved into an A-B-g fault. The fault is cleared from the system in 4.5 cycles (75 ms), as indicated at point c of the trace line L1-Vb-n. The plant was tripped 24 cycles (0.325 s) later from initiation of the fault at point d of the trace line L1-Ia (19.5 cycles from point c).

Analysis of the System Transient Stability *Transient stability* is normally defined as the ability of a power system to remain in synchronism by proper adjustment to sudden large changes caused by faults, load or generation changes, or switching. This can be taken as the ability of the CT gas-fired unit of a combined-cycle plant to remain in synchronism with the system during and after successful clearing of the A-B-g fault described above. In addition, the machine should adjust itself to a new steady-state condition. When a fault occurs, the power transmitted suddenly decreases, thus affecting the electrical output of the generators. The power transmitted by the machine is inversely proportional to the reactance between the machine and the system. This reactance value changes to a higher value at high speed during the fault.

Analysis of the Gas Turbine Undesired Trip Several theories can be postulated as to the source of the apparent trouble with the turbine control system. Electrical noise can be the source of generation of a false "drop load" signal in the control system, and the system's response in this condition is an upset state that it cannot recover. Alternatively, it can be speculated that the control system is merely responding to the grid electrical disturbance (transient) and then results in acceleration rates that lead to the "drop load" signal.

Corrective Actions The manufacturer of the CT gas unit was consulted to see if the speed control setting could be made less sensitive, to prevent similar occurrences. The acceleration trip setting (rpm/s) was raised by 25%. In addition, the causes of the problem need to be defined so that the machine behavior during and after system disturbances can be predicted with a machine-stable outcome condition following clearing of the fault. A scope-of-work proposal was therefore submitted to the manufacturer to perform the following:

1. Review the overall signal integrity of digital input and speed signals that might lead to a false "drop load" event, including control system transient immunity to substation surge influence according to ANSI/III C37.90.1. Adequacy of the machine sensor performance and adequacy and design performance of the instrumentation leads connecting the speed sensors to the control system need to be reviewed.
2. Develop logic diagrams for the control system to check that the settings of the load and speed control are within the proper design criteria for transient conditions, and recommend changes to ensure more reliable operation.

Case Study 4.21: Undesired Tripping of an Induction Generator by a Differential Relay Having a Capacitor Bank Within the Protection Zone

Abstract A small hydro induction generator unit rated at 1000 kVA and 4.16 kV is designed with a capacitor bank at the terminal of the generator. The capacitor bank, a three-phase delta-connected bank with a rating of 475 kVAR, is added for voltage support as shown in Fig. 4.128. The generator and its 4.16-kV lead are protected by an electromechanical differential relay with a 0.5-A tap setting and a slope of 30%. The capacitor bank and its leads are protected by external fuses. Upon completion of the project and on the first energization of the system, the generator differential relay operated falsely to trip the unit when the capacitor bank was energized by closing its contactor.

Analysis of the Undesired Trip Since the capacitor bank is included in the generator differential zone, the capacitor inrush current will appear as a differential current. The following calculation is performed to confirm the undesired operation of the differential relay.

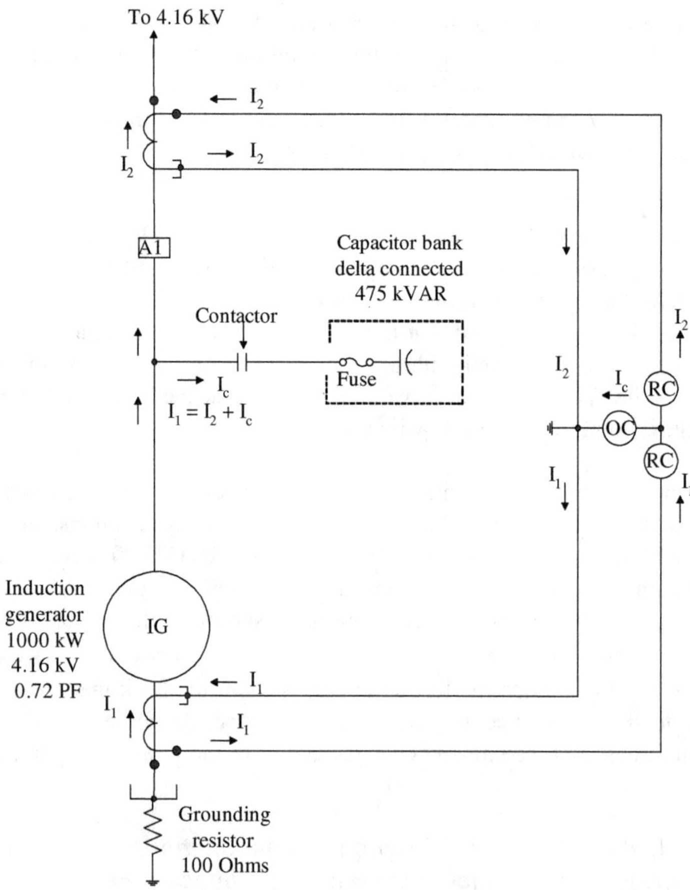

Fig. 4.128 One-line ac diagram showing the undesired trip of the differential relay during a capacitor inrush condition.

Generator full load current $= I_{FL}$:

$$I_{FL} = \frac{1000}{\sqrt{3} \times 4.16 \times 0.72} = 193\ \text{A}$$

Capacitor bank rating = 475 kVAR and capacitor bank rated current $= I_C$:

$$I_C = \frac{475}{\sqrt{3} \times 4.16} = 66\ \text{A}$$

Capacitor inrush current (assume $5 \times I_C$):

$$I_C(\text{inrush}) = 5 \times 66 = 330\ \text{A} \qquad \text{at natural frequency}$$

Differential relay restraining current $= I_{res}$: I_{res} = average magnitude of $(I_1 + I_2)$ during the capacitor inrush conditions of Fig. 4.128 (assume that $I_1 = I_2 = 193$ A):

$$I_{res} = \frac{193 + 193}{2} = 193 \text{ A}$$

Differential relay operating current $= I_{diff} = I_C$:

$$I_{diff} = \begin{cases} I_1 - I_2 = 66 \text{ A} & \text{for capacitor bank steady-state condition} \\ I_1 - I_2 = 330 \text{ A} & \text{for capacitor bank inrush condition} \end{cases}$$

The differential relay will operate if I_{diff} is greater than (slope $\times I_{res}$). The electromechanical differential relay has a tap setting = 0.5 A, and the slope $(S) = 30\%$. The secondary currents for inrush condition $= I_p$/CT; the CT ratio = 40 : 1:

$$I_{diff} = \frac{330}{40} = 8.25 \text{ A}$$

$$\text{slope} \times I_{res} = S \times I_{res} = 0.3\frac{193}{40} = 1.38 \text{ A}$$

since $I_{diff} = 8.25$ A is greater than $S \times I_{res} = 1.38$ A. This implies that the differential relay will operate when the capacitor bank is energized, as documented in this case study.

Corrective Action Add current transformers with a ratio of 40 : 1 at the terminals of the capacitor bank to exclude the capacitor bank from the generator differential zone, by connecting the added CTs to the generator differential system using the following procedure:

1. Assume an external fault outside the differential zone at the terminal of the capacitor bank at F as shown in Fig. 4.129.
2. Connect the capacitor bank CTs to the existing differential relay so that contributions I_X and I_Y are flowing through the restraint coils (with $I_{diff} = 0$). The same procedure can also be carried out using three-line ac, as shown in Fig. 4.130.

Lessons Learned

1. Capacitor banks should not be installed in a generator or transformer differential zone without connecting a current transformer dedicated to sensing capacitor phase currents in the differential circuit.
2. Capacitor bank added CTs for the differential relay connection should be installed on the high side of the bank to exclude the capacitor from the differential zone. This will eliminate undesired tripping of the differential relay during capacitor energization.

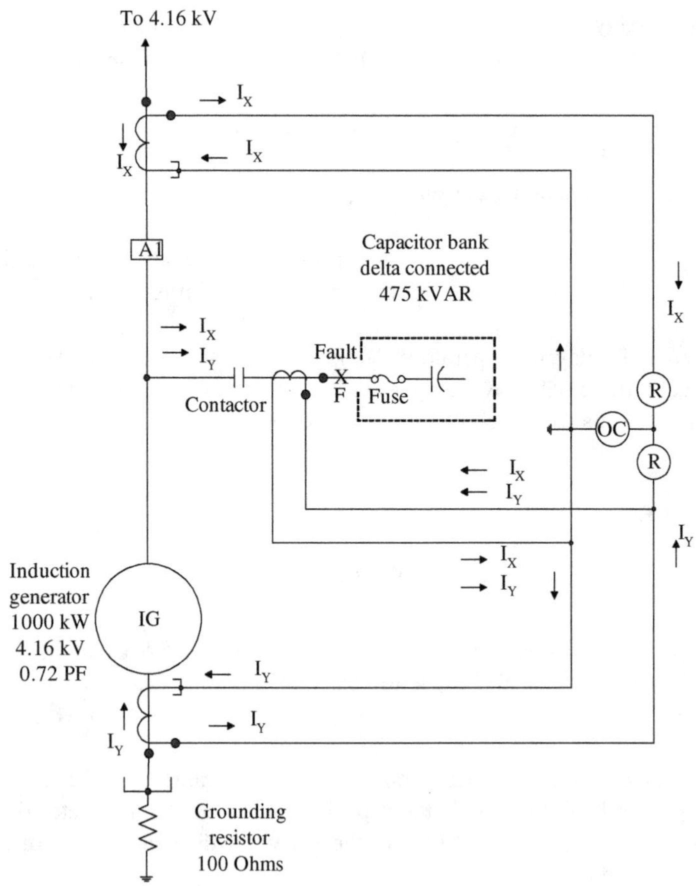

Fig. 4.129 One-line ac diagram showing a capacitor bank current connection to the differential relay during an external fault at F.

Case Study 4.22: Undesired Tripping of a Steam Unit Upon Its First Synchronization to the System During the Commissioning Phase of a Combined-Cycle Plant

Abstract Several undesired trips occurred during the commissioning phase of the steam unit of a combined-cycle plant. In this case study we analyze the causes of the undesired trip at the first synchronizing instant to the system. Corrective actions and lessons learned are also provided.

Description of the Incident Three undesired trips of the steam turbine unit (GS) were initiated by the generator differential elements of the numerical multifunction relays during the first synchronization attempts of the unit to the system. The first trip occurred immediately following the generator breaker closing by the automatic

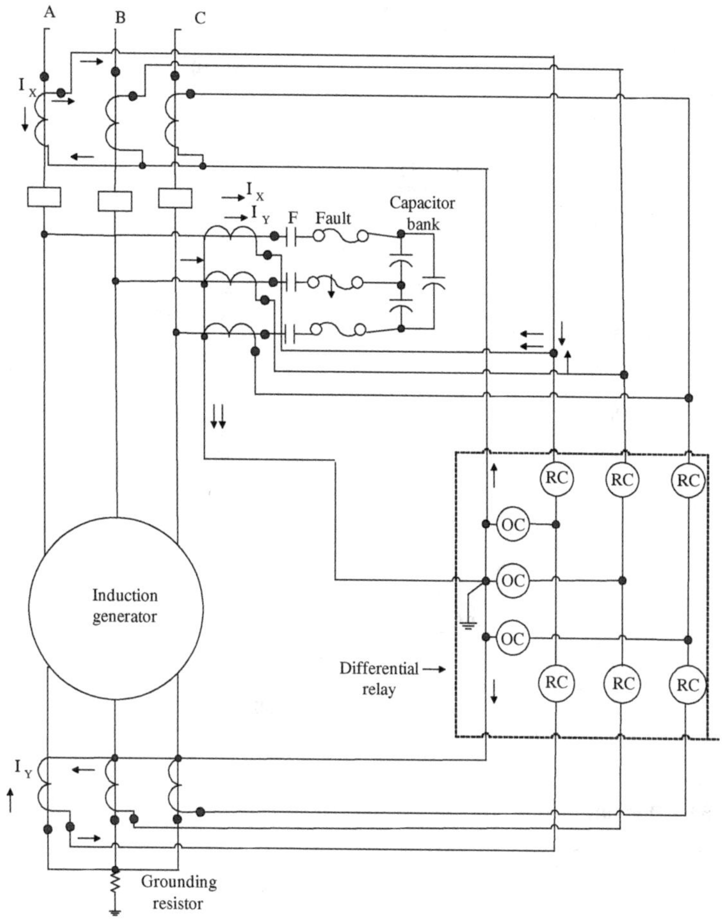

Fig. 4.130 Three-line ac diagram showing capacitor bank currents added to the differential relay circuit during an external L-g fault.

synchronizing equipment. The trip was initiated by the primary generator differential element of the primary numerical GRP relay on phases A and C.

Analysis of the Numerical Relay Oscillography Record for the Undesired Trip Figure 4.131 shows the numerical relay GRP-monitored currents and voltages as well as CT current inputs to the numerical multifunction relay for the GS unit. The numerical relay oscillography record in Fig. 4.132 shows that the phase A system side is in phase with the phase C generator side. Phase B is identical between the system and generator sides. The phase C system side is in phase with the phase A generator neutral side. Therefore, the cause of the trip of the unit upon the first synchronization instant to the system was the swapping of phase A and C current inputs from the system side to the differential relay.

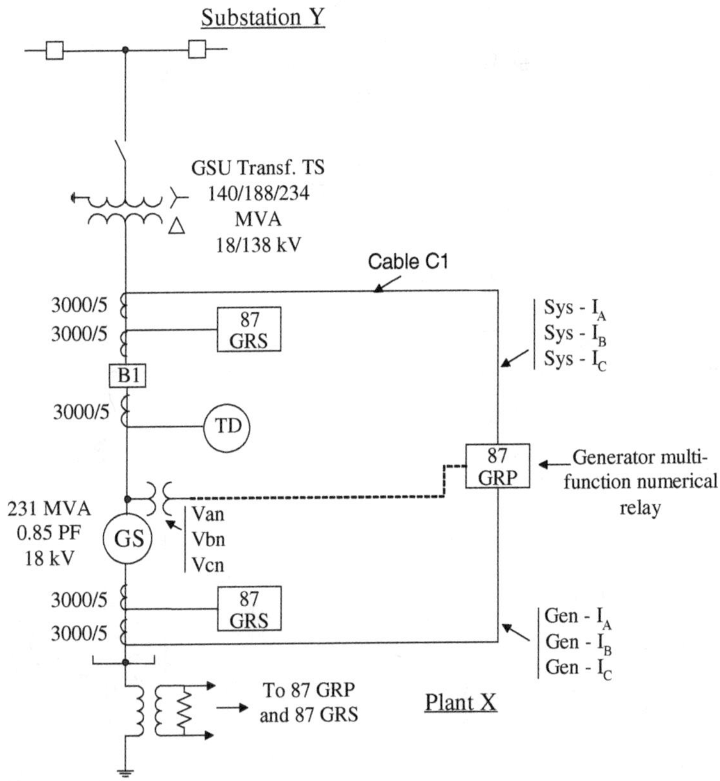

Fig. 4.131 One-line diagram for the steam unit GS with generator multifunction numerical relay CT current inputs.

The relay oscillography record for the generator numerical differential relay element of Fig. 4.132 shows that the system-side phase A current input at point a1 has a phase angle of 240° with the generator-side phase A current input corresponding point at a2. Figure 4.132 also shows that the system-side phase C current input at point c1 has a phase angle of 120° with the generator-side phase C current input corresponding point at c2. The relay record also shows the six currents obtained from the neutral and system sides of the generator to be slightly distorted due to the quantization errors of the low-level magnitudes of currents at the synchronizing instant.

The current transformer (CT) from the circuit breaker A bushing is carrying phase C system current and is being compared with the phase A breaker generator-side current. As a result, a phase angle error of 240° is obtained. Similarly, CT from the CB C bushing is carrying phase A system current and is being compared with the phase C breaker generator side current. As a result, a phase angle error of 120° is incurred. Phase B currents for the system and generator sides are identical, as revealed by the matching of peak points b1 and b2 for the currents shown in Fig. 4.132.

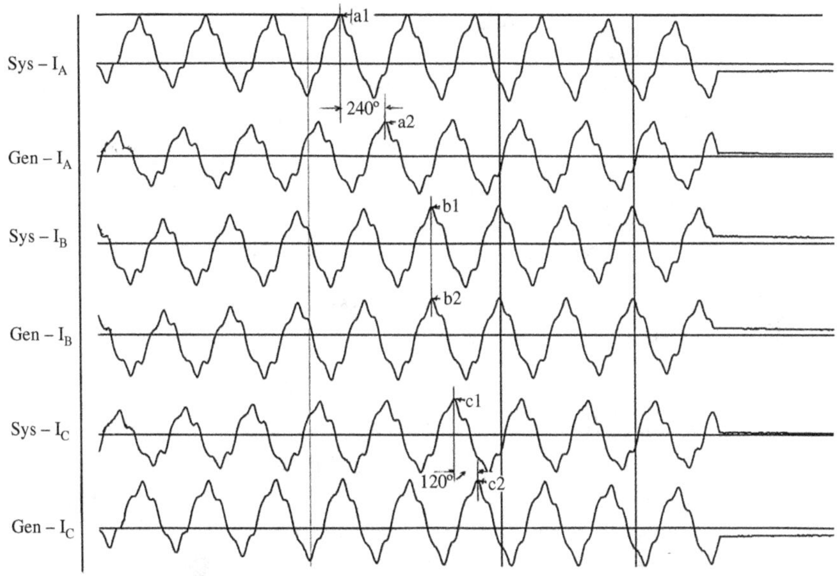

Fig. 4.132 Numerical relay oscillography record for the GS unit currents for the system and generator neutral sides during a synchronization attempt.

Corrective Actions

1. Swap phase A with phase C for the four-conductor cable C1 connecting the breaker CT input to the generator numerical relay. Figure 4.133 shows a comparison of the as-found condition between phasor diagrams for CT connections at the circuit breaker system and generator sides. Figure 4.134 documents additional steps in the solution by relabeling the CT conductor phase designations.
2. Leave the original labeling on the breaker compartment that identifies the three-phase system.

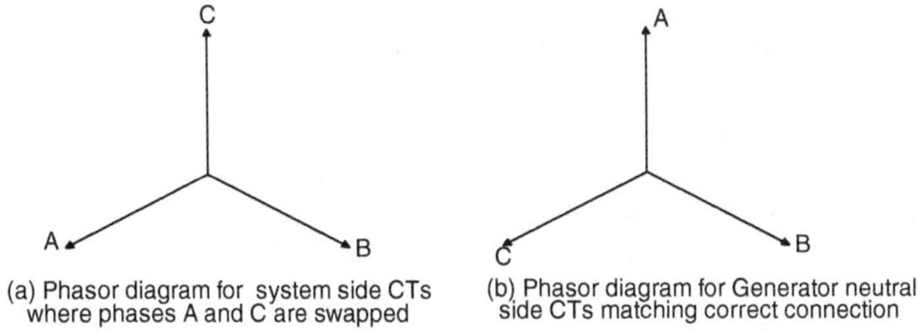

Fig. 4.133 Comparison of the CT phase connections at the CB system and generator sides.

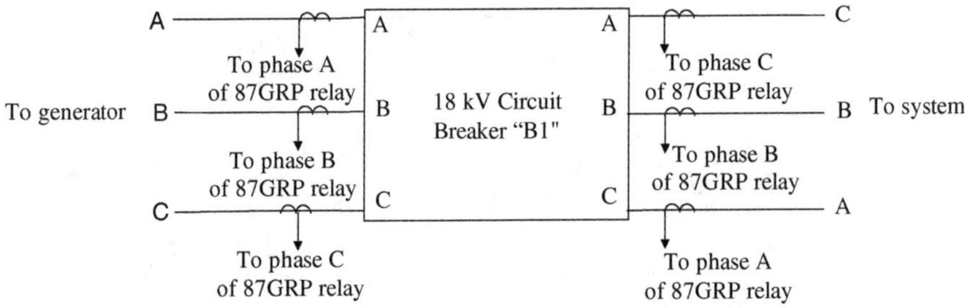

Fig. 4.134 Steam unit circuit breaker with the breaker bushing CT designations at the generator and system sides.

3. Add a note to the breaker housing compartment to read: "Breaker phase A bushing is lined up with system phase C, and breaker phase C bushing is lined up with system phase A." This is a simple solution that has the least impact on the design, including drawing revisions. Minimum work is required to switch the termination of two conductors with no more wiring changes.

Lessons Learned

1. Complete and thorough analysis of numerical relay oscillography records should always be carried out prior to restoring the system following any relay trip. If the first trip was analyzed properly as documented in this case study, the conclusion could have been reached that phase B is correct and the phase A and C currents are swapped between the system and neutral sides for the generator differential relay 87GRP. As a result, no more energization attempts will take place prior to error correction.
2. Equipment drawings should always be checked prior to integration with the rest of the system.

Case Study 4.23: Sequential Shutdown of a Steam-Driven Generating Unit as Part of a 500-MW Combined-Cycle Plant

Abstract The selection of appropriate generator tripping requires a complete understanding of generator protection and control devices, the capability of the generator/prime mover system, and unit operating and maintenance practices. The sequential tripping mode is employed on generators that are being driven by steam. This is done primarily to prevent overspeeding of the turbine when delayed tripping will not harm the generating unit, and therefore high-speed tripping is not a requirement. For turbine mechanical problems this is the preferred tripping mode since it prevents overspeeding of the machine. The scheme is implemented by first tripping of the prime mover, which will result in closing of the turbine valves.

The generator will revert to a motoring mode drawing its active power from the system. A reverse power relay will sense the power drawn by the generator and operate to close a contact output. The relay contact output is wired in series with the turbine valve closed position switches to trip the generator and the field breakers. The sequential tripping scheme provides protection against possible overspeed of the turbine by ensuring that steam flows to the turbine have been reduced to a safe level that will not produce an overspeeding condition when generators are tripped while supplying loads.

Description of the Steps Leading to Unit Sequential Shutdown Figure 4.135 shows the system one-line diagram for the combined-cycle plant. This case study provides a graphing of the steam generating unit GS power output and input in MW as a function of time during a sequential plant shutdown. As shown in Fig. 4.136, during the unit shutdown process, the first step is to trip the turbine at point a. When the steam valves are closed, the generator will start to draw power allowing the first reverse power relay element to operate at point b. The steam valves closed position in conjunction with the operation of the first reverse power relay element 32-1 will permit tripping of the generator breaker C1 after a time delay of 3 s at point c. The generator breaker is opened at point e, forcing the generator output to go to zero at point d. In the case of a failure of the generator breaker to open, the second-stage reverse power 32-2 element will operate to isolate the unit by tripping the 138-kV CBs A and A1. The reverse power relay elements are set based on the generator manufacturer's recommendation for the amount of power needed to motor the unit.

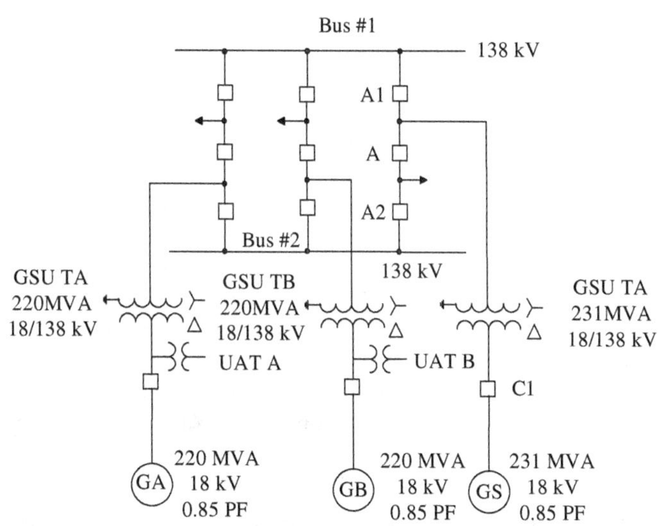

Fig. 4.135 Plant one-line diagram showing the combined-cycle plant.

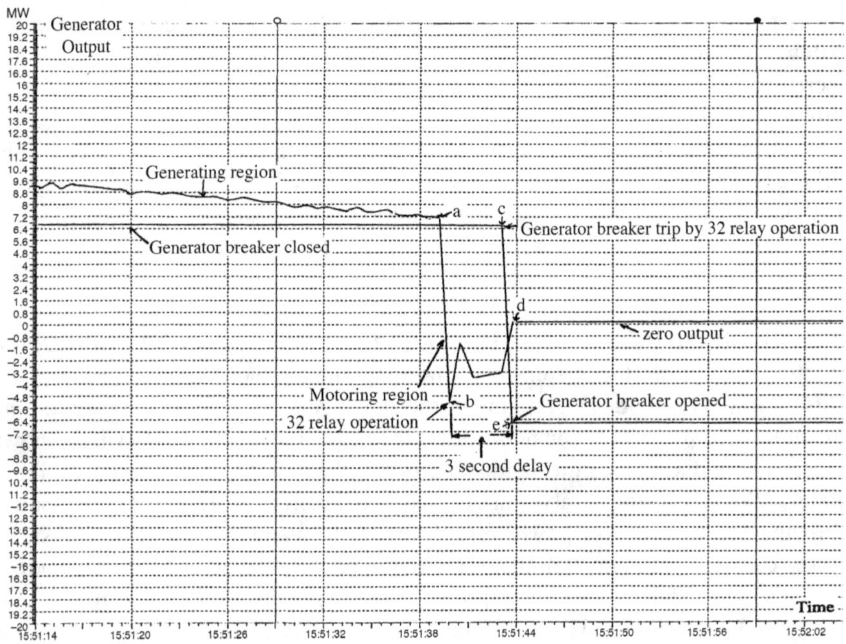

Fig. 4.136 Generator GS power output as a function of time during the unit sequential shutdown process.

Case Study 4.24: Wiring Errors Leading to Undesired Generator Numerical Differential Relay Operation During the Commissioning Phase of a New Unit

Abstract This case study documents some of the wiring errors that resulted in undesired trips of a 50-MW facility during the commissioning phase. It illustrates the benefits gained from the new multifunction numerical generator protection relays. The intelligent electronics device (IED) technology provides oscillography and fault records that were essential in quick troubleshooting and knowing the basic causes of the undesired trips.

Description of the Wiring Errors Figure 4.137 shows the plant one-line ac diagram and the CT connections to the generator differential circuit. The plant protection and control philosophy was based on a generator multifunction numerical relay, GRS. The original design was then modified by the utility. To enhance the dependability of the generator protection function, a second generator numerical relay, GRP, was added. Due to CT limitation in the generator neutral, the two relays were connected in series. CT1 was wired correctly to the GRP and CT2 was wired correctly to the GRP and GRS relays. As shown in Fig. 4.138, a four-conductor cable C1 completed the current differential circuit

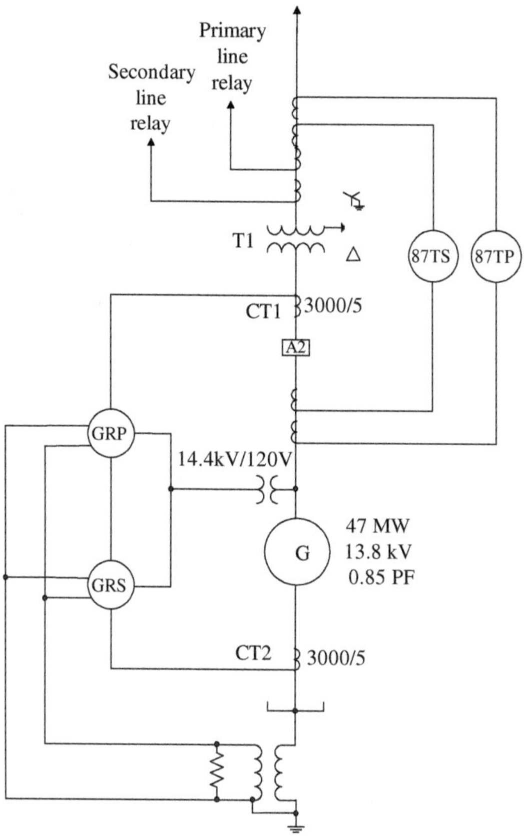

Fig. 4.137 One-line ac diagram showing series connection of the GRP and GRS relays.

connection between the two relays. It is this cable connection that was installed incorrectly and was not tested thoroughly for one of the generating sites. As a result, undesired trips were caused by the generator numerical relays, which is the subject of this case study.

Description of the Incident When the unit was first synchronized, the generator differential relay GRS tripped the generator breaker A2 and opened the breaker in 5.5 cycles. Since this is a through-loading condition for the differential circuit, secondary currents CT1 and CT2 should be identical. Analyzing the relay fault record shown in Fig. 4.139 reveals that the CT2 generator neutral-side current input is correct. However, CT1 system-side current input, which is established by running a four-conductor cable between the GRP and GRS relays, has problems. The three phase currents of CT2 are balanced with the phase sequence ABC. The CT1 phase A is missing, phase B current is swapped to phase A, and phase C is

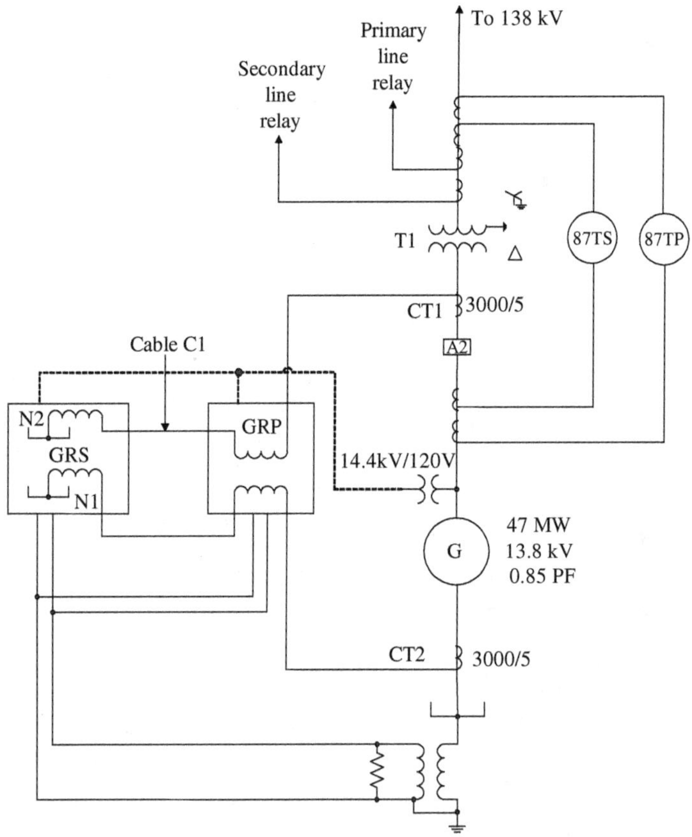

Fig. 4.138 One-line ac diagram showing the cable C1 connecting the two numerical generator relays in series.

swapped to phase B. Figure 4.140 illustrates the as-found condition with the wiring errors of cable C1.

Lessons Learned

1. Relay system wiring should always be checked and verified prior to commissioning.
2. The use of numerical relaying technology provides a wealth of information that is considered essential for quick troubleshooting and diagnosing of real system problems, especially during commissioning.
3. Manufacturers of IED generator multifunction relays should provide a DFR sample showing the relationship between the system and generator neutral currents for either an internal fault or through conditions for external fault or load flow conditions to be used as a reference to guide

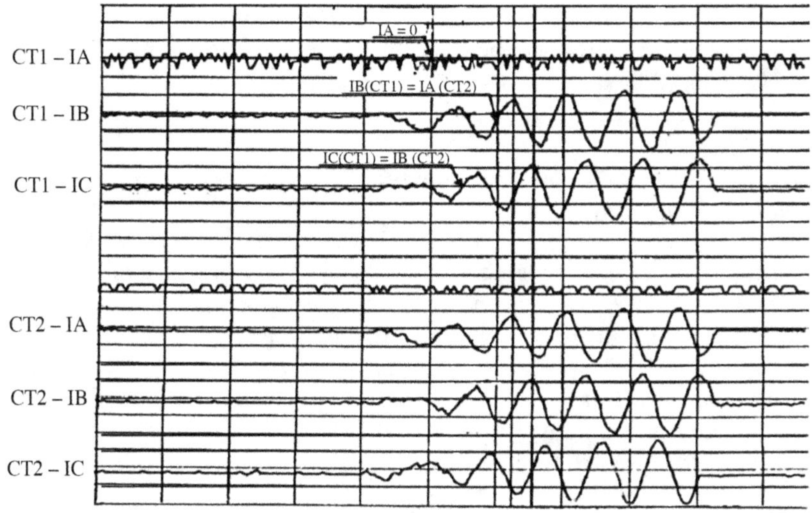

Fig. 4.139 Generator numerical relay fault record showing the currents from CT1 and CT2 for the undesired trip during the commissioning phase.

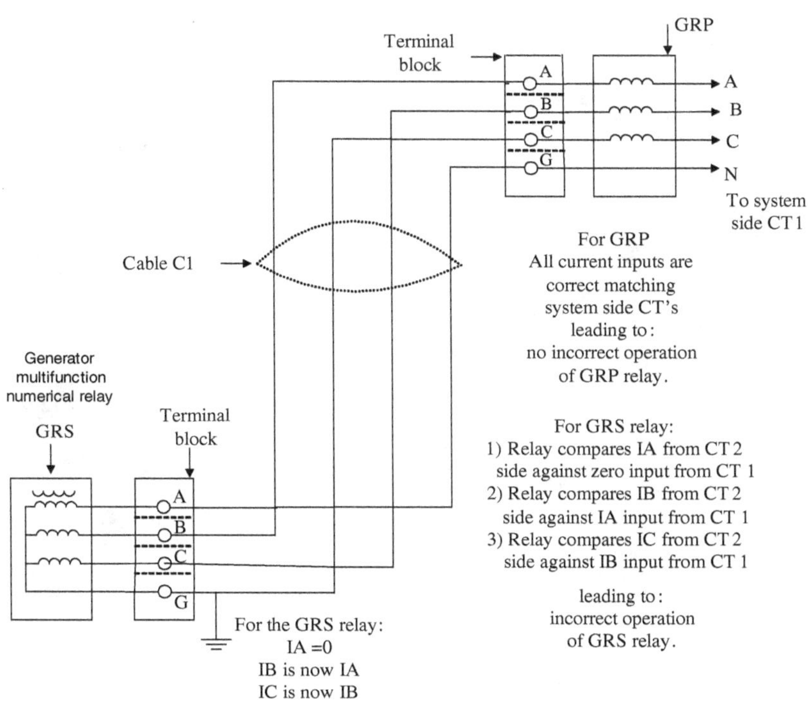

Fig. 4.140 As-found wiring condition for cable C1.

disturbance analysis. The users should not second guess the relationship between the currents to be either in phase or out of phase for any given fault location.

Case Study 4.25: Phasing a New Generator into the System Prior to Commissioning

Abstract This case study assumes a single-cycle combustion turbine (GT) unit rated at about 50 MW. The plant one-line diagram is shown in Fig. 4.141. A step- by-step approach is used to describe the tests that must be performed to connect the new plant to the system. The plant is synchronized to the power system using a 13.8-kV switchgear breaker. This procedure can be used as a guide and can be supplemented by tests specific to the utility, to provide additional assurances prior to connecting new units to the system.

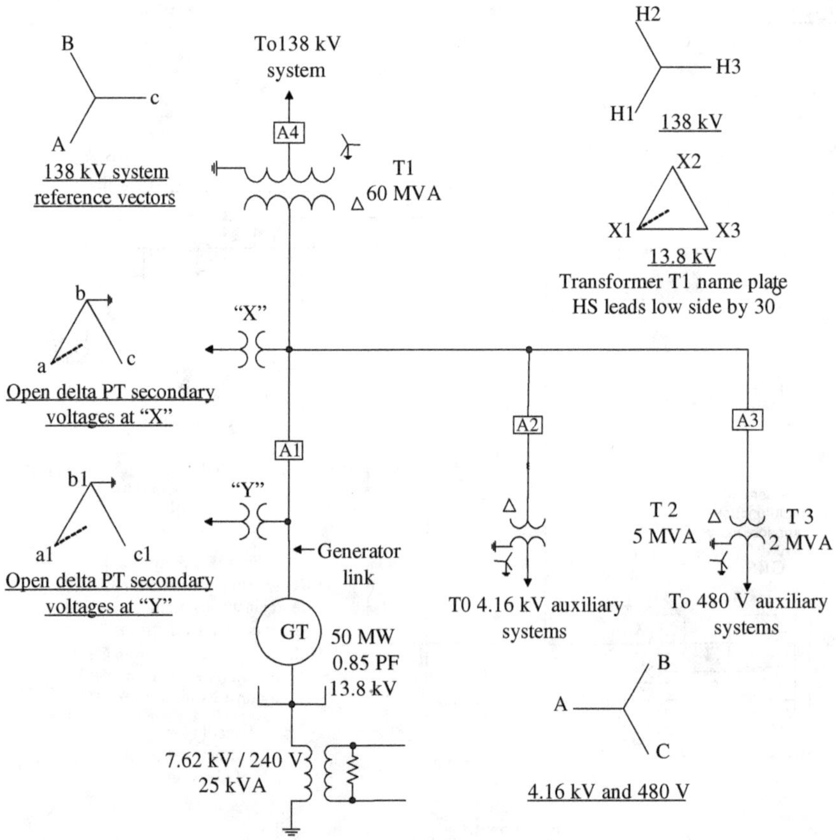

Fig. 4.141 One-line diagram of the synchronization procedure for a unit circuit breaker.

Plant Energization Steps Following are the major steps that must be performed (in order) to connect the machine to the system:

1. Backfeeding of plant station service
2. Confirming the PT circuits on both sides of the GT 13.8-kV circuit breaker
3. Closing the generator links
4. Firing the unit to bring it to a speed-no-load state
5. Carrying out the required machine tests at the speed-no-load state
6. Phasing the new machine
7. Synchronizing the unit to the system

These steps can be subgrouped into three primary categories, as explained in items 3, 4, and 5.

Backfeeding of the Plant from the High-Voltage System Connection The following procedure is recommended prior to backfeeding:

1. Verify the designation of the A,B,C 138-kV cable connection to the utility 138-kV reference system phasors.
2. Verify that the 138-kV phase A cable is connected to the H1 unit transformer terminal, the phase B cable to the H2 terminal, and the phase C cable to the H3 terminal.
3. Based on the unit transformer T1 nameplate phasing shown in Fig. 4.141, and to connect the transformer in accordance with the NEMA standard (so that the high side will lead the low side by 30°), verify that the 13.8-kV A phase is connected to the X1 unit transformer terminal, the B phase to the X2 terminal, and the C phase to the X3 terminal.
4. Verify the A,B,C phase designation for both ends of the 13.8-kV cables and bus connections.
5. Verify that the T2 and T3 station service transformers are connected in a manner similar to that of unit transformer T1. The 13.8-kV phase A cable should be connected to the H1 terminal, phase B to the H2 terminal, and phase C to the H3 terminal. The corresponding 4160- and 480-V cables should be connected as follows: phase A to the X1 terminal, phase B to the X2 terminal, and phase C to the X3 terminal.
6. Before connecting to the system and energizing the unit transformer for backfeeding, all protective relays should be in service and all work completed, including the removal of all temporary grounds and the return of all work permits. In addition, the 13.8-kV generator breaker A1 should be racked out and the breaker dc control fuses should be pulled out. With backfeeding, the 138-kV system phase A leads the 13.8-kV phase A by 30°and the 4160- and 480-V systems by 60°. This will confirm that the 13.8-kV system is now known in relation to the 138-kV reference system.

7. Check the phase sequence on the 13.8-kV, 4160-V, and 480-V systems using a phase rotation tester connected to the secondary side of potential transformers and confirm its counterclockwise ABC rotation.

Confirmation of PT Circuits on Both Sides of the Generator A1 Breaker
The following procedure is suggested

1. Open the 13.8-kV generator link shown in Fig. 4.141 and place insulation boots on the 13.8-kV bus stubs.
2. Apply breaker dc control fuses and rack-in the 13.8-kV generator breaker A1. This will energize both synchronizing PT circuits.
3. Perform the following voltage measurement and phasing tests between the system PT X and the generator PT Y.
   (a) Voltage magnitude measurements between Vab, Vbc, and Vca for PT X and Vab, Vbc, and Vca for PT Y. All voltage magnitude should be around the same value (either 115 or 120 V).
   (b) Confirm the phase sequence for PT X and PT Y. All should be counter-clockwise ABC.
   (c) Perform corner checks by doing voltage checks between a-a1, b-b1, and c-c1. All voltage measurements should be zero. Now they can provide valid voltage inputs to the machine synchronizer.
   (d) Confirm that during the phasing of the two PTs (with the open link), the sync scope should be at 12 o'clock when the two voltages are connected to the sync scope. This will imply correct phase inputs from the two PTs which are fed from the same system source.

These tests will confirm the correct connections of the two PTs X and Y. The only task left to do is to verify the machine phase sequence rotation and the voltage magnitudes generated.

Phasing the New Machine The following procedure is applicable to generator switchgear breakers and is suggested to check the phasing of a new generator prior to synchronization of the power system.

1. Remove the generator breaker A1 dc control fuses and rack-out the 13.8-kV breaker.
2. Bring the unit to the sync ideal state (speed-no-load).
3. When the machine reaches its rated speed, excitation is then applied automatically and rated voltage is reached.
4. Perform excitation system checks.
5. Perform a phase rotation check for the generator PT. Repeat voltage magnitude and corner checks between the system PT X and the generator PT Y.

6. Place the generator breaker in a rack-out position, confirmed by the open position of the "disconnected" indication on the switchgear breaker compartment door.
7. Jumper the track-operated contactor to allow the generator breaker to be closed in its racked-out position.
8. Exercise the automatic synchronizer to establish voltage, frequency, and slip matching between the unit and the system. Upon activation of the automatic synchronizer, the generator voltage will be raised to match the grid voltage, and the machine speed will be controlled until a speed match is accomplished. The synchronizer and the operation of sync check relays will close the generator breaker in the racked-out position.
9. Remove the breaker-closing jumper and install the jumper in the trip circuit.
10. Trip the generator breaker using breaker control switch device 52CS. Now the closing and tripping of the generator breaker have been verified during several dry runs with the generator breaker in its raked-out position, and we are ready for the final step.
11. Rack the generator breaker in position.
12. Synchronize the unit to the system.

4.1.1 Case Studies Related to Relay Setting Calculations

Two case studies are given, one related to the setting procedure of the third-harmonic undervoltage element for the application of 100% stator ground fault protection, and the second related to the philosophy of setting generator relaying elements to provide system backup protection.

Case Study 4.26: Third-Harmonic Undervoltage Element Setting Procedure for 100% Stator Ground Fault Protection

Abstract Use of the third-harmonic undervoltage conditions at the generator neutrals to provide 100% stator ground fault protection requires sufficient third-harmonic voltage generation. It also requires the development of curves for third-harmonic magnitudes as a function of the generator variation of active and reactive generator outputs to determine the minimum level generated. A hydro unit was upgraded from 170 to 220 MW by redesign of the waterwheel. Enhancement of the unit protection was also done using a combination of discrete electromechanical and solid-state relays. This upgrade predates the availability of generator numerical multifunction relays. Part of the enhancement was to add 100% stator ground fault protection instead of the existing 95% stator protection. In this case study we outline the setting procedure for the third-harmonic undervoltage element by first collecting the generator third-harmonic voltage as a function of the machine active and reactive output powers, followed by data analysis to set the relay reliably.

Methods of 100% Stator Ground Fault Protection The following methods are employed to provide 100% stator ground fault protection.

1. *Third-harmonic neutral undervoltage scheme.* The scheme monitors the third harmonic by an undervoltage element tuned to the third harmonic. When a generator stator ground fault occurs near the neutral or when the neutral connection disappears, the third harmonic monitoring the generator neutral goes to a very low value (near zero for neutral disappearance) below the element setting threshold.
2. *Third-harmonic comparison scheme.* This scheme compares the third harmonic at the neutral with the third harmonic at the generator terminal. For the stator ground fault near the neutral, the ratio will reach the tripping level.
3. *Subharmonic low-frequency injection scheme.* The subharmonic low-frequency injection method will detect a stator ground fault by monitoring the amount of signal received. A low frequency of 15 to 20 Hz (depending on the manufacturer) will show an increase in the return signal for the stator ground fault. This method has the advantage of detecting stator ground faults while the generator is on, turning gear prior to synchronization with the system. However, unlike methods 1 and 2 above, this injection method cannot detect an open generator neutral.

Setting Criteria for an Undervoltage Third-Harmonic Unit The third-harmonic machine-generated voltage varies as a function of unit outputs in terms of active and reactive power. The ideal approach is to determine three main curves:

1. Machine-generated third-harmonic voltage as a function of varying active power with zero reactive power.
2. Machine-generated third-harmonic voltage as a function of varying active power with maximum reactive power in value.
3. Machine-generated third-harmonic voltage as a function of varying active power with maximum reactive power out.

The objective is to find the minimum machine-generated third-harmonic voltage and to set the relay to drop out at a threshold equal to 50% of the minimum value generated.

Procedure Used to Collect Neutral Third-Harmonic Voltage Data The 100% generator stator ground fault protection is accomplished using an undervoltage element that is tuned to the third harmonic. The element requires a threshold setting that will cause the element to drop out when reached. A spectrum analyzer is connected across the secondary of the neutral distribution transformer shown in Fig. 4.142 to collect machine-generated third-harmonic neutral voltage. The active power output of the generator is then varied, while the reactive power is at zero value.

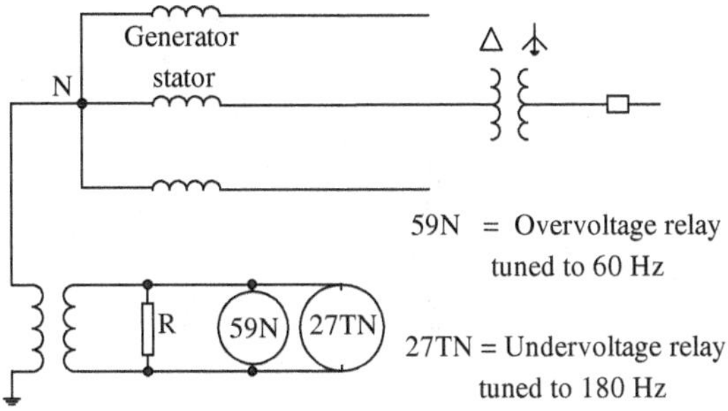

Fig. 4.142 100% stator ground fault protection using the third-harmonic undervoltage scheme.

The procedure for varying the active power is then repeated while the machine reactive power is at maximum in and maximum out. The machine output in megawatts and the neutral voltage in volts are then plotted for the three cases of machine reactive power outputs. A procedure can be established to coordinate the power ramping of the unit with the neutral voltage readings collections. Figure 4.143 illustrates the three curves that define the machine third-harmonic-generated voltage profile.

Setting Recommendation Analysis of all the third-harmonic measurements indicates that the lowest recorded third harmonic is 0.7 V. Based on these results, the setting of the third-harmonic undervoltage element should be set at half of the lowest recorded third harmonic, 0.35 V (50% of 0.7 V). This setting will provide better sensitivity to stator ground faults near the neutral end of the machine and greater security to prevent alarming on the intrinsic normally generated third harmonic of the machine.

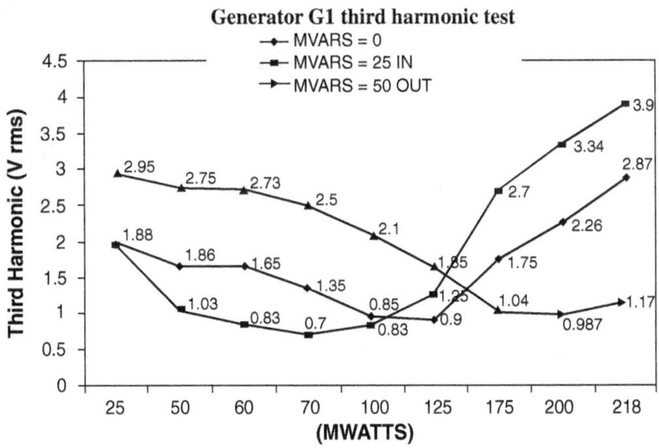

Fig. 4.143 Generator neutral third-harmonic plot as a function of active and reactive power.

Action Taken by the Undervoltage Third-Harmonic Relay Unit When the threshold setting value is reached, the undervoltage third-harmonic element will drop out, activating either a trip or an alarm output. For the alarm option, when the alarm is received, an independent assessment of the third harmonic in the generator neutral needs to be executed. The problem needs to be assessed and a schedule shutdown for the machine can be scheduled when replacement power with reasonable cost is arranged. Some utilities trip the generator for dropout of the third-harmonic undervoltage element.

Procedure to Confirm Operation of the Undervoltage Unit A separate spectrum analyzer is then connected across the secondary of the distribution transformer in the generator neutral. Absence of the third harmonic from the spectrum output will confirm the relay operation. The relay operation will be due to either a stator ground fault in the area (about 10%) near the neutral or to an open in the generator neutral connection.

The numerical multifunction generator relay can provide an assessment of the third harmonic through the meter function. In this case there is no need to connect a spectrum analyzer to meter the third-harmonic voltage.

Case Study 4.27: Basis for Setting the Generator Relaying Elements to Provide System Backup Protection

Abstract Figure 4.144 illustrates the application of system backup at the generator and the neutral of the generating step-up unit (GSU) transformer. For a unit-connected transformer the 21- or 51-V relays will provide backup for system phase faults. The negative-sequence relay device 46 provides backup protection for open phase, unbalanced loading and unbalanced faults. The negative-sequence relay needs to be set to plot under the generator negative-sequence capability curve $I_2^2 t$. The

$$\text{relay operating time} = \frac{\text{machine } I_2^2}{(I_2/I_{\text{nom}})^2}$$

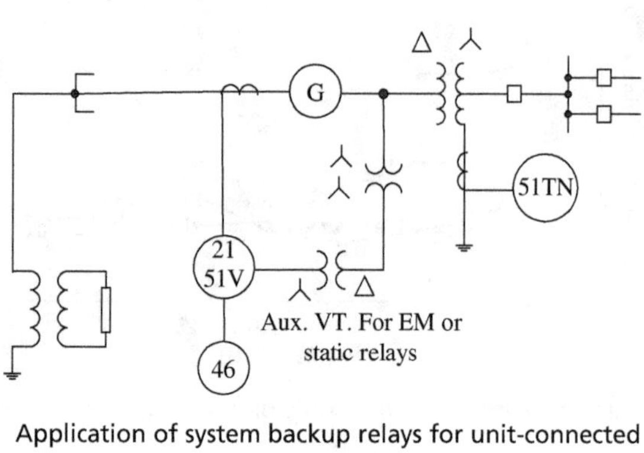

Fig. 4.144 Application of system backup relays for unit-connected generators.

where I_{nom} is the generator rated current (amperes) and I_2 is the relay calculated negative-sequence current (amperes).

The generator can be better protected using a numerical negative-sequence element that calculates I_2 using a symmetrical components equation. The old obsolete electromechanical relay has a sensitivity of 0.6 pu and was designed primarily to detect unbalanced, uncleared system faults, while the numerical element can sense negative-sequence current as low as 0.02 pu of the machine rating. Therefore, the numerical relay can, if required, detect series system unbalance for the open phase at the high side of the generator unit transformer.

The GSU neutral overcurrent relay 51TN will provide backup protection primarily for system high-resistance ground faults and open-phase conditions. Ground backup overcurrent relays in generator step-up neutrals should be set to coordinate with the system ground time overcurrent relays using the same relay curve type, to avoid loss of coordination and hence undesired tripping of the generator for ground faults outside the protected zone. For smaller units, a voltage-controlled overcurrent 51-V or voltage-restrained overcurrent relay is used to protect against uncleared system phase faults.

Basis for Setting the Generator Phase Fault Backup Distance Relay 21 For larger units, distance relay device 21 is preferred with voltage transformers (PTs) connected to the generator terminal and current transformers (CTs) connected to the generator neutral side. It is often applied with no offset so that the location of the generator PTs lies on the mho characteristic circle. The relay reach starts from a PT's connection point to the generator isolated-phase bus. The reach setting of the distance relay 21 must remain conservatively above the machine rating to prevent inadvertent trips on generator swings and severe voltage disturbances. The distance relay should be set to carry at least 200% of the generator rating at the rated power factor. At the same time, the relay 21 setting needs to be coordinated with backup distance relays on the system.

A generator phase backup distance relay device 21 is used to protect the generator from either three-phase or phase-to-phase system faults during extreme system conditions. Some of these extreme conditions may occur due to simultaneous breaker failure events occurring in cascade fashion, simultaneous failures of the transmission-line dual pilot protection systems, lack of redundancy for power element protection, or the occurrence of a single common failure that can disable entire protection systems.

Procedure Used to Set a Distance Relay Device 21 The philosophy behind setting these relays will depend on the type of protection application on the high-side power system and associated lines. The procedure therefore assumes that redundant line and bus relaying schemes as well as breaker failure protection are applied at substation X and the surrounding bulk power system as shown in Fig. 4.145. This protection philosophy will limit the need to extend the operation zone of the backup relay 21 beyond the lines emanating from the high side of the GSU transformer. It is also assumed that the relay 21 receives the proper 30° shift in its applied voltage.

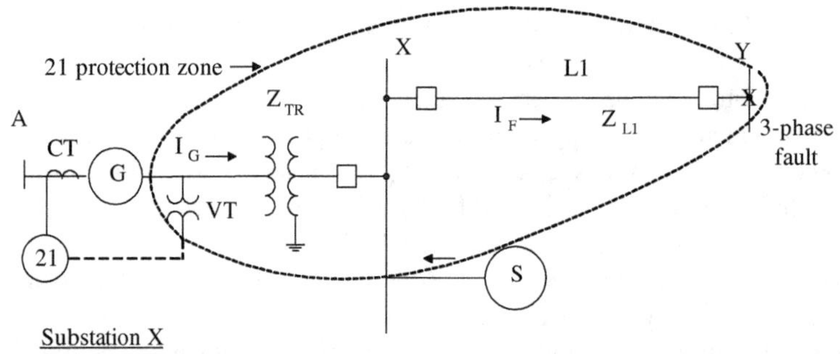

Fig. 4.145 One-line ac diagram with generator distance relay 21 protection zone.

For the application of dual (redundant) pilot line protection systems and breaker failure protection for all high-side circuit breakers, the following procedure to set the relays is recommended.

1. Select the longest line that emanates from substation X. Assume that it is L1.
2. Place a three-phase fault (stub) at the end of L1 at bus Y and calculate the apparent impedance seen by the generator 21 relay.

$$\begin{aligned} Z_{\text{app}} &= Z_{\text{tr}} + Z_{\text{L1}} \times (\text{in-feed ratio}) = Z_{\text{tr}} + Z_{\text{L1}}(I_{\text{F}}/I_{\text{G}}) \\ Z_{\text{set1}} &= Z_{\text{tr}} + 120\%[Z_{\text{L1}}(I_{\text{F}}/I_{\text{G}}) \end{aligned} \tag{1}$$

3. Calculate the impedance seen by relay 21 without in-feed using a 120% multiplier for the line impedance:

$$Z_{\text{set2}} = Z_{\text{tr}} + (120\% \times Z_{\text{L1}}) \tag{2}$$

4. Try to set the relay using the impedance calculated in equation (1), which is calculated in steps 1 and 2.
5. Check the loading ability of the setting to be above 200% of the generator maximum power output.
6. In case the criterion in steps 5 cannot be met, try to meet it by setting the relay using the impedance in equation (2), calculated in step 3.
7. Obtain the Z2 timer setting (T2) for line L1 phase backup distance relays.
8. Add a coordination margin of a minimum of 0.5 s to the Z2 timer setting value and set the relay 21 timer accordingly.
9. For high-voltage systems where three zone distance relays may be used, add a coordination margin of 0.4 s to the Z3 timer setting value and set the relay 21 timer accordingly.
10. For the application of two elements for generator phase backup protection, called 21-1 and 21-2, set the 21-1 element at 80% of the GSU transformer

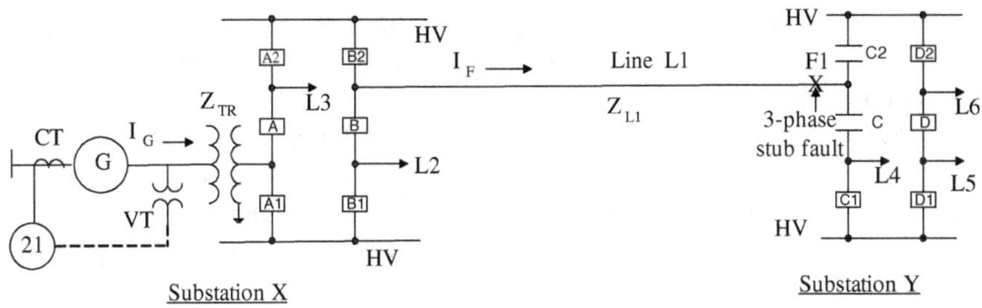

Fig. 4.146 One-line diagram showing a line-end fault scenario at substation Y.

impedance and the associated timer at 0.5 s (30 cycles). Set 21-2 using steps 1 through 8.

This philosophy is very conservative, as we demonstrate by going through an actual ring bus configuration with an end-of-line fault scenario.

Example 1: Simultaneous Failure of the Dual Pilot Relaying Systems Example 1 is illustrated in Fig. 4.146. If the three-phase stub fault at F1 at the end of the line cannot be cleared in primary or backup clearing times due to the extreme contingency (more than $N-1$) of a simultaneous failure of the dual pilot relaying systems at the substation X end of the line, the generator 21 distance backup relay will remove the generator by tripping high-voltage CBs A and A1. In this case, the assumed simultaneous failure of the dual pilot relaying systems will also result in no initiation of breaker failure protection.

Example 2: Simultaneous Breaker Failure Example 2 is illustrated in Fig. 4.147. If the three-phase stub fault at F1 at the end of line L1 cannot be cleared in primary or breaker failure (local back up) clearing times due to the failure of breaker B2 at substation X followed by a consecutive failure of CB A2 at the same substation (extreme contingency, beyond the N - 1 criterion), the generator distance backup relay 21 will remove the generator by tripping CBs A1 and A.

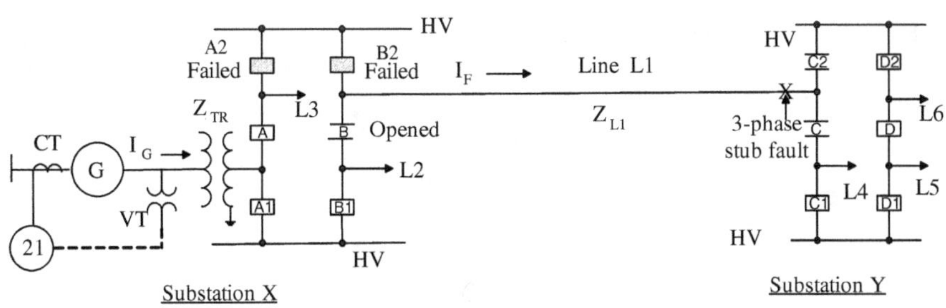

Fig. 4.147 One-line diagram showing a simultaneous breaker failure scenario at substation X.

Summary In view of Examples 1 and 2, operation of generator 21 relays occurs primarily during extreme system contingencies. It is obvious that the zone of protection of relay 21 is defined by the longest adjacent line, and therefore there is no need for a relay 21 to reach more than it should. If the line protection is applied using redundant relaying systems with diverse operating principles and the pilot protection is accomplished using diverse communication media, the Example 1 scenario can be avoided. The Example 2 scenario can be avoided by applying proper periodic breaker maintenance programs.

REFERENCES

Blackburn, J. L. *Applied Protective Relaying*. Pittsburgh, PA: Westinghouse Electric Corporation, 1979.

Blackburn, J. L. *Protective Relaying Principles and Applications*. New York: Marcel Dekker, 1987.

Blackburn, J. L. *Symmetrical Components for Power Systems Engineering*. New York: Marcel Dekker, 1993.

Disturbance Monitoring and Reliability Requirements. NERC Reliability Standard PRC-002-02.

Elmore, W. A., Ed. *Protective Relaying Theory and Applications*. New York: Marcel Dekker, 2000.

Fault and Disturbance Data Requirements for Automated Computer Analysis. IEEE Power System Relaying Committee Working Group 111, Special Publication 95 TP 107, 1995.

IEEE Guide for AC Generator Protection. IEEE Power System Relaying Committee Publication, 1986, and follow-up revisions.

Loss-of-Field Relay Operation During System Disturbances. IEEE Power System Relaying Committee Report. *IEEE Transactions on Power Apparatus and Systems*, Vol. PAS-94, 1975, pp. 1464–1472.

Out-of-Step Relaying for Generators. IEEE Power System Relaying Committee Report. *IEEE Transactions on Power Apparatus and Systems*, Vol. PAS-96, 1977, pp. 1556–1564.

Protective Relaying for Pumped Storage Hydro Units. IEEE Power System Relaying Committee Report. *IEEE Transactions on Power Apparatus and Systems*, Vol. PAS-94, 1975, pp. 899–907.

Schlake, R. L., G. W. Buckley, and G. McPherson. Performance of third harmonic ground fault protection schemes for generator stator windings. *IEEE Transactions on Power Apparatus and Systems*, Vol. PAS-100, 1981, pp. 3195–3202.

5

CASE STUDIES RELATED TO TRANSFORMER SYSTEM DISTURBANCES

The use of digital fault recorders with a limited number of analog channels at a large substation has affected the option of monitoring transformer high-side, low-side, and tertiary winding currents. Therefore, the use of transformer multifunction numerical relays has enhanced the monitoring and analysis of transformer disturbances. This is accomplished by providing oscillography records for the transformer current input at all voltage levels.

In this chapter we provide analyses and lessons learned from many disturbance case studies. Energization of a transformer with a high-voltage lead fault will document the generation of enough harmonics to prevent operation of the differential element and recommend reliance on different protection operating principles to clear these types of faults. Violation of the simple duality rule, where two or more elements, with a preference for diverse operating concepts, are used for transformer protection is explained. The failures of an autotransformer and a station service auxiliary transformer are analyzed. Recorded currents on the grounded-wye side are utilized to locate and classify faults on the delta winding of a two-winding transformer.

Disturbance analysis of many phase-to-phase and evolving three-phase faults around the leads of station service transformers and falling within transformer protection zones is also covered. The basics of successful implementation of

Disturbance Analysis for Power Systems, First Edition. Mohamed A. Ibrahim.

electromechanical and static differential relays are reiterated for the analysis of an undesired trip in an application of new numerical relaying concepts. Basic setting and hardwiring connections for the old technology are used as lessons learned for correct implementation of the new numerical relaying concepts.

5.1 TRANSFORMER BASICS

Basic transformer concepts essential for correct analysis of power transformer disturbances are covered in this section. In addition, examples describing transformer connection types, current flows, and coupling between windings are shown.

5.1.1 Most Widely Used Transformer Connection Types

Transformer winding connection types are normally based on system studies and how phase-to-ground faults will be detected. Most transmission systems are designed as solidly grounded systems to permit such parameters as currents and the derived impedance to be used to detect ground faults occurring on a system. Power systems are designed to include extrahigh-voltage (EHV) and high-voltage (HV) transmission levels. Most EHV and HV systems are designed to be in phase. For example, systems can be designed with 765-, 500-, 345-, 230-, 138-, and 115-kV voltage levels. As an example, a utility may have its 765-kV system designed to be in phase with the 345-, 230-, and 115-kV voltage levels. Therefore, an autotransformer with a delta tertiary or three-winding transformer YG/YG/delta application is ideal to maintain the in-phase relationship between the high- and low-side transmission voltage levels. In addition, the two-winding delta/YG transformer is employed as a generator step-up transformer and for distribution.

5.1.2 Autotransformers

To maintain that all these voltage levels are in phase, autotransformers or three-winding YG/YG/delta transformers are used. Autotransformers offer cost advantages compared with equivalent three-winding transformers; however, their through-fault withstanding capabilities must be checked and verified. This permits the connection of different parts of a system, which have different voltage levels, while maintaining the in-phase feature of these voltage levels. Therefore, most transformers used at substations are connected as autotransformers with a delta tertiary that can act as a zero-sequence current source. As shown in Fig. 5.1, the delta tertiary provides support to ground fault currents on either the low- or high-voltage sides. In addition, leads can be connected to the delta winding and used as a station service source. The delta leads can also be used to connect certain compensation or voltage control equipment if required.

When specifying autotransformers, there is an option of either adding or deleting the delta tertiary winding. The availability of sufficient ground fault currents for

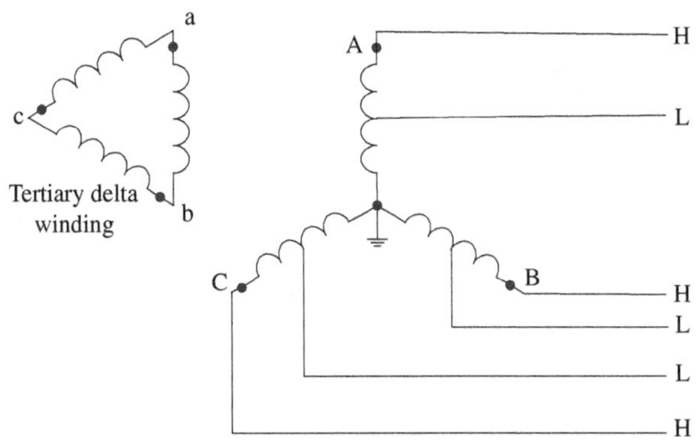

Fig. 5.1 Autotransformer with a delta winding tertiary.

normally solidly grounded transmission systems, as well as the absence of a need for substation station service, may eliminate the need for the delta tertiary winding. However, when specifying a delta tertiary winding, a preliminary short-circuit study should be run to examine the circuit breaker–interrupting capability and effects on ground time overcurrent relaying coordination. The added delta zero-sequence sources should not overstress the breaker-interrupting capability and hence should avoid the additional cost of breaker replacement.

For an embedded delta tertiary winding (no leads are brought out of the transformer), one corner of the delta should be grounded in accordance with ANSI standard requirements. This is normally done to reduce the likelihood of switching transient overvoltages and their associated insulation stresses on the transformer windings. Some applications specify autotransformers with embedded tertiary windings to improve the availability of the transformers by reducing the occurrence of tertiary lead faults. This is done by removing the lead exposures to faults caused primarily by animal contacts with the leads. In some applications the delta tertiary winding may be used as a station service supply, and the phase leads may be tapped by an insulated material to prevent faults caused by animal contacts or flying debris.

5.1.3 Analysis of the Fault Current Contribution from Autotransformers

Monitoring of autotransformer currents using a DFR can provide information that can be used to confirm power system modeling and behavior of the relay systems. Concentration is normally placed on the zero-sequence current contribution during ground faults, which is not affected by load flows. Four currents are of interest: high side, low side, neutral, and delta tertiary. These four currents are as shown in Fig. 5.2. However, due to the limited number of analog channels at a large substation, only a few or no autotransformer currents are monitored. Therefore, relationships between the autotransformer high side (I_{0H}), low side (I_{0L}), neutral (I_n), and tertiary

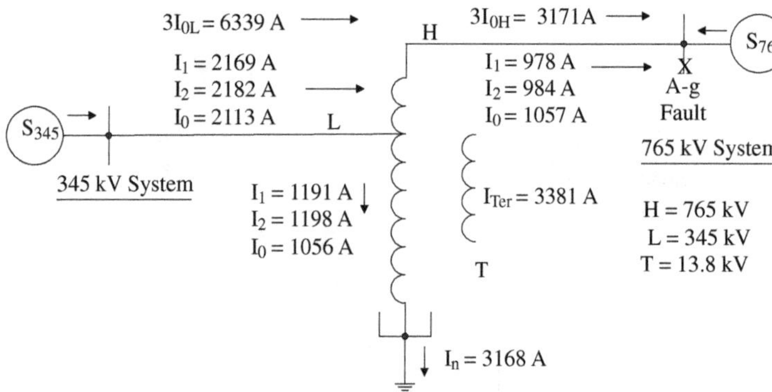

Fig. 5.2 Sequence current flow through an autotransformer for a high-side fault.

zero-sequence (I_{ter}) currents need to be established to be able to confirm the power system model. The following options are available for monitoring:

1. Direct recording of I_{0H} and I_{0L} can be used to deduce I_n and I_{ter}.
2. Direct recording of I_{0H} and I_n can be used to deduce I_{0L} and I_{ter}.
3. Direct recording of I_{0L} and I_n can be used to deduce I_{0H} and I_{ter}.
4. Direct recording of I_{0H}, I_{0L}, or I_n can be used directly to confirm the power system model.

5.1.4 Monitoring Delta Currents for Three-Winding or Autotransformers

Three-winding or autotransformers are normally specified with a current transformer for each phase of a delta tertiary winding. Traditionally, the delta current is used to polarize ground directional relaying for directional comparison blocking schemes and for ground backup relaying. It can also be used as a starting element for DFR record capturing during unbalanced ground faults. To obtain the ground current ($3I_0$), three sets of current transformers (one from each phase) are connected in parallel, similar to a zero-sequence current filter. This connection has the added benefit of not sensing balanced load current (sum of positive-sequence currents = 0) components when tertiary delta windings are used to supply station service to the associated substation. Monitoring the zero-sequence current by the DFR therefore consists of measuring $3I_0$ only, as desired.

5.1.5 Determining the Direction of a Delta-Winding Zero-Sequence Current for an Autotransformer During Faults

The ampere-turns balancing concept is used to find the direction of current flow inside the delta tertiary winding during low- and high-side faults. This direction is required for correct polarization of ground overcurrent primary and backup relaying. Tertiary delta current is reliable for polarizing of ground relaying provided that the negative branch of the transformer wye equivalent obtained from solving algebraic equations is not larger than the system equivalent impedance of the source behind the winding.

5.1.6 Per-Unit Basis for Delta Tertiary Zero-Sequence Current

In short-circuit study simulations, the tertiary current (which flows inside the delta winding) is calculated in pu using the following base current definition:

$$I_{\text{base}} = \text{MVA} \times 10^6/3 \times \text{L–L voltage (kV)} \times 10^3$$

Applying this base formula for current inside the delta winding will establish the ampere-turns current balance for all the windings of the autotransformer for either high- or low-side faults. For example, for a high-side fault, the high-side zero-sequence current in pu is equal to the low-side current in pu plus the tertiary current in pu based on the base current definition above.

5.1.7 Autotransformer Neutral Current Magnitude and Direction

The current flowing in an autotransformer neutral is equal to the difference between the high- and low-side ground ($3I_0$) currents. It is necessary to determine the magnitude of the neutral current for coordination with the system time overcurrent ground backup relaying. Normally, for high-side faults it will be flowing down the neutral, because the low-side zero-sequence current contribution (in amperes) is larger than the high-side current and flows toward the transformer. For low-side faults the current flows up the neutral, because the low-side zero-sequence current magnitude (in amperes) is larger than the high-side current and flows away from the transformer. In general, the transformer neutral current cannot be used for the polarization of ground relays, due to the change in current direction for high- and low-side faults.

Example 5.1: Verifying Autotransformer Current Flow Concepts

The following autotransformer example demonstrates the concepts stated above regarding current flows through an autotransformer bank. The example is lifted from a real power system to prove all the concepts associated with disturbance analysis involving autotransformers. The transformer described is rated as three single-phase 500-MVA 765/345/3.8-kV YG/YG/delta. The phase-to-ground fault is placed on the transformer high side (765 kV), and only current flows through the autotransformer are shown in Fig. 5.2. Using the sequence current flows in Fig. 5.2, the neutral current can be obtained as follows:

$$\begin{aligned} I_n &= 3I_{0L} - 3I_{0H} \\ &= 6339 - 3171 = 3168\ \text{A} \end{aligned} \tag{1}$$

Since the low-side contribution is greater than the high-side contribution, the direction of the neutral current will be down the neutral.

Figure 5.2 can also be used to verify the following relationship for zero-sequence currents:

$$I_{0H}(\text{pu}) = I_{0L}(\text{pu}) + I_{\text{ter}}(\text{pu}) \tag{2}$$

This is done using the following steps:

1. The base currents for the three voltage levels need to be calculated as follows:

$$I_{\text{base}} = \frac{\text{MVA} \times 10^6}{\sqrt{3} \times V_{\text{L-L}} \times 10^3}$$

   For 765 kV: $I_{\text{base}} = 100 \times 10^6/\sqrt{3} \times 765 \times 10^3 = 75.5\ \text{A}$

   For 345 kV: $I_{\text{base}} = 100 \times 10^6/\sqrt{3} \times 345 \times 10^3 = 167.4\ \text{A}$

   For 13.8 kV: $I_{\text{base}} = 100 \times 10^6/\sqrt{3} \times \sqrt{3} \times 13.8 \times 10^3 = 2415.5\ \text{A}$ ($I_{\text{phase}} = I_{\text{line}}/\sqrt{3}$)

2. The pu currents can be calculated as follows:

$$I_{0H}(\text{pu}) = \frac{I_{0H}}{I_{\text{base}}} = \frac{3171}{3 \times 75.5} = 14\ \text{pu}$$

$$I_{0L}(\text{pu}) = \frac{I_{0L}}{I_{\text{base}}} = \frac{6339}{3 \times 167.4} = 12.62\ \text{pu}$$

$$I_{\text{ter}}(\text{pu}) = \frac{I_{\text{ter}}}{I_{\text{base}}} = \frac{3381}{2415.5} = 1.40\ \text{pu}$$

   Substituting these values in equation (2) yields;

$$\text{RHS} = I_{0L}(\text{pu}) + I_{\text{ter}}(\text{pu}) = 12.62 + 1.40 = 14.02 = \text{LHS}$$

   thus verifying the relationship defined in equation (2).

5.1.8 Confirmation of Power System Modeling

Monitoring autotransformer currents using a DFR can provide information that can be used to confirm power system modeling and behavior of the relay system. One of the functions of system disturbance analysis is to verify the short-circuit model by trying to match DFR currents with currents obtained from short-circuit study simulations. Due to channel monitoring limitations, current flow for a ground fault on either side of an autotransformer must be determined.

Two of the four currents will be sufficient to deduce the remaining currents using relationships (1) and (2) above:

$$I_n = 3I_{0L} - 3I_{0H}$$
$$I_{0H}(\text{pu}) = I_{0L}(\text{pu}) + I_{ter}(\text{pu})$$

By DFR monitoring of the tertiary and neutral currents, we can use these proven relationships, which are two formulas with two unknowns, sufficient to obtain I_{0L} and I_{0H}.

Example 5.2: Determining the Direction of the Delta-Winding Zero-Sequence Current

For the 765-kV fault shown in Fig. 5.2, we can calculate the phase A current in the common and series winding portions of the transformer:

$$I_A = I_1 + I_2 + I_0$$

The phase-to-neutral voltage of the common winding is equal to 345 kV/$\sqrt{3}$ and can be selected as 1.0 pu turn, and the remaining windings can be scaled accordingly, as shown in Fig. 5.3. Assuming the current flow in the tertiary winding is in the down direction, and using the ampere-turns balance concept for transformer phase A for the up and down current directions, we have

$$\text{ampere-turns in the upward current direction} \\ = 3019 \times 1.217\ \text{pu} = 3674\ \text{pu-turn} \quad (1)$$

$$\text{ampere-turns in the downward current direction} \\ = 3445 \times 1\ \text{pu} + 3381 \times 0.0692\ \text{pu} = 3679\ \text{pu-turn} \quad (2)$$

Since equation (1) is nearly equal to (2), the solution is O.K. This confirms that the tertiary current is flowing down for a high-side L-g fault, and by using a transformer

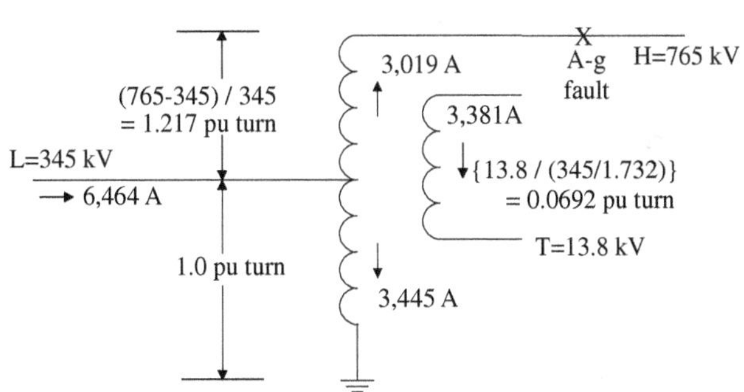

Fig. 5.3 Autotransformer phase A ampere-turns balance.

phasor diagram, the secondary CT current can be deduced accordingly, using the current polarity rule to provide correct polarization for ground overcurrent primary and backup relaying.

5.1.9 Delta/Grounded-Wye Transformers with the Grounded-Wye Side Leading Delta by 30° (Yd1 Transformer Connection)

For generator step-up unit (GSU) transformers the high-voltage (system) winding is connected as a grounded-wye and the low-voltage (generator) side as a delta, as shown in Fig. 5.4. The rational behind this connection is as follows:

1. Suppress the generated third and multiple harmonics and prevent them from flowing to the customer loads connected to the system by the delta winding connection.
2. Provide the needed zero-sequence currents (source) for the ground faults on the high-voltage solidly grounded transmission system.
3. Establish that the high-voltage system is operated as an effectively grounded (solid) system.
4. Establish selectivity in detecting ground faults on the high-voltage transmission system by relying on currents or the derived impedance to detect faults, thus creating relaying zones with effective coordination to isolate only the faulted element.
5. Isolate the generator high-resistance grounding system from the rest of the power system, thus protecting the generators from severe damage from stator ground faults.
6. Force the third-harmonic-generated currents to circulate between the generator high-resistance grounded neutral and the terminal capacitances around the generator isolated phase bus by the delta winding connection. This will allow the application of 100% stator ground fault protection. The use of third-harmonic flow or nonflow can be factored in the logic to cover the protection of the remaining 10% stator winding near the generator neutral.

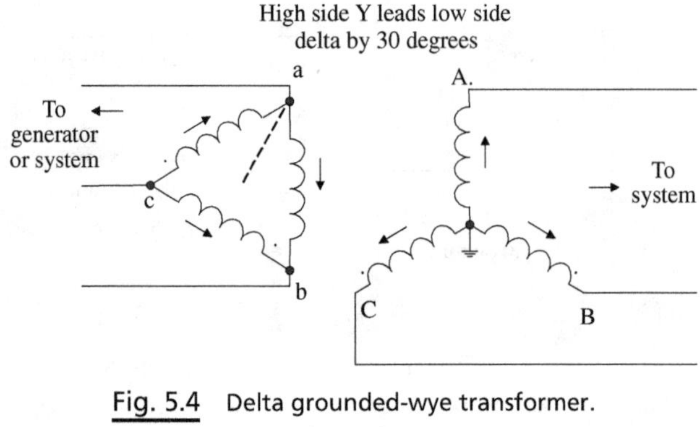

Fig. 5.4 Delta grounded-wye transformer.

7. Permit the ideal design of the generator grounding system by the matching of charging capacitive VAR around the generator with the neutral resistor loss. This will reduce transient overvoltages.

Steps to Find and Confirm the Phase Angle Through the Yd1 Transformer

The following steps can be used to phase the Yd1 transformer shown in Fig. 5.5.

1. Select a three-phase phasor rotation reference.
2. Apply the voltage drop rule for two of the winding magnetic flux couplings shown, using the assumed phasor rotation reference in step 1.
3. Combine the results of the windings in step 2 to deduce the open delta shown in the figure.
4. Complete the delta and show the phase-to-neutral side for the reference phase by connecting the median point (neutral) inside the delta to the corner of phase A.
5. Compare the phase A-to-neutral phasor on the wye side with the corresponding phase-to-neutral delta side. It can be shown that the grounded wye winding a-n leads the delta component A-n by 30°.

In accordance with the IEC standard, this transformer is a Yd1 and can be used as a generator step-up transformer, where according to the NEMA standard, the high side leads the low side by 30°.

5.1.10 Delta/Grounded-Wye Transformers Where the Delta Side Leads the Grounded-Wye Side by 30° (Yd11 Transformer Connection)

For distribution systems, the majority of step-down transformers are also connected as delta/YG to isolate the high-voltage ground relaying system from the low-side

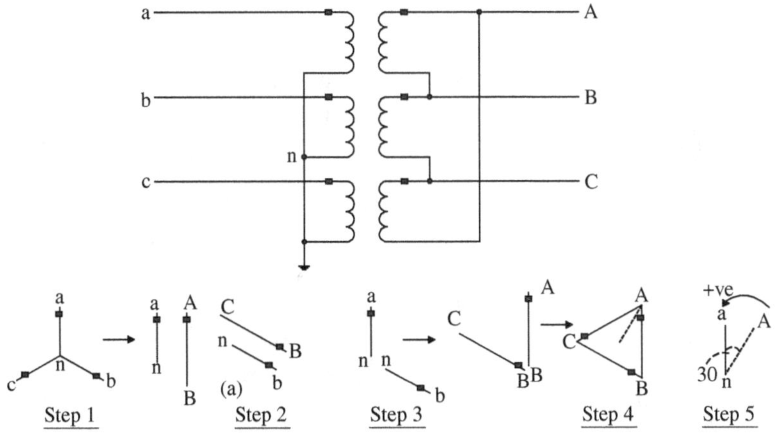

Fig. 5.5 Transformer phasing steps for a Yd1 connection.

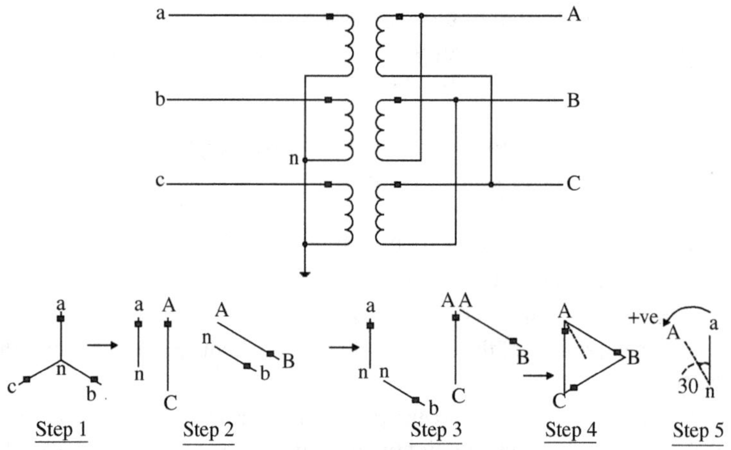

Fig. 5.6 Transformer phasing steps for a Yd11 connection.

relaying system and to provide a grounding source to the solidly grounded distribution system. This will enhance the coordination aspects of the protection systems. In addition, low-resistance grounding systems can be established by inserting a resistor in the neutral of the transformer low-voltage winding. Steps to find the phase angle across the Yd11 transformer shown in Fig. 5.6 are similar to those for the connection described in Fig. 5.5.

5.2 TRANSFORMER DIFFERENTIAL PROTECTION BASICS

5.2.1 Hardwired Differential Protection

Traditionally, transformers were protected using harmonic restraint electromechanical (EM) or solid-state discrete differential relays on a per-phase basis. Three relays are required to provide protection for three-phase transformers. For two-winding delta/grounded-wye transformers, connecting the CT at the wye side of the transformer in delta and at the delta side in wye allows the hardwired EM or static relays to correct for the 30° phase shift between the high and low sides and also to trap the zero-sequence current for external faults on the wye side. The traditional manufacturer of these relays was also the designer and manufacturer of the transformers. Therefore, complete knowledge of the transformer inrush phenomena was utilized in the design of the harmonic restraint feature of the differential relays. Relays are delivered to the users with a built-in circuitry to cope with inrush phenomena during transformer energization. Only two setting parameters are required to commission these relays. The parameters are relay taps (8.7, 5, 4.6, 4.2, 3.8, 3.5, 3.2, or 2.9 A) and relay slope (15, 25, or 40%).

5.2.2 Multifunction Numerical Differential Relaying

With the advent of numerical multifunction transformer differential relaying, manufacturers have designated many parameters to be settable by users. Several pages of setting are therefore required to commission microprocessor-based numerical relays, as compared with the few parameters needed traditionally for discrete electromechanical or static relays. The connection of all CT inputs in wye simplified fault analysis, made breaker failure fault detector elements (if required) easy to set, and allowed the formation of a single-point neutral grounding for the CT secondary wiring. Therefore, the numerical relay uses software to compensate for the phase angle between transformer windings and to block zero-sequence current flow as an operating quantity during external ground faults and to restrain the relay for inrush current generated during the energization of transformers.

The transformer multifunction numerical differential relay shown in Fig. 5.7 provides differential protection (87), instantaneous overcurrent protection (50), time overcurrent (TOC) protection (51), ground instantaneous (50N), ground time overcurrent (51N), undervoltage (27), overvoltage (59), overfrequency (81O), underfrequency (81U), and V/HZ (24) protection. The relay also provides a restricted transformer differential zone that offers sensitive protection for ground faults close to the transformer neutral formation point. In this case the differential element (87G) is fed from the ground current of the transformer neutral CT and the calculated ground current ($3I_0 = I_A + I_B + I_C$) derived from the phase CTs. In addition, the relay

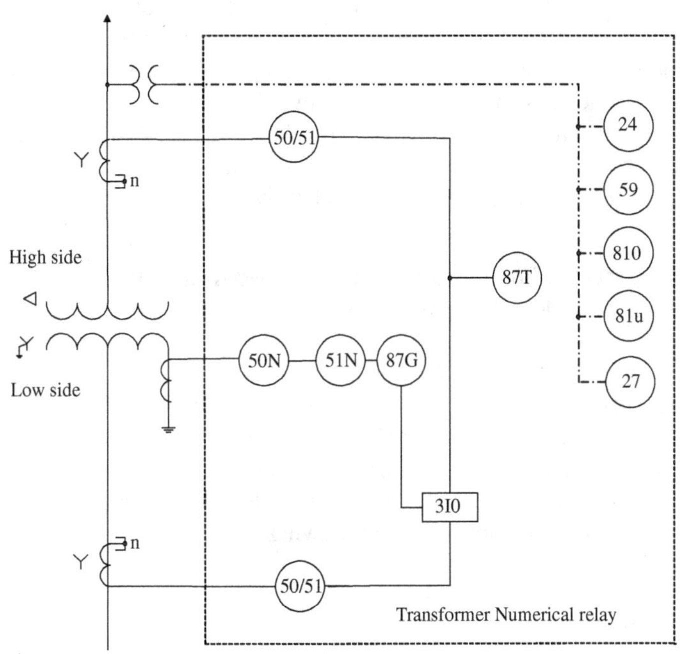

Fig. 5.7 Transformer multifunction numerical relay.

provides metering (MW, MVAR, *V*, *I*) quantities for the transformer as well as oscillography and fault records.

Parameters used for inrush restraint are among those required to set the relay. The rational behind user selection of the type of harmonics and their threshold ratios is unclear. Incorrect selection of harmonics and their ratios by users may result in undesired operation during transformer energization, as illustrated in Case Studies 5.10 and 5.11.

5.2.3 Applying Transformer Differential Protection

Four basic principles must be implemented to design and employ successful transformer differential protection systems:

1. Current magnitude balance
2. Phase angle compensation
3. Zero-sequence blocking for external L-g faults if the transformer is a source for zero-sequence currents (i.e., delta/grounded wye, autotransformer with delta tertiary, etc.)
4. Restraining relay operations from tripping the transformer during inrush energization

The importance of the basics associated with the invention and implementation of electromechanical and static relays should be emphasized. These basics are essential for correct implementation of the new numerical relaying technology and the avoidance of undesired relay operations. Electromechanical and static relaying schemes perfected through many years of operating experiences are sometimes being applied and replaced, with limited reviews, by less experienced engineers using the available logic offered by numerical relaying. This will highlight the importance of familiarity with the basic schemes, to apply (often done without checking) the new logic safely via the setting of the numerical relaying.

5.2.4 Differential Protection Basics for Two-Winding Delta/Grounded-Wye Transformers

Electromechanical and solid-state component relays were the main protection for transformers in the past. In the electromechanical and solid-state era, the formation of hardwired connections to the relay eliminated some of the confusion encountered when employing numerical relays. The wiring of CT as a delta on the grounded-wye side of a transformer and as a wye on the delta side of the transformer, as illustrated in Case Study 5.16, accomplished the following objectives:

1. Compensation for the 30° phase shift between the delta and grounded-wye sides of the transformer
2. Trapping of the zero-sequence current flow during external system ground faults on the grounded-wye side, away from the differential relay operating coil, thus preventing undesired relay trips for external ground faults

Most transformer multifunction numerical differential relays are applied using wye-connected CTs on both sides of the delta/YG transformer, as shown in Fig. 5.7. This is the preferred connection to obtain a relay oscillograph record on a per-phase basis, thus simplifying postmortem fault analysis. In addition, the wye-connected CTs permit single-point grounding of the neutral, thus enhancing personnel safety. Therefore, numerical relaying utilizes software routines both to cope with the 30° phase shift and to restraint the relay during external L-g faults and inrush conditions.

5.3 CASE STUDIES

Case Study 5.1: Energization of a 5-MVA 13.8/4.16-kV Station Service Transformer with a 13.8-kV Phase-to-Phase Bus Fault Within the Transformer Differential Protection Zone

Abstract A 5-MVA station service transformer T2 was energized into a phase-to-phase fault. Apparently, windblown water intruded into the outdoor station service transformer enclosure and resulted initially in a phase-to-phase fault, which evolved to a three-phase fault. The transformer T2 differential relay instantaneous unrestrained function and instantaneous overcurrent backup element of the unit transformer T1 numerical differential relay operated to clear the fault. In this case study we document the energization of a transformer into a fault within the differential relay protection zone. It reveals the generation of enough second-harmonic contents in the transformer inrush current to block the tripping output of the differential protection function of the numerical relay during the fault. We also highlight the importance of an instantaneous (nonrestrained) differential function and time overcurrent independent backup element for transformer protection. We also analyze the disturbance, and the second harmonic of the transformer faulted current and suggest corrective actions and lessons learned.

Description of the Protection System Station service transformer T2 is protected by dual relaying systems: one a dedicated transformer differential and the second a TOC element for the CT W3 input to the transformer T1 differential relay programmed as instantaneous and TOC elements. As shown in Fig. 5.8, the 50/51 TOC function of the 87T1 is fed from the parallel CTs that are associated with the 13.8-kV lead to station service transformers T2 and T3. The 50/51 programming function for the 13.8-kV side associated with 87T1 numerical differential relay input W3 provides an open protection zone that is controlled by the setting calculation of the overcurrent elements. In addition, the 87T2 numerical differential relay input W1 is programmed as an instantaneous and TOC element.

Zones of Protection As shown in Fig. 5.8, the 87T2 numerical differential relay provides a closed zone that covers the 13.8-kV breaker A3, the 13.8-kV cable connection to T2, transformer T2, and the 4.16-kV cable connection to the bank. The 50/51 programming function for the 13.8-kV side associated with numerical differential relay 87T2 provides an open protection zone that is controlled by the setting calculation. In addition, the 50/51 programming function associated with

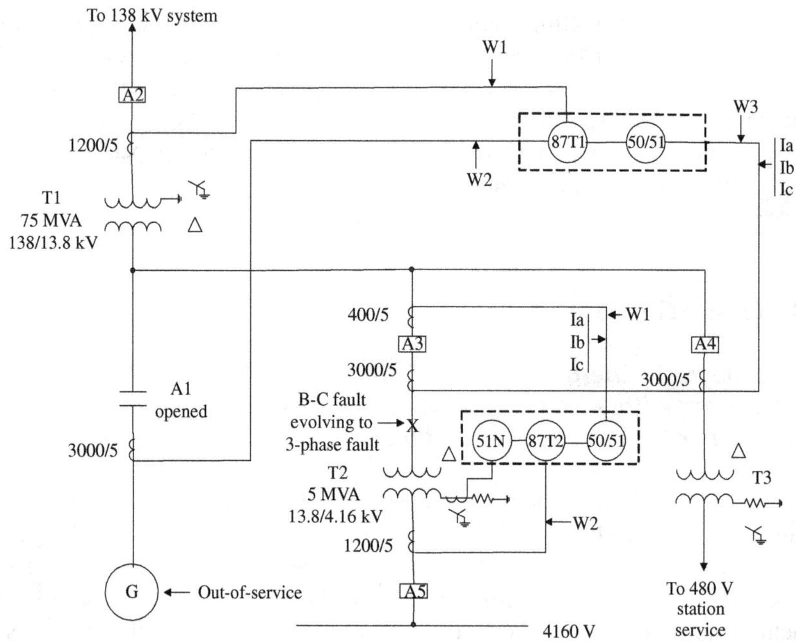

Fig. 5.8 One-line ac diagram showing the two independent relaying-systems for transformer T2 and the relay-monitored currents.

numerical differential relay 87T1 provides an open protection zone that is also controlled by the setting calculation. Element 50 is set to cover for faults within 80% of the transformer T2 impedance, and element 51 provides a coordinated trip with 4.16-kV switchgear feeder protection.

Relay Setting Criteria The overcurrent instantaneous element 50 of the numerical relay 87T1 is set above the transformer inrush and 125% of the low-side fault, whichever is larger. The instantaneous element 50 of numerical relay 87T1 is set at 125% of the low-side fault (4440 A). The TOC element 51 tap is set at 150% of the maximum transformer loading of 6667 kVA. The element 51 TOC time dial is selected to be plotted underneath the transformer damage curve and to coordinate with the 4.16-kV feeder TOC relays.

Description of the 13.8-kV Switchgear Enclosures The 13.8-kV bus is designed with a clearance of 7.5 inches between phases. The phases arced-over when the air between the bus metal barriers was saturated with moisture. The presence of moisture within a confined environment initiated a flashover across the top of the 13.8-kV phases.

Description of the Incident A phase B-C fault occurred on the transformer T2 13.8-kV connection lead when the 13.8-kV breaker A3 was closed to energize station service transformer T2. The fault may be attributed to windblown water and moisture, creating a humid environment inside the 5-MVA 13.8/4.16-kV transformer enclosure. The failure has occurred during heavy rain when water was trapped on the concrete slab. No water was leaking from the top of the transformer enclosure. The fault evolved to a three-phase fault when phase A got involved after 3.25 cycles from initiation of the B-C fault. The fault was cleared by removing the 138-kV source 4 cycles after initiation of the B-C fault.

Analysis of the T1 and T2 Numerical Relay Oscillograph and Fault Records

Transformer T2 Relay Figure 5.8 shows the system one-line diagram and numerical relay–monitored currents. As shown in Figs. 5.9 and 5.10, the incident

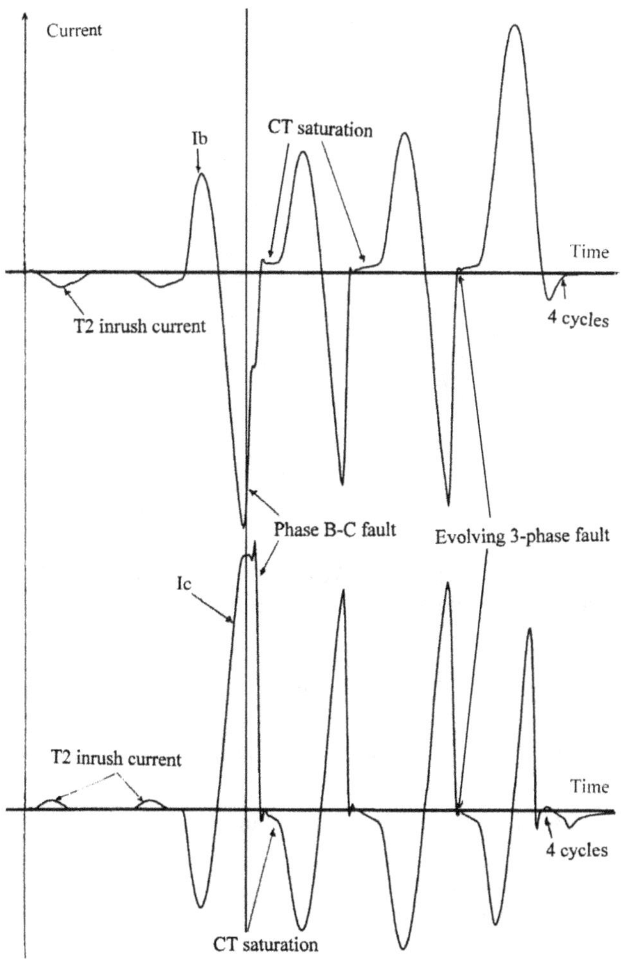

Fig. 5.9 Transformer T2 relay 87T2 fault record showing the W1 phase B and C 13.8-kV winding current during the fault.

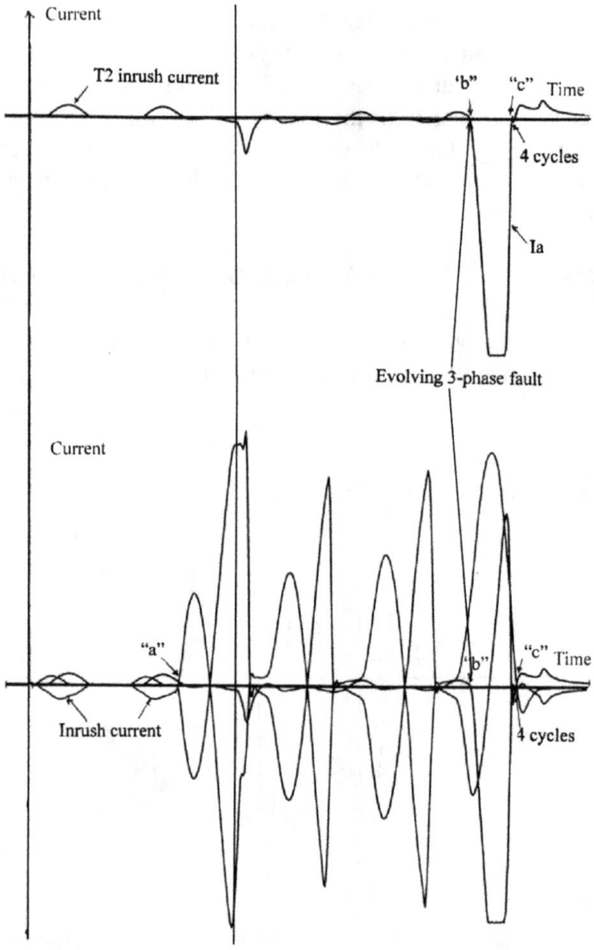

Fig. 5.10 Transformer T2 relay 87T2 fault record showing the W1 phase A and three-phase 13.8-kV winding current during the fault.

started with a few cycles of transformer T2 inrush currents followed by a 13.8-kV phase B-C fault that lasted for about 3.25 cycles, then evolved to a three-phase fault for an additional 0.75 cycle. The relay 87T2 fault record indicates that the RMS current phasors for the high-side (13.8-kV) winding W1 and low-side (4.16-kV) winding W2, which coincided with the operation of the instantaneous differential element, are as follows: Winding W1: $I_A = 356$ A at 0°, $I_B = 6504$ A at 67° lag, $I_C = 7435$ A at 247° lag, neutral (ground) current = 0. Winding W2 currents = 0; the low-side 4160 V has no fault current sources.

Differential currents (× CT rating of 5 A): $I_A = 1.12$ A, $I_B = 16.92$ A, $I_C = 17.92$ A

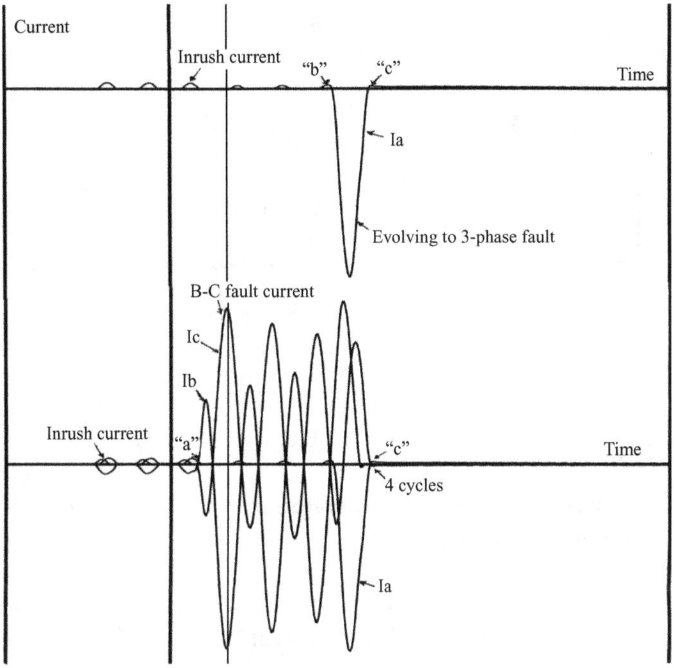

Fig. 5.11 Transformer T1 relay 87T1 fault record showing the W3 phase A and three-phase 13.8-kV winding current during the fault.

Restraint currents (× CT rating of 5 A): $I_A = 1.12$ A, $I_B = 16.92$ A, $I_C = 17.92$ A
Second harmonic as a percentage of 60 Hz: $I_A = 96\%$, $I_B = 99.9\%$, $I_C = 91.1\%$

Transformer T1 Relay Figure 5.11 shows phase A and three-phase currents of winding W3 of 87T1 during the 13.8-kV fault. During B-C fault phases B and C, currents were equal and opposite. The fault then evolved into a three-phase fault, confirming the T2 oscillography fault records shown in Figs. 5.9 and 5.10. Figure 5.11 also illustrates phase A current change from inrush current to fault current when the fault evolved to a three-phase fault for the remaining 0.75 cycle. Figure 5.12 reveals the three phase currents for the 138-kV (W1) side of numerical relay 87T1, where the phase B current is equal to the sum of phase A and C currents during the duration of the B-C fault. This relationship is also shown in Fig. 5.13 for the transformer T1 phasing diagram. The transformer T1 numerical fault record showing phase A and the three-phase relay 87T1 fault record indicates that the RMS current phasors for high-side (138 kV) winding W1 and low-side (13.8 kV) windings W2 and W3, which coincided with the operation of the phase instantaneous element of winding W3, are as follows:

Winding W1: $I_A = 259$ A at 0°, $I_B = 529$ A at 175° lag, $I_C = 269$ A at 348°, ground current = 0

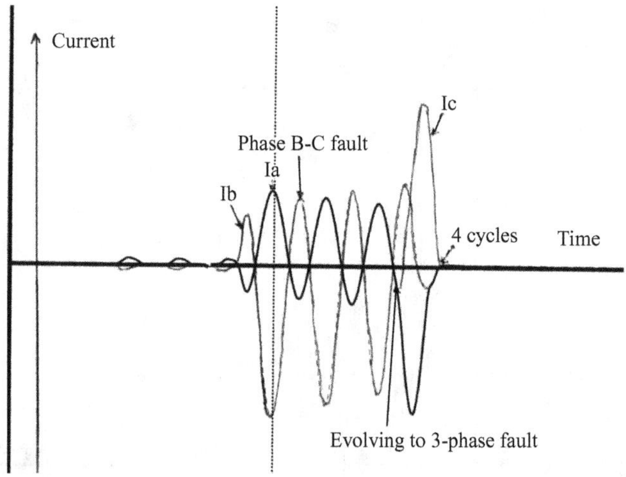

Fig. 5.12 Transformer T1 relay 87T1 fault record showing the W1 three- phase 138-kV winding currents during the fault.

Winding W2 : no current flow, due to the generator being out of service

Winding W3 : $I_A = 360$ A at 94°, $I_B = 4560$ A at 177° lag, $I_C = 4630$ A at 352° lag, ground current = 0

Analysis of the Numerical Relay 87T1 and 87T2 Fault Records The T2 oscillography fault record in Fig. 5.9 shows that the phase B current is about equal and opposite to the phase C current for a duration of 3 cycles. This will also confirm the currents shown in Fig. 5.10, which are recorded by the relay for the interval between a and b. In addition, since the relay W1 CT ratio is 80 : 1, the differential current and restraint current recorded by the relay can be correlated with the primary fault current. For example, for phase B the differential current = 16.92 × CT rating = 16.92 × 5 × 80 = 6768 A, which is very close to the recorded primary current of 6504 A.

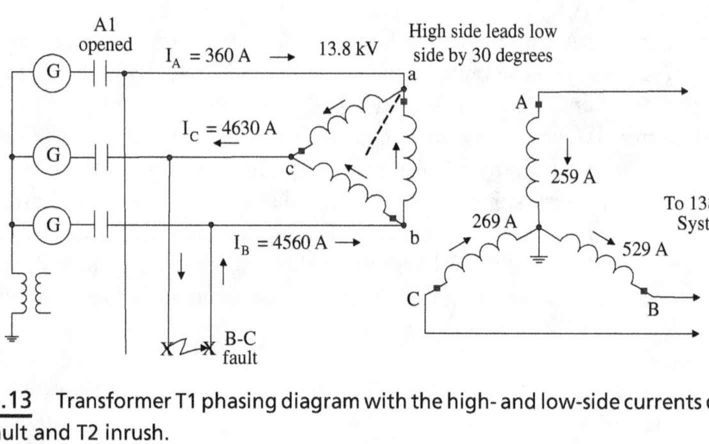

Fig. 5.13 Transformer T1 phasing diagram with the high- and low-side currents during the B-C fault and T2 inrush.

The T1 fault record is generated by an instantaneous trip of winding W3. The calculated RMS value of phase B of winding W3 is 6504 A, which exceeds the instantaneous setting threshold of 4440 A. The 138-kV winding W1 of record T1 confirms the fault record of Fig. 5.12, where $I_B = 529$ A at 175°, which is about equal to the sum of $I_A + I_C = 259$ A at 0° + 269 A at 348°. In addition, the phase A current is nearly equal to the phase C current. It can then be concluded that the relay fault records match the relay oscillography records. Figure 5.13 also confirms the ampere-turns coupling of transformer T1 as related to the relay fault record for the RMS currents calculated in the high (W1) and low (W3) windings.

Relay Targets Reported for the Operation The fault was cleared by the operation of instantaneous element 50 of the 5-MVA transformer numerical relay 87T2 and instantaneous element 50 of the multifunction GSU numerical differential relay 87T1. The relaying input as shown on the one-line diagram was programmed to provide the redundant element for the 5-MVA transformer protection via elements 50 and 51.

Response of the 87T2 Differential Element The relay fault record which coincided with operation of the 87T2 instantaneous differential element indicates that the second harmonic as a percentage of 60 Hz for phases A, B, and C is equal to 96, 99.9, and 91.1%, respectively. The values for all the phases at the energization of T2 and during the fault are above the second-harmonic restraint toehold of 20%. As a result, the percent differential function of T2 numerical relay was restraint from operation on all the phases during the fault.

Lessons Learned

1. Energization of a transformer with a fault within the differential zone presents a challenge to the harmonic restraint differential concepts and may result in generation of enough harmonic that may block operation of the relay, as demonstrated in this case study.
2. Transformer protection should be employed using diverse operating principles. Differential protection (current comparison) should be supplemented with instantaneous and TOC elements.

Case Study 5.2: Lack of Protection Redundancy for a Generator Step-up Transformer Leads to Interruption of a 230-kV Area

Abstract A surge arrester associated with the high side of a generator step-up (GSU) transformer failed, causing a line-to-ground fault. The fault was detected by transformer numerical differential relay device 87T1, which energized GSU lockout relay device 86T1. Device 86T1 failed to operate and isolate the fault in primary high-speed time by tripping and locking out CB A1 and shutting down the generating unit G1. Apparently, the independent power producer (IPP) facility was placed in service with a single lockout relay that interfaced to only one numerical transformer differential relay. As a result, the

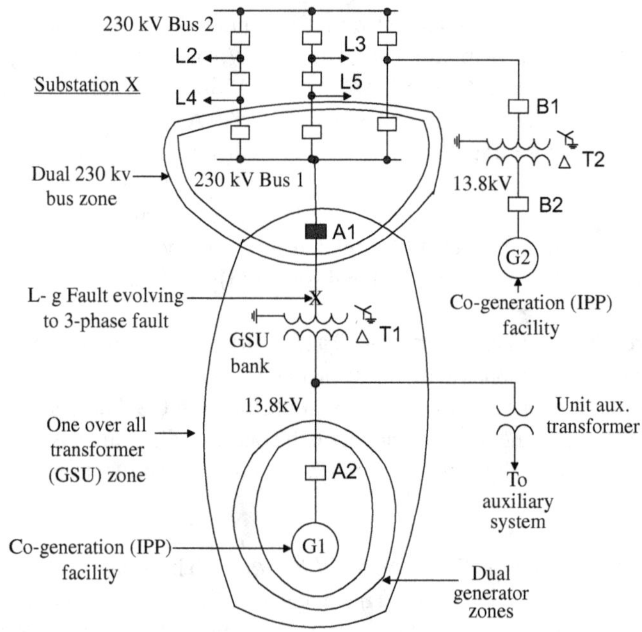

Fig. 5.14 Zones of protection for the IPP connection facility to substation X.

initial ionized cloud around the surge arrester grew, causing the fault to evolve into a double-phase-to-ground fault and eventually to a three-phase fault. The failure to isolate the fault locally led to exporting the event to the surrounding system. The failure of primary protection to trip CB A1 and to initiate its associated breaker failure can only be mitigated by operation of remote backup protection to clear the fault from the system. In this case study we describe the incident and provide corrective actions and lessons learned. In addition, we illustrate the violation of application of a dual-element principle for the protection of the bulk power system elements.

Description of the Plant Protection Systems Figure 5.14 illustrates the zones of protection for the IPP connection facility at substation X. Generating unit G1 is protected by dual multifunction numerical generator relays with separate dual lockout relays. The 230-kV bus 1 and the 230-kV lead connecting unit G1 are protected by dual high-impedance bus differential relays. However, the GSU T1 transformer bank is protected by only one overall transformer numerical differential relay.

As shown in Fig. 5.15, transformer T1 is protected by one numerical differential relay 87T1 with a single lockout relay 86T1. The numerical relay provides differential protection and inverse ground time overcurrent 51TN as a backup system protection. Transformer T1 sudden pressure relay 63T1 operation is routed through the numerical relay to be able to generate an event record during its operation. The 63T1 or 51TN operation will also energize the same lockout relay. It is therefore clear that failures of the transformer numerical relay or the lockout relay will disable T1 protection systems.

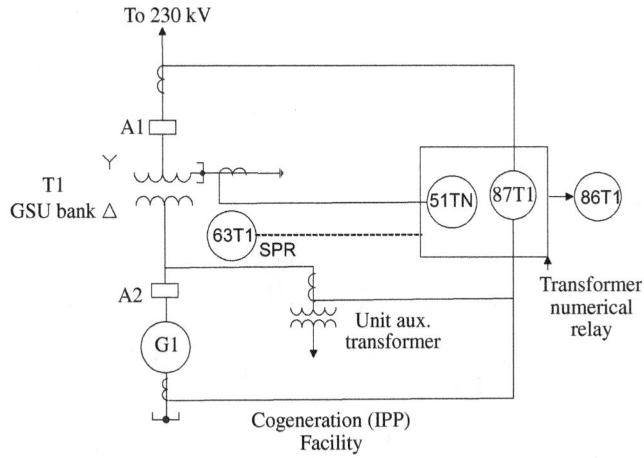

Fig. 5.15 One-line ac diagram showing the T1 GSU transformer protection systems.

Circuit breaker A1 is provided with a breaker failure protection function which is initiated by operation of the 86T1 lockout relay or the 230-kV bus 1 relay protection.

Analysis of the DFR Record Figure 5.16 shows the system one-line diagram and DFR-monitored voltages and currents for line L1 at substation W. The DFR record in Fig. 5.17 shows a C-g fault which was due to surge arrester failure on the high side of

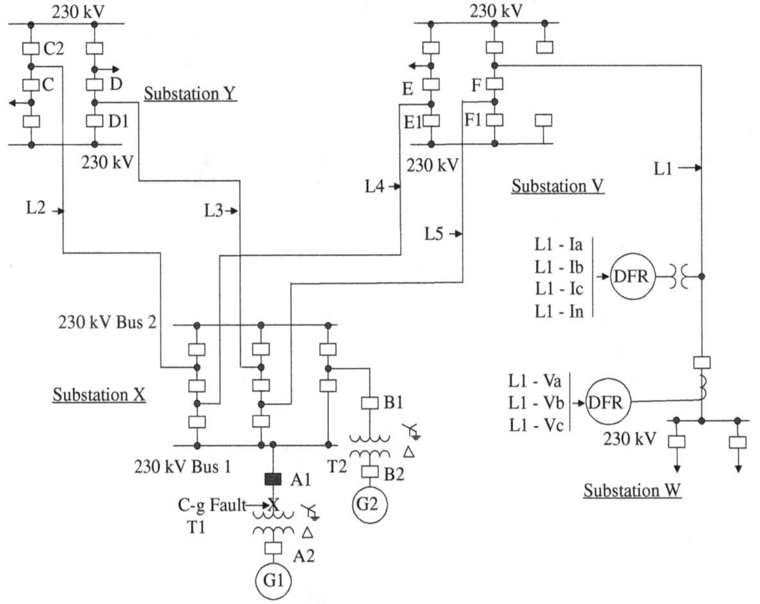

Fig. 5.16 System one-line diagram and line L1 DFR-monitored voltages and currents.

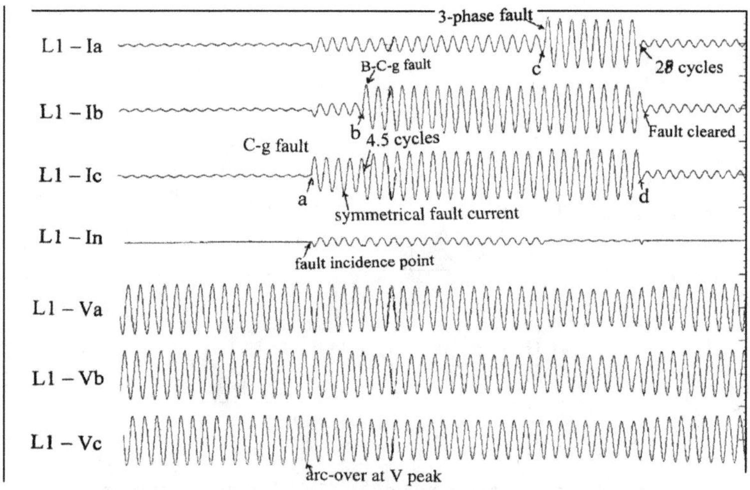

Fig. 5.17 Substation W DFR record for line L1 voltages and currents during the initial and subsequent evolving faults.

transformer T1. The fault incident point a occurred at the peak of the faulted phase C-n voltage, resulting in a symmetrical fault current that contains no dc current component. As mentioned earlier, the IPP facility was placed in service with a single lockout relay that interfaces to only one numerical transformer differential relay. The failure of lockout relay 86T1 to operate prolonged the presence of the fault, and as a result, the initial ionized cloud around the surge arrester grew, causing the C-g fault to evolve into a B-C-g fault at point b 4.5 cycles after point a ($t = 0$). The failure of lockout relay 86T1 also resulted in no breaker failure protection initiation for circuit breaker A1. Therefore, the single failure of relay 86T1 led to the failure of both primary relaying and breaker failure relaying (local backup). The fault can only be cleared by remote backup protection around substation X. The ionized cloud around the surge arrester grew further, eventually causing the fault to evolve into a three-phase-to-ground fault at point c 20 cycles after point a ($t = 0$).

At 28 cycles after $t = 0$, zone 2 (Z2) remote backup protection at substations Y and V operated to clear the fault at point d. Z2 relays at substation Y tripped 230-kV CBs C and C2 for line L2 and CBs D and D1 for line L3. Similarly, Z2 relays at substation V tripped 230-kV CBs E and E1 for line L4 and CBs F and F1 for line L5. The fault is now cleared from the 230-kV power system. Generating unit G1 and G2 contributions to the fault were stopped at 1.05 s from $t = 0$ by the operation of an undervoltage relay associated with each unit which is set to operate after a 1-s time delay. The undervoltage protection is being applied on most of the IPP interface to the connecting utilities.

Corrective Actions

1. Dedicate an output from numerical relay 87T1 to initiate CB A1 breaker failure directly to enhance the function and to cover for lockout relay 86T1 failure.

2. Add a redundant transformer numerical relay with a dedicated lockout relay. This addition will fulfill the dual-element requirement for the protection of a bulk power system.
3. Add a backup protection element at the 230-kV interface between the IPP generating facility and the connecting utility to provide independent backup protection for the IPP facility.

Lessons Learned

1. The occurrence of a single failure event that can disable the entire protection system for any power system element fault should be avoided. In this case study the numerical relay 87T1 output, SPR 63T1, and all related transformer backup protection is energizing a single lockout relay, which is a deficiency in meeting the avoidance of a single failure occurrence that can disable the entire transformer protection system.
2. Two high-speed protection systems are required for the 230-kV connection lead between the transformer and the interface breaker as well as the GSU transformer.
3. Each power system element should be protected by at least dual independent protection systems. The selection implementation will depend on the criticality of the power system and fault clearing speed requirements to keep the system stable during faults. Every fault occurring on any part of the power system should be detected by a minimum of two independent physically separated protection devices.
4. Remote backup protection is vital in isolating faults during extreme system conditions and when primary protection and breaker failure protection are disabled.

Case Study 5.3: Undesired Operation of a Numerical Transformer Differential Relay Due to a Relay Setting Error in the Winding Configuration

Abstract A GSU transformer was tripped by the primary transformer differential relay during an external 138-kV line-to-ground fault. In this case study we describe the protection system and explain the causes of this undesired trip. We demonstrate the difficulty and confusion sometimes encountered during application of the new numerical transformer protection technology. We also describe a numerical relay setting verification procedure, power system phenomena, corrective actions, and lessons learned.

Description of the Protection System The plant is protected using dual independent relays to sense and detect all faults in the relay protection zone. This criterion will provide two redundant relay systems to guarantee fault clearing. Primary GSU transformer differential relay 87TP is shown in Fig. 5.18 with the generator CT input not connected. The 138-kV system is used as a startup source to backfeed the plant auxiliary loads.

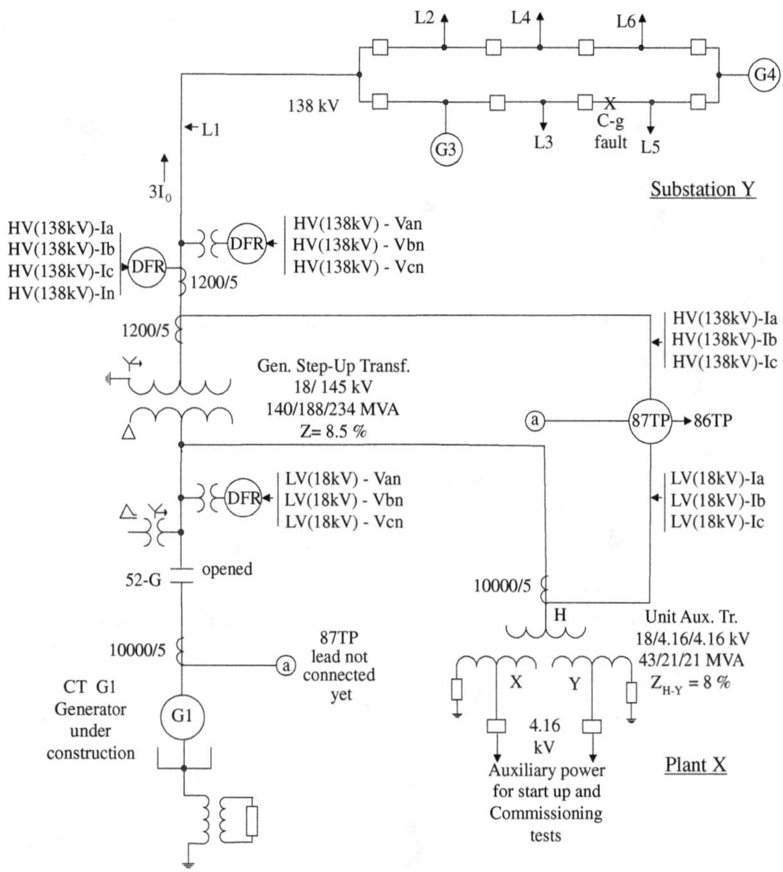

Fig. 5.18 One-line diagram showing DFR- and GSU numerical relay–monitored voltages and currents.

Basic Principles of Applying Transformer Differential Protection As stated earlier, four basic principles must be implemented to design and employ successful transformer protection systems:

1. Current magnitude balance
2. Phase angle compensation, depending on transformer winding connection types
3. Blocking of zero-sequence current for external L-g faults if the transformer is a source for zero-sequence currents (i.e., delta/grounded wye, autotransformer with delta tertiary, etc.)
4. Restraining relay operations from tripping the transformer during inrush energization

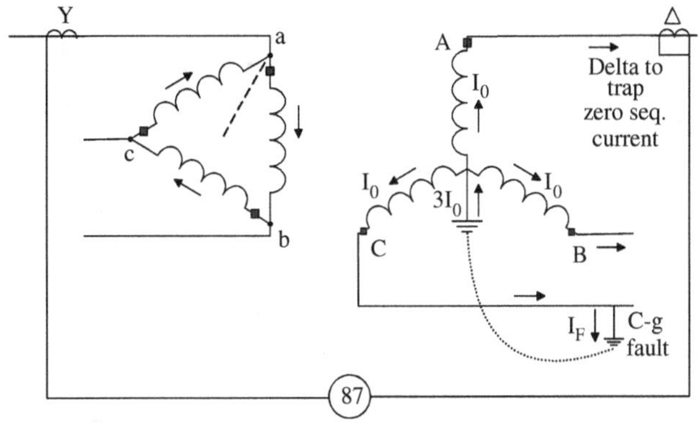

Fig. 5.19 Hardwired differential relay with delta-connected CTs on the wye side.

Differential Protection Basics for Two-Winding Delta/Grounded-Wye Transformers As stated earlier, electromechanical and solid-state component relays were the main protection for transformers in the past. In the electromechanical and solid-state era, the formation of hardwired connections to a relay eliminated some of the confusion encountered when employing numerical relays. The wiring of CT as delta on the grounded-wye side of the transformer and as wye on the delta side, as illustrated in Fig. 5.19, accomplished the following objectives:

1. Compensation for the 30° phase shift between the delta and grounded-wye sides of the transformer.
2. Trapping of the zero-sequence current flow during external ground faults by the delta-connected CTs as shown in Fig. 5.20 on the grounded-wye side away from the differential relay operating coil, thus preventing undesired relay trips for external ground faults

With the advent of numerical multifunction transformer differential relaying, manufacturers have designated many parameters to be settable by users. Several pages

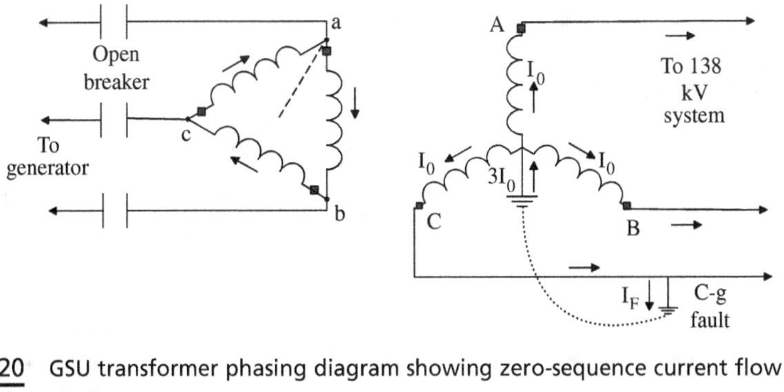

Fig. 5.20 GSU transformer phasing diagram showing zero-sequence current flow for an external C-g fault.

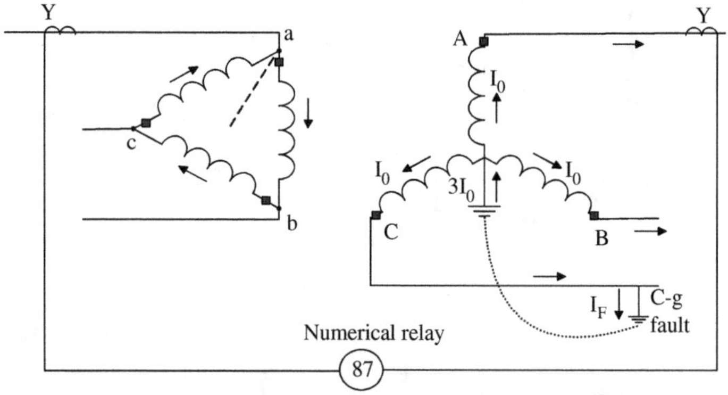

Fig. 5.21 Protection of the transformer using numerical relays.

of setting are therefore required to commission microprocessor-based numerical relays, compared with the few parameters needed traditionally for discrete electromechanical or static relays. As illustrated in Fig. 5.21, most transformer numerical relays are applied using wye-connected CTs on both sides of a delta/YG transformer. This is the preferred connection to obtain a relay oscillograph record on a per-phase basis, thus simplifying the postmortem fault analysis. In addition, the wye-connected CTs permit single-point grounding of the neutral, thus enhancing personnel safety. Therefore, the numerical relays employed in this case study must use software routines to cope with both the 30° phase shift and to restrain the relay during external L-g faults.

Numerical Differential Relay 87TP Setting Regarding Phase Shift Compensation and the Blocking of Zero-Sequence Current The numerical relay 87TP was set to provide phase shift compensation and zero-sequence current blocking during external ground faults internally via software.

Compensation for 30° The numerical relay is set by designating the delta winding as the reference and setting the angle as zero (0°) phase shift corrections and the HV grounded wye to be referenced to the delta in the negative (clockwise) direction. Based on the transformer nameplate, the HV (138 kV) leads LV (18 kV) by 30°. Therefore, the HV grounded-wye winding is set at −330° with respect to the reference delta winding.

Blocking Zero-Sequence Flow During External L-g Faults The relay is designed with software to block zero-sequence flow from the HV grounded-wye winding toward the system during external L-g faults. This is accomplished by setting the grounding parameter as "within zone." This will enable the setting mode to block zero-sequence current and prevent undesired relay operation during external ground faults. This can be implemented by allowing the software to calculate zero-sequence currents using symmetrical components and then subtracting zero-sequence current from the transformer grounded-wye-side phase current. Zero-sequence flow will

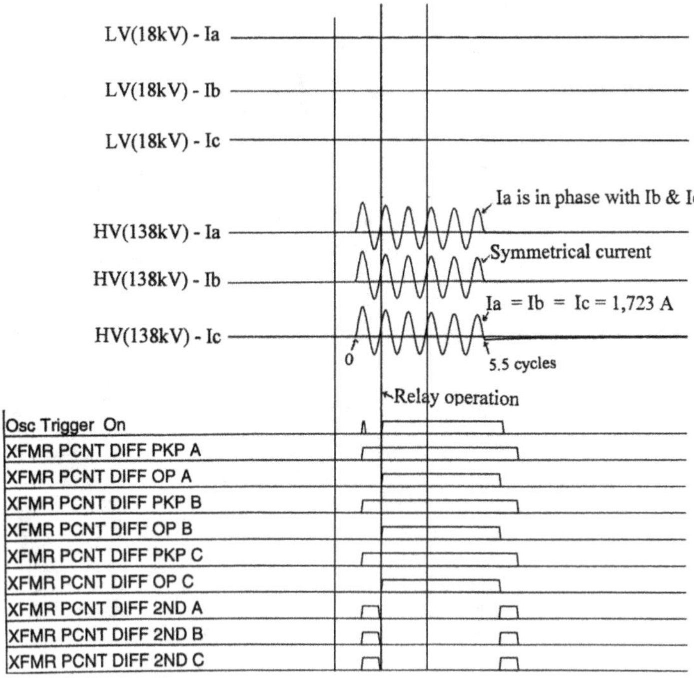

Fig. 5.22 Numerical relay 87TP oscillography record revealing the fault currents and relay responses.

not be removed from the phase currents of the relay if the setting of the winding parameter is made as "not within zone." The as-found grounding setting for the relay was done with "not within zone." Therefore, during the reported external L-g fault on the 138-kV bus, zero-sequence current flow was treated as a contribution to the fault, with flow only on the grounded-wye side without any blocking of the zero-sequence current. This is equivalent to having no delta CT connection for the electromechanical hardwired relay. This will cause zero-sequence differential current components to exist on phases A, B, and C. As shown in the relay fault record of Fig. 5.22, all three phase elements tripped, due to zero-sequence current flow during the external 138-kV C-g fault.

Analysis of the Plant DFR Record The DFR record shown in Fig. 5.23, reveals that the voltage trace HV (138 kV) V_{CN} dipped to zero, indicating a close-in solid phase C-g fault lasting for 5.5 cycles. The fault occurred when a high wind moved a metal wire to the 138-kV phase C bus. The delta/YG transformer is acting as a zero-sequence source supplying only zero-sequence current to the external C-g bus fault. HV (138 kV) currents I_A, I_B, and I_C are equal to 1723 A, and the in-phase confirming zero-sequence current flow to the bus fault with the I_n is equal to $3 \times I_A$, about 5200 A.

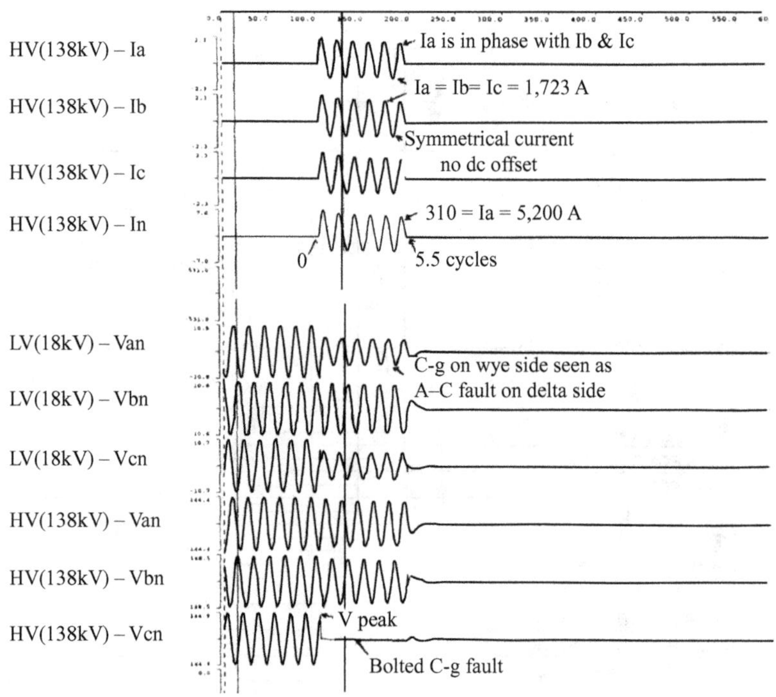

Fig. 5.23 Plant DFR record for the external C-g fault and subsequent false trip.

Analysis of the GSU Numerical Relay 87TP Oscillography Record The relay record shown in Fig. 5.22, reveals that HV (138-kV) currents Ia, Ib, and Ic are equal and in phase and thus represent zero-sequence current components. The record also shows an operating speed of about 1 cycle for all the phases (A, B, and C) as soon as the second-harmonic restraint signal drops out. The numerical differential relays speed of 1 cycle is far superior to the old electromechanical relaying technology speed of several cycles. The relay oscillography record shown in Fig. 5.22 matches the plant DFR record shown in Fig. 5.23.

Causes of the Transformer Differential Relay 87TP Undesired Trip The false operation was caused by a setting error for the "grounding" parameter for the HV GY winding that permitted the zero sequence to flow to the relay. The setting error was "winding = not within zone" instead of "winding = within zone."

Power System Phenomena

1. The delta/GY transformer acting as a zero-sequence source generating equal and in-phase zero-sequence currents, as shown in traces HV(138kV) phase and neutral currents in Figs. 5.22 and 5.23.

2. Phase C-g fault on the grounded-wye side appearing as a A-B fault on the delta side, as shown in the DFR record in the Fig. 5.23 traces LV(18kV)-Van and LV(18kV)-Vcn.
3. Bolted (solid) phase-to-ground fault where the voltage collapses to zero in high-speed time, as shown in Fig. 5.23 trace HV(138kV)-Vcn.
4. Arc-over at voltage peak, when a slow fault-creation mechanism will take place. The fault occurred when a high wind moved a metal wire to the 138-kV phase C bus, causing C-g to start at the voltage peak, thus producing symmetric current, as shown in Fig. 5.23 trace HV(138kV)-Vcn.

Corrective Actions Apply a setting procedure in which the central engineering office will receive the as-built field setting records to be compared with the original relay setting record. Software can be used to compare the two relay setting records automatically and report any deviation from the original record. Presently, most numerical relay manufacturers offer a comparison method for relay setting records as part of their software support.

Lessons Learned

1. The importance of the basics associated with the invention and implementation of electromechanical and static relays should be emphasized. These basics are essential for correct implementation of the new numerical relaying technology and the avoidance of undesired relay operations.
2. A site procedure for application of numerical relay setting during the construction phase should be established.
3. A testing procedure to test relay setting changes during the commissioning phase should be implemented.

Case Study 5.4: Location of a 13.8-kV Switchgear Phase-to-Phase Fault Using Transformer Differential Numerical Relay Fault Records

Abstract A station service supply for a 50-MW plant was tripped by the operation of unit transformer numerical differential relays during heavy rain. The trip was due to a phase-to-phase fault which was fed only from the 138-kV supply. The generating unit was off-line and the fault was first located by analysis of the numerical relay fault record only and confirmed by further analysis. In this case study we describe the procedure used to analyze and locate the fault. We illustrate how to deduce the 13.8-kV fault classification based on an analysis of the relay high-voltage 138-kV recorded currents during the fault. In addition, we describe the fault mechanism, fault analysis using the transformer phasing diagram, corrective actions, and lessons learned.

Description of the Power System and Associated Protection As shown in Fig. 5.24, the 50-MW gas-fired unit is connected to the 13.8-kV switchgear breakers

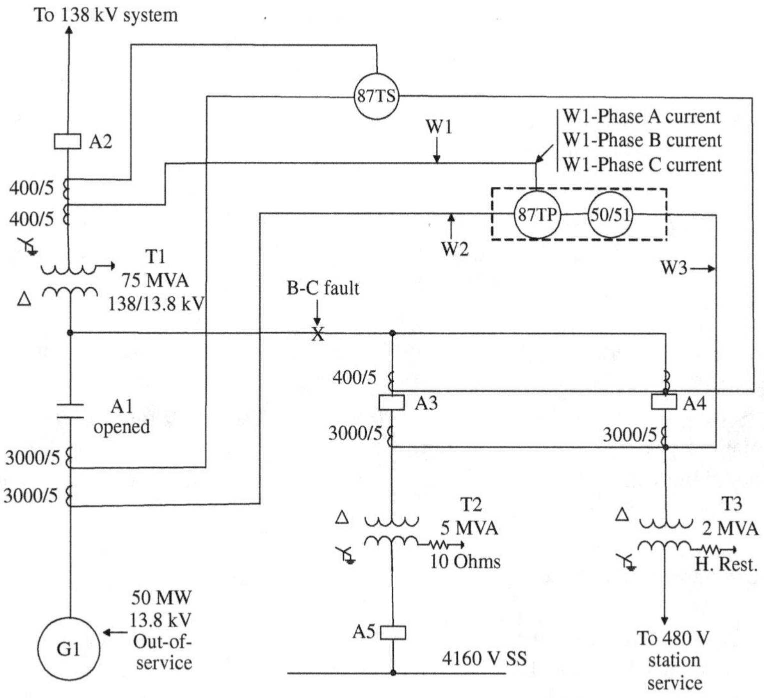

Fig. 5.24 One-line ac diagram showing the two numerical relays 87TP and 87TS and the transformer relay–monitored currents.

and the unit step-up transformer via solid dielectric cables. The transformer high side is connected to the utility system through 138-kV solid dielectric cable. The plant station service is fed from the high-voltage 138-kV system through the unit transformer via dedicated 13.8-kV switchgear breakers A3 and A4 for the G1 unit. The unit is synchronized with the system using 13.8-kV switchgear breaker A1. As shown in Fig. 5.24, the plant is protected by microprocessor-based intelligent electronic devices. Dual multifunction transformer numerical relays are used to protect the transformer, primary relay device 87TP, and secondary relay device 87TS. The transformer differential relay protection zone includes the 13.8-kV bus and the station service 13.8-kV switchgear breakers. The transformer protection zone overlaps the generator zone around the generator breaker. It also overlaps the protection zone of the station service transformer numerical relays.

Fault Mechanism Prior to the occurrence of the fault, the generator G1 was out of service, with its unit breaker opened. The 13.8-kV switchgear associated with unit G1 was designed for top entry by the generator cable leads. The original design was then altered to accommodate a lower entry of the leads to the switchgear. Concrete pads were completed for the original design, which made it difficult to drill holes for cable

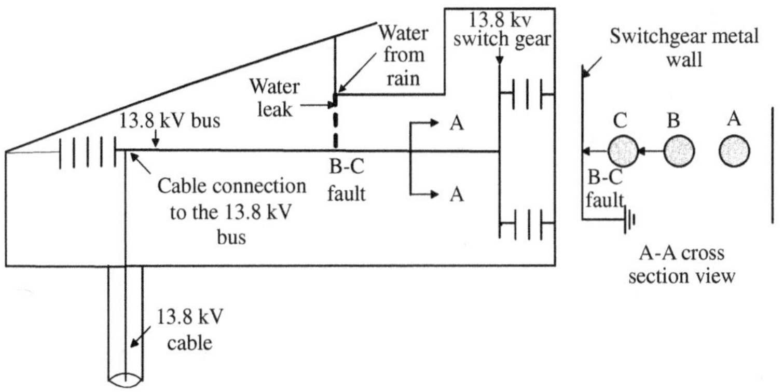

Fig. 5.25 Physical 13.8-kV cable routing to connect the generator to the switchgear breaker.

entry. As shown in Fig. 5.25, the design was modified to build an extension to the switchgear that would allow generator cable lead entry.

Figure 5.25 also shows a cross section of the switchgear extension enclosure showing the three-phase bus. During heavy rain, water apparently penetrated the enclosure, causing water dripping. As a result, phases b and c began to arc, causing a phase b-c-g fault on the 13.8-kVextension bus. The fault was cleared by operation of the dual numerical transformer differential relays in 4.5 cycles.

Transformer Numerical Relay Fault Records Transformer numerical relays 87TP and 87TS have produced the following fault records at the relay trip output for the 138-kV high-voltage transformer side. The primary relay 87TP fault record, shown in Table 5.1, indicates that the high-side (138-kV) RMS current phasors, which coincided with operation of the percent differential element, are as follows: $I_A = 919$ A at 0°, $I_B = 1838$ A at 180° lag and $I_C = 919$ A at 0°. The secondary relay 87TS fault record, shown in Table 5.2, indicates that the high-side (138-kV) RMS currents (magnitude only) at the time of the trip are as follows: $I_A = 933.4$ A, $I_B = 1872$ A, and $I_C = 931.7$ A. The relay fault records also included the differential operating and restraint currents. In

TABLE 5.1 Primary Relay 87TP Fault Record

Event Number Date of Event	Time of Event	Cause of Event	Winding 1 Phase A Current (A)	Winding 1 Phase B Current (A)	Winding 1 Phase C Current (A)
299 'June 17, 2001	'10:36:59.902 am	Trip 2 Off	16 A at 0°Lag	76 A at 183°Lag	15 A at 0°Lag
298 'June 17, 2001	'10:36:59.902 am	Solid Stats Trip Off	16 A at 0°Lag	76 A at 183°Lag	15 A at 0°Lag
297 'June 17, 2001	'10:36:59.900 am	Percent Differential Dropout	16 A at 0°Lag	76 A at 183°Lag	15 A at 0°Lag
296 'June 17, 2001	'10:36:59.836 am	Automatic Trace Trigger	919 A at 0°Lag	1838 A at 180°Lag	918 A at 0°Lag
295 'June 17, 2001	'10:36:59.836 am	Trip 2 On	919 A at 0°Lag	1838 A at 180°Lag	918 A at 0°Lag
294 'June 17, 2001	'10:36:56.836 am	Solid State Trip On	919 A at 0°Lag	1838 A at 180°Lag	918 A at 0°Lag
293 'June 17, 2001	'10:36:59.834 am	Percent Differential Oparate	919 A at 0°Lag	1838 A at 180°Lag	918 A at 0°Lag

TABLE 5.2 Secondary Relay 87TS Fault Record

```
Model Number:
Address: 001   Column: 01   Row: 14
Event Type: Fault record
HV Ia= 933.4 A Ib= 1872 A Ic= 931.7 A LV1 Ia= 0 A Ib= 0 A Ic= 0 A LV2 Ia=
  0 A Ib= 0 A Ic= 0 A Diff Ia= 135.1mA Ib= 25.77 A Ic= 25.77 A Bias Ia=
  81.06mA Ib= 12.90 A Ic= 12.87 A
```

addition, the fault record revealed that the currents for the 13.8-kV switchgear side are all zero (no fault sources).

Procedure to Locate and Classify the Fault The following steps are used to classify and locate the fault based on obtaining the numerical relay event records shown in Tables 5.1 and 5.2.

1. *Create the transformer phasing diagram.* From the transformer nameplate and the actual plant phasing diagram, a three-line ac diagram can be created. This diagram, shown in Fig. 5.26 for the unit transformer, confirms that the transformer is connected according to the NEMA standard, where the high-voltage (138-kV) side leads the low-voltage (13.8-kV) side by 30°.
2. *Assign the fault record currents on the wye winding.* The high-side phasors for I_A, I_B, and I_C of the faulted currents of the primary relay record can now be placed on the transformer primary wye-grounded winding. At the neutral point of the winding, phase B current is divided equally between phases A and C. Figure 5.30 shows the 87TP oscillograph relay record, where the only source for the B-C 13.8-kV fault is the 138-kV system (the generator was out of service with the generator breaker A1 opened). The record indicates initially that the phase B fault current has a phase shift of 180° with phases A and C. In addition, the phase A current is equal and in phase with the phase C current (identical). The B phase current is divided at the neutral point of the transformer primary wye winding into two equal components: $I_B = 2 \times I_A = 2 \times I_C$. This relationship confirms that the transformer high-voltage winding is not faulty. The

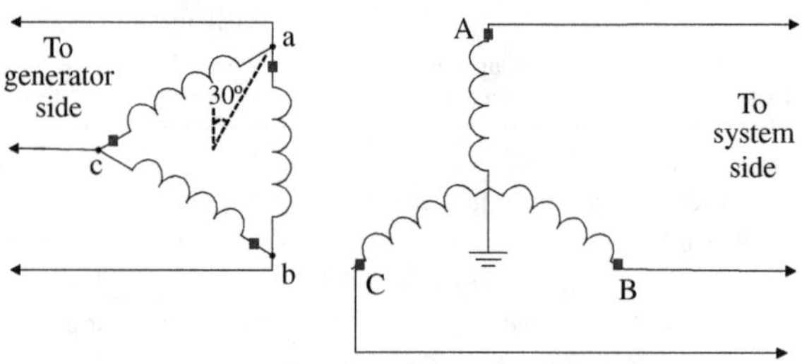

Fig. 5.26 Unit transformer T1 phasing diagram.

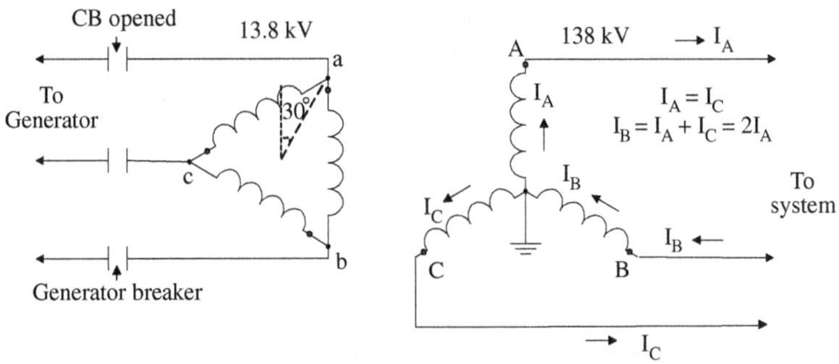

Fig. 5.27 Transformer 138-kV winding with normalized currents assigned in accordance with the numerical relay fault record.

normalized current values can now be placed on the transformer phasing diagram shown in Fig. 5.27 for the primary GY winding.

3. *Find the transformer low-side winding currents.* The transformer low-side currents can then be created from the relay high-side current record. The next step is to apply transformer ampere-turns coupling to the delta winding, as shown in Fig. 5.28. Transformer polarity rules can be used to obtain the low-side currents from the known high-side currents. The polarity rule states: "Currents flowing in at the polarity mark of one of the transformer windings is substantially in phase with current flowing out of the polarity mark of the other winding."

   For phase A on the high side, the I_A flows out of the polarity mark at the high side are in phase with current I_a flowing in the polarity mark of the low-side delta winding. For phase B on the high side, the I_B flows in polarity mark at the high side are in phase with current I_b flowing out of the polarity mark of the low-side delta winding. For phase C on the high side, the I_C flows out of polarity mark at the high side are in phase with current I_c flowing in the polarity mark of the low-side delta winding.

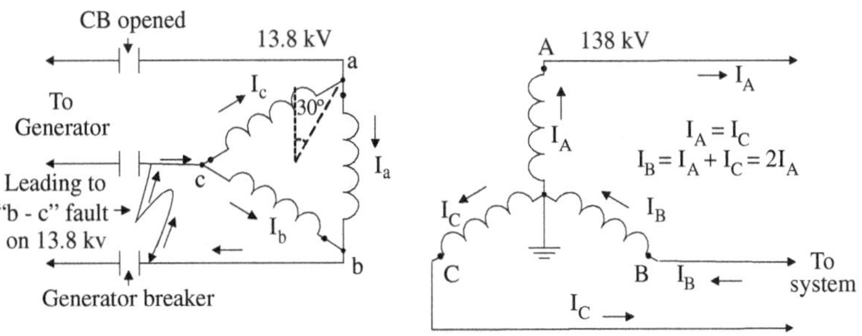

Fig. 5.28 Transformer low-side (13.8-kV) winding-deduced currents based on assigned primary currents and transformer polarity rules.

4. *Deduce the low-side line currents.* The low-side (13.8-kV) line currents can be obtained by applying Kirchhoff's first law to the delta currents, as shown in Fig. 5.28. It can then be concluded that the fault is on the 13.8-kV current as phase b-c, which lasted for 4.5 cycles.

Location and Classification of the Fault Using the Numerical Relay Fault Record Only By applying the procedure outlined above and using the relay fault record shown in Table 5.1, the fault can be classified as follows: The high-side 138-kV currents from Table 5.1 are $I_A = 918$ A at 0°, $I_B = 1838$ A at 180° lag, and $I_C = 919$ A at 0°. These currents can be superimposed on the wye winding of Fig. 5.29: $I_A = I_C = 919$ A and $I_B = 2I_A = 2I_C = 1838$ A.

It can be concluded from the foregoing current relationships that the fault is not internal to the transformer. This is based on application of Kirchhoff's first law at the neutral point of the transformer wye winding, where there are two equal current components flowing in the A and C winding currents. This confirms two equal impedances for the A and C windings, indicating an equal number of turns for each winding.

The low-side delta currents can then be obtained by applying the ampere-turns balance principles, yielding the following relationships for the delta currents shown in Fig. 5.29: The low-side line currents shown can then be obtained from the delta phase currents as follows: Since $I_a = I_c = I_A/N = I_C/N$, where N is the number of turns that are proportional to the winding voltage, we obtain

$$I_X = I_a - I_c = 0 \text{ (since } I_a = I_c)$$
$$I_Y = I_b + I_c$$
$$I_Z = -(I_b + I_c)$$

From these line currents we can now deduce that the low-side 13.8-kV fault is a phase b-to-phase c fault, with no involvement for phase a.

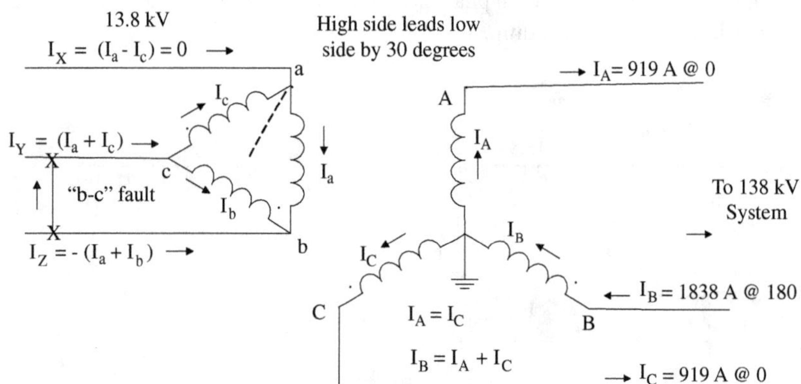

Fig. 5.29 Transformer T1 three-line ac diagram with fault location and fault currents as obtained from Table 5.1.

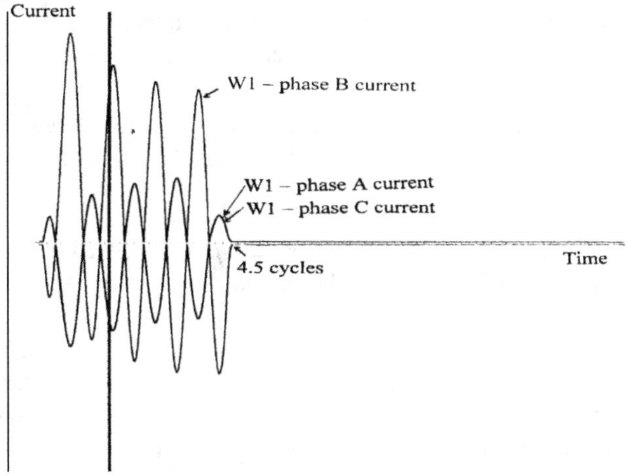

Fig. 5.30 Relay 87TP oscillography for the W1 138-kV current, indicating an a-c low-side fault.

The low-side currents can also be calculated as follows: Since $I_H N_H = I_L N_L$, where I_H is the high-side current, N_H the high-voltage winding turns, I_L the low-side current, and N_L the low-voltage winding turns, we can assume that for the high side, $N_H = 1$ pu turn $= 138/\sqrt{3}$, and for the low side,

$$N_L = \frac{13.8}{138/\sqrt{3}} = 0.173 \text{ pu}$$

For winding A,

$$I_L = I_a = \frac{I_H \times N_H}{N_L} = \frac{I_A \times N_H}{N_L} = \frac{919 \times 1}{0.173} = 5312 \text{ A}$$

The same can also be used to obtain currents for the b and c phases.

The relay fault records are also confirmed by the relay 87TP oscillograph record shown in Fig. 5.30, where the W1 phase A current is equal to and in phase with the current in phase C, while the W1 phase B current is twice the value of either A or C and out of phase.

Corrective Actions The failed switchgear bus was replaced and the outdoor 13.8-kV switchgear extension was sealed to prevent water penetration.

Lessons Learned

1. The procedure and analysis described above confirmed the faulted low-side phases of b and c. It is recommended that the analysis described in this case study be done using power system engineering basics.

2. Numerical relaying can provide fault record data that need to be applied with added basic knowledge about the power system to arrive at fault location and classification.
3. Basic relay fault data can be analyzed carefully to arrive at the correct conclusion without the need for an elaborate DFR record from a dedicated device.

Case Study 5.5: Operation of a Unit Step-Up Transformer with an Open Phase on the 13.8-kV Delta Winding

Abstract A generator step-up (GSU) transformer consists of three single-phase units separated by fire barriers, implemented to optimize the cost requirement for a spare transformer application. The GSU transformer is designed as a three-winding 13.8/13.8/230-kV transformer connected delta/delta/YG. When connecting the spare unit to substitute for one of the phases, the outside 13.8-kV bushings were not connected, leading to open delta connections on both 13.8-kV delta windings. The links used to connect the GSU to the generator ISO-phase bus were completed on all the phases, leading to an unbalanced condition due to the missing connections to form the two delta windings. The 13.8-kV normal delta winding is now converted to an open delta, leading to an open grounded wye on the high-voltage 230-kV side. As a result, neutral current began to flow on the 230-kV side of the GSU transformer toward the system side.

Description of the System Figure 5.31 shows the system one-line diagram with the three-winding 13.8/13.8/230-kV transformer connected delta/delta/YG. Dual hydro units are connected to each of the 13.8-kV delta windings. The transformer is connected to the 230-kV substation via a high-pressure fluid-filled (HPFF) cable feeder.

Description of the Incident Transformer bank T1, shown in Fig. 5.31, was removed from service by the operation of its lockout relay 86T1 associated with the bank protection system. The lockout relay was tripped by operation of the low oil pressure switch 63C associated with the 230-kV HPFF cable connection between the GSU transformer and the 230-kV substation. HPFF cable feeder L1 is located in the high-voltage cable tunnel. The lockout relay tripped the 230-kV breakers A and A1 and unit breaker B1. Units G2, G3, and G4 were out of service during the lockout relay trip. The field personnel were unable to reset the lockout relay, due to the standing permanent trip by the cable 63C device. Inspection of the oil pressure at the pumping station and in the HV cable in the power tunnel revealed normal readings.

The HV cable was inspected and found to be normal; accordingly, the 63C tripping circuit was isolated by opening the associated test blade. Due to the absence of any other relay targets, the 86T1 lockout relay was reset and the GSU T1 was restored successfully with generator G1 at full output at 54 MW. The remaining units, G2, G3, and G4, were also placed in service, with each generator output at 54 MW. A few hours later a high-temperature alarm was received on the HV cable. At this time the operator noticed no current flow on phase C for the L1 230-kV cable connection to the GSU T1.

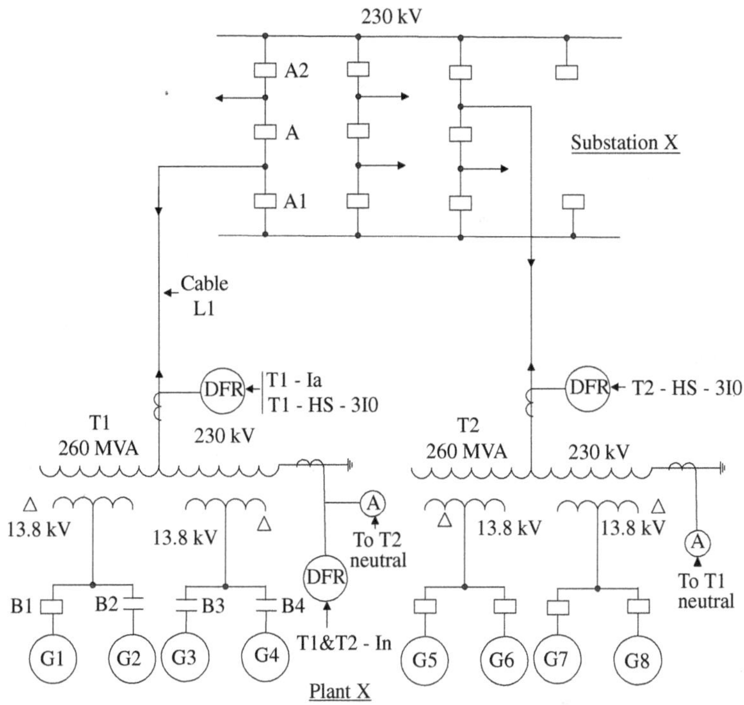

Fig. 5.31 Plant X one-line diagram showing DFR-monitored currents.

Early operation of the 63C cable pressure coupled with the cable high-temperature alarm prompted the immediate shutdown of the four units and their associated transformer T1.

Examination of the GSU T1 connections between the individual transformer phase units revealed that one of the phase connections that formed the delta winding for each of the 13.8-kV delta windings was left open, as shown in Fig. 5.32, which resulted in an open delta connection. The DFR disk was full, due to overwhelming data recording for the unit G1 connection, and hence no additional recording of signals was possible

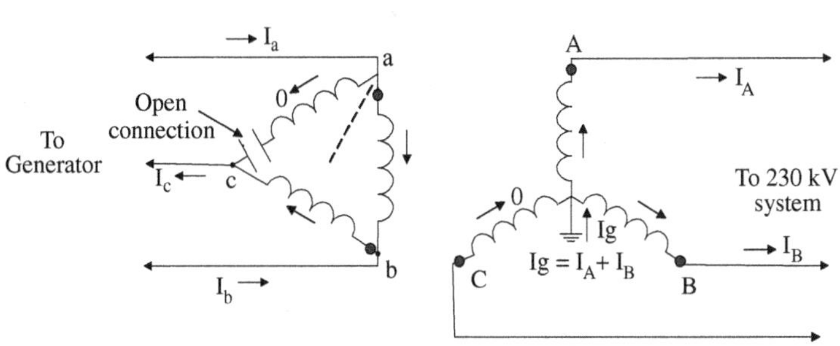

Fig. 5.32 GSU delta windings with a phase C open connection.

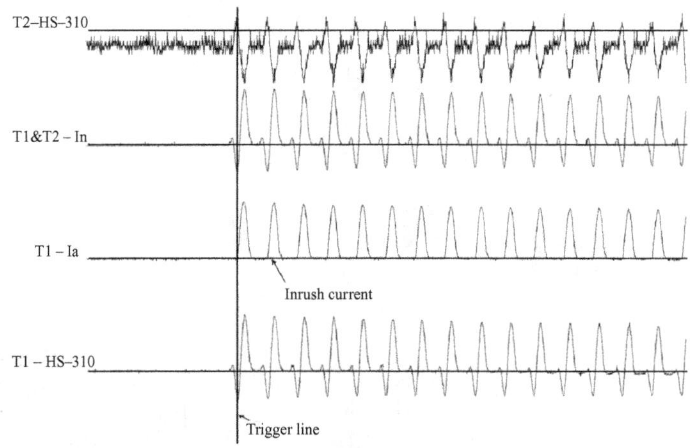

Fig. 5.33 DFR record with inrush currents during energization of transformer T1 from the high side with the 13.8-kV delta opened.

prior to connecting additional units to the transformer bank. The analysis was based only on data and DFR recordings obtained from the G1 unit.

Analysis of the DFR Records Figure 5.31 shows the system one-line diagram and the T1 and T2 DFR-monitored high-side and neutral currents. The substation X DFR record shown in Fig. 5.33 reveals successful energization of transformer T1 from the high side (230 kV) with the 13.8-kV deltas opened. The record reveals a transformer inrush current signature containing 60 Hz and harmonics. Figure 5.34 reveals the substation X DFR record for T1 and T2 residual current with one unit connected while the T1 delta windings are opened. The DFR record was captured by the presence of

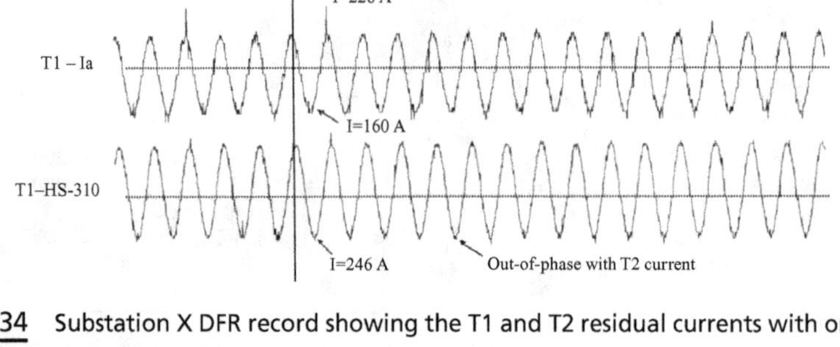

Fig. 5.34 Substation X DFR record showing the T1 and T2 residual currents with one unit connected while the T1 delta windings are open.

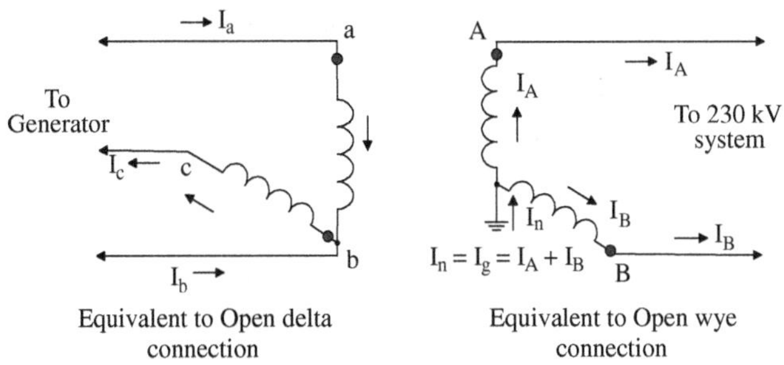

Fig. 5.35 Current flows through the equivalent open-delta low-side winding and open-wye high-side connection.

neutral (zero-sequence) current at the 230-kV high-voltage transformer side. Figure 5.35 illustrates the flow of neutral current on the high side due to the absence of phase C current flow. As shown in Fig. 5.34, it can be observed that the neutral current trace T1-HS-3I0 has a reverse polarity from that of the neutral current trace T2-HS-3I0. This will explain that the source of unbalance that allowed zero-sequence current to flow is the T1 neutral, with current returns coming from transformer neutrals other than the transformer T2 bank shown in Figs. 5.33 and 5.34. Some of the ground unbalanced currents returned via cable steel pipes, forcing the oil temperature and pressure to rise. The DFR disk was full of data prior to connecting additional units to the transformer bank, and therefore no DFR records were available for use to analyze the second energization of the four units.

Analysis of the Effect of the Open-Phase Conditions The missing connection for phase c has converted the delta/YG transformer connection to an open delta/open wye with a ground (grounded open wye). The current in the open delta side is obtained by allowing flow of the 60-MVA unit output through the two windings of the open delta. As illustrated in Fig. 5.35, the current flow from the T1 neutral to the ground will be equal to

$$I_n = I_g = I_a + I_b = -I_c$$

The opening of the braid connection of the delta winding resulted in no current flow through the delta-connected winding, designated c. This also resulted in no flow of current in the C winding of the grounded-wye high-side winding. The ampere-turns coupling principle cannot be violated, and as a result, neutral current flow was established for transformer T1.

Derivation of the Power Rating of an Open Delta Winding Connection Compared to a Delta Connection For a delta connection,

$$P_{\text{delta}} = \sqrt{3}\, V_{LL} \times I_L \times \cos\theta \tag{1}$$

where V_{LL} is the line-to-line voltage and I_L is the line current. For an open delta connection, the

$$\text{winding current} = I_P = \frac{I_L}{\sqrt{3}}$$

Therefore,

$$P_{\text{open delta}} = \sqrt{3}\, V_{LL} \times \frac{I_L}{\sqrt{3}} \times \cos\theta = V_{LL} \times I_L \times \cos\theta \qquad (2)$$

Dividing equation (2) by (1) yields

$$\frac{P_{\text{open delta}}}{P_{\text{delta}}} = \frac{V_{LL} \times I_L \times \cos\theta}{\sqrt{3} V_{LL} \times I_L \times \cos\theta} = \frac{1}{\sqrt{3}}$$

or

$$P_{\text{open delta}} = 58\%\, P_{\text{delta}}$$

It can be concluded that the open delta connection can supply a three-phase balanced load provided that the power served should not exceed 58% of a delta connection using the same phase-to-phase winding voltage.

Overloading of the GSU Transformer During the Incident The current flow in the open delta winding is the same as the line current since in a delta connection the ratio of the phase current to the line current is equal to $1/\sqrt{3} = 58\%$. Therefore, the ratio of the equivalent three-phase load for an open delta connection to a three-phase balanced delta load is 58%. It can also be stated that the open delta transformer rating is 58% of the three-phase capacity of the delta connection. The unit output of 58 MW was therefore flowing through the open delta winding and hence overloading the winding by a factor of $73\% = [(\sqrt{3}-1)/1] \times 100$. According to the transformer winding ampere-turns principle, the high-voltage winding is also overloaded by 73% of normal rating.

Corrective Actions

1. Install a ground time overcurrent in the neutral of the generator step-up transformer to provide protection against open-phase high-resistance ground faults and switching errors similar to this incident that can result in neutral current flow.
2. Provide a negative-sequence current protection to the generator to protect against uncleared unbalanced faults and unbalanced loading caused by an open phase or operating errors.

3. Wire an alarm for a full disk for the DFR, so that action can be taken to correct it.
4. Provide a clear procedure to avoid switching and connection mistakes.

Lessons Learned

1. Ammeter reading or current flow on individual phases should always be established following the successful synchronization of generating units. This is required to confirm that there is no open-phase condition or operating or connection error.
2. The use of ground time overcurrent in transformer neutrals provides valuable and unique protection against open-phase high-resistance ground faults and switching errors similar to this incident.
3. DFR records must be evaluated following any system trip. This will provide assurance that there will be correct switching and connections prior to restoration of the system.
4. Connection links for single-phase replacement must be reviewed prior to energizing the system.

Case Study 5.6: Using a Transformer Phasing Diagram, Digital Fault Recorder Record, and Relay Targets to Confirm the Damaged Phase of a Unit Auxiliary Transformer Failure

Abstract In this case study we explain how a unit auxiliary transformer failure can be analyzed using different information to arrive at the same conclusion. Analysis using a variety of information to find the failed transformer phase is described. We then show how a 345-kV phase-to-neutral voltage monitoring in conjunction with a transformer phasing diagram can be used to determine the transformer 6.9-kV winding faulted phase. We also describe power system phenomena, lessons learned, and corrective actions.

Description of the System and Associated Protection As shown in Fig. 5.36, the 1000-MVA 26-kV fossil unit is connected to the 345-kV system via three single-phase 333 MVA 26/345 kV (GSU) unit transformers (T1). The generator unit auxiliary transformer T2 steps down the generator voltage from 26 kV to 6.9 kV to supply the unit auxiliary loads. Transformer T2 is designed as a three-winding 26-kV delta winding and dual 6.9-kV grounded-wye windings. The 6.9-kV auxiliary supply ground faults are limited to a value of less than 1000 A by applying a 4-Ω 1000-A resistor to the neutral of each of the 6.9-kV windings.

The use of three single-phase units for the GSU transformer together with one spare unit, rather than one three-phase unit, makes it more economical to substitute the spare unit for any failures of one of the phases. Use of a fire barrier will normally limit the damage to the failing unit only (without involving the other nonfailed phases). This will enhance the availability of the generating unit by quick substitution of the spare unit. Normally in this situation the spare transformer will be designed with

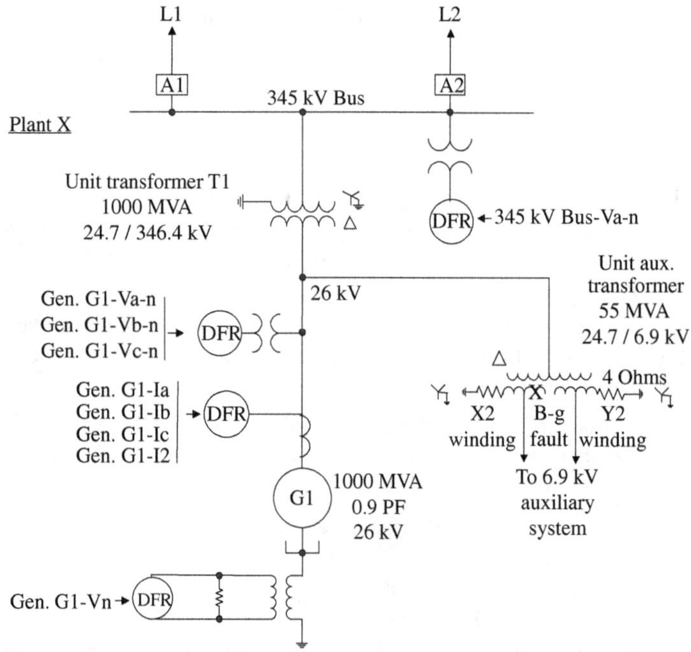

Fig. 5.36 Plant X one-line diagram with the failed transformer T2 winding X2 and DFR-monitored voltages and currents.

flexibility to reconfigure the unit protection through the sensing of associated disconnect switch positions.

Transformer T2 is protected by employing dedicated harmonic restraint percentage differential relay 87T2 and additional time overcurrent relays for backup protection. The sources for the 6.9-kV winding fault are the generator and the two 345-kV feeders. Therefore, the relay 87T2 operation energizes a lockout relay 86T2, which shuts the unit down and trips 345-kV feeder breakers A1 and A2.

Information Available for the Analysis The following information was available to analyze the failure and provide the needed sequence of events, and to answer basic relevant questions that follow every failure in the system.

1. Generating plant and substation one-line diagram
2. Three-line ac diagram for transformer differential relay connection
3. DFR records for the incident
4. Transformer phasing diagram (based on nameplate)
5. Transformer differential relay targets

Analysis of the DFR Record Figure 5.36 shows the plant one-line diagram and DFR-monitored voltages and currents that are used to analyze this disturbance. The

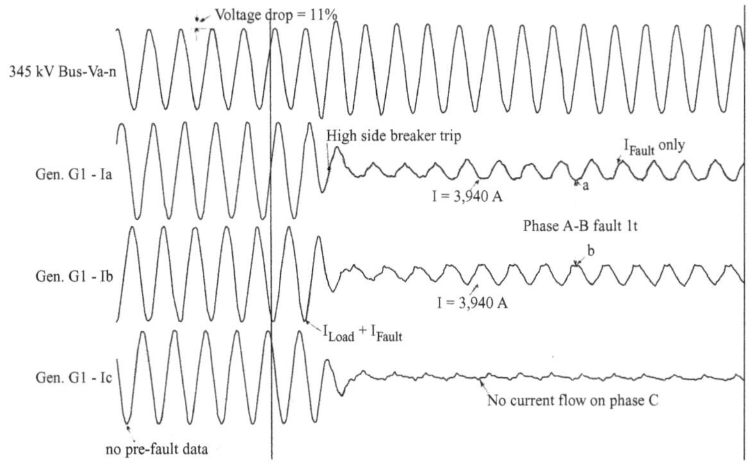

Fig. 5.37 Generator three-phase currents and 345-kV phase A-n voltage during T2 failure.

DFR record of Fig. 5.37 was not triggered by the transformer fault. The DFR capture of the fault record was started by a digital input from a breaker auxiliary contact, and as a result, all pre-fault data were lost. Based on the DFR record in Fig. 5.37, the 345-kV system phase A is down, indicating its involvement. From time 0 to about 6 cycles, the generator output indicates closed balanced phase currents with the presence of negative-sequence current. At 6 cycles from zero, the 345-kV CBs A1 and A2 at substation X were tripped and cleared the fault from the system. The generator was also shut down; however, the generator decayed energy (without the presence of a positive-sequence current contribution from the 345-kV system) illustrated clearly that phase A and B currents are close to being equal and 180° phase shift apart, and the phase C current is very small. The initial L-g fault is still supported by the presence of low-magnitude negative-sequence current. The generator decayed energy is now the only source feeding the 6.9-kV winding fault. Therefore, the generator continues to supply positive- and negative-sequence currents to the fault. Zero-sequence current is supplied from the auxiliary transformer winding connection (delta at 26 kV/grounded wye with a resistor on the 6.9-kV sides).

Determination of the Faulted Phase Using the DFR Record and Transformer Phasing Diagram According to the NEMA standard, delta/wye transformers are normally designed with the high-side line-to-neutral voltage leading the low side by 30°. As shown in Fig. 5.38, for transformer T2, the 26-kV winding phase-to-neutral voltage leads the 6.9-kV wye winding by 30°. This can be seen in Fig. 5.38, for positive rotation (counterclockwise), where phase A-n of the 26-kV winding leads the reference by 210° while phase A-n for the 6.9-kV winding leads the reference by 180°. Therefore, the difference between the two voltages is 30° (210° – 180°). For transformer T1, the 345-kV winding phase-to-neutral voltage leads the 26-kV wye winding by 30°. This can be seen in Fig. 5.39, where phase A-n of the 26-kV winding

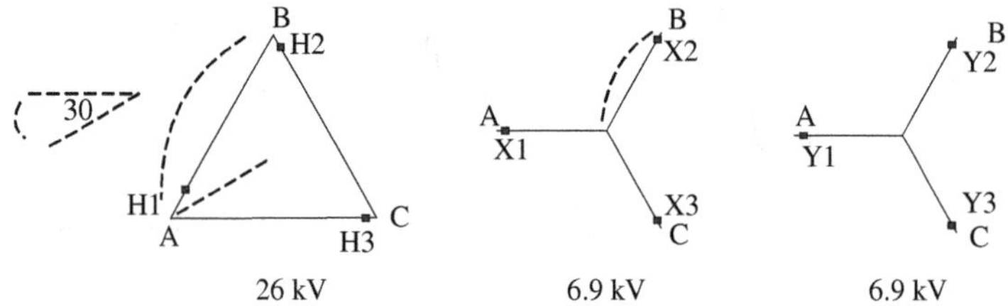

Fig. 5.38 Unit auxiliary transformer T2 phasing diagram.

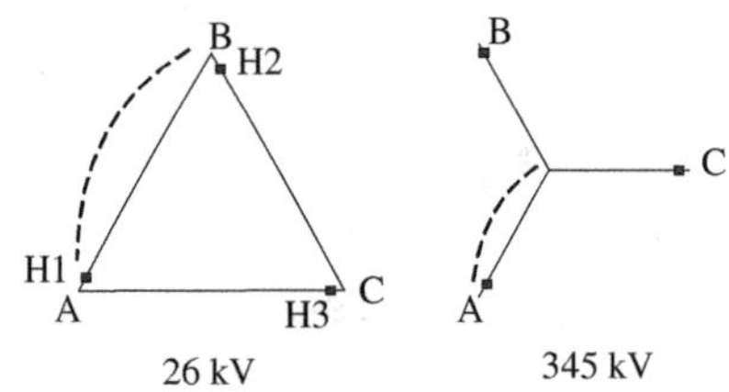

Fig. 5.39 Main unit transformer T1 phasing diagram.

leads the reference by 210°, while phase A-n for the 345-kV winding leads the reference by 240°. Based on the DFR record of trace 345 kV Bus-Va-n in Fig. 5.37, the 345-kV monitoring reveals that only phase A is depressed during the fault, with the other phases undisturbed. Using the T1 phasing diagram of Fig. 5.39, the 345-kV phase A is magnetically coupled to phases A and B of the 26-kV delta winding. Therefore, an A-g fault in the 345-kV system is seen as an A-B fault on the 26-kV side. Using the T2 phasing diagram of Fig. 5.38, the 26-kV phase A-B delta winding is coupled magnetically to phase B of the 6.9-kV wye winding. Therefore, the A-g fault on the 345-kV system, which is seen as the A-B fault on the 26-kV side, can now be confirmed as a ground fault on the 6.9-kVT2 phase B winding.

Procedure to Analyze and Determine the Failed Transformer Phase Due to the lack of monitoring of the 6.9-kV auxiliary voltages, the following steps can be substituted to confirm the fault location. The procedure starting point is the 345-kV side, where phase A voltage trace 345 kV Bus-Va-n shown in Fig. 5.37 is depressed at the beginning of the fault and can be summarized as follows:

1. By using the DFR record information in Fig. 5.37, the fault is seen as an A-g fault on the 345-kV side.
2. By using the T1 phasing diagram of Fig. 5.39, the 345-kV A-g fault can be used to deduce the corresponding faulted phase on the 26-kV delta side.
3. By using the T2 phasing diagram of Fig. 5.38, the corresponding fault on the 6.9-kV wye side can be deduced.
4. By using the transformer T2 relay targets, the faulted phase obtained in point 3 can be confirmed.

5. By using the DFR record for the generator voltages and currents, the conclusion arrived at by the analysis can be confirmed and the details of the sequence of events can be obtained.
6. By inspecting the failed transformer and damaged winding pictures, the analysis can be verified and the failed winding can be confirmed.

Analysis of the Failure Using the T2 Differential Relay Targets and Transformer Phasing Diagram The targets reported for the T2 failure are the phase A and B differential relays. Therefore, a secondary current will flow in a loop containing the A and B relays. This loop current is marked in Fig. 5.40, where the transformer

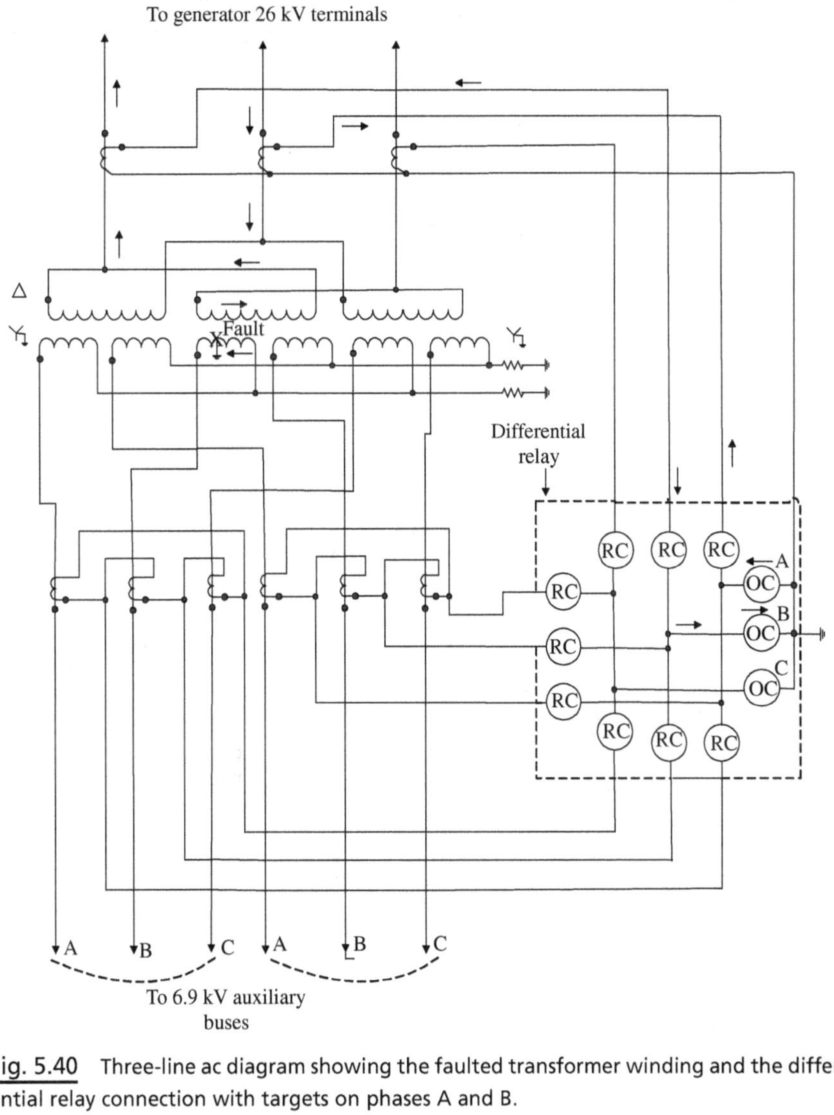

Fig. 5.40 Three-line ac diagram showing the faulted transformer winding and the differential relay connection with targets on phases A and B.

differential relay three-line AC is shown. A secondary current can then be circulated on the secondary wiring associated with the phase A and B relays. Using the current polarity rule, the primary current flow can be deduced for the A-B winding on the 26-kV delta side. According to the transformer T2 phasing diagram shown in Fig. 5.38, the A-B winding is coupled magnetically to the phase B winding of the grounded wye on the 6.9-kV side. This will also confirm the failure as a ground fault on the 6.9-kV T2 phase B winding.

Power System Phenomena

1. Variation of third-harmonic generation at the unit neutral as a function of machine active (MW) and reactive (MVAR) power outputs, as shown in trace Gen. G1-Vn in Fig. 5.41
2. Variation of negative-sequence generator current as it depends on the overall system equivalent impedance to the fault, as shown in trace Gen. G1-I2 in Fig. 5.42
3. Decayed unit trapped energy supply to the fault as shown in Fig. 5.42, where the unit is still feeding the fault after 1.67 s from mechanical unit shutdown

Corrective Action Since insufficient pre-fault analog data were captured by the DFR, the setting of triggering parameters was reviewed and the correct setting to provide enough pre-fault data was used.

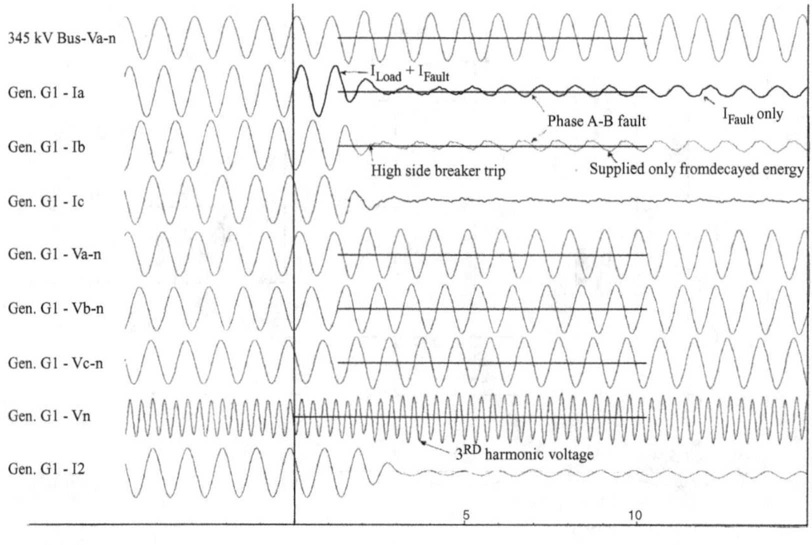

Fig. 5.41 Generator voltages and currents during the transformer T2 failure.

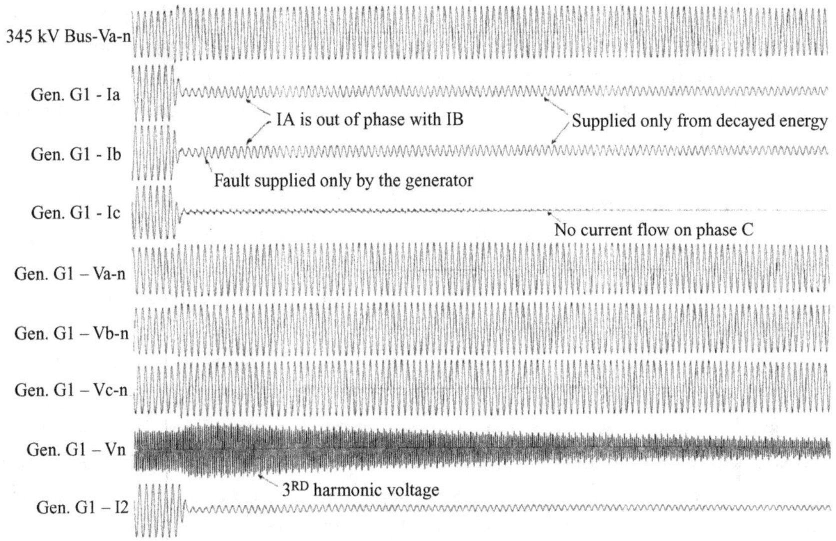

Fig. 5.42 Generator decayed energy feeding the T2 failure for up to 1.67 s from initiation of the fault.

Lessons Learned

1. The lack of monitoring on the 6.9-kV faulted transformer winding can be deduced by analysis of transformer phasing and monitoring of voltages at other transformer sides.
2. Understanding transformer phasing diagrams and how to move current from one winding to an other using the ampere-turns balance principle between windings, coupled with complete understanding of basic current polarity rules, is a useful tool for transformer relaying design and failure analysis.
3. Relay targets for transformer differential relays with the help of differential three-line ac circuit can provide confirmation of the transformer failed winding and faulted phase.
4. The tripping of the 345-kV breakers removed the 345-kV EHV strong positive-sequence source to the failed transformer. The only source left was the generating unit decayed energy, which continued to feed the 6.9-kV B-g fault. This decayed energy can be stopped only by employing a unit generator breaker. However, for the 1000-MW generator, the breaker is nonstandard and very costly to apply.

Case Study 5.7: Failure of a 450-MVA 345/138/13.2-kV Autotransformer

Abstract A 345/138-kV 450-MVA autotransformer with an embedded 13.2-kV tertiary failed, causing a transformer internal A-g fault. The fault was cleared in 4.5

cycles by both primary and secondary differential relaying systems. Successful high-speed operation of the relaying systems prevented a possible catastrophic failure of the bank, which could result in a serious fire. In this case study we reveal a step-by-step approach to confirm an internal transformer ground fault. We also describe the failure and analyze the DFR record.

Description of the Power System and Associated Protection The power system, shown in Fig. 5.43, consists of a 345-kV cable terminated in gas-insulated substation X. The substation includes two 345-kV circuit breakers, a 150-MVA shunt reactor bank, two 450-MVA 345-kV phase angle–regulating (PAR) transformers, and two 345/138-kV autotransformers connected to the 138-kV system. The autotransformer zones of protection extend from the PAR load-side bushing CTs to the 138-kV breaker bus-side bushing CTs. Transformers T1 and T2 are protected individually by a dual variable percentage slope harmonic-restraint differential relaying system fed from separate CTs and energizing separate lockout relays using separate dc sources. Ground inverse time overcurrent relays are used in the transformer neutrals to provide backup protection for ground faults and open-phase conditions. Phase overcurrent relays are employed on the 138-kV sides of the transformer to provide backup

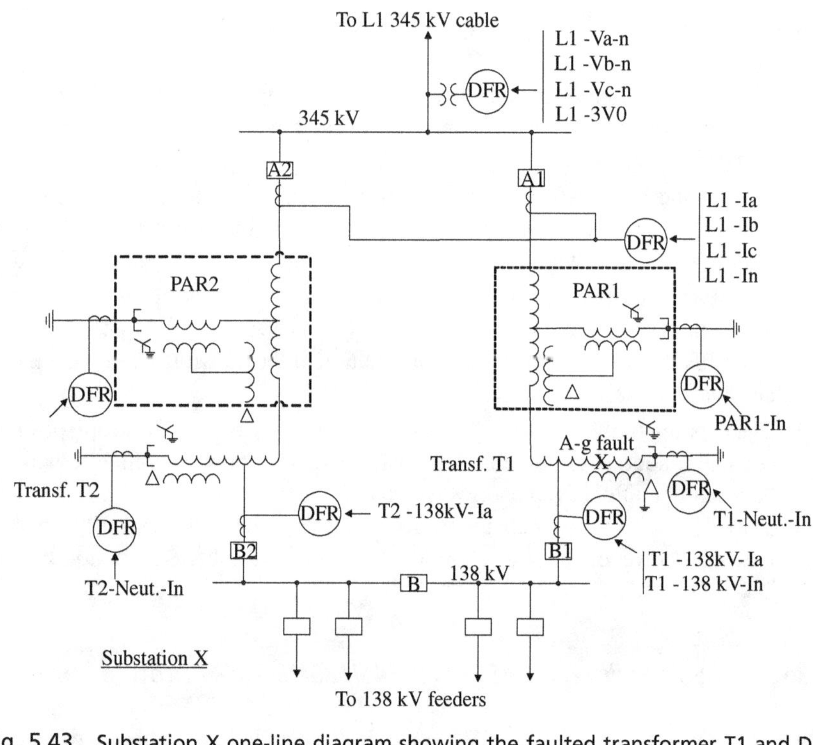

Fig. 5.43 Substation X one-line diagram showing the faulted transformer T1 and DFR-monitored voltages and current.

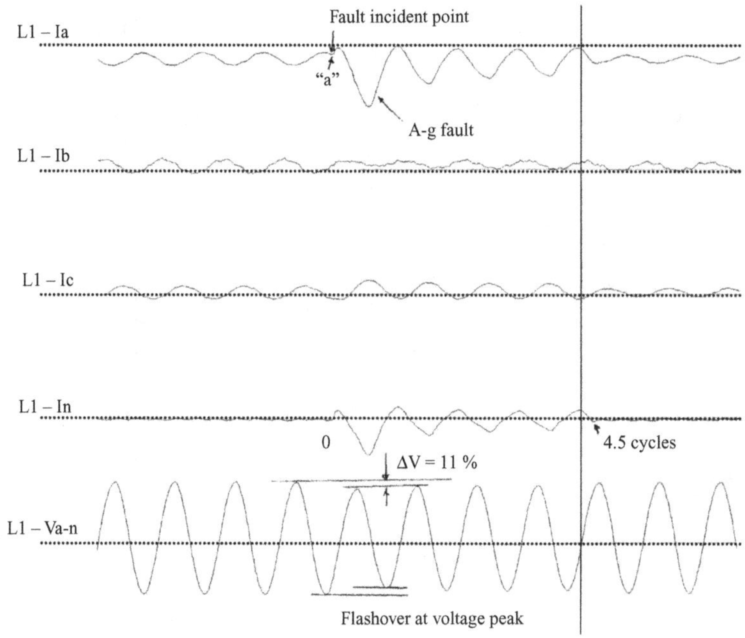

Fig. 5.44 DFR record showing the L1 currents and faulted phase voltage for transformer T1 failure.

protection for faults in the 138-kV system. Buchholz relays mounted in the pipe leading to the transformer conservator are used to alarm for gas conditions and to trip for increased oil flow between the main and conservator tanks. The relay will trip both primary and secondary lockout relays via separate contacts. In addition, rapid-pressure-rise relays are used to alarm for transformer faults.

Analysis of the DFR Record Autotransformer T1 was tripped off-line by the operation of several protective relays. These relays were the phase A and B differential protection, the Buchholz relay, and the mechanical relief device. The DFR record in Fig. 5.44 reveals a phase A-to-ground fault lasting for 4.5 cycles. The A-g fault is supported by the dip (11%) in phase A voltage, the increase in phase A current, and the presence of ground currents. The fault incident point a is at the voltage peak, indicating an insulation breakdown process leading to the flashover to ground. This resulted in symmetrical current with no presence of dc offset in the fault currents. The analysis confirmed that the phase-to-ground fault was inside transformer T1. The conclusion is based on the following facts, which were obtained by an analysis of DFR records and relay targets.

1. The operation of both relaying systems when currents entering the transformer are not balanced by currents leaving. This relay operation is indicative of a shunt unbalance (normally caused by a fault) rather than a series unbalance

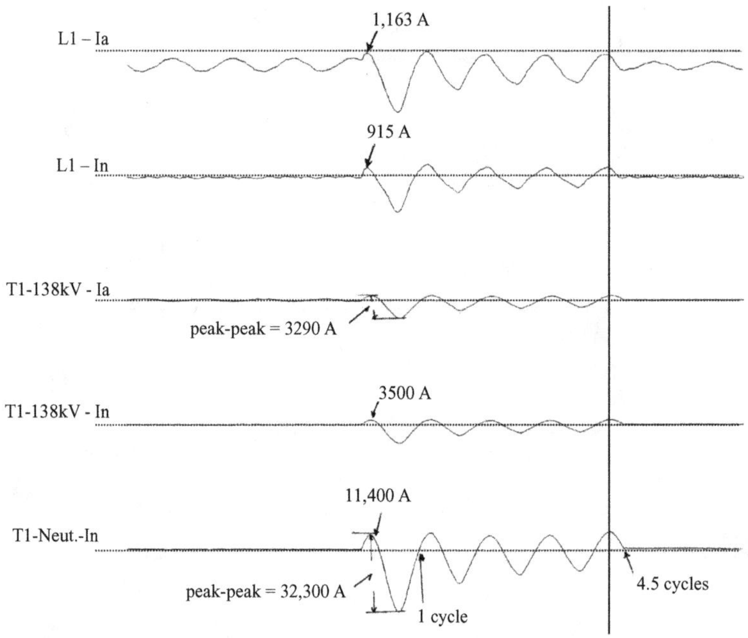

Fig. 5.45 Faulted phase and neutral currents for line L1 and transformer T1.

(normally caused by turn flashover or open phase). The differential relay operating current was consumed by the transformer shunt fault.

2. The appearance of ground currents for a duration of 4.5 cycles on the neutral of transformer T1, as shown in Fig. 5.45. In addition, ground current flows appear on the neutral of transformer T2, primary exciting windings of phase angle–regulating transformers PAR1 and PAR2, the 138-kV sides of T1 and T2, and the 345-kV L1 feeder, as shown in Figs. 5.45 and 5.46.
3. The appearance of zero-sequence voltage ($3V_0 = V_a + V_b + V_c$) on the L1 feeder, which is shown in trace L1-3V0 in Fig. 5.46. Normally, the presence of zero-sequence voltage is related to ground faults.
4. Postmortem analysis of a sampled DFR record for substation X was carried out. A symmetrical component transform operation was performed on the 138-kV system contribution to the ground fault, indicating the presence of positive-, negative-, and zero-sequence current components during the fault. Pre- and post-fault analysis revealed only positive-sequence currents. This matches the off-line simulation for the strength of the 138-kV system behind transformer T1.
5. Similar results were obtained for the L1 currents with the presence of negative- and zero-sequence currents only during the fault.
6. Figure 5.47 indicates that the pre-fault L1 phase A current trace L1-Ia and the T1 138-kV-side phase A current trace T1-138kV-Ia were out of phase. During

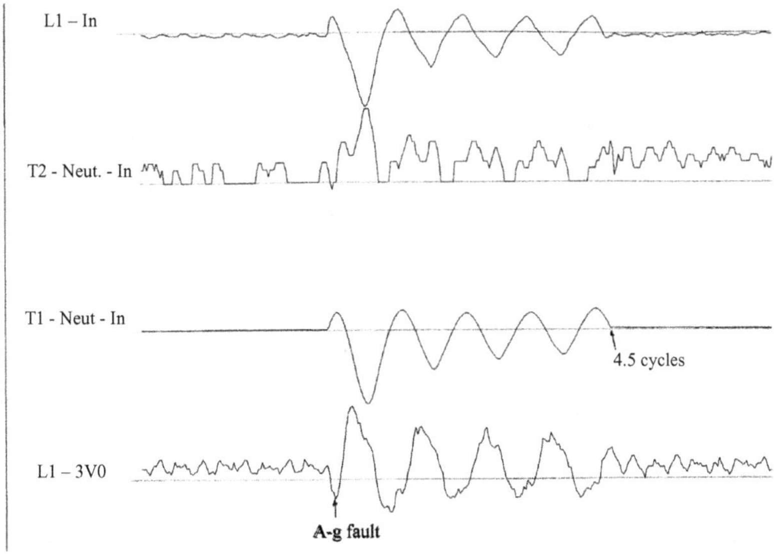

Fig. 5.46 DFR record for T1, T2, and L1 currents and line zero-sequence voltage.

the fault the T1 138-kV-side phase A current reversed to feed the T1 fault and is now in phase with the L1 phase A current. In addition, feeder L1 phase A current trace L1-Ia exhibits a phase angle shift from loading to fault currents at the fault incident point.

7. Zero- and positive-sequence currents ($I_0 - I_1$) on unfaulted phases of the L1 feeder in conjunction with load flow force I_B to go down and I_C to go up, as shown in Fig. 5.48.

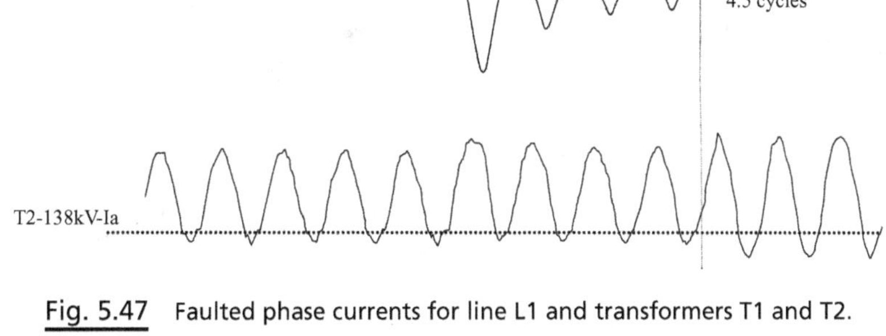

Fig. 5.47 Faulted phase currents for line L1 and transformers T1 and T2.

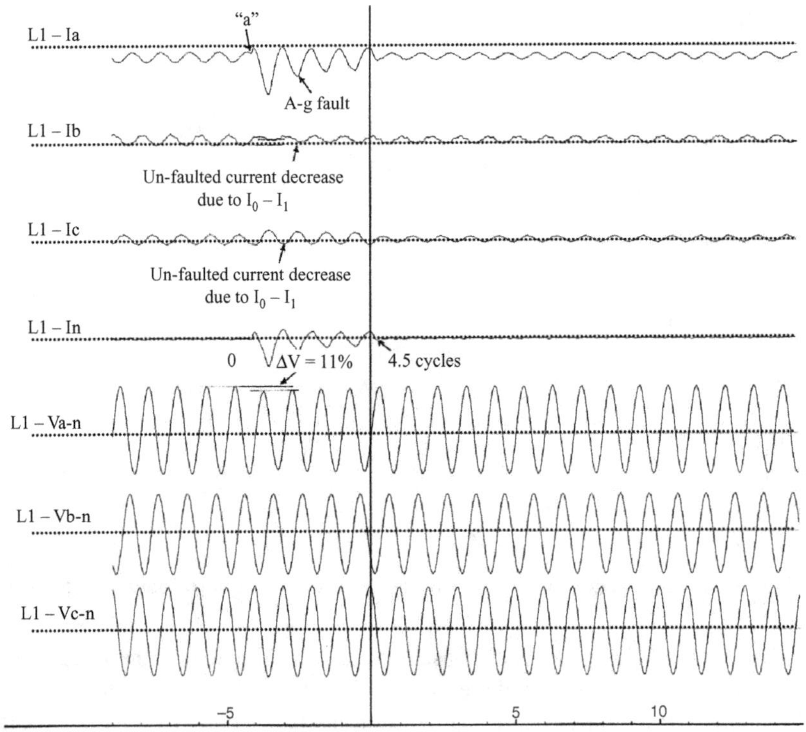

Fig. 5.48 Line L1 currents and voltages during the transformer T1 A-g fault.

Analysis of the Failure The fault-to-ground mechanism inside the transformer is not clear. However, since the fault was not solid to ground, a theory can be stated for some possible scenarios. It can be postulated that the A phase regulating winding has arced over to the associated core limb tertiary winding, allowing the ground fault current to flow through the existing ground of one of the corners of the delta tertiary winding. The embedded delta winding leads are not brought to the outside, and therefore one of the corners must be grounded in accordance with the standard to reduce electric stress. Other ground fault mechanism scenarios may also exist if the tertiary winding involvement can be ruled out. This can normally be confirmed by the transformer manufacturer during the detailed inspection phase. In this case, transformer design information can assist in confirming the theory of the failure process. The high fault current at the first cycle only could be explained by force generation that enlarged the early arc-over tracking path for the remaining fault-duration period.

Description of the Failure Mechanism Gas in oil analysis of a sample from the Buchholz relay showed 218 ppm of acetylene for a sample taken from the bottom valve after circulating the oil, compared to a previous value of 23 ppm. Previous samples taken a few years ago had shown no detectable hydrocarbons. The failure was initiated by a "flake" of copper or silver sulfide, which formed a bridge between a number of tap changer contacts. The sulfide layer had formed on all bare copper and silver surfaces, and had lose particles (probably caused by different expansion coefficients and temperature cycles). The resulting arc then caused a short-circuit current in a part of the regulating winding, giving rise to large forces in the winding and on the leads that altered the current flow path, resulting in lower currents after the first fault cycle.

Based on the transformer manufacturer's failure report, it was found that an electrical breakdown occurred in the selector of one of the tap changers. This was caused by a piece of silver sulfide that had fallen off a silver-plated contact. The sulfide layer was formed by corrosive sulfur, which must have been present some time before. The origin of this corrosive sulfur could not be determined. However, It can be assumed that it was in the components or materials from which the transformer was built, but not part of the insulating fluid that was used to fill the transformer.

Corrective Action The failed transformer was inspected by the manufacturer to confirm an internal phase A-to-ground fault and to make sure that no sulfur content exists inside the transformer tank.

Case Study 5.8: Failure of a 750-kVA 13.8/0.480-kV Station Service Transformer Due to a Possible Ferroresonance Condition

Abstract It is very well known that in a generator zone, it is possible to have a potential transformer ferroresonance condition. As shown in Fig. 5.49, this condition may occur when the generator is disconnected, leaving the unit transformer delta winding energized and essentially ungrounded. The grounded potential transformers are now connected to an ungrounded system with the possibility of a ferroresonance condition. As a result, a distorted 60-Hz or subharmonic voltage could overexcite the PT, forcing it into the saturated region. The PT variable nonlinear inductance may now resonate with the ungrounded system capacitance, thus causing extreme high voltage that may initiate the failure of system equipment in the generator zone if damping is not present.

This case study illustrates an actual occurrence of a ferroresonance condition which resulted in the failure of a station service transformer and 13.8-kV switchgear buses. This case study provides a sequence of events with an actual DFR record showing the three-phase fault that occurred when the station service transformer failed. We describe the power system and associated protection, the failure mechanism, the ferroresonance phenomenon, corrective actions, and lessons learned.

Description of the System One-Line Diagram Figure 5.49 shows a portion of the system one-line diagram where the generating units of the hydro plant are connected to the 230-kV transmission systems. A bank of four units is also shown connected to the 230-kV system via a three-winding delta/delta/YG 13.8/13.8/230-kV transformer. Dual units are bussed together and connected to each of the 13.8-kV delta windings. The transformer is protected by harmonic-restraint percentage differential relay device 87T. Other generating units, which are not shown, are connected to the

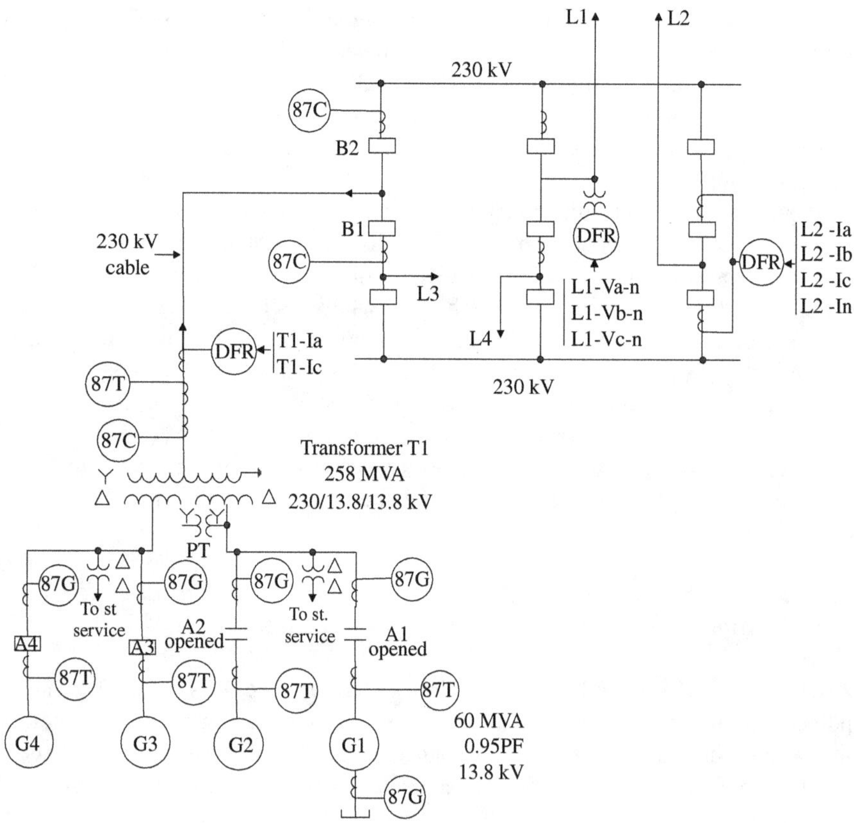

Fig. 5.49 System one-line diagram showing DFR-monitored voltages and currents.

115-kV system. To economize on the use of the spare transformer, the 115- and the 230-kV transformers are designed as single-phase units. One dual-voltage spare transformer is used to substitute for either the 115- or 230-kV single-phase transformers. The dual-voltage rating of the spare resulted in a capacitive reactance that is different from that of either the 115- or 230-kV transformer units. In the early application of the spare transformer, two failures occurred in the dry-type station service transformers. These failures prompted the use of loading resistors across the PTs during substitution of the spare unit. This eliminated further failures for many years. A new spare transformer was manufactured and connected to the system shown in Fig. 5.50 without the loading resistors. This case study documents an additional third failure for the system described. All the failures that coincided with the use of the spare transformer unit were attributed to the ferroresonance phenomenon.

Description of the Switching Sequence As shown in Fig. 5.49, units G2 and G1 are removed from service by opening 13.8-kV breakers A2 and A1, respectively. The following sequence of events, apparently triggered by the last breaker opening, is

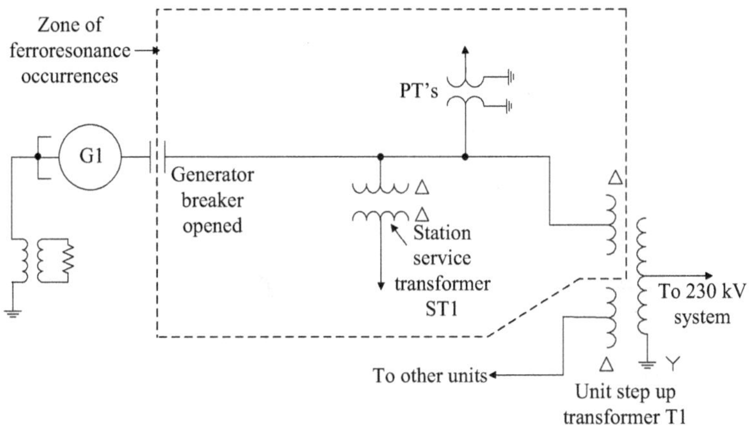

Fig. 5.50 Generator zone where ferroresonance may have occurred.

based on an analysis of the sequence-of-event recorder (SER) record and DFR record shown in Fig. 5.51.

At time 0 (reference point) the last breaker A1 was opened manually. Transformer bank T1 differential relay device 87 phase C picked up to energize the associated lockout relay at 0.75 s from the reference point. Eight milliseconds later, the transformer differential lockout relay operated to trip and lock out 13.8-kV breakers A3 and A4 and 230-kV breakers B1 and B2. The 87 differential relay apparently

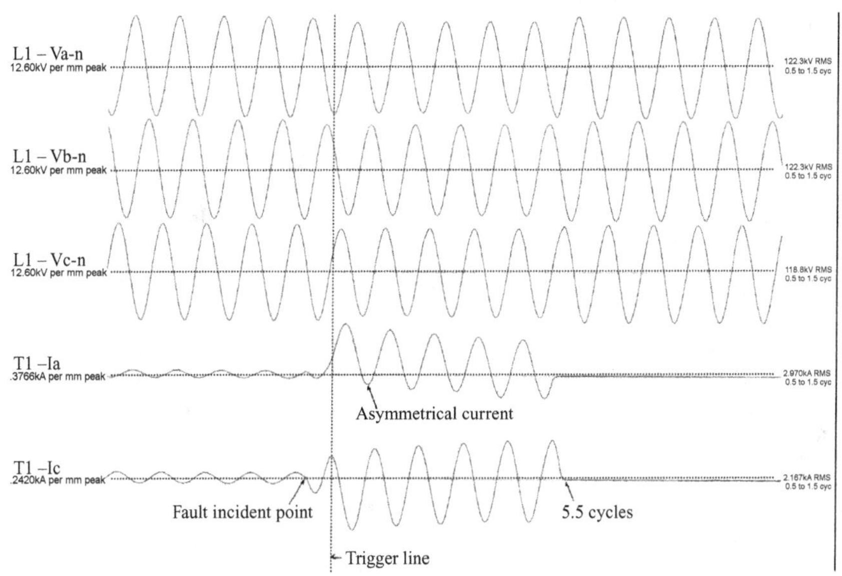

Fig. 5.51 Plant X DFR record for line L1 voltages and T1 phase A and C currents during the station service transformer ST1 fault.

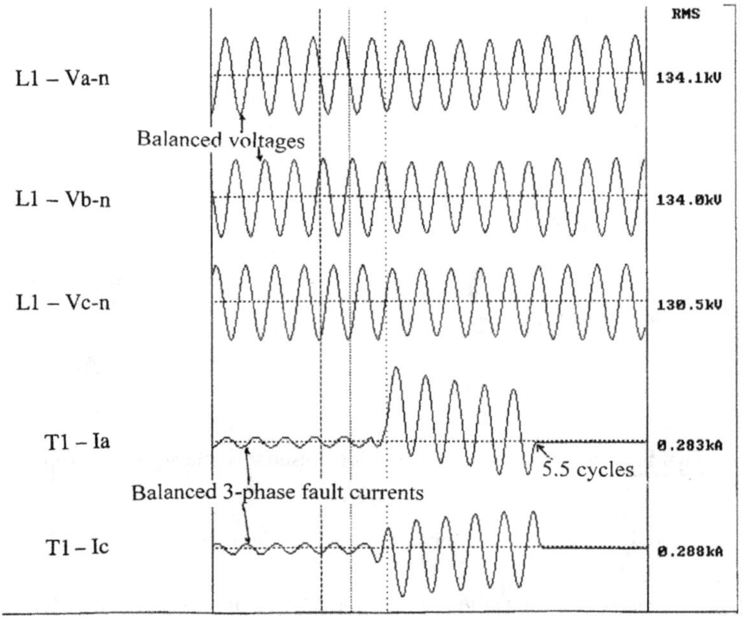

Fig. 5.52 Plant X DFR record showing RMS values for the pre-fault L1 voltages and T1 phase currents.

operated to clear failed station service transformer ST1 from the system. Figure 5.51 shows that the fault duration is 5.5 cycles, which can be broken into 2-cycle relay operating time, a half-cycle lockout relay operation, and a 3-cycle breaker interrupting time.

Analysis of the DFR Record Figure 5.49 shows the system one-line diagram and DFR-monitored voltages and currents that are used to analyze this disturbance. Less than a second (0.75 s) after the tripping of CB A1 and the isolation of generator G1 from the system, the associated station service transformer ST1 flashed over internally and to ground. The DFR record of Fig. 5.52 reveals a pre-fault loading current of 283 A on phase A and 288 A on phase C of the high-side (230-kV) transformer T1. This corresponds to unit 3 and 4 outputs of 50 MW each. The DFR in Fig. 5.51 indicates a total fault duration of 5.5 cycles. Figure 5.53 shows high currents on phases A, B, and C, confirming the occurrence of a phase A-B fault, evolving to a three-phase fault a few milliseconds later, confirmed by an increase in the phase C current, as shown in trace L2-Ic in Fig. 5.53. The three-phase fault was apparently caused by an insulation failure, and therefore a simulation for overvoltage study is required to affirm the failure mechanism.

The station service transformer represents a weak link from an equipment insulation viewpoint. The station service transformer (13.8-kV BIL) is 50 kV, which is lower than the normal 110-kV applied BIL class for the 13.8-kV winding of the main unit transformer T1 bank.

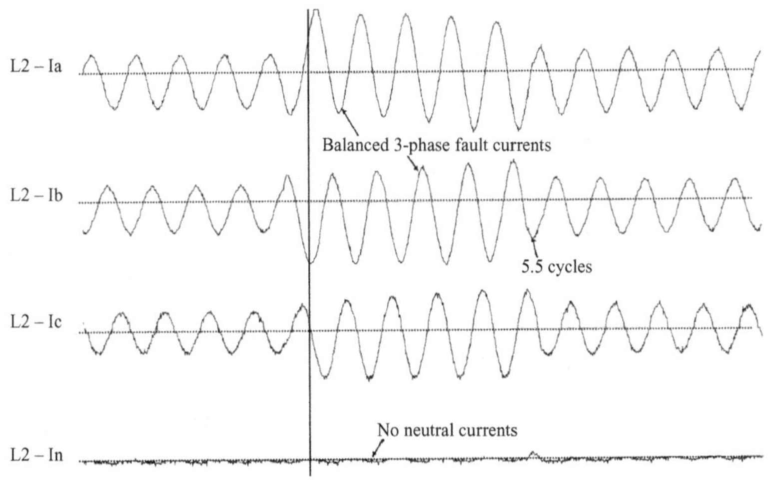

Fig. 5.53 Plant X DFR record showing the line L2 three-phase current contribution to the ST1 fault.

Analysis of the Station Service Transformer Failure As shown in Fig. 5.50, the opening of the second breaker has created an ungrounded 13.8-kV zone, which invited the occurrence of possible ferroresonance, thus producing relatively high phase-to-phase overvoltages. The ferroresonance phenomenon has caused the failure of two station service transformers in the past when the a dual-voltage spare single-phase transformer of different design was substituted. Apparently, this modified the overall equivalent capacitance enough to resonate with the PT nonlinear inductance. This failure occurred when a new spare single-phase unit was also added. It is postulated that the failure started inside the transformer winding and was followed by a flashover between the winding and ground. This resulted in excessive fault current through parts of the station service transformer windings, which in turn created significant electromagnetic forces in the windings and their terminal connections. The transformer mechanical damage observed is the result of high short-circuit currents. The 13.8-kV total asymmetrical fault current as derived from the recorded 230-kV transformer high side of 2970 A is 49,500 A $[= 2970 \times (230/\sqrt{3} \times 13.8) \times \sqrt{3}]$.

Fault Location Using Three-Phase Fault Calculation The DFR record shown in Fig. 5.51 reveals a three-phase fault current fed from the 230-kV side of the unit transformer with the RMS values calculated for a one-cycle window (0.5 to 1.5 cycles from $t = 0$). The phase C current has an RMS value of 2167 A, while phase A indicates a higher RMS value, 2970 A, due to the presence of dc offset.

Short-circuit study three-phase 13.8-kV bus fault simulation with all four units out of service gives a current of 2218 A fed from the 230-kV side, corresponding to a current of 36,966 A at the 13.8-kV generator side.

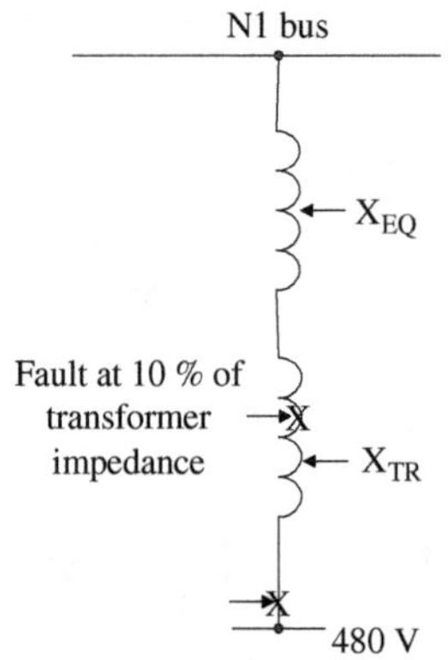

Fig. 5.54 Positive-sequence network for the phase fault.

For a 100-MVA base and 13.8-kV base voltage, the

$$\text{base current} = \frac{100 \times 10^6}{\sqrt{3} \times 13.8 \times 10^3} = 4183 \text{ A}$$

$$\text{per unit (pu) fault current} = I_{\text{pu}} = \frac{36,876}{4183} = 8.84 \text{ pu}$$

$$\text{system equivalent impedance} = \frac{1}{I_{\text{pu}}} = \frac{1}{8.84} = 0.113 \text{ pu}$$

The station service transformer rating is 750 kVA, 13.8 kV/480 V. The transformer impedance is 5.5% at 750 kVA and at the 100-MVA base is $5.5 \times (100/0.750) = 7.4$ pu. Simulate a fault at 10% of the station service transformer. The equivalent impedance to the fault $= 0.113 + 0.1 \times 7.4 = 0.85$ pu. For a three phase-fault at 10% as illustrated in Fig. 5.54, the fault current $= 1/0.853 = 1.17$ pu. The base current for the 230-kV side, 100 MVA/230 kV = 251 A. The 230-kV side contribution to the 10% fault location $= 1.17 \times 251 = 294$ A.

The recorded fault current of 2071 A is much greater than a fault current of 294 A. This will imply that the fault started near the terminal of the station service transformer, due to insulation failure caused by excessive line-to-line voltage generated by the ferroresonance phenomenon.

Explanation of the Ferroresonance Phenomenon Ferroresonance is generated by energy exchange between the equivalent system capacitances and the inductance of the voltage transformer. Figure 5.55 illustrates the basic equivalent circuit involved in the phenomenon. The oscillation frequency is determined by the PT inductance and the equivalent bus capacitance by the formula $f_s = 1/2\Pi\sqrt{LC}(\Pi = 3.14)$. The generator high-resistance grounding normally provides high-resistance grounding through the neutral grounding transformer. However, when the unit generator breakers A1 and A2 are opened, the 13.8-kV bus becomes ungrounded, with the ground for the bus area becoming very weak, and the path to ground will be only through the synchronizing potential transformers.

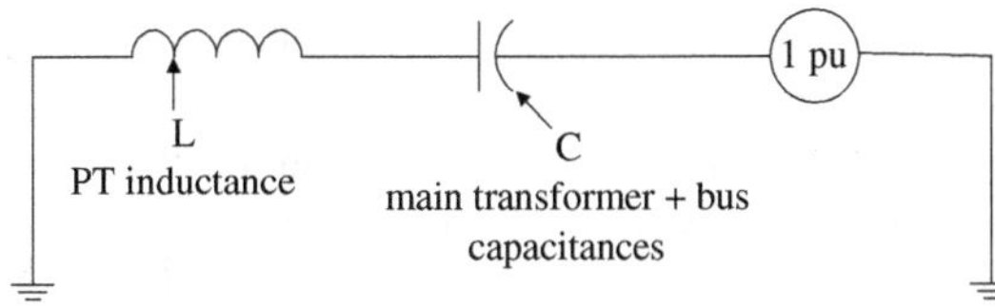

Fig. 5.55 Basic ferroresonance circuit representation.

The 115/13.8- and 230/13.8-kV unit step-up transformers for the hydro units are designed as single-phase units to optimize the use of a dual-voltage rating spare unit. Apparently, the dual-voltage winding of the spare unit has a different capacitance than that of either the 115- or 230-kV single units. The three regular transformers have a winding capacitance to ground that is less than that of two regulars and one spare. The ferroresonance phenomenon may therefore occur due to the change in the overall capacitance with the substitution of the spare coupled with the nonlinear inductance of the PT. Dry-type transformers apparently failed due to the overvoltage caused by the ferroresonance phenomenon. As mentioned earlier, ferroresonance occurs when the inductive reactance to ground equals the capacitive reactance to ground at the resonating frequency. Ferroresonance is a voltage oscillation which, if not damped, will continue to reach several times the value of the normal voltage.

In the generator zone, it is possible to have a PT ferroresonance problem. PT ferroresonance is a possibility if the generators are disconnected and the PT remain connected to the delta winding of the unit transformer, energizing the station service auxiliary loads without proper damping.

Corrective Actions

1. Load all the phases of the potential transformer by connecting a 100-W, 60-Ω resistor from phase to neutral as shown in Fig. 5.56, when the spare transformer is substituted for either the 115- or 230-kV transformer.

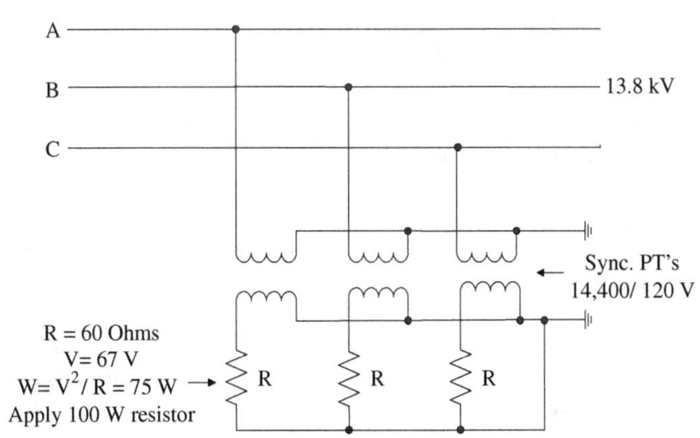

Fig. 5.56 Potential transformer secondary loading resistors.

2. Since the station service transformer has a delta/delta configuration, apply a four-legged surge arrester arrangement with three arresters connected between the transformer phases and the neutral, and the fourth connected between the neutral and ground. This arrester configuration provides adequate protection for both phase-to-phase and phase-to-ground insulation. For a 13.8-kV switchgear application, MOCV arresters rated at 8.4 kV can be employed to provide adequate protection. Voltage rating and arrester energy requirements should be confirmed using system transient studies.
3. Station service transformers should be specified using a higher BIL insulation level, similar to that of the unit GSU transformer.

Lessons Learned

1. Occurrence of ferroresonance is difficult to predict; however, precalculation should be used to design sufficient damping to cause it to die out. Recording voltages during system energization or deenergization to examine the phenomenon does not provide a full examination of the phenomenon.
2. Several failures arising from the use of old and new spare transformer units had to occur prior to learning the cause of the failure and understanding the ferroresonance phenomenon.
3. The ferroresonance phenomena suspected in the failure of the 13.8-kV bus and the station service of this case study can only be mitigated by damping the energy by loading all the phases of the potential transformer with resistors. This will prevent any further exchange of energy between the two storage elements, L and C. Therefore, resistance loading equal to the rated thermal capability of the PT may be required to suppress ferroresonance. The resistive PT loading can be removed when the generator is connected and the unit is operating in the normal mode. The ferroresonance effect can be nullified by adding burden resistors to the potential transformer secondary. If a 60-Ω resistor is used, the wattage can be calculated as follows

$$W = \frac{V^2}{R} = \frac{(67)^2}{60} = 75\ \text{W}$$

   Therefore, use of a 60-Ω, 100-W resistor will not overload the PT and will provide the needed damping.

Case Study 5.9: Undesired Tripping of a Numerical Transformer Differential Relay During an External Line-to-Ground Fault

Abstract A phase B-g fault that occurred on the 138-kV system was cleared in 4.5 cycles. An 80-MW plant, which is very remote from the fault location, was tripped undesirably by the operation of a unit transformer numerical differential relay. Due to the difficult access to the plant during the summer peaking load, no testing was permitted to find the basic cause of the undesired differential relay trip. As a corrective action, the relay was kept in service, for monitoring purposes, with its trip output removed by opening the flexi-test switch blades controlling the relay contact output.

During this time, another trip was issued by the relay, again during an external remote 138-kV C-g fault that lasted 6 cycles. The generating units were out of service with the unit transformer connected to the 138-kV system to supply the plant station service. A new case study was created for analysis of the same relay undesired trip. With the generating unit out of service, the delta/grounded-wye unit transformer connection is supplying zero-sequence current only during the external 138-kV C-g fault. The second undesired trip, due to the contribution of zero-sequence current only, added more mystery to the problem and hinted at a combination of errors with the secondary differential relay circuit. In the two case studies we analyze the numerical relay fault records, the relay in-service reading, and the as-found condition of the inspection for the differential relay circuit. We also provide an explanation of the undesired trips, corrective actions, and lessons learned.

Description of the System The 80-MW plant X consists of dual 40-MW combustion turbine (CT) units connected to the 138-kV system as shown in Fig. 5.57.

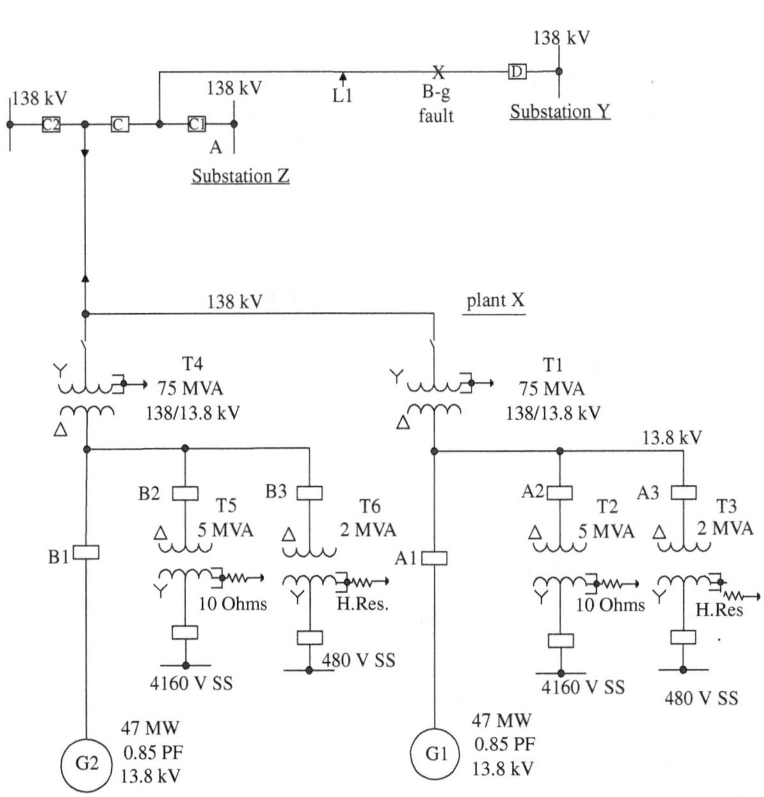

Fig. 5.57 Plant X one-line diagram and its connection to the 138-kV system.

TABLE 5.3 Differential Relay Setting Parameter Record[a]

Parameter	Setting	Parameter	Setting
Phase sequence	ABC	Percent differential	Enabled
Transformer type (name plate)	Y/d30/d30	Pickup	0.2 × CT
Winding (W1)		Slope 1	20%
$V_{L\text{-}L}$	138 kV	Knee point	1.0 × CT
T1 MVA rating (OA)	45 MVA	Slope 2	60%
Phase CT primary	600:5	Harmonic inhibit	
Winding (W2)		Function	Enabled
		Parameters	Second
$V_{L\text{-}L}$	13.8 kV	Harmonic averaging	Disabled
T1 MVA rating (OA)	45 MVA	Inhibit level	20% f_0
Phase CT primary	3000:5	Instantaneous differential	
Winding (W3)		Function	Enabled
$V_{L\text{-}L}$	13.8 kV	Pickup	8 × CT
T1 MVA rating (OA)	45 MVA		
Phase CT primary	3000:5		

[a]CT = 5 A secondary, as defined by the relay manufacturer.

The plant auxiliary starting power is derived from the 138-kV feeder connections to substation Z. Each GSU unit transformer is protected by dual numerical multifunction relays. Figure 5.57 also illustrates the GSU T1 transformer relay protection zones for the dual numerical relays.

Differential Relay Setting Record See Table 5.3. The numerical relay setting record offers:

1. A magnitude current balance between the 138- and 13.8-kV sides
2. The phase shift correction for the 30° by which the high-side 138-kV YG leads the low-side delta
3. Blocking the zero-sequence current for external ground faults on the system-grounded 138-kV side
4. Harmonic restraint to block the tripping of the numerical relay during the transformer inrush energization phase

Case Study 5.9A: Undesired Plant Trip During an External B-g Fault While the Generators Are in Service

Abstract Figure 5.58 shows the system one-line diagram with the dual numerical differential relays protecting transformer T1. The primary transformer numerical differential relay 87TP reacted to an external B-g fault occurring on line L1 as shown in Fig. 5.58 and produced an undesired trip. The 87TP produced an oscillography and event records that will be analyzed. As will be demonstrated, the analysis was complicated by the plant power generation in addition to fault contributions.

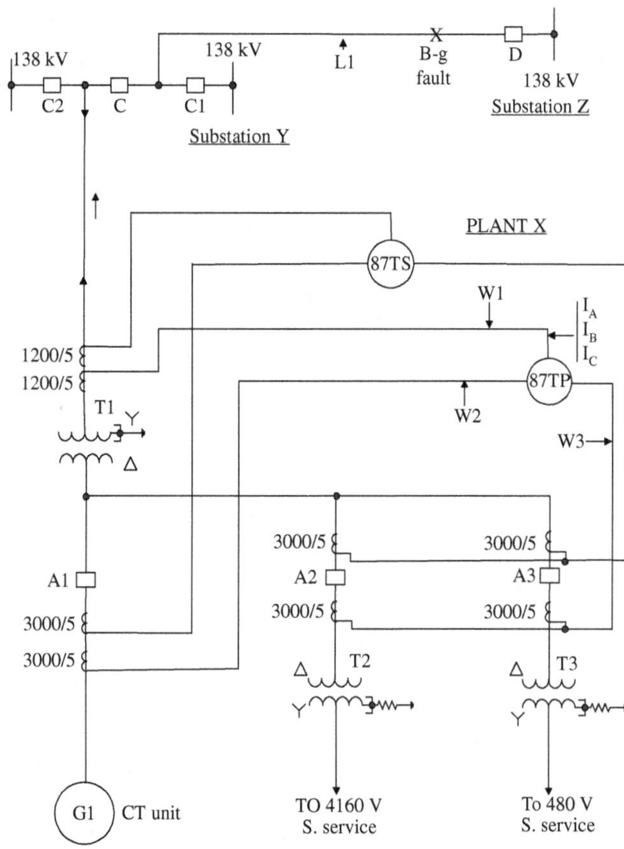

Fig. 5.58 Transformer T1 primary 87TP relay and its monitored phase currents.

Relay Fault Record The numerical differential relay generated the following undesired trip record:

1. 138-kV-side winding W1 primary currents with CT ratio $= 600:5$:

   Primary phase currents: $I_A = 154$ A at 0°; $I_B = 109$ A at 137° lag; $I_C = 101$ A at 225° lag.

   The relay recorded a ground current $3I_0 = 0$. The ground current magnitude can be verified by using symmetrical component transformation as follows: Since $3I_0 = I_a + I_b + I_c$,

$$I_a = 154 \text{ A at } 0^\circ = 154 + j0$$
$$I_b = 109 \text{ A at } 137^\circ \text{ lag} = -79.7 - j74.3$$
$$I_c = 101 \text{ A at } 225^\circ \text{ lag} = -71.4 + 71.4$$

   Therefore, $3I_0$ is approximately equal to zero.

2. 13.8-kV-side winding W2 primary currents with CT ratio = 3000 : 5:

   Primary phase currents: $I_A = 1440$ A at 257° lag; $I_B = 1530$ A at 343° lag; $I_C = 2130$ A at 120° lag

3. 13.8-kV-side winding W3 primary currents with CT ratio = 3000 : 5:

   Primary phase currents: $I_A = 70$ A at 238° lag; $I_B = 70$ A at 356° lag; $I_C = 50$ A at 109° lag

   Differential phase secondary currents: $I_A = 0.23 \times \text{CT} = 1.13$ A; $I_B = 0$; $I_C = 0.23 \times \text{CT} = 1.15$ A

   Restraint secondary currents: $I_A = I_B = I_C = 0.24 \times \text{CT} = 1.2$ A

   Here CT = 5 A. The 87TP relay tripped by the operation of the percent differential element by exceeding the setting threshold.

Analysis of the Relay 87TP Fault Record Figure 5.58 shows the system one-line diagram during the B-g fault and the 87TP relay–monitored currents. Winding W1 (138 kV)–recorded secondary currents contain no zero-sequence current components. On the primary side, the external B-g fault was fed by a zero-sequence current from the two unit transformer banks, which are connected as delta-grounded wye. The undesired trip occurred due to the absence of zero-sequence components in the relay W1 current. The lack of ampere-turns balance around the main transformer windings for the differential relay secondary currents produced sufficient differential current to cause the undesired trip.

Figure 5.59 shows the relay oscillography record for the three-phase currents. The predisturbance loading currents are balanced with equal magnitude. On the other hand, the combination of loading and fault currents during the fault duration is difficult to analyze, and therefore a need was established to perform in-service current measurements for the 87TP circuit.

In-Service Recorded Current Measurement Readings To check the integrity of the differential relay CT wiring circuits after the occurrence of the undesired trip,

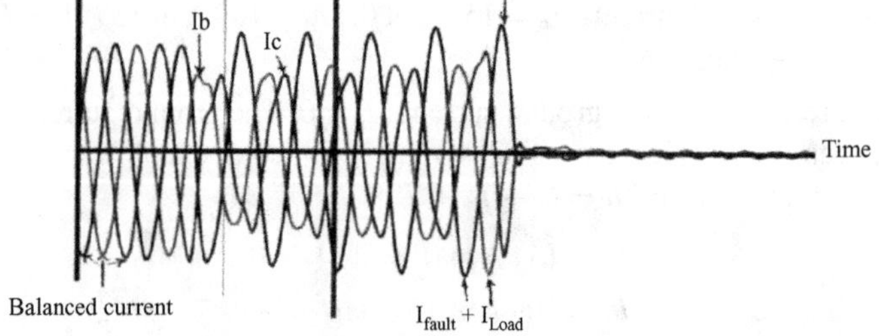

Fig. 5.59 Transformer numerical relay oscillography fault record showing the W1 138-kV three-phase currents for the initial B-g system fault.

in-service readings were taken for the plant loading and recorded as follows: The generator output = 39.9 MW.

The winding 1 (W1) primary current CT ratio 600 : 5 A:

$$I_a = 160\text{ A at }0^\circ\text{ lag}$$
$$I_b = 158\text{ A at }120^\circ\text{ lag}$$
$$I_c = 158\text{ A at }239^\circ\text{ lag}$$

The winding 2 (W2) primary current CT ratio 3000 : 5 A:

$$I_a = 1700\text{ A at }211^\circ$$
$$I_b = 1740\text{ A at }331^\circ$$
$$I_c = 1670\text{ A at }91^\circ$$

The winding 3 (W3) primary current CT ratio 3000 : 5 A:

$$I_a = 110\text{ A at }230^\circ$$
$$I_b = 110\text{ A at }352^\circ$$
$$I_c = 110\text{ A at }111^\circ$$

Differential currents: $I_a = I_b = I_c = 0$
Restraint secondary currents: $I_a = I_b = I_c = 0.29 \times \text{CT} = 1.45$ A secondary

Not knowing the exact cause of the undesired trips in both case studies, the in-service reading indicated a balanced three-phase current, which lead to the conclusion that the CT wiring circuit is correct. This balanced three-phase in-service reading from the HMI of numerical transformer differential relay 87TP made the cause of the false trip difficult to analyze. This dictated the need for field testing of the relay and associated CT wiring when the plant is taken out of service.

Corrective Actions The undesired trip occurred at a time where the unit was needed to meet the peak loading summer conditions. To prevent further undesired trips, the numerical relay trip output was isolated while the relay was left in service. This was done due to the protection redundancy left, which was offered by secondary numerical relay 87TS and a remote time overcurrent protection looking into the plant. The troubleshooting and problem investigation were postponed until an outage was scheduled. In addition, in-service measurements were taken to confirm the relay CT ratio and wiring connection.

Possible Causes of the Undesired Trips A review of all the information above leads to the following possible causes for the undesired operation of the 87TP:

1. The neutral of the unit transformer is not connected to ground. This was ruled out because only one of the differential relays operated falsely.
2. The neutral of the CT secondary circuit is not connected. This is a possibility provided that no zero-sequence current flows.

3. The numerical relay has software or hardware problems, causing a relay algorithm glitch. This is a remote possibility, because no alarms or error messages were generated.
4. Shorted or nonconnections of current transformers.
5. Combinations of more than one of the above.

The first trip was difficult to analyze, due to the fact that there was no access to the plant and the supply of positive-, negative-, and zero-sequence currents to the external fault. In addition, the generator is supplying load prior to the trip.

Case Study 5.9B: Undesired Plant Trip During an External C-g Fault While the Generators Are Out of Service

Abstract Figure 5.60 shows the system one-line diagram with the generator G1 out of service and only one numerical differential relay 87TS protecting transformer T1. Numerical relay 87TP was connected and energized while its trip output was removed from service after the first undesired trip that occurred in Case Study 5.9A. The primary transformer numerical differential relay 87TP reacted again to an external L-g fault and produced oscillography and event records similar to those of Case Study 5.9A. Since the plant was not generating, the second false trip covered in this case study with the generator out of service is easy to analyze, due to only the zero-sequence contribution from the delta/grounded-wye transformer to the external ground fault.

Relay Fault Record The numerical differential relay generated the following undesired trip record:

1. 138-kV-side winding W1 primary currents with CT ratio = 600 : 5:

   Primary phase currents: $I_A = 160$ A at 0°; $I_B = 160$ A at 0°; $I_C = 322$ A at 180° lag

   The relay recorded a ground current $3I_0 = 0$.
2. 13.8-kV-side winding W2 primary currents with CT ratio = 3000 : 5:

$$I_A = 0; I_B = 0; I_C = 0$$

3. 13.8-kV-side winding W3 primary currents with CT ratio = 3000 : 5:

$$I_A = 0; I_B = 0; I_C = 0$$

   Differential secondary phase currents: $I_A = 0.46 \times \mathrm{CT} = 2.3$ A; $I_B = 0$; $I_C = 0.46 \times \mathrm{CT} = 2.3$ A

   Restraint secondary currents: $I_A = I_B = I_C = 0.46 \times \mathrm{CT} = 2.3$ A, where CT = 5 A.

The 87TP relay tripped falsely when the percent differential element exceeded the setting threshold.

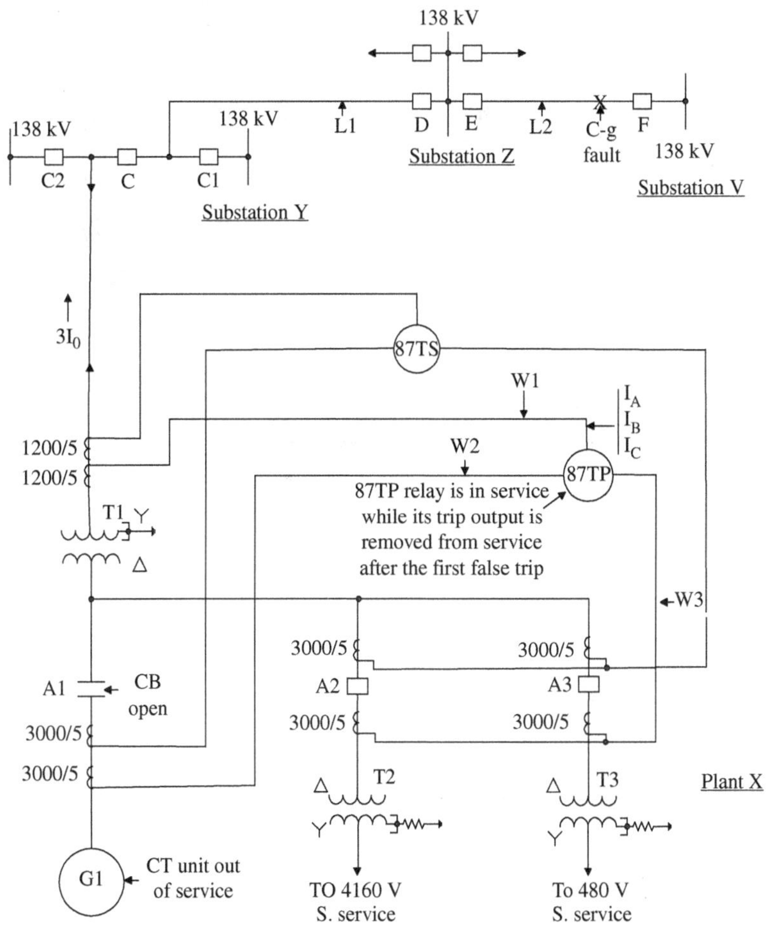

Fig. 5.60 System one-line diagram with relay–87TP monitored currents during the C-g external fault.

Analysis of the Fault Record Figure 5.60 shows the system one-line diagram with the C-g fault occurring on line L2 and the oscillograph-monitored 87TP currents. The numerical relay oscillography fault record shown in Fig. 5.61 reveals the W1 three-phase currents, and Fig. 5.62 reveals the individual W1 phase current during the second undesired trip. The oscillography record matches and confirms the relay fault record stated in item 2 above. The plant was shut down while the GSU unit transformer was supplying station service to the plant loads. As illustrated in Fig. 5.63, the delta/grounded-wye transformer is a zero-sequence source for high-voltage transmission system ground faults.

Analysis of the fault record reveals that the relay secondary current $I_a = I_b = 160$ A at 0° and $I_c = -(I_a + I_b) = 322$ A at 180°. Currents in phases A and B are in phase and have equal magnitudes and therefore represent zero-sequence currents. Since

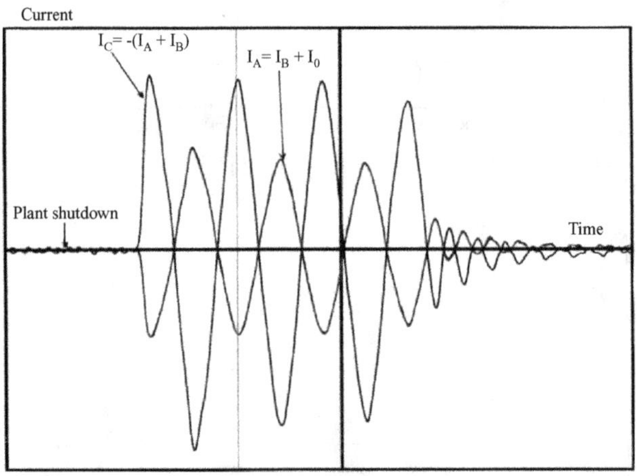

Fig. 5.61 Transformer numerical relay oscillography record showing the W1 138-kV three-phase currents for the second C-g system fault.

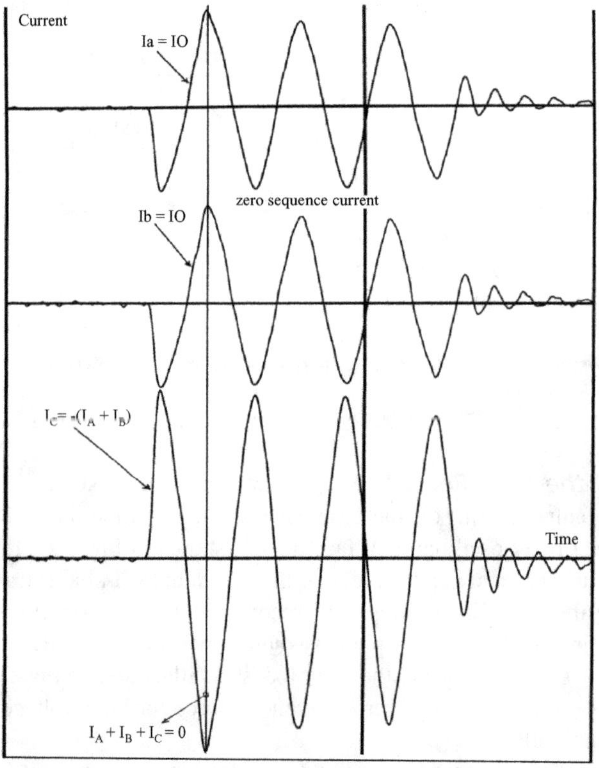

Fig. 5.62 Transformer numerical relay oscillography record showing the individual W1 phase currents during the C-g fault.

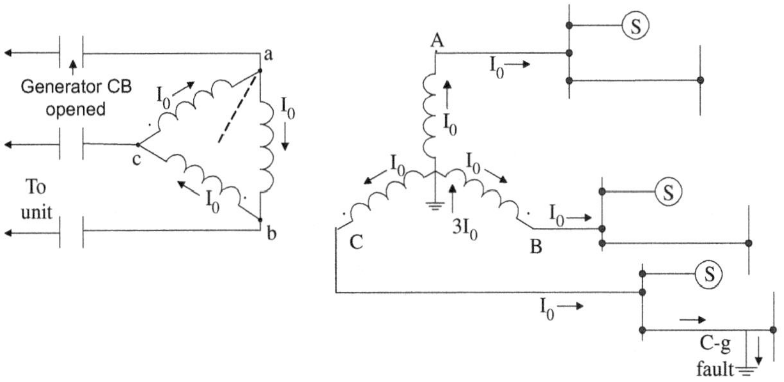

Fig. 5.63 Plant GSU T1 zero-sequence current contribution to the system C-g fault.

$I_c = -(I_a + I_b)$, the return of phases A and B zero-sequence currents was through phase C in the relay. Therefore, the sum of all currents sampled by the relay is equal to zero, and no zero-sequence current was extracted by the relay and its fault record since the relay from the delta current will filter the zero-sequence current, if present, by forming $I_a - I_c$, $I_b - I_a$, and $I_c - I_b$. Zero-sequence filtering is required to restraint the relay for external ground faults.

The differential currents were obtained by the relay as follows: The numerical relay creates delta currents for the wye winding based on the transformer connection and dividing by $\sqrt{3}$ to get the correct relay current. For example:

$$\text{phase A primary differential current} = I_a - I_c = 160 - (-322) = 482\ \text{A}$$
$$\text{phase A secondary differential current} = 482/(120 \times \sqrt{3} \times 5) = 0.46\ \text{CT}$$

As recorded by the relay fault record as stated in item 2 above,

$$I_{\text{restraint}} = \text{maximum current recorded by any of the windings for each phase}$$

Figure 5.64 shows the system one-line diagram together with the current transformer designation for numerical differential relay 87TP, which is used in Figs. 5.65 and 5.66. Since only zero-sequence primary current was supplied to the fault, a loss of neutral connection only should result in no secondary current flow, as shown in Fig. 5.65. As a result, no relay operation will occur. However, due to operation of the differential relay and the recorded three-phase current, an additional condition in the CT circuit must be present. Since the sum of the three phase currents is equal to zero, it can be concluded that one condition is a missing neutral connection in the CT secondary wiring. This dictated the need for additional field testing.

Additional Field Test The plant was out of service and the 138-kV systems were used as a station service. The outage coincided with a cold winter day where the

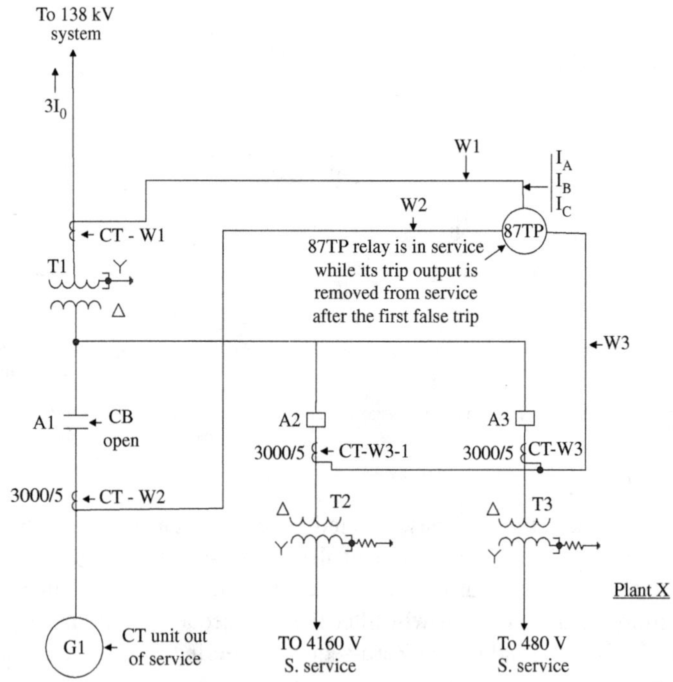

Fig. 5.64 One-line diagram showing the current transformer designations for numerical relay 87TP.

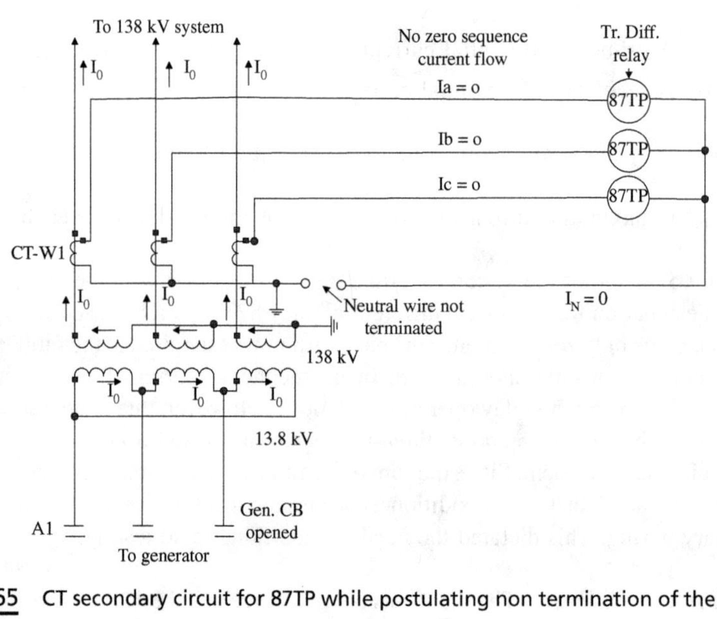

Fig. 5.65 CT secondary circuit for 87TP while postulating non termination of the neutral wire.

station service was also supplying the heating load. The unit transformer was energized, so no current injection testing was permitted. Field testing was restricted to a complete examination of the differential circuit secondary wiring diagram.

As-Found Condition of the Circuit Examination Inspecting the wiring system while the unit was out of service revealed the following:

1. The phase C current transformer was shorted by the shorting screw inserted in the CT shorting block.
2. The neutral wire for the CT cable was not terminated at the transformer CT cabinet termination.

The two as-found conditions are shown in Fig. 5.66.

Analysis of the Two Undesired Trips For the second trip, the primary current flows on phases A, B, and C are of zero-sequence nature. Therefore, I_a, I_b, and I_c are equal and in phase. The secondary phase C current was trapped in the shorted CT on phase C. The flow of current on phases A and B is summed up and flowed in the relay phase C and then back through the CT neutral connection. The resulting current was forced to flow on phase C due to disconnection of the neutral conductor.

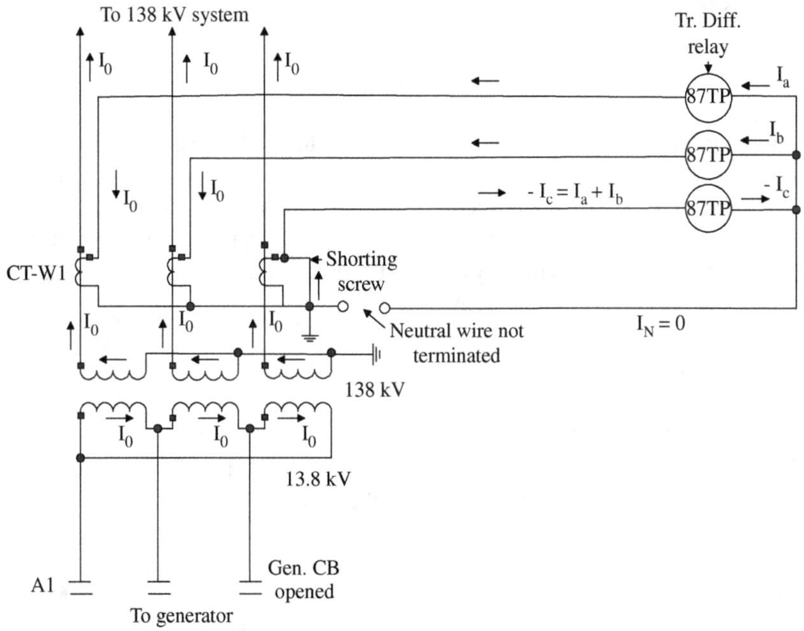

Fig. 5.66 As-found wiring condition of the 138-kV side CT-W1 secondary circuit indicating the short across phase C and the non termination of the neutral wire.

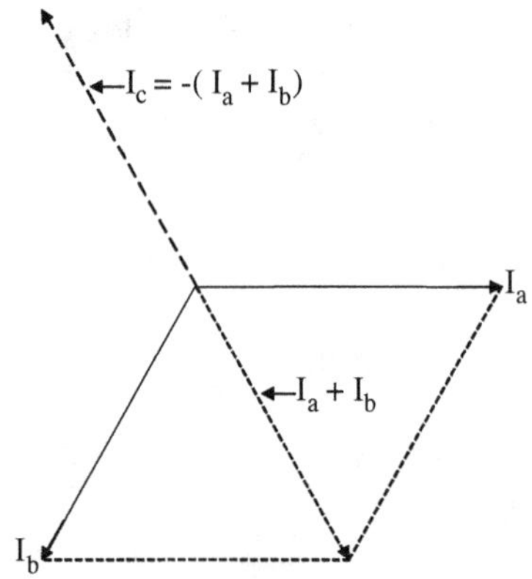

Fig. 5.67 Analysis of the in-service reading taken by the numerical transformer differential relay.

Analysis of the In-Service Reading in Light of the Finding of Wiring W1 The current in-service reading was taken while the unit was carrying a full load. Since for a balanced three-phase system the sum of the phase currents is equal to zero,

$$I_a + I_b + I_c = 0$$

or

$$I_c = -(I_a + I_b)$$

As shown in Fig. 5.67, the in-service reading produced the missing phase C current despite the shorting of the phase C current transformer. The inversion of the sum of $I_a + I_b$ produced the missing phase C current, and therefore a balanced set of three-phase currents was recorded by the numerical differential relay. As a result, the in-service reading did not provide any clues to the undesired trip and did not uncover the causes of the problem.

Corrective Actions

1. Remove the shorted screw at the phase C current transformer.
2. Terminate the neutral conductor of the 138-kV CT input at the CT neutral connection at the transformer terminal cabinet.

Lessons Learned

1. All possible causes of undesired trips should always be examined using a systematic approach.

2. The CT secondary circuits and associated cables should be tested thoroughly during commissioning prior to energization. Complete testing documentation should be a part of the commissioning phase.
3. Redundant relaying systems and a dual-element sensing philosophy can provide operating flexibility and options to isolate protective devices when they operate falsely. This case study highlights this fact by isolating the suspected relay while operating the plant in a normal mode.
4. As shown in Fig. 5.65, an open in the neutral conductor path connection will not permit the flow of zero-sequence current. Therefore, the flow of currents for the second operation had to be the combination of more than one error (namely, a shorted CT) and nontermination of the neutral wire.

Case Study 5.10: Undesired Operation of Numerical Transformer Differential Relays During Energization of Two 75-MVA 138/13.8-kV GSU Transformers

Abstract In this case study we illustrate the undesired operation of two numerical transformer differential relays. The undesired trip of the relays occurred by inrush currents generated by energization of the parallel transformers. Apparently, the second harmonic generated as a ratio of the fundamental 60 Hz was not sufficient to restrain the numerical relay as it should. In this case study we describe briefly the evolution of transformer protection, analyze the relay trip record, and provide corrective actions.

Description of the System As shown in Fig. 5.68, two 13.8-kV 50-MW combustion turbines are connected to the system via two 138/13.8-kV GSU unit transformers. Each transformer is rated at 75 MVA and is connected as a grounded wye on the high side and a delta on the generator side. The transformers are also used as backfeed sources to the plant station services. The two transformers are bussed together on the high side and connected to the 138-kV underground cable feeder.

Description of the Incident Circuit breaker A at plant X was closed to energize both transformers to provide startup sources for the plant auxiliary loads so as eventually to synchronize units G1 and G2 with the system. Energization of the transformers resulted in a trip by the numerical differential relays for each of the transformer lockout relays. The trip was generated due to the lack of the harmonic-restraint function needed to block the relay during transformer inrush conditions. The second harmonic-to-60 Hz ratio was below the setting threshold value of 20%.

Evolution of Transformer Protection Protection of transformers was traditionally supplied using harmonic-restraint electromechanical or solid-state discrete differential relays on a per-phase basis. Three relays are required to provide protection for three-phase transformers. The traditional manufacturer of these relays was also the designer and manufacturer of the transformers. Therefore, complete knowledge of the

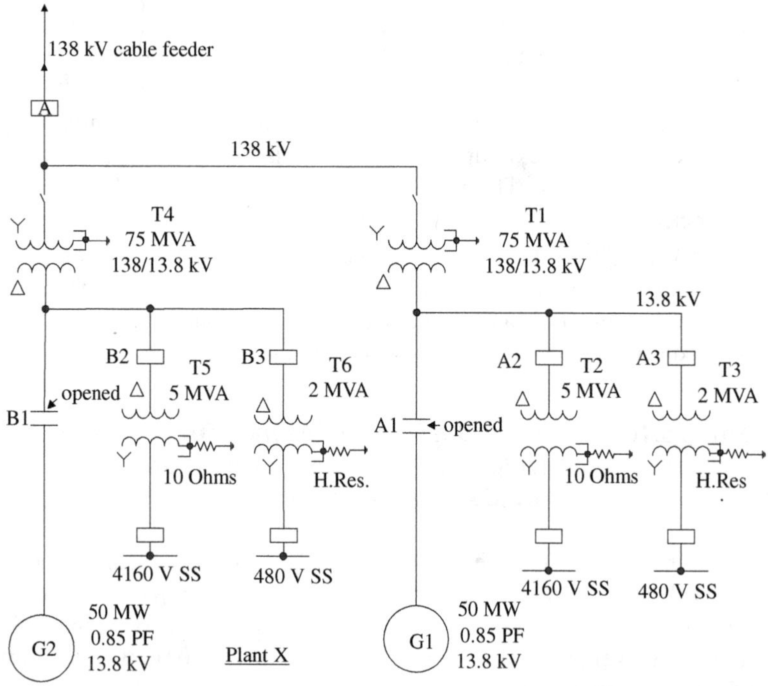

Fig. 5.68 One-line ac diagram showing the dual unit plant.

transformer inrush phenomena was utilized in the design of the harmonic-restraint feature of the differential relays. The relays are delivered to users with a built-in circuitry to cope with inrush phenomena during transformer energization. Only two setting parameters are required to commission these relays. The parameters are a relay tap, either (8.7, 5, 4.6, 4.2, 3.8, 3.5, 3.2, or 2.9 A) and a relay slope, either (15, 25, or 40%).

With the advent of numerical multifunction transformer differential relaying, manufacturers have designated many parameters to be settable by users. Several pages of settings are therefore required to commission microprocessor-based numerical relays compared with the traditional few parameters needed for discrete electromechanical or static relays. Parameters used for inrush restraint are among those required to set the relay. The rational behind user selection of the type of harmonics and their threshold ratios is unclear. Incorrect selection of harmonics and their ratios by users may result in undesired operation during transformer energization.

Numerical Differential Relay 87T1 Oscillography and Event Trip Records

Figure 5.69 shows the system one-line diagram with only transformer T1 differential numerical relay 87T1–monitored currents. The relay 87T1 oscillography record shown in Fig. 5.70 reveals the transformer high-side three-phase inrush currents. In

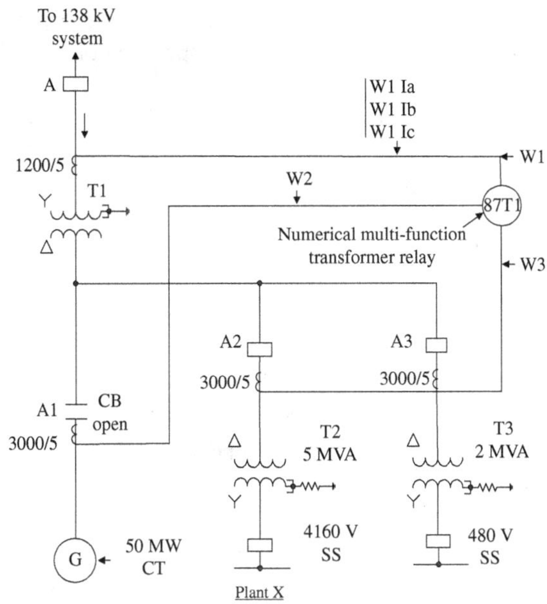

Fig. 5.69 One-line diagram showing the transformer relay 87T1–monitored currents.

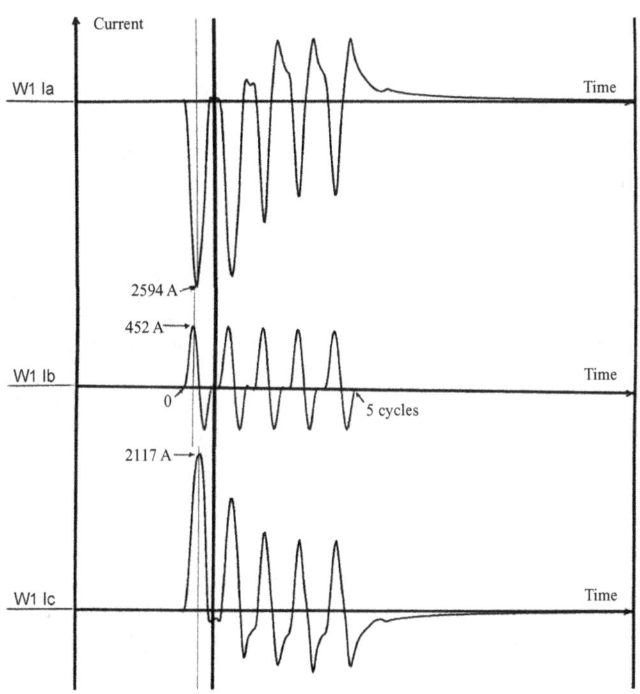

Fig. 5.70 Numerical relay 87T1 oscillography fault record showing the transformer T1 inrush currents that caused the undesired differential trip.

addition, the relay generated an event record at the differential relay element trip instant documenting the cause of the undesired operation of the 87T1.

1. *Relay trip event record.* The relay CT ratio for the 138-kV transformer side is 600 : 5.

   Differential operate secondary phase currents: $I_A = 4.1$ A; $I_B = 4.2$ A; $I_C = 1.6$ A

   Restraint secondary phase currents: $I_A = 4.1$ A; $I_B = 4.2$ A; $I_C = 1.6$ A

   Percent second harmonic/60 Hz: phase A = 15.9%; phase B = 29.6%; phase C = 29.4%

2. *Relay setting for harmonic restraint*

   Second harmonic inhibit: enabled

   Tripping inhibit level: 20% of fundamental 60 Hz

   Harmonic averaging: disabled

Analysis of the Relay Trip Event Records Since for phase A, the recorded ratio of 16% for percent second harmonic/60 Hz is less than the setting threshold of 20%, a trip output was generated by each of the numerical differential relays 87T1 for transformer T1 and 87T4 for transformer T4.

Options Available for the Use of Harmonic-Restraint Logic Functions Three options are available for second-harmonic detection logic, and their implementation depends on the specific relay manufacturer.

1. *Independent operation of each phase.* In most applications, similar to that of individual per-phase electromechanical or static relays, inrush restraint logic operates independently for each phase. This option enhances the dependability aspect of the protection, allowing relay trips for a faulted phase during inrush conditions, when the 60-Hz fault current will be much greater than that of the second harmonic. This option must be used for single-phase transformers that have separate tanks and are connected together to form a three-phase unit. This is normally the case for EHV application and for generator step-up (GSU) transformer single-phase units selected to optimize the use of a fourth single-phase spare unit rather than an expensive full three-phase spare unit.
2. *Cross blocking of tripping.* One phase exceeding the threshold will also issue a tripping block for the remaining two phases. This option enhances the security of the relay.
3. *Averaging technique.* A composite inrush detection threshold is calculated to be equal to one-third of the sum of the three individual phase thresholds. This single calculated composite threshold will be used to restrain all the phase, thus enhancing the security of the relay, especially for some of the new steel transformers, where one phase may be very low on second harmonics during energization, as demonstrated in this case study.

Permanent Corrective Action Since the percent harmonic restraint was lower on one phase only (and thus higher on the other two phases), options to building logic to restraint all phases upon the detection of a high-percentage harmonic threshold need to be explored. The percent harmonic restraint is defined as the ratio of the second harmonic to the fundamental (60 Hz). One of these options is the application of a cross restraint (if the option is provided by the manufacturer) to eliminate these types of false trips when the inrush current signature has insufficient harmonic restraint on one or two phases. However, the relay in question in this case study does not have cross restraint options. Instead, the relay has the feature of using a common averaging restraint value for all the phases instead of individually calculated restraints.

Therefore, the second available option is to use an averaging restraint threshold. The averaging threshold is defined to be equal to $\frac{1}{3}$ (% second "harmonic for A + % second "harmonic for B + % second "harmonic for C). In this case study, based on the relay record for this incident, the average threshold that can be used to restraint all the phases is calculated as

$$\text{average threshold} = \tfrac{1}{3}(16\% + 29.5\% + 28.9\%) = 24.8\%$$

This value is above the setting of 20%, resulting in no false trips. Therefore, use of the averaging technique rather than an individual phase will prevent similar future false trips during this random nature of transformer inrushes. This may affect the sensitivity of the relay, especially during energization of a faulty transformer. However, the programming of a time overcurrent 50/51 element on the high-side CT input to the numerical relay will add sufficient protection in this case. In addition, the Buchholz and remote backup are also providing additional protection in this case.

Another option is to lower the % second harmonic threshold from a 20% setting to 16%. This corrective action alone may not cover all inrush cases, especially if an energization case resulted in a percentage second harmonic of 15.9%. The average option was employed as a corrective action to eliminate future undesired trips during transformer energization.

Lessons Learned The use of numerical multifunction differential relaying offers an advantage during the commissioning phase of transformers by providing a learning mode for transformer inrush phenomena prior to finalization of the relay design. This, in addition to the options listed above, can provide secure selection for the harmonic-restraint logic.

Case Study 5.11: Undesired Operation of a Numerical Transformer Differential Relay During Energization of a 5-MVA 13.8/4.16-kV Station Service Transformer

Abstract A 13.8-kV transformer circuit breaker A2 was closed to energize transformer T2 and was tripped immediately, as were the plant 138- and 13.8-kV systems. The trip was activated by a phase A transformer differential numerical relay. The differential relay trip was due to the lack of a second-harmonic restraint for the phase A relay. The relay trip record indicated a recorded ratio of second harmonic to

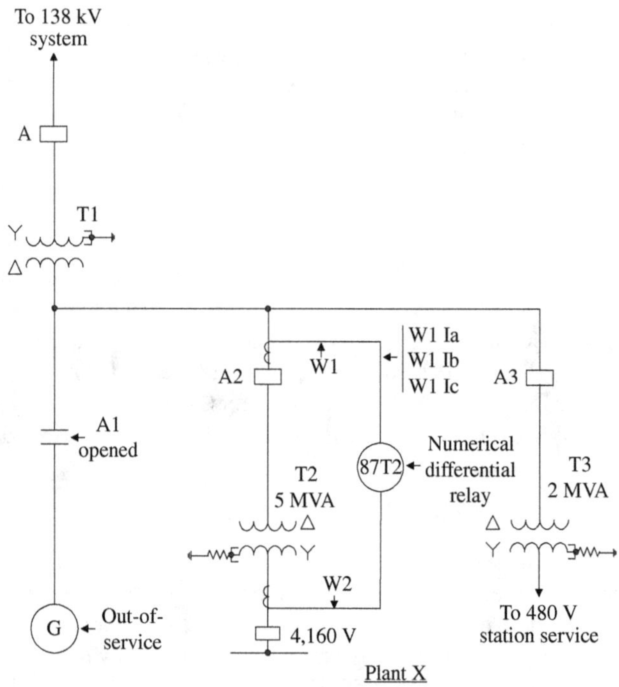

Fig. 5.71 One-line diagram showing the transformer 87T2 relay and its 13.8-kV W1-monitored currents.

fundamental (60 Hz) as 18.9%, which is lower than the relay setting threshold of 20%. Phases B and C of the differential relay recorded 59.9 and 30.2%, respectively, as shown in the relay record.

Description of the Incident Figure 5.71 shows the 5-MVA auxiliary transformer T2 of the 50-MW plant, which was energized by closing 13.8-kV switchgear breaker A2. Phase A of the transformer numerical differential relay operated to energize the plant lockout relays, which tripped the 13.8- and 138-kV breakers.

Analysis of the Relay Oscillography Record The numerical relay 87T2 oscillography record shown in Fig. 5.72 reveals the energization phase of T2, which lasted for only 5.5 cycles. Figure 5.72 also shows the inrush currents for phases A, B, and C as recorded by the oscillography function of the numerical relay. Since the energizing winding of T2 is connected as delta, the relay oscillography record shown in the figure confirms that the sum Ia + Ib + Ic for winding W1 must be equal to zero (there is no neutral return).

1. *Numerical relay 87T2 trip record.* In addition to the oscillography record, the relay generated an event record at the differential relay element trip instant, documenting the cause of the undesired operation.

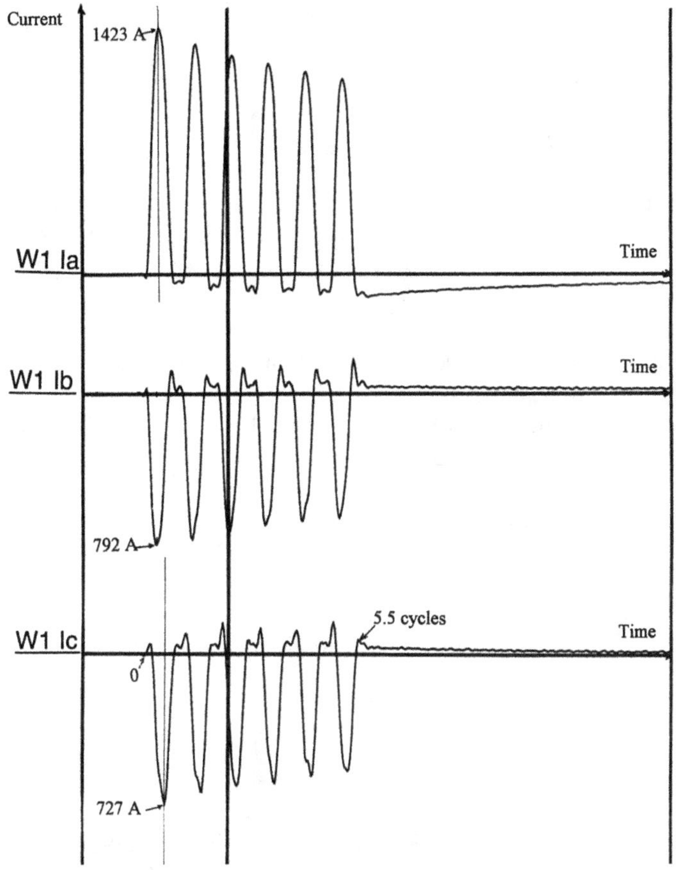

Fig. 5.72 Numerical relay oscillography fault record showing the transformer T1 inrush currents causing the differential relay trip.

Differential operate secondary phase currents: $I_A = 12\,\text{A}$; $I_B = 3.85\,\text{A}$; $I_C = 9.75\,\text{A}$

Restraint secondary phase currents: $I_A = 12\,\text{A}$; $I_B = 3.85\,\text{A}$; $I_C = 9.75\,\text{A}$

Percent second harmonic/60 Hz: phase A = 18.9%; phase B = 59.9%; phase C = 30.3%

2. *Relay setting for harmonic restraint*

   Second-harmonic inhibit: enabled

   Tripping inhibit level: 20% of fundamental 60 Hz

   Harmonic averaging: disabled

Analysis of the Relay Trip Event Record Based on the relay fault record, phase A of the numerical relay has calculated a second-harmonic ratio (percent second harmonic/60 Hz) of 18.9%, which is less than the setting inhibit threshold of 20%. This caused phase A of differential relay 87T2 to operate.

Corrective Action Employ the relay harmonic-restraint averaging feature value for all the phases instead of restraints calculated individually. The averaging threshold is defined to be equal to $\frac{1}{3}$(% second harmonic for A + % second harmonic for "B + % second harmonic for "C). Based on the relay fault record for this case study, the average option can be applied with the average setting threshold calculated as: average threshold value $= \frac{1}{3}(18.9\% + 59.9\% + 30.2\%) = 38\%$. The averaging threshold of 38% is above the setting of 20%, resulting in no future false trips for similar transformer inrush energization. Therefore, the use of an averaging technique rather than an individual phase will prevent similar future false trips during this random nature of transformer inrushes. This is preferred over the option of lowering an individual phase threshold, which could affect the relay sensitivity on energizing a faulty transformer.

Case Study 5.12: Phase-to-Phase Fault Evolving into a Three-Phase Fault at the High Side of a 5-MVA 13.8/4.16-kV Station Service Transformer

Abstract Windblown snow intruded into an outdoor station service transformer enclosure and resulted initially in a phase-to-phase fault, which evolved into a three-phase fault. The fault was cleared from the system successfully by correct operation of the designated relays. In this case study we describe the protection philosophy and analyze the disturbance using relay oscillography and event records for fault classification.

Description of the Protection Systems One of the basic protection criteria is to have two independent relays covering faults in any given protection zone. The first relaying system protecting station service transformer T2 is transformer numerical differential relay 87T2 with its closed protection zone from the 13.8-kV CB A2 to the T2 low-side (4.16-kV) CT. The second relay protecting T2 is provided by software programming of numerical differential relay 87T1 CT input from 13.8-kV CBs A2 and A3 to provide instantaneous and time overcurrent elements. Since the T2 and T3 leads are not fault sources, the two CTs are paralleled to form W3 input to relay, 87T1 as shown in Fig. 5.73. This is an open protection zone with an instantaneous element 50 device covering about 80% of T2 impedance and a TOC element that protects T2 thermally and coordinates with the transformer damage curve and low-side 4.16-kV and 480-V overcurrent relays. Two separate lockout relays are used to accomplish the tripping and shutdown of the generating plant. The two independent relaying systems for transformer T2 are shown in Fig. 5.74.

Description of the Disturbance The plant was not generating at the time of the incident, and as designed the plant station service is derived from the high-voltage 138-kV system. The windblown snow probably caused a phase-to-ground fault, either a B- or a C-to-ground fault. The initial phase-to-ground fault generated an ionized

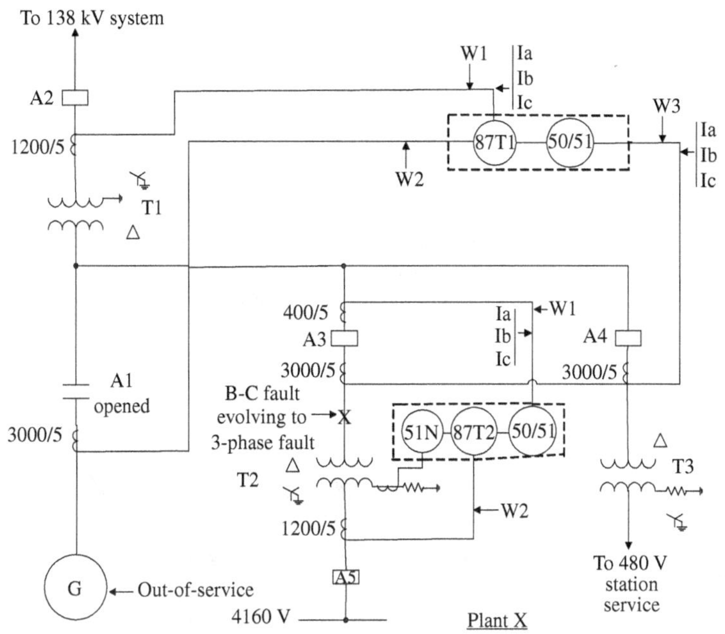

Fig. 5.73 One-line ac diagram showing transformer T2 relays and monitored currents.

cloud that caused the fault to evolve to a B-C fault. Several milliseconds later, the fault evolved into a three-phase fault. The transformer T2 differential relay and instantaneous backup element of the unit transformer T1 differential relay detected the fault and operated to clear the fault in 4 cycles, as confirmed by the numerical relay oscillograph records shown in Figs. 5.75 to 5.77.

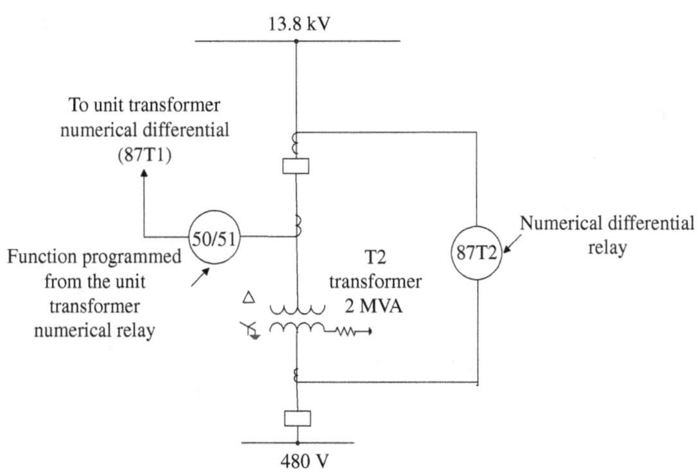

Fig. 5.74 Transformer T2 redundant relaying systems.

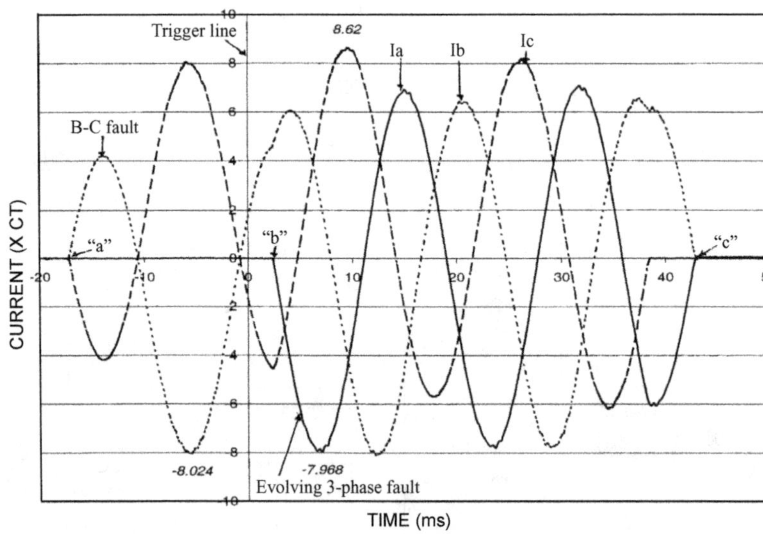

Fig. 5.75 Transformer T1 13.8-kV lead fault currents shown on W3 of the 87T1 relay oscillography record.

Analysis of the Numerical Relay Oscillography Records Figure 5.73 shows oscillography-monitored winding currents for W1 and W3 inputs for relay 87T1 and W1 input for relay 87T2. Figure 5.75 shows the T1 13.8-kV fault currents, where a phase B-C fault occurred at point a on the 13.8-kV bus between CB A3 and

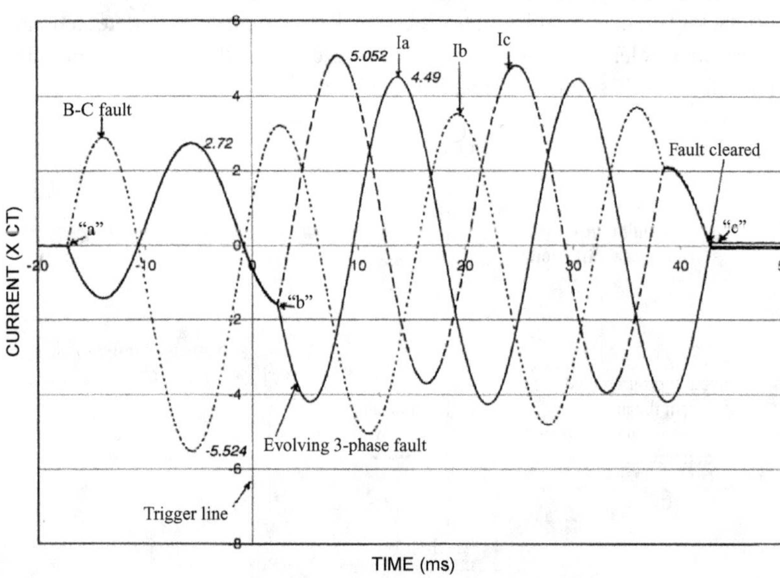

Fig. 5.76 Transformer T1 numerical relay 87T1 oscillography record showing the 138-kV W1 phase currents during the low-side fault.

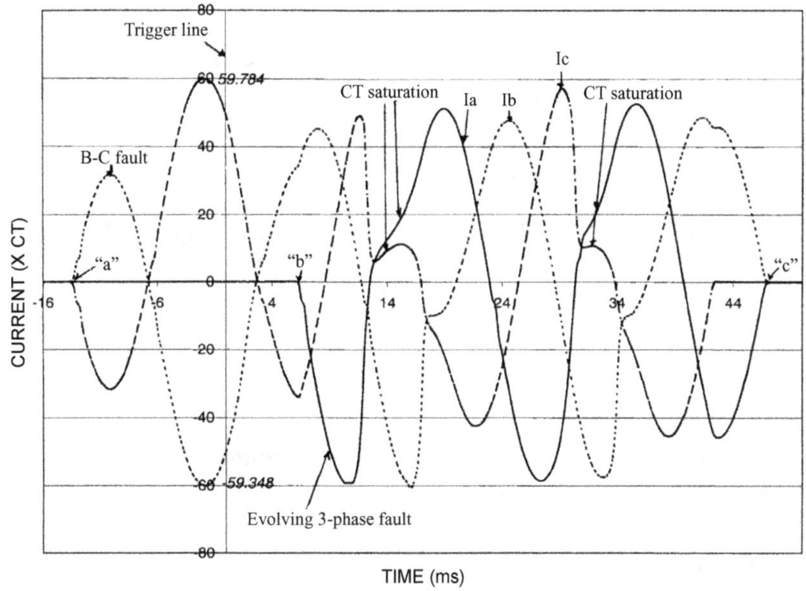

Fig. 5.77 Transformer T2 numerical relay 87T2 record showing the high-side W1 13.8-kV lead currents during the phase fault.

transformer T2. The fault was on the 13.8-kV bus inside the T2 enclosure, which feeds the T2 high side. Phase B current is equal and opposite to phase C current. The T2 13.8-kV fault currents shown in Fig. 5.77 reveal the same equal and opposite phase fault currents. Current transformer saturation during the three-phase fault is also shown as part of the T2 oscillography record. Figure 5.76 shows the T1 138-kV fault currents, which indicate that between points a and b, Ib = – (Ia + Ic) and Ia = Ic. Figure 5.78 confirms the 138-kV currents as derived from the B-C low-side 13.8-kV fault. Using ampere-turns coupling, it can be proven that at the 138-kV transformer T2 neutral point, Ib = Ia + Ic. At point b (1.25 cycles from point a), the initial B-C fault evolved into a three-phase fault. All oscillography records for T1 and T2 reflect the three-phase fault. Figure 5.75 reveals that the fault was cleared at point c (4 cycles from point a). The figure also shows that phase C of CB A2 opened 0.5 cycle earlier from point c. As a result, the three-phase fault evolved into a phase A-B fault (where Ia = –Ib).

Numerical Relay Fault Records

Transformer T2 Relay The relay 87T2 fault record indicates that the RMS current phasors for high-side (13.8-kV) windings W1 and W2, which coincided with operation of the percent differential element, are as follows:

Winding W1:

$$I_A = 0\,\text{A at } 0^\circ;\ I_B = 10,640\,\text{A at } 47^\circ \text{ lag};\ I_C = 10,599\,\text{A at } 227^\circ \text{ lag}$$

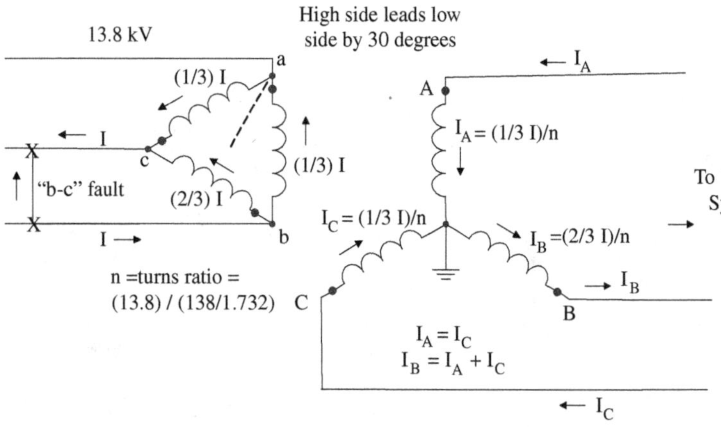

Fig. 5.78 Transformer T1 three-line ac diagram showing the low-side b-c fault at the beginning of the incident.

Winding W2:

currents = 0, low-side 480 V has no fault current sources

Differential currents (×CT rating of 5 A): $I_A = 0$, $I_B = 26.57$ A, $I_C = 26.53$ A

Restraint currents (× CT rating of 5 A): $I_A = 0$; $I_B = 26.57$ A; $I_C = 26.53$ A

Transformer T1 Relay The relay 87T1 fault record indicates that the RMS current phasors for the high-side 138-kV winding W1 and the low-side 13.8-kV windings W2 and W3, which coincided with operation of the instantaneous element of winding W3, are as follows:

Winding W1:

$I_A = 521$ A at 0°; $I_B = 1049$ A at 181° lag; $I_C = 519$ A at 0°

ground current = 0

Winding W2:

no current flow, due to the generator being out of service

Winding W3:

$I_A = 0$ A at 0°; $I_B = 9120$ A at 181° lag; $I_C = 9090$ A at 1° lag

Analysis of the Relay Fault Records The T2 fault record reveals that the phase B current is about equal and opposite to the phase C current. This will confirm the currents shown in Fig. 5.77, which are recorded by the relay for the interval between a and b.

In addition, since the relay W1 CT ratio is 80 : 1, the differential current and restraint current recorded by the relay can be correlated to the primary fault current. For example, for phase B the differential current = 26.57 × CT rating = 26.57 × 5 × 80 = 10,628 A, which is very close to the recorded primary current of 10,640 A.

The T1 fault record is generated by the instantaneous trip of winding W3. The RMS value calculated for phase B of winding W3 is 9120 A, which exceeds the instantaneous setting threshold of 4010 A. The W1 138-kV winding of the T1 record confirms the DFR record of Fig. 5.76, where $I_B = 1049$ A, which is about equal to the sum $I_A + I_C = 524$ A + 519 A. In addition, the phase A current is nearly equal to that of phase C. It can then be concluded that the relay fault records match the relay oscillography records.

Figure 5.78 also confirms the ampere-turn coupling of transformer T1 as related to the relay fault record for the RMS currents calculated for the high (W1) and low (W3) windings. From the figure, current flow in the bc loop of the 13.8-kV delta winding will split to one-third/two-thirds, as shown. Therefore, a current of 9105 A will split to 6070 A flow in the bc delta winding and 3035 A in the ab and ca windings, as shown. Using the ampere-turns coupling principle, $I_1N_1 = I_2N_2$, since the transformer turns ratio is 13.8 kV/(138/1.732) or 1/(10/1.732). Now we can check the accuracy of the recorded relay currents at 138-kV winding W1 and 13.8-kV winding W3. The 13.8-kV ab winding is linked to the 138-kV A winding; therefore, 3035 A = (10/1.732) × I_A or I_A = 3035 × (1.732/10) = 525 A, compared to the relay recorded value of 521 A. The 13.8 kV bc winding is linked to the 138-kV B winding; therefore, 6070 A = (10/1.732) × I_B or I_B = 6070 × (1.732/10) = 1051 A, compared to the relay recorded value of 1049 A. This implies that the relay calculates accurate values and that the ampere-turns coupling principle can be used to verify transformer winding currents and to classify and locate faults on a delta winding using current values on the grounded-wye transformer winding.

System Phenomena

1. Arc-over at the voltage peak when insulation is compromised by a conductive path (e.g., wet snow) in a non-high-speed (slow) fault-creation mechanism, as confirmed by the symmetric nature of the fault currents.
2. Evolving fault phenomena, caused by the growth of an ionized cloud in a confined space environment (transformer enclosure), as shown the DFR records.
3. CT saturation, as shown in Fig. 5.77 for the transformer T2 numerical relay DFR high-side currents. This is due to the use of a low-ratio (i.e., 80 : 1) CT on the 13.8-kV side for differential protection of the 5-MVA transformer T2.

Corrective Actions

1. Increase the spacing between the phases in the 13.8-kV bus connection to transformers T2 and T3 from 7 inches to 12 inches, thus increasing the striking distance between the phases within the enclosure.

2. Insulate the 13.8-kV phase connection metal strips to eliminate the likelihood of an arc-over between energized phases within a humid environment. This will also eliminate phase-to-ground or phase-to-phase arcing and the formation of ionized clouds, which cause initial faults to evolve.

Lessons Learned

1. Numerical multifunction transformer differential relays provide valuable information through their relay fault and oscillography records. Analysis of the information available provides the needed sequence of events and enhances the power system operation through optimization of the relay system performance.
2. The ampere-turns coupling principle is a very useful tool to analyze transformer winding currents for current magnitude verification and analysis for fault classification and location.

Case Study 5.13: Phase-to-Phase Fault Evolving into a Three-Phase Fault at the 13.8-kV Bus Connection of a 2-MVA 13.8/0.480-kV Station Service Enclosure

Abstract During a blizzard, windblown snow intruded into an outdoor station service transformer enclosure. The presence of snow and moisture inside the transformer enclosure caused a phase-to-phase fault, which evolved to a three-phase fault. The fault was cleared from the system successfully, and as a result, the generating unit was shut down. In this case study we describe the protection and setting philosophy for the station service transformer, illustrate the dual element criteria, and analyze the disturbance and the relay fault record.

Description of the Protection System Figure 5.79 shows the system one-line diagram and numerical relay 87T1–monitored currents. The 13.8-kV cable connection and station service transformer T3 are protected by dual relaying systems to fulfill the basic protection criteria for two independent systems, as shown in Fig. 5.80. One of the systems is numerical time overcurrent relay 50/51/T3, and the second system consists of instantaneous and time overcurrent (TOC) elements 50 and 51, derived from transformer T1 numerical differential relay 87T1. As shown in Fig. 5.80, one CT from the 13.8-kV cable lead connection to T2 is paralleled to the CTs from the 13.8-kV lead to T3. The added sum of the two currents is fed to the numerical relay 87T1 W3 winding.

Relay Setting Calculations The applied setting of the TOC 50/51/T3 is to protect T3, a 2-MVA transformer, and provide backup protection for the 480-V low-side system. The 50/51/T3 is set as follows: From the short-circuit study simulation using

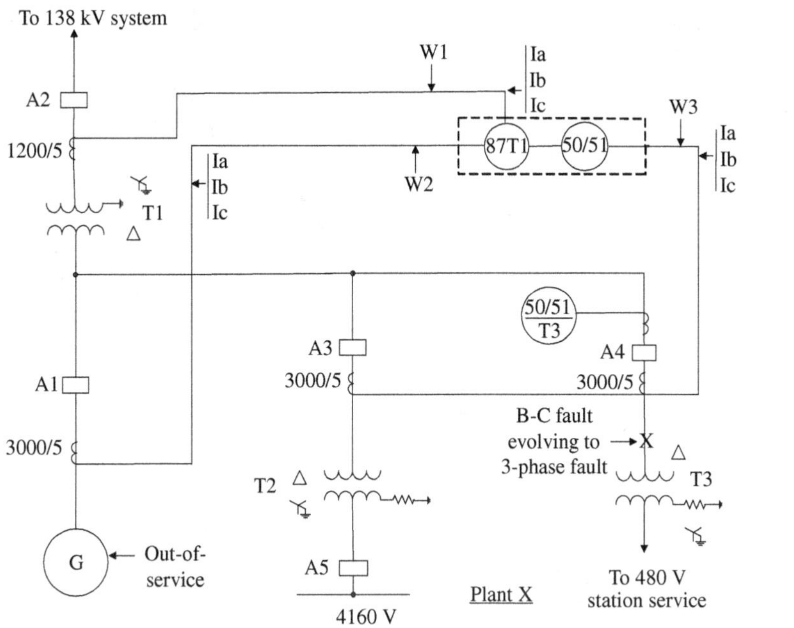

Fig. 5.79 One-line diagram showing transformer T1 numerical relay–monitored currents.

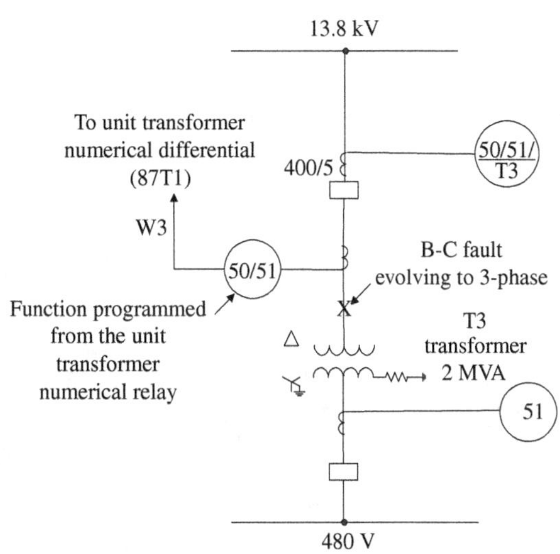

Fig. 5.80 One-line ac diagram showing the two independent time overcurrent relaying systems protecting the T3 transformer.

100 MVA as a base, the three-phase fault at the 13.8-kV generator bus = 40,208 A. The base current

$$I_{\text{base}} = \frac{100 \times 10^6}{\sqrt{3} \times 13.8 \times 10^3} = 4184\ \text{A}$$

$$\text{three-phase fault (pu)} = \frac{I_{\text{F}}}{I_{\text{base}}} = \frac{40,208}{4184} = 9.61\ \text{pu}$$

$$\text{system impedance} = X_S = \frac{1}{I_F} = \frac{1}{9.61} = 0.104\ \text{pu}$$

$$\text{T3 impedance} = Z_{\text{T3}} = X_{\text{T3}} = 5.75\%\ \text{on 2-MVA base}$$

$$Z_{\text{T3}} = X_{\text{T3}} = 5.75 \times \frac{100}{2} = 2.875\ \text{pu on 100 MVA}$$

For the low-side 480-V three-phase fault:

$$X_{\text{tot}} = X_S + X_{\text{T3}} = 0.104 + 2.875 = 2.98\ \text{pu}$$

$$I_{\text{pu}} = \frac{1}{2.98} = 0.34\ \text{pu}$$

$$\text{fault current} = I_{\text{pu}} \times I_{\text{base}} = 0.34 \times 4188 = 1405\ \text{A}$$

Set the instantaneous element 50 device above the transformer inrush and 125% of the low-side fault, whichever is larger.

$$I_{\text{set}} = 1.25 \times 1405 = 1756\ \text{A}$$

Now we can check for the T3 inrush currents. The T3 inrush can be assumed to be 12 × the transformer rating:

$$\text{T3 rating current} = \frac{2 \times 10^6}{1.732 \times 13.8 \times 10^3} = 84\ \text{A}$$

$$I_{\text{inrus}} = 12 \times 84\ \text{A} = 1008\ \text{A}$$

The element 50 setting of 1756 A will be above the T3 inrush current. Set TOC element 51 to protect T3 thermally below the damage curve and to give a 0.4-s. coordination margin with the slowest low-side (480-V) feeder relay.

Analysis of the Numerical Relay Oscillography Record The instantaneous overcurrent function associated with winding W3 of the T1 numerical differential relay 87T1 detected the 13.8-kV lead fault and generated a relay fault record. Figure 5.81 shows the 138-kV three-phase winding W1 currents during the fault, Fig. 5.82 the recorded numerical relay generator W2 currents during the fault, and

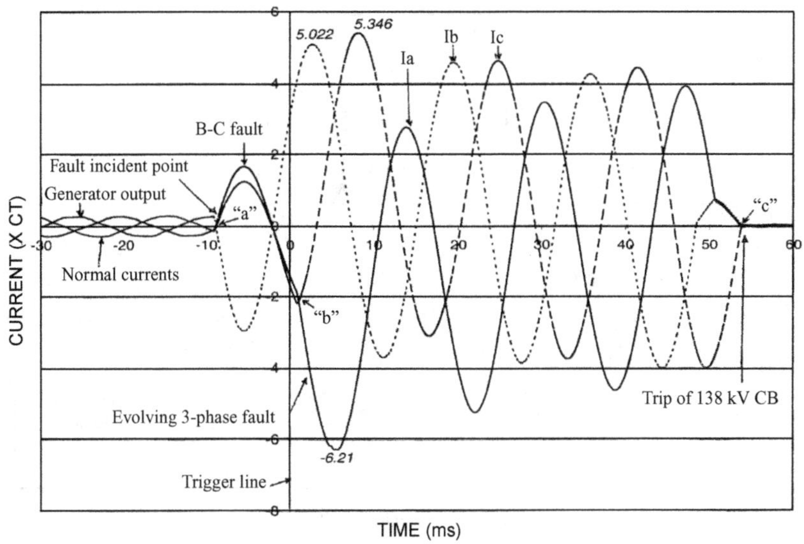

Fig. 5.81 Transformer numerical relay 87T1 oscillography record showing the 138-kV W1 phase currents during the low-side fault.

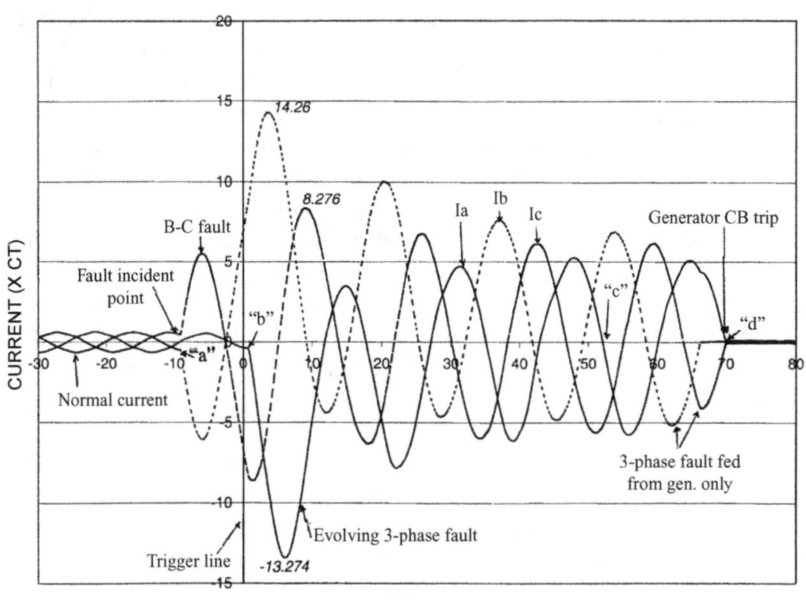

Fig. 5.82 Relay 87T1 oscillography record showing the generator 13.8-kV W2 output phase currents during the fault.

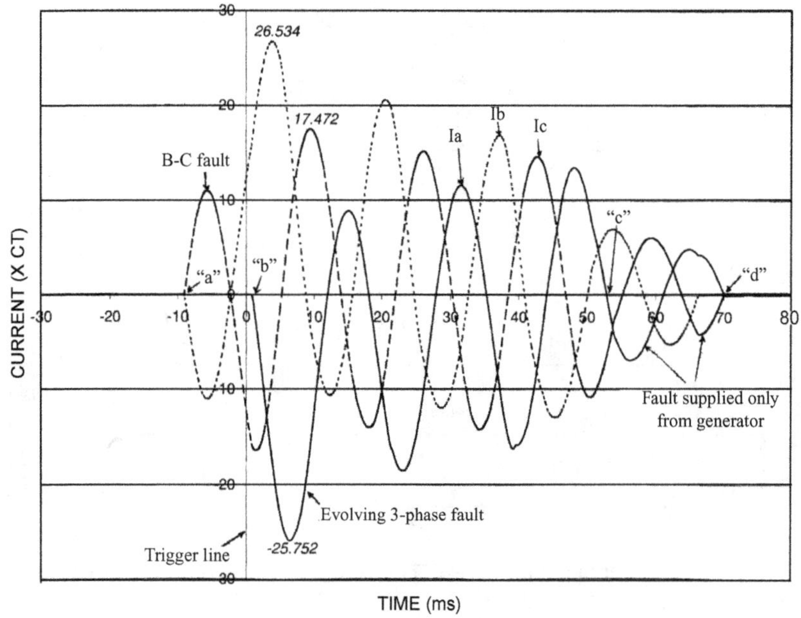

Fig. 5.83 Relay 87T1 oscillography record showing the W3 winding currents during the 13.8-kV fault.

Fig. 5.83 the W3 13.8-kV fault current. Figures 5.81 and 5.82 illustrate the generator balanced pre-fault three-phase output currents. A phase B-C fault occurred at point a, as shown in the W1 and W2 currents (with phase B equal and opposite to phase C). The phase-to-phase fault lasted for 12 ms, and at point b the initial fault evolved to a three-phase fault. Element 50 associated with winding W3 operated in about 1 cycle to shut down the plant and trip the 138- and 13.8-kV breakers. At 4 cycles from point a, 138-kV breaker A4 was tripped and cleared the fault from the system at point c. Generator G1 continues to feed fault F1 for an additional 1 cycle. The fault was finally cleared by tripping and opening CB A1 at point d. Figure 5.81 reveals the W1 current associated with the transformer T1 wye-winding connection between points a and b, where Ia = Ic and Ib = 2Ia as a reflection of the 13.8-kV fault currents through the transformer ampere-turns coupling principles and polarity rules.

Transformer T1 Numerical Relay Fault Record The relay 87T1 fault record indicates that the RMS current phasors for the high-side (138-kV) winding W1 and the low-side (13.8-kV) windings W2 and W3, which coincided with operation of the instantaneous overcurrent element of W3 winding, are as follows:

Winding W1:

$I_A = 64$ A at 0°; $I_B = 331$ A at 190° lag; $I_C = 263$A at 11° lag; ground current = 0

Winding W2 (generator G1 was in service):

$I_A = 1240$ A at 18° lag; $I_B = 4470$ A at 162° lag; $I_C = 3530$ A at 330° lag; ground current = 0

Winding W3:

$I_A = 70$ A at 35° lag; $I_B = 6670$ A at 171° lag; $I_C = 6640$ A at 350° lag
Differential currents (×CT rating of 5 A): $I_A = 0.01$; $I_B = 0.01$; $I_C = 0.02$
Restraint currents (×CT rating of 5 A): $I_A = 0.21$; $I_B = 1.11$; $I_C = 1.10$

Analysis of the T1 Numerical Relay Fault Record The fault was cleared by operation of W3 instantaneous overcurrent element 50 of the transformer T1 differential numerical relay. The T1 fault record is generated by the instantaneous trip of winding W3. The RMS value calculated for winding W3 B phase is 6670 A and for C phase is 6640 A, each of which exceeds the instantaneous setting threshold of 4440 A. Since the fault was external to the T1 differential zone, no significant differential operating current was generated, and hence a correct restraint operation occurred. The 50/51 T3 time overcurrent relay failed to produce a target, and as a result, no fault record was generated. The relay had a healthy light and the instantaneous element was set to see 70% of the transformer impedance, which is 1756 A of primary current. The T1 numerical relay fault record reveals a fault current of about 6600 A, which exceeds the T3 element 50 setting. Therefore, analyses of the 50/51 T3 relay failure to trip began by removing the relay from service and substituting a spare relay.

Corrective Actions The 50/51/T3 overcurrent relay was tested and shown to produce no output when subjected to a similar current. The relay had a healthy light with no indication of software problems. The relay setting file was then examined and relay setting group 2 (rather than group 1) was found to be enabled, with setting parameters not related to the transformer T3 setting record. The relay was designed with setting group 1 as the basis for T3 protection. Somehow an error occurred during commissioning of the relay in which the relay defaulted to setting group 2, initialized originally by the manufacturer with setting values that did not correspond to setting group 1 for T3 setting calculations.

Lessons Learned

1. Numerical relay setting files contain many parameters that need to be tested and verified prior to commissioning a relay system. In this incident the wrong relay setting group (group 2) was mistakenly enabled. The correct setting resided in

memory as group 1 in a disable mode, and the relay used group 2 for trip criteria. Group 2 had setting parameters not related to transformer T3 protection.

2. All relay system targets need to be analyzed to ensure that relay problems can be detected and analyzed prior to the occurrence of disturbances in which both relay systems can fail to clear the faults.
3. As explained above, the absence of a target for the 50/51/T3 overcurrent relay highlights the importance of the dual-element principles. Two separate TOC elements were used to protect transformer T3. Each system has a time overcurrent 50/51 function, with the fault being cleared by the element associated with the T1 differential relay.

Case Study 5.14: Phase-to-Phase Fault in a 13.8-kV Switchgear Caused by Heavy Rain Evolving into a Three-Phase Fault

Abstract In this case study we describe a 13.8-kV switchgear phase-to-phase fault that was cleared by transformer numerical relays. The generator was out of service and the fault was fed only from the 138-kV system. We illustrate how a relay oscillography record for the 138-kV transformer side can be used to develop the analysis for the classification of the fault and confirmation of the fault location.

Initial Troubleshooting of the Incident The initial assumption was for a fault inside the transformer, because field personnel claimed that the relays that operated were labeled as primary and secondary transformer protection. Transformer T1 is protected by dual numerical transformer differential relays 87TP and 87TS. Figure 5.84 shows only the connection for 87TP as one of the numerical relays that will be used to analyze the fault. The nonfamiliarity with the 87TP and 87TS zone of protection is the main reason for confusion on the field side. The relays have a protection zone that covers the transformer and the 13.8-kV cable bus from the generator breaker to the station service feeders as shown in Fig. 5.84.

Due to the incomplete and incomprehensive analysis of the incident, the transformer was tested and yielded good results. The doble test results almost match the factory testing data. The transformer was tested due to the lack of a real analysis of transformer ampere-turns coupling, which is used to prove that the turns ratio for all three-phase primary and secondary windings is correct.

Information Available to Classify and Locate the Fault Transformer numerical relay 87TP produced the oscillography record shown in Fig. 5.85. In addition, 87TP provided the following fault record at the relay trip output for the 138-kV high-voltage transformer W1 side: $I_A = 1771$ A at 0°; $I_B = 854$ A at 180° lag; and $I_C = 883$ A at 180° lag.

Classification and Location of the Fault Using the Information Above Determination of the fault can be started by an analysis of the relay 87TP

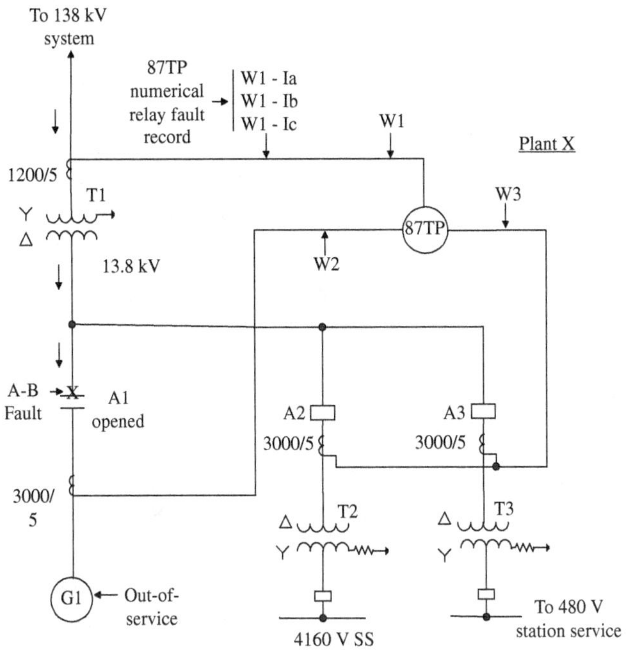

Fig. 5.84 One-line ac diagram showing the transformer 87TP relay and monitored currents with the location of the A-B 13.8-kV fault.

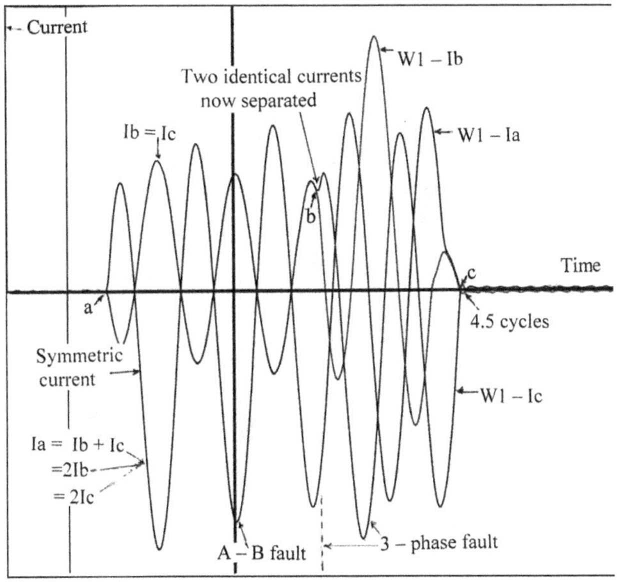

Fig. 5.85 Transformer T1 87TP relay oscillography record showing the high-side W1 phase currents feeding the A-B and evolving faults.

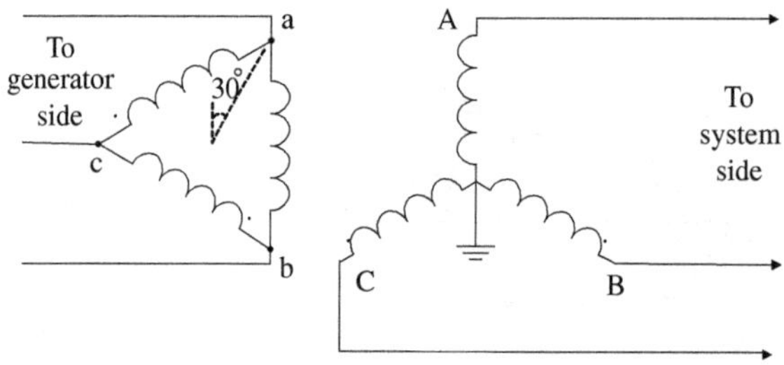

Fig. 5.86 Unit transformer nameplate phasing diagram.

oscillography and event fault records for the 138-kV high-voltage transformer W1 side. The analysis can proceed using the following steps:

1. Obtain the unit transformer phasing diagram, shown in Fig. 5.86, as defined on the transformer nameplate.
2. As illustrated in Fig. 5.85, the 87TP oscillography relay fault record displays the only source of the 13.8-kV fault as the 138-kV system (the generator was out of service with the generator breaker A1 open). The record indicates initially that the phase A fault current is a 180° phase shift with phase B and C currents. In addition, phase B current is equal and in phase with phase C current (identical). The phase A current is divided at the neutral point of the transformer primary Y winding into two equal components: $I_A = 2 \times I_B = 2 \times I_C$. These values can now be placed on the transformer phasing diagram shown in Fig. 5.87 for the primary GY winding. The same procedure can also be followed using the relay fault record currents outlined above. It can be concluded from this step that the transformer windings are healthy because they support Kirchhoff's first law at the neutral junction of the transformer high-voltage winding.
3. The next step is to apply transformer ampere-turns coupling to the delta winding and current polarity rules. These current relationships are applied

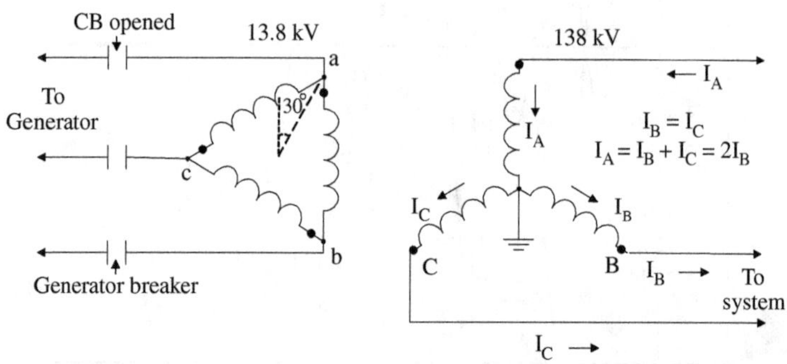

Fig. 5.87 Transformer primary 138-kV winding with currents assigned in accordance with the numerical relay fault record.

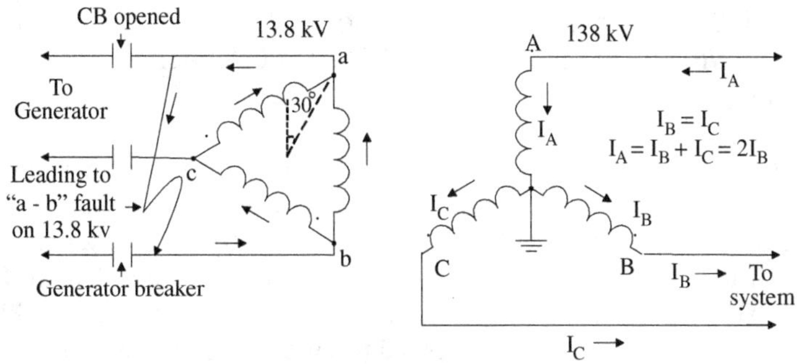

Fig. 5.88 Transformer secondary 13.8-kV winding-deduced currents based on the assigned primary currents and transformer polarity rules.

to the transformer phasing diagram shown in Fig. 5.88. The winding ampere-turns coupling principle is then applied to deduce the faulted phase on the 13.8-kV side. Figure 5.88 shows the analysis leading to fault classification as a phase a-b fault on the 13.8-kV lead. It can then be concluded that the fault is outside transformer T1 and located on the 13.8-kV bus as phase a-b. The fault started as a-b for a duration of less than 3 cycles prior to evolving to a three-phase fault that was cleared from the system in 4.5 cycles. This conclusion can also be reached by using the following information from the relay fault record: $I_A = 1771$ A at 0°; $I_B = 854$ A at 180° lag; $I_C = 883$ A at 180° lag. These current values can be placed on the transformer phasing diagram shown in Fig. 5.86 for the primary GY winding. The next step is to apply transformer ampere-turns coupling to the delta winding, as shown in Fig. 5.88. It can then be concluded that the fault is outside transformer T1 and located initially on the 13.8-kV bus as a phase a-b fault.

Lessons Learned

1. Protection and control drawings should be examined before troubleshooting any incident.
2. Familiarity with the protection scheme zone of operation is required to analyze disturbances.
3. The relay event and oscillography fault records should be examined prior to troubleshooting an incident.
4. Transformer ampere-turns coupling principles where $I_1N_1 = I_2N_2$ and current polarity rules should be employed to confirm whether a fault is external or internal to the transformer tank.
5. Available 138-kV fault record data should be used to deduce the fault type and its location within the transformer differential zone of protection.

Case Study 5.15: Undesired Operation of a Numerical Transformer Differential Relay Due to a Missing CT Cable Connection as an Input to the Relay Wiring

Abstract A 500-MW combined-cycle plant consists of two combustion turbine (CT) units and one-steam (ST) unit. The CT unit one-line ac diagram with the protection systems is shown in Fig. 5.89. During the construction phase of the plant, an undesired trip of the transformer primary differential numerical relay 87TP/T1 occurred. The trip coincided with the energization of one of the 4.16-kV/480-V station service transformers. The undesired trip was caused by an error shorting the LV CT input to the 87TP/T1. As a result, the through inrush current flow during the energization of transformer T3 generated differential operating current for relay 87TP/T1, which produced a trip output. In this case study we describe the incident,

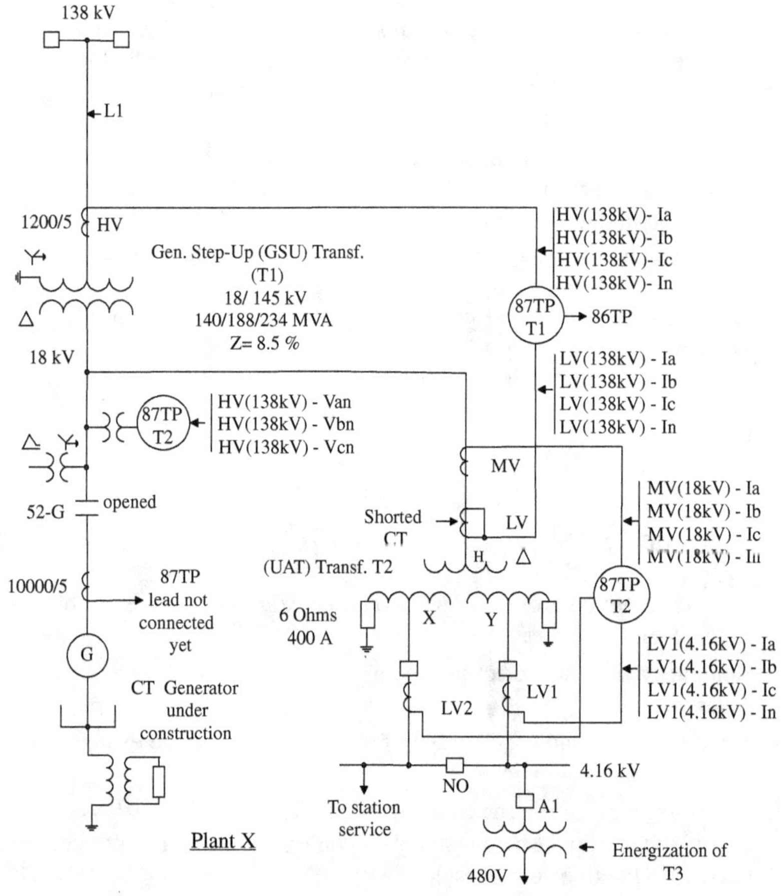

Fig. 5.89 Plant CT unit one-line diagram with T1 and T2 numerical relay–monitored currents and voltages.

state the cause of the trip through analysis of the numerical relay fault records associated with relays 87TP/T1 and 87TP/T2, and provide lessons learned.

Description of the Protection System The plant is protected using dual independent relays to sense and detect all faults in the relay protection zones. This criterion will provide two redundant relay systems to guarantee fault clearing. The only exception to this criterion is the clearing of ground faults occurring in the 18-kV ISO-phase bus area during the backfeeding phase. The 18-kV areas will be ungrounded when the generator breaker is opened and the 138-kV systems are used as a source for the plant auxiliary systems. The ungrounded 18-kV bus, which connects the GSU and UAT transformers to the generator, will only be protected by the 60-Hz tuned overvoltage 59N relay, which will detect ground faults through the broken delta PT connection at the 18-kV generator lead. However, once the unit is synchronized to the system, the dual relay element principle will be fulfilled by the generator stator ground fault protection system. Figure 5.89 illustrates the overlapping protection zones of the CT unit with the GSU and UAT transformer connections.

Description of the Incident During the construction phase of the CT unit, while energizing one of the 4.16/.480-kV station service transformer T3, an undesired operation of the differential element of the 87TP/T1 occurred. The 87TP/T1 relays energized the lockout relay 86TP to trip and lock out the 138-kV feeder L1-associated circuit breakers. Figure 5.89 illustrates the connected 18-kV system at the construction phase. The generator CT circuit cables were not connected to the 87TP/T1, and its wiring was restricted to the final construction phase. Oscillography and event records were generated for the numerical relays of both T1 and T2.

Analyses of the T1 and T2 Numerical Relay Oscillography and Event Records The 87TP transformer numerical relay is designed with a minimum differential current value, below which the relay does not register the differential calculation or the second harmonic. As shown in Fig. 5.90, the oscillography record from 87TP/T1 on the GSU (T1) transformer shows current only on the high-side 138 kV. The event lasted for 9.5 cycles and Fig. 5.90 shows no current on the low-side 18-kV traces LV(18kV)-Ia, LV(18kV)-Ib, and LV(18kV)-Ic, which is fed from CTs on the 18-kV side of the UAT (T2) transformer. The currents shown on the high side is the inrush current for energization of transformer T3 on the 4.16-kV bus. Since only the high-side currents were fed to the 87TP/T1 relay, it saw differential current that caused the relay to trip falsely upon the disappearance of the second-harmonic-restraint signal when the second harmonic-to-60 Hz ratio dropped below the setting threshold of 20%.

As illustrated in Fig. 5.90, the phase B percent differential element operated 6 cycles from the energization of T3. The B phase values are above the cutoff during the entire event. The average second harmonic drops below the restraint setting of 20% and the B phase differential (the only active one) trips. The differential current exists because the CT connection from the 18-kV system is missing. This analysis confirms the error of not connecting the low-voltage 18-kV current transformer leads to the 87TP/T1. The connection of the CT leads to the 87TP/T1 will balance the relay,

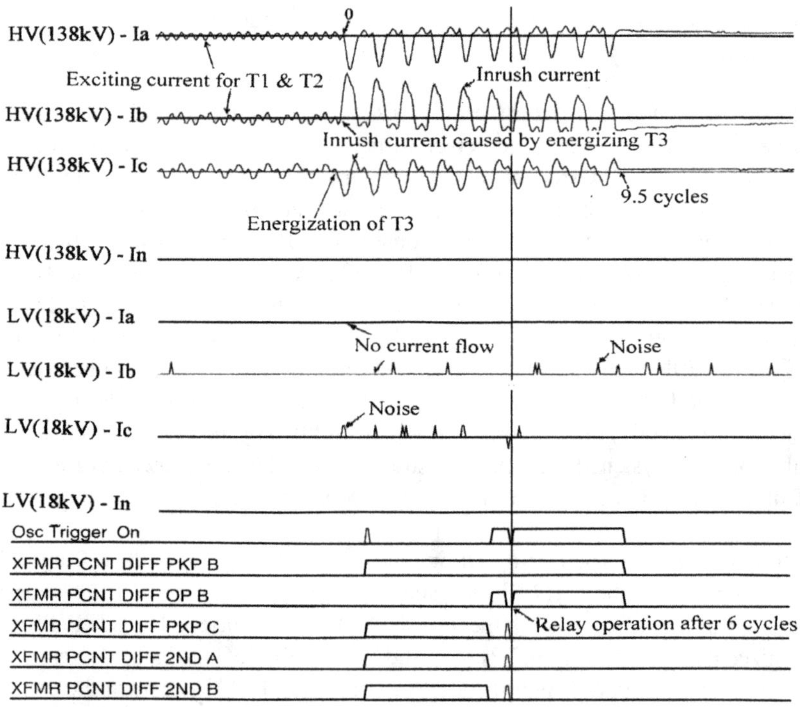

Fig. 5.90 T1 numerical relay 87TP oscillography record showing the missing 13.8-kV CT input to the differential relay.

producing no differential current, and the energization of T3 will appear as a through-flow condition for the relay.

Figure 5.91 shows the current on the high side of the T2 (UAT) as captured by the 87TP/T2 on transformer T2. This is the current that should also be seen by the 87TP/T1 relay for transformer T1. If these currents had been fed into the 87TP/T1 relay, the differential currents would have been close to zero (nonlinear currents will never be fully canceled out). The figure also illustrates (light traces) nonoperation of the relay due to low differential currents compared to the high restraint currents associated with the generation of harmonic restraint.

Causes of the Transformer Differential Relay Undesired Trip The sudden application of voltage to the T3 transformer generated a current that appeared on one side only. The pulsating current is rich with second harmonic. The energization of transformer T3 generated inrush currents that are considered a through-current condition for 87TP protecting transformer T1. However, the undesired trip occurred due to the generation of differential current at the time where the second-harmonic restraint fell below the restraint setting of 20%, as illustrated in Fig. 5.90. The differential current was due to the missing 87TP/T1 relay current CT lead connections associated with the 18-kV numerical relay input. This is considered as an error during

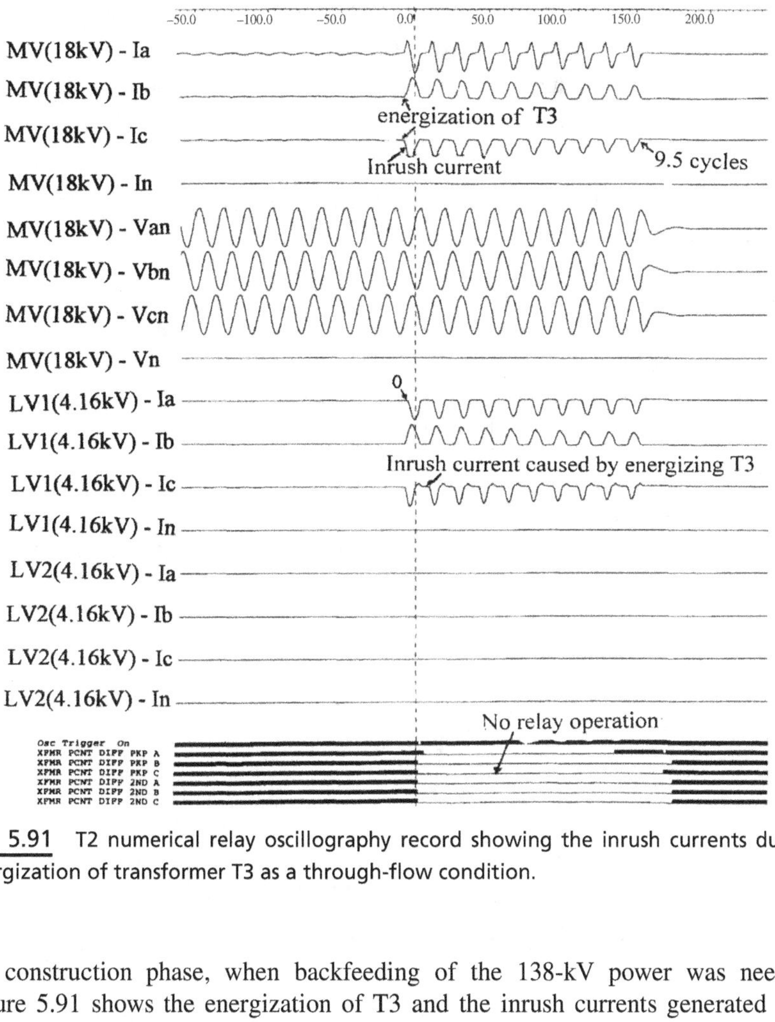

Fig. 5.91 T2 numerical relay oscillography record showing the inrush currents during energization of transformer T3 as a through-flow condition.

the construction phase, when backfeeding of the 138-kV power was needed. Figure 5.91 shows the energization of T3 and the inrush currents generated as a through condition for the 87TP/T2 relay protecting transformer T2. No differential currents were generated to cause any undesired 87TP/T2 numerical relay trip.

Corrective Actions Add the missing current to the 87TP/T1 relay CT input by wiring the current transformer secondary windings leads to the relay.

Lessons Learned

1. Thorough relay testing should be performed during the commissioning phase of new protection systems. This will eliminate undesired relay trips.
2. Although the second-harmonic setting was not the cause of the incident, the second-harmonic mode, the "per phase" option, should be avoided on new three-phase transformer units, as experience shows that at least one phase has a

low harmonic content in inrush currents during transformer energization for the new transformer steel core design. The "average" setting mode is recommended for this case study and normally works well because two phases are usually high at the same time.

Case Study 5.16: Phase-to-Ground Fault Caused by Flashover of a Transformer 115-kV Bushing Due to a Bird Droppings

Abstract A flashover occurred across a 115-kV transformer bushing, causing a phase B-to-ground fault. The fault was cleared from the system successfully by operation of the transformer differential relays. In this case study we explain the cause of the flashover, highlight fault resistance as recorded by the DFR record, and provide fault study simulation to calculate the fault resistance and confirm the system model.

Description of the Power System As shown in Fig. 5.92, the transformer is a three-winding grounded wye/delta/delta 115/13.8/13.8 kV rated as 260 MVA. Four hydro units rated 50 MW each are shown, each two of which are paralleled and connected to one of the delta windings via an isolated phase bus. The units are synchronized via dedicated 13.8-kV generator breakers.

Description of the Protection System As shown in Fig. 5.92, transformer T1 is protected by a harmonic-restraint differential relay with three inputs feeding the

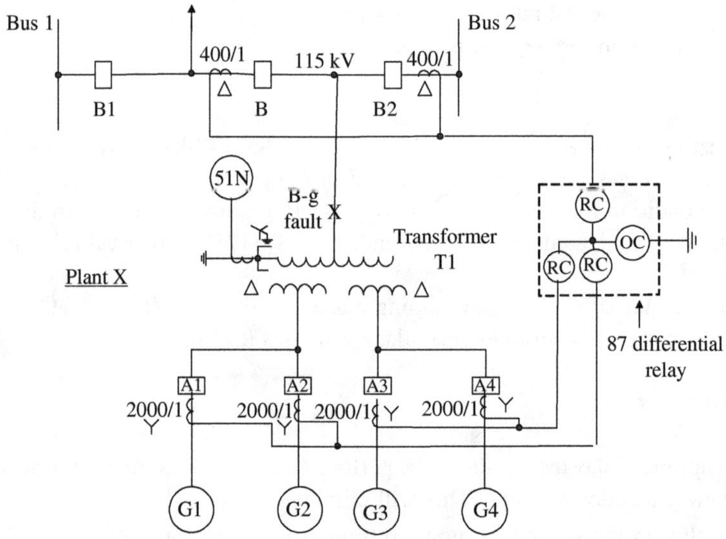

Fig. 5.92 One-line ac diagram showing the transformer T1 overall differential protection zone around the B-g fault.

restraint circuits. CTs from each two units are paralleled to feed the transformer differential relay. CTs from the high-side 115-kV side are also paralleled and connected to one of the differential relay restraint circuits. It should be noted that this paralleling of CTs for elements defined as fault sources is not recommended by present practice. The protection system was designed in the late 1950s with a three-winding input transformer differential relay. A ground backup relay was installed in the transformer neutral to provide ground fault backup protection to the 115-kV cable feeder and the surrounding system. Generator backup relay device 21 is also set to provide system phase backup protection.

Protection Zones As illustrated in Fig. 5.92, the transformer closed protection zone includes the 13.8-kV generator, the unit breaker, the 115-kV feeder cable, and the 115-kV circuit breakers. Overlapping zones are provided for generator protection and 115-kV bus and line protection. A transformer neutral overcurrent relay provides an open zone for the 115-kV cable and system L-g faults. Overvoltage relay tuned to 60 Hz and connected to broken delta PTs on the 13.8-kV generator provides ground detection for the ungrounded system area when units are out of service. The generator phase backup and negative sequence provides an open zone for system phase faults.

Design and Setting of the Transformer Differential Relay The electromechanical harmonic restraint differential relay protecting the transformer has three-restraint winding. The transformer protection zone has six inputs, two CTs from the 115-kV side and four CTs from the 13.8-kV side. Therefore, paralleling of CTs must be done despite the fact that all six inputs are sources of faults within or outside the protection zone. The relay had a limit of three restraint inputs during the time of the design, as was done in the late 1950s. As shown in Fig. 5.92, the 115-kV side will be one input, and every two units are the remaining relay inputs. The differential relay setting calculations are shown in Table 5.4, where for the 115-kV side transformer rating is 258 MVA and is used to balance the differential.

TABLE 5.4 Differential Relay Settings

	115-kV Side	13.8-kV Side	13.8-kV Side
Transformer rating	258 MVA	130 MVA	130 MVA
Maximum current (MVA $\times 10^3/\sqrt{3} \times$ kV)	1295 A	5439 A	5439 A
CT connection	Delta	Wye	Wye
CT ratio (maximum current = 448)	$(1295 \times \sqrt{3}):5$	$5439:5 = 1088$	1088
Actual CT ratio	400 : 1	2000 : 1	2000 : 1
258 MVA to balance the differential	1295 A	10,878 A	10,878 A
Secondary current	$(1295 \times \sqrt{3})/400 = 5.61$ A	$10{,}878/2000 = 5.44$ A	$10{,}878/2000 = 5.44$ A
Relay tap	5 A	5 A	5 A

Check the mismatch:

$$\text{current ratio} = \frac{5.61}{5.44} = 1.022 \qquad \text{tap ratio} = \frac{5}{5} = 1.0$$

$$\text{mismatch }\% = \frac{\text{current ratio} - \text{tap ratio}}{\text{the smaller of the two}} \times 100$$

$$= \frac{1.022 - 1.0}{1.0} \times 100 = 2.2\% < 5 \qquad \text{O.K.}$$

Description of the Disturbance The 115-kV bushing integrity was compromised when a seagull defecated on the high-voltage bushing. This caused the bushing to track down and flash to the top of the transformer tank. The B-g fault was detected by the transformer differential relay and cleared by tripping the associated 115- and 13.8-kV circuit breakers. The 115-kV high-voltage bushing, which is mounted on the top of transformer T1, flashed over through the neutral bushing when it arced-over to ground.

Analysis of the DFR Record Figure 5.93 shows the DFR-monitored voltages and currents used to analyze this disturbance. The DFR record in Fig. 5.94 reveals that the flashover across the transformer B phase bushing occurred at the voltage peak. The phase B-g fault is supported by phase B voltage drop and current increase for phase B and a presence of neutral current. The fault was cleared from the system in 5 cycles by tripping of 115-kV breakers A1 and A2. The hydro generators continued to feed the fault for an additional 3.5 cycles. The fault was detected by the transformer differential relays, which energized the associated lockout relays to trip and lock out the circuit breakers.

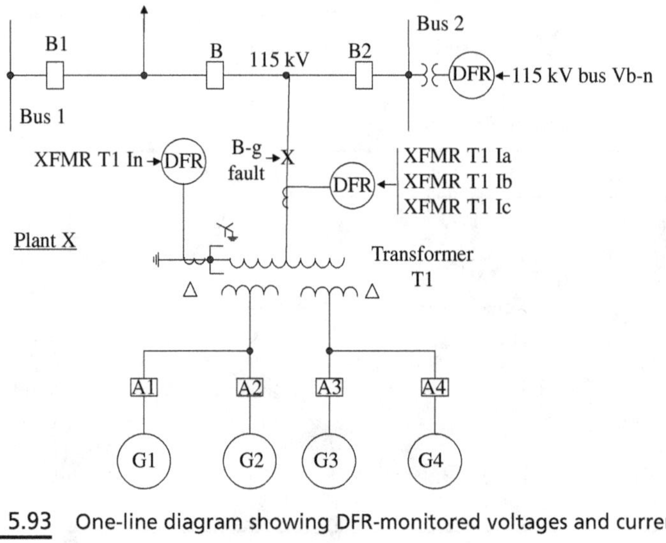

Fig. 5.93 One-line diagram showing DFR-monitored voltages and currents.

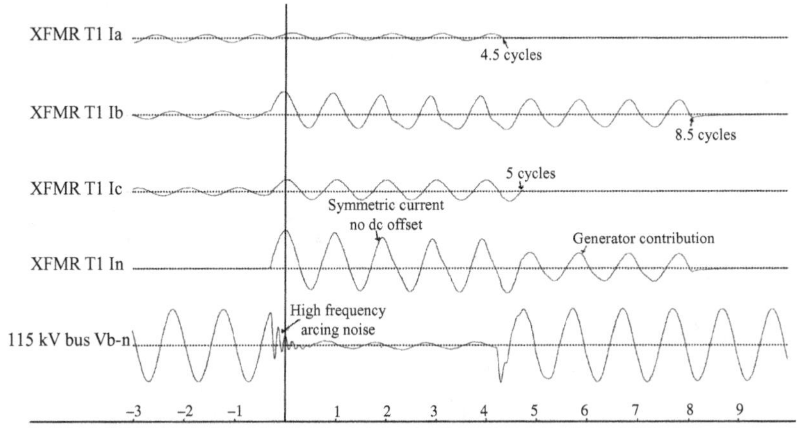

Fig. 5.94 Transformer T1 phase and neutral currents during the fault.

Analysis of the Cause of the Ground Fault The flashover of the phase B bushing on the top of transformer T1 was through a fault resistance. The fault resistance path was established raising the top of the transformer tank voltage above ground by an amount that caused arc-over of the neutral 15-kV bushing. The neutral bushing is located on the top of the transformer tank, and its flashover has allowed the fault current to flow from the HV bushing through the top of the tank, to neutral, and then to ground.

Another fault mechanism could be that tracking down the HV bushing has raised the potential of the top cover of the transformer, which is connected to the grounded tank via a gasket and some steel conduits. The high potential of the top of the transformer caused the neutral bushing to flash over, thus establishing a ground path for the fault current.

The fault mechanism theory is supported by the following:

1. The HV bushing had a glaze burned off the porcelain on the top three skirts.
2. The top of the transformer had a large area of white seagull excrement near the base of the HV bushing, and the excrement was splattered up onto the bottom skirts of the HV bushing.
3. A conduit on the top of the transformer, which runs down the side of the transformer, showed marks of arc-over. The conduit support bracket on the side of the transformer had carbon tracking, indicating heavy current flow down the conduit to the ground cables attached to the transformer base.
4. The arcing stud positioned near the neutral bushing had been melted to the neutral bushing by an arc.

Verification of the Power System Model and Estimation of Fault Resistance Since the 13.8-kV generator breakers are slower than the 115-kV breakers, the initial fault was fed for an additional 3.5 cycles from the hydro generators. This is an isolated model that can be used to verify the representation of the generators and transformer

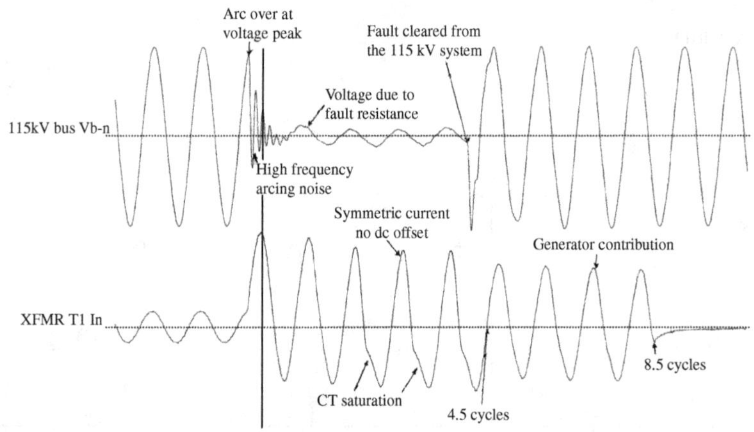

Fig. 5.95 Transformer T1 faulted phase current and 115-kV faulted phase voltage.

parameters in short-circuit study simulation. Since the fault incident point is at the voltage peak, the fault current is symmetrical, containing no dc offset. This pure 60-Hz symmetrical current is the same component obtained from the quasi-steady-state fault calculations using symmetrical components. The ground currents (neutral) can therefore be used to verify the power system model. However, the presence of fault resistance makes this power system model verification difficult to achieve. It can be concluded that the fault current will be higher, with no fault resistance simulated. The fault current for a solid 115-kV bus L-g fault is 3348 A, while the DFR record shows 3000 A. Using the short-circuit study for the 115-kV bus fault with varying the fault resistance, a match is established at $R_F = 10\,\Omega$ for a fault current of 3004 A.

Power System Phenomena

1. Flashover (arc-over) at the voltage peak as a slow fault-creation mechanism, as shown in trace 115 kV bus Vb-n in Fig. 3.95.
2. Arc-over noise appearing as the high-frequency noise shown in the trace 115 kV bus Vb-n in Fig. 5.95.
3. Fault resistance as shown in trace 115 kV bus Vb-n in Fig. 5.95, where the faulted phase voltage Vb-n is about 7.1 kV at the fault point.
4. An increase in the unfaulted phase currents when the fault is fed from more than one source, while the contributions from individual sources for I_1 are not equal to I_0. This is confirmed by phases C and A of transformer T1 in Fig. 5.94.
5. CT saturation as evident in trace XFMR T1 In in Fig. 5.95.

Lessons Learned

1. The paralleling of CT inputs to connect the differential relay restraint input should be done only for non-fault sources. In this case study the design of the

relay system goes back to the late 1950s, when differential relays were only available for a maximum of three inputs, and thus paralleling of CTs associated with fault sources was done to a common restraint winding.

2. Case studies having symmetrical L-g faults containing no dc offset can be used to verify a power system model.

Case Study 5.17: Using a Transformer Numerical Relay Oscillography Record to Analyze Phase-to-Ground Faults in a 4.16-kV Low-Resistance Grounding Supply

Abstract A combustion turbine (CT) plant, shown in Fig. 5.96, consists of two units with power generation at the 80-MW level. A phase-to-ground fault occurred in the 4.16-kV leads to the gas compressor motor of one of the units. The fault resulted in the shutdown of both generating units. Based on the numerical relay fault record, the faulted supply was tested by energizing the suspected gas compressor feeder, which was tripped again by the same relay. In this case study we describe how a frayed cable in the gas compressor junction box caused both units to trip. In addition, we describe the protection philosophy of the 4.16-kV auxiliary systems and its shortcoming (deficiency), analyze the fault record cases of the plant shutdown, and provide corrective actions and lessons learned.

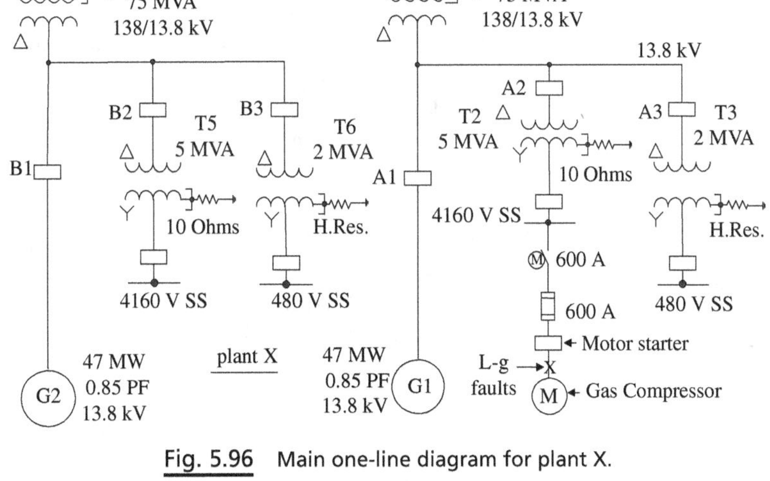

Fig. 5.96 Main one-line diagram for plant X.

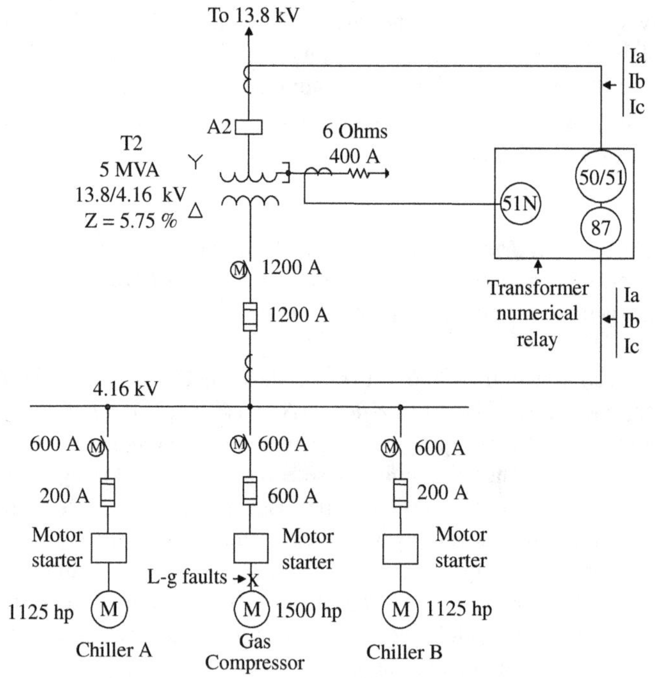

Fig. 5.97 One-line diagram showing the numerical relay–monitored currents and ground fault protection for motor feeders.

Description of the 4.16-kV Station Service Supply The medium-voltage station service is designed as a low-resistance grounding system with a maximum ground fault current of 400 A. The three-phase fault current available at the 4.16-kV bus is 11 kA. This type of system reduces ground fault currents to values lower than normal loading currents and hence eliminates excessive forces and stresses that can be caused by ground faults on other solidly grounded systems. The low-resistance system supply is designed as a fuse disconnect for all the motor feeders. This option is selected to reduce the plant cost by designing feeders to the 4.16-kV auxiliary loads as a fuse disconnect, compared to switchgear breakers having shunt trip coils. Figure 5.97 shows a one-line diagram for the main plant 4.16-kV motors.

Motor Protection Philosophy The medium-voltage (4.16-kV) motor circuits are protected by a combination of high-voltage motor starters and current-limiting fuses. The starters utilize overload relays and backup current-limiting fuses to provide complete overcurrent protection. The fuses operate to interrupt high values of fault current that exceed the interrupting rating of the contactor, and the overload relay operates to open the contactor before the fuse blows for lesser, yet abnormal currents due to motor overloads, locked rotors, repeated starts, extended accelerating time, or low-value ground fault currents.

The motor starter package shown in Fig. 5.97 was specified without ground fault protection elements. No ground fault current transformer was installed in the 4.16-kV motor leads. Fuse-disconnect application in a medium-voltage motor feeder requires permitting motor starting conditions and normally is sized at 1.5 × the rating of the motor. Therefore, the fuse rating of 300 A may not detect the maximum ground fault current of 400 A. As a result, neither the motor starter nor the 300-A fuse will offer guaranteed ground fault protection even for the maximum value of 400 A.

Ground Protection Philosophy As mentioned earlier, the 4.16-kV system is designed as a low-resistance grounding system with the ground fault current limited to a value of 400 A. Therefore, detection of ground faults in the medium-voltage auxiliary supply will depend on sensing zero-sequence current. The detection is normally carried out by inserting sensitive overcurrent relays in the neutral connections or by a donut CT where all phases pass through a window-type CT. In this case study a mistake was made by relying only on the 51N element of the station service transformer numerical relay, which monitored the current flowing in the neutral of the transformer T1 grounded wye winding.

As shown in Fig. 5.97, station service transformer T2 is protected by a numerical multifunction differential relay. The relay-enabled elements are: differential protection 87T2, high-side (13.8-kV) instantaneous and time overcurrent 50/51, and transformer neutral time overcurrent (51N). The neutral time overcurrent element setting is: CT ratio = 200 : 5; pickup = 4.25 A secondary; curve shape is extremely inverse; curve multiplier = 1.

Phase-to Ground Fault Occurring on the Gas Compressor Feeder A 4.16-kV cable feeding a gas compressor for a single-cycle 50-MW CT unit failed to ground, first on phase B while the plant was in service and then on phase C during testing of the feeder. The faults were seen only by the station service transformer neutral relay. The first fault was cleared by the 51N in 1.3 s and resulted in the shutdown of both gas units, which is not desirable, as explained below. The second C-g occurred when the gas compressor feeder was isolated and then energized alone from the 4.16-kV supply. The fault was also cleared in 1.3 s since the ground fault is limited by low-resistance grounding of the transformer neutral using a 6-Ω resistor.

Transformer Relay Fault Record Two relay fault records were obtained. The first relay record, for the B-g motor lead fault, was triggered by relay trip output by the operation of neutral time overcurrent 51N, which is an element of the numerical transformer T2 differential relay, while the plant was generating 79 MW. The recorded fault currents are as follows:

Primary currents for the 13.8-kV-side winding W1:

phase A = 100 A at 0°; phase B = 110 A at 150° lag; phase C = 52 A at 265° lag

Primary currents for the 4.16-kV-side winding W2:

phase A = 176 A at 239°; phase B = 520 A at 342° lag; phase C = 169 A at 113° lag

The relay recorded a ground current $3I_0 = 371$ A at 156° lag.

Differential phase currents: phase A = 0; phase B = 0; phase C = 0

Restraint currents: phase A = 0.25 A; phase B = 0.28 A; phase C = 0.13 A

The fault was outside the T2 differential protection zone, and therefore no differential operating current was recorded.

The second relay record, for a C-g fault, was triggered by relay trip output by the operation of neutral time overcurrent 51N while the plant was shut down during the energization (testing) of the suspected gas compressor feeder.

Primary currents for the 13.8-kV side winding W1:

phase A = 221 A at 0°; phase B = 266 A at 127° lag; phase C = 223 A at 254° lag

Primary currents for the 4.16 kV side winding W2:

phase A = 768 A at 226°; phase B = 928 A at 326° lag;
phase C = 764 A at 105° lag

The relay recorded a ground current $3I_0 = 356$ A at 95° lag.

Analysis of the Relay Fault Record The relay oscillography record shown in Fig. 5.98 reveals that prior to the trigger line a, the winding W2 three-phase currents

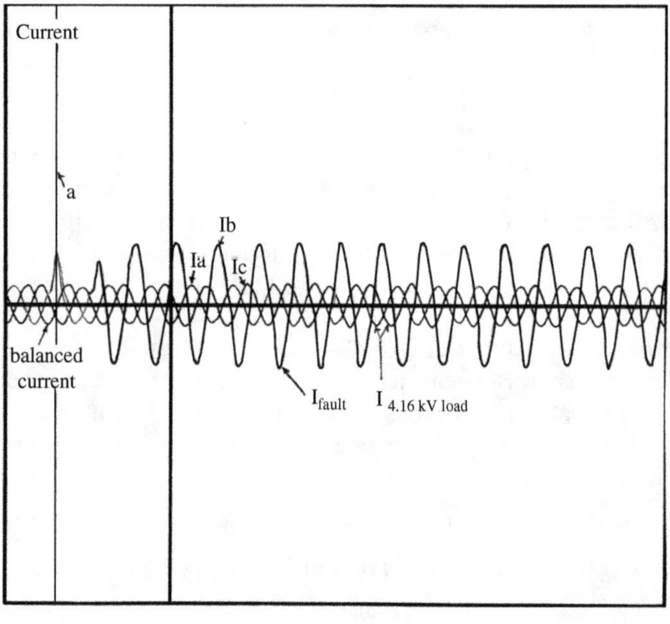

Fig. 5.98 Transformer numerical relay oscillography record showing the 4.16-kV phase currents for the initial B-g 4.16-kV fault.

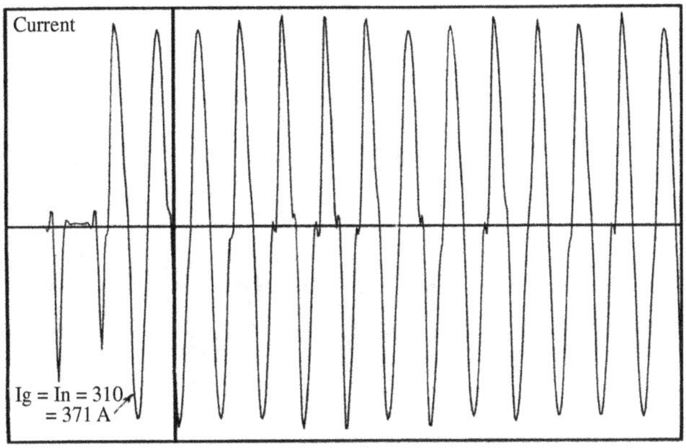

Fig. 5.99 Transformer numerical relay oscillography record showing the T2 neutral current for the initial B-g 4.16-kV fault.

were equal and balanced. The phase B current reveals an increase due to the B-g fault. Figure 5.99 shows the ground current that flows from phase B to ground, then up from the transformer grounded neutral. The numerical relay recorded only 16 cycles of fault current. The relay fault record confirmed a symmetrical ground fault current with a magnitude of 371 A. As shown in Fig. 5.100, there is no pre-fault current and the energization of the gas compressor motor resulted in a locked rotor current with higher current on phase C due to the C-g fault. Figure 5.100 also reveals the 4.16-kV transformer side for the resulting current of the motor inrush (starting) and the 4.16-kV motor lead C-g fault. The relay oscillography fault record shown in Fig. 5.101 shows that the sum of the three currents is zero, due to the delta-connected transformer 13.8-kV winding. The ground current shown in Fig. 5.102 is symmetrical having a magnitude of 356 A. It should be noted that the ground fault current is symmetrical, due to arc-over of phase C at the incident point of the voltage peak.

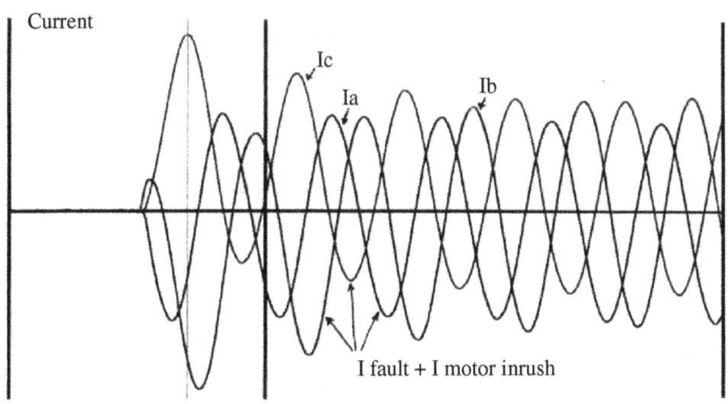

Fig. 5.100 Differential relay oscillography record showing the 4.16-kV phase currents for the second C-g 4.16-kV motor lead fault.

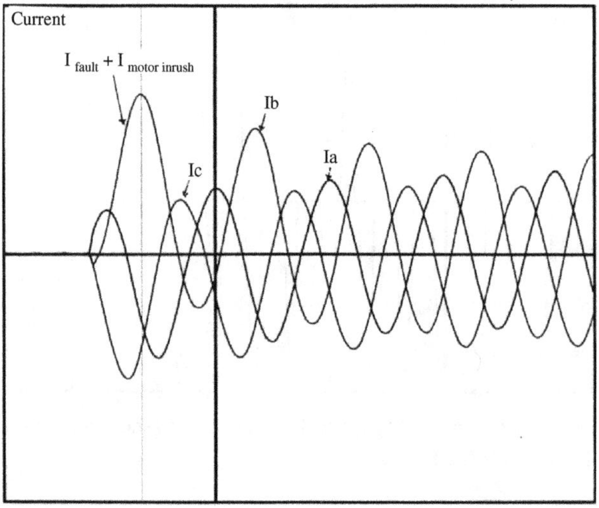

Fig. 5.101 Differential relay oscillography record showing the 13.8-kV phase currents for the second C-g motor lead fault.

As-Found Condition The damage to the motor leads occurred at the point where they penetrated through the rear wall of the terminal box, as they were able to rub against the sheet metal. The fuse characteristic shown in Fig. 5.103, sized at 300 A, which is 1.5 × the normal compressor motor current, may not sense the maximum L-g fault current. There is a lack of compatibility in coordination between the low-resistance grounding system and the fuse-disconnect protection concept. The motor starter package has an overload protection function only. No ground fault current transformer is included in the starter package. Therefore, the only ground fault protection for the 4.16-kV system is the time overcurrent element of the transformer numerical multifunction relay. Use of a sole ground protection element is considered a protection deficiency and violates the basic protection rule of having more than one relay element responding to any given system fault.

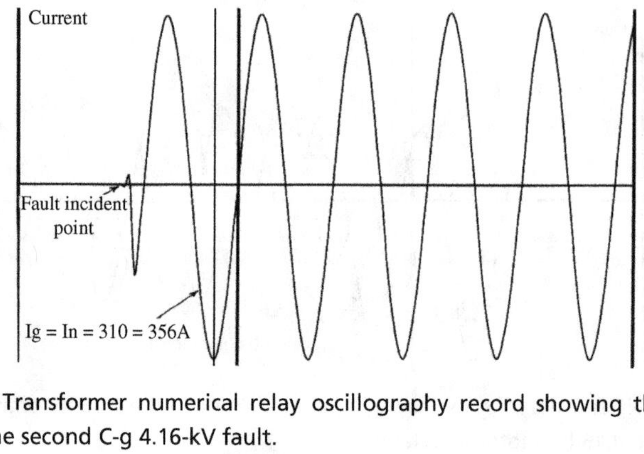

Fig. 5.102 Transformer numerical relay oscillography record showing the T2 neutral current for the second C-g 4.16-kV fault.

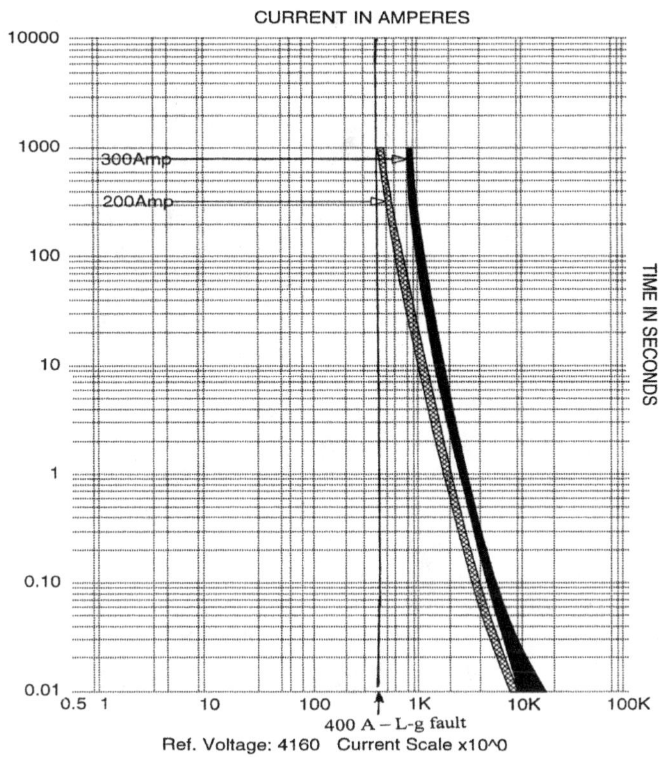

Fig. 5.103 Characteristics of the fuses used for the protection of 4.16-kV motors.

Explanation of the Complete Shutdown of Both Units Following the Motor Lead Ground Fault As shown in Fig. 5.97, the T2 differential relay has an area from the 13.8-kV bus in its protection zone. The plant has dual units connected together without individual high-side 138-kV breakers. Therefore, for 13.8-kV faults at any of the plant buses, shutdown of both units is required. To optimize the project cost and to reduce the numbers of lockout relays used for the plant protection system, a few common lockout relays were used. For this reason, any T2-associated protection will trip a common lockout relay, which will eliminate fault sources. As a result, both units and the 138-kV feeder were tripped by the operation of the T2 neutral time overcurrent relay. The 51N trip activated the main lockout relays, which removed the three sources (two units and the 138-kV system) of the fault.

Corrective Actions

1. Insulate around the cable entry opening to soften the sharp edges of the conduit. Wrap the motor leads with several layers of 69-kV insulating tape, and then wrap this material with regular electrical tape for physical protection.

2. Add ground protection sensing to the motor starter package to enhance fault detection.
3. Add a tripping logic to prevent the shutdown of both units for 4.16-kV motor lead faults and add a new lockout relay to trip only the unit affected and build a breaker failure function with the numerical differential relay as explained below.

Description of the New Added Logic to Eliminate Future Trips of the Entire Dual-Unit Plant for a Similar Fault As shown in Fig. 5.104, a separate new lockout relay, device 86T2, is added and will be dedicated to each 13.8/4.16-kV station service transformer protection. The lockout relay will be tripped by the W2 4.16-kV winding neutral time overcurrent 51N element. The 86T2 relay will trip and lock out all 13.8-kV breakers associated only with each plant, thus eliminating the undesired trip of the second connected unit. In addition, an instantaneous neutral current element 50N, associated with the transformer numerical relay, will be enabled. A setting threshold of 150 A (based on a maximum ground fault of 400 A) will be applied to the 50N element. A 4.16-kV ground fault will be seen by the 51N and 50N elements. An output from 86T2 will then be added with operation of the 50N element and nonoperation of the 86TS plant lockout relay. A coordination timer set at 1.5 s will then provide an output to trip both units. This may occur if the 13.8-kV breaker A2 has failed to clear the 4.16-kV ground fault. Therefore, the new added logic will provide protection coverage for the failure of 13.8-kV circuit breaker A2.

Lessons Learned

1. Applying a fuse disconnect instead of breakers with fault-sensing relays requires sufficient fault currents. Applying a fuse disconnect with low-resistance grounding

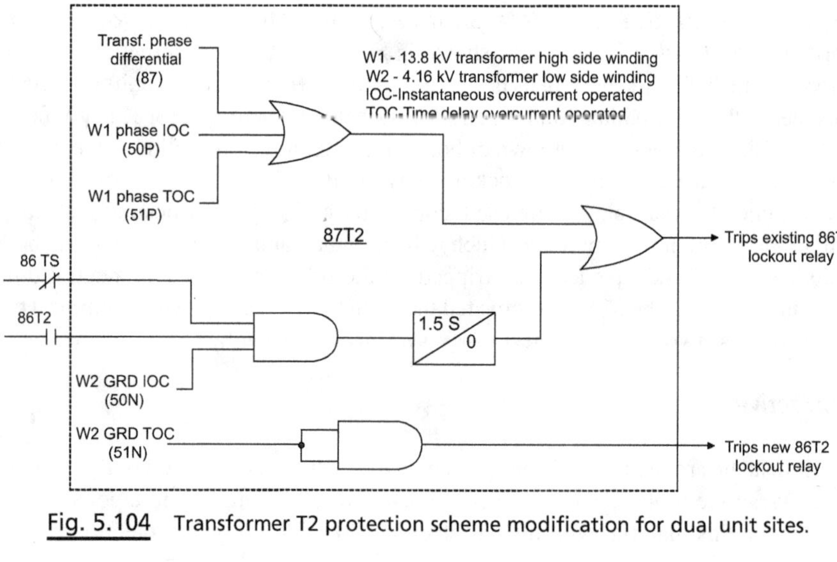

Fig. 5.104 Transformer T2 protection scheme modification for dual unit sites.

system (400 to 1200 A) should be selected carefully, due to the lack of sensitivity to ground faults. Normally, a fuse can withstand 1.25 to 1.5 × its rating. The speed of fuse operation for low- and medium-magnitude fault current should be checked to determine whether supplementary protection (such as ground fault protection) is needed to clear arcing-type ground faults. Fuse sizing and characteristics may not coordinate with the limited ground fault current level.

2. There is a need for another backup protection (separate overcurrent) function to act as backup protection in the event that the overcurrent devices closer to the fault either fails to operate due to a malfunction or operates too slowly due to incorrect (higher) ratings or settings.
3. The motor starter should include a ground fault detection option.
4. Installation of cables and conduits has to be checked for situations in which motor leads are rubbing against sharp edges of the conduits. Motor vibration can cause the lead insulation to break at the sharp edge, causing ground faults.
5. Low-voltage protective device settings should be selected to provide time coordination. Maximum protection against ground faults can be obtained by applying ground protection on every feeder circuit from source to load. Normally, three protection steps need to be provided for ground fault protection:
   (a) At the motor or the feeder breaker
   (b) At the incoming feeder (the transformer low side)
   (c) At the neutral of the station service transformer grounded-wye winding

Case Study 5.18: Phase-to-Phase Fault Caused by a Squirrel in a 13.8-kV Cable Bus Which Evolves into a Three-Phase Fault

Abstract A phase B-C fault occurred in the 13.8-kV cable bus of plant X. Figure 5.105 shows the plant one-line ac with the location of the fault. As shown in the plant cable bus of Fig. 5.106, the fault was caused by a squirrel contacting the B and C phases. The fault evolved to a three-phase fault by the spread (growth) of the ionized cloud. The fault was cleared from the system successfully. In this case study we describe the incident, analyze the numerical relay fault records, and correlate currents on the grounded-wye side to the B-C fault on the delta side using the transformer ampere-turns principle.

Description of the Incident and the Associated Sequence of Events The B-C fault was fed from the 138-kV system and the 50-MW combustion turbine (CT) generator. As shown in Fig. 5.107, the initial B-C fault on the delta side was supported by current flow on phase B, which divides and flows to the other phases, A and C (in accordance with winding ampere-turns principles). The unit current output supply to the system load is superimposed on the fault currents. The two current components have affected the 138-kV system and the unit current flow contributions during the fault. The ionized cloud grew, causing the fault to evolve to a three-phase fault which was cleared from the 138-kV side in 3.5 cycles, as shown in Fig. 5.108 for the numerical

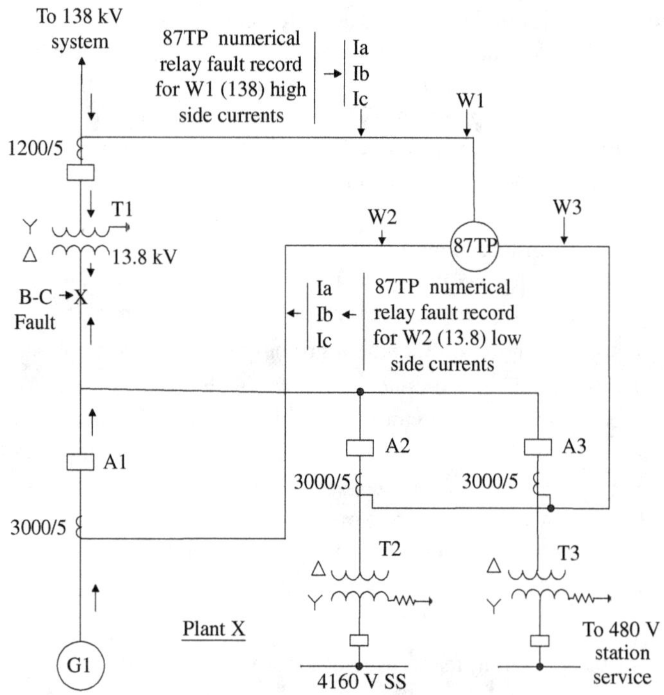

Fig. 5.105 One-line ac diagram showing transformer numerical relay–monitored currents and the location of the B-C 13.8-kV fault.

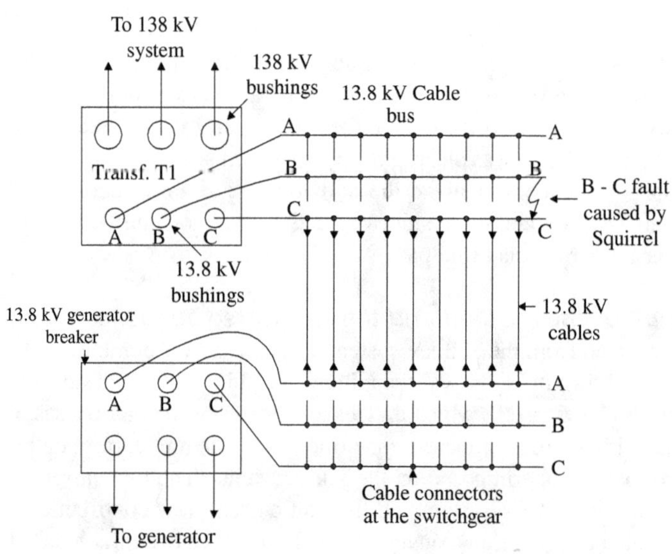

Fig. 5.106 13.8-kV cable bus with the phase-to-phase fault shown.

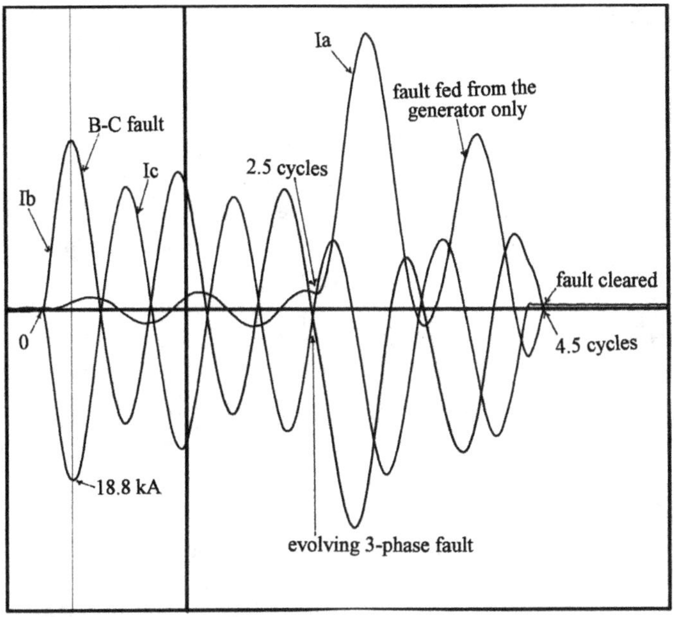

Fig. 5.107 Transformer T1 numerical relay oscillography record showing the low-side (W2) 13.8-kV phase currents during the bus fault.

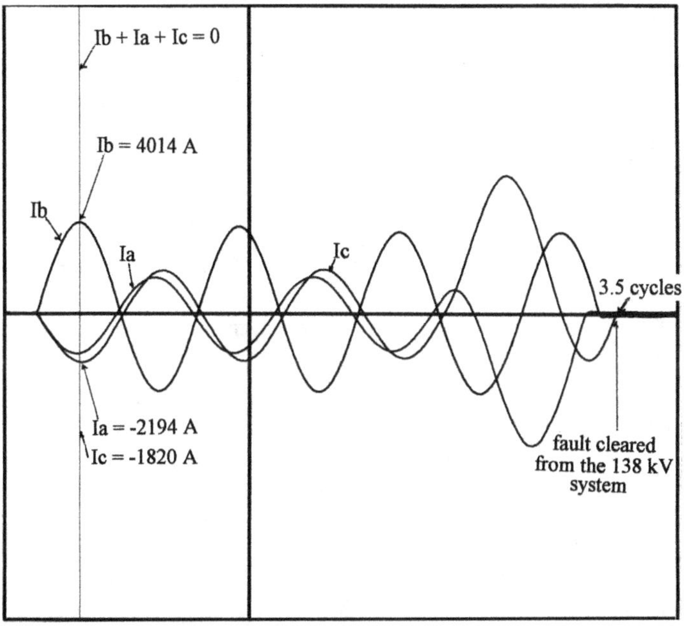

Fig. 5.108 Transformer T1 numerical relay oscillography record showing the high-side (W1) 138-kV phase currents during the bus fault.

relay 138-kV W1-side oscillography record. The relay fault record reveals close to a 180° phase shift between Ib and either Ia or Ic. Figure 5.107 shows the 13.8-kV relay current input side from the generator when the initial magnitude of I_A is very close to I_C with a phase angle close to 180° (phase-to-phase) and a slight current imbalance flow on generator phase A. The initial B-C fault lasted for 2.5 cycles and then evolved to a three-phase fault which was cleared by tripping the generator breaker in 4.5 cycles. Figure 5.107 shows transformer phase currents during the initial B-C fault, where the ampere-turns coupling analysis confirms the relay fault record for 2.5 cycles.

Numerical Relay Fault Record Captured During the Phase-to-Phase Fault Figure 5.108 shows the numerical differential relay currents that are part of the relay oscillography records during the fault.

The high-side 138-kV W1 primary currents are

$$I_a = 723 \text{ A at } 0^\circ;\ I_b = 1318 \text{ A at } 173^\circ \text{ lag};\ I_c = 594 \text{ A at } 344^\circ \text{ lag}$$

The low-side 13.8-kV generator W2 primary currents are

$$I_A = 1160 \text{ A at } 228^\circ \text{ lag};\ I_B = 13{,}410 \text{ A at } 169^\circ \text{ lag};\ I_C = 13{,}980 \text{ A at } 353^\circ \text{ lag}$$

The low-side 13.8-kV station service load W3 primary currents are

$$I_A = 90 \text{ A at } 28^\circ \text{ lag};\ I_B = 130 \text{ A at } 356^\circ \text{ lag};\ I_C = 250 \text{ A at } 185^\circ \text{ lag}$$

Classification of the Fault The fault can be classified and located by using the numerical relay oscillography record for the 138-kV transformer side or the numerical relay oscillography record for the 13.8-kV generator side. The 138-kV numerical relay fault record shown in Fig. 5.108 can be used to classify the fault as described below.

According to the transformer phasing as documented in the nameplate, the transformer is connected in accordance with the NEMA standard, where the 138-kV (HS) system leads the 13.8-kV (LS) system by 30°. The 138 kV three phase current values can now be placed on the transformer phasing diagram shown in Fig. 5.109 for the primary GY winding. The next step is to apply transformer ampere-turns coupling to the delta winding, as shown in Fig. 5.109. It can then be concluded that the fault is located on the 13.8-kV side and is classified initially as a phase B-C fault, lasting for 2.5 cycles prior to evolving to a three-phase fault. The generator low-side 13.8-kV W2 currents display $I_B = 13{,}410$ A at 169° lag and $I_C = 13{,}980$ A at 353° lag. The magnitude of I_B and I_C currents are nearly equal, and the phase angle between them is very close to 180°, this will lead to the classification of the fault initially as a B-C fault on the 13.8-kV side.

Corrective Action Insulate the 13.8-kV 50-MW generator cable bus to avoid similar interruptions. Normally, insulating tape is used on generator buses for medium-size units to avoid generator interruption by animal contact or conducting debris.

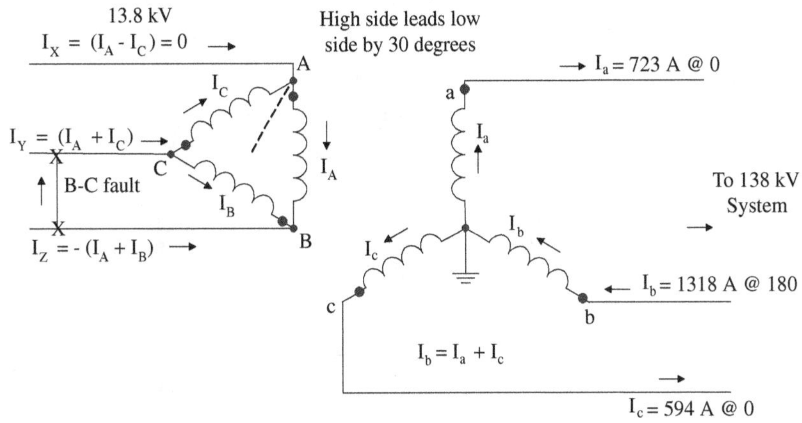

Fig. 5.109 Transformer T1 three-line AC diagram with fault location and fault currents derived from the numerical relay oscillography record.

Case Study 5.19: 13.8-kV Transformer Lead Phase-to-Phase Fault Due to Animal Contact, Evolving into a 115-kV Transformer Bushing Fault

Abstract In this case study we describe a transformer low-side 13.8-kV phase-to-phase-to-ground fault caused by a raccoon contacting the phase c bushing connection to the surge arrester. We show how the 115-kV DFR-recorded currents and voltages are used to analyze initial and evolving faults on the 13.8-kV bushings of the 13.8-kV delta winding. We also provide power system phenomena, corrective actions, and lessons learned.

Description of the System Involved in the Incident Transformer T4 is rated as 360/180/180 MVA and designed with three windings 115/13.8/13.8 kV connected YG/delta/delta. As shown in Fig. 5.110, one of the delta windings is connected to a synchronous condenser. The second delta is not loaded and can be considered as an embedded delta tertiary winding, where the 13.8-kV bushings are connected only to the surge arresters (no other leads are connected to the bushings). In this type of application, one corner of the delta is grounded in accordance with the ANSI standard requirements, as shown in Fig. 5.111. The grounding of the delta b corner is normally done to reduce the likelihood of switching transient overvoltages and their associated insulation stresses on the transformer windings. Therefore, delta winding leads are only terminated to the 13.8-kV bushings, with no outside connection to the system. The 115-kV substation has several generators and is connected to the 230-kV system via autotransformers T1, T2, and T3.

Description of the Protection Systems Transformer T4 is protected by a dedicated harmonic restraint percentage differential relay device 87TP, while the 115-kV transformer lead is protected by a high-impedance bus differential relay 87BP. Overall harmonic restraint transformer differential device 87TBS provides the

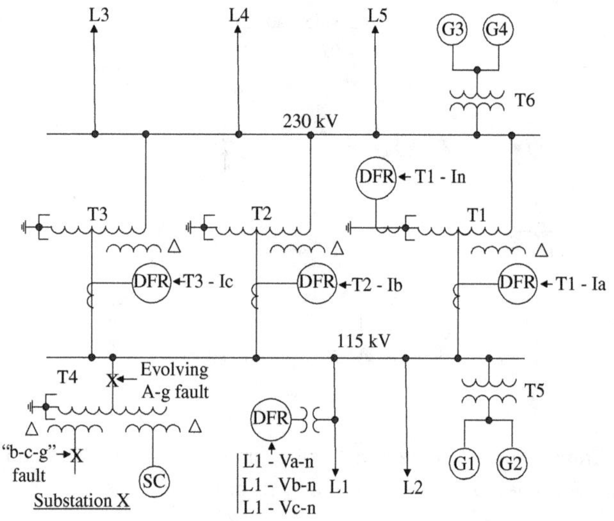

Fig. 5.110 One-line ac diagram showing DFR-monitored currents and voltages and the locations of the evolving fault around transformer T4.

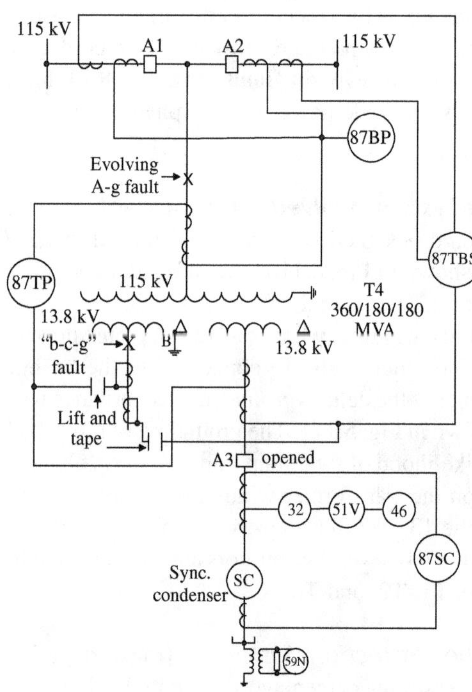

Fig. 5.111 Transformer T4 one-line ac diagram showing the protection systems for T4 and the sync condenser.

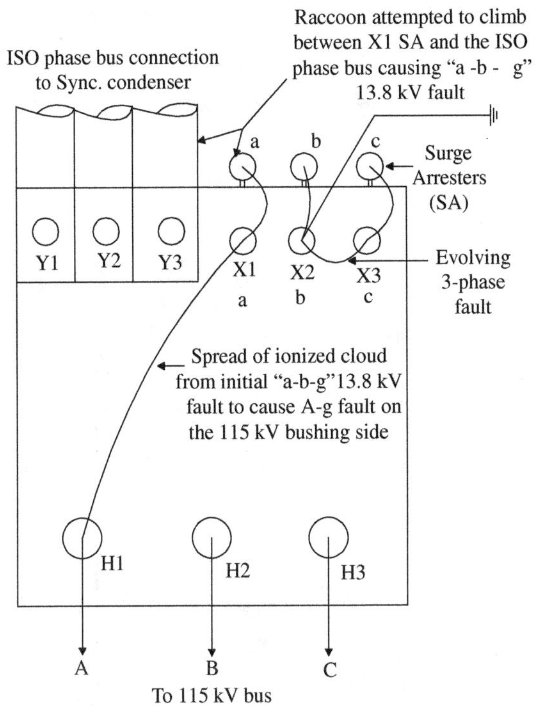

Fig. 5.112 Top view of the transformer T4 bushings, showing the initial fault mechanism and the spread of an ionized cloud to involve the high side.

second protection system for the combined zones of T4 and its 115-kV associated lead. The current transformer wiring associated with the 13.8-kV leads for the inputs to the differential relays is lifted to protect the tertiary bushings and associated surge arrestors by T4 differential relays. The one-line ac diagram shown in Fig. 5.111 illustrates the protection systems for transformer T4 and the synchronous condenser.

Description of the Incident As shown in Fig. 5.112, the incident started on the low side with a raccoon contacting phase a of the 13.8-kV bushing, causing a phase a-b-g fault through the ground. The transformer embedded delta tertiary winding is not brought to the outside, and therefore one of the delta corners (phase b) must be grounded as shown. The ionized cloud grew, causing the fault to evolve to a three-phase fault. The ionized cloud began to spread toward the high-side bushings and caused an A-g fault on the 115-kV bushing. Due to the large size of the substation, transformer T4 phase currents are not monitored directly by the DFR device. Therefore, scattered phase current traces from the remaining autotransformers, T1, T2, and T3, are substituted and used for the fault analysis.

Classification of the Low-Side 13.8-kV (Delta) Fault Using DFR Information from the 115-kV System As illustrated in Fig. 5.113, phase A current from

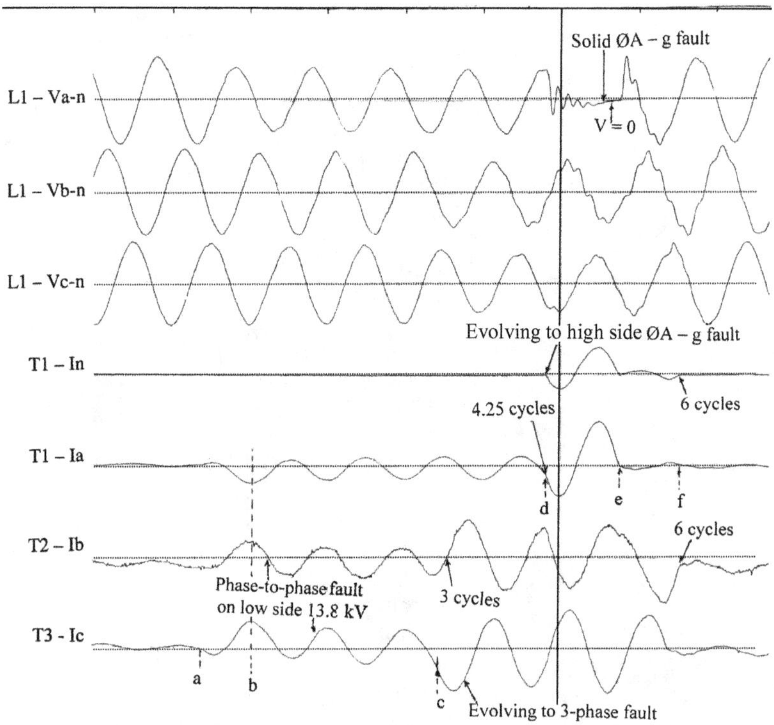

Fig. 5.113 Substation X DFR record showing the line L1 voltages and transformer phase and neutral currents during the initial low-side and evolving high-side faults.

transformer T1, phase B current from transformer T2, and phase C from transformer T3 will be used to confirm the analysis of the 13.8-kV fault caused by an animal climbing on the transformer. All these currents are monitoring the 115-kV high sides for transformers T1, T2, and T3. At cursor b of Fig. 5.113, phase B current trace T2-Ib is in phase with phase C current trace T3-Ic, and both are out of phase with phase A current trace T1-Ia. The DFR calibration for the T1 phase A current trace is higher than that for the other currents shown for transformers T2 and T3. The low-side fault can be classified using the basic relationships between the transformer HV and LV sides, as shown in Fig. 5.114, while applying the transformer ampere-turns coupling principles.

The three currents' phasor relationships are therefore superimposed on the grounded-wye 115-kV HV side of the T4 phasor diagram, as shown in Fig. 5.114. Applying T4 ampere-turns coupling principles between the high side and the delta low side, the primary 115-kV currents can be moved to the delta side. Applying Kirchhoff's first law, the 13.8-kV lead currents can be deduced and the initial fault can be classified as an a-b-g fault when the raccoon climbed to connect phase a to ground.

Analysis of the DFR Record Figure 5.108 shows the DFR-monitored currents and voltages used to analyze the disturbance. As shown in the DFR record in Fig. 5.111,

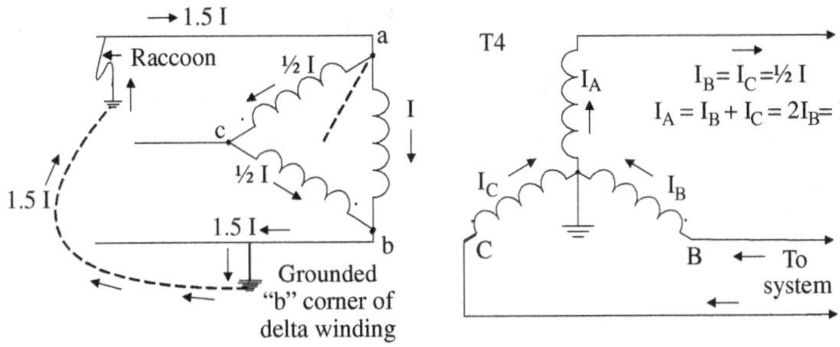

Fig. 5.114 Initial 13.8-kV a-b-g fault caused when a raccoon tried to climb between the phase a X1 bushing and the ISO-phase bus connecting the sync condenser.

the a-b-g fault occurred at point a, lasted for 3 cycles, and then at point c evolved into a three-phase fault which lasted for an additional 1.25 cycles by the spread of the ionized cloud to the 13.8-kV phase c bushing. The spread of the ionized cloud continues to involve the 115-kV phase A bushing, causing a high-side phase A-g fault at point d which lasted for an additional 1.75 cycles.

The DFR trace L1-Va-n reveals a solid phase A-g fault, and trace T1-In illustrates the reduction of fault current when the 115-kV oil circuit breakers tripped and cleared phase A at point e. Clearing a three-phase fault at current zero with 120° between the phases will result in the creation of a non-60-Hz neutral current, which in this case lasted for a duration of less than 1 cycle between points e and f. Phases B and C eventually cleared the simultaneous faults at point f. The simultaneous occurrence of the 115-kV L-g fault with the three-phase 13.8-kV fault was cleared in 6 cycles at point f by the operation of the transformer T4 primary differential relay 87TP and the secondary overall differential relay 87TBS. All phases on both relay systems operated to energize their respective lockout relays to trip and lock out 115-kV circuit breakers A1 and A2. Circuit breaker (13.8-kV) A3 associated with synchronous condenser was opened prior to the incident.

Conclusions By knowing the transformer phasing diagram from its nameplate and using the basic analysis of the currents on the high side, the line-to-line fault on the low side can be derived from the high-side currents via application of transformer ampere-turns balance and polarity rules. It can then be concluded that the raccoon started the fault by contacting phase a, causing a phase a-b-g fault (due to the grounding of the delta corner at b). The ionized cloud grew and the fault evolved to include the 13.8-kV phase c and became a three-phase fault, which evolved again to include the 115-kV bushing, causing a simultaneous phase-A-to-ground fault on the 115-kV transformer high side.

Power System Phenomena

1. Evolving of a double-phase-to-ground fault to a three-phase fault on the delta 13.8-kV side, then evolving to a simultaneous phase-to-ground on the 115-kV high side, as shown in Fig. 5.113.

2. Basic relationship between delta and grounded wye for a phase-to-phase fault on the delta side.
3. Clearing of a three-phase fault at current zero with 120° between the phases will result in the creation of a non-60-Hz neutral current as shown in trace T1-In in Fig. 5.113.

Corrective Actions Insulate the 13.8-kV bushings and the connections to the 13.8-kV surge arresters to eliminate any future exposure to faults caused by animals.

Lessons Learned Basic knowledge of transformer theory, including winding ampere-turns balance, current and voltage, and current polarity rules, is essential for correct analysis to arrive at the correct sequence of events in the incident.

Case Study 5.20: Undesired Tripping of a Numerical Multifunction Transformer Relay by Assertion of a Digital Input Wired to the Buchholz Relay Trip Output

Abstract A 138-kV feeder connecting the combustion turbine (CT) unit G1 of a 500-MW combined cycle at plant X tripped via the operation of the GSU transformer

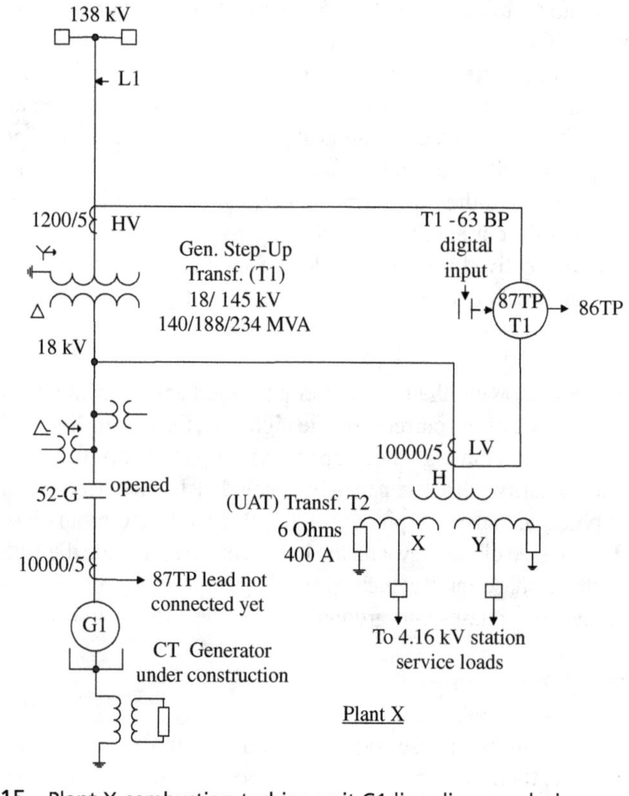

Fig. 5.115 Plant X combustion turbine unit G1 line diagram during construction.

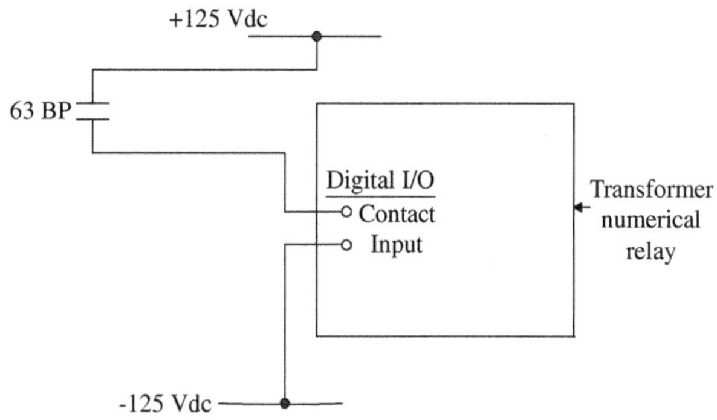

Fig. 5.116 Basic digital inputs to the transformer numerical relay using a wet contact.

T1 lockout relay 86TP, which keyed the direct transfer trip to the connecting utility. The undesired trip was initiated by an assertion of a digital input dedicated to the operation of a transformer Buchholz relay (63BP), as shown in Fig. 5.115. In this case study we analyze the cause of the trip, provide corrective actions, and state the lessons learned. In addition, we highlight the benefits of using the Buchholz trip via 87TP output relay to generate an event record. Wiring of inputs from transformer protection devices such as the Buchholz relay, pressure relays, and low-low oil levels as digital input to 87TP relay can provide event records that will enhance postmortem analysis functions. This is one of the advantages that the new numerical technology offers.

Description of the Protection Systems The unit GSU transformer T1 is protected by dual numerical multifunction transformer relays applied with individual lockout relays. The transformer unit is designed with a conservator tank, allowing the use of a Buchholz relay for sensitive fault detection. It is always assumed that a transformer turn-to-turn fault can only be detected by the Buchholz and pressure devices. As shown in Fig. 5.116, the 63BP tripping contact is interfaced to the numerical relay to trip the 86TP lockout relay output with an event record generated.

Description of the Incident The numerical relay event record is shown in Fig. 5.117. The 87TP numerical relay has a digital input scanning rate of 0.5 ms,

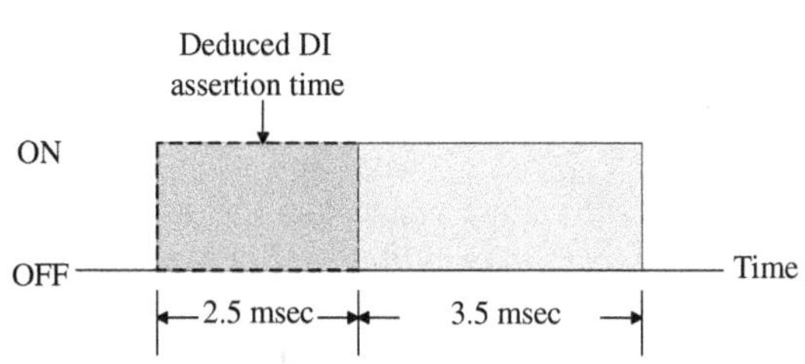

Fig. 5.117 Buchholz 63BP digital input duration during the relay false trip.

and it asserts the digital input based on the setting of the debounce time. With a relay-commissioned debounce time setting of 2 ms, it is postulated that the relay asserted the 63BP in 2.5 ms. Based on the fine resolution of the numerical transformer relay event time stamp (to the nearest microsecond), the event lasted for only 3.5 ms. This implies that the entire event lasted for 6.0 ms when the 63BP digital input is up.

Analysis of the Performance of the Buchholz Relay The Buchholz relay detects gas and the flow of oil between the transformer main tank and the conservator. Gas detection is normally used to alarm, whereas the flow of oil is to trip the transformer. The device must be employed on transformers with a conservator tank design in which the relay is mounted in the throat between the two tanks. The 63BP device used in T1 has magnet-operated reed switches, which makes it immune to vibration. Due to the undesired trip, which was initiated by the assertion of a digital input dedicated to the operation of a transformer Buchholz relay 63BP, the relay was inspected and compared with other transformer devices at the site and found to be normal. In addition, according to the numerical relay record, the 63BP trip contact was closed for only 3 to 3.5 ms. By adding the numerical relay assertion time, the overall contact closing time will be between 5.5 and 6 ms. This type of high-speed phenomenon, which represents a complete cycle of the Buchholz magnetic structure rotation in the make, and then in the break, direction, is difficult to explain and therefore cannot be attributed to the relay. The Buchholz trip contact is also wired to the plant computer, and the log shows the presence for no relay tripping output. In addition, transformer oil sample analysis reveals normal readings. In view of the above, it can be concluded that the Buchholz relay did not operate undesirably to cause a trip input to the T1 numerical relay binary (digital) input.

Cause of the Undesired Trip The sensitive setting for the 63BP contact as a digital input to the transformer numerical relay of 2 ms invited a possible reaction to noise. This noise can be generated in the control cabling of the plant possibly during energization of big motors at the plant construction site. The recognition time of 2 ms is short and does not offer security to ensure that the input assertion is related to actual operation of the oil flow trip of the 63BP device and not to noise sources. It is postulated that the trip was due to noise pulses caused by electromagnetic interference.

Corrective Action Options The following options are available as remedies to the undesired trip of the Buchholz 63BP:

1. Connect the Buchholz tripping contact leads (dry contact) to energize the 86TP primary transformer lockout relay directly through a flexi-test switch blade. This will eliminate the use of a relay digital input circuit and hence eliminate the effect of noise pulses on the 63BP trip input. The loss of the event record

provided by the use of a numerical relay digital input can be compensated by the use of a plant computer alarm log for the 63BP operation.

2. Revise the numerical relay programmed logic to change the present 63BP function from tripping to alarming. The remaining dual differential multifunction numerical relays provide adequate protection for transformer T1; however, sensitive turn-to-turn winding protection will be lost.
3. Increase the debounce time for the 63BP digital input assertion time from 2 ms to 8 ms. This added delay will allow the T1 numerical relay digital sampling logic to ignore a noise pulse of 5 to 6 ms duration and thus prevent the 63BP input from being falsely asserted.

Evaluation and Selection of the Ideal Option for the Plant Corrective action option 1 requires wiring changes that will impose additional testing, and option 2 will alarm for 63BP but not trip. Therefore, option 3, which only requires using software to revise the relay setting, was selected.

Investigation of the Source of the Noise Pulses To speed up the identification of the source of the noise pulses, the present digital input of the 63BP was also wired to a separate 87TP relay digital input group with a sensitive debounce time setting of 2 ms. The assertion of 63BP input triggers the relay oscillograph, and both digital inputs, 2 milliseconds and 8 milliseconds settings, can be included in the numerical relay oscillography list. The recorded event time can be compared with the on/off switching events of the plant auxiliary loads to correlate the noise pulses with their corresponding sources.

Lesson Learned For numerical relay binary (digital) inputs, the debounce (assertion) time should be set at 8 ms, above any noise burst duration, to avoid undesired operation.

REFERENCES

Bean, R. L., N. Chackan, H. R. Moore, and E. C. Wentz,. *Transformers for the Electric Power Industry*. New York: McGraw-Hill, for the Westinghouse Electric Corporation, 1959.

Blackburn, J. L. *Applied Protective Relaying*. Pittsburgh PA: Westinghouse Electric Corporation, 1979.

Blackburn, J. L. *Protective Relaying Principles and Applications*. New York: Marcel Dekker, 1987.

Blackburn, J. L. *Symmetrical Components for Power Systems Engineering*. New York: Marcel Dekker, 1993.

Elmore, W. A., Ed. *Protective Relaying Theory and Applications*. New York: Marcel Dekker, 2000.

Greenwood, A. *Electrical Transients in Power Systems*. New York: Wiley-Interscience, 1971.

Ibrahim, M. A., and F. Stacom. *Phase angle regulating transformer protection.* Presented at the IEEE Power Engineering Society Winter Meeting, January 31–February 5, 1993, Columbus, OH.

Khalifa, M. *High-Voltage Engineering*. New York: Marcel Dekker, 1990.

Neuenswander, J. R. *Modern Power Systems*. Scranton, PA: International Textbook Company, 1971.

Sonnemann, W. K., C. L. Wagner, and G. D. Rockefeller. Magnetizing inrush phenomena in transformer banks. *Transactions of the IEEE*, Vol. 77, No. 3, 1958, pp. 884–892.

6

CASE STUDIES RELATED TO OVERHEAD TRANSMISSION-LINE SYSTEM DISTURBANCES

Case Studies 6.1 and 6.2 illustrate how the sequence of events can be derived from analysis using one DFR record or oscillogram record from only one end of the line, provided that the line voltages are monitored. Case Study 6.3 deals with the analysis of a three-phase fault caused by lightning, and the use of DFR technology that permits expansion of the recorded traces in the directions of both the time and magnitude axes, to determine that the fault began as a phase-to-phase-to-ground fault that evolved sequentially rather than occurring simultaneously as a three-phase-to-ground fault. Case Study 6.4 deals with analysis of an EHV horizontal line configuration double-phase-to-ground fault caused by lightning and the utilization of DFR technology, which permits expansion of the recorded traces in the direction of both the time and magnitude axes, to confirm the sequential nature of the fault as a direct hit on one phase followed by a voltage buildup to ground, causing involvement of the second phase. Case Study 6.5 deals with the analysis of a three-phase fault caused by lightning. The DFR software package is used to determine the nature of the occurrence of the fault. The analysis confirmed the sequential nature of the fault and flagged a problem with the transmission tower footing resistance.

Disturbance Analysis for Power Systems, First Edition. Mohamed A. Ibrahim.

Case Study 6.6 documents the occurrence of a ground fault caused by vandals shooting at insulators on a 115-kV transmission structure. It describes an initial phase-to-ground fault on a solidly grounded system, which was cleared first from the grounded system successfully and then, due to sequential clearing from a tap station on the line, was connected to an ungrounded system through the tap station transformer. Case Study 6.7 displays the occurrence of a phase-to-ground fault on a 345-kV line by an act of vandalism. Vandals removed 28 nuts from a 160-foot transmission pole base, and as a result, the double-circuit tower swung into a hill, causing a phase C-to-ground fault which was cleared from the system successfully. The line reclosed unsuccessfully by automatic reclosing, followed by successful reclosing a few minutes later via SCADA. The line was then removed from service after notification of the tower problem. Case Study 6.8 documents a phase A-to-ground fault occurring on a 345-kV circuit. The fault was caused by an accident along the line right-of-way. The high-resistance fault was cleared successfully at both ends of the line.

Case Study 6.9 deals with an undesired trip of a numerical current differential relaying system applied on a 138-kV tie for a combined-cycle plant. The trip occurred during an external phase-to-ground fault. Analysis of transformer current performance during the external fault using relay oscillograph fault records and DFR records is described. Case Study 6.10 describes the failure of an outdoor cubicle housing 13.8-kV disconnect switches and surge arresters during heavy rain. The metal enclosure apparently permitted water to accumulate and to cause a three-phase fault, which was cleared in 5.5 cycles by operation of the primary and secondary transformer differential relays. A design error left an extra CT's ground, resulting in the undesired operation of the remaining feeders to the station service supply. Case Study 6.11 illustrates how analysis of a mundane system operation with a successfully cleared ground fault on a 115-kV system can lead to the correction of a system modeling error. In addition, it describes the validation process of the short-circuit model and power system phenomena obtained from an analysis of the fault. Case Study 6.12 describes a procedure followed to locate a 345-kV phase-to-ground fault for a line protected by electromechanical distance relays. Fault location and fault type for EHV systems are important and must be determined as quickly as possible so that system inspection and corrective actions can be taken to restore the system. A DFR record from only one of the 345-kV line ends was used for the analysis. Power system modeling and DFR analysis are also described. Fault location calculation accuracy is examined and system phenomena associated with this case study are described.

Case Study 6.13 describes the use of a DFR record in conjunction with relaying target information to analyze and locate the fault at a 100-MW combined-cycle independent power-producing (IPP) facility. The plant has no fault recording and monitoring devices. This case study shows how a remote DFR record and short-circuit study simulation can be used to locate the fault at the IPP switchgear equipment. Case Study 6.14 describes a high-resistance tree phase A-g fault occurring on a 345-kV line during an ice and snow storm coupled with high wind. The initial phase-to-ground fault was cleared successfully and followed by two automatic reclosing attempts that reestablished the initial fault. During the last fault, which lasted about 3.5 s, one subconductor of the wire bundle broke and fell to the ground. This case study illustrates the importance of a ground time overcurrent element for the protection of HV and EHV transmission lines. Normally, the ground time overcurrent element provides protection against high-resistance (tree) ground faults and open-phase conditions. Case Study 6.15 describes an initial phase-to-phase fault and an unsuccessful high-speed reclosing that energized a phase-to-ground fault. It deals with the application of opening and closing resistors for 115-kV oil circuit breakers and transients associated with energization of a line-connected capacitor bank. Case Study 6.16 documents an improper setting that was applied to the ground time overcurrent function of a numerical distance relay protecting a 115-kV line. This resulted in an undesired trip of the line during an external L-g fault on an adjacent line. The setting did not coordinate with downstream relays in the system. This was an error created by the fact that numerical relays were sometimes given the same electromechanical relay curve designation.

Case Study 6.17 begins with the derivation of ground distance relay formulas for a variety of compensation factors. It also illustrates how the mutual coupling phenomena can cause the ground distance relay to either overreach or underreach, depending on the relative directions of current flows in the mutually coupled lines. It provides an analysis of how the reach of a ground distance relay is influenced on mutually coupled lines if one of the lines is out for maintenance with three-phase ground chains applied at both ends. The modeling of this maintenance condition in short-circuit study simulations is described for the first time. Mitigation of the effects of zero-sequence mutual coupling phenomena on the impedance measured by ground distance relays is also covered. The approach recommended reveals a step-by-step calculation approach to cope with the effect of zero-sequence mutual coupling on the setting of ground distance relaying. It also describes an actual example for the application of relays for the protection of an EHV 345-kV transmission system. In addition, it describes another example for the protection of two parallel 345-kV lines. It then concludes that the best solution is through the correct setting of the relay based on an understanding of the phenomena and by simulation via short-circuit studies of all relevant system outage conditions. This solution is recommended instead of attempting to hardwire current transformers from unfaulted lines to the protected line CT circuit.

6.1 LINE PROTECTION BASICS

The main function of a relay system is to detect faults and isolate them by tripping the associated circuit breakers. This is accomplished primarily by producing a contact closure, regardless of whether the relay is numerical, static, or electromechanical. Relaying *reliability* is defined as dependability plus security. *Dependability* is an assurance that the relay will detect and isolate faults within the relay operating zone. *Security* is the degree of certainty that the relay will not react to external faults.

Selectivity is defined as isolating only the faulted element, while *sensitivity* is defined as enough relay pickup for the minimum fault condition.

6.1.1 Duality Principle for Protection and Control of Power System Elements

Every power system element should be protected by at least two independent protection systems to guarantee fault clearing despite the failure of any of the systems. The selection implementation will depend on the criticality of the power system and fault-clearing speed requirements as defined by system planning studies. Therefore, the correct option must be selected for two independent relaying systems having either two high-speed (pilot) clearings, one high-speed and one delayed clearing, or two delayed clearings. Examples are two high-speed pilot relaying systems for bulk power systems, one pilot and one stepped distance (slow) clearing if the power system can tolerate slow clearing, a combination of a relay and a fuse, or two separate fuses for distribution and subtransmission systems.

6.1.2 Historical Evolution of Transmission-Line Relaying Technology

The evolution of transmission-line protective relaying products began around 1912 with the availability of electromechanical distance relays on the market. These relays have performed an adequate job in the protection of transmission lines. A total of six relays (three for phase and three for ground) are used to provide protection for all types of phase and ground faults in a system. The discrete nature of these relays has afforded users the flexibility of in-service testing of these component relays simply by substituting spare relays. In addition, electromechanical relays have offered better redundancy features for multiphase faults. Electromechanical distance relays with sufficient fault currents can operate in high-speed times to provide fast clearing, and they have also survived the harsh substation environment, however, they are quiescent and their failures can only be detected during either routine testing or operational problems. It is the author's opinion that with sufficient fault currents the induction cup or cylinder used for the design of electromechanical relays can turn to operate in few milliseconds (4 to 8) with a speed that is difficult to match with other technologies.

In the early 1960s, solid-state (static) distance relaying systems came into use. The static relay systems were used in the same environment as that of electromechanical relaying systems. The initial static relaying systems have experienced numerous failures that resulted in either false operations or failure to operate. The failures were attributed to the noisy and harsh environment of the substations and to the use of individual transistors and components that resulted in many unreliable connections. In the course of time, the performance of static relays has improved, due to the hardening of the interface to the relay by adding surge protection and by employing medium- and large-scale integrated (LSI) circuit concepts as a replacement for the discrete components. The use of a hybrid design that mixes digital and analog concepts has further improved the performance of static relays.

Electromechanical and solid-state phase distance relays are energized by the difference in phase voltages and currents which are known as "delta" voltages and currents, to respond to the positive-sequence impedance to multiphase faults. Ground distance relays apply zero-sequence compensation using a portion (K_0) of $3I_0$ to tune the ground distance relay to measure an impedance equal to the positive-sequence impedance of the line to the point of the fault (similar to phase distance relays).

In 1969, the first substation computer was installed in the field on a utility system to evaluate the performance of an experimental hardware/software computer system that functions as one terminal of a transmission-line fault protection device. Similar efforts were also ongoing to develop digital algorithms for symmetrical components and impedance calculations at several locations in different countries.

Around 1984, a totally integrated substation protection and control system became available for field trials. The concept was difficult to implement and raised many issues regarding reliability and maintainability of the system and the higher cost. In the late 1980s, microprocessor-based (numerical) relay systems became available commercially for use as either a protective or a fault locator device for transmission lines.

Around 1988, integration of substation communication systems for several intelligent electronic devices began to emerge for practical implementation. The new numerical technology has benefited from the proven concepts adopted by electromechanical relaying. The surge immunity for the new technology is also based on very well established and proven surge and filtering interface techniques for solid-state relaying.

6.1.3 Bulk Power System Protection Criteria

A bulk power system is defined as a system where uncleared faults will have an impact on systems connected outside the faulted area. The criterion used for power system protection in the northeastern part of the United States that emerged as one of the corrective actions of the major 1965 Northeast Blackout will be used as a sample.

The sample criterion is translated into two separate relaying systems from input to output, meeting the following requirements:

1. Two primary line relaying systems
   (a) Different manufacturers (preferable)
   (b) Different operating principles (preferable)
   (c) Mix of numerical, static (existing), and some electromechanical elements (existing)
   (d) Avoidance of common-mode failure
2. Two different communication media
   (a) Power line carrier (PLC) and leased telephone circuit (LTC)
   (b) or PLC and microwave

   (c) or microwave and LTC
   (d) or fiber optic with any one of the media above
   (e) or dual fiber optic with diverse routing (a single failure should not disable both fibers)
3. Dual dc batteries
4. Dual circuit breaker trip coils
5. Dual CTs and separate secondary voltage windings
6. Physical barrier between the two relaying systems
7. Secure CB failure systems

6.1.4 Applying Pilot Relaying for Line Protection

Pilot relaying covers 100% of the line by instantaneous protection and requires the use of a communication link with a transmitter and receiver employed at each end of the line. Following are some of the advantages of pilot relay protection use:

1. Provides 100% protection coverage of the transmission line on an instantaneous basis
2. Minimizes damage by providing high-speed tripping
3. Allows high-speed reclosing
4. Improves system stability
5. Improves safety by minimizing public exposure during faults

Line relaying protection systems are affected by the following system configuration: bus arrangement, line arrangement (long line, short line, tap station, or three terminals), system impedance ratio, mutual coupling, loading restriction, and in-feed and out-feed conditions.

Ground time overcurrent relays for line protection play an important role in providing protection to the power system against high-resistance (tree) faults and open-phase operation. It is the only element that provides most of the challenge for relay setting calculation and coordination. In EHV and HV systems, two redundant ground time overcurrent relays are frequently employed.

6.2 CASE STUDIES

Case Study 6.1: Using a DFR Record From One End Only to Determine Local and Remote-End Clearing Times for a Line-to-Ground Fault

Abstract A phase C-g fault occurred on 115-kV line L1 due to a lightning strike during a thunderstorm. The line fault was cleared successfully from both ends of the line. In this case study we illustrate how the sequence of events can be derived from using one DFR record only from one end of the line provided that line voltages are monitored. We also cover power system phenomena and lessons learned.

Description of the Line L1 Protection Systems The 115-kV line L1 is protected by step-distance relays for phase protection and instantaneous and time overcurrent relays for ground fault protection. Therefore, only 80% of the line faults will be cleared by instantaneous protection. The remaining 20% of the line faults will be cleared by Z2 backup time and ground time overcurrent.

Sequence of Events Obtained from DFR Analysis Figure 6.1 shows the system one-line diagram with DFR-monitored voltages and currents for line L1 used to analyze the fault. The DFR record shown in Fig. 6.2 reveals the line L1 voltages and currents recorded during the fault. The C-g fault was identified from the DFR analysis by a voltage dip in phase C, high currents in phase C, and the presence of *neutral* (also referred to as *ground* or *residual*) current. The C-g fault, which apparently was caused by a lightning strike, was cleared from substation X at the end of line L1 in 5 cycles by the operation of the Z1 ground distance relay and ground instantaneous time overcurrent element. The fault was then cleared from the remote substation Y end of line L1 in 22 cycles by ground time overcurrent element.

Using the Local DFR Record to Deduce the Remote-End Clearing Time Figure 6.2 illustrates fault current flow from substation X toward line L1 fault for 5 cycles on phase C current trace L1-Ic and neutral trace L1-In. Remote clearing

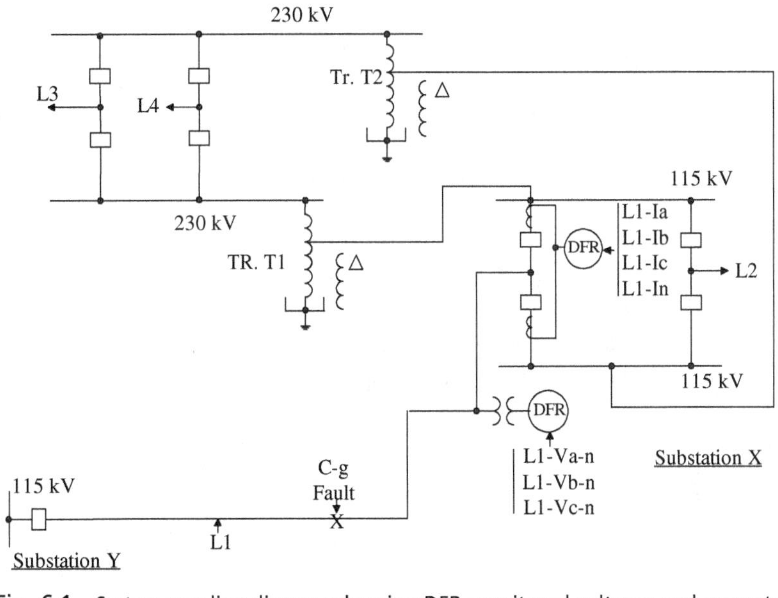

Fig. 6.1 System one-line diagram showing DFR-monitored voltages and currents.

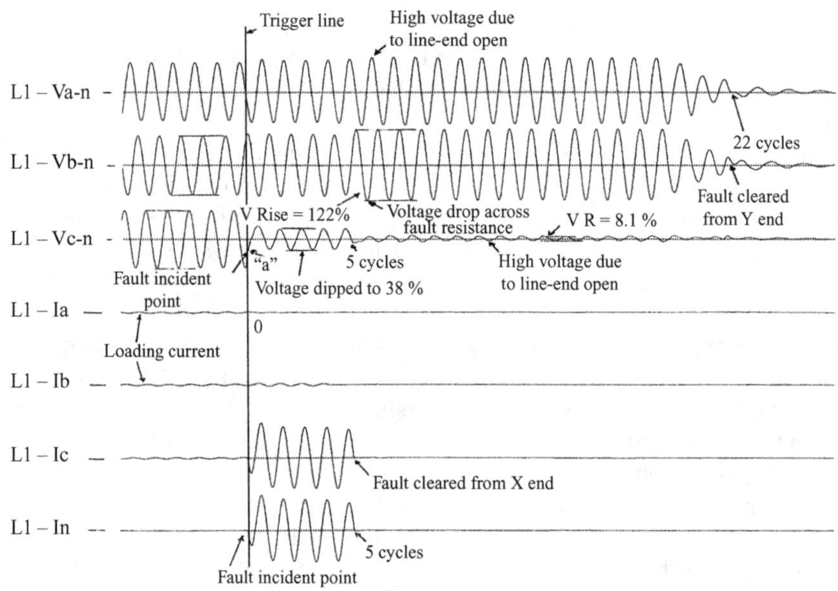

Fig. 6.2 Substation X DFR record showing local and remote clearing of the L1 C-g fault.

is obtained by analyzing line L1 voltage traces. The trace L1-Vc-n illustrates healthy pre-fault voltage, then a fault incident at point a with a voltage dip to 38% for 5 cycles, followed by an 8.1% voltage drop for an additional 17 cycles. The voltage drop is caused by remote substation Y feeding fault current through fault resistance. For solid (bolted) faults this voltage will decrease to zero value. Line L1 voltage traces for phases A and B reflect the remote clearing at 22 cycles from the fault incident point.

Power System Phenomena

1. The Ferranti rise phenomenon in which a transmission line, with one end closed and one end open, is exposed to higher voltage at the open end of the line, as shown in the DFR records of traces L1-Vb-n and L1-Vc-n in Fig. 6.2.
2. Fault resistance as indicated by the voltage drop seen at the line end at substation X. The fault resistance may be caused by a high value of the transmission tower footing resistance.
3. Sequential clearing of line faults when nonpilot step-distance relaying schemes are used to protect the line.

Lessons Learned

1. Monitoring of three-phase line voltages by DFR at one end of the transmission line can be used to derive the clearing time of the remote end of the line for line faults.

2. The Ferranti rise voltage phenomenon illustrates the danger of high voltage during line-end-open conditions and the need to add shunt reactors for compensation and overvoltage protection and control for EHV lines.

Case Study 6.2: Analysis of Clearing Times for a Phase-to-Ground Fault from Both Ends of a 345-kV Transmission Line Using Oscillograms from One End Only

Abstract A phase-to-ground fault occurred on a 345-kV line L1 due to a lightning strike. The line fault was cleared successfully from both ends of the line. In this case study we demonstrate how the entire sequence of events can be derived from using oscillogram records from only one end of the line. We also illustrate power system phenomena obtained from the fault analysis.

Description of the Line L1 Protection Systems The 345-kV line is protected by two redundant distance-based pilot relaying systems using diverse communication media, designed in the early 1980s. The first pilot relaying system is a solid-state (static) permissive overreaching transfer trip (POTT) using a power line carrier. The second pilot relaying system is also a POTT using electromechanical relays and a leased telephone circuit. Each scheme is applied using two zones of phase and ground distance relays. In addition, each system is supplemented with ground time overcurrent inverse relays to protect against open-phase and high-resistance ground faults and a high set phase instantaneous for protection against switch into a three-phase (zero voltage protection) fault.

Description of the Fault An apparent lightning strike caused a C-g fault on a 345-kV circuit shown in Fig. 6.3. The substation X sequence-of-events recorder (SER) printout indicated that the instantaneous high set electromechanical phase overcurrent relay operated in 10 ms, to pick up an auxiliary tripping relay that energized the secondary trip coils of CBs A1 and A2 in 6 ms. The ground instantaneous electromechanical overcurrent relay operated in 16 ms. The SER also indicated that the primary Z1 solid-state distance relaying system operated in 15 ms. to pick up the auxiliary tripping relay, which energized the primary trip coils of 345-kV CBs A1 and A2.

Using the Oscillogram from Substation X to Provide Clearing Times for Substation Y As shown in traces L1-Ic and L1-In in Fig. 6.4, the substation X oscillogram reveal that the fault currents were interrupted at the substation X end of the line 3 cycles (50 ms) from the initiation of the fault. When the line-side voltages for all three phases are monitored by a DFR or an oscillograph at one end of the line only, the clearing time for the other end of the line can be deduced. Therefore, the clearing time for the substation Y end of circuit L1 can be obtained by analyzing the line voltages of substation X oscillogram shown in Fig. 6.4, where trace L1-Vcn for the phase C line voltage reveals that from $t = 0$ to $t = 3$ cycles, the phase C-n voltage

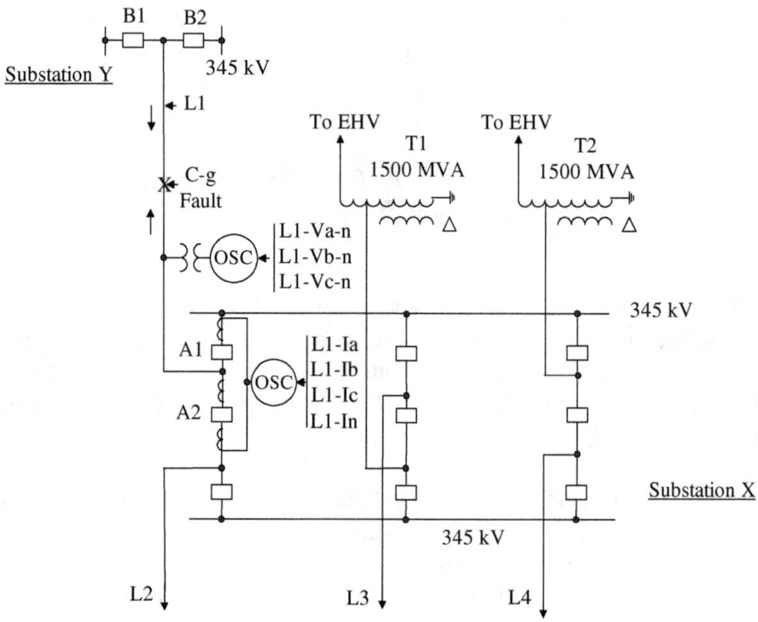

Fig. 6.3 One-line diagram showing oscillograph-monitored voltages and currents.

dipped to 60% due to the ground fault, and from $t = 3$ cycles to $t = 5$ cycles the voltage dipped to zero. In addition, from $t = 3$ cycles to $t = 5$ cycles, traces L1-Van and L1-Vbn reveal that slightly higher than normal voltages appear at substation X for the unfaulted A and B phases. The appearance of voltages on the unfaulted A and B phases and the zero voltage value on the faulted phase C indicate that line L1 Y end was still feeding the initial fault. As a result, the clearing time of the Y end as determined by analysis of the L1 X end is 5 cycles.

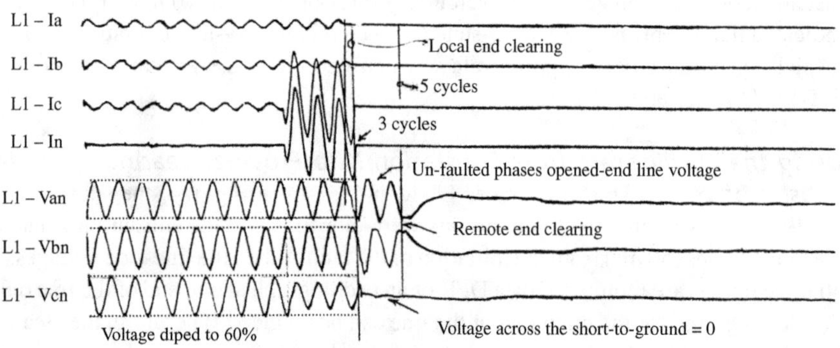

Fig. 6.4 Substation X oscillograph record showing local and remote clearing of line L1 C-g fault.

Power System Phenomena

1. Since lightning is a random phenomenon, a fault incident point can occur at any point on the voltage sine wave. Trace L1-Ic in Fig. 6.4 reveals an asymmetric fault current with dc offset, due to the fault incident point occurring near the voltage zero crossing.
2. Higher-than-normal voltages on unfaulted opened phases A and B due to line-end-open conditions, as shown in traces L1-Va-n and L1-Vb-n in Fig. 6.4. This phenomenon, known as the Ferranti effect, occurs due to capacitive current flowing through inductive line reactance.
3. A solid phase-to-ground fault is indicated by zero voltage across the short to ground as shown in trace L1-Vcn in Fig. 6.4, between 3 and 5 cycles, following by opening of the X end of the line.

Case Study 6.3: Analysis of a Three-Phase Fault Caused by Lightning

Abstract In this case study we deal with analysis of a three-phase fault caused by lightning. The DFR technology permits the expanding of the recorded traces in the direction of both the time and magnitude axes, which permits further detailed analysis to determine the nature of the occurrence of the fault. The analysis confirmed that the three-phase fault occurred sequentially rather than simultaneously.

Description of the Fault Figure 6.5 illustrates the system simplified one-line diagram and the DFR-monitored voltages and currents that are used to analyze the disturbance. Apparently, lightning caused a three-phase fault on the L1 115-kV circuit. The DFR record in Fig. 6.6 suggests that the three-phase fault is simultaneous. However, a lightning creation of a simultaneous three-phase fault requires further analysis. The expanded traces of currents of the DFR record shown in Fig. 6.7 reveal that the fault was cleared in 4.5 cycles, with the phase B trace L1-Ib being the last phase to clear the fault. Phase C trace L1-Ic opened $\frac{1}{3}$ cycle (120°) earlier, and phase A

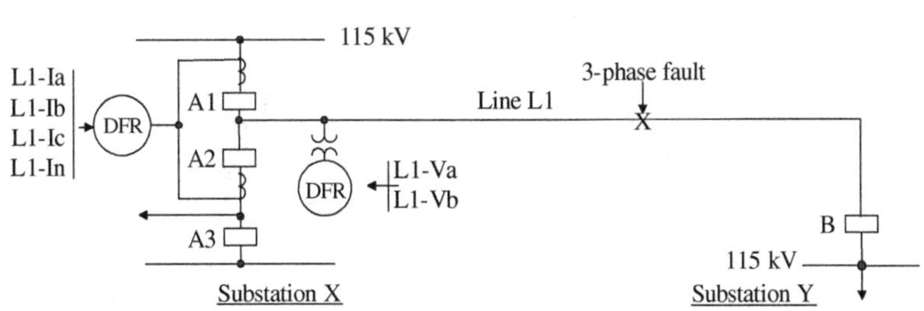

Fig. 6.5 Line L1 DFR-monitored voltages and currents.

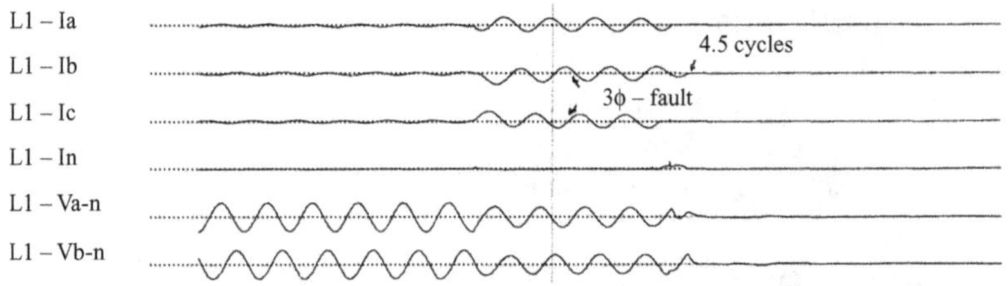

Fig. 6.6 Substation X DFR record showing the three-phase fault.

trace L1-Ia opened $\frac{2}{3}$ cycle (240°) before phase B. As a result of the interruption process of the tree-phase fault, current component not related to 60Hz will flow in the neutral, as exhibited by trace L1-In.

Analysis of the Fault The DFR technology permits expansion of the recorded traces in the direction of both the time and magnitude axes. As shown in Fig. 6.8, expanding the three-phase currents of line L1 reveal that the fault started as a direct hit on phases A and C, causing an A-C fault, as indicated by the intersection of cursor 1 to phase A current at point a and phase C current at point c. The current of the lighting stroke went to the ground via the tower and built enough voltage across the tower footing resistance. The voltage buildup caused a backflash from the ground to the phase B conductor 0.6 ms later, causing the phase-to-phase fault to evolve to a three-phase-to-ground fault, as indicated by the intersection of cursor 2 to phase B current at point b. This analysis confirms the nonsimultaneous nature of the three-phase fault that was caused by lightning.

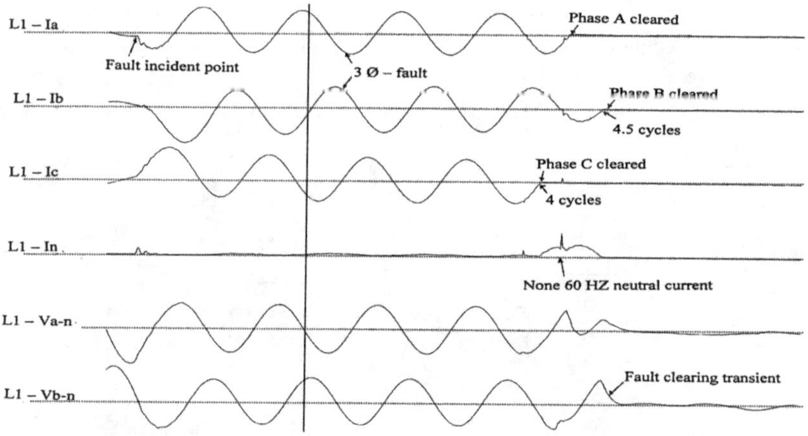

Fig. 6.7 Substation X DFR record showing expanded traces for line L1 currents and voltages during the three-phase fault.

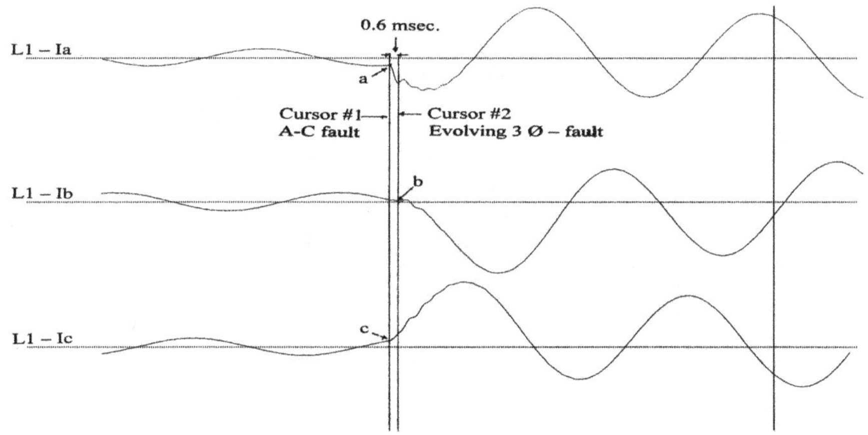

Fig. 6.8 Time expansion of the three-phase recorded currents.

Case Study 6.4: Analysis of a Double-Phase-to-Ground 765-kV Fault Caused by Lightning

Abstract This case study deals with an analysis of a double-phase-to-ground fault caused by lightning. The lightning strike at fault location point was confirmed by the lightning system monitor. Since the 765-kV transmission lines are normally designed as a horizontal configuration, further analysis was required to confirm the possible evolving process of the fault. The DFR technology permits expansion of the recorded traces in the direction of both the time and magnitude axes, which allows further detailed analysis to determine the nature of the occurrence of the fault. The analysis confirmed the sequential nature of the fault as a direct hit on one phase followed by a voltage buildup to ground, causing involvement of the second phase.

Description of the Fault An apparent lightning caused a double-phase-to-ground fault on 765-kV line L1. Figure 6.9 shows the system simplified one-line diagram with line L1 DFR-monitored voltages and currents. As shown in the DFR record of Fig. 6.10, the fault was cleared in 2.5 cycles, with phase A trace L1-Ia being the last phase to clear. Trace L1-Ic reveals that phase C opened in 2 cycles. Figure 6.10 also reveals that it appears that the phase-to-phase-to-ground fault was simultaneous. However, a lightning creation of a simultaneous double-phase-to-ground fault between the two outer phases that are shown in Fig. 6.11, for the cross section of the horizontal configuration of 765-kV line L1 requires further analysis.

Analysis of the Fault The fixed DFR trigger line will be used as a reference to analyze the nature of the fault. Figure 6.12 shows the time expansion substation X DFR record for L1 voltages and currents and confirms that the phase A voltage and current began to change simultaneously, indicating a phase A-to-ground fault. The increase in phase A current, trace L1-Ia, at −0.6 ms from the fixed cursor can lead to the conclusion that phase A had a direct lightning hit. The fault incident point started at −0.6 ms from the DFR record trigger line. Figure 6.12 also shows that the phase C

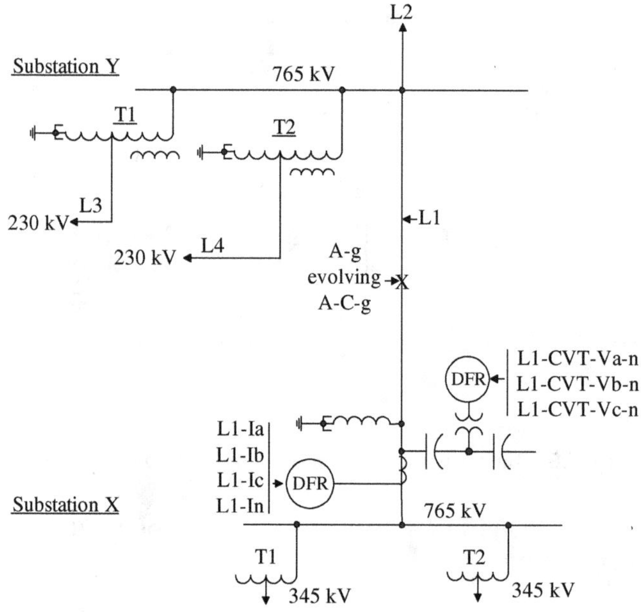

Fig. 6.9 System one-line diagram with line L1 DFR-monitored voltages and currents.

voltage and current started to change simultaneously, indicating an evolving A-C-g fault. The fault incident point started at the DFR record trigger reference line. The current of the lighting stroke went to the ground via the tower and built enough voltage across the tower footing resistance.

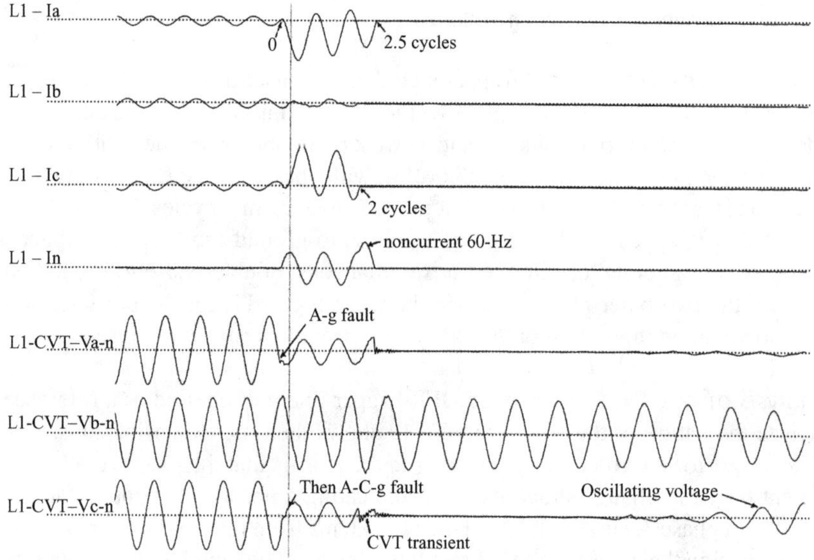

Fig. 6.10 Substation X DFR record showing line L1 voltages and currents during the fault.

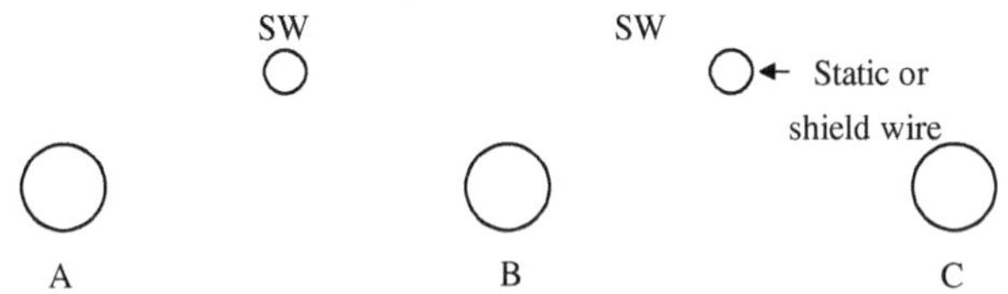

Fig. 6.11 Cross section for the horizontal configuration of the 765-kV line L1.

The voltage buildup caused a backflash from the ground to phase C after 0.6 ms (the difference between the two cursors shown in Fig. 6.12) from the initial phase A-g fault. This analysis will confirm the nonsimultaneous nature of the double-phase-to-ground fault. The fault started as a phase-A-g, due to a direct lightning hit to one of the outer phase, and evolved into a double-phase-to-ground (A-C-g) after 0.6 ms. The evolving of the fault is caused by a voltage buildup from the flow of current from the initial direct stroke through the transmission tower footing resistance, resulting in a backflash from the ground-to-phase C conductor.

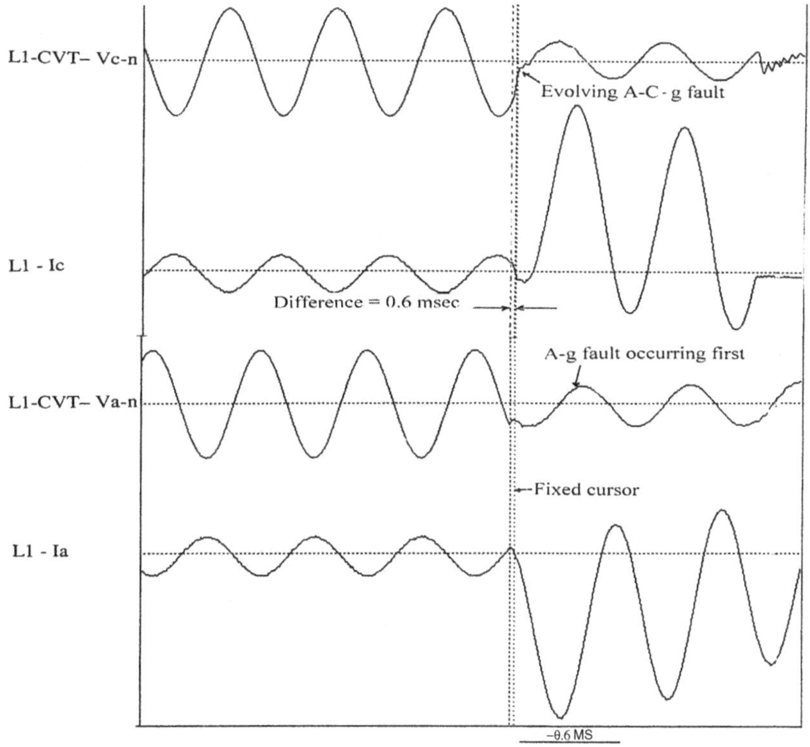

Fig. 6.12 Substation X DFR record showing time expansion of L1 voltages and currents during the initial A-g fault and the evolving A-C-g fault.

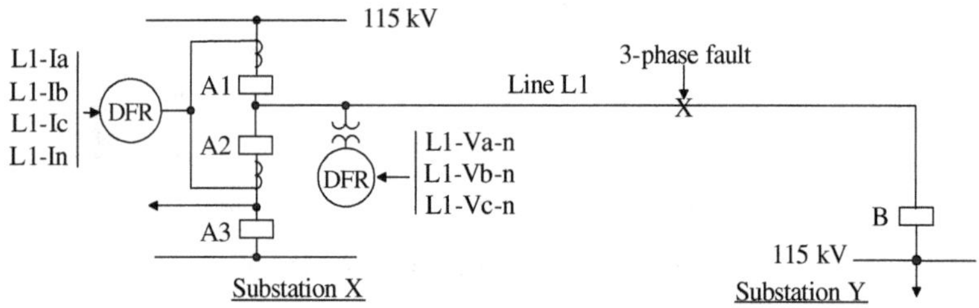

Fig. 6.13 Line L1 DFR-monitored voltages and currents.

Case Study 6.5: Assessment of Transmission Tower Footing Resistance by Analyzing a Three-Phase-to-Ground Fault Caused by Lightning

Abstract This case study deals with an analysis of a three-phase-to-ground fault caused by lightning. The DFR technology permits expansion of the recorded traces in the direction of both the time and magnitude axes, which permits further detailed analysis to determine the nature of the occurrence of the fault. The analysis confirmed the sequential nature of the fault and flagged a problem with the transmission tower footing resistance.

Description of the Fault Figure 6.13 shows a simplified system one-line diagram and DFR-monitored voltages and currents. An apparent lightning strike caused a three-phase-to-ground fault on a 115-kV L1 circuit. The DFR record shown in Fig. 6.14 revealed that the fault was cleared in 5.5 cycles, with phase C trace L1-Ic being the last phase to clear. Phase B trace L1-Ib opened $\frac{1}{3}$ cycle (120°) earlier, while phase A trace L1-Ia opened $\frac{2}{3}$ cycle (240°) before phase C. Circuit breaker interruption of three-phase faults, which are normally balanced, leaves an odd-looking nonsinusoidal signal on the neutral trace L1-In as shown, following interruption of the first phase. This signal may cause some ground relays to give

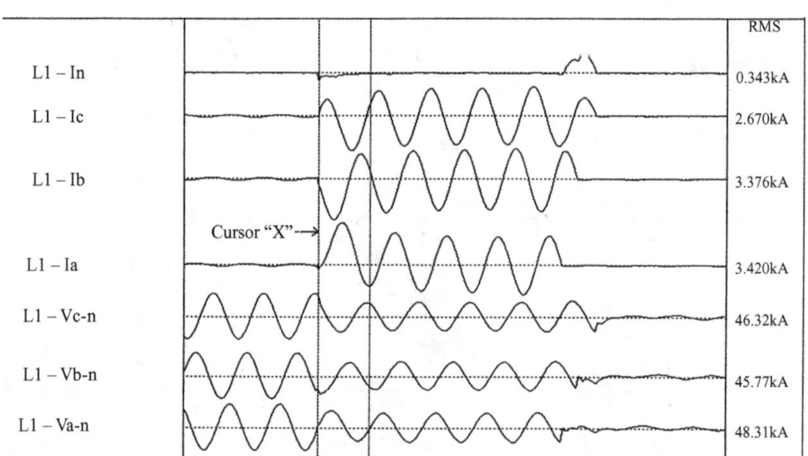

Fig. 6.14 Substation X DFR record for the three-phase fault.

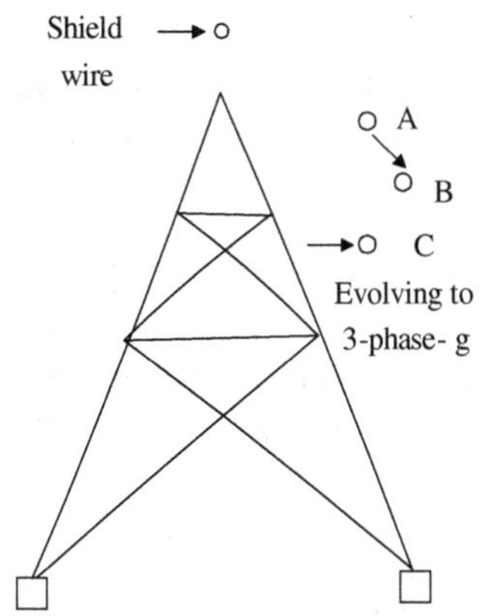

Fig. 6.15 115-kV transmission tower configuration.

targets by reacting to the three-phase fault. Figure 6.14 also reveals that it appears that the three-phase fault was simultaneous, as indicated by cursor X. However, a lightning creation of a simultaneous three-phase fault for the transmission tower configuration shown in Fig. 6.15 needs to be confirmed and analyzed further.

Analysis of the Fault The DFR technology permits the expansion of the recorded traces in the directions of both the time and magnitude axes. As shown in Fig. 6.16, expanding the three-phase currents of line L1 reveals that the fault started as a direct hit on phases A and B, as indicated by cursor a. The current of the lighting stroke went to the ground via the tower and built enough voltage across the tower footing resistance. The voltage buildup caused a backflash from the ground to phase C 0.7 ms later, as indicated by cursor b. Phases A and B are located at the top of the

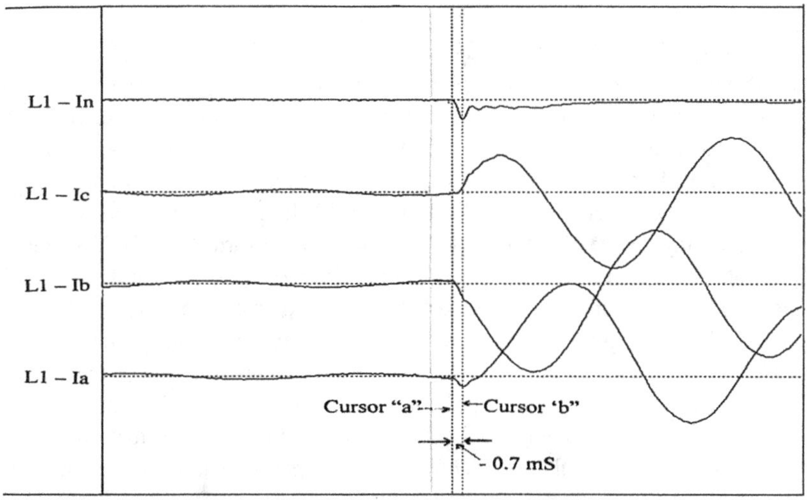

Fig. 6.16 Time expansion of the DFR-recorded currents at the fault incident point.

transmission tower shown in Fig. 6.15, thus confirming that phases A and B can have a direct lightning hit followed by a flashover to phase C, causing the phase A-B-g fault to evolve to a three-phase-to-ground fault. Manipulation of DFR records can give many clues to analysis of disturbances on the system.

Corrective Action The fault was located by using a short-circuit study by sliding the fault on the line until a match is established between the study results and the DFR-measured currents and voltages. Measurements of tower footing resistance around the fault location revealed a higher value of about 60 Ω, which was corrected to a value below 10 Ω.

Case Study 6.6: 115-kV Phase-to-Ground Fault Cleared First from a Solidly Grounded System, Then Connected and Cleared from an Ungrounded System

Abstract In this case study we describe an initial phase-to-ground fault on a solidly grounded system which was caused by vandals shooting at insulators mounted on the 115-kV transmission structure. The fault was cleared successfully first from the grounded system and then, due to sequential clearing, from a tap station on the line. The fault was then connected to an ungrounded system through the tap transformer. It will also provide analysis of overvoltages on a faulted ungrounded 115-kV system. In addition, it will cover the analysis of DFR and numerical relay fault records and power system phenomena.

Description of the 115-kV System As shown in Fig. 6.17, the 115-kV line L1 between substations X and Z is tapped to feed the 46-kV subtransmission loop. At substation Y, two parallel transformers, each rated as 20 MVA, 115/46 kV, are connected to the tap substation via a 115-kV circuit switcher. The transformers are connected delta on the 115-kV side and grounded wye on the 46-kV side. The 46-kV loop is fed from another 115-kV source at substation W.

Description of the Protection System The 115-kV line L1 is protected by dual numerical distance relays. One of the numerical relays forms permissive overreaching pilot protection over power line carrier single-sideband equipment. The second relay is a step distance using Z1, Z2, and Z3 elements. Each relay offers a ground time overcurrent backup function for open-phase and high-resistance ground faults. As shown in Fig. 6.17, backup protection for the 46-kV loop consists of time overcurrent directional relay device 67L for phase faults looking toward the 115-kV system and overvoltage relay device 59G for ground faults. The 67L is installed at the 46-kV sides of the transformers, while 59G is at the 115-kV side to protect the ungrounded system created by the tap supply being isolated on the 115-kV system during ground faults. The PTs used are connected as YG on the 115-kV side and as broken delta on the secondary side to act as a zero-sequence voltage filter detecting ground faults when the supply becomes ungrounded. A direct transfer trip is sent from the tap station to the remote ends of the line whenever the fault level is above the interrupting capability of circuit switcher B, by operation of instantaneous relay device 50CS.

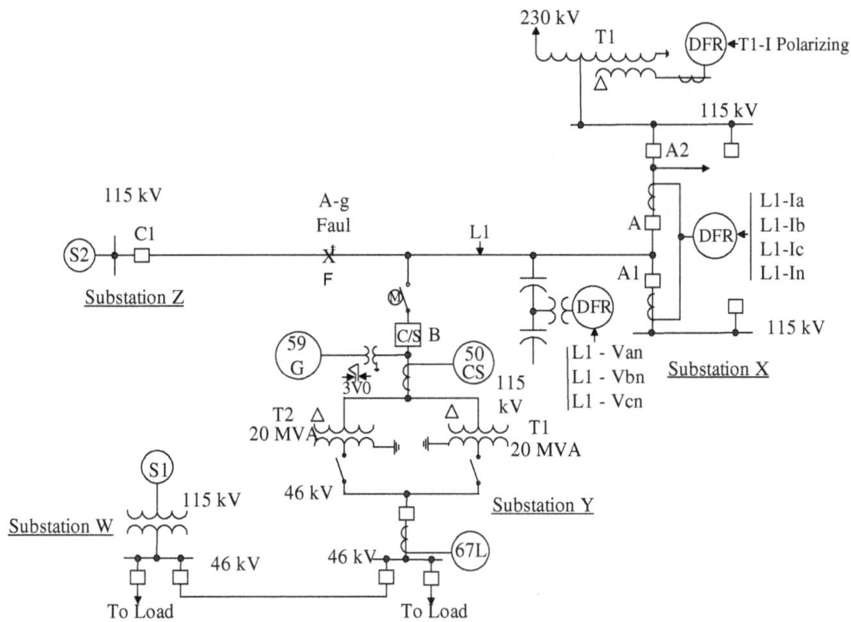

Fig. 6.17 One-line diagram showing monitored DFR voltages and currents and tap station backup protection devices.

Description of the Incident During an act of vandalism, insulators for the 115-kV line L1 were shot at, damaging several insulators. Empty cartrage shells were found nearby the damaged line insulators. This resulted initially in an A-g fault on a grounded system and then an A-g fault on an ungrounded system, evolving into A-C-g on the ungrounded portion of the system. Line crew reported that all 18 insulators on a suspension structure were damaged at the fault location.

Relay Setting Criteria for the Tap Station Backup Relays The 67L relay is set to have adequate pickup for a phase-to-phase fault at substation Z, with the fault being fed only from the 46-kV system (stub). For the 59G, the PT ratio is 1000 : 1, giving a secondary voltage of 69 V for each phase. During the ground fault, the neutral shifts fully, with unfaulted phases rising to 120 V. The broken delta voltage $3V_0 = 3 \times$ phase to neutral $= 3 \times 69 = 207$ V. The 59G relay is set at 70 V to have $3 \times$ pickup with a timer set at 0.5 s.

Analysis of the Fault Records Figure 6.17 shows the system one-line diagram and the line L1 DFR-monitored currents and voltages used to analyze the A-g line faults. The DFR record shown in Fig. 6.18 reveals an A-g fault occurring on line L1 at point a, which is the peak of phase A voltage. At substation X, the 115-kV CBs A and A1 tripped and cleared in 4 cycles at point b. Circuit breaker C1 at substation Y tripped and cleared the fault from the solidly grounded system in 5 cycles.

The fault is still connected to the system via the 46-kV loop subtransmission system. As shown in Fig. 6.17, transformers T1 and T2 are connected delta on the 115-kV side and

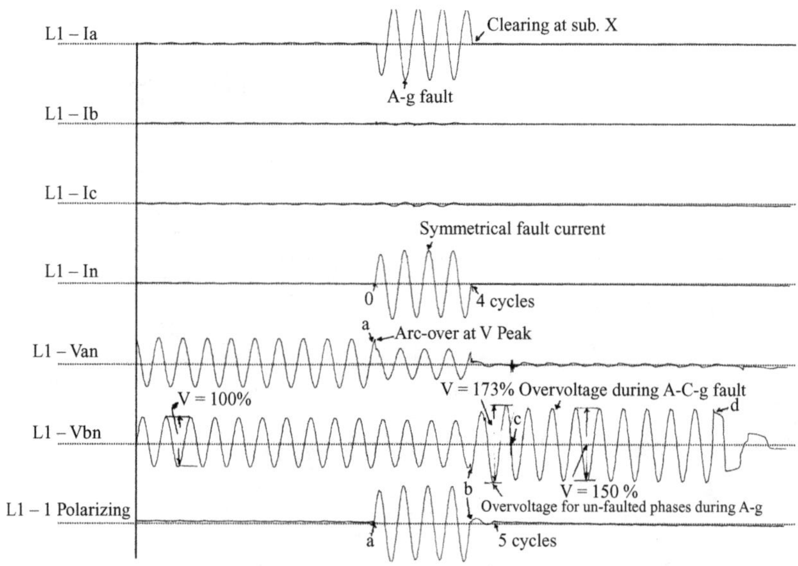

Fig. 6.18 Substation X DFR record showing line L1 currents and voltages during the initial A-g line fault.

grounded wye on the 46-kV side. At this time, the initial L-g fault is on an ungrounded system for a duration of 1.5 cycles. Therefore, the voltages on the unfaulted phases B and C rise to a full L-L value (= 173% nominal value) as shown in the substation X DFR of Fig. 6.18 trace L1-Vbn for phase B, and the numerical relay oscillography record of Fig. 6.19 traces VB and VC for phases B and C. At 6.5 cycles, the A-g fault on the ungrounded system evolved to an A-C-g fault, probably due to the spread of the ionized

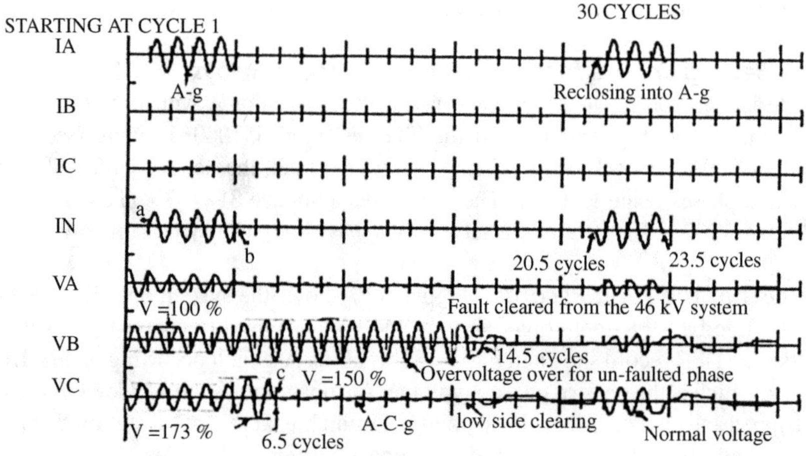

Fig. 6.19 Substation X numerical distance relay fault record for line L1 currents and voltages during the initial A-g line fault.

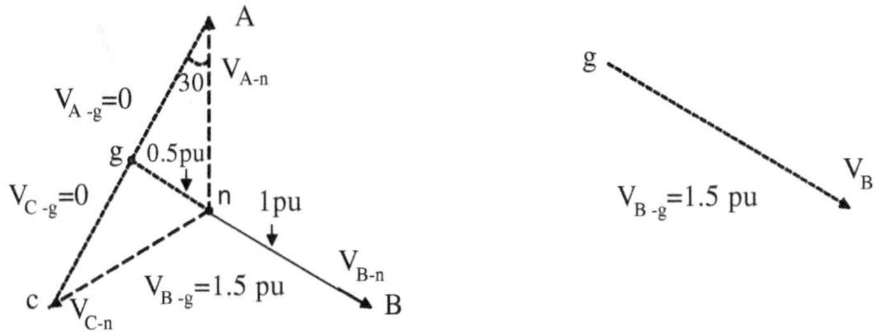

Fig. 6.20 A-C-g fault on an ungrounded system at F.

cloud. For a duration between the 6.5 cycles at point c to 14.5 cycles at point d, phase B voltage has an overvoltage of 150% nominal, as supported by the analysis shown in Fig. 6.20. Phases A and C directional backup time overcurrent device 67 shown in Fig. 6.17 operated to clear the fault from the 46-kV system after 14.5 cycles from $t = 0$. Tuned overvoltage relay 59G is set at 70 V and 0.5 s and therefore did not have sufficient time to operate and produce a target. The 46-kV source was locked out, waiting for the 115-kV transmission system to be energized. Apparently, the involvement of phase C from 6.5 to 14.5 cycles could also be by the spread of the ionized cloud.

At 20.5 cycles, CB A2 reclosed undesirably onto the A-g fault. The reclosing was initiated by a sneak circuit that had a shorted diode forcing CB A2 to reclose immediately following the tripping of the breaker. At 23.5 cycles, CB A2 tripped and cleared the fault. Figure 6.21 documents the substation X DFR record showing line L1 currents and voltages during unprogrammed breaker reclosing into the initial A-g fault. The numerical relay oscillograph record of traces IA and IN in Fig. 6.19 confirms the reclosing onto a fault for a duration of 3 cycles.

At 6.5 s the CB A reclosed unsuccessfully by autoenergizing a C-g fault. The DFR record of line L1 at substation X shown in traces L1-Ic and l1-In in Fig. 6.22 reveal an increase in the phase C current and the presence of the neutral current, respectively, for a duration of 5 cycles. It appears that the C-g fault occurred at the same initial location by vandals shooting at all line insulators.

Analysis of Fault Clearing from the Solidly Grounded System As shown in Fig. 6.18, the DFR record indicates ground current flow from the ground sources at substations X and Z. The fault was cleared from the X end of the line in 4 cycles. The T1 polarizing current trace reveals 4 cycles of high-zero-sequence current, followed by an additional 1 cycle of reduced zero-sequence current flow, indicating that substation Z still feeds the initial fault for an additional 1 cycle. Figure 6.23 illustrates the three phase voltages for a balanced power system. The voltages are equal in magnitude and separated by 120°. For the A-g fault at F the voltage will dip as shown in Fig. 6.24 and the zero-sequence voltage at the relay location can be calculated as shown. The DFR record in Fig. 6.18 confirms a phase A voltage dip of 0.65 pu and therefore the zero-sequence voltage will be calculated as $3V_0 = 1\text{ pu} - 0.65\text{ pu} = 0.35$ pu for an A-g fault on the solidly grounded 115-kV system. At 5 cycles from initiation

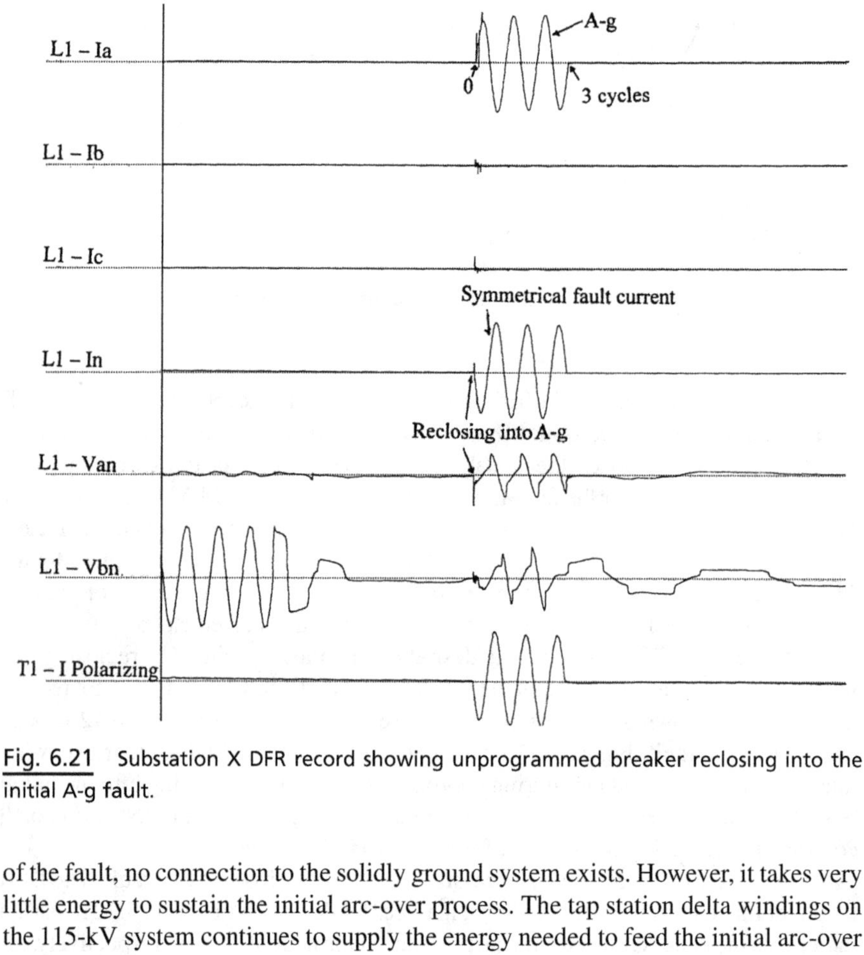

Fig. 6.21 Substation X DFR record showing unprogrammed breaker reclosing into the initial A-g fault.

of the fault, no connection to the solidly ground system exists. However, it takes very little energy to sustain the initial arc-over process. The tap station delta windings on the 115-kV system continues to supply the energy needed to feed the initial arc-over process.

Analysis of Fault Clearing from the Ungrounded System From 5 to 6.5 cycles, the ungrounded system sees a phase A-g fault. This is supported by the DFR voltage of trace L1-Vbn in Fig. 6.18 for the unfaulted phase B to rise to the value of the line-to-line voltage (173% of nominal voltage). The numerical relay record of Fig. 6.19 at substation X confirms the same amount of nominal voltage (i.e., 173%) for phases B and C. The voltage phasor diagram of Fig. 6.25 shows that the neutral point has shifted to A.

At 6.5 cycles from $t = 0$ the ionized cloud grew and the A-g fault evolved to an A-C-g fault for an additional 8 cycles. The fault is now being fed from the 46-kV system with sufficient current to operate the 67 directional phase overcurrent relay. Figure 6.20 shows the vector analysis used to calculate the voltage for the unfaulted phases during the A-C-g fault and reveals the voltage seen by the unfaulted phase B to be 150% of normal. The DFR record shown for trace L1-Vbn in Fig. 6.18 confirms a

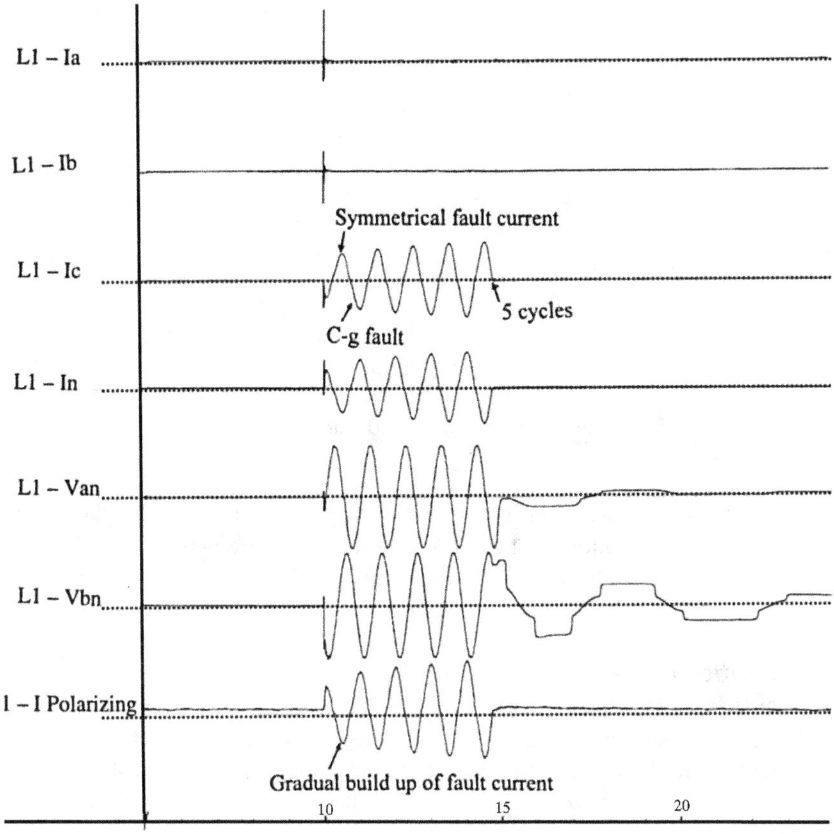

Fig. 6.22 Substation X DFR record showing automatic reclosing by one of the 115-kV line L1 breakers.

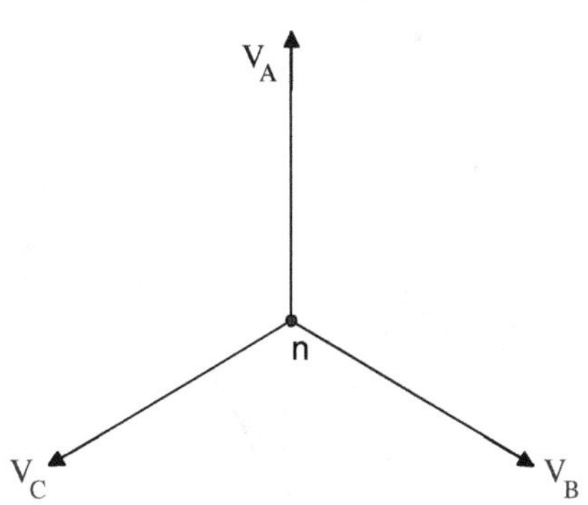

Fig. 6.23 Pre-fault three-phase voltages.

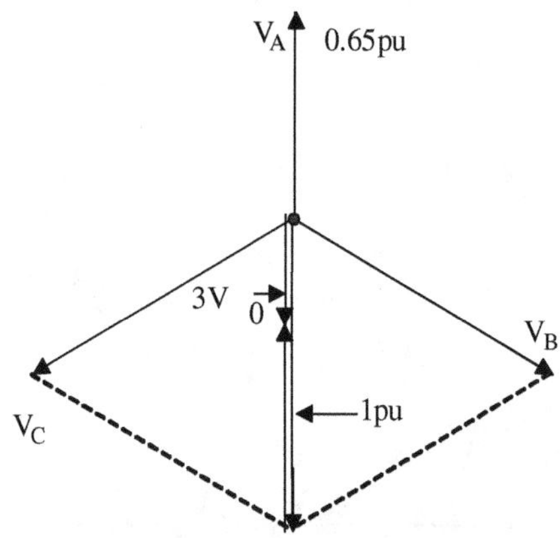

Fig. 6.24 A-g fault on a solidly ground system at F.

measured phase B voltage value of 150% of normal, while trace L1-Van shows a zero value for phase A. In addition, the numerical relay oscillograph record of Fig. 6.19 indicates a zero value for phase C voltage and confirms the recorded values cited above for phases A and B.

Confirmation of the DFR-Measured Voltage During the A-C-g Fault on the Ungrounded 115-kV System As in Fig. 6.20, points A and C will move to the median point g, forcing the $V_{A\text{-}g}$ and $V_{C\text{-}g}$ voltages to zero. The ground point is now at g and the neutral point n now moves to ground point g. As a result, voltages $V_{A\text{-}g} = V_{A\text{-}n} = V_{C\text{-}g} = V_{C\text{-}n}$ will be zero. The phase B-g voltage now consists of two components:

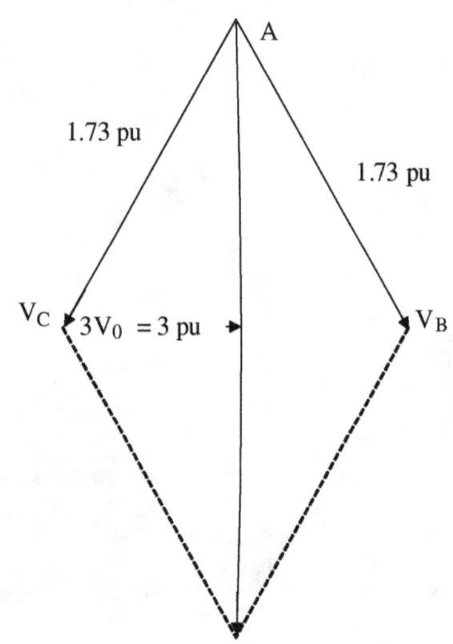

Fig. 6.25 A-g fault on an ungrounded system at F.

$V_{B\text{-}n}$ and the neutral-to-ground voltage, V_{ng}. $V_{B\text{-}n} = 1$ pu, and $V_{ng} = V_{A\text{-}n} \times \sin 30° = 0.5$ pu. The resulting value for phase B voltage $= 1.0 + 0.5 = 1.5$ pu, as confirmed by the DFR voltage, recorded as 150% of the phase-to-neutral value for phase B and zero for phases A and C.

System Phenomena

1. Arc-over at the voltage peak, as is always the case with a slow arc-over process of insulator flashovers due to mechanical failures, as shown in trace L1-Van in Fig. 6.18.
2. A ground fault starting on a solidly grounded system, and then cleared from the grounded transmission system, followed by slow clearing from a tap station fed from a delta/GY transformer, forcing the fault to be fed from the ungrounded portion of the system via the delta high-side winding as shown trace L1-Vbn in Fig. 6.18 and trace VB in Fig. 6.19.
3. Overvoltages on the unfaulted phases due to ground faults on the ungrounded systems as shown in trace L1-Vbn in Fig. 6.18. Additional supportive analysis is shown in Fig. 6.25 for confirmation of an overvoltage value of 173% of nominal value on phase B (unfaulted phase) for an initial A-g fault and in Fig. 6.20 for a value of 150% of nominal for in A-C-g fault.
4. An evolving fault from A-g to A-C-g, due to spread of the ionized cloud, as shown in trace VC in Fig. 6.19.

Case Study 6.7: 345-kV Phase-to-Ground Fault (C-g) Caused by an Act of Vandalism

Abstract A phase-to-ground fault occurred on a 345-kV line by an act of vandalism. The fault was cleared from the system successfully. The line reclosed unsuccessfully by automatic reclosing, followed by successful reclosing a few minutes later via SCADA. The line was then removed from service after notification of a circuit problem. In this case study we describe the incident, sequence of events, power system phenomena, and corrective actions.

Description of the 345-kV and Associated Protection Systems The 345-kV system shown in Fig. 6.26 consists of double-circuit steel pole towers. The protection of the 345-kV line was designed in the early 1980s and consists of dual pilot relaying systems using diverse communication media. System S1 is a permissive overreaching transfer trip (POTT) system using distance-based static relays applied over a power line carrier (PLC). The POTT scheme has an unblocking feature that generates a channel trip of 150-ms duration for a PLC signal loss. This feature enhances the relay dependability of tripping for L-G faults occurring on the top of the capacitive voltage transformer, where the relay voltage signal is derived. System S2 is also a POTT scheme, using phase and ground electromechanical distance relays applied on leased

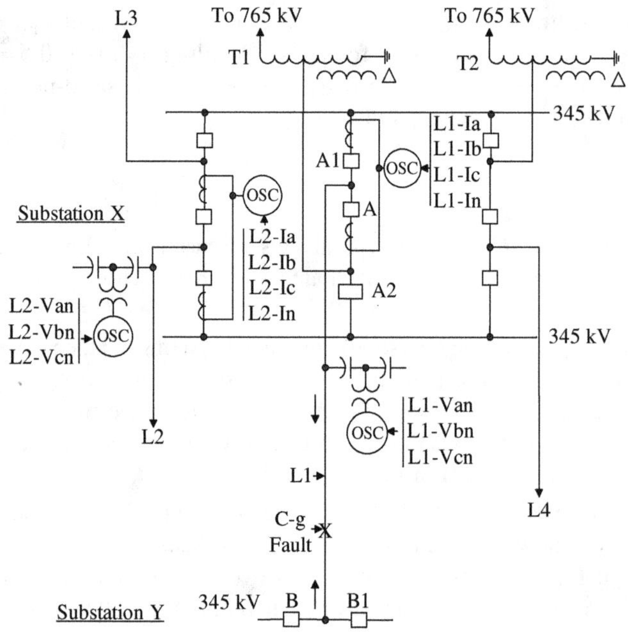

Fig. 6.26 System one-line diagram and oscillograph-monitored voltages and currents at substation X.

telephone circuits. Both relay systems have Z1 as a direct trip and Z2 as a pilot element, also providing remote backup protection.

Description of the Incident Vandals removed 28 nuts from the base of a 345-kV 160-foot steel transmission pole. As a result, the double-circuit tower swung into a hill, causing a phase C-to-ground fault on one of the lines. Simulation of the C-g fault by the titled tower is shown in Fig. 6.27.

Sequence of Events The sequence of events was constructed from an analysis of oscillograms and the sequence of events at substation X. Figure 6.26 shows the line L1

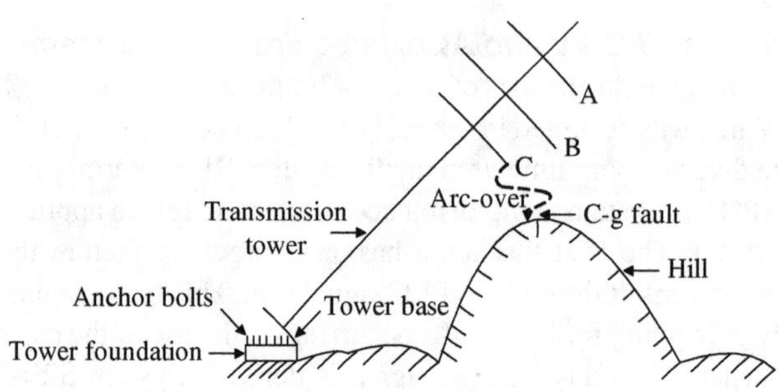

Fig. 6.27 Simulation of the C-g fault by the titled tower.

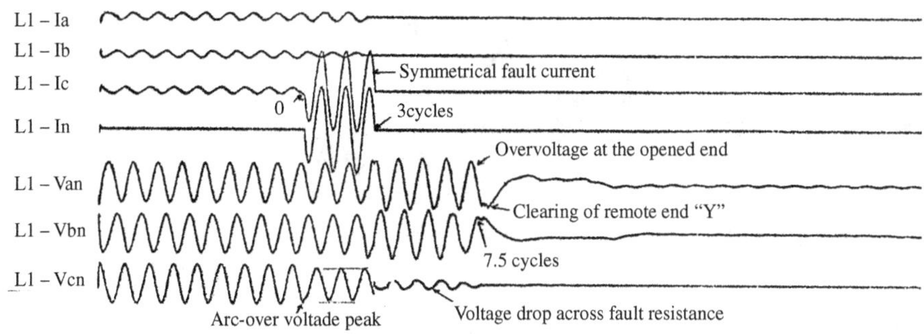

Fig. 6.28 Substation X oscillogram illustrating local and remote clearing of the C-g fault.

and L2 oscillograph-monitored voltages and currents at substation X. The oscillogram record shown in Fig. 6.28 reveals a phase C-g fault occurring on line L1. Three cycles later, 345-kV CBs A and A1 at substation X tripped by the operation of zone 1 instantaneous elements of the line relaying systems. From 3 to 7.5 cycles, the oscillogram record of Fig. 6.28 reveals a lower voltage on phase C and higher voltages on phases A and B. This indicates that the initial fault was still being fed from the remote end at substation Y. Seven and a half cycles later, 345-kV CBs B and B1 at substation Y tripped sequentially to clear the fault.

Ten seconds from $t = 0$, the CB B at substation Y reclosed automatically, energizing line L1 with the initial C-g fault. For a duration of 3 cycles, the oscillograph record in Fig. 6.29, which monitors line L1 voltages at substation X, reveals the lower

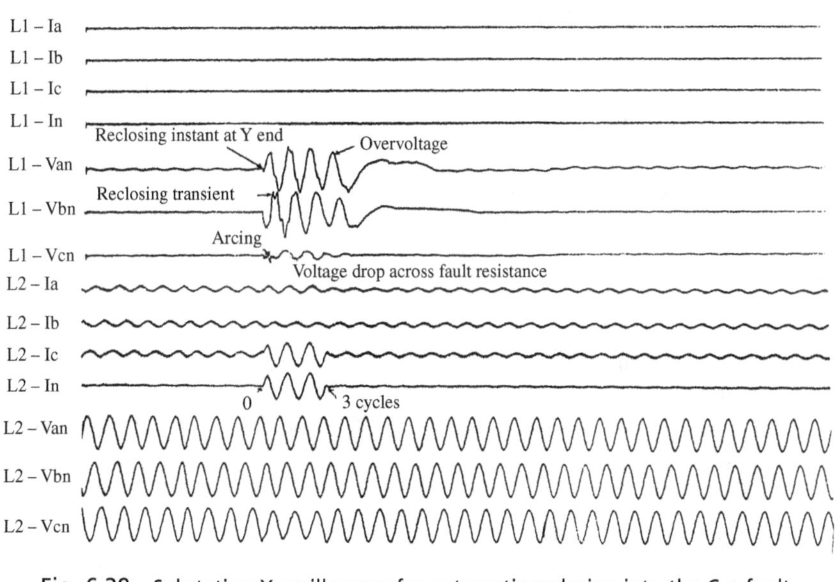

Fig. 6.29 Substation X oscillogram for automatic reclosing into the C-g fault.

voltage on phase C and higher voltages on phases A and B. This indicates that the initial fault is again being fed from substation Y and was cleared from the system by tripping CB B at substation Y. Eight minutes later, 345-kV breakers at both ends were reclosed successfully by SCADA energizing line L1. Apparently, the leaning tower bounced with the help of phase conductors and static wires to increase the clearance at the hill where initially the phase C conductor arced to ground. Two hours later the line was removed from service by tripping all associated 345-kV CBs via SCADA following notification regarding the leaning 345-kV transmission tower.

System Simulation of the Incident System disturbance analysis requires running short-circuit study simulations to verify the power system model. Oscillograms of this case study should be compared with the results of the study simulations to validate the power system model. As shown in Fig. 6.28, the L1-Vcn voltage trace reveals that the fault has a resistance to ground. It was also reported that the fault left burning marks on the grass and vegetation of the hill. This is always the case for a live conductor that moves near a ground surface and then touches the ground, where ground currents are seeking a return to ground sources via the water table in moist ground soil. Simulation of this incident in the short-circuit study will require knowing the fault location and the value of the fault resistance. The fault location is known in this incident, leaving only the fault resistance as the remaining unknown to be defined.

The distance to the fault is now known and can be input to the short-circuit analysis study. Estimation of fault resistance in these cases is based on knowing the location of the fault. A sensitivity analysis can now be performed by simulating the known fault location, varying the fault resistance, and examining the result for comparison, then matching the results obtained by measuring the oscillograph current and voltage outputs. Simulation is needed to estimate the fault resistance where current is seeking a path at the surface of the Earth for a return to ground sources. The analysis produced a resistance value of 3.5 Ω. The oscillogram record reveals a ground current ($3I_0$) of 5078 A as a contribution from substation X to the fault as compared with a value of 5063 A from the short-circuit output shown in Fig. 6.30. The recorded phase A voltage is 0.8 pu, compared with the value of 157.7 kV ($= 157.7/200 = 0.78$) calculated in the short-circuit study simulation shown in Fig. 6.30.

Power System Phenomena

1. Flashover at the voltage peak due to a slow fault-creation mechanism when the tower swung and phase C arced-over to ground, as shown in trace L1-Vcn in Fig. 6.28.
2. Fault resistance as the voltage appears at substation X when breakers A and A1 cleared the substation X contribution to the fault, as shown in trace L1-Vcn in Fig. 6.28.
3. Line open-end voltage rise (Ferranti effect) due to capacitive current flow of the line through the line inductive reactance, as shown in traces L1-Van and L1-Vbn in Fig. 6.28.

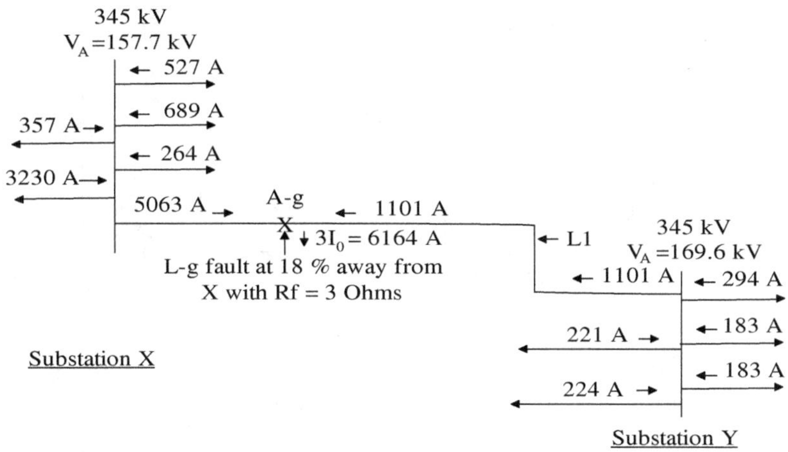

Fig. 6.30 Simulation of the ground fault using a short-circuit study.

4. Sequential clearing when one end clears first, forcing the opposite-end fault current to increase, allowing relays to operate as displayed in the oscillogram shown in Fig. 6.28.

Corrective Action All bolts used to anchor the towers on the 200-mile transmission corridor were sealed using an epoxy material to prevent a similar incident in the future.

Lesson Learned Monitoring the line voltage side by an oscillograph, DFR equipment, or a numerical relay at one end is sufficient to provide remote end-line clearing times without the need to analyze the remote end fault record.

Case Study 6.8: 345-kV Phase-to-Ground (A-g) Fault Due to an Accident Along the Line Right-of-Way

Abstract A phase A-to-ground fault occurred on a line L1 345-kV circuit. The fault was caused by an accident along the line right-of-way. The high-resistance fault was cleared successfully at both ends of the line. In this case study we describe the incident, analyze it, and provide corrective action, calculation of fault resistance, and lessons learned.

Description of the System and Associated Protection Figure 6.31 shows the system one-line diagram and DFR-monitored currents and voltages used to analyze the disturbance. A protection system for the 345-kV line, designed in the early 1980s, consists of dual pilot relaying systems using diverse communication media. System A is a permissive overreaching transfer trip system using phase and ground electromechanical distance relays employed on leased telephone circuits. System B is a

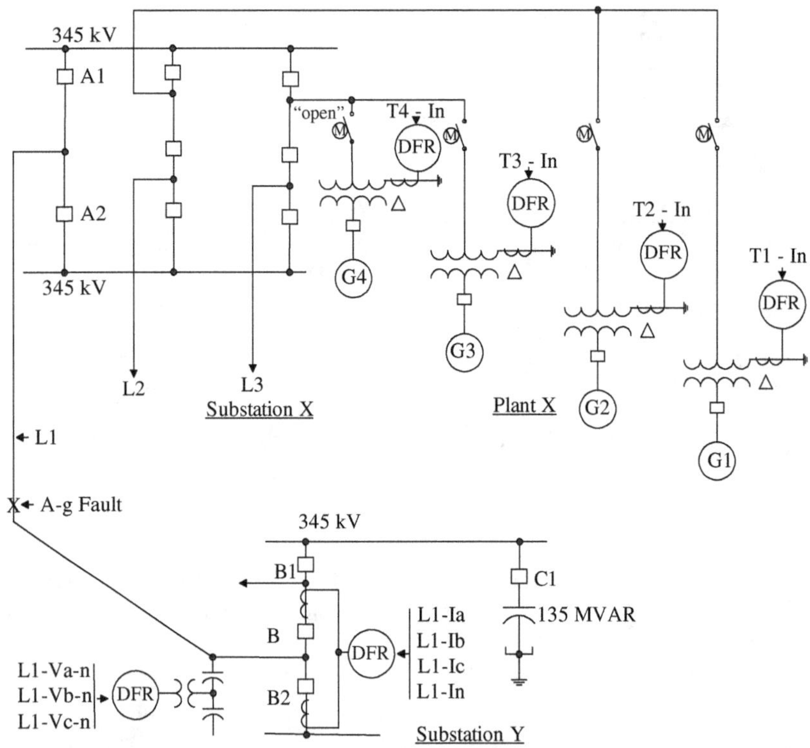

Fig. 6.31 One-line diagram showing line L1 DFR-monitored voltages and currents.

directional comparison blocking scheme system using distance-based static relays applied over a power line carrier.

Description of the Incident As shown in Fig. 6.32, a utility worker was moving a 40-foot electrical pole outfitted with a ground cable. The pole was being installed to provide electricity to a new home. The worker was driving a truck equipped with jaws to move utility poles. He selected his unloading area near the right-of-way of a 345-kV line. He began to unload one of the poles and moved it underneath the EHV 345-kV line L1. The 345-kV line arced-over to ground via the pole-installed ground cable when the pole was lifted in the air, thus causing a phase A-g fault. The worker had probably been touching the lifting truck, and as a result he received a very serious electric shock.

Analysis of the Incident The DFR record shown in Fig. 6.33 reveals a phase-to-ground fault occurring at the voltage peak, which is typical for an arc-over of a live conductor to a moving grounded object. The fault is classified as A-g, due to the phase A voltage drop and the presence of high current on phase A and the neutral current. The fault current is gradually increasing, indicating an initial high-resistance L-g fault

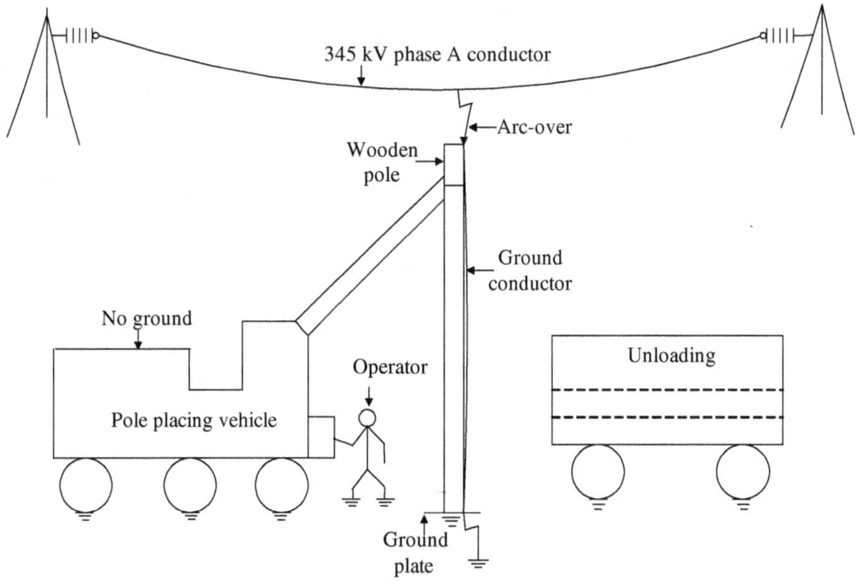

Fig. 6.32 Occurrence of the 345-kV A-g fault along the right-of-way.

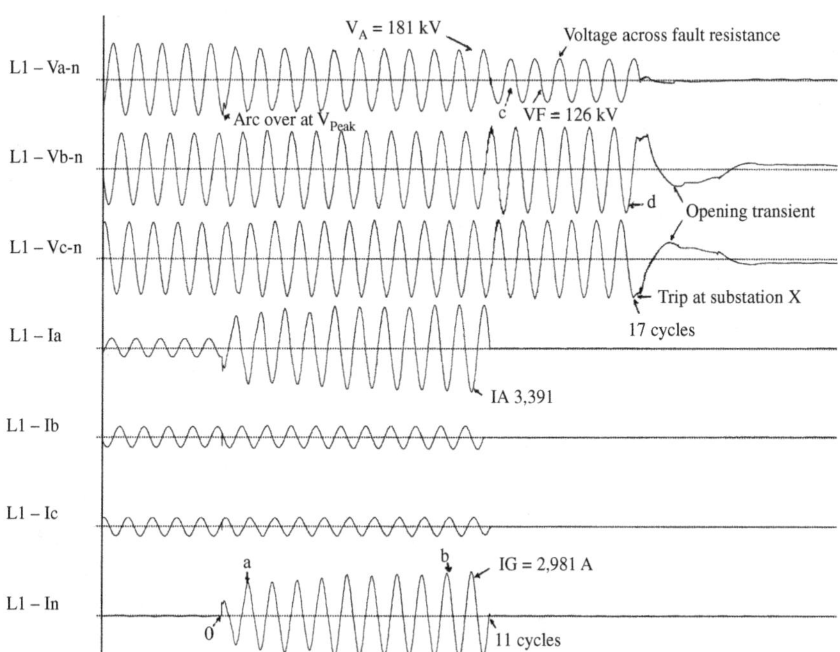

Fig. 6.33 Substation Y DFR record showing line L1 voltages and currents during the A-g fault.

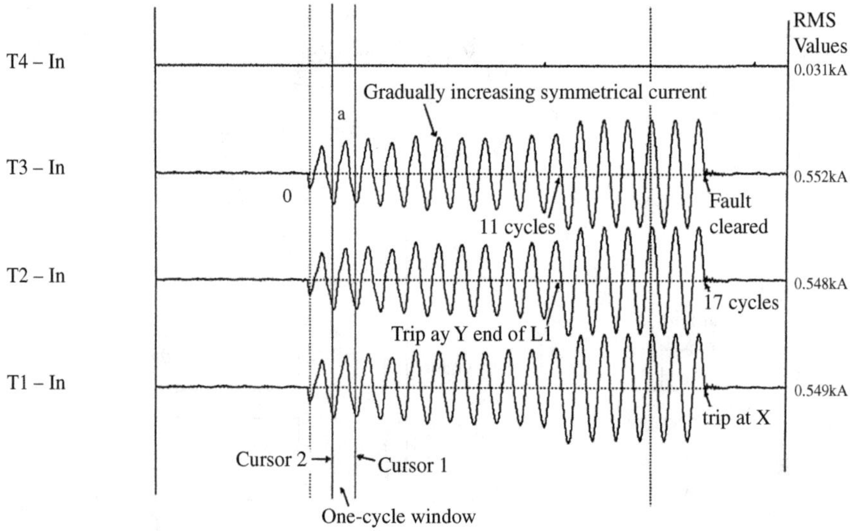

Fig. 6.34 RMS transformer neutral currents for plant X 1 cycle after fault initiation.

whose resistive value decreased while the ground currents were seeking a ground return to the neutrals of the ground sources. The fault was cleared from substation Y in 11 cycles by a zone 1 ground distance element and tripped sequentially from substation X to clear in 17 cycles by a ground distance mho tripping element of the directional blocking carrier pilot scheme.

Analysis of the Plant DFR Record The DFR record in Fig. 6.34 shows the ground currents flow in the neutrals of generator step-up transformers for units G1, G2, and G3. Unit G4 has no ground current contribution to the fault during the accident because of the isolation of the unit opening disconnect switch shown in Fig. 6.31. The symmetrical fault current started as about 550 A in Fig. 6.34 with a gradual increase. Figure 6.35 illustrates that the current reached a value of about 695 A in 9 cycles. Figure 6.36 reveals that the ground current reached a value of about 998 A after 12 cycles.

Performance of the Protection System A sequence-of-event recorder, which has a 1-ms resolution, has confirmed that both the primary and secondary ground distance elements reacted to the fault when the substation Y terminal of the line tripped first. The tripping of the line L1 end at substation Y resulted in a 50% increase in the ground fault current from substation X and subsequent operation of the secondary carrier blocking system and primary unblock system pilot element. The directional ground time overcurrent backup element at plant X would have operated in 0.6 s for this A-g fault contribution. The fault was cleared in 0.27 s by the ground distance relay used in the secondary carrier blocking protection system.

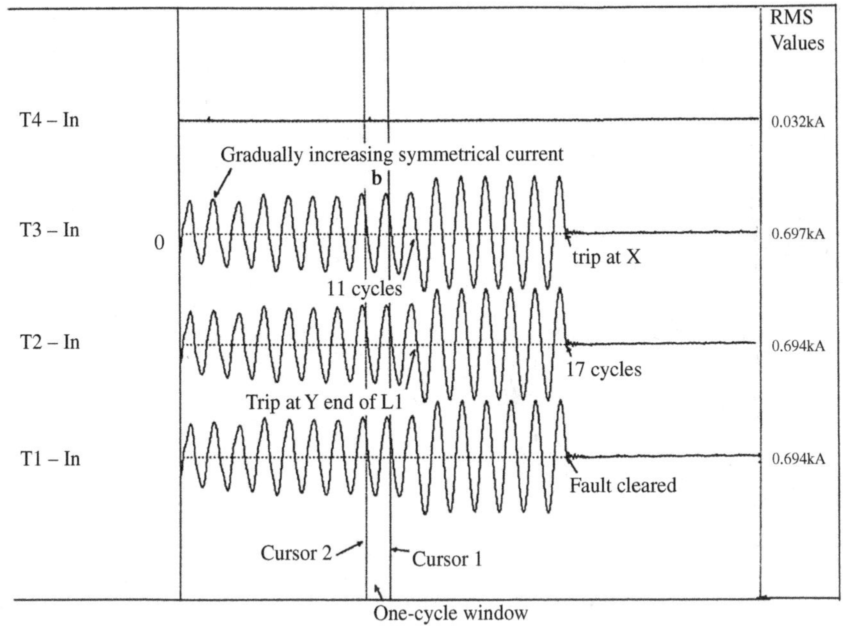

Fig. 6.35 RMS transformer neutral currents for plant X 9 cycles after fault initiation.

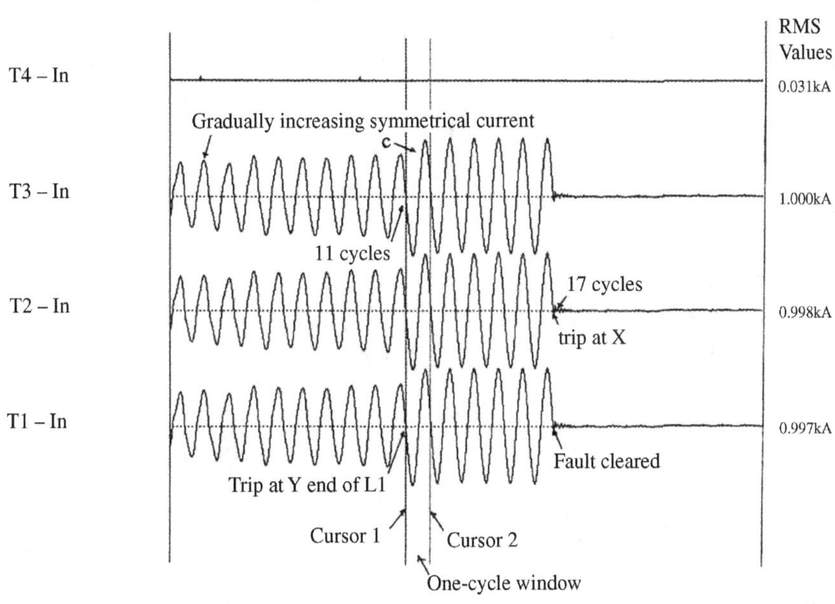

Fig. 6.36 RMS transformer neutral currents for plant X 11 cycles after fault initiation.

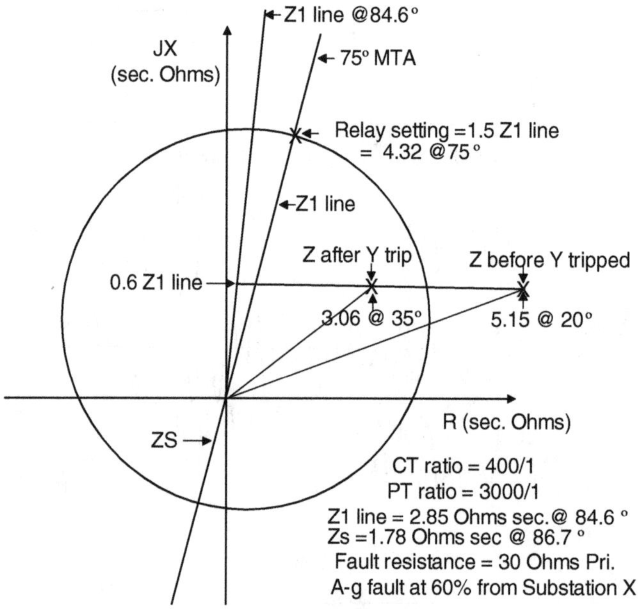

Fig. 6.37 Substation X ground distance relay dynamic characteristic during the clearing process of the A-g fault.

Figure 6.37 shows the dynamic characteristics of the primary ground pilot distance mho relay, illustrating the sequential clearing of the fault. When initially the fault was being fed from both ends, the element reach in the *R*–*X* diagram was plotted outside the mho element dynamic characteristic, as illustrated. When the fault was cleared from the remote end of line L1 at substation Y, the impedance vector measurement moved inside the mho relay tripping zone, as shown.

Protection Against High-Resistance (Tree) Faults and Open-Phase Conditions The ground time overcurrent relays play an important role in providing protection to the power system against high-resistance faults and open-phase conditions. The ground TOC element provides most of the challenge for relay setting calculation and coordination. In EHV and HV systems, two redundant ground time overcurrent relays are often used to fulfill the duality relaying principle (more than one element responding to a given fault).

Power System Phenomena

1. Arc-over at the voltage peak, which is typical for a fault caused by arc-over of a live conductor to a moving grounded object, as shown in trace L1-Va-n in Fig. 6.33.
2. A high-resistance fault with a varying magnitude of fault resistance, as shown in traces L1-Ia and L1-In in Fig. 6.33.

Lessons Learned

1. High-resistance faults can occur on EHV systems, and normally such faults are cleared from the system by backup directional ground time overcurrent relays. These types of faults offer a challenge to relaying systems that are based on operating principles using impedance (distance) measurements. In this case, the delayed clearing is acceptable since these high-resistance faults have less impact on the transient stability of the system.
2. Protection of EHV lines should be based on applying redundant ground directional overcurrent protection to provide a guarantee to clear high-resistance ground faults and to sense open-phase conditions.

Case Study 6.9: False Tripping of a 138-kV Current Differential Relaying System During an External Phase-to-Ground Fault

Abstract In this case study we deal with the undesired trip of a numerical current differential relaying system applied on a 138-kV tie for a combined-cycle plant. The trip occurred during an external phase-to-ground fault. An analysis of current transformer performance during the external fault using relay fault records and DFR records is described. We also cover corrective actions, lessons learned, and a description of system phenomena.

Description of the Undesired Operation A technician accidentally bumped a transformer lockout relay, which closed high-side ground switch V, causing a 138-kV C-g fault at substation Z shown in Fig. 6.38. Based on an analysis of substation X DFR record shown in Figs. 6.39 and 6.40, the plant was tripped 7 cycles from the start of the incident. The tripping of the 138-kV tie between substations X and Y was done by the current differential relays at both ends of the line. The relay tripped CB A at substation X and CB B at substation Y. Loss of the 138-kV tie resulted in shutdown of the combined-cycle 150-MW plant which is connected to substation X.

Analysis of the Undesired Trip of the 138-kV Line L1 Undesired operation of current differential relaying systems could be attributed to one or more of the following: poor current transformer (CT) transient performance at one of the terminals, failure of the communication link, or failure of the relay. The DFR equipment at substation Y failed and no fault record was generated for the incident. The lack of documentation for the secondary current input to the differential relay at Y end of the line made the analysis of the false trip difficult. The availability of CT secondary current waveshapes for both ends will certainly help in isolating the CT as the source of the false trip. Analysis of the event therefore had to be based on the DFR record at substation X and relay fault records from substations X and Y.

A process of elimination can be used to provide an explanation for relay system false trips and to pinpoint the main causes. First, an analysis of the relay records reveals identical local and remote phase current quantities, as shown in Tables 6.1

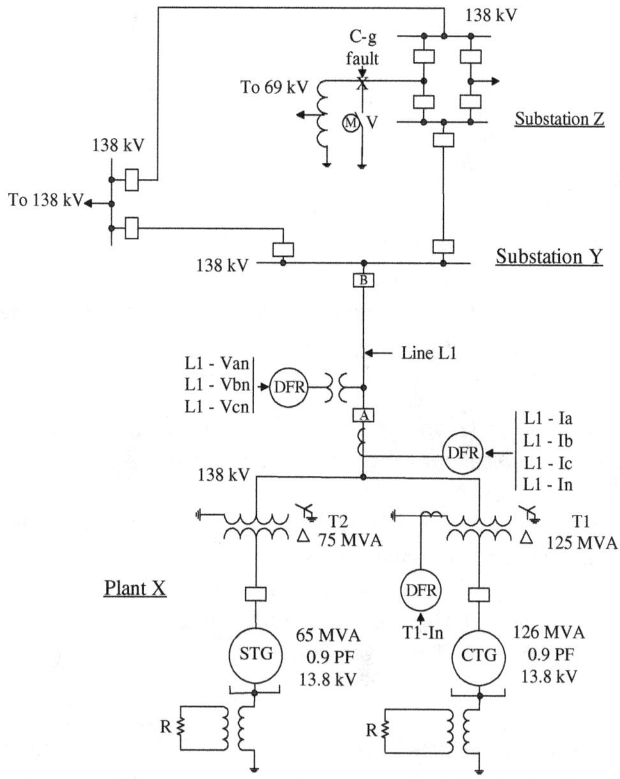

Fig. 6.38 System one-line diagram with DFR-monitored voltages and currents.

and 6.2. The records indicate that phase C current inputs into the differential relays do not match during the duration of the fault. However, when the 138-kV line L1 was later energized and returned to service, the phase C load current readings from the two relays were very closely matched. This implies that the current differential relays and their associated fiber optic communication link are functioning normally, as designed, with no obvious failure. Now the analysis can concentrate on the CT performance during the external phase-to-ground fault. Poor transient performance of CT at substation Y can be a main suspect for the false trip of the differential relay. These hypotheses can be supported by the following observations:

1. The utility that owns substation Y uses one common specification for its 138-kV circuit breaker procurement with all CTs specified with a multiratio 3000 : 5 and an accuracy of C800 at the full CT tap. The utility that owns substation X specified its 138-kV circuit breaker with CTs having a ratio of 2000 : 5 with an accuracy of C800 at the full CT tap. The use of a 1200 : 5 tap at both ends of

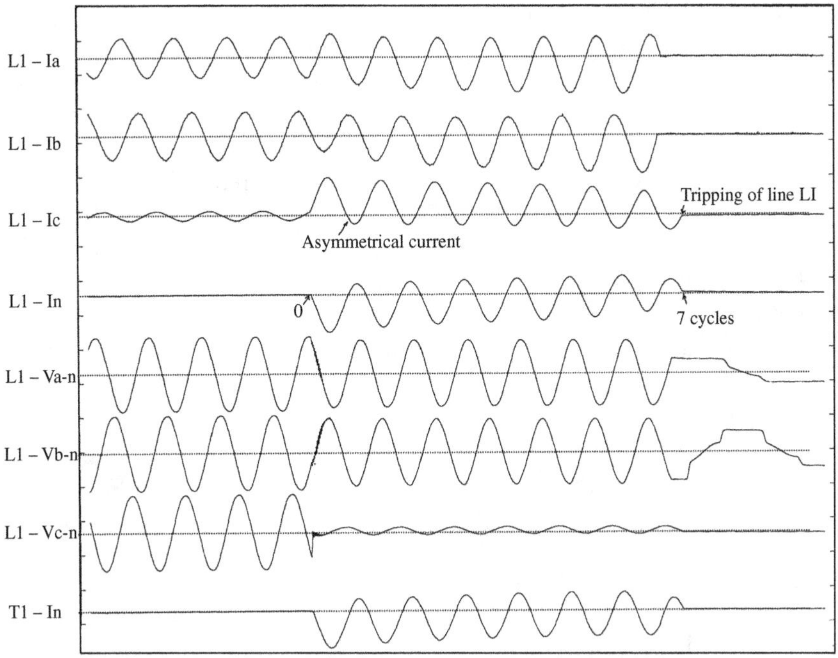

Fig. 6.39 Plant X DFR record confirming line L1 undesired trip during the external C-g fault.

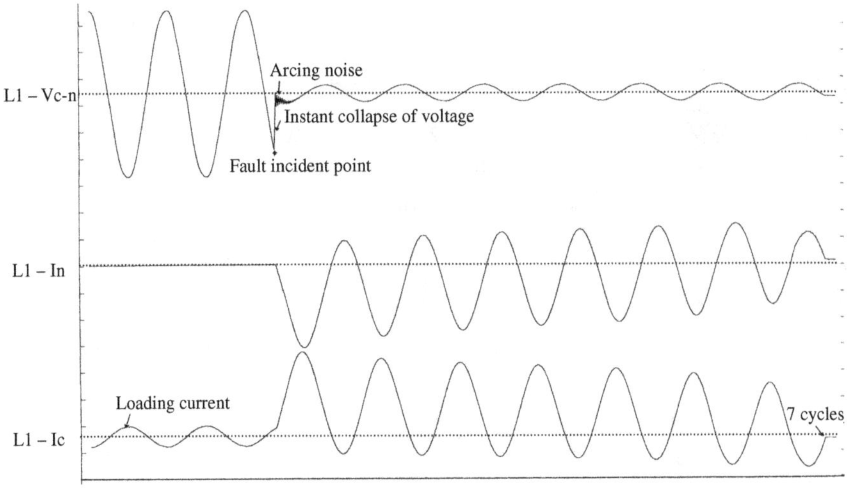

Fig. 6.40 Plant X DFR record for line L1 faulted phase voltage and currents.

the tie to connect the relay has made the CT at substation Y less accurate than the CT at substation X.

2. The current differential relay at substation X was connected using dedicated CTs, while the relay at substation Y had additional phase and ground electromechanical backup relays. This resulted in different burdens connected to the CTs at both ends, which may lead to different CT responses for the same transient input currents.
3. The presence of the asymmetrical current shown in trace L1-Ic in Fig. 6.40, which contains slow decaying dc offsets (normally present near generating stations) as inputs to the differential relays.
4. The possible presence of a significant level of remnant flux of an adverse polarity with respect to the polarity of fault current waveform offset. This can occur if CTs are subjected to dc resistance tests during maintenance outages without the proper demagnetization afterward.
5. The relay fault records shown in Tables 6.1 and 6.2 reveal different phasor magnitudes, with the substation Y relay calculated phasor magnitude being less than that of the X relay. Since these phasors are calculated from samples obtained from the original secondary currents and stored in digital form, it may imply that the Y end had some lower magnitude samples. This could happen only if the Y end had partial CT transient saturation. The degree of current waveform distortion at only one relay terminal (Y) appears to have produced sufficient differential current to operate the relay falsely.

It can now be concluded that the undesired trip occurred due to partial saturation of the substation Y end phase C current transformer input to the differential relay. This resulted in enough differential current to satisfy the tripping criteria of the K1 slope percentage bias characteristic shown in Fig. 6.41.

Analysis of the Relay Fault Records The generation of relay fault records during system disturbances is considered as one of the important advantages of applying microprocessor-based (numerical) relaying systems. Tables 6.1 and 6.2 comprise the relay fault records obtained from substation X and Y, respectively, for the undesired trip of the 138-kV tie. In the tables, DIFF $= I_{\text{diff}} =$ the vectorial sum $I_{\text{R}} + I_{\text{L}}$ and BIAS $= I_{\text{bias}} = \frac{1}{2}$ (the sum of magnitudes of I_{R} and I_{L}, where I_{R} is the remote input to the differential relay and I_{L} is the local input to the differential relay.

The relay setting data, defined in Fig. 6.41, are

$$
\begin{aligned}
I_{\text{S1}} &= \text{relay pickup} = 0.2\,\text{pu} \\
I_{\text{S2}} &= \text{bias current} = 2.0\,\text{pu} \\
K_1 &= \text{slope of first segment} = 30\% = 0.3 \\
K_2 &= \text{slope of second segment} = 100\% = 1.0
\end{aligned}
$$

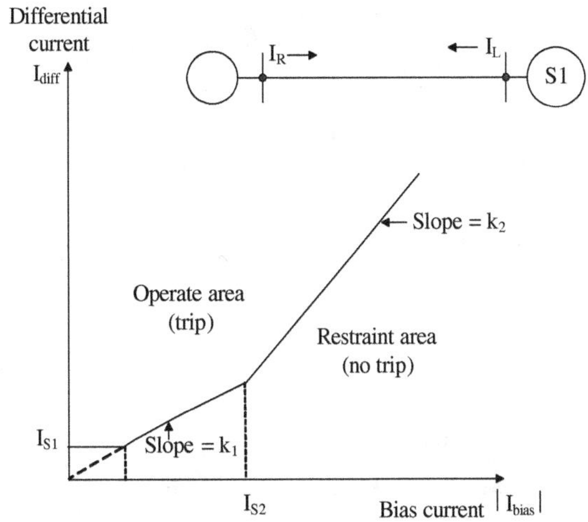

Fig. 6.41 Current differential relay operating characteristic.

The relay operates as follows: for $I_{\text{bias}} < I_{S2}$,

$$I_{\text{diff}} \geq K_1 \times I_{\text{bias}} + I_{S1} \tag{1}$$

and for $I_{\text{bias}} > I_{S2}$,

$$I_{\text{diff}} \geq K_2 \times I_{\text{bias}} - (K_2 - K_1)I_{S2} + I_{S1} \tag{2}$$

The fault records indicate a perfect match of local and remote currents for the unfaulted phases A and B during the C-g external fault. This resulted in close-to-zero

TABLE 6.1 Substation X Current Differential Relay Fault Record

Date	Time	Alarm Type	Details
	Alarm Record		
1999 Sept. 16	13:18:42	Diff Prot Trip	Phase C
	Fault Records		
1999 Sept. 16	13:18:42	Differential	Phase C
	Phase A	Phase B	Phase C
Local	0.64 pu	0.51 pu	1.33 pu
Remote	0.64 pu	0.51 pu	2.08 pu
DIFF	0.00 pu	0.01 pu	1.13 pu
BIAS	0.64 pu	0.51 pu	1.70 pu

TABLE 6.2 Substation Y Current Differential Relay Fault Record

Date	Time	Alarm Type	Details
	Alarm Record		
1999 Sep 16	13:30:26	Diff Prot Trip	Phase C
	Fault Record		
1999 Sept. 16	13:30:26	Differential	Phase C
	Phase A	Phase B	Phase C
Local	0.64 pu	0.51 pu	2.07 pu
Remote	0.64 pu	0.49 pu	1.32 pu
DIFF	0.02 pu	0.02 pu	1.14 pu
BIAS	0.64 pu	0.50 pu	1.70 pu

differential currents and sufficient bias current to restrain the tripping of the relay for phases A and B.

However, phase C has a differential current of about 1.14 pu (where 1 pu = 5 A secondary) and a bias current of 1.7 pu $\left[=\frac{1}{2}(2.08+1.13)\right]$. Based on the relay characteristic shown in Fig. 6.41 and since the bias current of 1.7 pu is less than the I_{S2} setting of 2 pu, the relay will generate a trip if equation (1) is satisfied, given $I_{diff} = 1.14$ pu, $K_1 = 30\% = 0.3$, $I_{S1} = 0.2$ pu, and $I_{bias} = 1.7$ pu.

Substituting these values in equation (1) will satisfy the relation

$$1.14 > [(0.3 \times 1.7) + 0.2] > 0.71$$

This result will confirm the tripping of the 138-kV tie by phase C of the relay, as indicated in both relay alarm and fault records.

Transient Performance of the Current Transformer During External Faults

Apparently, transient current transformer saturation has occurred at one terminal only, which depressed the output current for phase C at terminal Y and gave an effective phase shift to the output current. Through fault current saturation at a low level of fault current rarely occurs. It is only near a generating plant where conditions may lead to partial CT saturation. This would be where CT saturation occurs due to a very slow rate of decay of the transient dc offset in the through-fault current (a high X/R ratio). In addition, the nonmatching of CT accuracies at both ends of line L1 contributed to the CT transformation differential current. The differential relay at substation Y uses a multiratio 3000 : 5 CT and an accuracy of C800 at the full tap, while the relay at substation X uses a multiratio 2000 : 5 with an accuracy of C800 at the full tap. The use of the 1200 : 5 tap at both ends produces an equivalent connected CT accuracy of C200 at substation Y versus C400 at substation X. Connecting electromechanical backup relays in series with the differential relay at substation Y has also contributed to the burden mismatch between the two terminals. The differential current created by the CT error caused an undesired trip due to the differential relay operation on the sensitive segment of the operating curve at a slope of 30%. The operating point was due to the through-fault current being low.

System Phenomena

1. Instant collapse of the voltage at fault occurrence and the presence of arcing noise, which accompanies initiation of the fault, as indicated in trace L1-Vc-n in Fig. 6.40.
2. Asymmetrical fault current containing slowly decaying dc offset for generator contribution to nearby faults, as indicated in trace L1-Ic in Fig. 6.40.

Corrective Actions

1. Increase the CT ratio for the current differential relaying system from 1200 : 5 to 2000 : 5, which is the maximum available ratio for terminal X. This increase will improve the performance of CTs and enhance the security of the relay. The change in the CT ratio will not affect the sensitivity of the relay, due to the availability of sufficient fault current (i.e., 36,000 A) during internal faults. Phase and ground fault magnitudes for faults at substation X are close to faults at substation Y, due to the short length of the 138-kV cable ties (1000 feet).
2. Increase relay slope K1 from 30% to 50% to provide more security during external faults. This setting still provides adequate protection for minimum internal fault levels fed from the generating plant at 2400 A.
3. Relocate the electromechanical backup phase and ground relays at the Y end to a different CT circuit or replace them with equivalent lower-burden numerical relays.
4. Find a way to monitor the availability of DFR continuously either by alarming or by generating a periodic trigger for a record to be sent or faxed remotely to an operating center.

Lessons Learned

1. Current differential relaying systems require matching the CT performance at both ends of the protected line. This will require that the CT ratio, accuracy, and burden connected be identical at each end of the transmission line.
2. Current differential relays are particularly susceptible to the effects of asymmetric CT saturation at low levels of through-fault current if CT transient saturation can occur.
3. CT accuracy specified at full multiratio is downgraded when only part of the winding is connected to the relay system. For example, C800 at 3000 : 5 becomes equivalent to C200 [$= 800 \times (1200/3000)$] at a ratio of 1200 : 5, and

C800 at 2000 : 5 becomes an equivalent to C400 [= 800 × (1200/2000)] at a ratio of 1200 : 5.

4. Care should be taken when testing the CTs used for differential relaying near generating plants to remove any remnant flux left in the CT magnetic circuit.
5. Numerical relaying provides valuable fault records that facilitate the analysis of undesired relay system operations.
6. The DFR status should be monitored continuously either by periodic software or by timer triggering to generate a record to be faxed to remote locations or by producing an alarm to record a failure.
7. Technicians should be educated as to the consequences of inadvertent bumping of relays or auxiliary relays at substations.

Case Study 6.10: Undesired Operation of a 13.8-kV Feeder Ground Relay During a Three-Phase Fault Due to an Extra CT Circuit Ground

Abstract An outdoor cubicle housing 13.8-kV disconnect switches and surge arresters failed during a heavy rain. The 13.8-kV outdoor switchgear metal enclosures apparently permitted water to accumulate and to cause a three-phase fault, which was cleared by operation of the primary and secondary transformer T1 differential relays in 5.5 cycles. A design error left an extra current transformer ground, which resulted in undesired operation of the remaining feed to the station service supply. In this case study we describe the system, CT switching scheme, undesired operation of the T2 feeder relay, calculations of current split, confirmation of the short-circuit study model, system phenomena, corrective actions, and lessons learned.

Description of the Station Service System and Associated Protection As shown in Fig. 6.42, the station service system supplies power to a hydro facility that produces more than 900 MW of power and is connected to the grid through 230- and 115-kV substations. The station service uses two sources from the 13.8-kV tertiary windings of two autotransformers, T1 and T2. An outage of either transformer is substituted by a third tertiary from autotransformer T3 through the use of an extensive CT switching scheme. A grounding transformer is added to each of the main 13.8-kV buses, as shown in Fig. 6.42. A 10-Ω resistor is added in each of the grounding transformer neutrals to limit the ground faults to about 800 A. The autotransformers are protected by dual differential relays, labeled as primary and secondary. Figure 6.43 shows the protection system for transformer T1, which is also typical of those for transformers T2 and T3. The 13.8-kV tertiary lead is protected by instantaneous and time overcurrent relays. The grounding transformer neutral has a time overcurrent relay for ground backup protection.

Description of the CT Switching Scheme A two-position multideck control switch is used for the CT switching scheme. One switch is needed for each of the 13.8-kV feeders L1 to CB C1 and L2 to CB C2. The switch for L1 is designated as

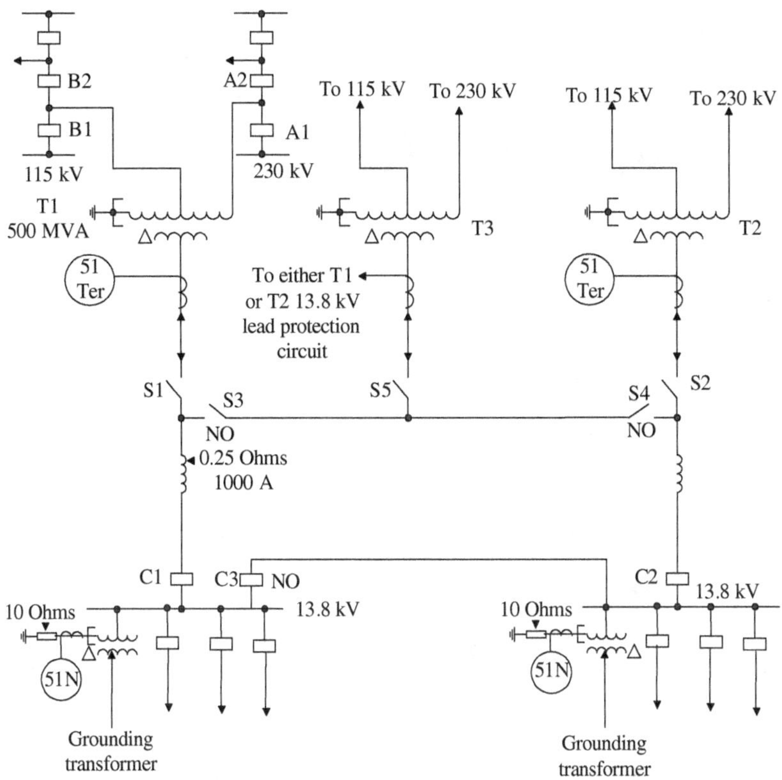

Fig. 6.42 One-line diagram for the 13.8-kV supply.

43-C1, with two positions, designated normal T1 and transfer T3, and for L2 as 43-C2, with two positions designated as normal T2 and transfer T3. As shown in Fig. 6.43, the normal position will also allow the 6000 : 5 CTs associated with 13.8-kV CB C1 to be connected to the T1 primary and secondary differential relays. As shown in Fig. 6.44, at the normal position for T1, the switch contacts close to connect the T1 13.8-kV lead 600 : 5 CTs to the 50/51/T1 overcurrent relays. The normal position for T2 will similarly connect the T2 13.8-kV CTs to 50/51/T2 relays and connect the CTs associated with CB C2 to T2 differential relays. Both switches in their normal position will short the T3 13.8-kV lead CTs through the dedicated ground shown in Fig. 6.44.

When transformer T1 is not available as a source to the plant 13.8-kV station service, the switch will be position T3. As shown in Fig. 6.44, the T1 13.8-kV lead CTs will be shorted through a dedicated ground and the T3 13.8-kV lead CTs short will be removed to connect them to the 50/51/T1 overcurrent protection. In addition, the 6000 : 5 CTs associated with 13.8-kV CB C1 will be disconnected from the T1 differential relay circuits and connected to the T3 differential protection circuits. A similar transfer can take place between the T2 and T3 13.8-kV feeders. The transfer

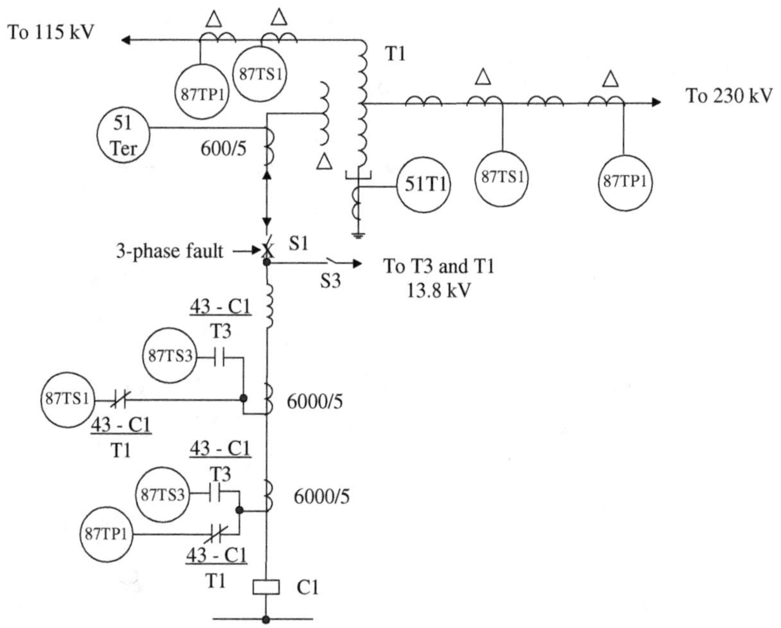

Fig. 6.43 Transformer T1 one-line ac diagram with differential protection using the CT switching scheme.

from either T1 or T2 (normal position) to T3 (transfer position) is normally done manually following a forced outage or scheduled maintenance outage of the respective transformer. For transfer from T1 to T3, the following switching order sequence is used: close CB C3; open CB C1; open disconnect switch S1; close disconnect switches S3 and S5; close CB C1; and finally, open CB C3.

Description of the Fault Due to heavy rain, the 13.8-kV disconnect switch S1 structure, associated with autotransformer T1, failed and created a three-phase fault. The fault-generated forces destroyed the disconnect switch and surge arrester cubicles. Transformer T1's primary and secondary differential relaying operated to energize the primary and secondary lockout relays to clear the fault.

Analysis of the DFR Records Due to the large size of the substation, scattered phase current traces from autotransformers T1, T2, T3, and T4 are monitored and will be used for the 13.8-kV fault analysis. Figure 6.45 shows the monitored currents for transformers T1, T2, T3, and T4 and 115-kV bus voltages at substation X. The DFR record shown in Fig. 6.46 reveals that the phase A current for T1 cleared in 5.5 cycles, confirming the tripping of the 230- and 115-kV circuit breakers. Figures 6.47 and 6.48 indicating the involvement of phases A, B, and C in supplying the 13.8-kV fault. This confirms the occurrence of a three-phase fault with symmetrical phase A current and asymmetrical currents for phases B and C. The presence of dc offset on phases B and C

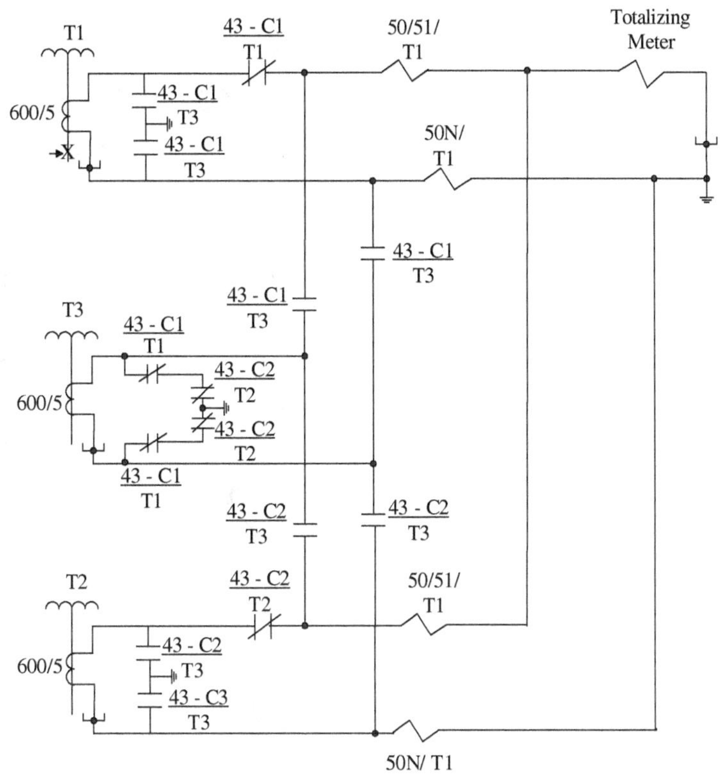

Fig. 6.44 CT switching scheme with three additional grounding points that will be used only during the shorting of CTs.

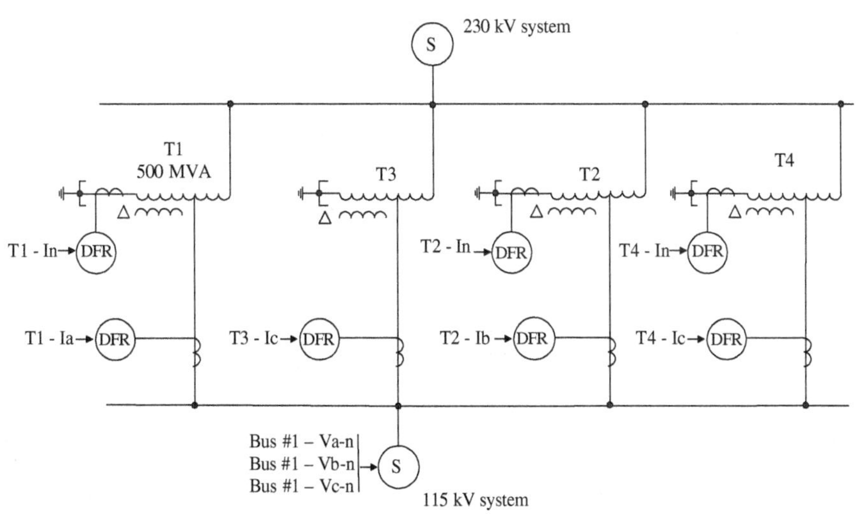

Fig. 6.45 Substation X one-line diagram showing 115-kV DFR-monitored voltages and currents.

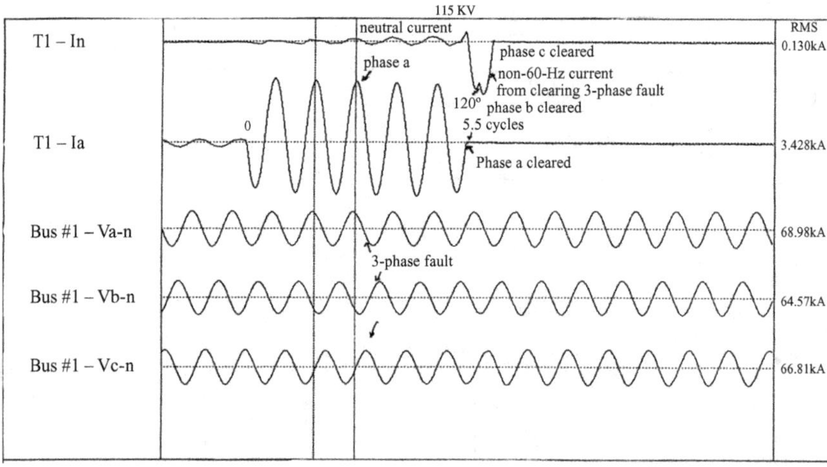

Fig. 6.46 Substation X DFR record showing transformer T1 115-kV phase A current feeding the 13.8-kV fault.

resulted in a neutral current flow on the 115-kV side of transformer T1 as shown in the DFR record of trace T1-In in Fig. 6.46. The DFR record of Fig. 6.47 illustrates trace T2-Ia feeding the three-phase fault for a duration of 5.5 cycles. The DFR record of Fig. 6.48 illustrates trace T4-Ic feeding the three-phase fault for an additional 11 ms. (240°) beyond the 5.5-cycle clearing of phase A. The DFR record illustrates no visible voltage drop during the 13.8-kV fault, implying the high-resistance nature of the fault.

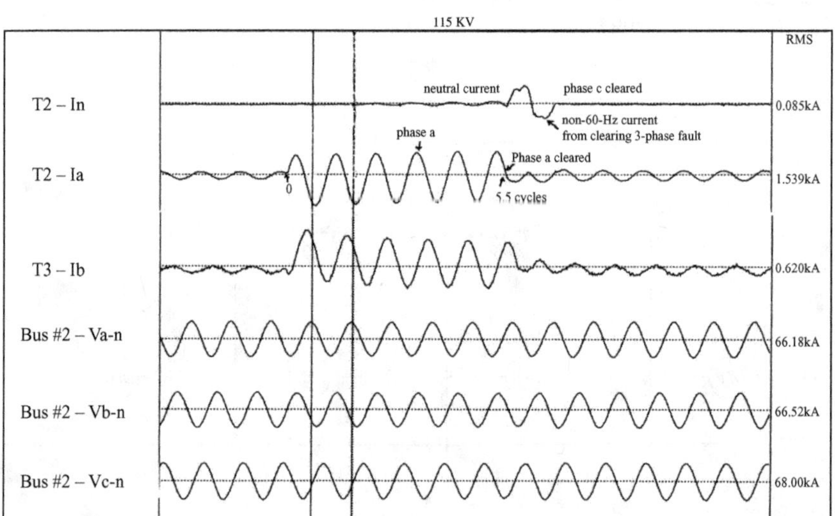

Fig. 6.47 Substation X DFR record for transformer T2 and T3 currents feeding the 13.8-kV fault.

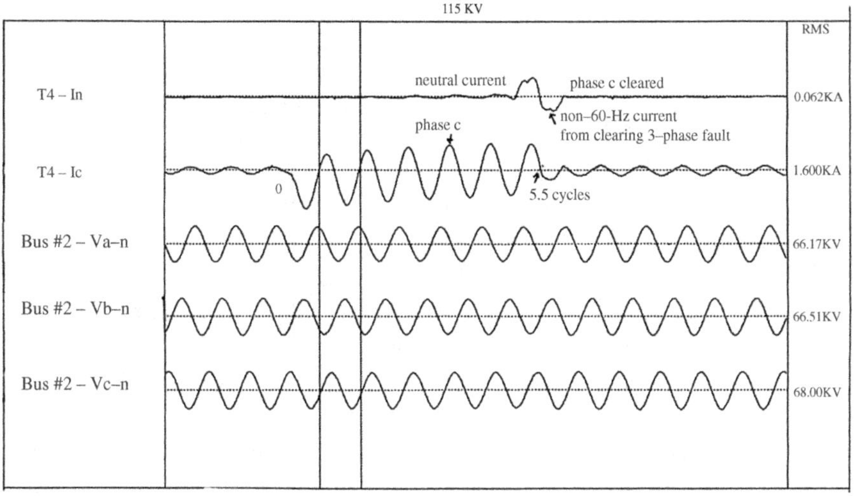

Fig. 6.48 Substation X DFR record showing transformer T4 115-kV phase C current feeding the 13.8-kV fault.

The relaying targets indicated the operation of all three primary differential relays 87TP1-A, B, C and all three secondary differential relays 87TS1-A, B, C. The differential relays energized their respective lockout relays to trip and lock out the T1 high-side 230-kV circuit breakers A1 and A2 and the T1 low-side 115-kV CBs B1 and B2 and 13.8-kV CB C1.

Undesired Operation of the T2 13.8-kV Feeder Ground Relay Switchgear ACB C2 in the switchyard service building that houses 13.8-kV switchgear over-tripped during the fault. The undesired operation of the ground relay associated with ACB C2 was due to its CT being grounded, both in the relay room and at the CTs of the 13.8-kV tertiary lead of transformer T2. The additional ground at point c in Figs. 6.49 and 6.50 has caused a current split and provided an alternate return for ground current back through the ACB C2 ground relay 50N/T2. The residual current was caused by the clearing of the three-phase fault where 120° exists between the zero crossings of the phases, as shown in trace T1-In in Fig. 6.46. In addition, the heavy three-phase fault current that occurred ahead of the current-limiting reactor has apparently caused an uneven saturation of 13.8-kV CTs, resulting in a residual neutral current flow. The residual current has also caused the operation of instantaneous neutral relay device 50N/T1 associated with the T1 13.8-kV lead.

Neutral Current Distribution (Split) Calculation for the T2 13.8-kV Feeder Ground Relay The ground element 50/51 used for protection of the 13.8-kV feeders L1 for T1 and L2 for T2 is an electromechanical relay with an inverse time overcurrent (TOC) and instantaneous elements.

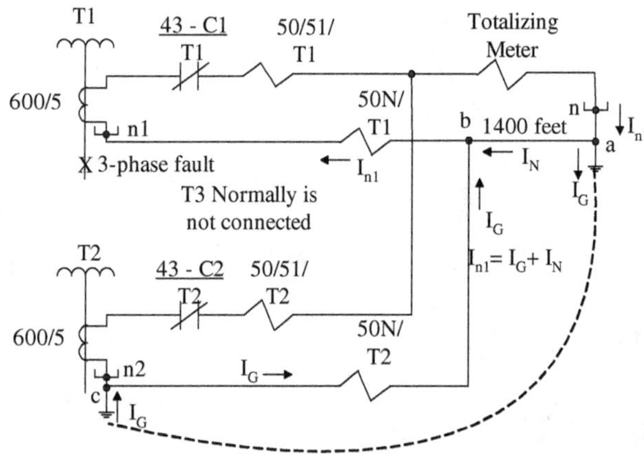

Fig. 6.49 Neutral current split due to an additional ground point at the T2-associated CT.

The time overcurrent (TOC) element is set at 1 A and time dial at 0.5, and the instantaneous element is set at 10 A. The relay burden for the TOC element is 2.38 VA at 0.5 A. The burden for the instantaneous IOC is 4.5 VA at 10 A. Since $S = I^2 \times Z = V^2/Z$,

$$Z = \frac{S}{I^2} = \frac{2.38}{0.5 \times 0.5} + \frac{4.5}{10 \times 10} = 9.55\,\Omega$$

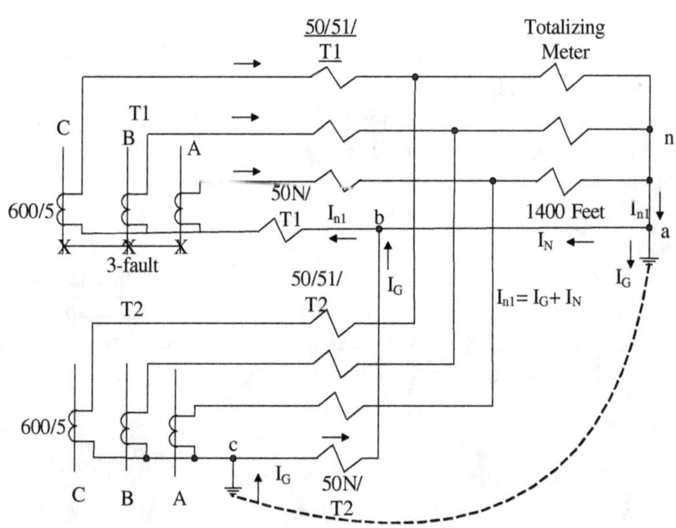

Fig. 6.50 Three-line ac diagram showing current split due to extra ground at CTs associated with transformer T2.

Assume that the relay burden is resistive. The TOC relay is located at the 13.8-kV switchgear room, while the totaling meter device is also located at the control room. The distance between the two devices is about 1400 feet. This distance is shown in Figs. 6.49 and 6.50 between points a and b. Assuming a CT cable resistance of 1 Ω per 1000 feet, the cable resistance is 1.4 Ω.

Now the neutral current I_{n1} flowing out of the neutral n will split at point a to the 1.4-Ω branch as I_n and to the other branch, which consists of current I_g going to the ground at a and then out of ground at the extra ground at point c. The current continues to flow through the 50/51/T2 relay element and then back to point b. The distribution factor for the relay current = 1.4/(1.4 + 9.55) = 0.128 pu. The neutral current I_n is equal to 11,000 A as calculated from the residual trace, and the relay current = 0.128 × 11,000 = 1406 A. The relay CT ratio is 120, So the instantaneous relay primary pickup is 120 × 10 = 1200 A. The calculated current is 1406 A, which is greater than the instantaneous relay pickup setting of 1200 A. This calculation confirms the undesired operation of 50/51N/T2, which was due primarily to the extra ground applied at the CT associated with the T2 13.8-kV tertiary lead.

Fault Calculation to Confirm Fault Location The 13.8-kV tertiary lead three-phase fault is fed mainly from the 230- and the 115-kV sides of autotransformer T1. The 13.8-kV tertiary lead current is not monitored by the DFR, due to the priority of monitoring the numerous more important high-voltage analog channels. Therefore, the 13.8-kV lead current must be calculated. Prior to performing fault current calculations, the system model needs to be verified using DFR-monitored voltages and currents.

Using a short-circuit study, a lead three-phase fault gives a current of 52,474 A on the 13.8-kV side. The study gives the 230-kV contribution as 4764 A and the 115-kV contribution as 3505 A. Converting this current to source impedance yields

$$I_{\text{base}} = \frac{\text{MVA} \times 10^6}{(\sqrt{3} \times V_{\text{L-L}} \times 10^3)} = 4183.8\ \text{A on 13.8-kV and 100-MVA bases}$$

$$I_{\text{pu}} = \frac{I_{\text{actual}}}{I_{\text{base}}} = \frac{52{,}474}{4183.8} = 12.54\ \text{pu}$$

$$X_{\text{pu}} = \frac{1}{I_{\text{pu}}} = \frac{1}{12.54} = j0.08\ \text{pu}$$

Using a 0.25-Ω current-limiting reactor, we first get the base impedance:

$$X_{\text{base}} = \frac{(\text{kV})^2}{\text{MVA}} = \frac{(13.8)^2}{100} = 1.904\ \Omega$$

$$X = \frac{j0.25}{1.904} = j0.131\ \text{pu}$$

Place a three-phase fault beyond the reactor; then

$$X_{\text{tot}} = j0.131 + j0.0797 = j0.2107 \text{ pu}$$

$$I_{\text{pu}} = \frac{1}{0.2107} = 4.745 \text{ pu}$$

$$= 4.745 \times 4183.8 = 19{,}854 \text{ A}$$

The

$$115\text{-kV contribution} = 19.854 \times \text{distribution factor}$$

$$= 19{,}854 \times \frac{3505 \text{ A}}{52{,}474 \text{ A}} = 1326 \text{ A}$$

Since the recorded T1 115-kV contribution on phase A (shown in the DFR record of Fig. 6.46) is 3428 A, the fault cannot be located after the current-limiting reactor. Therefore, the fault occurred in the disconnect switch enclosure between the 0.25-Ω current-limiting reactor and the T1 transformer tertiary winding.

Confirmation of the Short-Circuit Model Analysis of the DFR records should always be performed to match the recorded currents and voltages with outputs of fault simulation short-circuit studies. The simulation was used to locate the switchgear three-phase fault. The heavy fault current of 52.5 kA has provided an explanation of the violent switchgear damage. A short-circuit study simulation is normally carried out to compare recorded (measured) fault currents versus calculated values. For this case study the voltages and current values obtained from Figs. 6.47, 6.48, and 6.51 as well as the values calculated using short-circuit study simulation are tabulated and analyzed in Table 6.3. The percentage errors shown in the table are all less than 10%, which confirms that the power system model is adequate for power system design studies and relay setting calculations.

System Phenomena

1. Nonsinusoidal neutral currents as a result of clearing the three-phase fault, as in Figs. 6.46, 6.47 and 6.48 for all neutral current traces. The nonsinusoidal residual currents are caused by clearing the three-phase fault where 120° exists between the zero crossings of the phase currents.
2. Heavy three-phase fault currents on the delta tertiary lead where uneven saturation of three CTs associated with the lead may cause a false neutral current as shown in trace T1-In in Fig. 6.46.

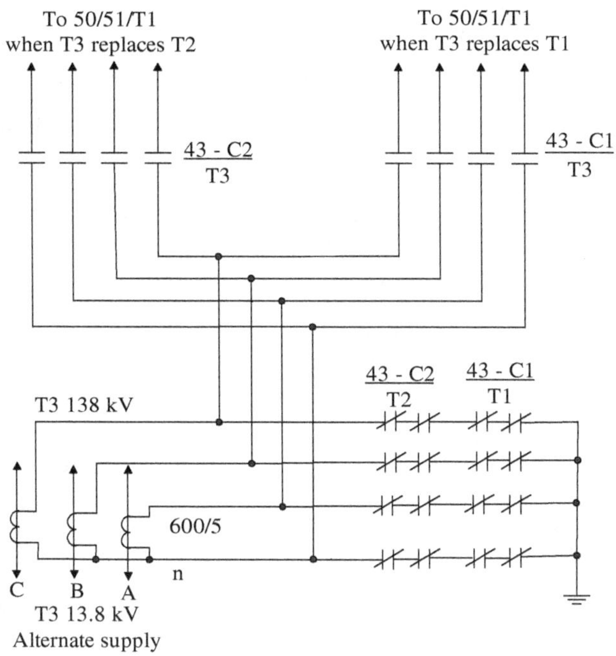

Fig. 6.51 CT switching scheme to substitute T3 for either transformer T1 or T2.

Corrective Actions

1. Remove the extra ground from the CT neutral for the autotransformer 2 13.8-kV tertiary lead as shown in Fig. 6.52.
2. Remove the instantaneous neutral relay from both T1 and T2 13.8-kV leads. As shown in Fig. 6.42, the zone of protection of this relay is very limited, due to the addition of a grounding transformer to the 13.8-kV delta supply for each of the 13.8-kV station service buses with a 10-Ω neutral resistor. The availability of maximum ground fault current is therefore limited to the value $V_{\text{l-n}}/R = 13{,}800/\sqrt{3} \times 10 = 800$ A. The relay setting of 1200 A was also not proper, as it exceeds the available ground fault current of 800 A.

TABLE 6.3 Confirmation of the Short-Circuit Model

Monitored Analog Trace	Measured	Calculated	Percent Error[a]
T1 : 115-kV phase A current	3428 A	3505 A	2.2%
T3 : 115-kV phase C current	620 A	635 A	2.4%
T2 : 115-kV phase A current	1539 A	1618 A	5.1%
T4 : 115-kV phase C current	1600 A	1630 A	1.9%
230-kV phase b-n bus voltage	116.4 kV	115.2 kV	1.0%

[a]% Error = [(measured − calculated)/measured] × 100.

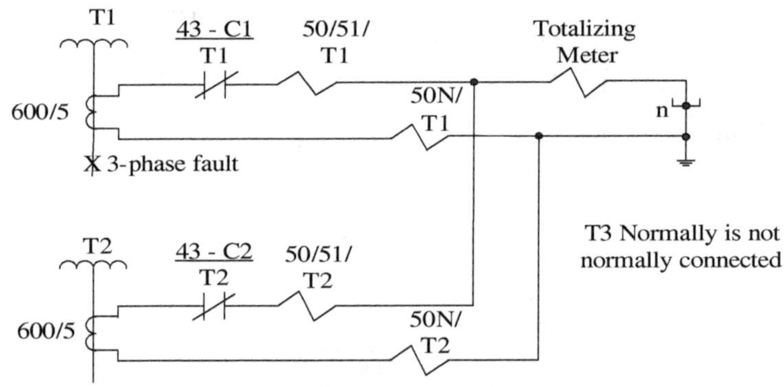

Fig. 6.52 13.8-kV normal feed from transformers T1 and T2 with correct single grounding point at the relay panel.

3. Evaluate the entire station service supply system to remove the CT switching scheme and correct all the existing deficiencies.

Lessons Learned

1. Avoid switching current transformer inputs to relay systems. The extra ground missed originally due to the sharing of relays between two separate CT circuits. Transformer 3 is used as a substitute station service supply to either transformer 1 or 2. In addition, the three-line ac connection was complicated by paralleling T1 and T2 CT circuits to sum their inputs to a common revenue metering device, as shown in Figs. 6.49 and 6.50.
2. Review the application of protective devices for their zones of protection and settings prior to their use. Instantaneous relay devices 50N/T1 and 50N/T2 were set above the maximum available L-g fault current of 800 A, and their zones of protection was limited.
3. Review the application of 13.8-kV CTs and their connected burdens to ensure that the available fault current will not cause CT saturation.

Case Study 6.11: Correction of a System Model Error from Analysis of a Failure of a Post Insulator Associated with a 115-kV Disconnect Switch

Abstract In this case study we illustrate how analysis of a mundane system operation with a successfully cleared ground fault on the 115-kV system can lead to a correction of a system modeling error. In addition, we describe the validation process of the short-circuit model and power system phenomena obtained from an analysis of the fault.

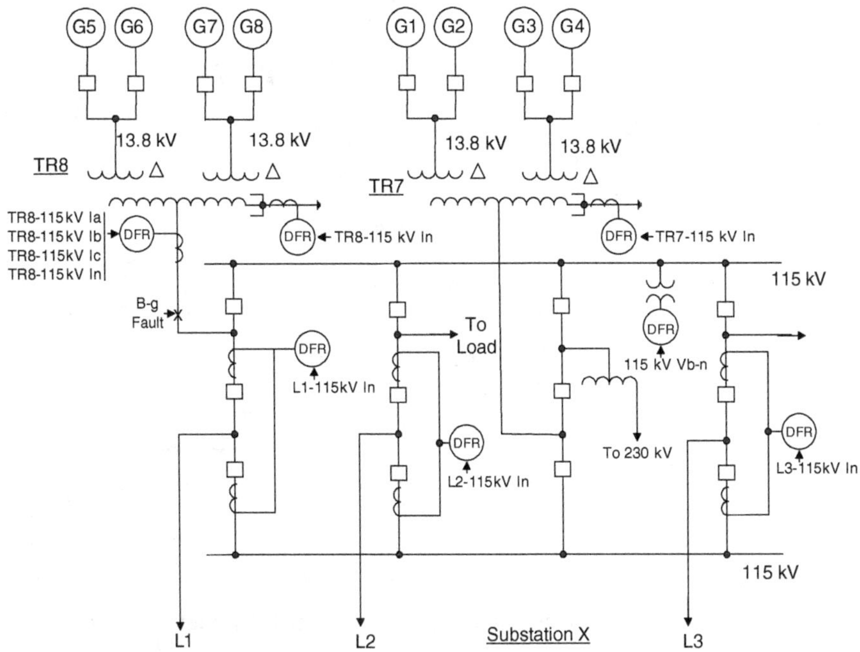

Fig. 6.53 Substation X one-line diagram showing DFR-monitored voltages and currents.

Description of the Power System and Associated Protection Systems As shown in Fig. 6.53, substation X is connected to the power system through 115-kV lines and autotransformers to the 230-kV system. In addition, eight hydro units of 50 MW each are connected to the substation through two three-winding transformers TR7 and TR8. Each transformer is a three-winding connected delta/delta/YG, where two hydro generating units are connected to each delta winding. A DFR is monitoring the transformer TR8 phase and neutral currents, 115-kV lines neutral currents, and 115-kV bus voltages. The DFR was triggered by the 115-kV ground fault and the DFR records are shown as part of the case study.

As shown in Fig. 6.54, the transformer differential zone includes the generator 13.8-kV breakers, the 13.8-kV ISO-phase bus, and the 115-kV cable connection to substation X. This facility was built in the early 1960s and only one overall differential relay was used for high-speed protection in conjunction with a neutral TOC backup system. In addition, sudden pressure relays are used to protect the transformer and trip for a bank internal fault.

Description of the Fault A post insulator supporting 115-kV phase B disconnect switch W, shown in Fig. 6.54, failed and flashed over, causing a solid phase-to-ground fault. Phases B and C of the harmonic-restraint electromechanical overall differential relays shown in Fig. 6.55 for a portion (two-CT input) of the simplified three-line ac diagram of transformer differential TR8 connection operated to clear the fault in

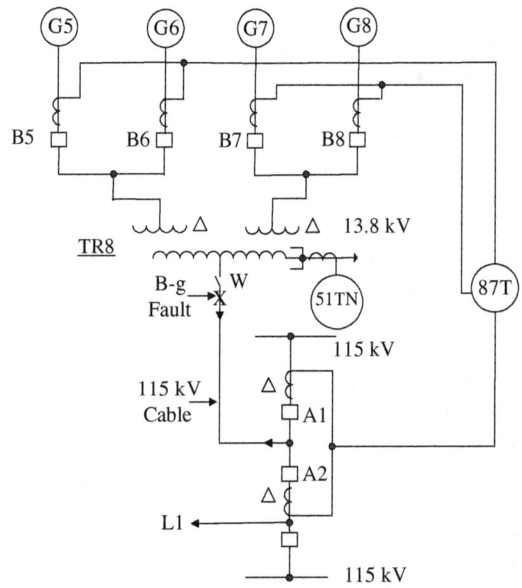

Fig. 6.54 One-line ac diagram showing transformer TR8 protection systems.

1 cycle. The differential relays energized the associated lockout relay to trip the 115-kV circuit breakers to clear the fault from the system in 4.5 cycles and the 13.8-kV circuit breakers to shut down the four hydro units and clear the fault in 8.5 cycles. The DFR record in Fig. 6.56 reveals that trace 115-kV Vb-n, representing phase B-g of the 115-kV bus voltage, collapsed to zero and the ground fault was cleared from the 115-kV system in 4.5 cycles at point b. Trace TR8-115-kV In of Fig. 6.56 indicates that between points b and c the fault is being fed only from the hydro units, due to their slower-opening breakers.

Sequence of Events As shown by the DFR record in Fig. 6.56, the B-g fault was cleared from the 115-kV system by tripping CBs E1 and E2 in 4.5 cycles, and the 13.8-kV generator breakers F1, F2, F3, and F4 tripped and cleared the fault in 8.5 cycles.

Classification of the Fault Using Differential Relay Targets For cases having no DFR records, the fault can be classified using differential relay targets only. Referring to Fig. 6.55, where a three-line ac diagram is shown for a portion (two-CT input) of the simplified unit transformer differential protection, the targets obtained for phases B and C can be traced back to a secondary loop that involves only the high-side (115-kV) CT of phase B only. The flow of secondary current on the high-side differential CT circuit is coupled only to the phase B primary. This concludes that the fault can be classified as B-g fault on the high-side grounded-wye 115-kV system. With the absence of transformer pressure relay targets, the fault has to be external to transformer TR8 and on the 115-kV high side within the differential zone.

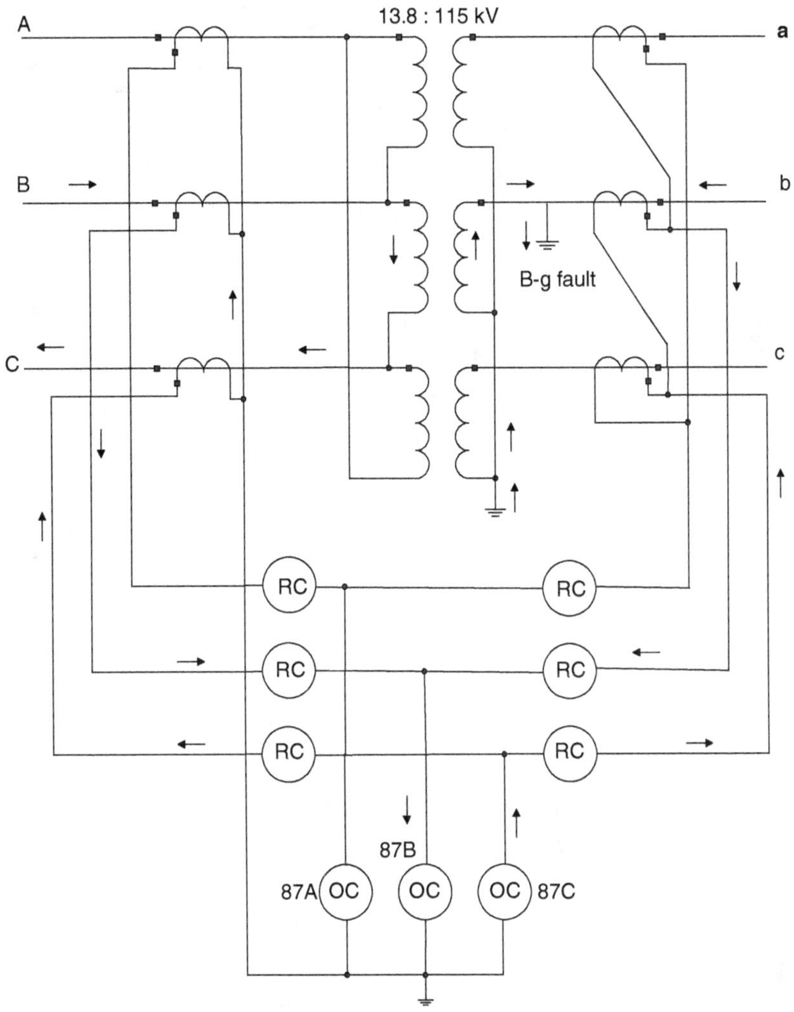

Fig. 6.55 Three-line ac diagram for a portion (two-CT input) of a simplified transformer differential TR8 connection.

Validation of the Power System Model for Short-Circuit Study The presence of symmetric fault current provides an ideal condition to validate the power system around the 115-kV faulted bus. Two power system conditions can be examined in this case study. The first condition is the initial fault being fed from all sources connected to the 115-kV system at the beginning of the fault, as shown in Fig. 6.56. The second condition occurred after removal of the fault from the 115-kV system after 4.5 cycles and continued feeding of the fault from the four hydro units only through TR8 for an additional 4 cycles, as shown in Fig. 6.57. The second configuration presents a simple stub model for four units and a three-winding transformer, as shown in Fig. 6.58.

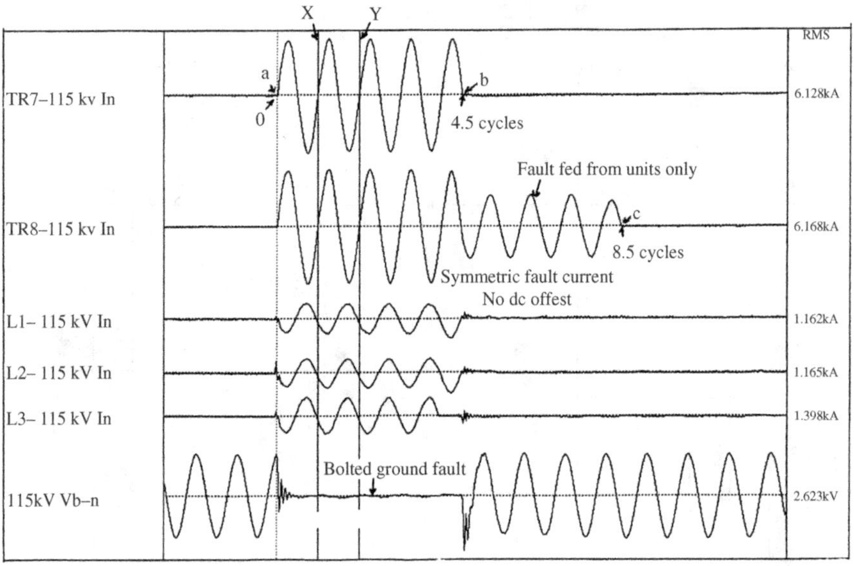

Fig. 6.56 Transformer TR8 neutral current feeding the 115-kV B-g bus fault.

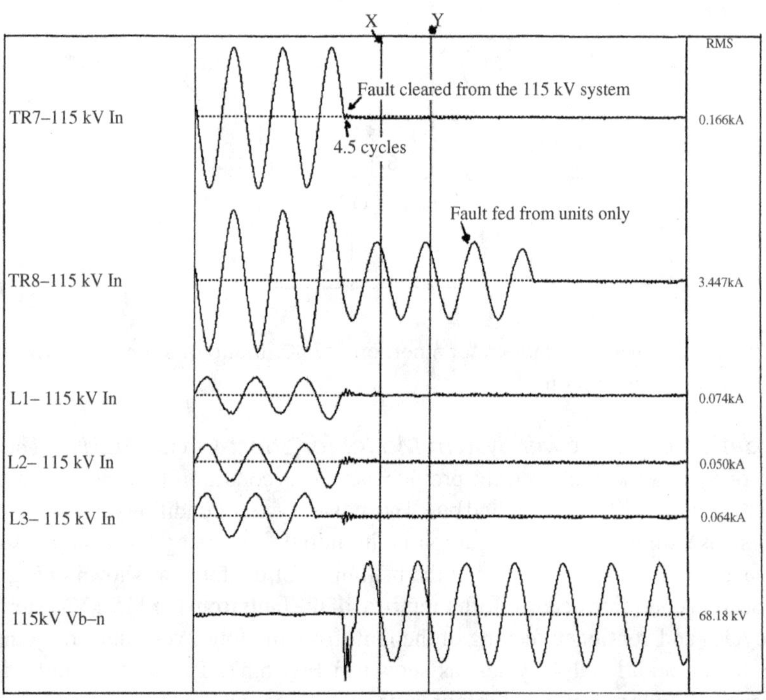

Fig. 6.57 Transformer TR8 neutral current RMS magnitude during the stub bus fault.

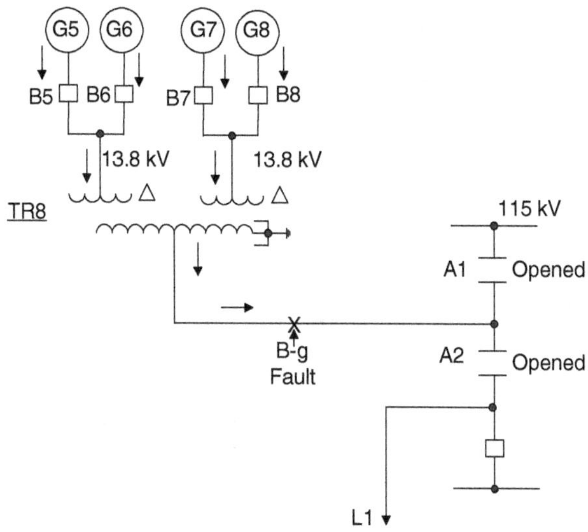

Fig. 6.58 B-g fault being fed from the units only after tripping of the 115-kV CBs.

The DFR records in Figs. 6.56 and 6.57 show two cursors, labeled X and Y, separated by a single fixed cycle (16.66 ms). These cursors move together freely along the DFR record to provide RMS calculation. The DFR software calculates an accurate RMS using discrete Fourier transform analysis for the samples contained within the 1-cycle window and stamping the right side of the DFR record with the current values calculated, producing RMS results as shown. These results are the measured RMS currents that will be used to validate the short-circuit power system model. Figure 6.56 shows the TR8 measured RMS neutral current as 6160 A when the B-g fault is being fed from the system and the four units are connected to transformer TR8. Figure 6.57 shows the TR8-measured RMS neutral current as 3447 A for the B-g fault being fed only (stub bus) from the four units connected to transformer TR8 after removal of the system source.

Comparison of measured and calculated currents obtained from simulating the power system involved in the fault reveals errors exceeding the allowable error of $\pm 10\%$. Table 6.4 indicates that the zero-sequence contribution from unit transformers TR7 and TR8 has errors, as do all the other 115-kV fault sources. However, comparison of the unit stub model also revealed errors, which eventually hinted at the modeling of the unit equivalent being the main source of errors. The equivalent provides an impedance lump for four hydro units and a three-winding transformer. The transformer has a dual delta low-voltage winding, with two hydro units bused together and connected to each of delta winding and a grounded-wye high-voltage winding. Upon reviewing the transformer model used in the study, an error was detected that caused the discrepancy in the results. The transformer nameplate had an error that was the main source of the difference between the currents. This resulted in an error of 591 A (17%), shown in Table 6.5, between the measured DFR

TABLE 6.4 Confirmation of the Short-Circuit Model for an Initial Fault Fed from All Sources[a]

Monitored Analog Trace	Measured	Calculated	Percent Error[b]
T7 : 115-kV neutral current	6128 A	4646 A	24
T8 : 115-kV neutral current	6160 A	4656 A	24
L1 : 115-kV neutral current	1162 A	716 A	38
L2 : 115-kV neutral current	1105 A	683 A	38
L3 : 115-kV neutral current	1390 A	809 A	41

[a]A short-circuit study simulation is normally done to compare recorded (measured) fault currents versus calculated values. For this case study the current values obtained from Fig. 6.57 and the results of short-circuit study are tabulated and analyzed.
[b]% Error = [(measured − calculated)/measured] × 100. The percentage errors are all above 10, which confirms that the power system model is inadequate for power system design studies and relay setting calculations.

of 3447 A and the calculated value of 2856 A for the stub model. The same fault condition was simulated after the model was corrected, producing a calculated current of 3390 A with an acceptable error of 57 A (1.7%). Table 6.6, also shows acceptable errors between measured and calculated currents after the system modeling, was corrected.

The error in modeling occurred when a patch-mode short-circuit study data format was converted to a new format suitable for a PC-based short-circuit program.

System Phenomena

1. Flashover of a failed insulator at the voltage peak that produces symmetrical fault currents with no dc offset as shown in the DFR record for trace L1-Vb in Fig. 6.57, which reveals an arc-over at the peak of the 115-kV phase B voltage.
2. Capacitive voltage transformer transients that occur when voltages collapse to near zero, forcing energy storage elements contained in the CVT device to

TABLE 6.5 Confirmation of the Short-Circuit Model for a Stub Fault Fed Only from the Four Generators Through Their Unit Transformer, Prior to Error Correction[a]

Monitored Analog Trace	Measured	Calculated	Percent Error[b]
TR8 : 115-kV neutral current	3447 A	2856 A	17

[a]For this model the RMS measured current value obtained from the DFR record in Fig. 6.58 for the transformer bank T8 stub configuration is compared with the value calculated from the results of a short-circuit study.
[b]% Error = [(measured − calculated)/measured] × 100. The percentage errors are all above 10, which confirms that the power system model is inadequate for power system design studies and relay setting calculations.

TABLE 6.6 Confirmation of the Short-Circuit Model After Correction of the Modeling Error

Monitored Analog Trace	Measured	Calculated	Percent Error[a]
T7 : 115-kV neutral current	6128 A	6173 A	−0.7
T8 : 115-kV neutral current	6160 A	6194 A	−0.6
L1 : 115-kV neutral current	1162 A	1190 A	−2.4
L2 : 115-kV neutral current	1105 A	1128 A	−2.1
L3 : 115-kV neutral current	1390 A	1420 A	−2.2

[a]The percentage error is within the allowable ±10%, indicating that the power system model is adequate for power system design studies and relay setting calculations.

generate a transient in the secondary voltage that does not mimic the primary system voltage. The DFR record shown for trace 115 kV Vb-n in Fig. 6.57 reveals a high-frequency transient on the phase B voltage on the order of about 400 Hz that lasted for about 6 ms.

Corrective Action The error encountered in the unit transformer model was corrected and a comprehensive analysis was done as shown in Table 6.6, with good matching results, confirming the validity of the model.

Lessons Learned

1. Review of mundane operations that results in successful fault clearing can reveal important information for a power system and associated protection systems. The feedback process can enhance the overall design and operation of the protection system.
2. Short-circuit data must be reviewed thoroughly, especially when software conversion is done.

Case Study 6.12: Location of a 345-kV Line Fault Protected by Electromechanical Distance Relays Using Information from a DFR Record

Abstract Fault location for the EHV systems is an important function and must be done as quickly as possible so that system inspection and corrective actions are taken to restore the system. Replacing existing electromechanical and static relays with numerical relays enhances the operation of the system by providing accurate fault location. In this case study we describe a procedure that was followed to locate a 345-kV phase-to-ground fault for a line that is protected by electromechanical distance relays. A DFR record from only one of the 345-kV line ends was used for the analysis. In addition, we examine fault location calculation accuracy and describe power system modeling, DFR analysis, and system phenomena.

Description of the Line Protection The line is protected with dual redundant pilot relaying systems using diverse communication media, dual dc sources, and separate CTs and VTs. One system is a permissive overreaching transfer trip (POTT) system using electromechanical distance relays and a power line carrier for communication; the second system is also a POTT pilot scheme using electromechanical distance relays over leased telephone circuits. As of the late 1990s, most EHV (345-kV) transmission lines are protected using electromechanical and/or solid-state relaying systems. In the course of time, all obsolete relays will be replaced by numerical protection systems. The lack of a fault location option (now available with numerical microprocessor-based relaying systems) for some of the present EHV relaying systems may slow down the system restoration task and make fault location using other methods a challenging task.

Description of the Fault Figure 6.59 shows the system one-line diagram and DFR-monitored currents and voltages. The DFR record in Fig. 6.60 reveals a phase B-g fault that occurred on the 345-kV line L1. The fault was caused by a tree split near the line right-of-way, resulting in some of the branches falling on the outside (B phase) conductor of line L1. The fault incident point is at the voltage peak, resulting in symmetric current with no dc offset. The fault was cleared at both ends of the line in 4 cycles. Terminal X of line L1 reclosed successfully at high speed in 33 cycles, while the other end, at substation Y, reclosed successfully in 10 s using the sync check option. The DFR records shown in Figs. 6.60 and 6.61, available only from substation X, were used to direct a helicopter to the fault to determine its cause, thus expediting the replacement of failed insulators.

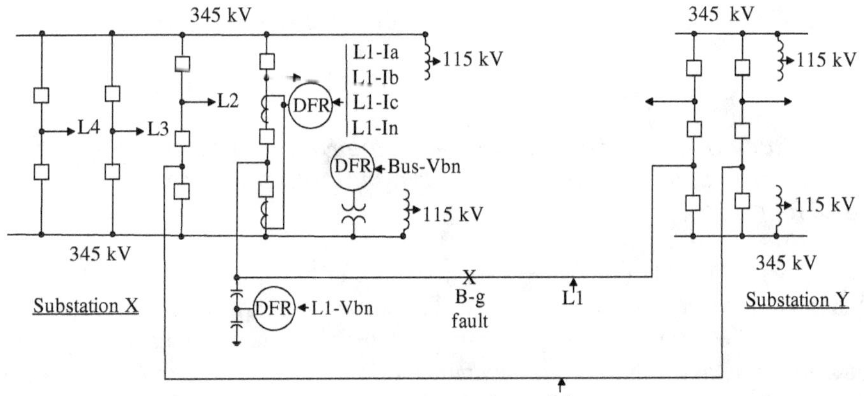

Fig. 6.59 345-kV system one-line diagram for substation X DFR-monitored voltages and currents.

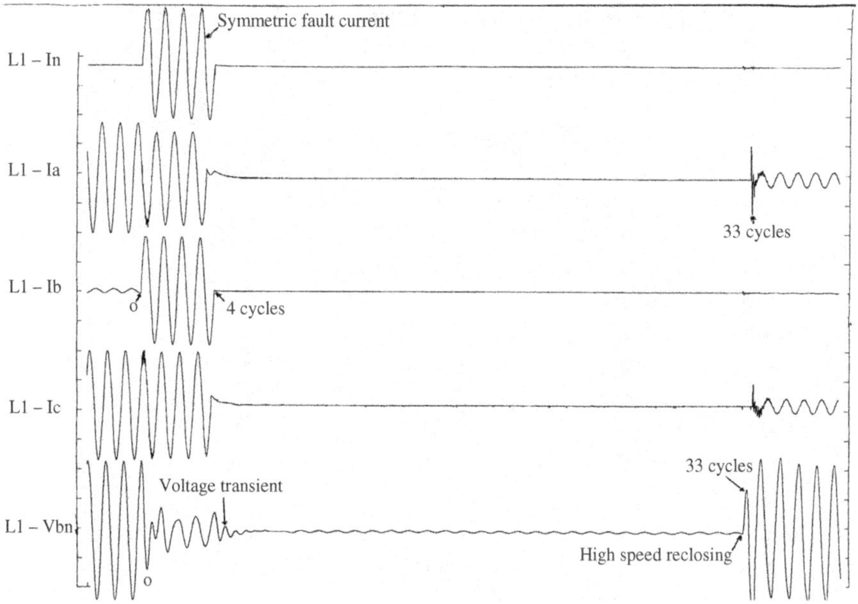

Fig. 6.60 Faulted line L1 currents and voltage during the initial L-g fault.

Location of the Fault Using DFR Information from One End Only Quite often, the location of 345-kV faults becomes important when restoring a line and to avoid the reoccurrence of tree faults, which may result in generation rejection to avoid line thermal overloads. This made the location of this fault important in

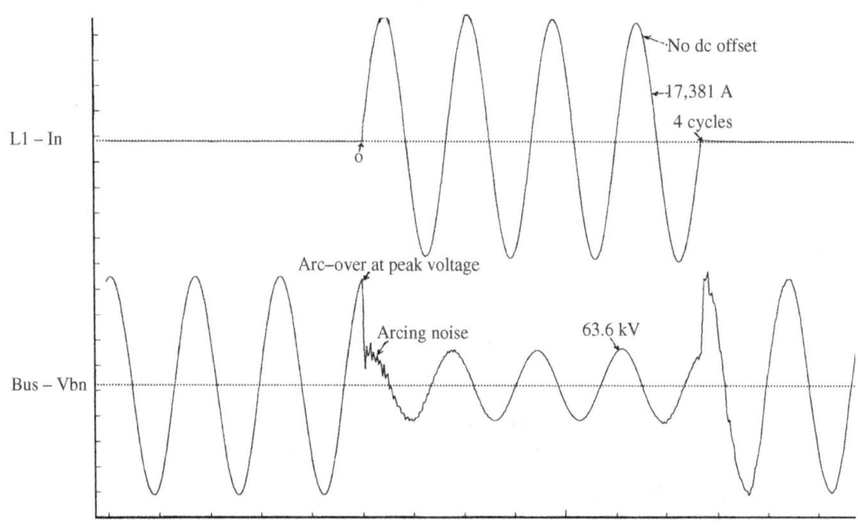

Fig. 6.61 Fault incident point occurring at the voltage peak.

assessment of the right-of-way tree trimming program. The following approach was used to first find the cause of the fault and then to locate it using DFR record from substation X:

1. Analysis of the DFR record reveals that the fault incident point coincided with the phase B voltage peak, resulting in a symmetric fault current with no dc offset. This may lead to the conclusion that a fault has occurred by a slow arc-over mechanism, due either to a grounded object contacting the line (tree) or to an insulator flashover. Contamination on line insulators combined with dew or fog can form a conductive surface, permitting flashover to the grounded tower. Since the fault occurred very early in the morning (5:00 A.M.) and fog was reported in the vicinity of the faulted line, the insulator flashover story also had to be considered in conjunction with the tree as the main cause of the fault.
2. A symmetric fault current is easy to simulate in short-circuit studies, which are based on steady-state 60-Hz modeling.
3. Using the DFR record, the recorded ground currents and faulted phase voltage for the faulted line L1 had to be compared with the results obtained from short-circuit study simulation.
4. The targets reported indicated operation of zone 1 (set at 80% of Z_{L1}) relays only at substation X of the L1 end, which places the fault within the 20% zone near substation X (area of nonoperation of zone 1 at the substation Y end of L1).
5. DFR phase currents should not be used for the matching process, due to the effect of load currents that cannot be matched easily by short-circuit simulations.
6. Using a short-circuit study that matches the system prior to the fault, the fault location on line L1 can be found by sliding the fault point within 20% of substation X until a close match is obtained when the errors are minimized between the actual recording and the simulation study results.
7. A small fault resistance in the path to ground can be modeled (if required) to account for arcing phenomena.
8. An average pre-fault voltage $[(V_a + V_b + V_c)/3]$ can be obtained from the DFR record and used as a base voltage in the short-circuit simulation study.
9. When all line neutral currents and ground sources at the substation are monitored by the DFR, a check can be made to see that all sources feeding the L1 B-g fault can add up to the total fault current. This will provide confirmation of the accuracy of the DFR-recorded currents.

Analysis of Results and Overall System Accuracy Locating the fault begins by obtaining RMS values for line L1 ground current and the faulted voltage. The RMS was calculated for the DFR-recorded currents and voltages using analysis software: moving a cursor on the screen to locate the peak value. The RMS results were confirmed using software that searches for the maximum sampled value within the

1-cycle interval window (from 0.5 to 1.5 cycles) of stored fault data and then divides the result by 1.414. DFR records were used before accurate Windows-based RMS calculations were available. The primary RMS results, shown in Fig. 6.61, are as follows:

1. The DFR record in Fig. 6.61 shows that the primary current of trace L1-In = 17,381 A.
2. The DFR record in Fig. 6.61 shows that the primary voltage of trace Bus-Vbn = 63,690 V.

The primary RMS results can be compared with the short-circuit study output.

Using short-circuit simulation and sliding the fault, the ground current match was established at a location 7% distant from terminal X. However, the faulted phase voltage calculated was 66,300 V. This will result in an error that is defined as follows:

$$\text{error (\%)} = \frac{\text{measured} - \text{calculated}}{\text{the smaller of the two} \times 100}$$

$$= \frac{(66{,}300 - 63{,}690) \times 100}{63{,}690} = 4.1\%$$

By sliding the fault, a match was established for the recorded phase B faulted voltage of 63,690 V 6.6% distant from terminal X. However, the calculated ground current was 17,683 A, which will result in

$$\text{error (\%)} = \frac{(17{,}683 - 17{,}381) \times 100}{17{,}381} = 1.7\%$$

Since the entire process is an educated guess to guide a helicopter in locating the fault, a range has to be established around the results. Due to the DFR and system simulation parameter inaccuracies, an additional error of +5% and −5% of line length will be added to the range above. This process will provide a zone that will include the actual fault location point. The lower end of the range can be calculated by subtracting 5% from the smaller result obtained above. The higher end of the range can be obtained by adding 5% of the line distance to the larger result. The lower end of the range will be 6.6 −5 = 1.6%, and the upper end of the range is 7 + 5 = 12%. The search for the location of the fault was directed for a zone of 1.6 to 12% of the line length.

The helicopter located the fault at a distance of 6.6 miles from terminal X of L1 line, where the fault was caused by a tree split near the line right-of-way. The line total length is 61.6 miles, which locates the fault at 6.6 × 100/61.6 = 10.7%, near the high end of the search range of 12%. This confirms that the fault location may be given as a range regardless of the method used to obtain the result.

An actual error in fault location can now be calculated using the result above that produces the least error. The normalized error is thus 10.7% − 6.6% = 4.1%, which is within the assumed 5% error.

Power System Phenomena

1. Flashover to ground at the voltage peak, resulting in a symmetrical fault current with no dc offset, as shown in trace Bus-Vbn Fig. 6.61
2. Successful high-speed reclosing in 33 cycles (0.55 s) as shown in trace L1-Vbn in Fig. 6.60

Corrective Actions

1. A helicopter was flown to patrol the line based on faulted phase identification and preliminary analysis of the fault location.
2. System modeling, including zero-sequence mutual coupling and data from neighboring utilities, was reviewed and corrections were made as needed.

Lessons Learned

1. Information such as relay targets (in this case, study Z1 operation at X only) provide needed input for fault location function to the slide fault simulation point within a defined range (0 to 20% from substation X).
2. System phenomena (fault occurrence at the voltage peak) may provide clues to the causes of the faults on the system (in this case the study fault was caused either by an insulator flashover or by a grounded object contacting the line).
3. High-speed reclosings should be implemented at both terminals of the line to help to keep the system together following successful high-speed clearing of temporary system faults. The use of high-speed reclosing at both ends of the line (when justified) will enhance the transient stability of the system.
4. Due to system modeling and DFR record inaccuracies, fault location calculation may be given as a range.
5. Use of numerical relays as a replacement for obsolete electromechanical and static relays will enhance operation of the system by providing accurate fault location.

Case Study 6.13: Location of an Outdoor 13.8-kV Switchgear Fault at a Cogeneration Facility Using a DFR Fault Record from a Remote Substation

Abstract A 100-MW combined-cycle plant owned by an independent power producer (IPP) is connected to substation X through a 115-kV transmission line, as shown in Fig. 6.62. The IPP plant was on cold shutdown and manned by a few

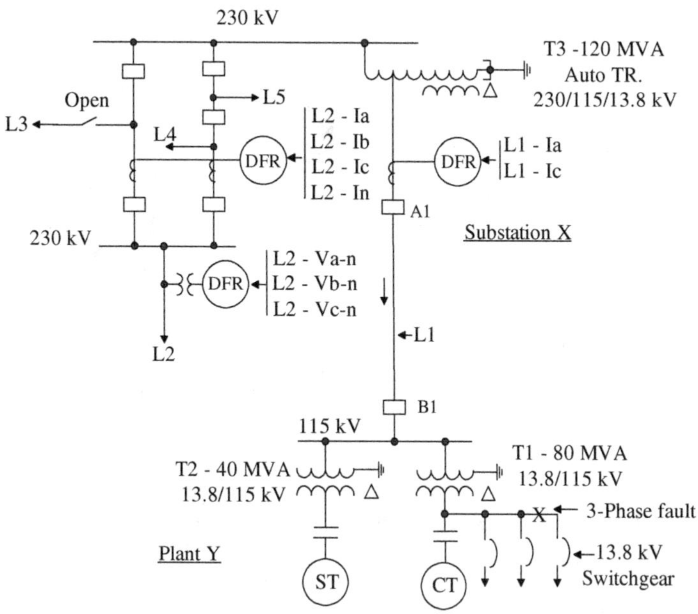

Fig. 6.62 System one-line diagram showing substation X DFR-monitored voltages and currents.

lightly experienced personnel. The plant has no fault recording or monitoring devices. Substation X is connected to the bulk power system and is equipped with a DFR device. The 115-kV connection to the IPP is used to backfeed the IPP facility supplying the station service loads. The station service source was lost and the IPP claimed that the loss of its station service source was caused by the connecting utility at substation X. The DFR at substation X was triggered and a record was captured showing a three-phase fault. The DFR record, in conjunction with no relaying targets or operation, especially for the 115-kV tie line L1, was used to analyze and locate the fault at the IPP facility. In this case study we show how a remote DFR record and short-circuit study simulation can be used to locate the fault at the IPP 13.8-kV switchgear.

Description of the 13.8-kV Fault After Confirmation of Its Location The three-phase fault apparently was caused by the failure of a seal on one of the cable entries to the outdoor 13.8-kV switchgear. This permitted rain to get into the switchgear cubicle and caused the three-phase fault. Based on the DFR record shown in Fig. 6.63 at substation X, the fault lasted for 4.5 cycles. Inspection of the switchgear revealed burn marks as evidence of arcing from the fault.

Information Available for Analysis of the Fault The IPP facility is equipped with no real-time monitoring equipment. In addition, when problems occur at the IPP facility, no technical staff exists to troubleshoot the problem, so the connecting

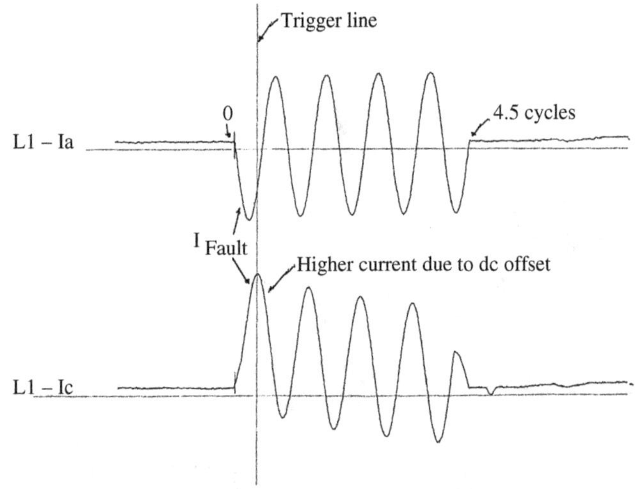

Fig. 6.63 Substation X DFR record for line L1 phase A and C current contributions to the 13.8-kV bus fault at plant Y.

utility is obligated to help the IPP. Therefore, the DFR record at substation X of the connecting utility, where as shown in Fig. 6.62, line L2 three-phase voltages and currents as well as phase A and C currents for line L1 will be utilized to assist the IPP in locating the fault at its facility. The connecting utility analysis was also needed to suppress the IPP allegation that loss of the station service power is caused by the operation of protection and control equipment owned by the connecting utility.

Procedure Used to Locate the 13.8-kV Fault at the IPP Facility The following procedure is used to locate the fault at the IPP facility:

1. Since 115-kV circuit breaker A1 at substation X did not trip during the incident, it can be assumed that clearing the fault had to be done by the circuit breakers at the IPP facility.
2. The DFR recorded fault current that is close to being symmetrical and that contains the fewest dc offset components should be selected for analysis.
3. Since load current components affect the phase current during three-phase and phase-to-phase faults, it can be assumed that complete matching cannot be expected between DFR measured versus calculated short-circuit study outputs.
4. DFR feeder currents that are symmetrical and have no dc offset components or no load flow should be selected for analysis.
5. The DFR-monitored currents and voltages in Figs. 6.63, 6.64, and 6.65 should be analyzed to classify the fault type; in our case study it is a three-phase fault.

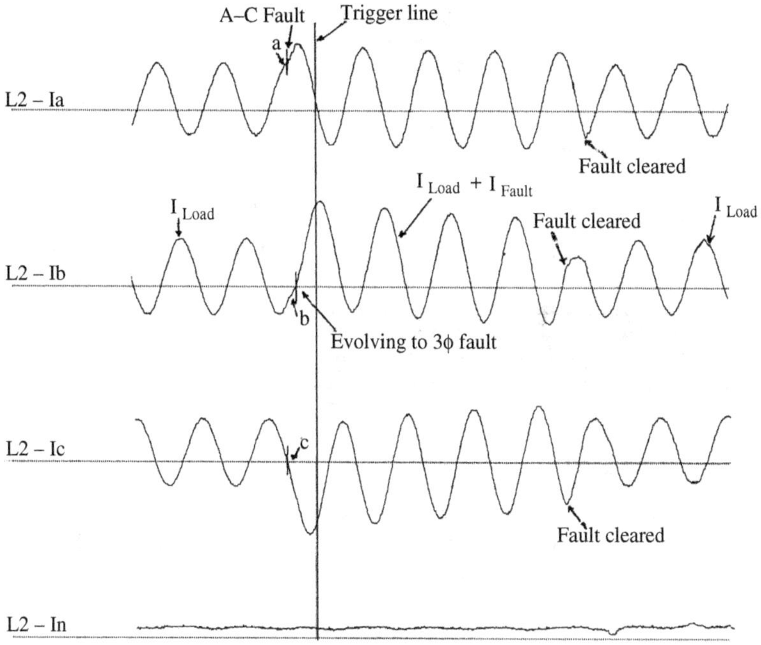

Fig. 6.64 Substation X DFR record for line L2 currents during the 13.8-kV bus fault at plant Y.

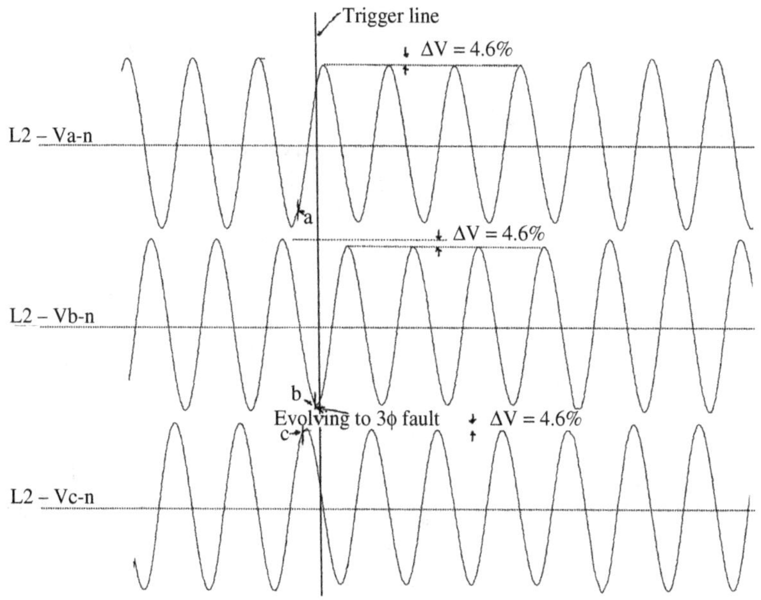

Fig. 6.65 Substation X DFR record for line L2 voltages during the 13.8-kV bus fault at plant Y.

TABLE 6.7 Comparison Between Measured and Calculated Fault Currents

Monitored Analog Trace	Measured	Calculated	Percent Error[a]
115-kV L1 phase A current	980.5 A	944 A	3.9% < 10%—O.K.
115-kV L1 phase C current	1220 A	944 A	29% > 10%
230-kV L2 phase A current	173 A	172 A	0%—O.K.
230-kV L2 phase B current	206 A	172 A	19% > 10%
230-kV L2 phase C current	179 A	172 A	4% < 10%—O.K.
L2 phase A-n voltage	123.7 kV	128 kV	0.35% < 10%—O.K.

[a]The asymmetrical phase B and C currents provide high errors due to the presence of dc offset. The percent error is defined as [(measured − calculated)/the smaller of the two] × 100. Based on the error results, the values calculated, which are symmetrical, provide a complete match with the DFR-measured symmetrical currents and voltages.

6. A short-circuit study with the deduced fault type of step 5 and fault location should be simulated to compare the measured versus the calculated currents and voltages using an error range of ±10%.

Comparison of the DFR Record and the Simulated Short-Circuit Study

Using a PC-based short-circuit study, a three-phase fault was simulated at the IPP 13.8-kV bus. A short-circuit study simulation is normally carried out to compare recorded (measured) fault currents versus calculated values. Table 6.7 compares the measured (DFR) records obtained from Figs. 6.63, 6.64, and 6.65 versus currents and voltages calculated from the short-circuit study results in Fig. 6.66.

Confirmation of the Fault Location The 13.8-kV switchgear three-phase fault location was confirmed by the IPP personnel, and it was attributed to water penetration of the outdoor 13.8-kV switchgear during heavy rain.

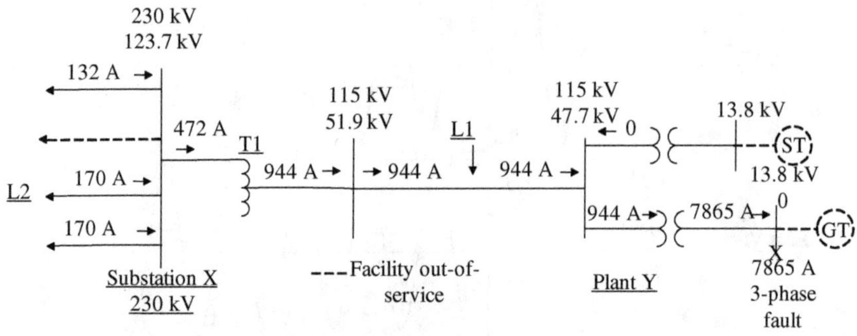

Fig. 6.66 Short-circuit study results for a three-phase fault at the 13.8-kV switchgear bus of plant Y.

Lessons Learned

1. Analysis of a mundane DFR record can provide a wealth of information that can be used to locate faults on the system, especially for plants and substations where no monitoring equipment is installed.
2. Partial DFR coverage of the power system is important and can enhance system disturbance analysis.
3. DFR traces should be adjusted to have their traces coincide with the axis of symmetry.

Case Study 6.14: Breakage (Failure) of a 345-kV Subconductor Bundle During a High-Resistance Tree Fault, Due to the Heavily Loaded Line Sagging to a Tree

Abstract A high-resistance tree phase A-g fault occurred on the L1 345-kV circuit during an ice or snow storm, coupled with high wind. The initial phase-to-ground fault was cleared successfully and followed by two automatic reclosing attempts which reestablished the initial fault. During the last fault, which lasted about 3.5 s, one of the subconductors of the wire bundle broke and fell to the ground. In this case study we illustrate the importance of a ground time overcurrent element for the protection of HV and EHV transmission lines. Normally, the ground time overcurrent element provides protection against high-resistance (tree) ground faults and open-phase conditions. We also highlight system phenomena, describe a procedure to estimate the value of fault resistance based on the reported location of the fault, and provide an analysis of conductor breakage during the fault.

Description of the 345-kV Line Protection Systems The 345-kV transmission line L1, shown in Fig. 6.67, is protected by dual pilot protection systems applied using diverse communication media. The design of the 345-kV relaying system was done in the early 1980s, when microprocessor-based relays were not commercially available. Therefore, the dual relaying systems were designed based on electromechanical and solid-state distance relays. The first system is a permissive overreaching transfer trip pilot system (POTT) using a solid-state distance-based relaying package applied on power line carrier communication media. The second system is also a permissive overreaching transfer trip system utilizing electromechanical distance-based relays applied on leased telephone circuits. The application of dual POTT schemes using diverse communication media offers the best compromise between dependability and security features of the line relaying systems.

Sequence of Events Figure 6.67 shows the system one-line diagram and DFR-monitored voltages and currents. The sequence of events is constructed from an analysis of the DFR records shown in Figs. 6.68, 6.69, 6.70, 6.71, and 6.72 and the sequence-of-events systems at substation X and the energy management system operator report at substation Y.

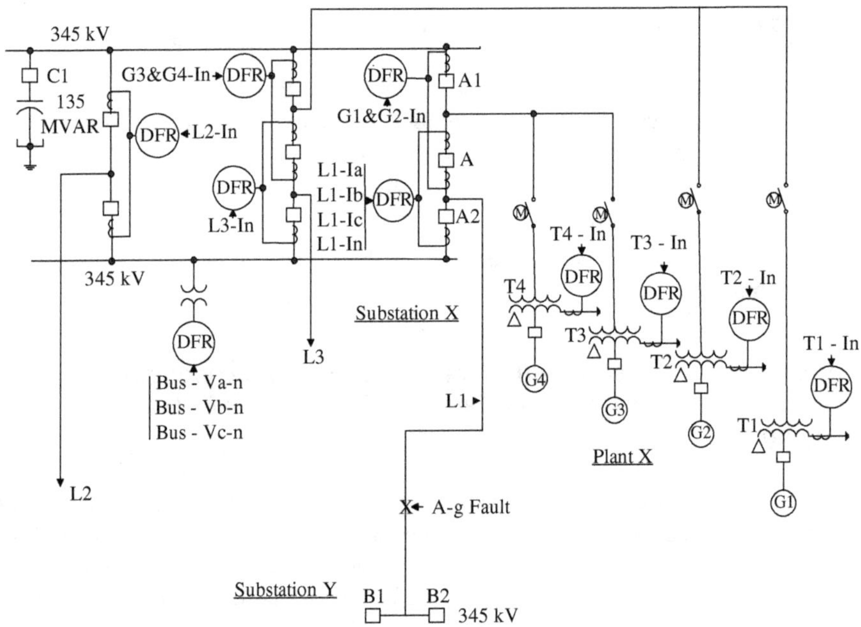

Fig. 6.67 One-line diagram showing DFR-monitored voltages and currents for substation X.

Fault 1, Occurring at $t_0 = 0$ The DFR record in Fig. 6.68 shows a phase A-g fault occurring on the 345-kV line L1. The fault was due to a tree that contacted the line under high wind conditions. The voltage trace Bus-Va-n reveals that the fault incident point is at the voltage peak, indicating a slow fault-creation mechanism with arc-over at the highest electric stress instant. This incident point will be marked as the reference $t_0 = 0$. Six and a half cycles later, CBs B1 and B2 at substation Y tripped, as indicated by an interruption of the phase B loading current trace L1-Ib and an increase in the neutral current L1-In. Ten cycles after t_0, the CBs A and A2 at substation X tripped and cleared the fault. The clearing at substation X can be considered sequential when it occurs after a time delay of several cycles from the clearing of the remote end of the line.

The DFR record shown in trace L1-In in Fig. 6.69 reveals an increase in ground current from an initial value of 1673 A to a value of 4006 A when the remote end of line L1 clears at substation Y. All ground current traces shown in Figs. 6.69 and 6.70 reveal a gradually increasing current at the beginning of the fault, thus indicating the high-resistance nature of the tree fault when ground currents are seeking ground returns to their sources. Twenty seconds after t_0, circuit breaker B1 at substation Y reclosed successfully automatically. Thirty-seven seconds after t_0, circuit breaker A at substation X reclosed successfully automatically via a sync check relay.

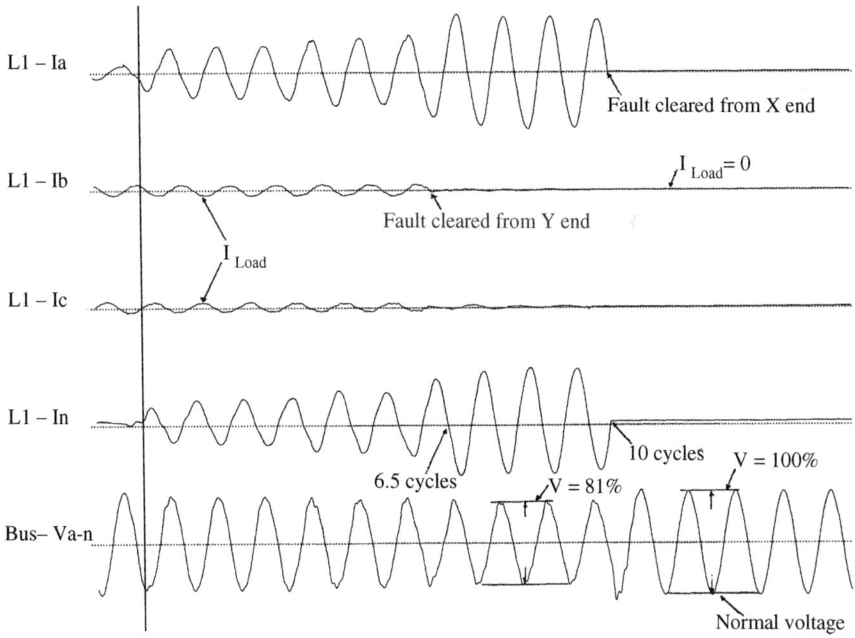

Fig. 6.68 Substation X DFR record showing local and remote clearing during line L1 first A-g fault.

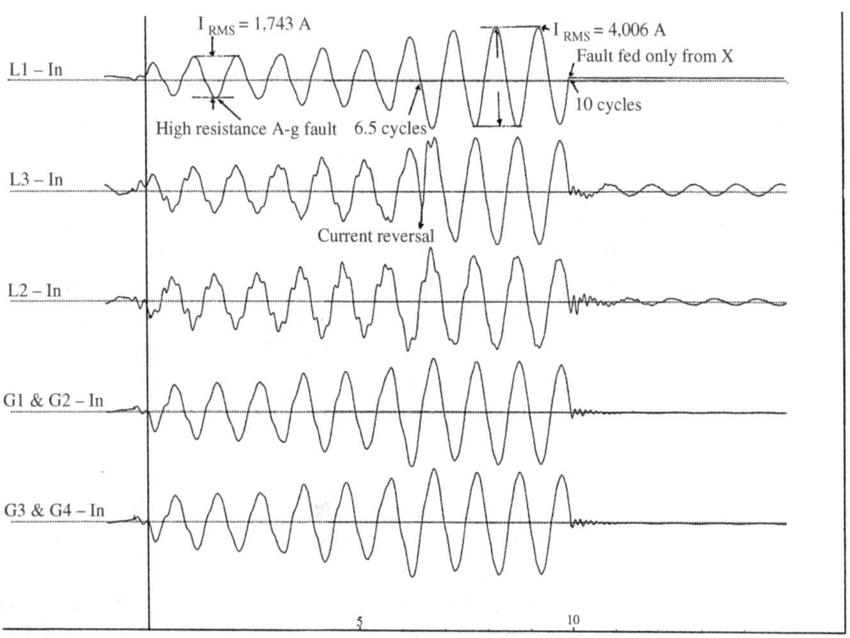

Fig. 6.69 Substation X DFR record showing neutral currents feeding the first A-g fault.

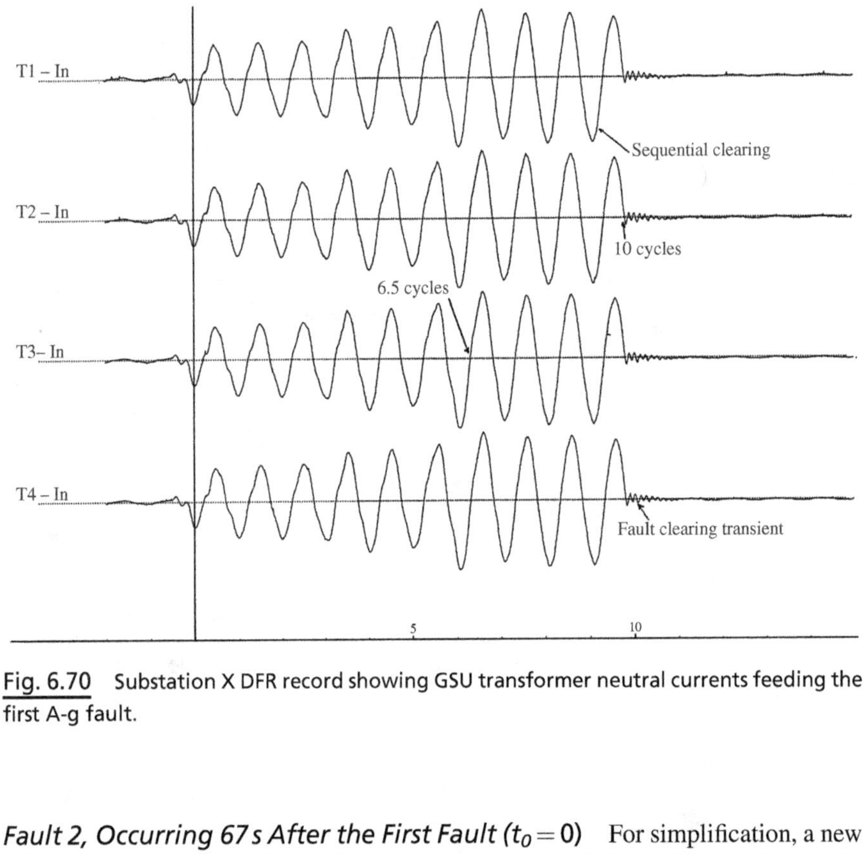

Fig. 6.70 Substation X DFR record showing GSU transformer neutral currents feeding the first A-g fault.

Fault 2, Occurring 67 s After the First Fault ($t_0 = 0$) For simplification, a new reference time point $t_1 = 67$ s will be used to analyze fault 2. The DFR record shown in Fig. 6.71 illustrates that the early phase A-g is reestablished again at the voltage peak, resulting in symmetrical fault currents with no dc offset components. Ground current trace L1-In shown in Fig. 6.71 reveals a gradually increasing current at the beginning of the fault, again indicating the high-resistance nature of the tree fault. Circuit breaker B1 at substation Y tripped 12.5 cycles after reference point t_1. Trace L1-In shown in Fig. 6.71 also reveals an increase in ground current from an initial value of 1680 A to a value of 3067 A when the remote end of line L1 clears at substation Y. At 43 cycles after t_1, circuit breaker A at substation X tripped and cleared the fault. The high-resistance fault was cleared only by a ground time overcurrent relay element in 39 cycles.

Twenty seconds after t_1, circuit breaker B1 at substation Y reclosed successfully automatically. The reclosing sequence was reset after the first fault when the breaker stayed in the closed position beyond the reclosing reset time. The high-resistance fault was cleared only by ground time overcurrent relay element in 39 cycles. Circuit breaker A at substation X reclosed successfully automatically via the sync check 37 seconds after t_1. The automatic reclosing circuit was also reset when the CB A stayed in the closed position longer than the reset time.

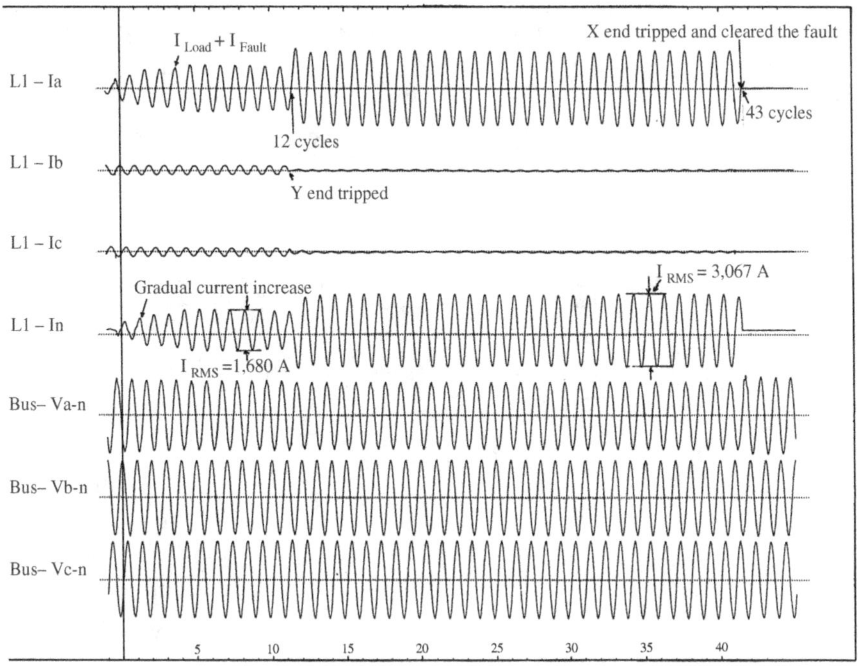

Fig. 6.71 Substation X DFR record showing local and remote clearing during the second line L1 A-g fault.

Fault 3, Occurring at t_2 496 s After the First Fault ($T_0 = 0$) The DFR record in Fig. 6.72 shows that the early phase A-g is reestablished with symmetrical fault currents with no dc offset components. The ground current trace L1-In shown in Fig. 6.72 reveals a gradually increasing current at the beginning of the fault, again indicating the high-resistance nature of the tree fault. The high-resistance fault was cleared from substation X by tripping CB A and from substation Y by tripping CB B after 3.5 s from the T_2 reference point. The solid-state primary relay system operated to clear the fault at both ends. The fault was cleared when one of the subconductors of the two-wire-bundle broke and fell, causing a new low-resistance (perhaps solid) line-to-ground fault. The fault clearing was confirmed by the substation X sequence-of-events recorder by the operation of primary and secondary permissive overreaching pilot schemes at both ends. The substation X DFR record in Fig. 6.72 illustrates the start of the high-resistance fault for a duration of 1.8 s, where recording was stopped without showing the clearing of the fault caused by subconductor fall. Apparently, the arc caused by the flow of low fault current for several seconds during the high-resistance fault acted as a torch to melt the conductor and cause it to fall to ground. It is postulated that the DFR recorded 1.8 s and the flow of low ground current for several seconds acted as a torch to melt one of the subconductors and cause it to fall, creating high current ground fault flow, due to the new low-resistance path to ground.

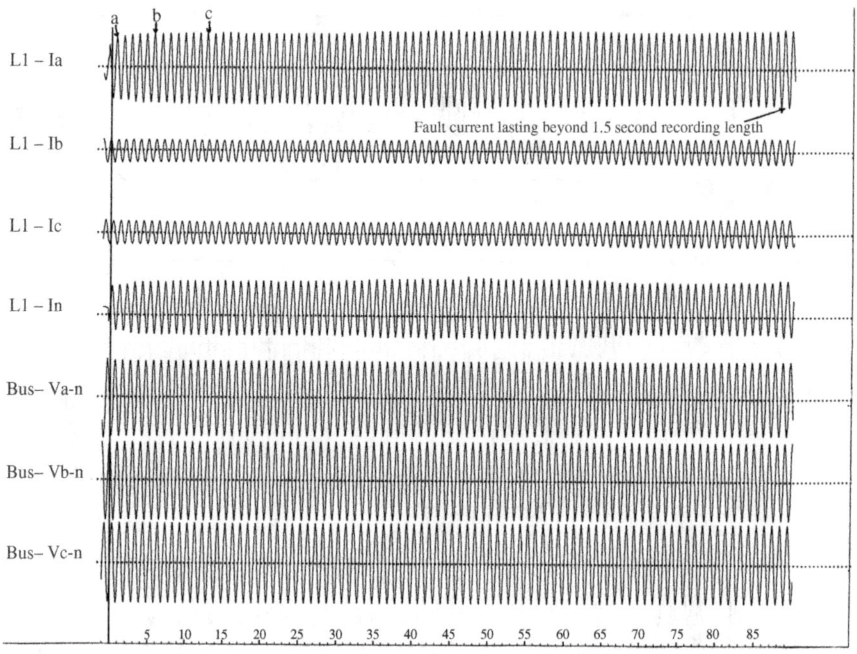

Fig. 6.72 Substation X DFR record showing line L1 currents feeding the third A-g fault.

Fault 4, Occurring 20 s After the Third Fault Reference Point T_2 Twenty seconds from T_2, circuit breaker B1 at substation Y reclosed unsuccessfully into the permanent A-g fault caused by the falling of the subconductor and tripped immediately. The automatic reclosing relay at substation Y locked out, preventing further testing of the permanent L-g fault. The exact trip time is not known, due to the lack of a DFR recording device at substation Y.

Procedure Used to Estimate the Tree Fault Resistance When the DFR Record Is Obtained from Only One End of the Line Normally conducting simulation studies require securing two unknowns, fault location and fault resistance, in addition to fault type. The resistance of the fault can be calculated by sequential clearing (nonsimulation tripping of the two ends of the line) using the following procedure:

1. Estimate the ground fault current contributions from the substation X end of the line at the beginning of the fault and when the substation Y end of the line tripped.
2. Using short-circuit study simulation, fault cases can be run at a tree location.
3. The simulated fault resistance magnitude can be varied until a match is established between the study result and the recorded DFR values. The fault

records confirm that ground currents varied at the beginning of the fault and became stable near the end of the record.

4. The resistance value calculated can be prorated using the current ratio to estimate the initial value of R_F.
5. Simulation studies with both ends of the lines need to be run to confirm the value of R_F.

Analysis of the 345-kV Subconductor Breakages Due to a low loading flow (about 233 MVA) on the L1 circuit prior to the occurrence of the three separate L-g faults, it can be postulated that the fault was caused by either the heavy weight of the snow and ice, by a high wind randomly causing the tree to come closer to the live conductor, or by both snow and a high wind. Since two separate successful automatic reclosing attempts were established for the L1 circuit followed by a fault after 40 s in one case and about 7 minutes for the other fault, it can be concluded that high wind has moved the tree branches close enough for the phase A live conductor to arc-over to ground via the tree. Based on short-circuit study simulations, the three L-g faults are of high resistance. The fault resistance values vary from about 24 Ω at the first fault to 60 Ω at the last fault.

The two successful automatic reclosings at both ends of the lines and the nonsuccessful last automatic reclosing from the substation Y end indicated that the conductor broke at the last fault. Since the last fault lasted for 3.5 s, the prolonging of arcing duration supported by a neutral current of about 1200 A has contributed to conductor failure, similar to a torchlike phenomenon.

System Phenomena

1. The high-resistance tree fault shown in trace L1-In in Figs. 6.68, 6.71, and 6.72
2. Arc-over at the voltage peak as shown in trace Bus-Va-n in Fig. 6.68
3. The low fault current torching effect on the conductor, leading to failure of the subconductor wire

Corrective Actions

1. A tree-trimming program should always be executed to remove endangered trees from the right-of-way of transmission lines.
2. Relay setting calculations for ground time overcurrent relays should be reviewed periodically (especially for power system changes) to enhance fault clearing times for high-resistance ground faults. This will also minimize conductor failure and breakage for slow-clearing low-magnitude fault currents.
3. The pre-fault recording time for the DFR should be increased from 1 cycle to 5 cycles.

Lessons Learned

1. High-voltage and extrahigh-voltage transmission line protection systems should always include a ground time overcurrent element to provide protection against high-resistance tree faults and open-phase conditions. This will also improve the aspects of personnel safety issues.
2. The importance of fast clearing of high-resistance tree faults is illustrated in this case study. The prolonged clearing (in this case, of about 2 to 3 s) has resulted in the breakage of one of the subconductor bundles.
3. The importance of clearing high-resistance ground faults only by ground backup relays and the impact of slow clearing times on conductor melting and breakage is illustrated in this operation.

Case Study 6.15: 115-kV Phase-to-Phase Fault Caused by Failure of a Circuit Switcher

Abstract In this case study we describe an initial phase-to-phase fault and an unsuccessful high-speed reclosing that energized a phase-to-ground fault. We deal with the use of opening and closing resistors for 115-kV oil circuit breakers and the transients associated with energization of a line-connected capacitor bank. Power system phenomena associated with both faults are also described.

Description of the Fault Circuit switcher D was closed to energize a 115-kV 14-MVAR capacitor bank as shown in Fig. 6.73. A few minutes later, the phase B interrupter of circuit switcher D failed, causing a phase B-to-C fault. The fault was cleared successfully from both ends of the line and followed by unsuccessful high-speed automatic reclosing from the Y end of the line. The automatic reclosing energized a phase C-g fault at the same location, which was also cleared from the system.

Sequence of Events Figure 6.73 shows the system one-line diagram and DFR-monitored voltages and currents that are used to analyze this fault. The sequence of events as confirmed by the analysis of DFR records of substation X is as follows. The DFR record in Fig. 6.74 shows that at $t = 0$ (reference point) a phase B-C fault occurred on the L1 115-kV circuit. The phase-to-phase fault is classified by the presence of phase B current being equal and opposite to phase C current and voltage dips on phases B and C. The fault was first cleared from substation X by tripping CBs A1 and A2 after 3.5 cycles from $t = 0$. The initial fault evolved to a double-phase-to-ground fault (B-C-g) upon clearing of the X end of line L1 at 3.5 cycles from $t = 0$. The evolution to a double-phase-to-ground fault is supported by the decreasing of phase B and C voltages to zero in the DFR record in Fig. 6.74 and the line L2 ground current contribution shown in Fig. 6.75. Substation Y tripped and opened the 115-kV CB B1 main interrupter 7 cycles after $t = 0$. From 7 to 9 cycles, the preinsertion opening

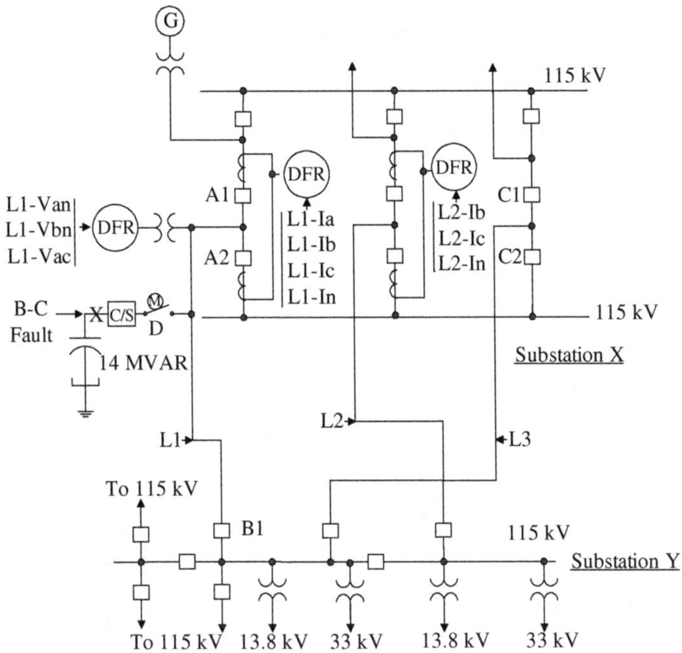

Fig. 6.73 System one-line diagram showing DFR-monitored voltages and currents.

resistor at substation Y is conducting. At 9 cycles the fault current is interrupted and the fault was cleared from the 115-kV system.

At 18 cycles from $t = 0$, CB B1 at substation Y reclosed automatically by high-speed reclosing, inserting closing resistors on all phases. At 20 cycles the 115-kV CB B1 closing resistors are removed from the circuit by the circuit breaker mechanical movement. A phase C-g fault is established at the circuit switcher (the same location as the early B-C fault). At 26 cycles the 115-kV CB B1 at substation Y tripped. At this time, CB B1 opening resistors are inserted for all the phases. At 28 cycles the 115-kV CB B1 cleared the fault by removing opening resistors from the system via the breaker auxiliary interrupter contacts.

Using the DFR Information from One End Only to Develop a Complete Sequence of Events for the Entire Disturbance When line-side three-phase voltages are monitored with a DFR, tripping and reclosing times for the remote end of the faulted line can be obtained. In addition, when more than one circuit is connected between the two substations during the fault, more confirmation about the sequence of events can also be obtained.

In this case study, traces L1-Van, L1-Vbn, and L1-Vcn of line L1 voltages of the DFR record shown in Fig. 6.74 provide information about the following events: insertion of opening resistors at 7 cycles, the clearing of the Y end of L1 in 9 cycles, high-speed reclosing in 18 cycles, removal of closing resistors in 20 cycles followed

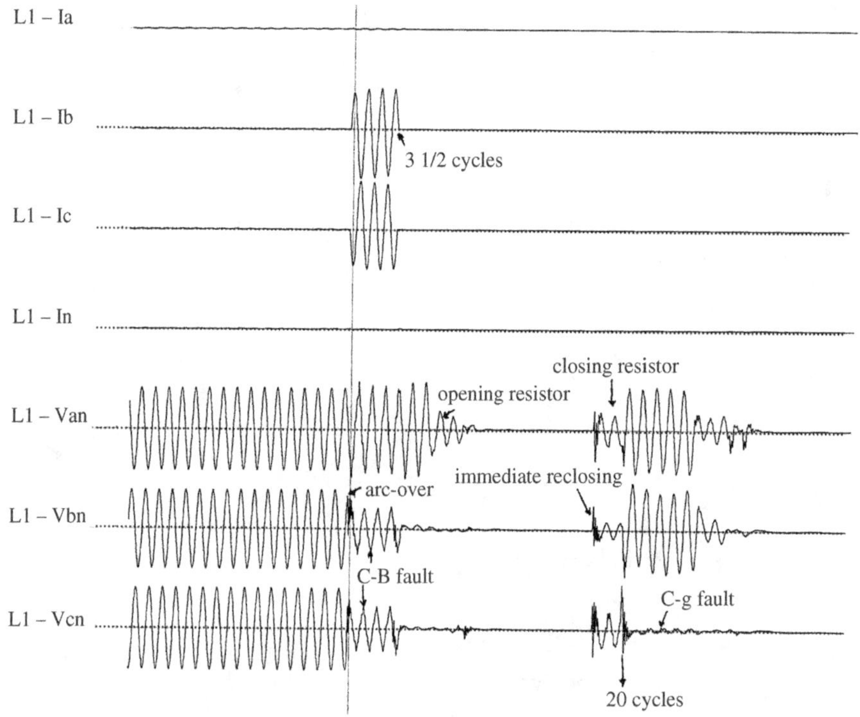

Fig. 6.74 Substation X DFR record showing the phase-to-phase fault and subsequent immediate reclosing of line L1 at the end of substation Y.

by re-fault at the same time, and finally, clearing the C-g fault in 28 cycles. The timing sequence above is confirmed further by the analysis of L2 monitored phase currents shown in Fig. 6.75. The initial B-C fault evolved to ground when the X end of the line tripped at 3.5 cycles. The C-g fault that occurred following the high-speed reclosing occurred at the voltage peak, resulting in a symmetrical fault current, which makes this a good case for system modeling verification.

Corrective Actions High-speed reclosing for one end only by CB B1 at substation Y was disabled. High-speed reclosing practice provides benefits to a power system when it is applied at both ends of the line. Since substation X is a generating plant, line testing by automatic reclosing should always be carried out from the remote end of the line (away from the generator). Therefore, the high-speed reclosing option should be disabled and a 10-s automatic reclosing shot should be used instead for CB B1 using hot bus–dead line reclosing logic.

System Phenomena

1. Switching of circuit breaker opening resistors as indicated in trace L1-Van in Fig. 6.75 at the initial phase-to-phase fault.

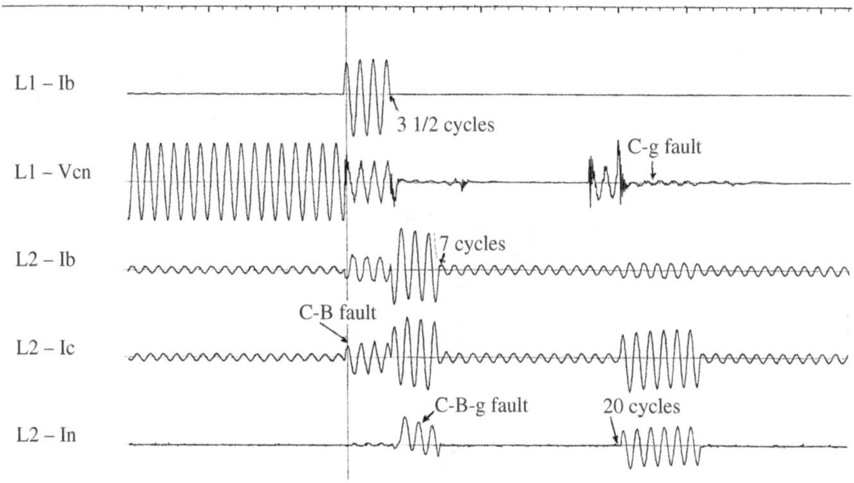

Fig. 6.75 Substation X DFR record showing the line L2 contribution to the phase-to-phase fault and subsequent phase-to-ground fault at reclosing.

2. Switching of circuit breaker closing resistors at the instant of immediate reclosing, as shown in traces L1-Van, L1-Vbn, and L1-Vcn in Fig. 6.75.
3. Sequential fault clearing as indicated by the clearing of the Y end of the line 3.5 cycles following the clearing of the other end at substation X. Sequential clearing is not considered a good relaying practice and should be avoided.
4. Fault current reversal associated with parallel circuits when one end of the faulted line clears faster than the other. As indicated in Fig. 6.75, current flow on circuits L1 and L2 was toward substation X for the initial 3.5 cycles. When end X of the line cleared at 3.5 cycles, the current flows reverse, flowing from substation X to substation Y.
5. Evolving faults, due to the spread of an ionized cloud and gases as a by-product of an early arc-over (flashover). As shown in Fig. 6.76, the symmetrical B-C fault evolved into an asymmetrical B-C-g when the end X of the line tripped at 3.5 cycles, as shown in traces L2-Ib and L2 –Ic in Fig. 6.75

Case Study 6.16: Undesired Tripping of a 115-kV Feeder Due to a Setting Application Error in the Time Overcurrent Element for a Numerical Line Protection Relay

Abstract An improper setting was applied to a ground time overcurrent function of a numerical distance relay protecting 115-kV line L1. This resulted in an undesired trip of line L1 during an external L-g fault at F1 on line L2. The setting did not coordinate with downstream relays at substation Y. This was an error

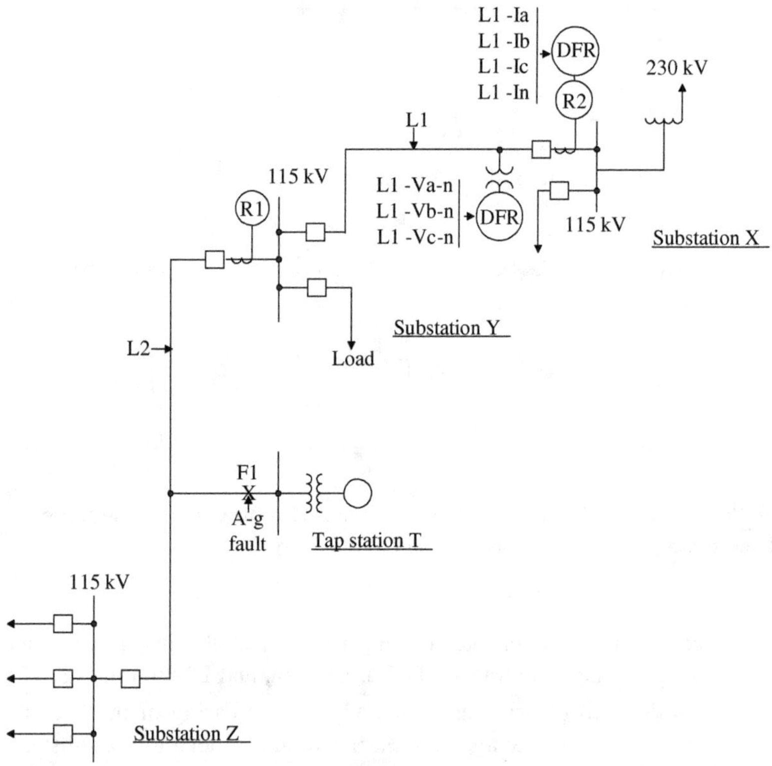

Fig. 6.76 System one-line diagram with substation X DFR-monitored currents and voltages.

created by the fact that numerical relays were sometimes given the same electromechanical relay curve shape designation. In this case study we provide an analysis of the DFR record, power system phenomena, corrective actions, and lessons learned.

Description of the Protection System The 115-kV line L1, shown in Fig. 6.76, is protected by a pilot relaying system using a directional comparison blocking scheme employed in power line carrier communication. The second protection system consists of step distance and ground time overcurrent as a backup scheme using numerical relays. The ground TOC is implemented as a function offered by the numerical distance relay.

Ground Time Overcurrent Relay Setting Calculations The instantaneous element of R2 is set using stub fault (line L1 end fault, out of substation X). From short-circuit study simulation, the stub fault current is 2944 A. To avoid overreaching of the instantaneous element for external fault, a margin of 125% is used. The instantaneous element setting is therefore $= 125\% \times 2944 = 3680$ A. The R1

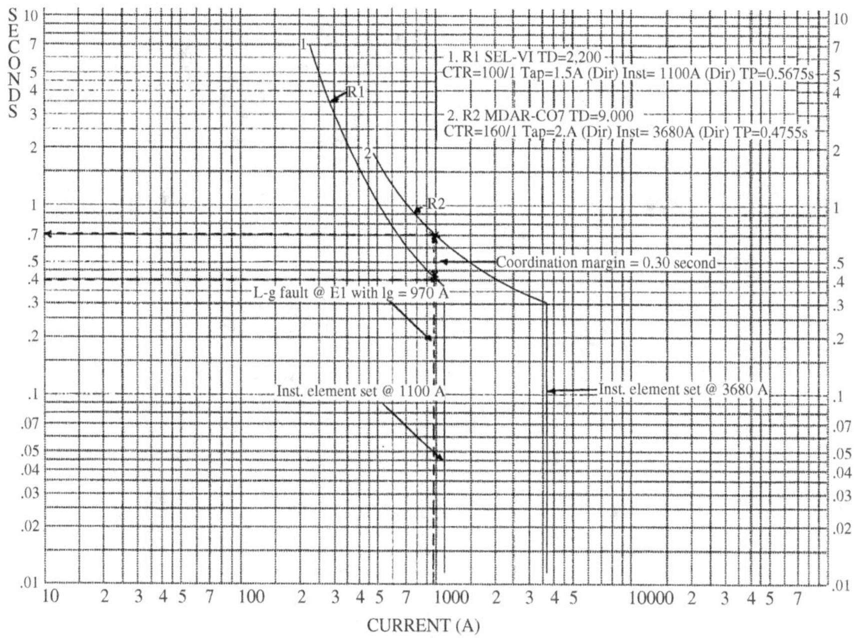

Fig. 6.77 Coordination curves for R1 and R2 relays with the correct setting applied to the R2 ground element.

instantaneous element is set similarly for the stub condition at 1100 A. The R1 TOC element setting is shown in Fig. 6.77, with CT ratio = 100:1 and relay tap = 1.5 A, with a time dial at 2.2 very inverse curve. The ground time overcurrent relay R2 was set to coordinate with R1, the slowest ground relay at substation Y. Figure 6.77 documents the application and coordination between R1 and R2, with the correct setting applied to the R2 ground element as CT ratio = 160:1 and relay tap = 2 A, with a time dial at 9.0 very inverse curve. The study result yielded a coordination margin of 0.3 s, as shown in Fig. 6.77.

Using automated relay setting programs, the electromechanical relay time dial 2 was used mistakenly for R2 to check the coordination between ground relays R2 and R1. The correct R2 numerical relay was not used in the automated relay setting programs, and hence its corresponding correct time dial needed to produce a coordination margin of 0.3 s between R2 and R1 was not identified. The numerical relay time dial 9 is equivalent to the electromechanical relay time dial 2. As a result, the numerical relay setting was issued with a minimum time dial number of 3 instead of 9. Figure 6.78 shows the coordination curves for R1 and R2 relays with the wrong setting applied to the R2 ground element.

Description of the Undesired Relay Operation A phase-to-ground (A-g) fault occurred on 115-kV line L2 due to an insulator failure at F1. The fault was closer to substation Z and was cleared by the zone 1 (Z1) ground distance element. Due to the

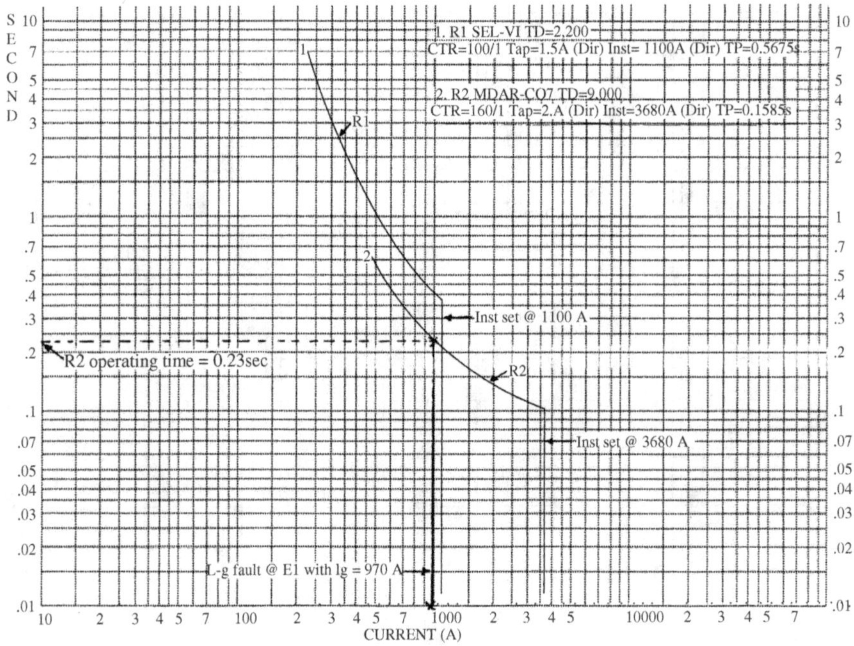

Fig. 6.78 Coordination curves for relays R1 and R2 with the wrong setting applied to the R2 ground element.

relay setting application error mentioned above, coordination was lost between ground time overcurrent elements R1 and R2, resulting in the undesired tripping of line L1 at substation X by relay R2. The ground time overcurrent element of numerical relay R2 generates 64 coordination steps, starting at 3, while the equivalent electromechanical time overcurrent relay has only 11 time dial steps, from 0.5 to 10. The coordination level was done initially using time dial (TD) 2 for an electromechanical relay, which is equivalent to TD curve 9 for the numerical relay. A computer-based coordination program was used to verify coordination based on TD curve 2 for an electromechanical relay. Then an error was made during development of the numerical relay setting file. The same TD was used, TD2, which was translated by the coordination software as the minimum curve of the numerical relay, which is TD 3 rather than the correct one, which is TD 9.

Analysis of the DFR Record The DFR record shown in Fig. 6.79 reveals an arc-over at the peak of phase A voltage, which later was confirmed as an insulation breakdown on the phase A insulator, resulting in a phase A-to-ground fault at F1. The fault was first cleared by the operation of Z1 ground distance at substation Z in 3.5 cycles. The opening of the relatively strong ground source at substation Z boosted the fault current fed from substation X from 685 A to a value of 970 A. The flow of this

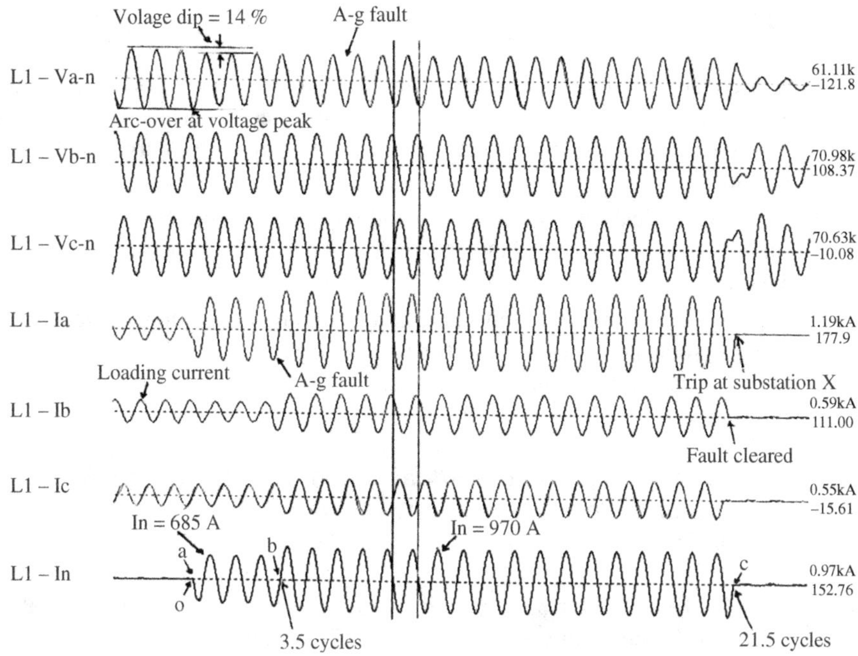

Fig. 6.79 DFR record for substation X showing the undesired trip of line L1 during the line L2 A-g fault at F1.

current through both ground relays R1 and R2, with R2 having an incorrect low time dial setting, forced R2 to operate faster than R1 and trip line L1 in 21.5 cycles from the initiation of fault F1, as documented in Fig. 6.78.

Figure 6.77 illustrates the DFR record for a line L1 undesired trip during the L2 fault at F1. The appearance of neutral (ground $3I_0$) current on L1 at point a, the voltage dip on phase A, and an increase in the phase A current confirm the A-g fault. The symmetric nature of the neutral current (no dc offset) confirms an arc-over of an insulator that has failed exactly at peak voltage. The line L2 end at substation Z cleared at point b after 3.5 cycles, while the undesired trip of line L1 at substation X cleared at point c after 21.5 cycles.

Power System Phenomenon An arc-over occurs at the voltage peak, as shown in the DFR record of trace L1-Va-n in Fig. 6.77 when insulation fails at the peak point of phase A voltage. As a result, the fault current will be symmetric, with no dc offset component present in the fault current.

Corrective Action Download a new setting file to the numerical relay at substation X for line L1 with the correct time dial setting of 9 as a preliminary quick fix to the loss-of-coordination problem between ground TOC elements R1 and R2. The correct use and coordination between R1 and R2 are shown in Fig. 6.78.

Lessons Learned

1. Numerical relay setting documentation should always be reviewed thoroughly to understand the basic differences between the electromechanical published relay curves and the numerical relay equivalent coordination curve numbering system.
2. Proper numerical relay designations and the correct equivalent coordination curve numbering system should be reviewed prior to conducting relay coordination checks.
3. The use of the same electromechanical TOC relay designation for documentation of some numerical relay setting files, which provides equivalent curves, can be misleading and should be avoided.

Case Study 6.17: Mitigation of Mutual Coupling Effects on the Reach of Ground Distance Relays Protecting High- and Extrahigh-Voltage Transmission Lines

Abstract We begin this case study with a derivation of the ground distance relay formulas for different compensation factors. Zero-sequence mutual coupling phenomena using an actual DFR record to illustrate mutual coupling–induced zero-sequence voltage in one circuit as a result of zero-sequence current flow in another circuit is explained in Section 3.38. We also illustrate how the phenomenon can cause the ground distance relay to either overreach or underreach, depending on the relative directions of current flows in the mutually coupled lines. We provide an analysis of how the reach of a ground distance relay is influenced on mutually coupled lines if one of the lines is out for maintenance, with three-phase ground chains applied at both ends. The modeling of this maintenance condition in short-circuit study simulations is described for the first time. Mitigation of the effects of zero-sequence mutual coupling phenomena on the measured impedance by ground distance relays is also covered. The approach recommended reveals a step-by-step calculation approach to cope with the effect of zero-sequence mutual coupling on the setting of ground distance relaying. We also describe an actual example of the use of relays for the protection of an EHV 345-kV transmission system. In addition, we describe another example for the protection of two parallel 345-kV lines. We then conclude that the best solution is through the correct setting of the relay based on an understanding of the phenomena and by the simulation via short-circuit studies of all relevant system outage conditions. This solution is recommended instead of attempting to hardwire current transformers from unfaulted lines to the protected line CT circuit.

Derivation of the Ground Distance Relay Formula Assume that a phase a-to-ground fault occurred at location F at the end of the line for the system shown in Fig. 6.80. Ignoring the effects of mutual coupling and fault resistance, the

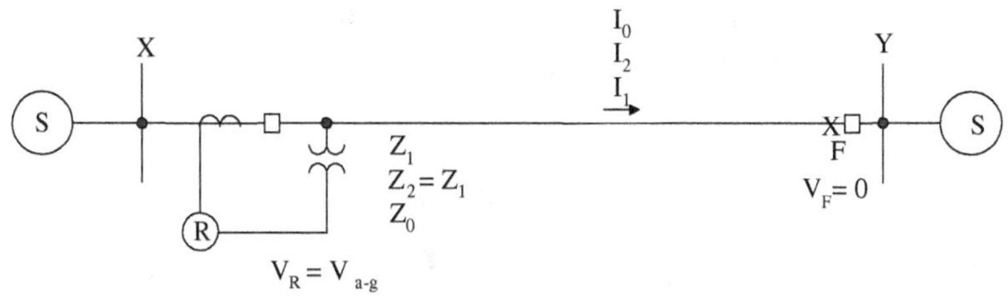

Fig. 6.80 Simplified system model.

voltage drop across the line will be

$$V_R - V_F = I_1 Z_1 + I_2 Z_2 + I_0 Z_0$$

Since $V_F = 0$ and assuming that $V_R = V_{\text{a-g}}$, we obtain

$$V_{\text{a-g}} = I_1 Z_1 + I_2 Z_2 + I_0 Z_0 \tag{1}$$

where $V_{\text{a-g}}$ is the faulted phase voltage at relay location and V_F is the voltage at the fault, I_1, I_2, and I_0 are the symmetrical components of the currents fed from the end of X, and Z_1, Z_2, and Z_0 are the positive-, negative-, and zero-sequence impedances of the line, respectively.

The symmetrical component current I_1 and I_0 contributions from the X end of the faulted line are functions of the positive- and zero-sequence distribution factors as related to the total fault current. The values of the distribution factors can vary due to generators or transformers being out of service. These noncontrolled variations of distribution factors result in a varying measure of apparent impedances by the relay for the same fault point. It is certainly desirable to design ground distance relays to be set based on the positive-sequence impedance to the ground fault point. To solve this problem, a zero-sequence compensation method can be used to adjust the measured impedance to reflect the actual positive-sequence impedance. First we derive the correct compensation factor to use.

Adding and subtracting $I_0 Z_1$ in equation (1) and assuming that $Z_1 = Z_2$ yields the following:

$$V_{\text{a-g}} = I_1 Z_1 + I_2 Z_1 + I_0 Z_0 + I_0 Z_1 - I_0 Z_1$$

Rearranging yields

$$V_{\text{a-g}} = (I_1 + I_2 + I_0) Z_1 + I_0 (Z_0 - Z_1)$$

Substituting $I_a = I_1 + I_2 + I_0$ and factoring Z_1 gives us

$$V_{\text{a-g}} = Z_1 \left(I_a + I_0 \frac{Z_0 - Z_1}{Z_1} \right)$$

where I_a is the faulted phase current. Substituting $Z_1 = Z_f =$ measured relay positive-sequence impedance to the fault, we have

$$Z_f = \left(\frac{V_{a\text{-}g}}{I_a + I_0(Z_0 - Z_1)/Z_1} \right) \tag{2}$$

For electromechanical and static ground distance relays, the neutral (or residual) current $3I_0$ can easily be injected for impedance measurement compensation rather than I_0. Therefore, equation (2) has to be modified as follows [some numerical distance relays are also designed using equation (3)]:

$$Z_f = \frac{V_{a\text{-}g}}{I_a + 3I_0(Z_0 - Z_1)/3Z_1}$$

or

$$Z_f = \frac{V_{a\text{-}g}}{I_a + K_0(3I_0)} \tag{3}$$

where K_0 is the zero-sequence compensation factor $= (Z_0 - Z_1)/3Z_1$ and Z_0 and Z_1 are the zero- and positive-sequence impedances of the protected line. In this formula, K_0 is less than 1 for transmission lines having a zero-sequence impedance/positive-sequence impedance ratio of less than 4. In addition, for overhead lines, having a zero-sequence/positive-sequence impedance ratio value of 3 will produce a K_0 value of 0.67. Zero-sequence compensation using the K_0 portion of $3I_0$ will tune the ground distance relay to measure apparent impedance equal to the positive-sequence impedance of the line to the point of the fault.

Some numerical distance relays can obtain symmetrical components for the current during ground faults. Formula (3) can be modified to reflect this concept. From equation (1),

$$V_{a\text{-}g} = I_1 Z_1 + I_2 Z_2 + I_0 Z_0$$

By assuming that $Z_1 = Z_2$ and factoring Z_1 for the RHS, we obtain

$$V_{a\text{-}g} = Z_1 \left(I_1 + I_2 + I_0 \frac{Z_0}{Z_1} \right)$$

or

$$Z_f = \frac{V_{a\text{-}g}}{I_1 + I_2 + K'_0(I_0)} \tag{4}$$

where K'_0 is the zero-sequence compensation factor Z_0/Z_1. For this type of relay compensation, the value of K'_0 for overhead lines is greater than 1. Other numerical relays, where the zero-sequence current I_0 can be calculated from the sampled current, are designed in accordance with equation (2):

$$Z_f = \frac{V_{\text{a-g}}}{I_a + I_0(Z_0 - Z_1)/Z_1}$$

or

$$Z_f = \frac{V_{\text{a-g}}}{I_a + K''_0(I_0)} \tag{5}$$

where K''_0 is the zero-sequence compensation factor $(Z_0/Z_1) - 1 = (Z_0 - Z_1)/Z_1$, which is three times greater than the K_0 factor used in equation (3). (*Caution*: The correct K_0 expression must be used when setting the ground distance relay to avoid undesired operations.) K''_0 will be greater than 1 for Z_0/Z_1 ratio values greater than 2. In summary, some ground distance relays are designed using three different compensation factors, K_0, K'_0, and K''_0, as illustrated in equations (3), (4), and (5).

Description of the Zero-Sequence Mutual Coupling Phenomena Refer to Section 3.38 for a description of mutual coupling phenomena.

Effects of Mutual Coupling on the Reach of Ground Distance Relays The numerators in equations (3), (4), and (5) are the sum of positive-, negative-, and zero-sequence voltages. Therefore, during ground faults the relay reach can be influenced by zero-sequence voltages generated by mutual coupling. Since this mutually coupled zero-sequence voltage to the protected line cannot be correlated with the actual fault current flow, an impedance relay measurement type of concept is used. The impedance relay can either overreach or underreach, depending on the relative directions of fault currents in the parallel lines. This is explained below.

Case of Overreach As shown in Fig. 6.81, the fault currents in the two parallel lines are flowing in opposite directions. Figure 6.82 shows the equivalent zero-sequence mutual network. An ideal auxiliary current transformer with a ratio of 1 : 1 is assumed and added to simplify the analysis and to study the impact of mutual coupling

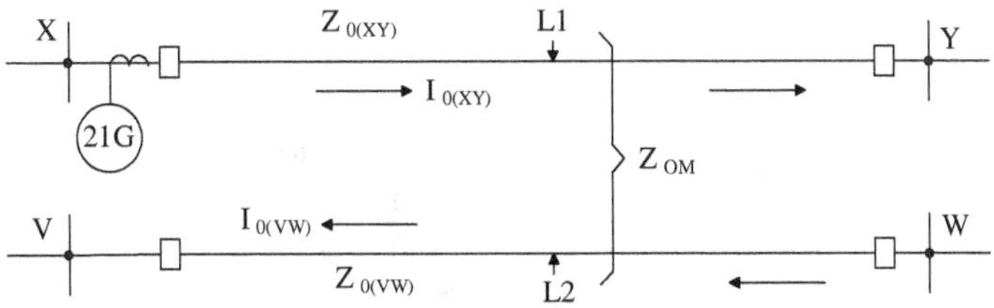

Fig. 6.81 Zero-sequence current flow in the opposite direction for two mutually coupled lines.

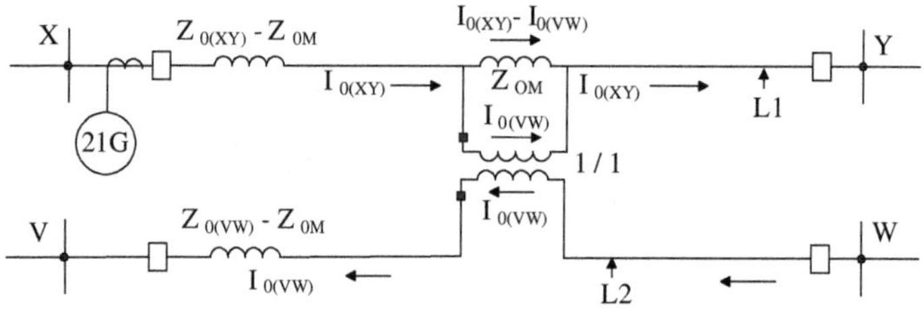

Fig. 6.82 Zero-sequence network equivalent for Fig. 6.81.

on the ground distance relay reach. A line-to-ground fault or a line-end (stub) fault is assumed near terminal Y. The voltage seen by the ground distance relay at X is composed of positive-, negative-, and zero-sequence components:

$$V_a = V_1 + V_2 + V_0 \tag{6}$$

Original zero-sequence fault current flow in line XY sets zero-sequence voltage, which forces current flow on line VW (provided that connections to ground exist between the two terminals). Using an ideal transformer with a turns ratio of 1 : 1, the current $I_{0(\mathrm{VW})}$ can be moved to the faulted line to study the impact on the line zero-sequence voltage.

Current $I_{0(\mathrm{vw})}$ flows out of the auxiliary CT polarity in line VW, which will force the same current to flow in polarity for line XY. This will result in current flow through $Z_{0\mathrm{M}}$ to be $I_{0\mathrm{XY}} - I_{0\mathrm{VW}}$. The voltage drop can then be written as

$$V_{0(\mathrm{XY})} = I_{0(\mathrm{XY})}(Z_{0(\mathrm{XY})} - Z_{0\mathrm{M}}) + (I_{0(\mathrm{XY})} - I_{0(\mathrm{VW})})Z_{0\mathrm{M}} \tag{7}$$

$$= I_{0(\mathrm{XY})}Z_{0(\mathrm{XY})} - I_{0(\mathrm{VW})}Z_{0\mathrm{M}} \tag{8}$$

As indicated in equation (8), the subtracted zero-sequence voltage component due to $Z_{0\mathrm{M}}$ will decrease the relay voltage in equation (6), thus decreasing the apparent distance reach of the ground distance relay at X. As a result, the relay will overreach ($Z_{\mathrm{app}} < Z_{\mathrm{actual}}$). For a fault located at the end of the line, the apparent impedance measured by the relay will be smaller than the actual line impedance, and it may cause false operation of Z1. Therefore, the Z1 element of the ground distance relay (if not set properly) will react to bus faults and faults external to the Z1 zone of detection.

Case of Underreach As shown in Fig. 6.83, fault currents in the two parallel lines L1 and L2 are now flowing in the same direction. Figure 6.84 shows the equivalent zero-sequence mutual networks. An ideal auxiliary current transformer with a ratio of 1 : 1 is assumed and added to simplify the analysis and to study the impact of mutual coupling on the ground distance relay reach. A bus fault or a line-to-ground fault is assumed near terminal Y.

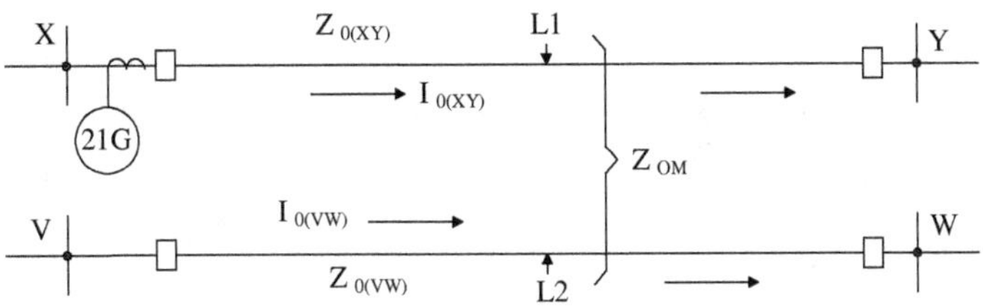

Fig. 6.83 Zero-sequence current flow in the same direction for two mutually coupled lines.

The voltage seen by the ground distance relay is composed of positive-, negative-, and zero-sequence components:

$$V_a = V_1 + V_2 + V_0$$

The original zero-sequence fault current flow in line XY sets a zero-sequence voltage that forces current flow on line VW (provided that connections to ground exist between the two terminals). Using an ideal transformer with a turns ratio of 1 : 1, the current $I_{0(VW)}$ can be moved to the faulted line to study the impact on the line zero-sequence voltage. The voltage drop can then be written

$$V_{0(XY)} = I_{0(XY)}(Z_{0(XY)} - Z_{0M}) + (I_{0(XY)} + I_{0(VW)})Z_{0M} \tag{9}$$

$$= I_{0(XY)}Z_{0(XY)} + I_{0(VW)}Z_{0M} \tag{10}$$

As indicated in equation (10), the added voltage component due to Z_{0M} will increase the zero-sequence voltage at the relay location and hence increase the apparent reach of the ground distance relay at X. This increase in the relay reach will cause the relay to underreach ($Z_{apparent} > Z_{actual}$). For faults at the balance point of Z2, the apparent impedance will be greater than the setting value and may cause no relay operation for this fault. Therefore, the relay will not cover part of the relay protection zone near the balance point. In this case of relay underreach, the system is being underprotected.

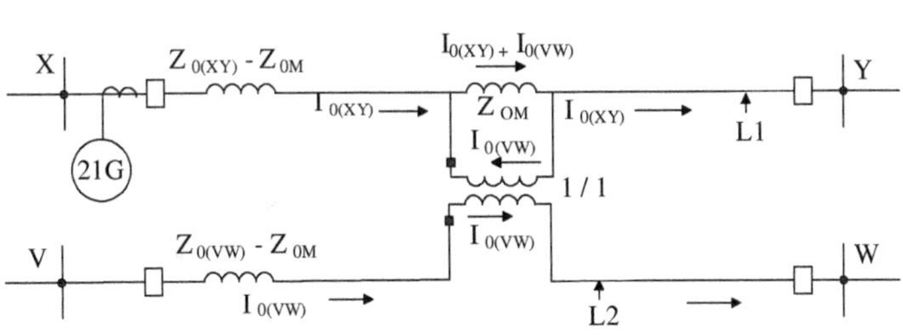

Fig. 6.84 Zero-sequence network equivalent for Fig. 6.83.

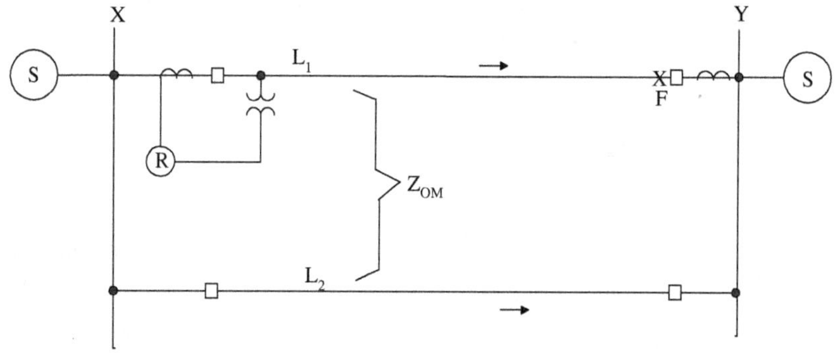

Fig. 6.85 Simplified model for two parallel lines sharing the same right-of-way with zero-sequence current flows in the same direction.

System Configurations Leading to the Existence of Mutual Coupling Phenomena

Full Sharing of the Right-of-Way Between Lines (Transmission Corridor) As shown in Fig. 6.85, full sharing can occur when the two transmission lines L1 and L2 terminate at both ends, in common substations. In this configuration it is possible to gain access to the adjacent line zero-sequence current, provided that no other line mutual couplings exist, to introduce it (if required) for compensation for the line that is affected by its presence.

Partial Sharing Between Transmission Lines As shown in Fig. 6.86, consider a transmission line L1 from substation X to substation Y. Assume that line L2 from substation X runs parallel with line L1 for a considerable distance and then branches off to terminate at substation Z. It is not possible to compensate ground distance relays

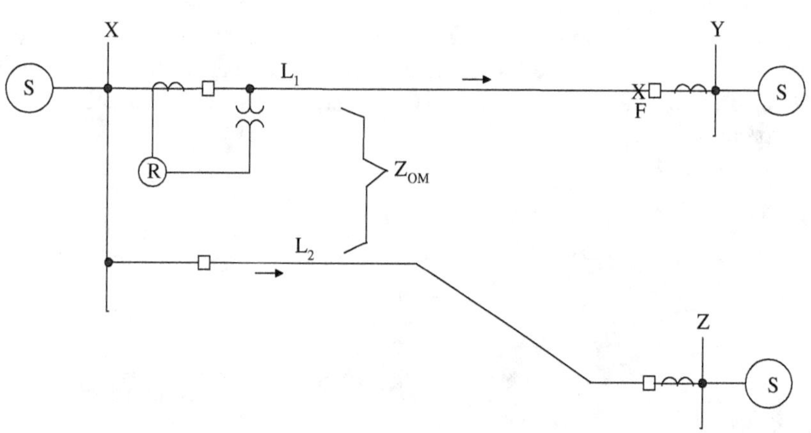

Fig. 6.86 Simplified model for partial sharing of the right-of-way for transmission lines.

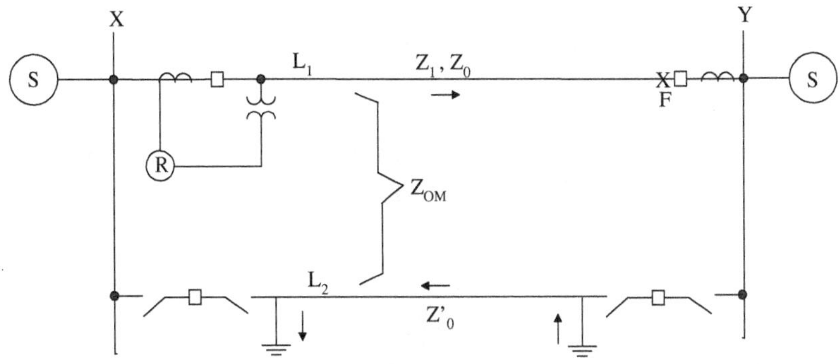

Fig. 6.87 Line out of service with three-phase grounding straps applied at both ends of L2.

at either substation Y or Z (if required) for zero-sequence mutual coupling between the two lines.

Analysis of the Effects of a Line Out of Service with a Three-Phase Ground Chain Applied at Both Ends on the Reach of a Ground Distance Relay Applied on Mutually Coupled Lines The application of three-phase grounding chains at both ends of a transmission line after its removal from service is required for the safety of field personnel. As shown in Fig. 6.87, an occurrence of ground faults on the system where zero-sequence current I_0 flow on line L1 will induce zero-sequence voltage V'_0 on line L2 due to mutual coupling Z_{0M} between lines L1 and L2. At substation Y, for a fault at F on line L1, an induced zero-sequence voltage will occur on line L2 due to the mutual coupling effect. This voltage will force zero-sequence current flow in the L2 loop through ground at both ends of the line. The current will be

$$I'_0 = \frac{I_0 Z_{0M}}{Z'_0}$$

The currents in lines L1 and L2 during the fault will be in opposite directions. The zero-sequence current I''_0 flow in line L2 will induce zero-sequence voltage in line L1. This voltage will be

$$V_{0M} = \frac{I_0 Z_{0M}}{Z'_0} \times Z_{0M} = \frac{I_0 Z_{0M}^2}{Z'_0}$$

This voltage will be subtracted from the voltage measured at location X causing the relay to overreach. Therefore, this maintenance condition is required to be simulated prior to setting Z1 to prevent Z1 overreach. The voltage measured by the relay at X,

$$V_{a\text{-}g} = I_a Z_1 + I_0(Z_0 - Z_1) - \frac{I_0 Z_{0M}^2}{Z'_0}$$

This condition can only be mitigated by relay setting due to the unavailability of access to a zero-sequence ground current contribution from line L2 to be added to line L1 with a multiplier Z_{0M}.

$$\%\text{ overreach} = 100\left[\frac{I_a Z_1 + I_0(Z_0 - Z_1)}{I_a Z_1 + I_0(Z_0 - Z_1) - I_0 Z_{0M}^2/Z'_0} - 1\right]$$

Short-Circuit Study Simulation of a Line Out of Service with Grounding Chains Applied at Both Ends As shown in Fig. 6.88, the simulation of a line L2 out-of-service condition can be done by splitting the buses where L2 is terminated and forming two new buses, C and D. Line L2 is modeled in the study with its positive- and zero-sequence self- and mutual coupling impedance to line L1 as Z'_1, Z'_0, and Z'_{0M}, respectively. The grounding at both ends is simulated by shunt branches with a very large positive-sequence impedance of *j*9999 pu and a very small zero-sequence impedance of *j*0.0001 pu. To link the isolated line L2 with the rest of the power system, a very high impedance across one of the split buses is inserted with a positive-sequence impedance = zero-sequence impedance = *j*9999 pu. Placing ground faults on the system will force induced ground currents to flow in the L2 loop to ground, thus simulating the effects of applying grounding chains at both ends of the line on the line L1 ground distance relays. Techniques can be used with any arbitrary number of mutually coupled lines, although mutual modeling can be very tedious and cumbersome for some short-circuit programs.

Mitigation of Mutual Coupling Effects Effects of zero-sequence mutual coupling phenomena on the measured impedance by ground distance relays can be corrected using one of the methods described below.

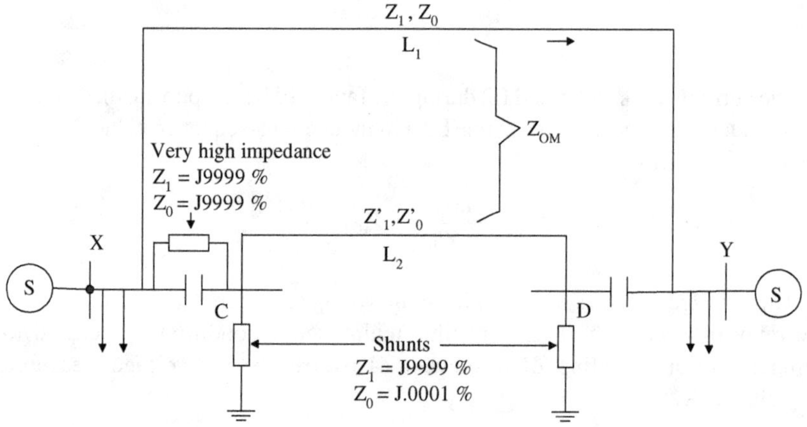

Fig. 6.88 Modeling of a line out of service with three-phase grounding straps applied at both ends.

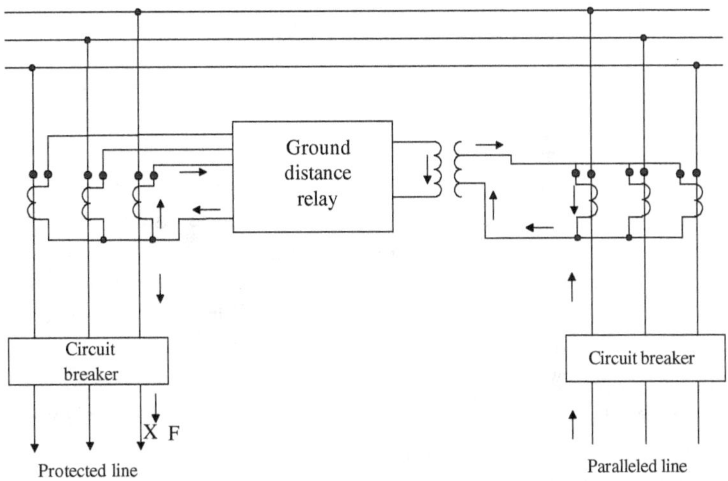

Fig. 6.89 Mutual coupling compensation using hardwiring.

Compensation of Zero-Sequence Current from the Mutually Coupled Unfaulted Line For the system shown in Fig. 6.89, it is possible to gain access to the adjacent parallel line zero-sequence current to introduce it to the relay associated with the protected line, provided that both lines emanate from the same substation and there are no additional line mutual couplings. Based on analysis of the overreach and underreach mutual coupling cases, it is evident that the significance of this current in a relaying measurement varies depending on fault locations and hence current direction. To use it correctly, an evaluation of all these cases must be done to compensate carefully for the mutual effects. Mutual compensation can then be accomplished by injecting a portion of the $3I_0$ current from the parallel line into the protected line relay to offset the decrease or increase in the zero-sequence component of the voltage measured by the relay. The magnitude of the compensated portion of the current is proportional to the zero-sequence mutual impedance Z_{0M}. The correct polarity connection of the compensation current is essential and is normally based on assumed directions for fault currents in the two parallel lines. One way of ensuring correct connections is to study the case for a ground fault in the protected line and assume that the parallel line contribution is flowing opposite the protected line. In this case, the two secondary currents are flowing opposite to each other in the compensation transformer, as shown in Fig. 6.89. As a result, the net compensated current will be adjusted to nullify the effect of induced mutual zero sequence voltage on the distance relay reach.

To use mutual compensation, the parallel lines must terminate in the same substations to have their current available. However, this method cannot provide compensation for the following cases:

1. *Partial sharing*. This occurs when two lines run parallel for a portion of their total length without termination at the same substation, as shown in Fig. 6.86.

Due to the physical separation of substations, zero-sequence current cannot be obtained from the parallel line to be used for mutual compensation.

2. *One transmission line sharing the right-of-way with multiple lines connected either in parallel or in series.* In the parallel case, it is not practical to use compensation when several transmission lines are mutually coupled due to sharing the same right-of-way, thus forming a transmission corridor or when multiple circuits (more than two) are sharing the same tower. The cross connections of more than one line are complex and increase the possibility of an incorrect connection or testing mistake. In the multiple series connection case, currents from remote sections are not available for local connections in the compensation circuit, as explained for Case Study 6.17B.
3. *Line out of service with three-phase ground chains applied at both ends.* The case where the parallel line is opened and grounded at both ends was covered earlier. The zero-sequence current from the parallel line is not available for use in the compensation circuit. This is due to application of the grounding point away from the current transformer location and isolation of the line by opening the line disconnect switches.

Factoring Mutual Coupling Effects in Ground Distance Setting Calculations Using a short-circuit study, several line-to-ground fault cases for system normal and outage conditions can be simulated. For conventional short-circuit studies, a step-by-step approach using symmetrical components can then be used to obtain sequence voltages and currents. The sequence quantities can then be substituted in symmetrical component equations to get faulted currents and voltages. Fault impedance measured by the ground distance relay can then be calculated in accordance with the applicable relay reach formula. As an example, equation (3), defined earlier in this case study, which is applicable for electromechanical, solid, state, and some numerical relays, will be used:

$$Z_f = \frac{V_a}{I_a + K_0(3I_0)}$$

Most available PC-based short-circuit study simulation programs provide faulted phase voltages and currents and apparent impedance calculations. However, for older conventional short-circuit studies, the following simple procedure can be used to calculate V_a, I_a, and $3I_0$. We begin with

$$V_a = V_1 + V_2 + V_0$$

and

$$V_1 = E_s - I_1 \times Z_{1s}$$

$$V_2 = -I_2 \times Z_{2s}$$

$$V_0 = -I_0 \times Z_{0s}$$

Here Z_{1s}, Z_{2s}, and Z_{0s} are, respectively, the positive-, negative-, and zero-sequence impedances of the power system equivalent behind the protected line at the relay location and E_s is the source voltage, which is normally assumed to be 1.0 pu. Substituting V_1, V_2, and V_0 produces V_a. Similarly, $I_{a.} = I_1 + I_2 + I_0$ (normally, $I_1 = I_2$). These values and $3I_0$ can then be substituted in equation (3) to obtain the ground distance relay reach for any ground fault case simulation of the system.

Recommended Practice to Cope with the Effect of Mutual Coupling on a Ground Distance Relay Reach Compensation of mutual effects by hardwiring of current circuits from unfaulted lines is complex and can lead to incorrect operations and testing difficulties. In addition, as indicated earlier, there are many system configurations that will defeat the purpose of the compensation. Therefore, mitigation of the effects of zero-sequence mutual coupling on the ground distance relaying reach should be implemented using relay setting calculation rather than wiring CTs from mutually coupled lines. This recommended method requires complete knowledge of the applied ground distance relay reach formula and its zero-sequence compensation factor.

Analysis of the power system using legitimate outages to calculate the apparent ground distance measured impedance as affected by mutual coupling phenomena will lead to a simpler solution to mitigate the effects on the relay reach. Complete understanding of the sequence quantities and how the applied relay measures the ground fault impedance will lead to a secure Z1 setting and dependable Z2 setting. Grounding of one of the line sections when removing it from service using ground chains on the phases can be modeled to study the effect on the relay reach.

A calculation procedure that utilizes all possible contingencies on the system for ground fault simulations at the end of the protected line will produce an apparent fault impedance range. The maximum overreach value can then be subtracted from the impedance of the line to derive a Z1 setting. The goal is to find the maximum overreach that can be subtracted from the actual transmission line to obtain the line length required to set Z1 to cover 80% of the new line length. This will define the correct line impedance that will prevent the overreaching of Z1 due to zero-sequence mutual coupling phenomena, thus enhancing the security of the relaying system. The maximum underreach value can then be added to the impedance of the line to derive Z2 element setting. The Z2 element can then be set at 125% of the new defined impedance of the line to be used as a permissive overreaching pilot element. This will define the correct line impedance that will guarantee covering of the line by the pilot element during line ground faults, thus enhancing the dependability of the protection system. Figure 6.90 defines mutual coupling with zero-sequence flows in the same direction for two parallel lines during a bus fault, causing the distance relay Z2 to underreach. Figure 6.91 defines mutual coupling with zero-sequence flows in the opposite direction for two parallel lines during a line-end fault, causing the distance relay Z1 to overreach.

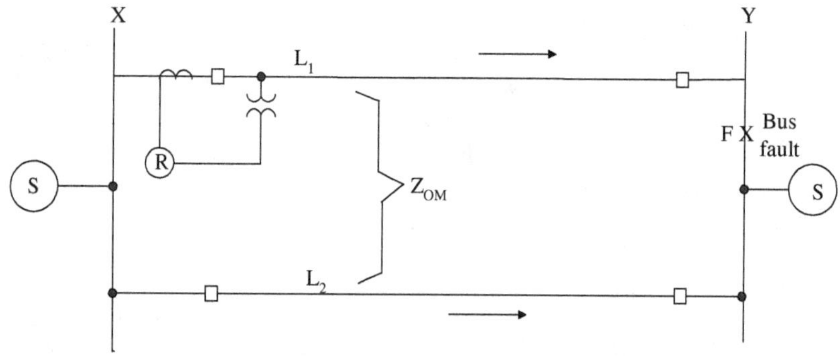

Fig. 6.90 Mutual coupling with zero-sequence flows in the same direction for two parallel lines during a bus fault.

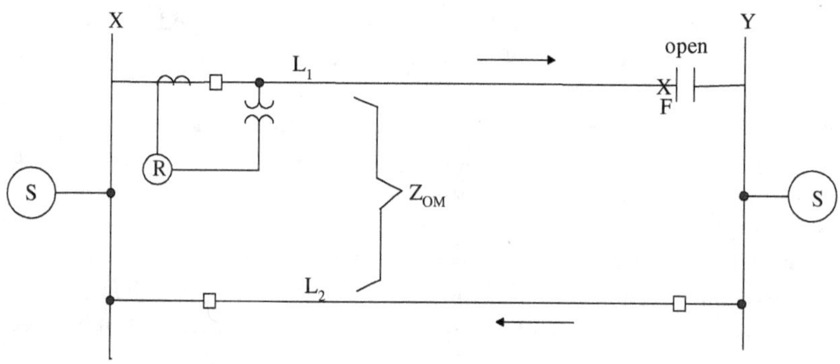

Fig. 6.91 Mutual coupling with zero-sequence flows in the opposite direction for two parallel lines during a line-end fault.

Case Study 6.17A: Application of the Procedure Recommended for the Protection of Two Parallel Lines

Abstract In this case study we describe the use of the procedure recommended, and it is analyzed using the very simple system shown in Fig. 6.92, where two parallel 345-kV lines have $X_1 = j0.1$ pu and $X_0 = j0.3$ pu with a simple system equivalent at each end. Zero-sequence mutual coupling between the two lines is assumed to be 60% of individual line zero-sequence impedance = 0.18 pu. All impedances are given on 100-MVA and 345-kV bases.

Objective Calculate the Z1 and Z2 settings of a permissive overreaching transfer trip (POTT) application for the ground distance relay (21G), R1 of line L1 at the bus A terminal.

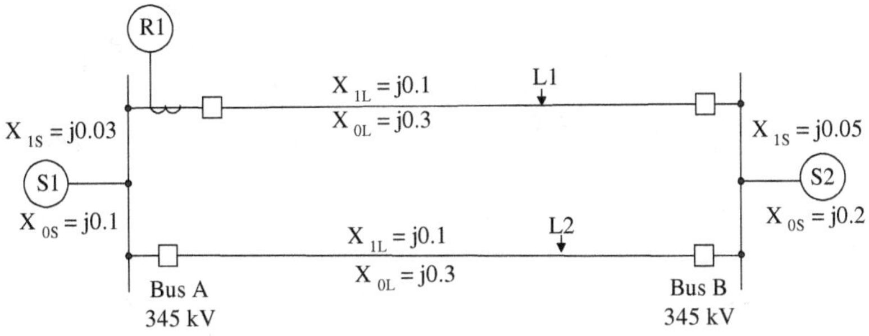

Fig. 6.92 Two parallel 345-kV lines with simple power system equivalent sources.

Relay-Setting Calculation Steps

1. Simulate a line-to-ground fault at bus B with line L2 out of service as illustrated in Fig. 6.93. The opening of line L2 (with no ground chains applied at both ends) will remove zero-sequence mutual coupling impact on the R1 relay reach for ground faults. Using the relay correct ground distance reach formula, R1 will measure line L1 positive-sequence impedance, and this value should be used as a reference to evaluate the effect of mutual coupling on the R1 relay; or

$$Z_F = \frac{V_{\text{a-g}}}{I_a + K_0(3I_0)} \tag{1}$$

From Fig. 6.93,

$$V_{\text{a-g}} = 150.7\,\text{kV}$$

$$I_a = 0.724\,\text{kA}$$

$$K_0 = \frac{Z_0 - Z_1}{3Z_1} = \frac{0.3 - 0.1}{3 \times 0.1} = 0.666$$

$$3I_0 = 0.814\,\text{kA}$$

By substituting all these values in equation (1), and since all impedances are pure reactances, to simplify the calculations, we write

$$X_F = \frac{150.7}{0.724 + (0.666 \times 0.814)} = 119\,\Omega \text{ primary} = 0.1\text{ pu}$$

To check the calculation:

$$X_{L1} = X_{pu} \times X_{base} = 0.1 \times \frac{(\text{kV})^2}{\text{MVA}} = 0.1 \times \frac{(345)^2}{100} = 119\,\Omega\,\text{primary} \qquad \text{O.K.}$$

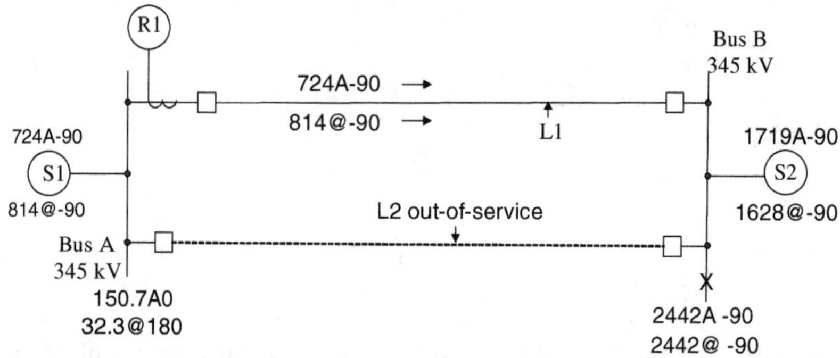

Fig. 6.93 RMS phase and zero-sequence currents and voltages for an L-g fault at bus B with L2 out of service.

2. Simulate the line-end L-g (stub) fault. As shown in Fig. 6.94, zero-sequence currents flow in opposite directions in lines L1 and L2, producing an apparent impedance overreach for relay R1. We use equation (1) for the L-g fault:

$$Z_F = \frac{V_{\text{a-g}}}{I_a + K_0(3I_0)}$$

From Fig. 6.94,

$$V_{\text{a-g}} = 159.5\ \text{kV}$$

$$I_a = 0.967\ \text{kA}$$

$$K_0 = \frac{Z_0 - Z_1}{3Z_1} = \frac{0.3 - 0.1}{3 \times 0.1} = 0.666$$

$$3I_0 = 0.967\ \text{kA}$$

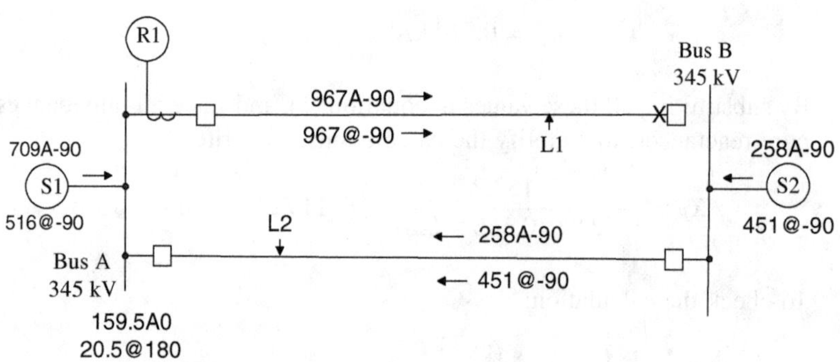

Fig. 6.94 RMS phase and zero-sequence currents and voltages for the L1 line-end L-g fault.

By substituting all these values in equation (1), and since all impedances are pure reactances to simplify calculations, we have

$$X_F = \frac{159.5}{0.967 + (0.666 \times 0.0.967)} = 99\,\Omega\,\text{primary}$$

$$\%\,\text{overreach} = \frac{119 - 99}{119} \times 100 = 16.8\%$$

3. Simulate line L2 out with ground chains applied at each end of the line. As shown in Fig. 6.95, zero-sequence currents flow in opposite directions in lines L1 and L2, producing an apparent impedance overreach for relay R1. Again using equation (1) for the L-g fault, we have

$$Z_F = \frac{V_{\text{a-g}}}{I_{\text{a}} + K_0(3I_0)}$$

From Fig. 6.95,

$$V_{\text{a-g}} = 139.4\,\text{kV}$$

$$I_{\text{a}} = 0.843\,\text{kA}$$

$$K_0 = \frac{Z_0 - Z_1}{3Z_1} = \frac{0.3 - 0.1}{3 \times 0.1} = 0.666$$

$$3I_0 = 1.069\,\text{kA}$$

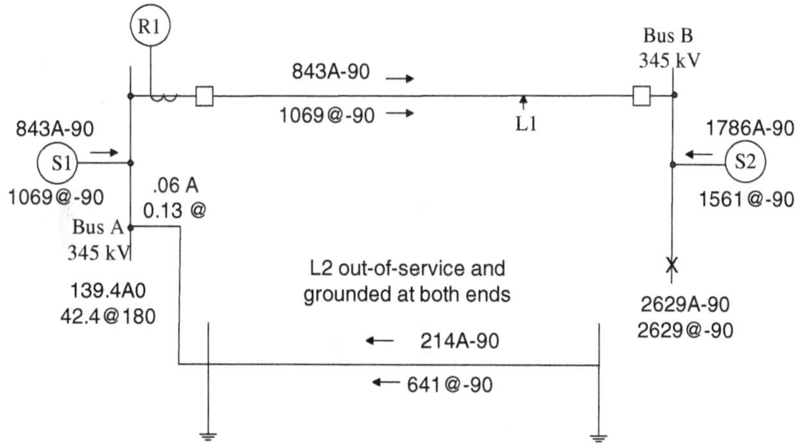

Fig. 6.95 RMS phase and zero-sequence currents and voltages for the L-g fault at bus B with L2 out of service and ground chains applied at both ends of the line.

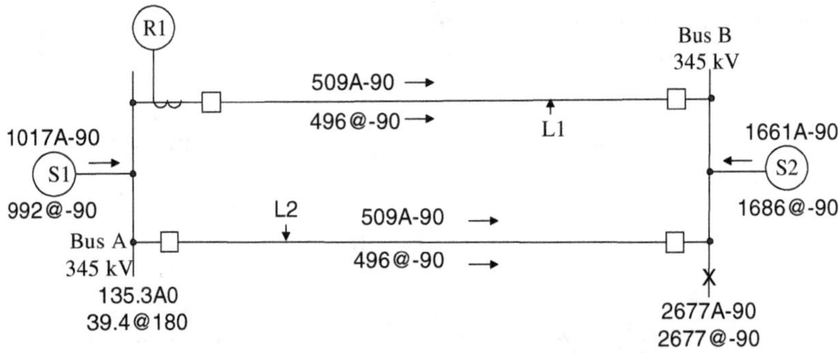

Fig. 6.96 RMS phase and zero-sequence currents and voltages for the L-g fault at bus B.

By substituting all these values in equation (1), and since all impedances are pure reactances, to simplify calculations, we have

$$X_F = \frac{139.4}{0.843 + (0.666 \times 1.069)} = 89.6\ \Omega \text{ primary}$$

$$\%\ \text{overreach} = \frac{119 - 89.6}{119} \times 100 = 24.7\%$$

4. Simulate an L-g fault at the 345-kV bus B. As shown in Fig. 6.96, zero-sequence currents flow in the same direction in lines L1 and L2, producing an apparent impedance underreach for relay R1. We use equation (1) for the L-g fault:

$$Z_F = \frac{V_{\text{a-g}}}{I_a + K_0(3I_0)}$$

From Fig. 6.96,

$$\begin{aligned} V_{\text{a-g}} &= 135.3\ \text{kV} \\ I_a &= 0.509\ \text{kA} \\ K_0 &= \frac{Z_0 - Z_1}{3Z_1} = \frac{0.3 - 0.1}{3 \times 0.1} = 0.666 \\ 3I_0 &= 0.496\ \text{kA} \end{aligned}$$

By substituting all these values in equation (1), and since all impedances are pure reactances, to simplify calculations, we have

$$X_F = \frac{135.3}{0.509 + (0.666 \times 0.496)} = 161\ \Omega \text{ primary}$$

$$\%\ \text{underreach} = \frac{161 - 119}{119} \times 100 = 35\%$$

Analysis of the Calculation Results The line positive-sequence actual measured impedance by ground distance relay during phase-to-ground faults in step 1 is 119-Ω primary and will be used as a reference representing 100% of Z_L with zero measurement error (due to line L2 being open). As shown in steps 2 and 3, the overreach range is 16.8 to 24.7% of Z_L and in step 4 the underreach value is 35% of Z_L.

Relay Setting Recommendation Zone 1 (Z1) for the ground distance relay should not be set to overreach for the worse zero-sequence mutual coupling effect. The maximum overreach is 24.7%, corresponding to the fault case simulating line L2 out with ground chains applied at each end of the line. From step 3, the correct apparent line length for the Z1 setting is therefore equal to 100 – 24.7 = 75.3 % of Z_L. Z1 should be set at 80% of the apparent line length. The Z1 setting = 0.8 × 75.3 = 60% of Z_L.

Zone 2 (Z2) for the ground distance relay should be set to cover the worse underreach case due to mutual coupling effects. Since the element is used for tripping for POTT scheme, it should be set with a margin for the apparent line impedance. From step 4, the apparent line length = 100% + maximum under reach = 100% + 35% = 135% Z_L Z2 should be set at125% of the newly defined apparent line impedance. The Z2 setting = 1.25 × 135% = 169% of Z_L:

$$\text{line base impedance} = \frac{(\text{kV})^2}{\text{MVA}} = \frac{345^2}{100} = 1190\,\Omega\,\text{primary}$$

The primary impedance of Z1 = 0.6 × 1190 = 72 Ω primary. If the CT ratio is 400 : 1 and the PT ratio is 3000 : 1, then

$$\text{Z1} = 72 \times \frac{400}{3000} = 9.6\,\Omega\text{-s}$$

The primary Z2 impedance = 1.69 × 0.1 × 1190 = 201 Ω primary:

$$\text{Z2} = 201 \times \frac{400}{3000} = 26.8\,\Omega\text{-s}$$

Case Study 6.17B: Protection of a 345-kV System That Has Several Lines Connected in Series and Mutually Coupled to a Longer Line

Abstract In this case study we begin describe the use of the procedure recommended. Figure 6.97 illustrates the 345-kV L1 transmission line between substations X and Y, which is mutually coupled to the 345-kV transmission lines; L2 between

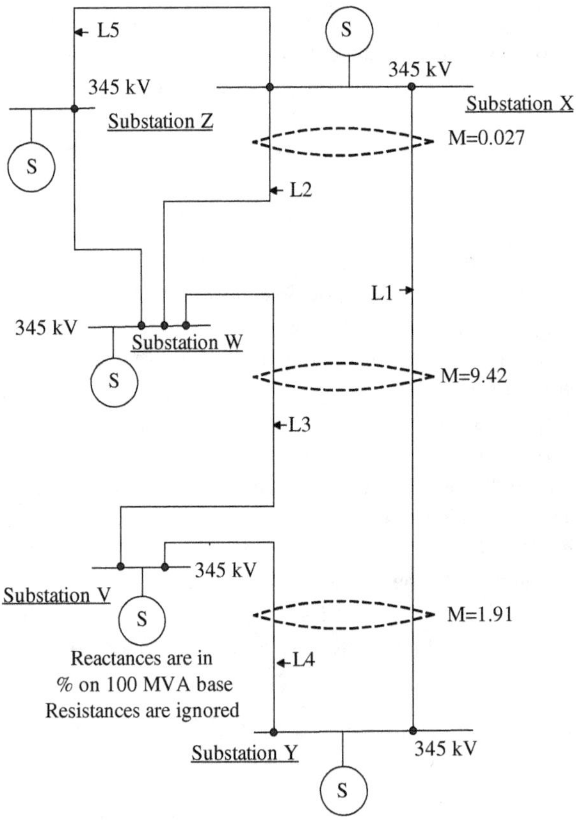

Fig. 6.97 EHV system used for the mutual coupling case study.

substations X and W; L3 between substations W and V; and L4 between substations V and Y. The L1 line is protected using phase and ground distance relays at each end. The pilot relaying scheme used for L1 protection is a permissive overreaching transfer trip scheme. The ground distance relays of line L1 terminal at substation X need to be set with due consideration to zero-sequence mutual coupling. The objective is to calculate the ground distance relay settings for zone 1 (Z1) and zone 2 (Z2) for the X end of the 345-kV L1 circuit.

Description of the 345-kV EHV System The 345-kV system is shown in Fig. 6.98 with system impedance values. The system consists of a double-circuit 345-kV system between substations X, W, V, and Y. Transmission line L1 shares the right-of-way for a distance of 135 miles with three other lines: L2, L3, and L4. Transmission line L5 connects substations X and Z.

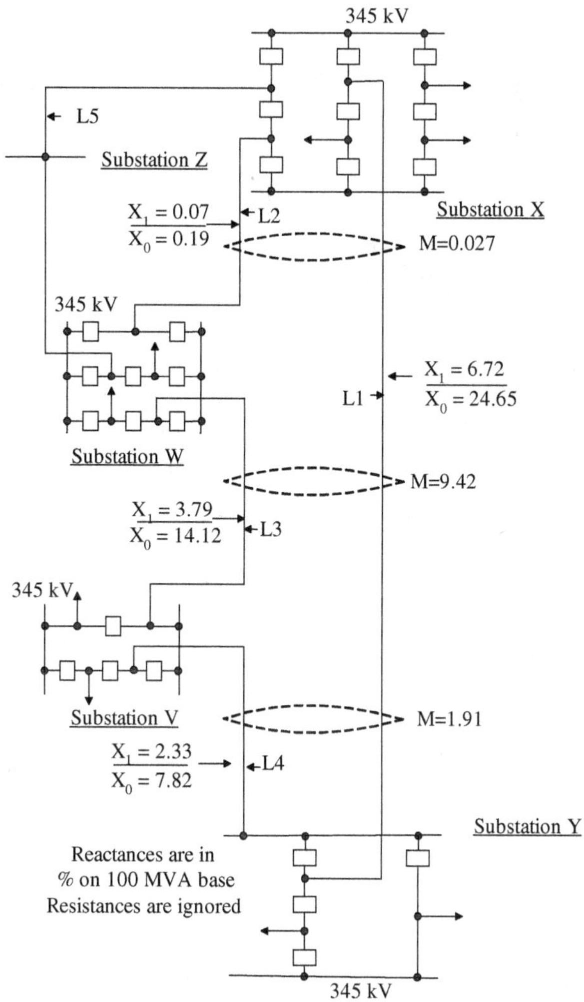

Fig. 6.98 345-kV system impedance diagram.

Description of Bulk Power System Protection The 345-kV transmission line L1 is protected by dual pilot protection systems utilizing diverse communication media. The design of the 345-kV relaying system was done in the early 1980s, when microprocessor-based relays were not commercially available. Therefore, the dual relaying systems were designed based on electromechanical and solid-state distance relays. The first system is a permissive overreaching transfer trip pilot system using a solid-state distance-based relaying package applied on power line carrier communication media. The second system is also a permissive overreaching transfer trip system utilizing electromechanical distance-based relays applied on leased telephone circuits.

Transmission-Line L1 Data

$$Z_{1L1} = 0.0039 + j0.0672 \text{ pu} \qquad \text{on a 100-MVA basis}$$

$$Z_{0L1} = 0.0365 + j0.2465 \text{ pu} \qquad \text{on a 100-MVA basis}$$

$$Z_{\text{base}} = \frac{(\text{kV})^2}{\text{MVA}} = \frac{(345)^2}{100} = 1190.25\,\Omega$$

$$Z_{1L1(\text{pri})} = (0.0039 + j0.9672)1190.25$$

$$Z_{1L1(\text{pri})} = 4.64 + j79.985 = 80.12\,\Omega \text{ at } 86.6^{\circ}$$

$$Z_{0L1(\text{pri})} = (0.0365 + j0.2465)1190.25$$

$$Z_{0L1(\text{pri})} = 43.44 + j293.4 = 296.6\,\Omega \text{ at } 81.6^{\circ}$$

$$\text{potential transformer ratio (PT)} = 3000 : 1$$

$$\text{current transformer ratio (CT)} = 1500 : 5$$

Calculation of the zero-sequence compensation factor K_0 will be done using transmission line L1 impedance parameters:

$$K_0 = \frac{Z_{0L1} - Z_{1L1}}{3Z_{1L1}}$$

Electromechanical relays use a simple autotransformer to obtain the zero-sequence compensation, and hence

$$K_0 = \frac{296.6 - 80.12}{3 \times 80.12} = 0.9$$

Calculation Procedure Using a conventional short-circuit study simulation, a simple procedure can be used to calculate V_a, I_a, and $3I_0$, as outlined earlier. These values and the K_0 calculated earlier can then be substituted in equation (3), which is applicable to the electromechanical and static distance relaying systems used for the protection of line L1. The ground distance relaying reach for ground faults case simulation can then be calculated. This procedure was employed for this example due to the unavailability of the-state-of-the-art PC-based short-circuit and relay coordination programs in the early 1980s.

Simulated Study Cases to Arrive at the Optimum Ground Distance Relay Setting for Relays at Substation X Protecting Line L1 Eleven cases were simulated to arrive at the voltages and currents calculated. The first simulated case is for a three-phase fault at the end of line L1. The three-phase fault positive-sequence impedance will be used as a reference to establish the maximum and minimum relay reach errors when the mutual coupling phenomena are considered. Since line L1 shown in Fig. 6.98 is mutually coupled with three separate lines (L2, L3, and L4) along the transmission corridor (right-of-way), these lines will be placed out of service individually when a phase-to-ground remote bus fault is simulated at substation Y. Normal system and line-end faults are also simulated. In addition, some double-contingency cases are used. Lines out of service with a grounding chain applied at both ends of the line is also simulated.

The following cases simulating faults at substation Y are studied to provide the optimum settings for line L1 zone 1 and zone 2 ground distance relays to cope with mutual coupling effects:

1. Three-phase fault at substation Y
2. Normal system L-g fault at substation Y, as shown in Fig. 6.99
3. End-of-line (stub) L-g fault at substation Y
4. Line L2 out of service with L-g fault at substation Y
5. Line L3 out of service with L-g fault at substation Y
6. Line L4 out of service with L-g fault at substation Y
7. Lines L3 and L5 out of service with L-g fault at substation Y
8. Lines L2 and L5 out of service with L-g fault at substation Y
9. Lines L4 and L5 out of service with L-g fault at substation Y
10. Line L3 out of service and three-phase ground chain applied at each end coincident with a L-g fault at substation Y
11. Line L4 out of service and three-phase ground chain applied at each end coincident with a L-g fault at substation Y as shown in Fig. 6.100.

The results for fault calculations for all 11 simulated short-circuit study cases are shown in Table 6.8.

Analysis of the Calculation Results The three-phase fault measured impedance by phase relays during phase faults of 8.016 Ω secondary will be used as a reference representing 100% of Z_L with zero measurement error. As shown in Table 6.9, the overreach range is 13.8 to 16.8% of Z_L and the underreach range is 3 to 11.8% of Z_L.

Relay Setting Recommendations Zone 1 (Z1) for the ground distance relay should be set not to overreach for the worst zero-sequence mutual coupling effect. The maximum overreach is 16.8%, corresponding to cases 3 and 11.

The correct apparent line length for the Z1 setting is therefore equal to $100 - 16.8 = 83.2\%$ of Z_L. Z1 should be set at 80% of the apparent line length. The Z1 setting $= 0.8 \times 83.2 = 70\%$ of Z_L. Zone 2 (Z2) for the ground distance relay should be set to cover the worst underreach due to mutual coupling effects. Since the element

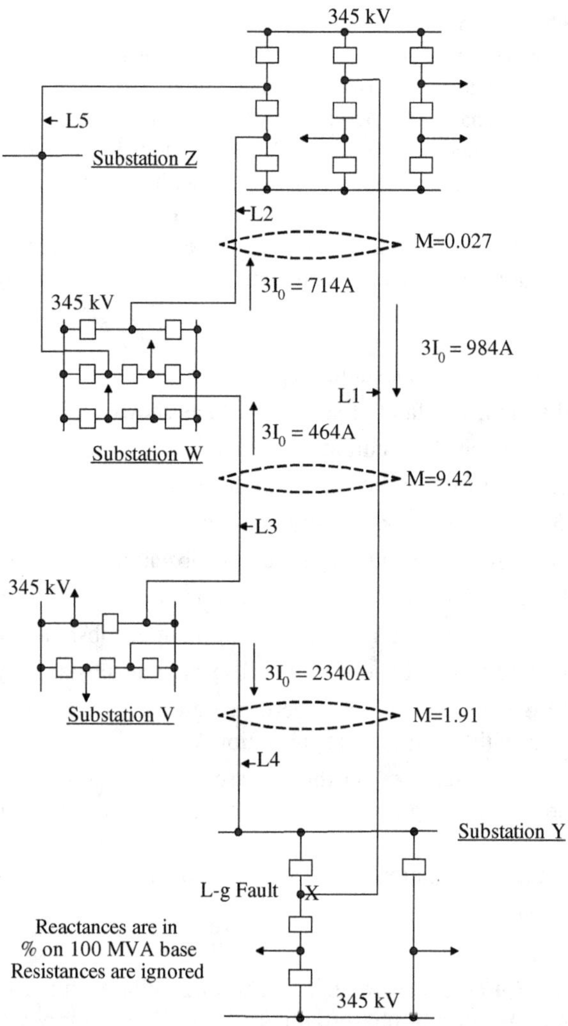

Fig. 6.99 Zero-sequence current flows during a simulation of the normal system L-g fault at substation Y.

is used for tripping in the POTT scheme, it should be set with a margin for the apparent line impedance. The maximum underreach is 11.8%, corresponding to cases 5 and 7. The apparent line length = 100% + maximum underreach = 100% + 11.8% = 111.8% Z_L. Z2 should be set at 125% of the apparent line impedance. The Z2 setting = 1.25 × 111.8% = 150% of Z_L.

Newer numerical relays will allow the use of different setting groups to accommodate different system outage conditions (if desired). This may permit an optimum

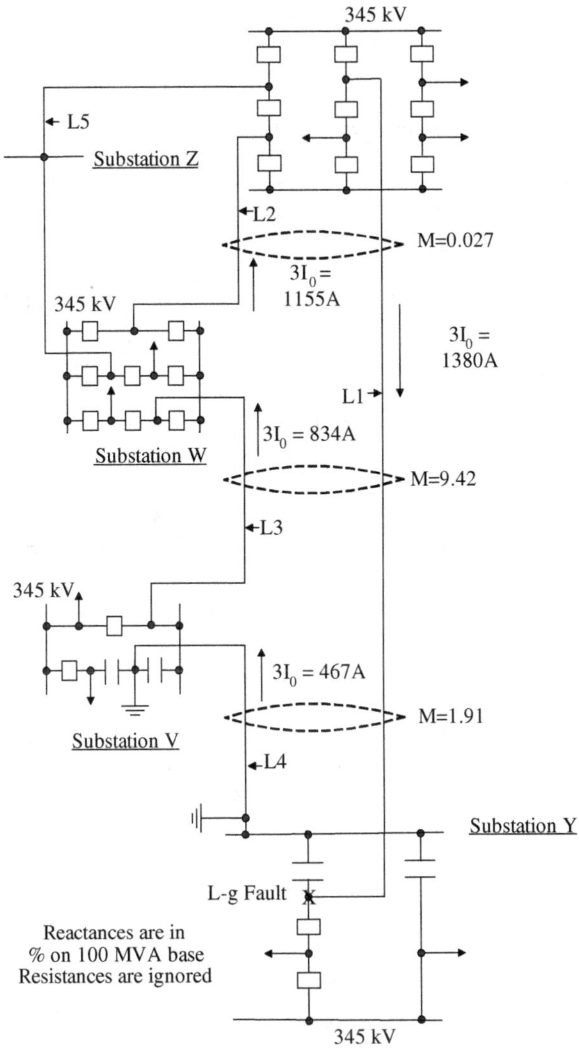

Fig. 6.100 Zero-sequence current flows during an L-g fault at bus Y with line L4 out and ground chains applied at both ends of L4.

setting for zone 1 application. However, it is rather important to keep the relay system simple and not be dependent on several inputs to derive the correct setting for zone 1. It is therefore recommended that only one setting be used for each of the distance elements, based on the procedure outlined in this case study, and that it can be tested and verified prior to commissioning of the relay.

TABLE 6.8 Results for Fault Calculations

L-g Fault at Substation Y with Specific System Conditions	Faulted Voltages and Currents at Substation Y	Calculated Apparent Impedance, $Z = V_a/[I_a + K_0\,(3I_0)]$
Case 1	$V_a = 168.1$ kV	$Z = V_a/I_a = 168.1/2.097$
Normal system three-phase fault	$I_a = 2097$ A (−87)	= 80.16 Ω primary
Case 2	$V_a = 182.3$ kV; $I_a = 1263$A (−84)	$Z = 182.3/(1.263 + 0.9 \times 0.964)$
Normal system L-g fault	$3I_0 = 984$ A (−83)	= 84.85 Ω primary
Case 3	$V_a = 189.8$kV; $I_a = 1497$ A (−85)	
End of L1 L-g fault at Y	$3I_0 = 1497$ A (−85)	= 66.7 Ω primary
Case 4	$V_a = 173.4$ kV; $I_a = 1184$ A (−84)	
Line L2 out of service	$3I_0 = 899$A (−83)	= 87 Ω primary
Case 5	$V_a = 184.5$ kV; $I_a = 1271$ A (−84)	
Line L3 out of service	$3I_0 = 876$ A (−82)	= 89.6 Ω primary
Case 6	$V_a = 187.5$ kV; $I_a = 1503$ A (−85)	
Line L4 out of service	$3I_0 = 1344$A (−84)	= 69.1 Ω primary
Case 7	$V_a = 184.7$ kV; $I_a = 1274$A (−84)	
Lines L3 and L5 out of service	$3I_0 = 875$A (−82)	= 89.6 Ω primary
Case 8	$V_a = 170.2$ kV; $I_a = 1156$ A (−84)	
Lines L2 and L5 out of service	$3I_0 = 875$ A (−83)	= 82.6 Ω primary
Case 9	$V_a = 187.2$ kV; $I_a = 1500$ A (−85)	
Lines L4 and L5 out of service	$3I_0 = 1343$ A (−84)	= 69.1 Ω primary
Case 10	$V_a = 187.9$ kV	
Lines L3 out of service and ground chains applied at both ends	$I_a = 151$ A (−84) $3I_0 = 13$ A (−82)	= 68.2 Ω primary
Case 11	$V_a = 186.8$ kV	
Lines L4 out of service and ground chains applied at both ends	$I_a = 1475$ A (−83) $3I_0 = 1475$ A (−82)	= 66.7 Ω primary

TABLE 6.9 Summary of Calculation Results

L-g Fault at Substation Y with Specific System Conditions	Primary Apparent Impedance	Secondary Apparent Impedance	Percent Overreach	Percent Underreach
Case 1 Normal system three-phase fault	80.16 Ω	(CT/PT = 0.1) 80.16 × 0.1 = 8.02	Zero	Zero
Case 2 Normal system L-g fault	84.85 Ω	8.49 Ω	—	5.90
Case 3 End of L1 L-g fault at Y	66.7 Ω	6.67 Ω	16.80	
Case 4 Line L2 out of service	87 Ω	8.7 Ω	—	8.50
Case 5 Line L3 out of service	89.6 Ω	8.96 Ω	—	11.80
Case 6 Line L4 out of service	69.1 Ω	6.91 Ω	13.80	
Case 7 Lines L3 and L5 out of service	82.6 Ω	8.96 Ω	—	11.80
Case 8 Lines L2 and L5 out of service	89.6 Ω	8.26 Ω	—	3.0
Case 9 Lines L4 and L5 out of service	69.1 Ω	6.91 Ω	13.80	
Case 10 Lines L3 out of service and ground chains applied at both ends	68.2 Ω	6.82 Ω	14.92	
Case 11 Lines L4 out of service and ground chains applied at both ends	66.7 Ω	6.67 Ω	16.80	

REFERENCES

Blackburn, J. L. *Applied Protective Relaying*. Pittsburgh, PA: Westinghouse Electric Corporation.

Blackburn, J. L. *Protective Relaying Principles and Applications*. New York: Marcel Dekker, 1987.

Blackburn, J. L. *Symmetrical Components for Power Systems Engineering*. New York: Marcel Dekker, 1993.

Disturbance Monitoring and Reliability Requirements. NERC Reliability Standard PRC-002-02.

Elmore, W. A. Current differential and phase comparison relaying compared with pilot distance schemes. Presented at the 25th Annual Western Protective Relay Conference, October 13–15, 1998, Spokane, WA.

Elmore, W. A., Ed. *Pilot Protective Relaying*. New York: Marcel Dekker, 2000.

Elmore, W. A., Ed. *Protective Relaying Theory and Applications*. New York: Marcel Dekker, 2000.

Guide for Protective Relaying Application to Transmission Lines. IEEE Power System Relaying Committee Report, IEEE C37.113, 1998.

Pilot Relaying Performance Analysis. IEEE Power System Relaying Committee Report. *IEEE Transactions on Power Delivery*, Vol. 5, No. 1, January 1990, pp. 85–102.

Sun, S. C., and R. E. Ray. A current differential relay system using fiber optics communication. *IEEE Transactions on Power Apparatus and Systems*, Vol. PAS-102, 1978, pp. 410–419.

7

CASE STUDIES RELATED TO CABLE TRANSMISSION FEEDER SYSTEM DISTURBANCES

Analysis of cable disturbance case studies provides a description of the power system affected and its associated protection systems, sequence of events, analysis of the failure, power system phenomena, corrective actions, and lessons learned. The cases can be summarized as follows.

Case Study 7.1 describes the design of the relay system protection zones and demonstrates how optimum design of relay protection zones can lead to a quick identification of the faulted cable section to provide information for quick service restoration. Case Study 7.2 provides an analysis of an undesired trip of an underground cable feeder during the soaking process of a generator step- up (GSU) transformer, which was due to a design error that resulted in reverse polarity connections at one end for the numerical current differential relaying systems. Case Study 7.3 describes the failure of a high-pressure fluid-filled cable connecting the high side of a GSU transformer and the substation, and how the use of relay targets coupled with a complete knowledge of the protection zones can lead to correct analysis and determination of the faulted area. The correct determination is achieved despite the mechanical failure of the cable pot head porcelain, which can give a false indication to the faulted area. Case Study 7.4 documents a 345-kV high-pressure fluid-filled cable trip by the operation of the relay systems, and how

Disturbance Analysis for Power Systems, First Edition. Mohamed A. Ibrahim.

the troubleshooting and restoration of the system can be done using relay targets only. It then introduces more fault data, which became available later, to confirm the initial analysis using very basic data: namely, relay targets only. Case Study 7.5 documents a failure of high-pressure fluid-filled 345-kV cable connecting the unit GSU transformer and the substation, which occurred at the peak voltage, thus supporting a cable insulation failure scenario and producing symmetrical fault currents. Case 7.6 covers a violent 138-kV cable pot head failure, causing a phase A-to-ground fault. The force generated by the fault current caused the pot head to explode and damage the adjacent equipment.

CASE STUDIES

Case Study 7.1: Optimum Design of Relaying Protection Zones Leads to Quick Identification of a Faulted 345-kV Submarine Cable Section

Abstract A submarine cable, which is a portion of a 345-KV cable circuit shown in Fig. 7.1, was faulted to ground. In this case study we describe the design of the relay system protection zones and demonstrate that the optimum design of relay protection zones can lead to a quick identification of the faulted cable section to provide information for quick service restoration. In addition, we describe the incident, provide DFR records analysis, and discuss power system phenomena, corrective actions, and lessons learned.

Description of the 345-kV Cable Systems As shown in Fig. 7.1, the 345-kV cable consists of 9 miles of high-pressure fluid-filled (HPFF) pipe-type cable, 8 miles of underwater self-contained fluid-filled (SCFF) cable, and 10 miles of HPFF pipe-type cable. At substation X the 345-kV cable terminates in a gas-insulated substation (GIS). Substation X includes two 345-kV circuit breakers A1 and A2, a 150-MVA shunt reactor bank, two 450-MVA phase angle–regulating (PAR) transformers, and two 345/138-kV autotransformers. The autotransformers connect to the 138-kV system. Substation Y is a conventional air-insulated breaker-and-a-half switching arrangement with two added 345-kV SF6 live tank breakers. The cable is fully compensated with 600-MVAR shunt reactors, two 150-MVAR shunt reactor banks at substation Y, one 150-MVAR reactor at the transition from submarine cable to land cable at transition station V, and one 150-MVAR shunt reactor at substation X.

The transitions stations V and W are SF-6 gas-insulated stations which include terminations for the connection of the HPFF cable systems to the underwater SCFF cable system. The underwater SCFF cable system includes four single-phase cables, one of them designated as a spare, separated by 500 feet. Disconnect switches are included at the V and W transition stations to facilitate the switching of the spare SCFF underwater cable in the event of the failure of any of the connected phases.

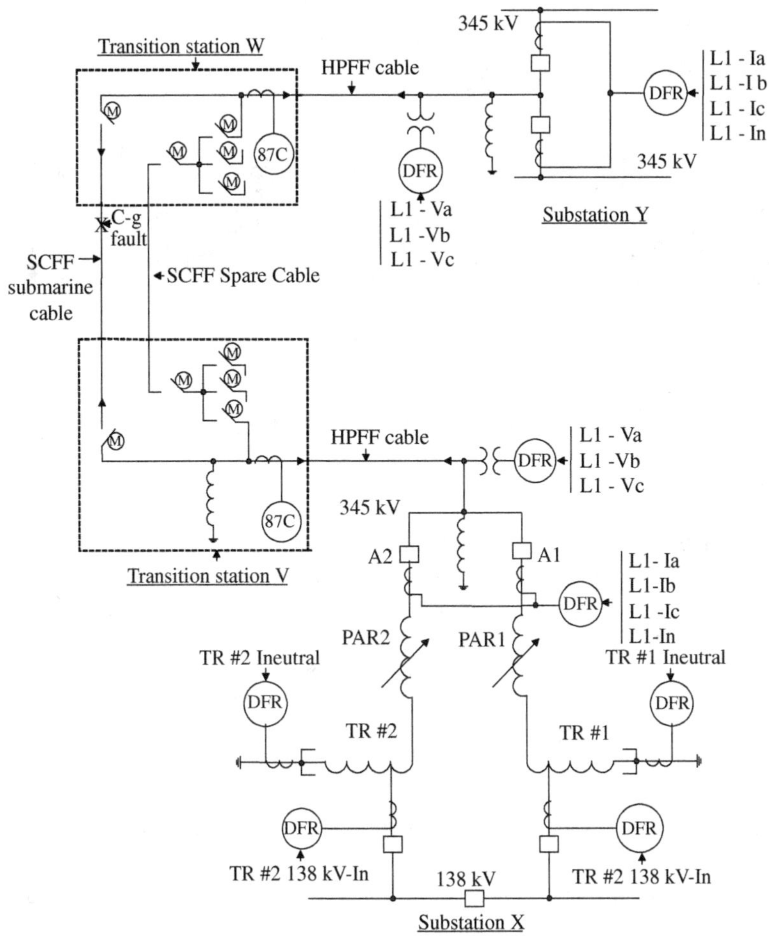

Fig. 7.1 System one-line diagram showing the DFR-monitored voltages and currents at substations X and Y.

Description of the Protection Systems The 345-kV cable is protected by three relaying systems. System 1 is a solid-state permissive overreaching transfer trip (POTT) utilizing a dedicated pair of fiber optics, system 2A is a POTT electromechanical relay system with audio tone equipment applied on leased telephone circuits, and the third system 2B is a solid-state current differential relay with an integral fiber optic interface utilizing a dedicated fiber optic pair on a separate cable. System 2B shares the secondary current transformer circuits and a dc system with system 2A. Three 3000 : 5 A multiratio C800 accuracy current transformers are included with the GIS at both transition stations for use with the SCFF faulted-phase targeting system.

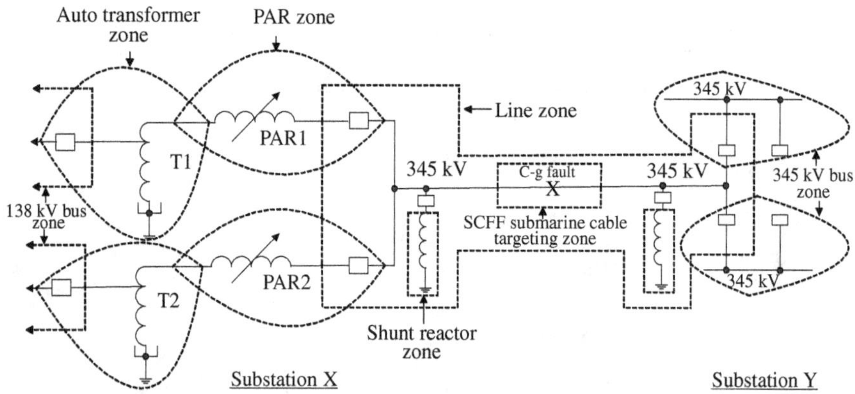

Fig. 7.2 Zones of protection for the 345-kV cable and associated termination equipment.

Underwater SCFF Cable Faulted Phase Targeting System For the purpose of targeting the faulted phase and to facilitate the switching of the submarine cable, the current differential relaying system 87C shown in Fig. 7.1 is added to the self-contained cable between transition stations V and W. During the early planning stage for the protection systems, no single-phase current differential relay system using a fiber optic communication link was readily available. Hence, a three-phase current differential relaying system with phase instantaneous overcurrent relays was substituted. The relay system is interfaced to the fiber optic link via a voice channel of the secondary fiber optic communication systems.

The phase instantaneous relays are set above maximum load currents and the currents in the unfaulted phase $(I_0 - I_1)$ during internal phase-to-ground faults. Operation of the current differential relay and the proper phase instantaneous relay will provide a target for the faulted phase identification of the SCFF submarine cable.

Zones of Protection The zones of protection shown in Fig. 7.2 are designed to provide quick restoration of the cable following the occurrence of faults. The 345-kV cable circuit is protected by the following overlapping redundant zones of protection:

1. *Line zone.* This extends from substation X breakers A1 and A2 PAR side bushing CTs to substation Y bay 2 free-standing CTs.
2. *SCFF cable zone.* This extends from current transformers included with the GIS at transition stations V to CTs included in transition station W.
3. *Shunt reactor zones.* Although protected by dedicated redundant relaying systems, all shunt reactors are also included in the line relaying zone.
4. *Phase angle regulating (PAR) zone.* This extends from the line-side bushing CTs of substation X to the autotransformer high-side bushing CTs.

5. *Auto transformer zone.* This extends from the PAR load-side bushing CTs to substation X 138-kV breaker bus-side bushing CTs.
6. 345- *and* 138-*kV bus zones.*

Description of the Incident The cable was damaged during the burying of a gas pipeline extension in the water. The ground fault was initiated when an anchor from a barge hit the submarine cable and began to drag it. The fault was cleared successfully by operation of the relays at both ends of the circuit to trip the associated CBs. Divers were used to assess the extent of the damage before repairs. Physical and electrical tests of the remaining phases were done to ensure that they were unaffected. During the incident, insulating fluid, which also cools the SCFF cable, was leaking into the water at a very slow rate (as designed). The slow flow rate of the cable dielectric fluid to the water is designed to prevent water penetration to the failed cable. The insulating fluid is also known as a dielectric fluid, designed to be nontoxic and lightweight. The small insulating fluid leak will therefore reach the water surface and immediately begin to disperse and evaporate.

Sequence of Events Obtained from Analysis of the DFR Records The DFR record of trace L1-In in Fig. 7.3 reveals that the fault incident point is occurring at a. At this point, instantaneous collapse of the C phase voltage occurred coupled with an increase in the C phase current and the presence of ground current. This will lead to the conclusion that the fault is C-g, lasting for 2.5 cycles, and that the arc-over occurred near the voltage peak. The fault was in the SCFF submarine cable section as determined by the targeting of the differential relay associated with the SCFF cable.

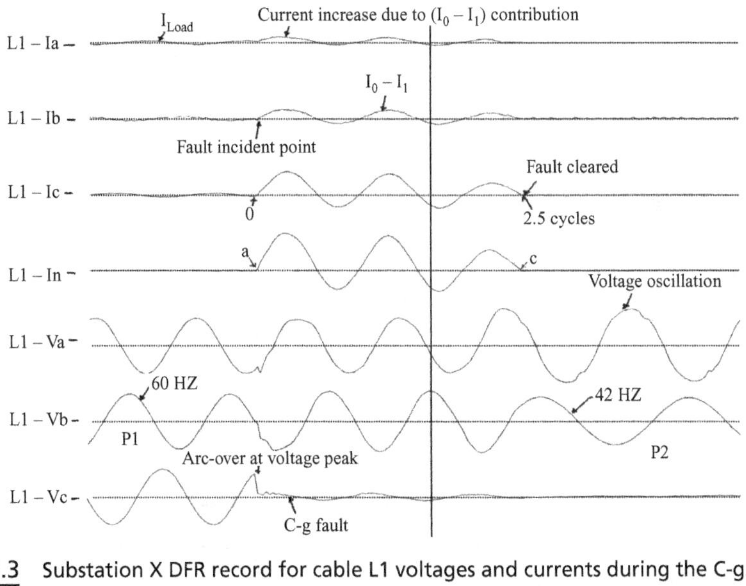

Fig. 7.3 Substation X DFR record for cable L1 voltages and currents during the C-g fault.

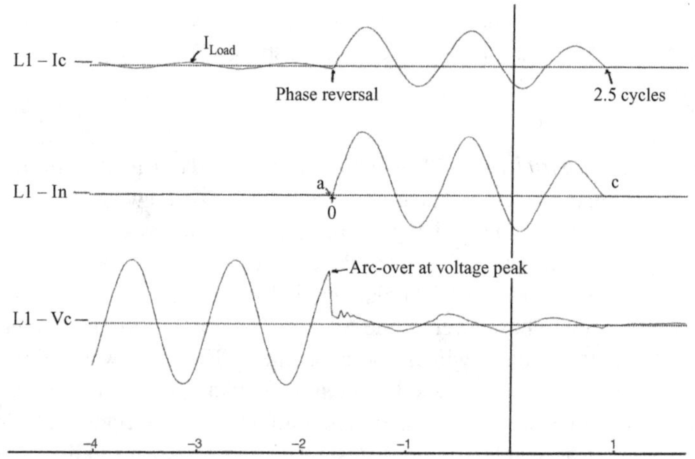

Fig. 7.4 Substation X expanded DFR record for cable L1 faulted phase voltage and current.

The faulted phase was confirmed by the phase C instantaneous fault detector relay target and DFR analysis. The DFR record in Fig. 7.4 shows a time expansion of the faulted phase currents and voltage. Figure 7.5 reveals that one of the two parallel branches, which consist of PAR2 in series with autotransformer TR2, tripped its associated CB A2 to clear the fault in two cycles (0.5 cycle earlier than the CB A1 fault

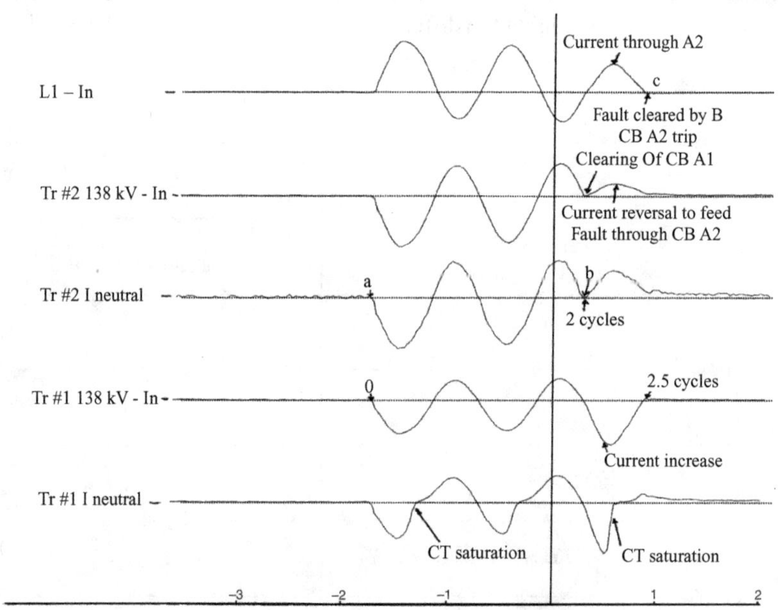

Fig. 7.5 Zero-sequence current flows for the circuit breaker A1 and A2 branches at substation X during the C-g fault.

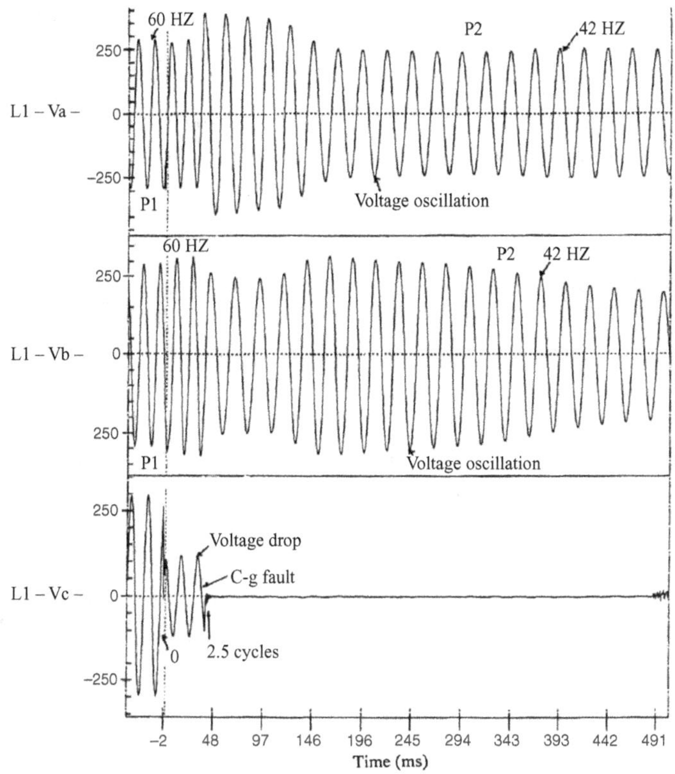

Fig. 7.6 Substation Y DFR record for cable L1 voltages during and after the C-g fault.

clearing). The autotransformer is now connected to the 138-kV system and is still acting as a zero-sequence source for an additional 0.5 cycle. Current is then forced to reverse and continue to supply the fault. The DFR record Fig. 7.6 reveals that substation Y cleared the fault in 2.5 cycles.

Power System Phenomena

1. Arc-over near the voltage peak, indicating an insulation failure as shown in trace L1-Vc in Fig. 7.3. The hitting and lifting of the cable by the anchor is considered a slow movement process for a fault-creation mechanism, resulting in the breakdown of insulation near the voltage peak.
2. Oscillation of energy between the cable charging system and the shunt reactors used for cable compensation, as shown in trace L1-Vb in Fig. 7.6.
3. Current flow in the unfaulted phases, where ground fault current is fed from two sources, which may cause the current distribution factors for the zero-sequence current to differ from the positive-sequence current distribution factor, as shown in traces L1-Ia and L1-Ib in Fig. 7.3.

4. Current reversal occurring on phases B and C of line L1 when the initial loading current is flowing from substation. X to substation. Y is altered by the contribution of the unfaulted phases of $I_0 - I_1$, as shown in Figs. 7.3 and 7.4.
5. Current reversal occurring when CB A1 clears the C-g fault half a cycle faster than CB A2. Autotransformer T1 with its tertiary winding continues to supply zero-sequence current to the fault via CB A2 branch. The $3I_0$ flow reversed their course of flow for an additional 0.5 cycle, as shown in traces Tr #2 138 kV-In and Tr #2 I neutral in Fig. 7.5.
6. CT saturation, shown in trace Tr #1-I neutral in Fig. 7.5. The CT connected in the neutral of auto transformer T2 is exhibiting saturation, as indicated in the secondary current.

Corrective Actions

1. Following the determination of the cable possible faulted area, which is the SCFF, the system was reconfigured to isolate the faulted cable as shown in Fig. 7.1. The spare cable was substituted in place of phase C to restore the cable using the SF6 disconnect switches.
2. The faulted submarine cable was cut and a cap was applied to stop insulating fluid flow in the water. Later, two new splices were installed to repair the failed cable.

Lessons Learned

1. Optimum design of protection zones can lead to quick identification of the faulted area of the power system, which enhances power system operation and restoration.
2. Numerical current differential relays presently available, used in fiber optic communication links, provide elaborate fault record information, including that on the faulted phase. Use of the new systems results in simplification of relaying schemes compared with the SCFF targeting system described in this case study.

Case Study 7.2: Undesired Operation of a 138-kV Cable Feeder Differential Relay During the Commissioning Phase of a 500-MW Plant

Abstract In this case study we analyze the undesired trips of an underground cable feeder during the soaking process of a generator step-up (GSU) transformer. The undesired trips were due to a design error that resulted in reverse CT cable differential connections at plant X for both primary and secondary numerical current differential relaying systems. We describe the power system and associated protection and highlight how the error occurred that caused the undesired trips. In addition, we provide

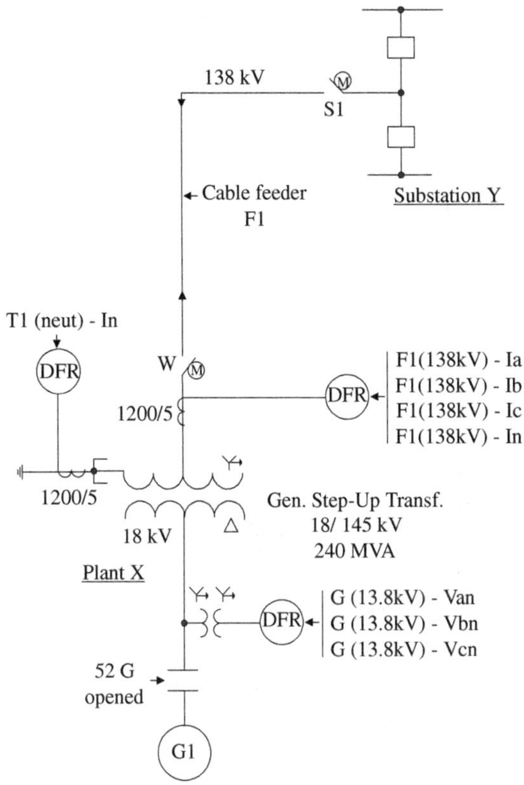

Fig. 7.7 Plant X displaying DFR-monitored voltage and current signals.

the basic CT polarity concepts, an analysis of relay oscillography records, and lessons learned.

Description of the System and Associated Protection A 240-MVA generator step-up transformer is connected to the 138-kV transmission system via a short (less than a mile) solid dielectric cable L1. The feeder F1 is protected by dual numerical current differential relaying systems from diverse manufacturers. Two separate fiber optic cables are used for protection communication between plant X and 138-kV substation Y (Fig. 7.7) As shown in Fig. 7.8, numerical current differential relays 87FP and 87FS are fed from separate CTs, communicating using physical separate fiber optic cables and tripping separate breaker trip coils.

Description of the Undesired Trip The 138-kV system was being prepared to energize an underground cable feeder F1 from substation Y, for the purpose of soaking transformer T1 at plant X. Upon closing disconnect switch S1 to prepare for the soaking, a cable differential relay operation took place for feeder F1. In addition, an arc was observed with the closing of the disconnect switch. Apparently, CB A1 had

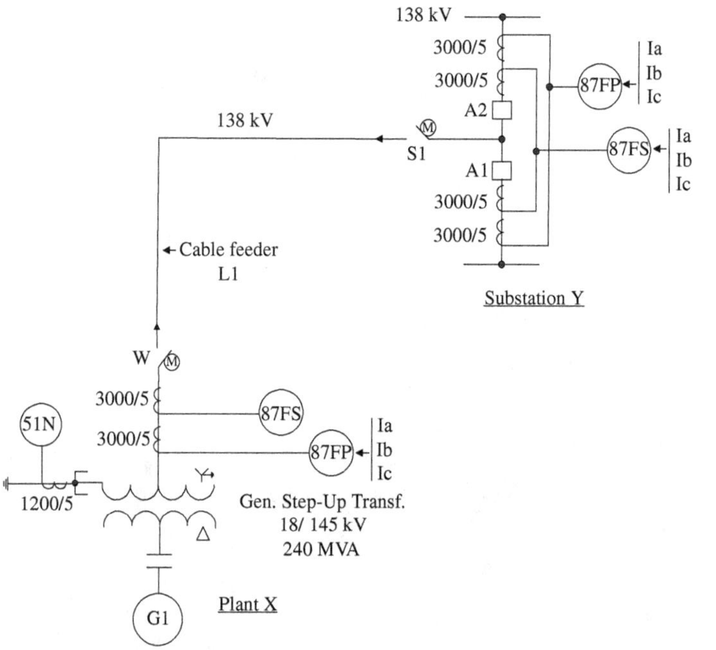

Fig. 7.8 System one-line diagram with the cable protection system and the numerical relay oscillography record–monitored currents.

unequal opening operation, prior to closing disconnect switch S1, that resulted in phase B remaining in the closed position while phases A and C were opened. As a result, the GSU transformer was energized when the phase B disconnect switch S1 was closed. The breaker was isolated and a soaking process was attempted by using breaker A2 only. No attempt was made to investigate the differential relay undesired operation that took place earlier when disconnect switch S1 was closed. Breaker A2 was also tripped as soon as it was closed to energize transformer T1. Subsequent analysis for both trips revealed a design error that resulted in CT secondary connection errors (reverse) at plant X for both primary and secondary numerical current differential relaying systems (Fig. 7.9).

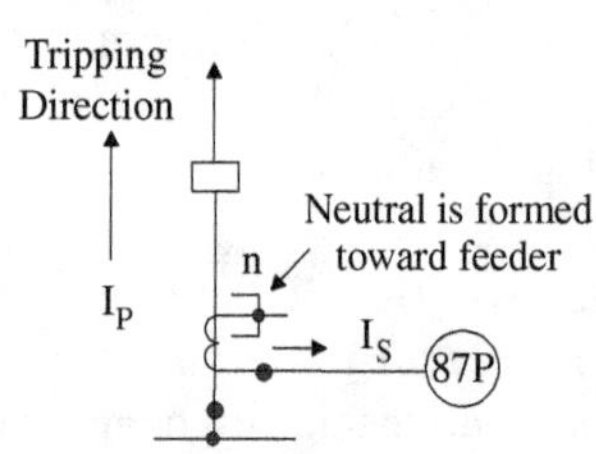

Fig. 7.9 Determination of the side where the CT neutral is formed according to the relay manufacturer's requirements.

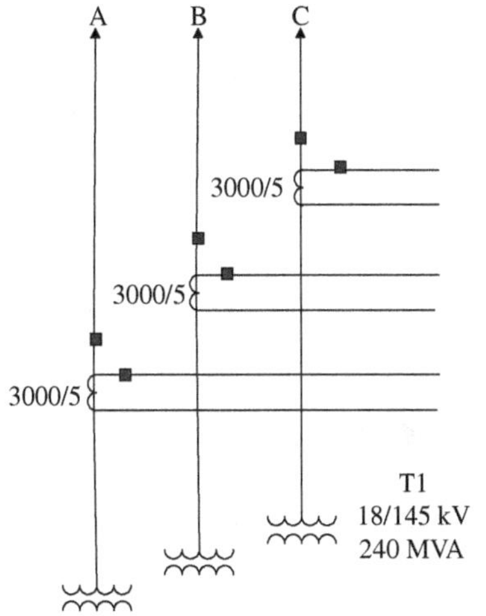

Fig. 7.10 Transformer T1 nameplate showing CT polarity away from the transformer.

Analysis of the Causes of the Undesired Trips During the process of verification of CT polarity lead designation, an error was made by field engineering personnel. The transformer manufacturer's drawing shown in Fig. 7.10 indicates that the 138-kV-side CT lead polarity markings are away from the associated transformer, which is different from the original drawing, where polarity markings are toward the transformer used to connect the differential relay. Field engineering personnel apparently reacted to the new information and committed a mistake by changing the CT neutral formation from the earlier checked correct system side of the relay to the transformer side, which disagrees with the relay manufacturer's drawing shown in Fig. 7.11. This resulted in a violation of the manufacturer's typical required connections and subsequent undesired operation of both relay systems by transformer T1 inrush currents during the energization phase for the soaking process. Analysis of the first feeder trip due to the energization of one phase only could have prevented the second unsuccessful energization attempt. Figure 7.12 illustrates the commissioning of the differential relay with the wrong three-line relay CT connections, where the neutral is formed toward the transformer. The wrong connection resulted in reversal of the CT connection, which forced the through-inrush-current condition to be a cable differential relay operating quantity.

Connection of the Current Differential Relay According to the Numerical Relay Manufacturer's Drawings The CT neutral formation side will control how the current will flow to the relay. The current transformer neutral connection side to the numerical differential relay must match the CT neutral in accordance with the numerical relay manufacturer's published connection diagram, regardless of whether the CT marked polarity is toward or away from the associated power equipment. This will guarantee correct performance of the differential relay connection in restraining for through current flow conditions and tripping only for internal faults.

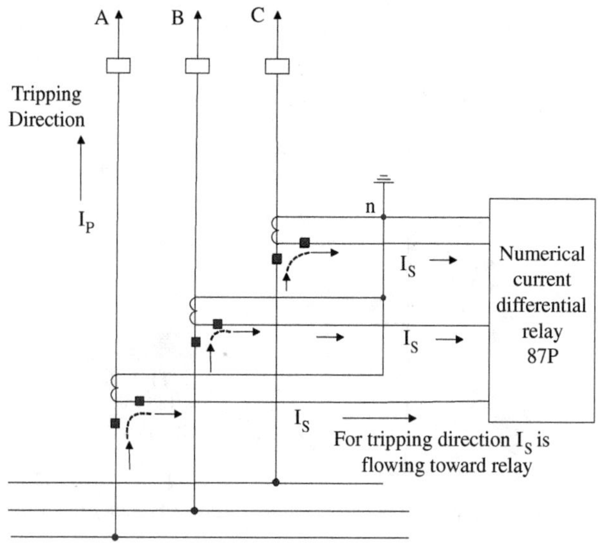

Fig. 7.11 Relay manufacturer three-line ac diagram connection to the numerical differential relay to sense internal faults.

Analysis of the DFR and Relay Event Records Figure 7.8 illustrates the feeder dual differential relay connections and the relay oscillograph-monitored currents. Figure 7.13 shows a numerical relay 87FP oscillograph record during single-phase energization, when transformer inrush current is present only on phase B, while

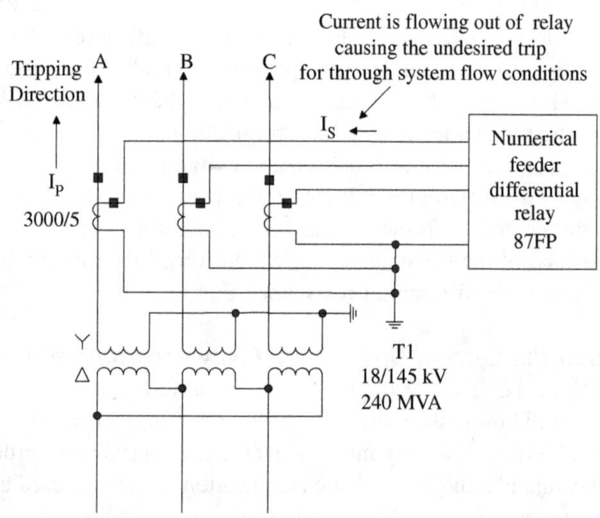

Fig. 7.12 Commissioning of the differential relay with the wrong three-line relay CT connections.

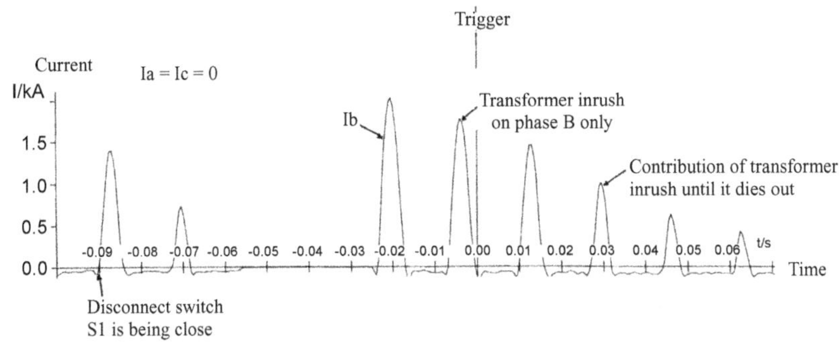

Fig. 7.13 Substation Y relay 87FP oscillography record showing the phase B inrush current.

$I_a = I_c = 0$. This occurred when disconnect switch S1 was closed, drawing an arc and accidentally energizing transformer T1. Circuit breaker A1 had unequal opening operation prior to closing disconnect switch S1, resulting in phase B of the breaker remaining in the closed position while phases A and C were opened. The phase B transformer inrush current generated relay trip outputs for both the primary and secondary numerical relays protecting the 138-kV cable feeder L1. The 87FP relay also generated an oscillograph record showing a current trip level of 505 A. Breaker A1 was isolated and an attempt was made to energize the 138-kV feeder L1 using breaker A2 at substation Y. No field investigation was conducted to determine the causes of the relay 87FS and 87FP trip targets. The relay fault record shown in Fig. 7.13 reveals only 11 cycles for the phase B inrush current.

Figure 7.14 shows a numerical differential relay 87FS oscillograph record for substation Y during energization by CB A2. The three-phase inrush current is represented by primary current when the breaker tripped after 4 cycles. The inrush peak current recorded is about 2700 A.

Figure 7.15 shows a numerical differential relay 87P oscillograph record for substation Y during energization by circuit breaker A2. The three-phase inrush current is represented by secondary current interruption after 4 cycles. The inrush peak current is about 4.5 A secondary, which when multiplied with the CT ratio of 600:1 will produce the same 2700-A primary current recorded by the relay 87FS.

Figure 7.16 shows a numerical differential relay 87FP oscillograph record for plant X during energization by CB A2. The record at plant X is identical to the record of Fig. 7.15 at substation Y, as it should be. The differential remote and local relay records for the through-inrush-current condition is identical, as proved by Figs. 7.15 and 7.16.

Figure 7.8 shows the plant one-line diagram and DFR-monitored currents and voltages. The plant X DFR record shown in Fig. 7.17 reveals the cable feeder L1 voltages and currents during the GSU transformer inrush energization. In addition, Fig. 7.17 shows the feeder F1 neutral current, which is similar and opposite to the transformer T1 neutral current. The DFR record is identical to the local and remote oscillograph records of both numerical relays 87P and 87S.

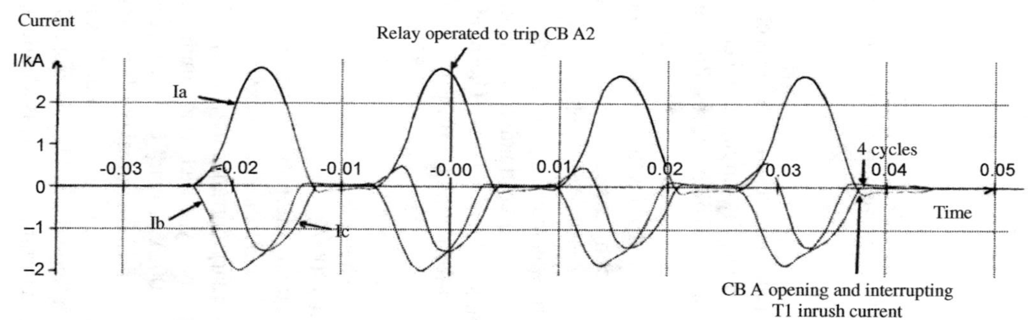

Fig. 7.14 Substation Y relay 87FS oscillography record with four-cycle three-phase transformer inrush currents.

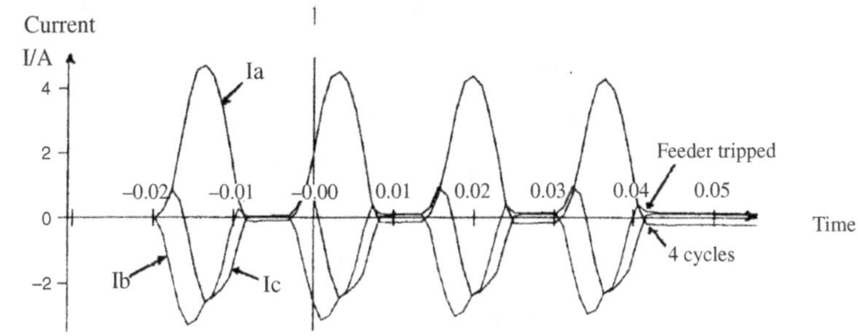

Fig. 7.15 Substation Y relay 87FP oscillography record with four-cycle three-phase transformer inrush currents.

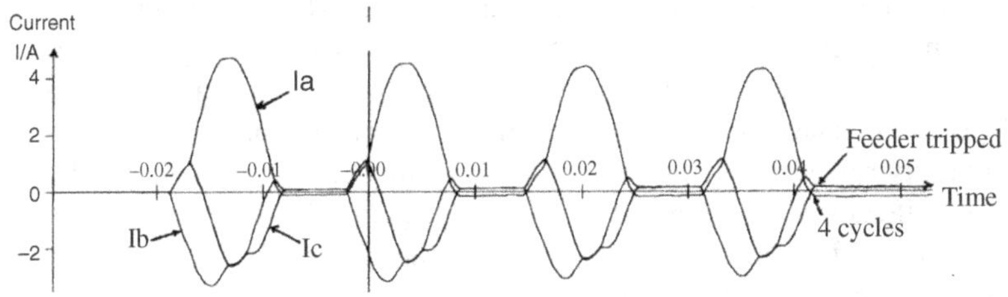

Fig. 7.16 Plant X relay 87FP oscillography record showing the same feeder inrush current as fed from the Y end.

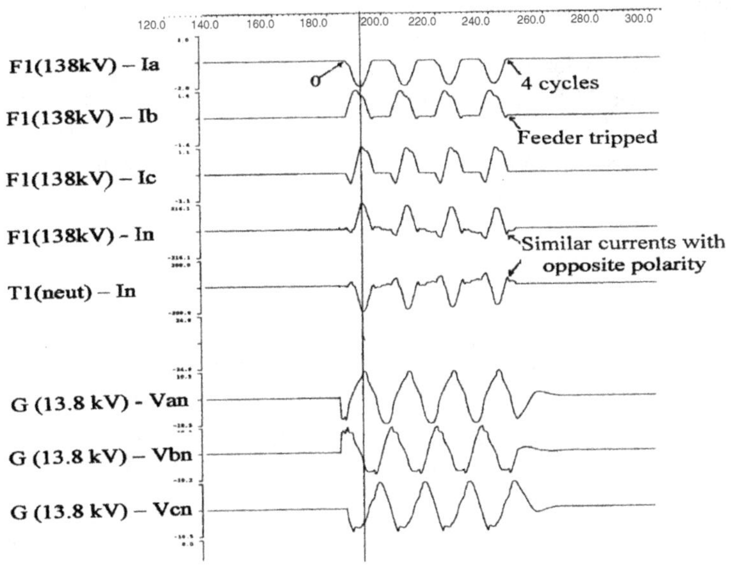

Fig. 7.17 Plant X DFR record showing 138-kV cable feeder three-phase voltages and currents during transformer inrush energization.

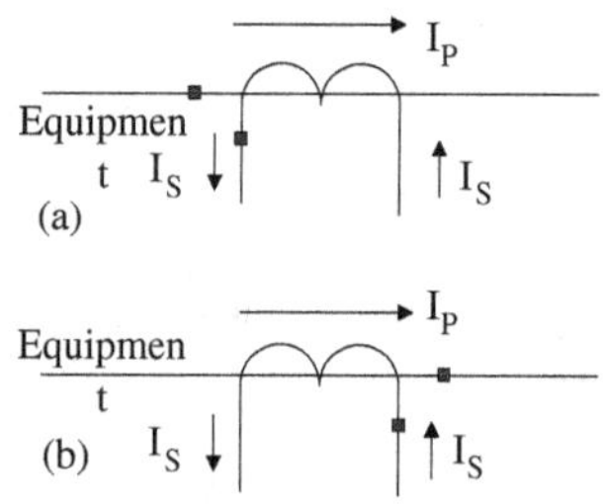

Fig. 7.18 Polarity markings for current transformers.

Basic Definition of Polarity CT polarity is normally away from the associated equipment (transformer or breaker); however, equipment nameplates should always be used to confirm polarity markings. Figure 7.18 shows the two options of having CT polarity markings either toward or away from the associated equipment. Figure 7.18 also represents the polarity rule for current transformers. Knowing the primary current flow, the secondary currents can be deduced by the polarity rule principles. The polarity rule states that current in polarity for the primary side is substantially in phase with current out of polarity on the secondary side. It should be observed that the direction of the secondary current is the same regardless of whether the polarity marks are together on one side or on the other.

Current Direction as Affected by Neutral Formation For a given primary direction (toward the feeder), the current in the secondary lead will maintain the same direction regardless of whether the polarity of the CT is toward or away from the associated equipment. In other words, there is only one correct way to form the neutral at the CT secondary leads for a given unique primary current flow toward the relay tripping zone. The CT neutral formation side will control how the current will flow to the relay. Figure 7.19 shows that formation of the neutral at the CT polarity marking and with the assumed primary current I_P flow that is shown in Fig. 7.18, the

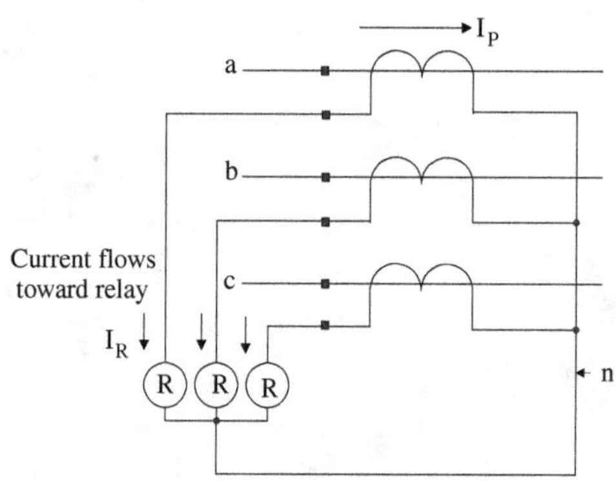

Fig. 7.19 Formation of neutral connection on the nonpolarity side of the CTs.

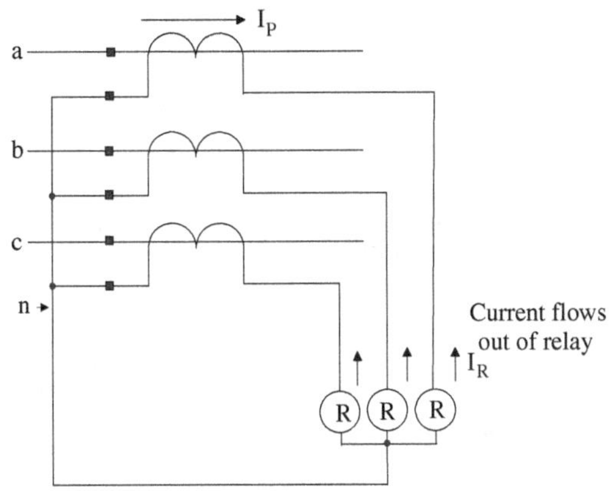

Fig. 7.20 Formation of neutral connection on the polarity side of the CTs.

corresponding secondary current I_R flow will be going into the numerical relay R. However, by forming the neutral on the nonpolarity side of the CT, the secondary current I_R will flow out of the relay, as shown in Fig. 7.20.

Simple Rule to Connect CTs Correctly to Numerical Differential Relays

The CT neutral formation side will control how the current will flow to the relay. The numerical differential relay current transformer neutral connection has two options: to be formed either toward the cable feeder or toward the transformer bank. The following rule will guarantee correct performance of the numerical differential relay: The current transformer neutral connection side must match the CT neutral, in accordance with the numerical relay manufacturer's published typical three-line ac connection diagram, regardless of whether the CT marked polarity is toward or away from the associated power equipment. The rule was applied to connect the differential relay correctly, as shown in Fig. 7.21.

Lessons Learned

1. Undesired trips should always be investigated to arrive at the causes of the trip prior to reenergization of the system. If field analysis had been carried out on the first feeder trip due to the energization of one phase only, the second unsuccessful energization attempt could have been avoided.
2. The CT neutral formation side will control how the current will flow in the secondary CT connection to the relay for a given current flow in the primary CT circuit.
3. Correct connection of this type of current differential relaying systems requires that their design and wiring be done in accordance with the numerical relay manufacturer's published typical three-line ac diagram. By applying this rule at each end of the feeder for the numerical current differential relay, the relay will

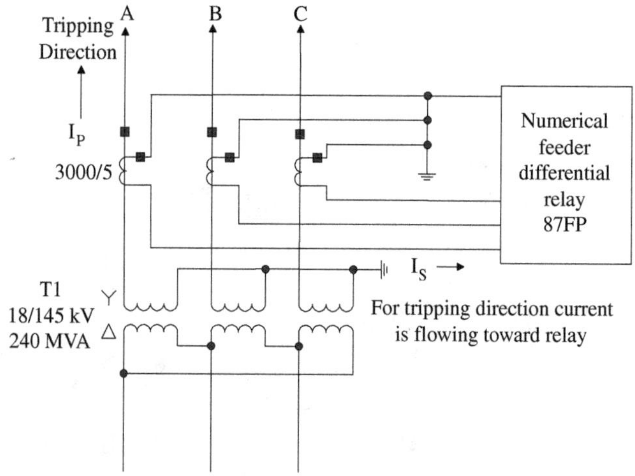

Fig. 7.21 Correction of the differential relay connection to match the manufacturer's drawing.

be guaranteed to function correctly as intended during external faults or through-current-flow situations.

4. For a given primary current flow, the secondary differential or directional relay current flows must be deduced using the polarity rule: Current flowing in at the polarity mark of one winding flows out of the polarity mark of the other winding. Both currents are substantially in phase.
5. Overall, thorough review of current differential schemes should always be performed prior to the first energization of the feeder.

Case Study 7.3: Phase-to-Ground Fault Caused by Failure of a 345-kV Cable Connection Between the Generator and the Switchyard, Accompanied by Mechanical Failure of One of the Cable Pot Head Phases

Abstract A high-pressure fluid-filled (HPFF) cable connecting the high side of a generator step-up (GSU) transformer to a 345-kV substation failed, causing an L-g fault. The fault was cleared from the system successfully. In this case study we show that the use of relay targets coupled with complete knowledge of the protection zones can lead to the correct analysis and determination of the faulted area. Correct determination is achieved despite the mechanical failure of the cable pot head porcelain, which can give a false indication of the faulted area. Non-operation (no targets) of the complete relaying system which overlaps the circuit breaker rules out the possibility that the solid A-g fault was not within the bushing and that the fault was due to cable failure when only cable protection systems targets were obtained.

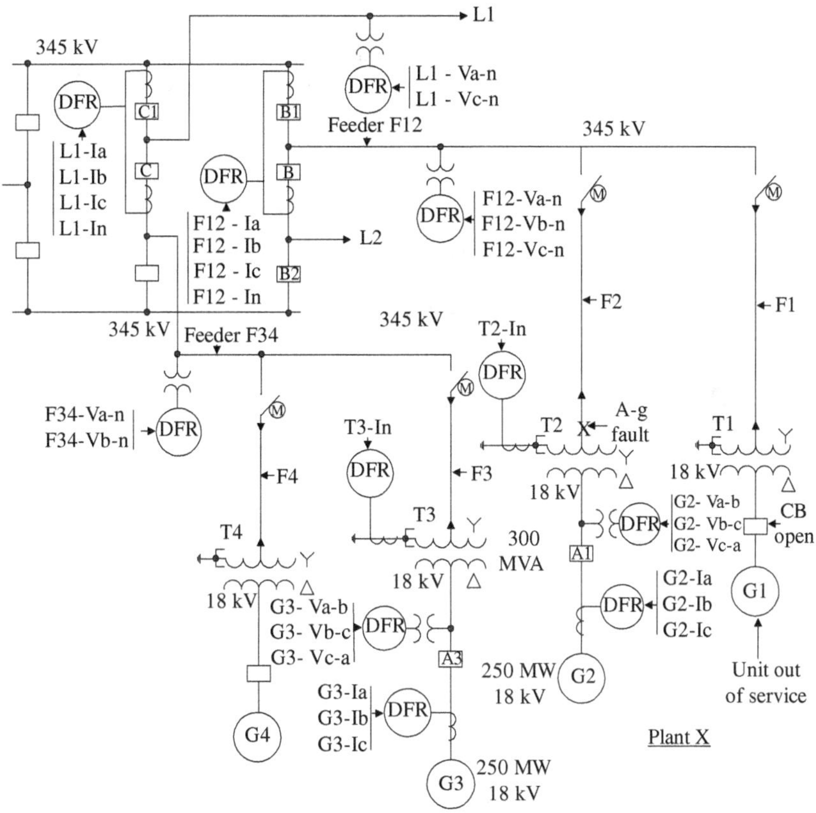

Fig. 7.22 DFR-monitored signals for plant X and the associated 345-kV substation.

Description of the Protection Systems The system shown in Figure 7.22 is protected by dual protection systems. As illustrated in Fig. 7.23, the generator protection zone is covered by dual differential relays 87GP and 87GS. Transformer T2 (GSU) is also protected by dual harmonic restraint differential relays 87TP and 87TS. The 345-kV HPFF cable lead connection between the GSU transformer and the 345-kV switchyard is protected by a pilot wire current differential relay 87C and a backup relaying system located at the switchyard end of the cable. The phase backup protection is a phase distance relay 21S, and the ground backup protection is a high-speed directional time overcurrent 67GS relay polarized from the zero-sequence voltage. The 345-kV substation lead that connects the unit 1 and 2 cables is protected by high-impedance bus protection device 87BP. The 21S and 67GS relays provide backup protection for 345-kV cable leads for units G1 and G2 as well as the substation lead connection.

Description of the Cable Failure A high-pressure fluid-filled cable connecting the high side of a generator step-up transformer to a 345-kV substation failed, causing

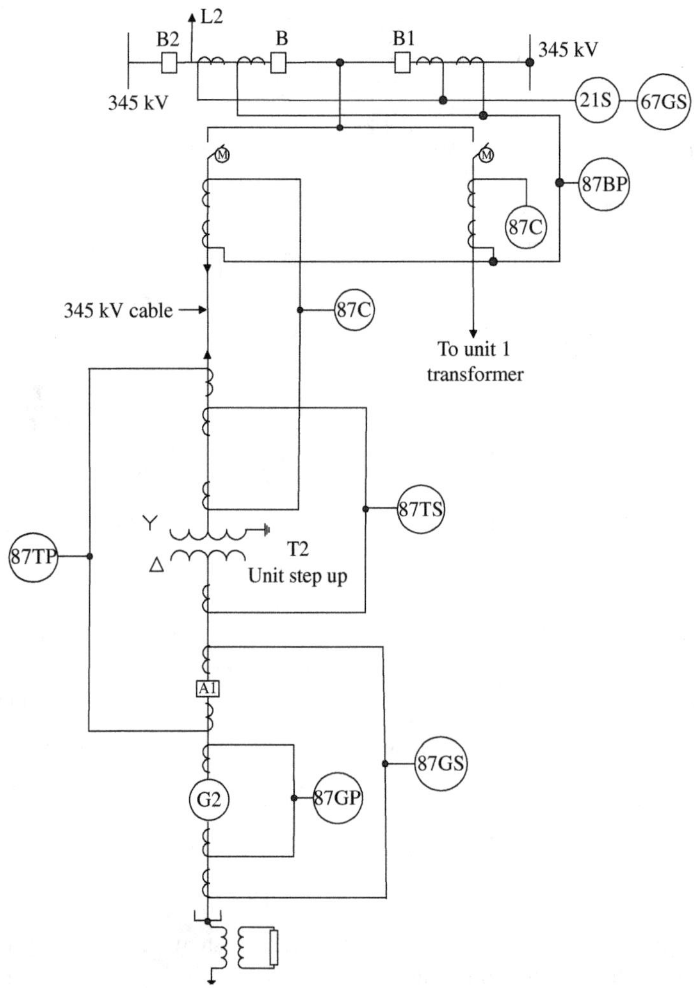

Fig. 7.23 Protection systems and zones for the unit connection to the 345-kV substation.

a phase A-to-ground fault. The fault incident point is at the voltage peak, as illustrated in trace F34-Va-n in Fig. 7.24. This is an indication of cable insulation failure, leading to the ground fault. Examination of the failed cable area reveals burn marks and a fault opening area about 8 inches wide that appeared to have opened as a result of the insulating fluid pressure surge during the failure.

Location of the Fault and the Failed Power Element In this case study we demonstrate the use of relaying targets and the overlapping feature of protection application to determine the faulted element. Figure 7.25 shows the overlapping protection zone created by transformer T2 differential protection and 345-kV cable pilot wire current differential relay 87CP. Since no targets were obtained for the

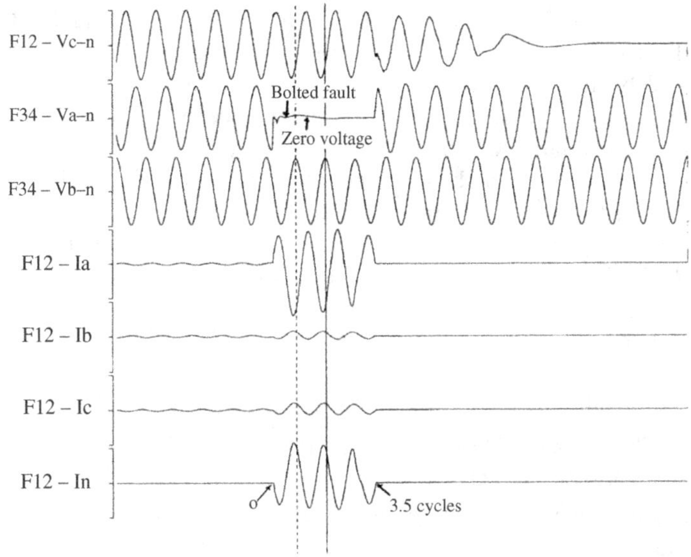

Fig. 7.24 Substation Y DFR record showing the 345-kV feeder F12 current contribution to the A-g fault.

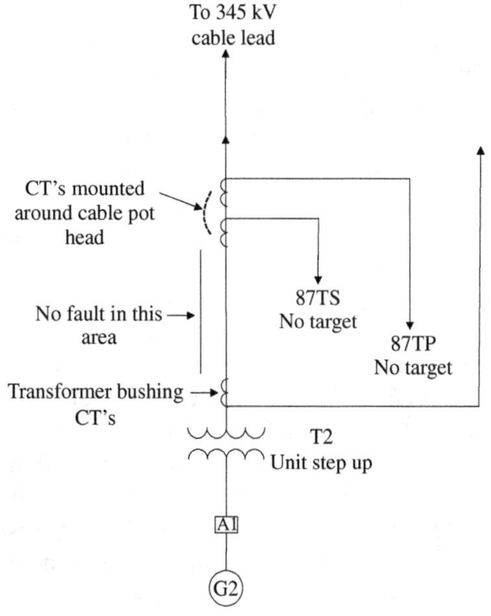

Fig. 7.25 One-line ac diagram showing the protection overlap and the bushing area clear of faults.

T2 differential relays 87TP and 87TS, the pot head area is free of faults. This is true despite the damage reported for the pot head during the fault. The pot head damage was caused by the excessive mechanical forces created by fault currents of an approximate current value of 20 kA fed by the generator positive- and negative-sequence currents and GSU unit transformer zero-sequence current source. Based on obtaining a relay target for the cable pilot wire current differential (only) and no targets for transformer differential relays, the failure must be in the cable. As stated earlier, the failure is located in the 345-kV cable lead and not in the pot head, because of the 87CS cable target, without obtaining any targets for 87TP and 87TS.

Crucial Factors Involved in Determining Fault Location

1. Understanding of the relay systems.
2. Understanding of the relay zones.
3. Examination of the relay targets.
4. Examination of the overlapping relay elements around the adjacent protection zones.
5. Faults, can sometimes, as in this case, be located without the need to analyze DFR records.
6. DFR records are then used to classify the fault and confirm the conclusion arrived at by the analysis documented in steps 1 to 5.

Analysis of the DFR Records Figure 7.22 shows the system one-line diagram and DFR-monitored voltages and currents. The DFR record shown in Fig. 7.24 reveals a cable failure when the fault incident point occurred at the voltage peak. The DFR record reveals an undervoltage on phase A, an overcurrent on phase A, and the presence of neutral current, thus confirming an A-to-ground fault. The A-g fault is bolted (zero voltage) and fed from the system as well as unit G2. Positive- and negative-sequence currents are fed from unit G2, while zero-sequence currents are supplied from the unit G2 GSU transformer T2.

The fault was cleared from the system side in 3.5 cycles by tripping 345-kV CBs B and B1, as illustrated in Fig. 7.24, by current traces of feeder F12. Finally, The fault was cleared from the generating side in 5.5 cycles by tripping 18-kV CB A1, as illustrated in Fig. 7.26, by current traces of unit G2. According to the GSU transformer T2 phasor diagram, the 345-kV high-side A-g fault on the grounded-wye winding will be coupled as a phase-A-to-phase-B fault on the 18-kV low-side delta winding. Traces G2-Ia and G2-Ib confirm the phase-to-phase fault current flow on the generator side. In addition, trace G2-Ic confirms a clearing time of 3.5 cycles when the system side cleared and no source is available for the unit in its pumping mode. The time duration from 3.5 to 5.5 cycles shows a reduction of zero-sequence currents for trace T2-In; this is due to isolation of the system's strong zero-sequence source. During the same period, current trace G2-Ia in Fig. 7.26 shows an increase in magnitude due to the disappearance of the loading current component.

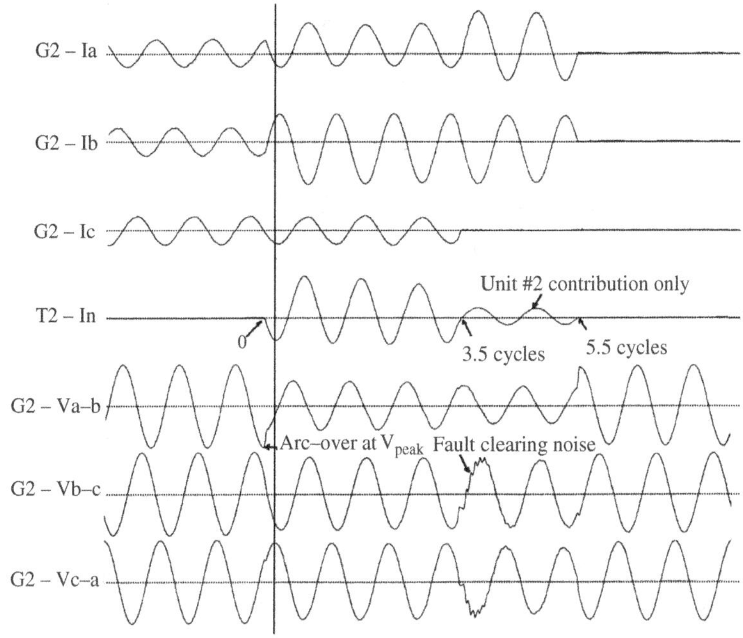

Fig. 7.26 Plant X DFR record showing the generator G2 contribution to the A-g feeder fault.

Undesired Trip of 345-kV Line L1 The DFR record in Fig. 7.27 shows the disappearance of line L1 currents on all the phases around 3 cycles from initiation of the A-g fault in feeder F12. The DFR record also reveals the presence of line L1 voltages on phases A and C after the false line trip. This voltage presence will confirm the trip of line L1 at one end only, which is the 345-kV substation of plant X by tripping 345-kV CBs C and C1. The undesired trip was initiated by the Z1 distance-based solid-state terminal upon the loss of directionality during the close-in solid phase-to-ground fault in the reverse direction. These types of false trips of static relay terminals during reverse close-in faults have occurred infrequently throughout EHV transmission systems.

Undesired Trip of Unit 3 Figure 7.28 illustrates unit G3 currents and voltages during the A-g fault in feeder F12 cable. Trace G3-Ia reveals first current reversal at 2 cycles from the start of the record. This indicates that G3 was in a pumping mode and had to feed positive- and negative-sequence currents to the F12 ground fault. The same trace reveals the second current reversal 3.5 cycles from initiation of the fault. This point coincides with the interruption of zero-sequence current in trace T3-In, when the fault was cleared from the power system by tripping 345-kV CBs B1 and B2. The G3 current traces for phases A, B, and C reveal the tripping of unit G3 8.5 cycles after the fault incident point (5 cycles after system fault clearing) by tripping 18-kV CB A3. The undesired trip of the unit was caused by a

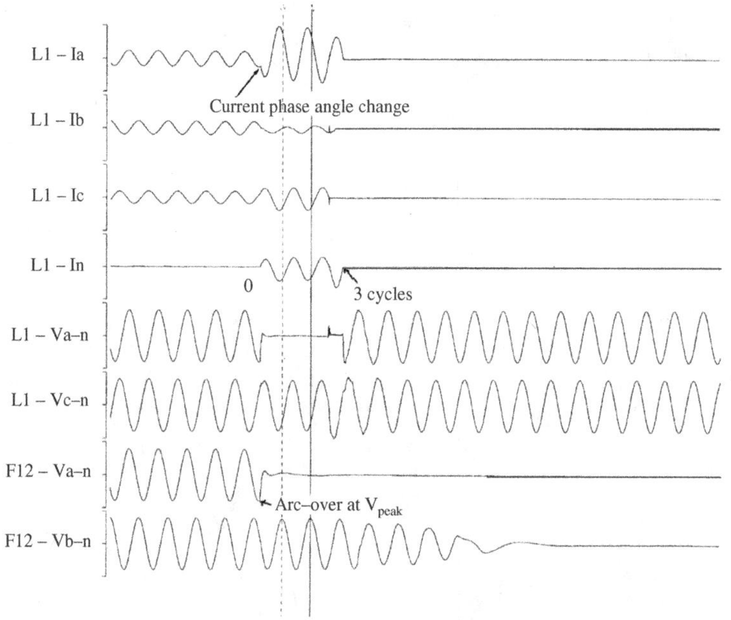

Fig. 7.27 Substation Y DFR record confirming the 345-kV line L1 undesired trip at the beginning of the feeder F12 A-g fault.

failure of the underfrequency relay, which is normally enabled during the unit pumping mode. The instantaneous relay output should integrate a timer instead of direct tripping of the unit.

Power System Phenomena

1. Insulation failure at the voltage peak, as shown in trace F12-Va-n in Fig. 7.27
2. Solid (bolted) phase to ground fault, as shown in trace F31 Va n in Fig. 7.25
3. Current reversal, as shown in trace G3-Ia in Fig. 7.28
4. Phase-to-ground fault on transformer grounded-wye side seen as a phase-to-phase fault on the delta side, as shown in Fig. 7.26
5. Loss of directionality for static line distance relay during close-in reverse faults

Corrective Actions

1. Plan replacement of the line L1 obsolete static relay terminal by a numerical relay.
2. Add a time delay for the underfrequency relay outputs during the unit pumping mode.

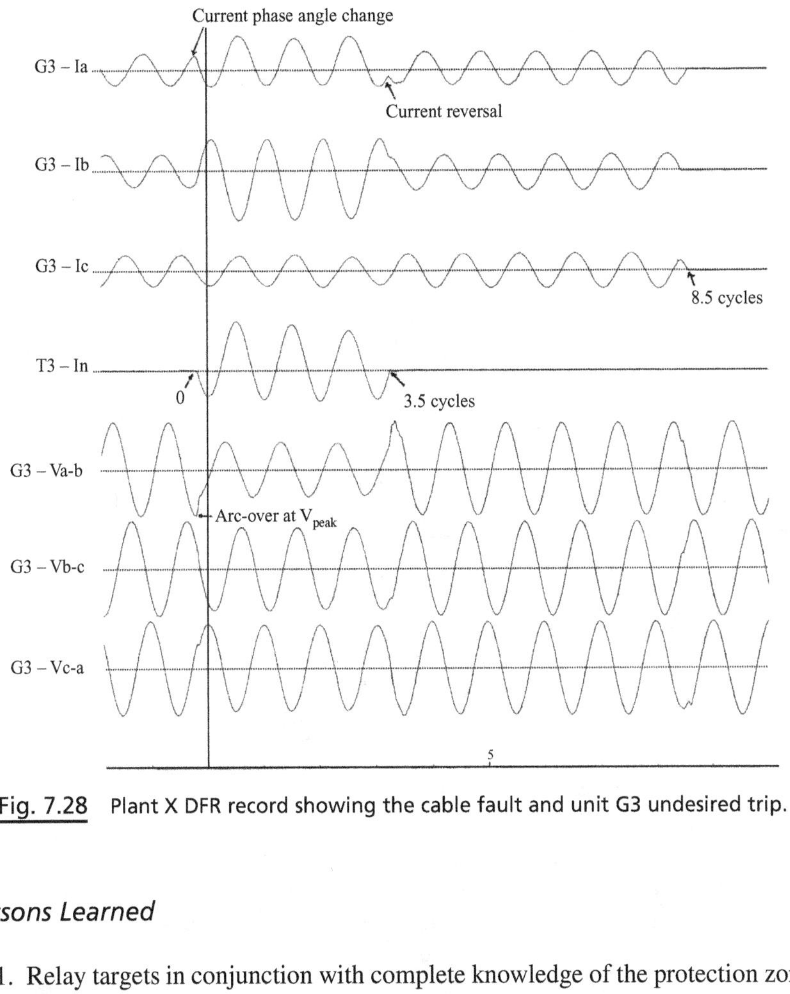

Fig. 7.28 Plant X DFR record showing the cable fault and unit G3 undesired trip.

Lessons Learned

1. Relay targets in conjunction with complete knowledge of the protection zones can lead to correct analysis and determination of the faulted area.
2. Optimum design of protection zones can lead to correct isolation of the faulted area and can enhance the process of restoration by identifying the faulted system elements.

Case Study 7.4: Troubleshooting a 345-kV Phase-to-Ground Fault Using Relay Targets Only

Abstract A 345-kV cable system, shown in Fig. 7.29, was out of service to perform routine maintenance. The system was restored and the phase angle regulators were tapped to control the power flow, allowing only 15-MW power flow through the cable. Nine minutes later the cable was tripped by operation of the relay systems. One of the 345-kV dead-tank breakers had failed, causing a line-to-ground fault. In this case

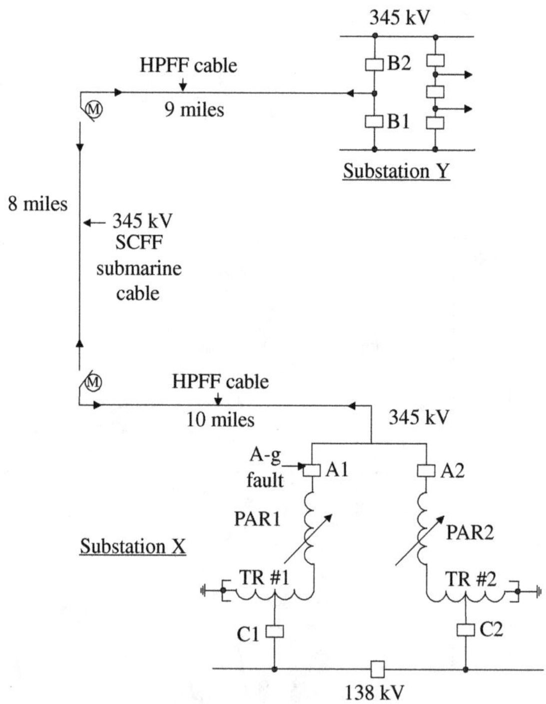

Fig. 7.29 System one-line diagram.

study we illustrate the troubleshooting and restoration of the system using relay targets only. we then introduce more fault data, which became available later, to confirm the initial analysis using very basic data: namely, relay targets only. We also cover verification of the power system short-circuit model, system phenomena, corrective actions, and lessons learned.

Description of the System As shown in Fig. 7.29, the 345-kV cable consists of 9 mile-high-pressure fluid-filled (HPFF) pipe-type cable, 8 miles of submarine self-contained fluid-filled (SCFF) cable, and 10 miles of HPFF pipe-type cable. At substation X the 345-kV cable terminates in a gas-insulated substation. Substation X also includes two 345-kV circuit breakers, two 450-MVA phase–angle regulating transformers and two 345/138-kV autotransformers. Substation Y is a conventional air-insulated breaker-and-a-half switching arrangement with 345-kV SF6 live tank breakers.

Description of the Protection Systems The 345-kV cable is protected by three relaying systems; system 1 is a static permissive overreaching transfer trip (POTT) utilizing a dedicated pair of fiber optic, system 2 is a POTT electromechanical relay system with audio tone equipment applied on leased telephone circuits, and system 3 is a static current differential relay with an integral fiber optic interface. The phase angle–regulating transformers are protected individually with dual electromechanical harmonic-restraint percentage differential relays.

Relay Targets for Cable Feeder L1 Tripping Operation

1. The 345-kV CBs B1 and B2 at substation Y tripped and the following 345-kV cable feeder L1 targets were reported:
   (a) Current differential relay device 87 L1 trip LEDs
   (b) Static distance package circuit breaker trip LEDs
   (c) POTT audio-tone permissive trip sent signal
2. The 345-kV CBs A1 and A2 and the 138-kV CB C1 at substation X tripped with the following phase angle–regulating transformer (PAR1) targets reported:
   (a) Phase B secondary differential electromechanical relay
   (b) Phase A secondary differential electromechanical relay
   (c) PAR lockout 86S
   (d) Phase A primary differential electromechanical relay
   (e) PAR lockout 86P
3. 345-kV cable feeder L1 targets reported:
   (a) Current differential relay device 87 L1 trip LED
   (b) Primary solid-state relay zone 1 ground
   (c) POTT audio-tone permissive trip sent signal

Analysis of the PAR1 Targets The PAR transformer is protected by an overall dual differential relaying systems. The primary differential relay is designed with CTs connected in wye. For a CB A1 fault, a phase A target was obtained only for the PAR primary differential relaying. The secondary differential relay is designed with CTs from the 345-kV PAR source and load side to be connected in delta. Figure 7.30 shows that the A relay is fed from the delta current $I_A - I_C$, while the B relay is fed from the delta current $I_B - I_A$. Therefore, two targets on phases A, B for the secondary differential relay were obtained for the common phase A and thus confirming the occurrence of the A-g fault. The correct PAR1 primary phase A target and the secondary phases A and B targets have confirmed the faulted phase as A. The PAR1 relay targets in conjunction with the 345-kV cable feeder protection targets have also confirmed the faulted overlapping zone around the 345-kV CB A1.

Confirmation of the Initial Analysis Using Relay Targets Only As shown in Fig. 7.31, overlapping of the cable protection zone and the overall PAR differential zone provides a definite suspected faulted area around the SF6 breaker A1 of the

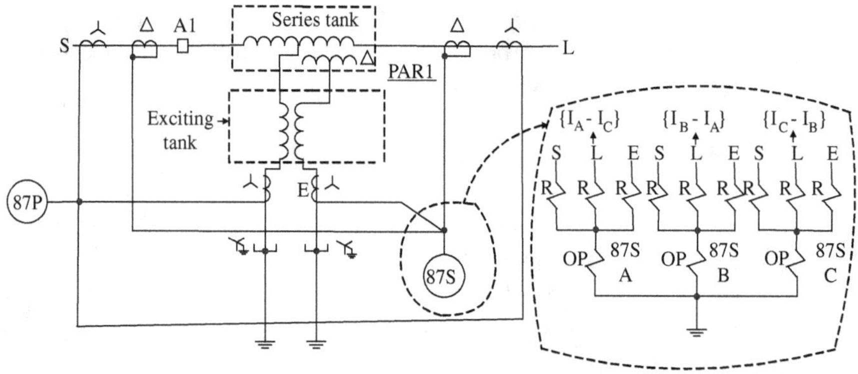

Fig. 7.30 One-line ac connection for PAR1 87P and 87S differential relays with the three-line ac for the 87S relay current input.

gas-insulated substation. Several SF6 samples were taken from six different compartments of breakers and disconnect switches for analysis. The suspected breaker zone exhibited 15 ppm, while the remaining gas zones showed undetectable ppm. This implies that gas analysis confirmed the initial A-g fault finding. In addition, DFR records also became available from substations X, Y, and Z. All also lead to a conclusion similar to that of relay target analysis: namely, an A-g fault inside the CB A1 compartment. The DFR record reveals that the fault

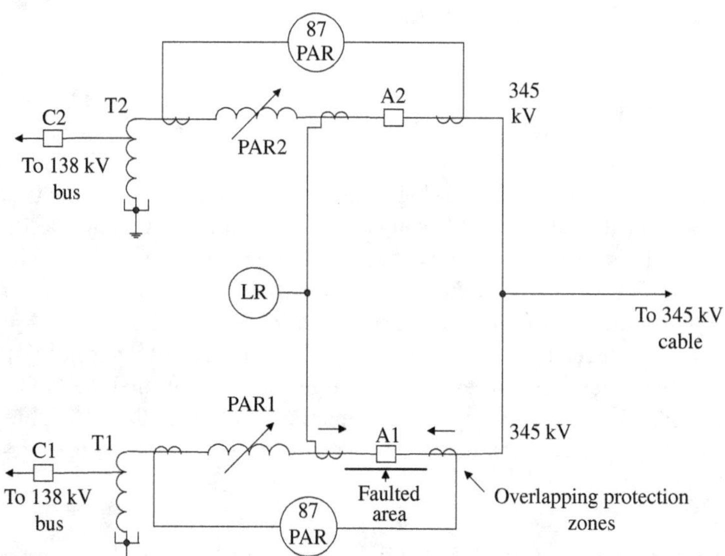

Fig. 7.31 One-line diagram showing overlapping protection zones around the faulted area.

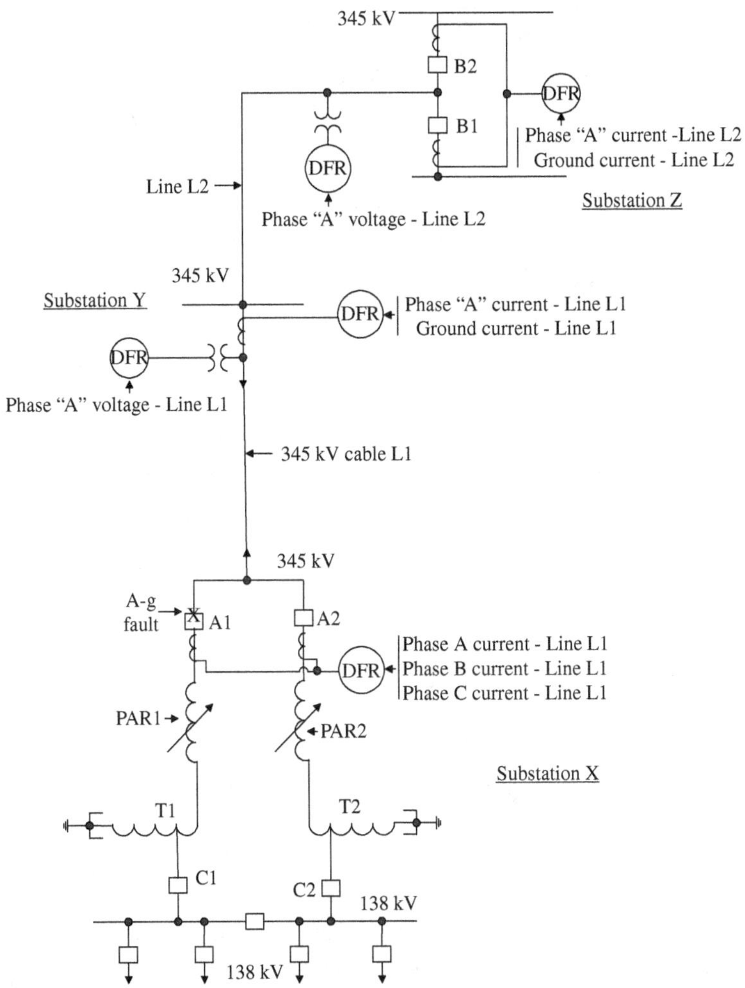

Fig. 7.32 System one-line diagram showing the DFR-monitored currents and voltages for substations X, Y, and Z.

incident point occurred at the peak voltage. This indicates a slow fault-creation mechanism that is caused by an insulation failure. Figure 7.32 shows the system one-line diagram and DFR-monitored voltages and currents. The DFR record in Fig. 7.33 reveals that the A-g fault was cleared in 2.5 cycles from the substation Y end of the cable.

The DFR record shown in Fig. 7.34 for line L2 at substation Z confirms flashover at the voltage peak and the clearing of the fault in 2.5 cycles from the substation Y end of the cable. The DFR record shown in Fig. 7.35 reveals that the A-g fault was cleared in 3.5 cycles from the substation X end of the cable. The DFR records show fault current contributions matching a solid L-g fault simulation in the short-circuit study for a fault

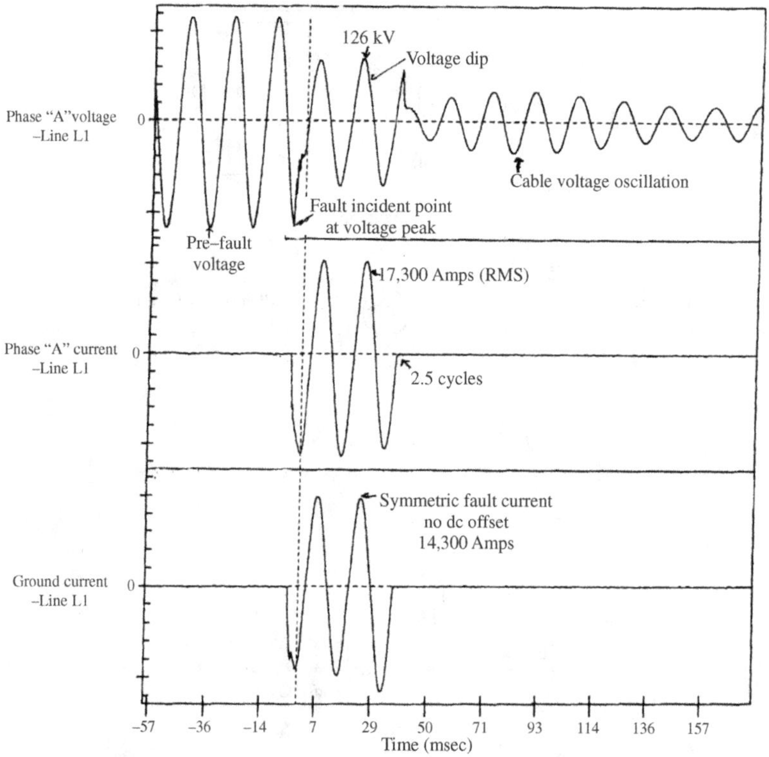

Fig. 7.33 Substation Y DFR record for line L1 currents and voltages confirming the A-g fault.

at point F. In addition, the SF6 breaker A1 phase A was disassembled, and evidence of arcing and burning was observed at several elements of the breaker.

Confirmation of the Power System Model A short-circuit study simulation is carried out to compare DFR-recorded (measured) fault currents versus calculated values. Figure 7.36 illustrates the short-circuit study simulation results for L-g fault at bus X with actual phase current contributions shown. Figure 7.37 illustrates the short-circuit study simulation for L-g fault at bus X with zero-sequence ($3I_0$) current contribution shown.

For this case study the voltages and current values obtained from DFR figures, and the calculated values shown in Figs. 7.36 and 7.37, are tabulated and analyzed in Table 7.1. It should be noted that line L1 has only 15 MW of load, allowing the phase A current measured to be very close to the value calculated. The percentage error values shown in the table are all less than 10%, which confirms that the power system model is adequate for power system design studies and relay setting calculations.

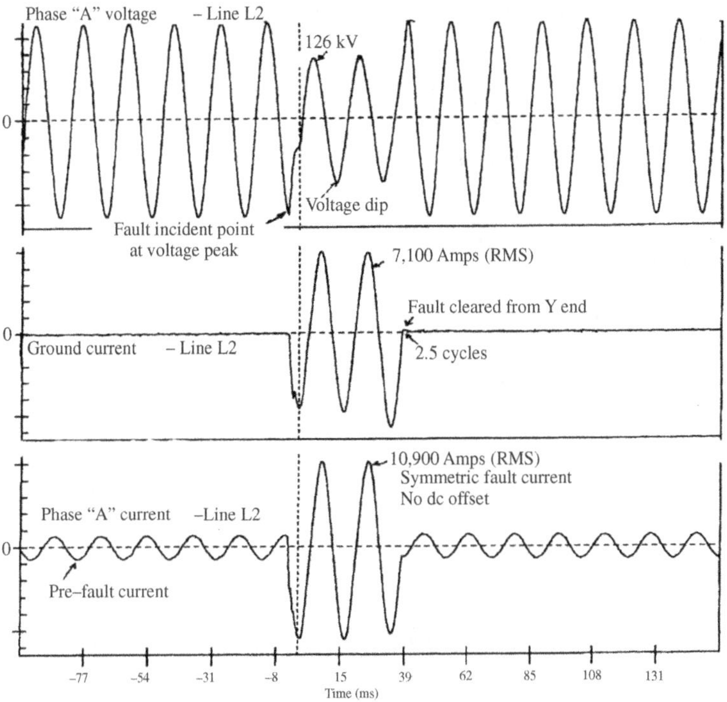

Fig. 7.34 DFR record for line L2 at substation Z, confirming the A-g fault.

Power System Phenomena

1. Arc-over at the voltage peak, indicating an insulation failure as shown in trace phase A voltage—line L1 in Fig. 7.31.
2. Oscillation of energy between the cable charging system and the shunt reactors used for compensation. This produces a beat with a frequency that is normally

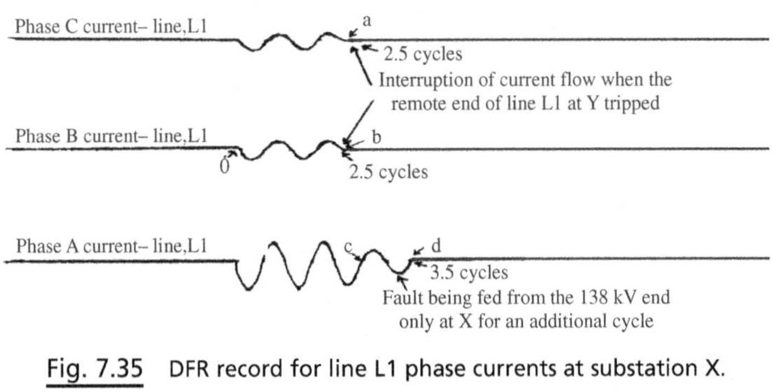

Fig. 7.35 DFR record for line L1 phase currents at substation X.

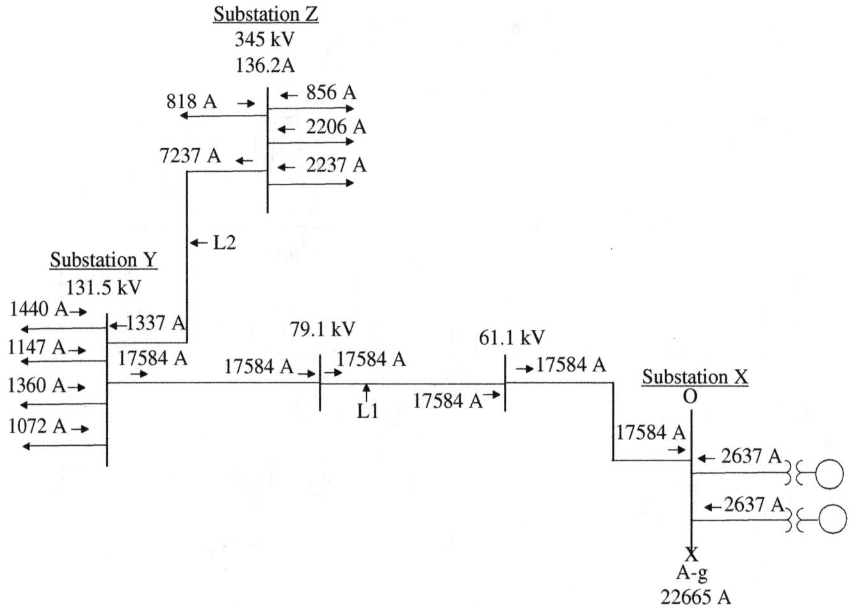

Fig. 7.36 Short-circuit study simulation for the L-g fault at bus X with actual phase current contributions shown.

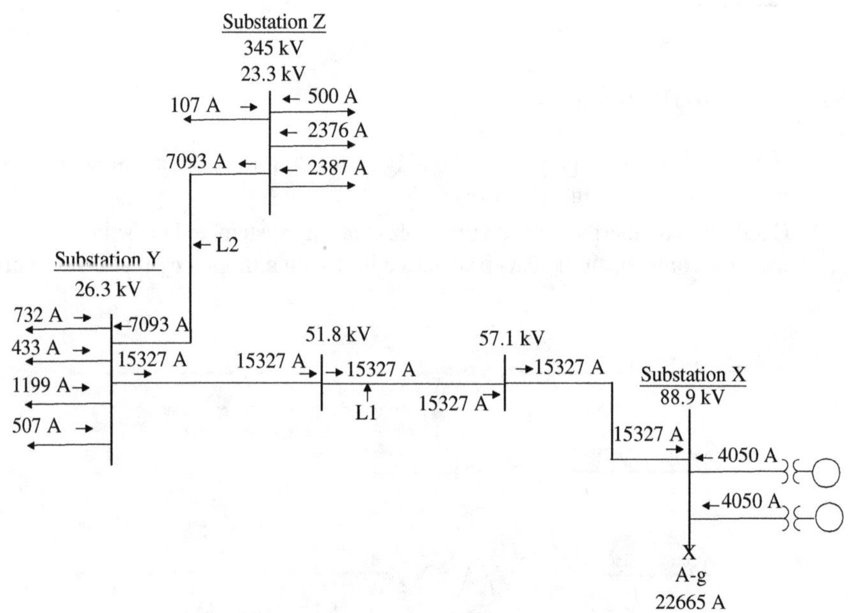

Fig. 7.37 Short-circuit study simulation for the L-g fault at bus X with zero- sequence ($3I_0$) current contributions shown.

TABLE 7.1 Comparison Between Measured and Calculated Currents

Monitored Analog Trace	Measured	Calculated	Percent Error[a]
Faulted phase (A) voltage at Z bus	128 kV	136 kV	1.4
Faulted phase (A) voltage at Y bus	126 kV	131.5 kV	4.4
Ig ($3I_0$)) contribution of Z to Y (L2)	7,100 A	7,093 A	0
Ig ($3I_0$)) contribution of Y to X (L1)	14,300 A	15,327 A	7.2
IA contribution of Y to X (L1)	17,300 A	17,584 A	1.6

[a]% Error = [(measured − calculated)/measured] × 100.

less than 60 Hz. Figure 7.33 reveals voltage oscillation for the phase A voltage upon clearing of the fault.

Corrective Actions The system was reconfigured after determination of the possible faulted area, to isolate the faulted SF6 area according to Fig. 7.31. The SF6 gas zones were tested for six compartments: four disconnect switches that surround the two 345-kV CBs A1 and A2. All SF6 testing for five compartments showed nondetectable PPM. The CB A1 compartment has a value of 15 ppm, indicating a breakdown in the SF6 gas. The faulted SF6 circuit breaker was disassembled for further analysis, and the circuit breaker was repaired and placed in service.

Lessons Learned

1. Relay targets can provide sufficient information to analyze disturbances and proffer recommendations to restore the system. Additional DFR data which can then become available can provide confirmation of the early analysis.
2. Experienced relay engineers need to know which data to analyzed first to provide a quick and accurate rough analysis and not get bogged down in the sea of data being produced by several IEDs.
3. Relay engineers should be aware of the phenomena of fault current flows in the unfaulted phases during ground faults so that correct fault classification analysis is carried out, with the correct faulted phases identified.

Case Study 7.5: Failure of a 345-kV Cable Connection Between a 300-MW Generator and a 345-kV Switchyard, Causing a Phase-to-Ground Fault

Abstract A high-pressure fluid-filled 345-kV cable connecting the unit GSU transformer at plant X to the 345-kV substation Y failed, causing a phase-to-ground fault. The failure occurred at peak voltage, thus supporting a cable insulation failure scenario. This is a slow fault-creation mechanism which results in

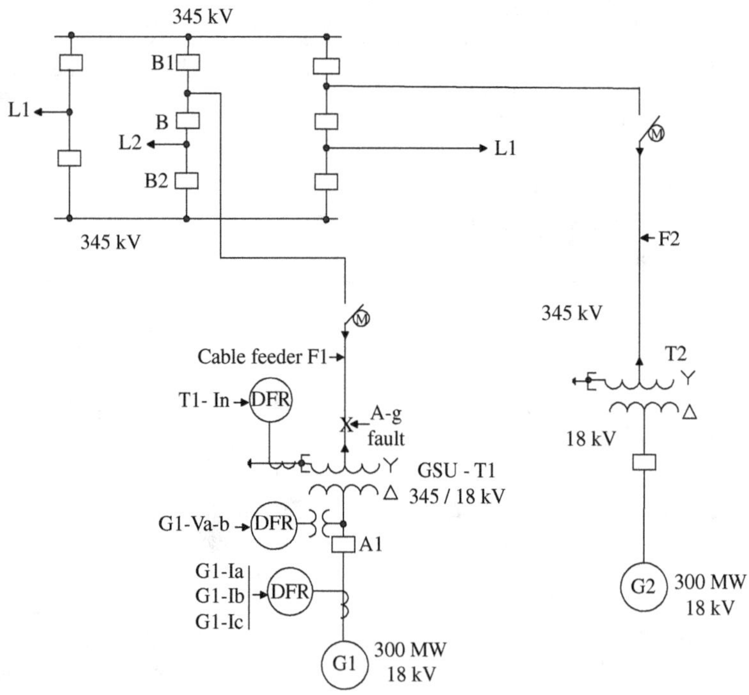

Fig. 7.38 One-line diagram showing the unit DFR-monitored voltages and currents.

symmetrical fault currents with no dc offset components. In this case study we describe the system and associated protection, display a redundant protection zone, and analyze the plant low-side DFR fault record to determine details for the high-side 345-kV fault.

Description of the System and Associated Protection As shown in Fig. 7.38, the generating unit GSU transformers are connected to the 345-kV system via dedicated high-pressure fluid-filled cables. The zones of protection shown in Fig. 7.39 consist of dual relays dedicated to each power system element zone with complete protection overlap. The generator and GSU transformer are protected with dual differential relays. The 345-kV cable is protected by a dedicated pilot wire current differential relay and a distance relay added at the substation. Therefore, the duality protection principle rule is fulfilled throughout this system

Description of the Cable Failure An HPFF cable connecting the high side of a generator step-up transformer to a 345-kV substation failed, causing phase A-g fault. The fault incident point is at the voltage peak, as illustrated in trace G1-Va-b in Fig. 7.40. This is also supported by the symmetrical nature (containing no dc component) of the T1 neutral current trace T1-In. This is an indication of cable insulation failure leading to the phase A-g fault.

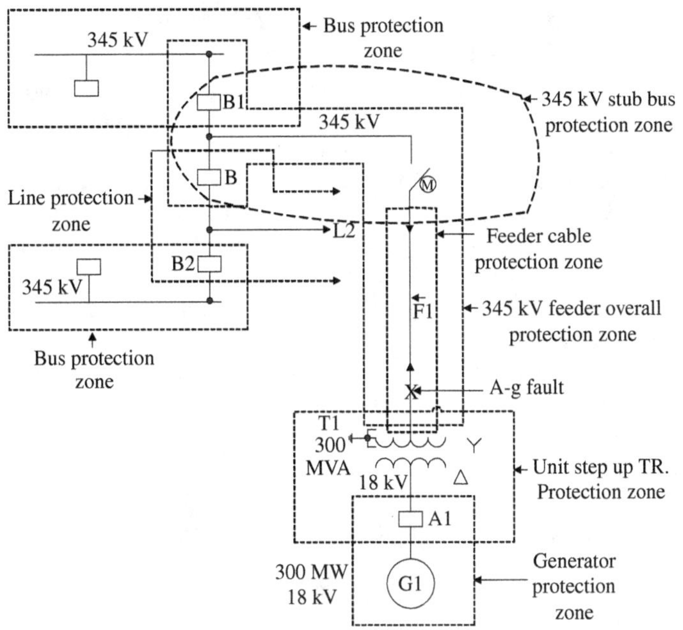

Fig. 7.39 Protection zones for the unit connection to the 345-kV substation.

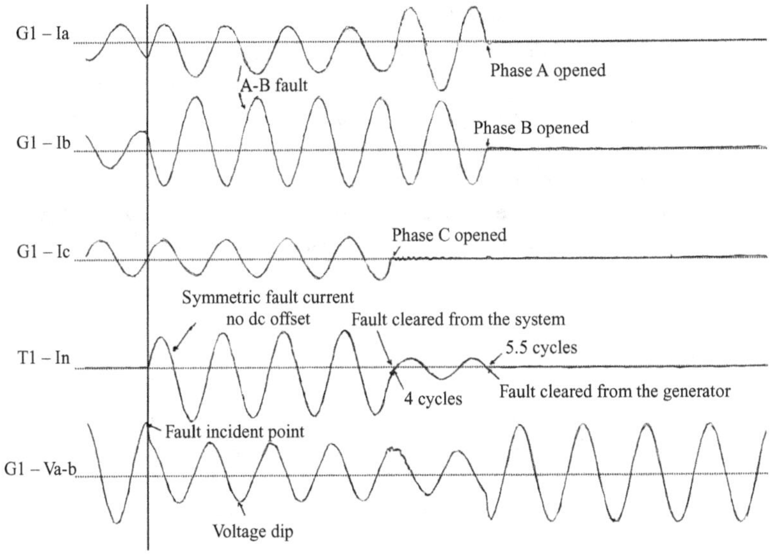

Fig. 7.40 Substation X DFR record confirming a phase-to-phase fault coupled through the T1 transformer from the high-side A-g fault.

Analysis of the High-Side Fault Type Using DFR Information from the GSU Low Side Figure 7.38 shows the system one-line diagram and DFR-monitored voltages and currents. The switchyard DFR device did not trigger to generate a fault record that can be used to analyze the 345-kV cable fault, and therefore the low-side recording by the powerhouse will be substituted to complete the analysis. First, the fault was located by obtaining a relay target for the pilot wire current differential relay covering the cable feeder F1 protection zone. The second step was to classify the fault using the plant DFR record covering unit G1 low-side 18-kV voltages and currents and GSU transformer T1 neutral current.

Figure 7.40 shows the DFR record for the cable failure where the fault incident point occurred at the voltage peak. The DFR record reveals an undervoltage on the line-to-line (V_{ab}) voltage, overcurrent on phases a and b, and the presence of neutral current, which confirms a line-to-ground fault. According to the transformer phasing diagram shown in Fig. 7.41, the a-b winding for the T1 18-kV delta winding is coupled to the A winding on the 345-kV high side, thus indicating an A-g fault. The F1 cable feeder A-g fault is fed from the system as well as unit G1. Positive- and negative-sequence currents are fed from unit G1, while zero-sequence current is supplied from the unit G2 GSU transformer T1.

The fault was cleared from the system side in 4 cycles by tripping 345-kV CBs B and B1. This is illustrated in by trace G1-Ic in Fig. 7.40, confirming a clearing time of 4 cycles when the system side cleared. The fault was finally cleared from the generating side in 5.5 cycles by tripping the 18-kV circuit breaker A1, as illustrated in Fig. 7.40 by current traces G1-Ia and G1-Ib of unit G1. Phase a and phase b currents are equal and opposite for a duration of 1.5 cycles when the fault is fed only from unit G1. According to the GSU transformer T1 phasor diagram shown in Fig. 7.41, the 345-kV high-side A-g fault on the grounded-wye winding will be coupled as a phase a-b fault on the 18-kV low-side delta winding.

Therefore using the low-side DFR recordings, traces G1-Ia and G1-Ib indicate current flow for an a-b fault on the generator side, which when coupled to the high side will confirm the 345-kV A-g fault. The time duration from 4 to 5.5 cycles shows a reduction of zero-sequence currents for trace T1-In, this is due to isolation of the system's strong zero-sequence source. During the same period, current trace G1-Ia shows an increase in magnitude due to the disappearance of the loading current

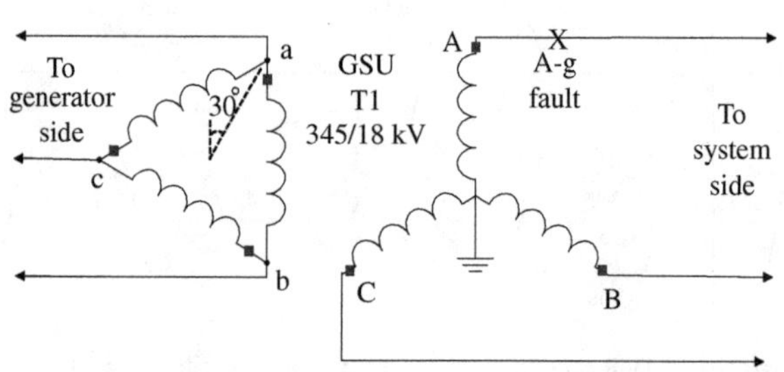

Fig. 7.41 Transformer T1 phasing diagram.

component that was flowing opposite to the fault current contribution from unit G1 during the fault.

Case Study 7.6: 138-kV Cable Pot Head Failure Analysis Using Numerical Current Differential Relay Oscillography and Event Records

Abstract A 138-kV cable pot head failed violently, causing a phase A-to-ground fault. The force generated by the fault current caused the pot head to explode and damage adjacent equipment. The ground fault was cleared from the system successfully. In this case study we describe cable protection systems and provide a sequence of events, an analysis of the failure, power system phenomena, and lessons learned.

Description of the 138-kV Cable Protection Systems As shown in Fig. 7.42, the 1-mile 138-kV solid dielectric cable feeder L1 is protected by dual numerical current differential relaying systems 87LP and 87LS. Each numerical differential relay is using a dedicated fiber optic link. Two separate fiber optic cables with physical separation are utilized for a differential relay communication interface between plant X and substation Y. This will fulfill the protection requirements of having more than one element responding to any given fault in the protection zone. This criterion enhances the personnel safety requirement, limits the impact on the stability of the network, and limits equipment damage by guaranteeing fault clearing by at least one of the independent protection schemes.

Figure 7.43 shows the overlapping protection zones for the cable feeder L1 and the combustion turbine units and their associated generator step-up transformers. The protection zones consist of a generator-dedicated zone, a GSU zone, and a cable feeder zone. Unit faults or GSU faults are cleared by tripping local CBs A2 and A3 at plant X directly and the remote CB A1 at substation Y by using two independent direct transfer trip systems, which are part of the numerical differential relays.

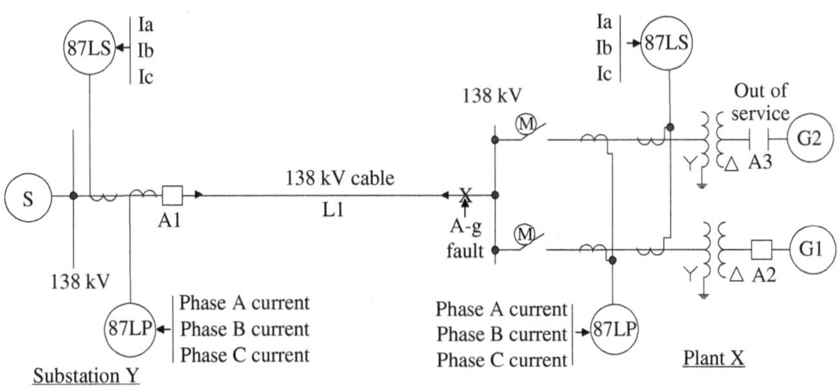

Fig. 7.42 One-line diagram with the cable protection systems and numerical relay oscillography–monitored current designations.

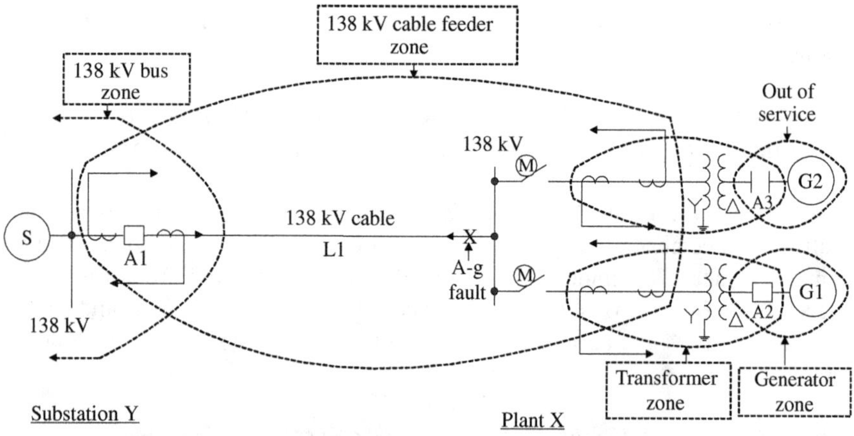

Fig. 7.43 Relay system zones of protection for plant X and substation Y.

Description of the Failure Incident A 138-kV cable pot head failed violently, causing a phase A-to-ground fault. The force generated by the fault current caused the pot head to explode and damage adjacent equipment. The ground fault was cleared from the system successfully. Figure 7.44 shows the system symmetrical component current contribution to the A-g fault when fed from both ends of the feeder. Figure 7.45 shows the system symmetrical component current contribution to the A-g fault from plant X after tripping of the 138-kV system at substation Y.

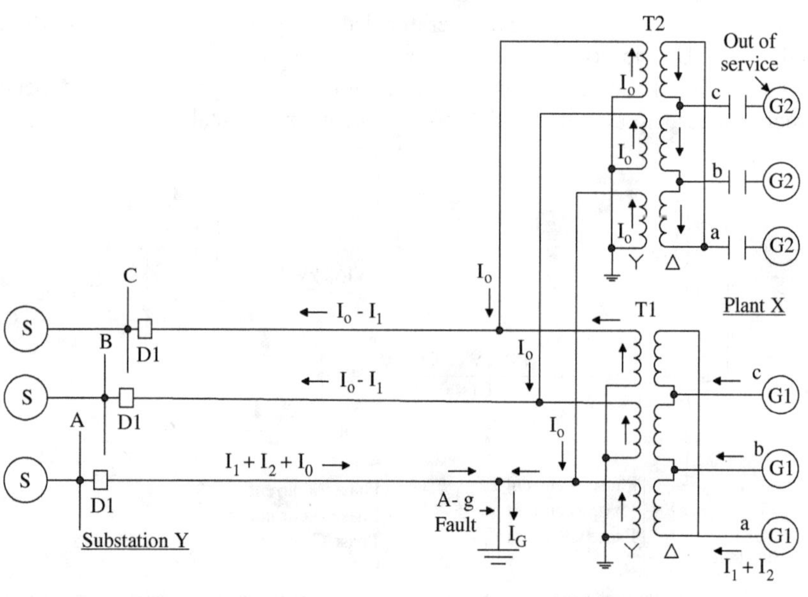

Fig. 7.44 System symmetrical component current contributions to the A-g fault.

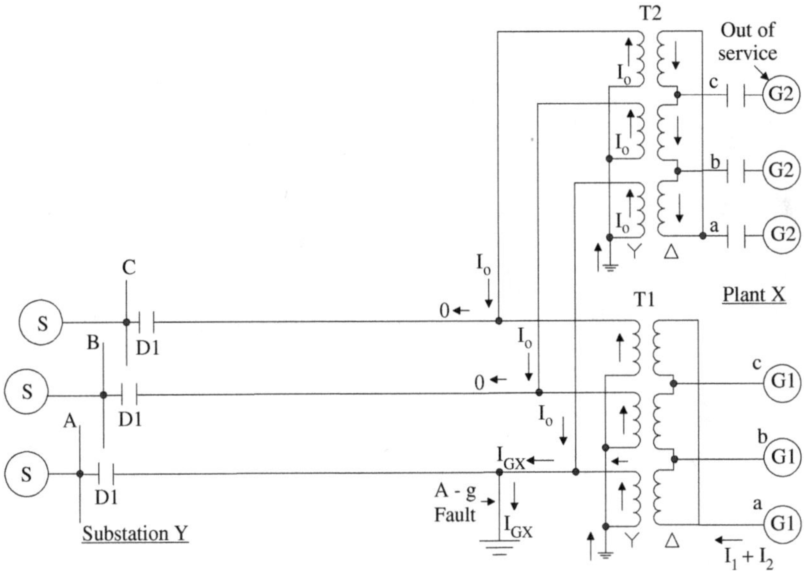

Fig. 7.45 Plant X current contribution to the A-g fault after the tripping at substation Y.

Analysis of the Relay Fault Records The 50-MW CT units and the 138-kV cable feeder are protected by 100% numerical relays, which provide oscillography and event records during the fault, with sufficient details for postmortem analysis. Therefore, digital fault reordering equipment is not required to be installed at either plant X or substation Y. Figure 7.42 shows the monitored three-phase currents of cable feeder L1 for the primary and secondary numerical differential relays at plant X and substation Y. As shown in Fig. 7.43, the fault occurred at plant X when the pot head of phase A of cable feeder L1 failed. Figure 7.46 reveals the numerical cable differential relay oscillography record at plant X. The pre-fault generator output balanced three-phase current is also shown.

Figure 7.46 illustrates the fault incident point at a, then clearing of phases B and C at point b after 3.5 cycles. The interruption of currents on phases B and C at substation Y forced the unfaulted phase current flow ($I_0 - I_1$) at plant X to drop to zero at 3.5 cycles, as shown. Tripping of the remote end at substation Y at 1.5 cycles earlier than substation X resulted in interruption of the unfaulted phases and a drop in magnitude in fault current from the power plant. The substation Y relay fault record shown in Fig. 7.47 confirms that phases B and C tripped and interrupted the unfaulted phase B and C current flow. This forces the fault current being fed from plant X to go down, representing positive-, negative-, and zero-sequence currents from plant X only. From 3.5 to 5 cycles the fault current contribution from plant X dropped, due to the tripping of the system strong zero- and positive-sequence sources at substation Y. Faulted phase A at substation Y interrupted and cleared the fault, which was caused by the failed pot head at

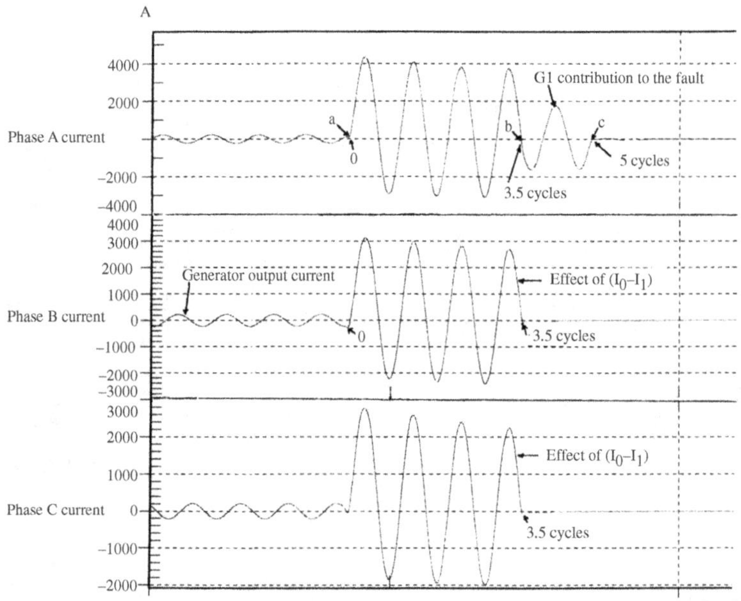

Fig. 7.46 Plant X primary differential relay 87LP fault record showing phase currents during the phase A pot head failure.

plant X, from the system at 4 cycles. Plant X finally cleared the fault at point c after 5 cycles.

Figures 7.48 and 7.49 illustrate the secondary differential relay 87LS oscillography records for plant X and substation Y. Analysis of these records reveals the same

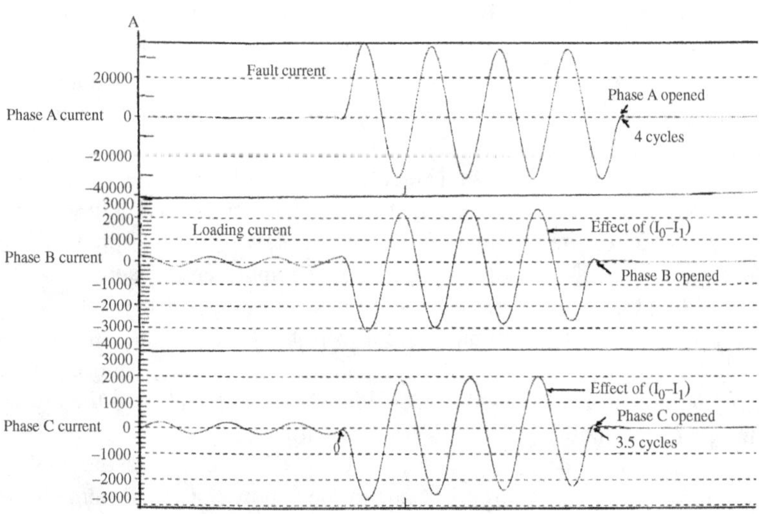

Fig. 7.47 Substation Y primary differential 87LP relay fault record showing the three-phase currents during the phase A pot head failure.

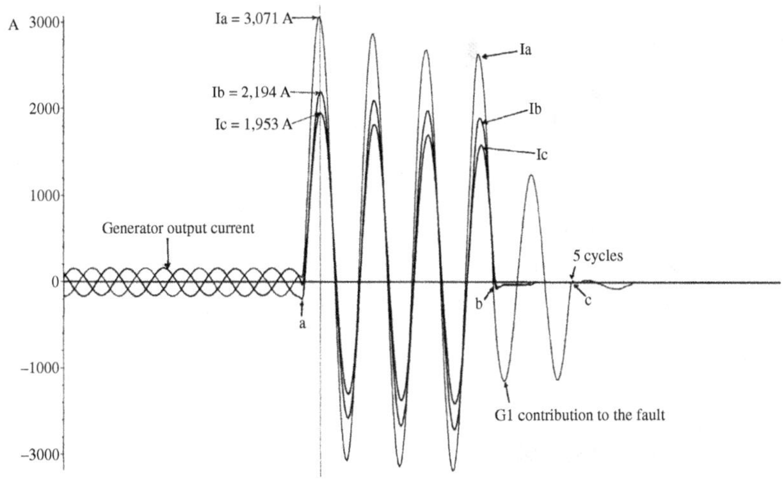

Fig. 7.48 Plant X secondary 87LS relay fault record for the phase A pot head failure.

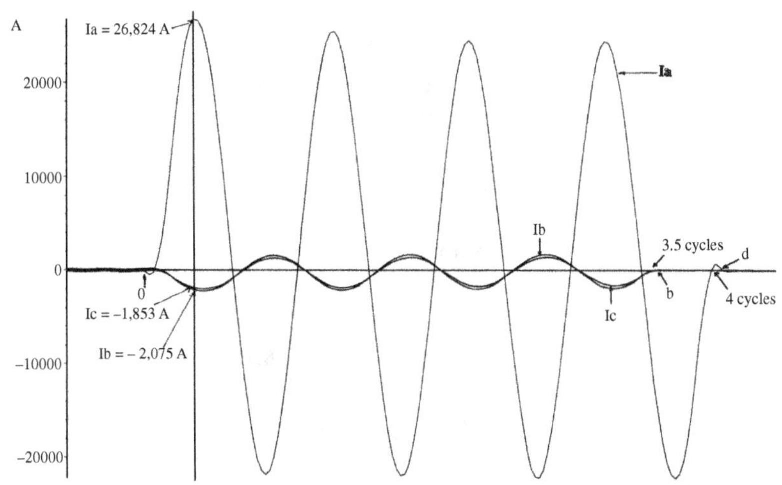

Fig. 7.49 Substation Y secondary differential 87LS relay oscillography fault record for the phase A pot head failure.

sequence of events as obtained from previous analysis of the primary differential relay 87LP oscillography records.

Differential Relay Event Record The primary relay 87LP fault record at substation Y reveals the following data:

Fault duration	67.00 ms
CB operating time	58.00 ms

Relay trip time	4.00 ms
I_A local	19.49 kA
I_B local	1604 A
I_C local	1351 A
I_A remote	2144 A
I_B remote	1602 A
I_C remote	1353 A
I_A differential	21.58 kA
I_B differential	5 A
I_C differential	13 A
I_A bias	10.82 kA
I_B bias	1603 A
I_C bias	1352 A

The primary relay fault record at Y can be analyzed to confirm the numerical relay differential and biasing currents calculated for faulted phase A and unfaulted phases B and C. Let I_L be the local current, I_R be the remote current, and, I_{diff} = the vectorial difference between I_L and I_R (as defined by the relay manufacturer). For faulted phase A;

$$I_{diff} = 19{,}490-(-2144) = 21{,}634\text{ A}$$

which is very close to the calculated value of 21,580 A.

$$\begin{aligned} I_{bias} &= \text{average sum of the magnitude of } I_L \text{ and } I_R \\ &= \tfrac{1}{2}(I_R + I_L) = \tfrac{1}{2}(19{,}490 + 2144) = 10{,}817\text{ A} \end{aligned}$$

which is very close to the calculated value of 10,820 A.

Analysis of the Failure Possible cracks in the bushing base may have resulted in loss of the insulating fluid. The possible presence of voids in the bushing paper insulation may have led to the amplification of electrical stress and eventual failure. An increase in the internal pressure of insulating fluid may also have contributed to the failure, causing the bushing base to be blown. In addition, mechanical causes such as improper base plate mounting, improper cable positioning and grips, improper installation, and possible lateral loading may also have contributed to the 138-kV bushing failure.

Power System Phenomena

1. Current flow in unfaulted phases due to the difference between zero- and positive-sequence current (I_0–I_1) flow, as shown in the trace phase B current in Fig. 7.47.

2. Simultaneous interruption of nonfaulted phases b and c at substation Y, due to removal of the positive- and zero-sequence current at remote sources followed by interruption of the faulted phase current half a cycle later

Lessons Learned

1. Symmetrical fault currents (with no dc offset) obtained from numerical differential relay oscillograph records can be used to deduce that the fault incident point has occurred at the voltage peak, implying that the phase A pot head failure has occurred due to insulation failure. This can be stated without the need to monitor the faulted-phase voltage by the numerical relay.
2. Numerical relay oscillography fault records provide vital information to provide the sequence of events for the fault and can substitute for the function normally provided by traditional digital fault reordering equipment.
3. Symmetrical component principles provide understanding and justification of fault current flows.
4. The current magnitudes for the relay record and for faulted and unfaulted phases can be confirmed using PC-based short-circuit program simulations.

REFERENCES

Blackburn, J. L. *Applied Protective Relaying*. Westinghouse Electric Corporation.

Blackburn, J. L. *Protective Relaying Principles and Applications*. New York: Marcel Dekker, 1987.

Blackburn, J. L. *Symmetrical Components for Power Systems Engineering*. New York: Marcel Dekker, 1993.

Elmore, W. A. Current differential and phase comparison relaying compared with pilot distance schemes. Presented at the 25th Annual Western Protective Relay Conference, October 13–15, 1998, Spokane, WA.

Elmore, W. A., ed. *Pilot Protective Relaying*. New York: Marcel Dekker, 2000.

Elmore, W. A., ed. *Protective Relaying Theory and Applications*. New York: Marcel Dekker, 2000.

Guide for Protective Relaying Application to Transmission Lines. IEEE Power System Relaying Committee Report, IEEE C37.113, 1998.

Ibrahim, M. A., and F. Stacom. Phase angle regulating transformer protection. Presented at the IEEE Power Engineering Society Winter Meeting, January 31–February 5, 1993, Columbus, OH.

Pilot Relaying Performance Analysis. IEEE Power System Relaying Committee Report. *IEEE Transactions on Power Delivery,* Vol. 5, No. 1, January 1990, pp. 85–102.

Sun, S. C., and R. E. Ray. A current differential relay system using fiber optics communication. *IEEE Transactions on Power Apparatus and Systems*, Vol. PAS-102, 1978, pp. 410–419.

8

CASE STUDIES RELATED TO BREAKER FAILURE PROTECTION SYSTEM DISTURBANCES

In this chapter we cover a variety of breaker failure case studies. Case Study 8.1 documents the undesired operation of a breaker failure due to an initiation relay welded contact which was determined by playing back the DFR record using the COMTRADE format to test the relay and confirm its operation. Case Study 8.2 emphasizes the importance of generator phase backup devices as well as remote backup protection for substations to clear faults during extreme contingencies that can occur as a result of simultaneous breaker failures. Case Study 8.3 displays magnetic flux cancellation phenomenon associated with a circuit breaker dual trip coil assembly as the main cause of substation isolation from the bulk power system. Case Study 8.4 describes an incident involving numerous breakers tripping as well as generation rejection initiation by a fault that grew out of proportion due to the undesired operation of a breaker failure relaying system, by backfeeding to the fault detector CT circuit, thus causing additional breaker tripping. Case Study 8.5 describes a unit breaker mechanical problem during an attempt to synchronize a unit to the system while in a pumping mode, resulting in an out-of-step condition, with the unit oscillating three times against the system.

Case Study 8.6 covers re-striking phenomena of oil circuit breakers while clearing simultaneous B-g faults on two 230-kV parallel circuits. Case Study 8.7 documents a

Disturbance Analysis for Power Systems, First Edition. Mohamed A. Ibrahim.

phase-to-ground fault occurring inside a capacitor breaker with another breaker failing subsequently during the process of initial fault isolation. Case Study 8.8 documents multiple faults that represented a challenge to the performance of breaker failure protection systems, resulting in isolation of the failed breaker. Case Study 8.9 illustrates how a remote breaker (located at a different substation) problem can be detected from an analysis of the local substation oscillograms, where the three-phase line voltages and currents are monitored. Case Study 8.10 describes the complete isolation of a 138-kV area around a substation from all adjacent line terminals due to a primary relay system failure and the lack of a separate dc control power for the secondary relay system and breaker failure circuits. Case Study 8.11 describes the installation of two 345-kV breakers in series within a ring substation to mitigate the loss of critical lines during breaker failure events. Case Study 8.12 also describes the installation of two circuit breakers in series to provide breaker failure protection, due to the unavailability of new communication media and the marginal performance of an existing power line carrier system over high-voltage cable.

8.1 BREAKER FAILURE PROTECTION BASICS

Breaker failure protection was recommended and its application was mandated following the 1965 Northeast Blackout. Breaker failure protection is considered to be a local backup function that will clear the failed circuit breaker locally. At any given time, only one circuit breaker can be assumed to fail during the clearing of power system faults. This will conform to the $N-1$ system planning contingency criterion, which is analogous to a fault and the failure of only one of the circuit breakers assigned to clear the fault within the protection zone of relay operation. Breaker failure schemes respond only when protective relays operate to clear faults. Traditionally, a timer was energized by fault detectors, which made the scheme dependent on consistently fast dropout of the overcurrent units used to sense breaker current flow and the associated timer initiating auxiliary relays. The application of some static and numerical relaying systems removes this traditional dropout problem. Circuit breakers are always surrounded by two or more overlapping protection zones. Therefore, a circuit breaker can fail to interrupt faults that occur in any of the two overlapping zones of protection around the breaker. Circuit breaker failure fault detectors should be set to sense fault levels within the overlapping zones of protection. Figure 8.1 is a flowchart of the fault-clearing process from a power system as well as a traditional breaker failure scheme. Breaker failure protection is mandated on all bulk power systems. Only one circuit breaker should be assumed to fail to interrupt a fault occurring in the system at any given time. One breaker failure protection system is sufficient to meet the design criteria of a bulk power system, so there is no need for redundancy. Breaker failure protection requires initiation by a protective relay and the operation of a breaker failure fault detector, confirming that fault current still flows through the breaker. This will confirm that the breaker has failed to interrupt the fault current.

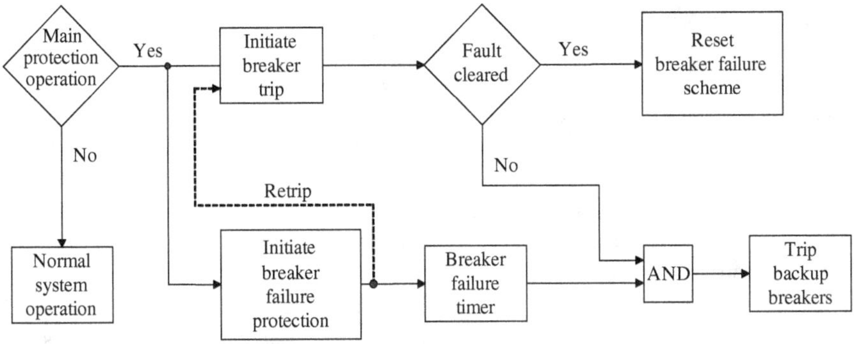

Fig. 8.1 Flowchart of a fault clearing process and a traditional breaker failure scheme.

Some bulk power systems are designed to be capable of withstanding normally cleared three-phase faults and breaker failure for ground faults. Some HV and EHV circuit breakers are designed with three independent pole mechanisms, which is advantageous in downgrading breaker failure to clear a three-phase fault to a phase-to-ground fault, with the assumption that only one pole (phase) will fail to interrupt. Implementation of this feature will downgrade a three-phase fault to a phase-to-ground fault and hence will comply with the design criteria calling for the power system to withstand a phase-to-ground fault coupled with a breaker failure.

Backup relaying is defined as a remote backup or local backup system. Remote backup relaying clears faults one substation away where the failure to clear faults has occurred. Remote backup relaying trips more system than necessary to clear a fault, and it has long backup operation times, 1 to 3 s. It is difficult to set remote backup relays due to the in-feed effect on distance relay setting and the requirement to carry line emergency loading.

Local backup relaying (also known as breaker failure relaying) clears a fault in the substation where the failure has occurred and therefore trips the least number of system elements in clearing a fault. The local backup clearing times range from 9 to 20 cycles, depending on the scheme timer setting to meet the system requirement for critical switching time as defined by stability studies. High-speed primary fault clearing times for bulk power systems range from 3 to 6 cycles. Causes that contribute to breaker failure are loss of dc supply, an open trip coil, a short-circuited trip coil, mechanical failure of the tripping mechanism, and failure of main contacts to interrupt.

8.1.1 Breaker Failure Application Options

Dedicated Stand-Alone Protection Traditionally, breaker failure protection was implemented using dedicated electromechanical (discrete) component relays consisting of fault detectors, timers, and auxiliary tripping or lockout relays. In early 1960s, integrated solid-state (static) packages were employed for breaker failure protection where the fault detector and timer functions are integrated. For some solid-state breaker failure packages, the power supply is activated by protective relay breaker failure initiation output. For some low-grade fault current magnitude in transformer tanks that may be cleared by pressure or Buchholz relays, type a breaker auxiliary contacts may be used to supplement fault detectors. The re-trip feature of the failed circuit breaker is used without a time delay, to avoid isolating breakers by tripping breaker failure lockout relays during scheme testing errors. Breaker failure application on breaker-and-a-half and ring bus substation configurations will require clearing of remote breakers via the application of dedicated direct transfer trip communication systems.

For a bulk power system, total separations and independence should be established between communication systems for line pilot protection schemes and breaker failure protection systems. Dedicated numerical breaker failure relays are used as a replacement for old electromechanical and static breaker failure relaying systems and should be used to comply with breaker failure separation and independence criteria.

Element Included with the Numerical Line Protection Package Application of new numerical line relaying with individual breaker current sampling opened the door for the use of breaker failure relays as an element of the line relaying package. Figure 8.2 shows the use of a breaker failure protection function as an element of numerical line relaying firmware. This can be implemented when the line relay samples individual breaker currents for the ring bus, double-bus double-breaker, and breaker-and-a-half substation configurations. For bulk power systems, breaker failure protection separation criteria must not be violated. This will require duplication of the breaker failure protection, perhaps as an element within each of the dual redundant

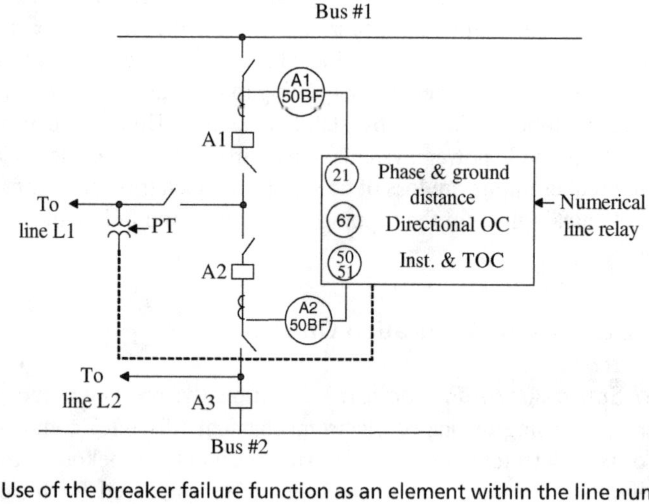

Fig. 8.2 Use of the breaker failure function as an element within the line numerical relay protection package.

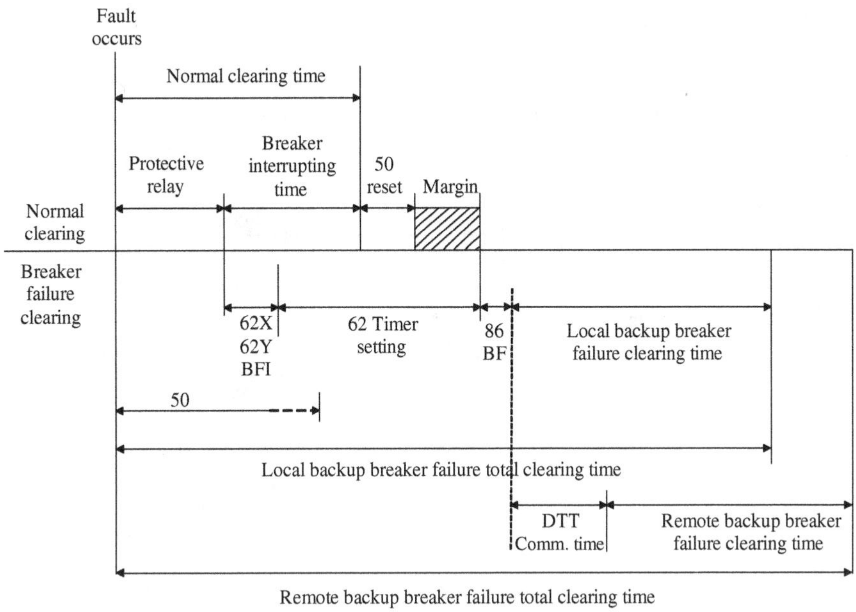

Fig. 8.3 Time chart of a traditional breaker failure scheme.

line numerical relays. Caution must be exercised when testing a line relaying system to avoid false operation of the breaker failure element.

8.1.2 Time Chart for Traditional Breaker Failure

Figure 8.3 shows the main breaker failure time coordination chart, with normal clearing and breaker failure local backup clearing and remote breaker failure clearing, where a DTT signal is sent to trip the remote-end circuit breakers of the line.

8.1.3 Conventional Breaker Failure DC Schematic

Figure 8.4 shows a typical complete dc schematic for a live tank gas (SF6)-insulated circuit breaker with an independent pole mechanism design using discrete relay elements. Breaker and CT column flashover protection device 50 trips and locks out breaker failure lockout relay 86BF directly. Pole disagreement 94X coupled with loss of SF6 gas may trip 86BF directly. Protective relay initiation in a series with either phase or ground breaker failure fault detectors will energize the timer 62BF, which will output to trip 86BF. The re-trip relay 62Z is designed to minimize the impact of undesired operation during testing. The lockout relay 86BF coil operation is monitored by a light that will go out upon energization of the lockout relay coil. New lockout relays are designed with LED lights for monitoring and relay operation indications.

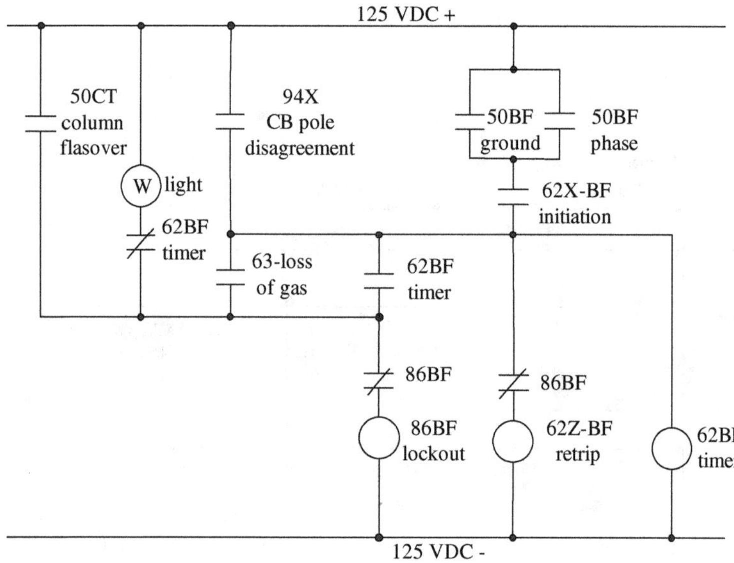

Fig. 8.4 Conventional breaker failure dc scheme.

8.1.4 Primary Protection Blind Spots Cleared by Breaker Failure Protection

Use of a dead tank breaker with CTs available at each side of the breaker permits complete clearing of faults within the overlapping zones in primary clearing times. However, the use of live tank breakers with a CT column applied at only one side of the breaker provides primary protection blind spots where certain faulted areas will require more time to clear. In these cases, faults will be cleared by breaker failure relaying.

Figure 8.5 shows blind spots that are cleared with a time delay using breaker failure protection systems. For Fig. 8.5(a), the CT column is toward the line; with a fault between the CB and CT columns, the fault falls in the bus B1 protection zone. However, opening breaker A1 does not clear the fault in primary high-speed time. The fault F1 is external to the line L1 primary protection zone and therefore remains fed from the system. However, bus B1 relaying will initiate CB A1 failure protection, and with operation of the 50BF fault detector, the breaker failure timer will start. The fault F1 will then be cleared when the breaker failure timer expires.

For Fig. 8.5(b), the CT column is toward the bus, with a fault between the CB and CT columns the fault falls in the line protection zone. However, tripping and opening breaker A1 does not clear the fault in primary high-speed time. The fault F1 is external to the bus B1 primary protection zone and therefore remains fed from the system. However, line L1 relaying will initiate CB A1 failure protection, and with operation of the 50BF fault detector, the breaker failure timer will start. The fault F1 will be then cleared when the breaker failure timer expires.

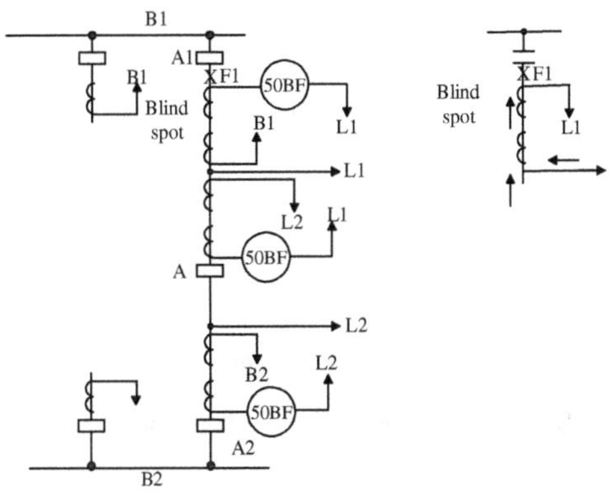

(a) Current transformer (CT) column is toward the line

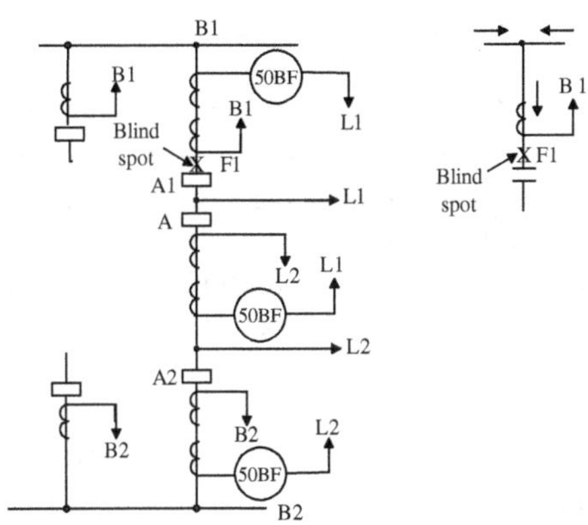

(b) Current transformer (CT) column is toward the bus

Fig. 8.5 Blind protection spots created by a CT column applied to live-tank circuit breakers.

8.1.5 Effect of Breaker Failure Action on the Power System as a Function of Substation Configurations

The effect of breaker failure operation will depend on the type of substation configuration.

Breaker-and-a-Half Substations Any power element fault with one breaker failing to clear results in the loss of an additional power element. Loss of the element is a function of the type of elements connected to the substation. The element loss could be either two lines, one bus and one line, one bus and one generator, one bus and one transformer, one line and one transformer, or one line and one generator. No substation split buses or separation of the remaining line connection at the substation will occur as a result of a breaker failure operation. Figure 8.6 illustrates one case for a line L1 fault with breaker A failing that will result in the additional loss of line L2.

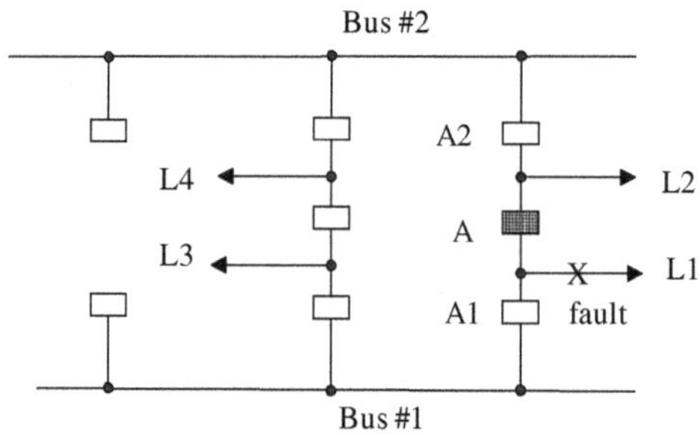

Fig. 8.6 Breaker-and-a-half substation configuration.

Ring Bus Substations Any power element fault with one breaker failing to clear results in the loss of an additional power element. The element loss is a function of the type of element connected to the substation ring and is similar to breaker-and-a-half applications. The element loss could be either two lines, or one line and one transformer, or one line and one generator. No separation of the remaining line connection at the substation will occur as a result of breaker failure operation. Figure 8.7 illustrates one case for a line L1 fault with breaker A1 failing that will result in the additional loss of line L2. This may present a challenge to the power system, especially when the two lost lines connect sensitive areas in the system. This situation could justify the use of two breakers in series to mitigate loss of the second line.

Double-Bus Double-Breaker Substation Configuration Any power element fault with one breaker failing to clear results in the loss of the faulted power element only. In addition, one of the buses will be lost; however, the remaining lines will stay connected at the other bus as a result of breaker failure operations. Figure 8.8 illustrates one case for a line L1 fault with breaker A1 failing that will result in the isolation of bus 1 by tripping three additional breakers: B1, C1, and D1. The remaining lines, L2, L3, and L4, will remain connected to bus 2. This highlights one of the main advantages that could justify the investment in designing this type of substation configuration.

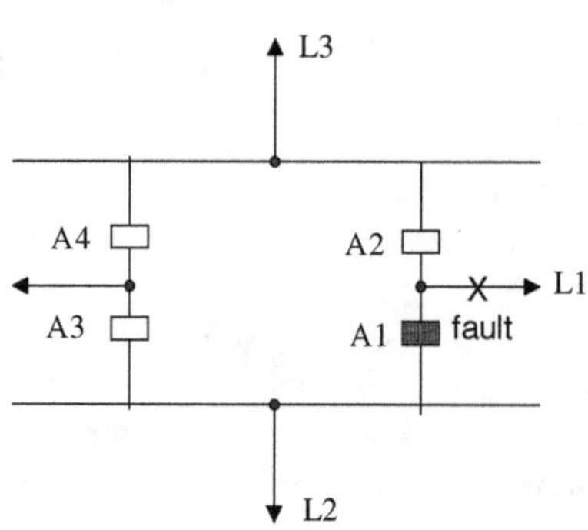

Fig. 8.7 Ring bus substation configuration.

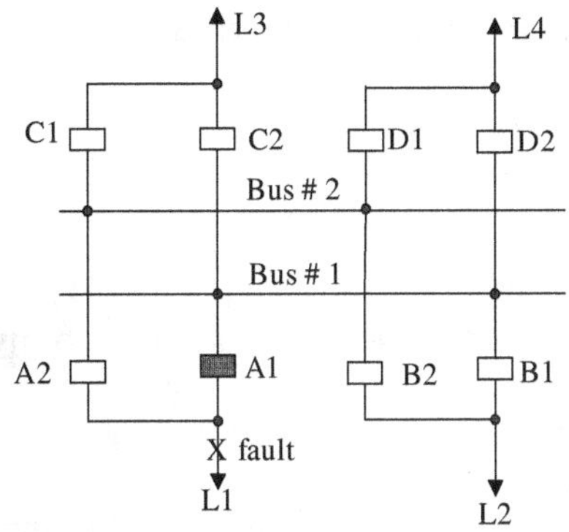

Fig. 8.8 Double-bus double-breaker substation configuration.

Example 8.1: Breaker Failure Tripping Functions

An example of the required tripping functions of a breaker failure protection scheme is illustrated in Fig. 8.9. For a fault at the end of the line L1 zone of protection with CB A assumed to fail, the following action will carried out by a breaker A failure lockout relay or auxiliary (no lockout) relay:

1. Trip and lock out A.
2. Trip and lock out CBs A1 and A2.

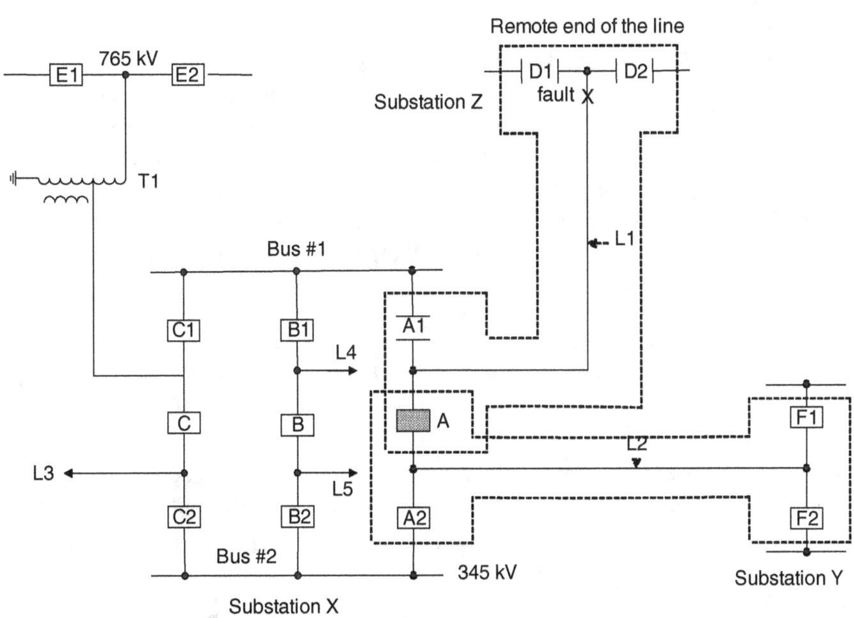

Fig. 8.9 Breaker failure function for a failed middle breaker in a conventional breaker-and-a-half substation.

3. Initiate direct transfer trip on line L1 to trip and lock out remote CBs D1 and D2 at substation Z.
4. Initiate direct transfer trip on line L2 to trip and lock out remote CBs F1 and F2 at substation Y.

8.1.6 Breaker Failure Protection Setting Philosophy

Circuit breakers are always surrounded by overlapping protection zones. Therefore, a CB can fail to interrupt faults that occur in either of the two zones of protection. CB failure fault detectors should be set to sense fault levels within the zones of protection. The closed and open protection zones of a circuit breaker should first be defined to set the breaker failure fault detectors optimally. The end of the zone point will yield the minimum fault current required to set the breaker failure fault detectors. In addition, the setting margin should be included by taking system outages, which will result in lowering fault currents. It should be noted that the setting of breaker failure fault detectors is normally done using a stub (one end of the system is opened) system condition. This is justified on the basis that at any given time, one circuit breaker is only assumed to be failing. This is ideal for schemes where the timer is started only with fault initiation and the fault detectors are checked upon expiration of the timer to accomplish the tripping. Consideration should be given to examining fault currents during normal system condition with the two ends of the line closed. This will allow the breaker failure timer to start at the beginning of the fault (for some older schemes), thus avoiding longer clearing times that may occur if the fault detector is set at system stub (one end of the faulted line opens first) conditions.

The setting of fault detectors must be at least two times pickup under minimum fault conditions. If possible, the phase unit should be set above maximum loading on the associated element. Ground elements for fault detectors should be set very sensitive (since no loading restriction is required). The ground element should be set above the loading imbalance (20% of maximum loading). The CB failure timer setting is based on the system critical switching time, which is obtained from system stability study simulations.

Circuit Breaker Failure Phase Fault Detector Setting Procedure The setting procedure for a breaker failure phase fault detector is as follows:

1. Determine the breaker overlapping primary zones of protection to determine the fault locations to be simulated for CB failure phase fault detector setting.
2. Simulate a line-end (stub) phase-to-phase fault for each of the primary protection zones. This is ideal for schemes where the timer is started only with fault initiation.
3. Set the phase element to have 2× pickup for the smaller of the two phase-to-phase line-end faults in step 2.

4. For breaker failure schemes where the timer is started by the fault detector and fault initiation, examine normal system phase faults and try to see if the fault detector will operate using a 50% current split between the two breakers, for ring bus, breaker-and-a-half, and double-bus double-breaker substation configurations.

Circuit Breaker Failure Ground Fault Detector Setting Procedure The setting procedure for a breaker failure ground fault detector is as follows:

1. Determine the breaker overlapping primary zones of protection to determine the fault locations to be simulated for CB failure ground fault detector setting.
2. Simulate a line-end (stub) phase-to-ground fault for each of the primary protection zones.
3. Set the element above the maximum loading unbalance of 20% maximum line loading.
4. Set the ground element to have plenty of pickup (4 to 6) to cover for high-resistance ground faults.
5. For breaker failure schemes where the timer is started by the fault detector and fault initiation, examine system normal ground faults and try to see if the fault detector will operate using a 50% current split between the two breakers, for ring bus, breaker-and-a-half, and double-bus double-brear substation configurations.

Example 8.2: Setting a Ground Fault Detector

A portion of a 345-kV system is shown in Fig. 8.10, where CB A breaker failure fault detectors need to be set. The CB A overlapping zones of protection are determined from short-circuit study simulation as lines L1 and L2, as shown.

Contribution from substation X for system normal faults:

Fault Locations	*Three-Phase* ($I_{3\emptyset}$)	*Phase-to-Ground* (I_g)
F1	3,344 A	2,256 A
F2	12,362 A	8,392 A

Contribution from substation X for line-end three-phase faults (stub):

L1-end fault (out of substation X; open at substation Y):	4760 A
L2-end fault (out of substation X; open at substation Z):	14,676 A

Contribution from substation X for line-end phase-to-ground faults (stub):

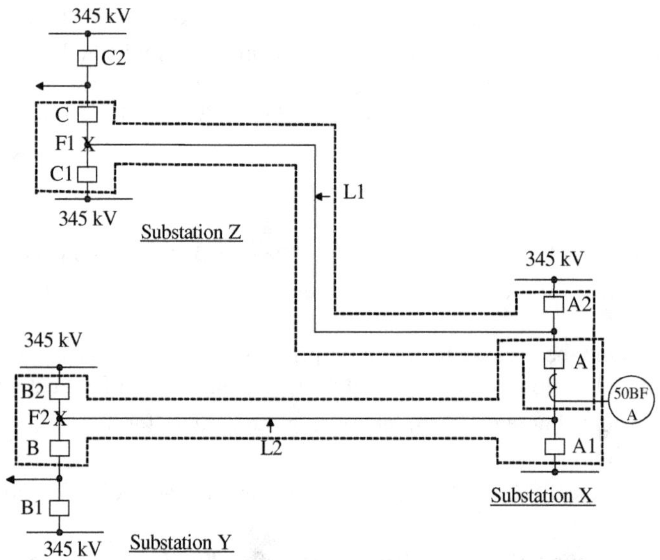

Fig. 8.10 One-line diagram of a portion of a 345-kV system.

L1-end fault (out of substation X; open at substation Y):	3282 A
L2-end fault (out of substation X; open at substation Z):	11,503 A

The maximum loading for L1 = 2393 A. Since the current for fault F1 is smaller than that for F2, use line L1 zone of protection to set the CB A failure fault detectors.

Set the phase element using a phase-to-phase line-end fault. Set the phase at (4760 × 0.866)/2 = 2061 A. Use the CT ratio 400 : 1.

$$I_{set} = \frac{2061}{400} = 5.15\text{ A}, \qquad 400:1 \text{ set at } 5\text{ A}$$

Set the ground element using a phase-to-ground line-end fault. Set the ground above 20% of the load = 479 A. Set the ground at 3282/5X = 656 greater than 479.

$$I_{set} = \frac{656}{400} = 1.6\text{ A}, \qquad 400:1 \text{ set at } 1.5\text{ A}$$

CASE STUDIES

Case Study 8.1: Tripping of a Combined-Cycle 150-MW Plant by Undesired Operation of a Solid-State Breaker Failure Relaying System

Abstract In this case study we describe an incident that is triggered by a switching transient, which was recorded by a DFR. The transient activated an undesired breaker failure operation, which resulted in the tripping of a 150-MW combined-cycle plant. The DFR record for the incident was first converted to COMTRADE format and then played back to test the breaker failure relay, after its isolation, using an automated test set. The cause of the undesired operation was determined by the DFR playback testing option. In this case study we describe the power system, the 138-kV breaker failure protection system, troubleshooting of the incident, analysis of the DFR record, relevant power system phenomena, corrective actions, and lessons learned.

Description of the Power System Figure 8.11 shows the combined-cycle plant, which consists of a 100-MW combustion turbine (CT) unit and a 50-MW steam turbine (ST) unit. The plant is connected to the 138-kV system via a gas-filled 138-kV cable. The CT generator is connected to the grid via a 125-MVA 13.8/138-kV transformer. The steam turbine unit is connected to the system via an 80-MVA 13.8/138-kV transformer. Both transformers are connected delta on the 13.8-kV side and grounded wye on the 138-kV side. The plant is designed with 13.8-kV generator breakers used for the synchronization of generators to the system, thus allowing the 138-kV system to be the starting source for the plant auxiliary system. Figure 8.12 shows a portion of the 138-kV system near the combined-cycle plant X.

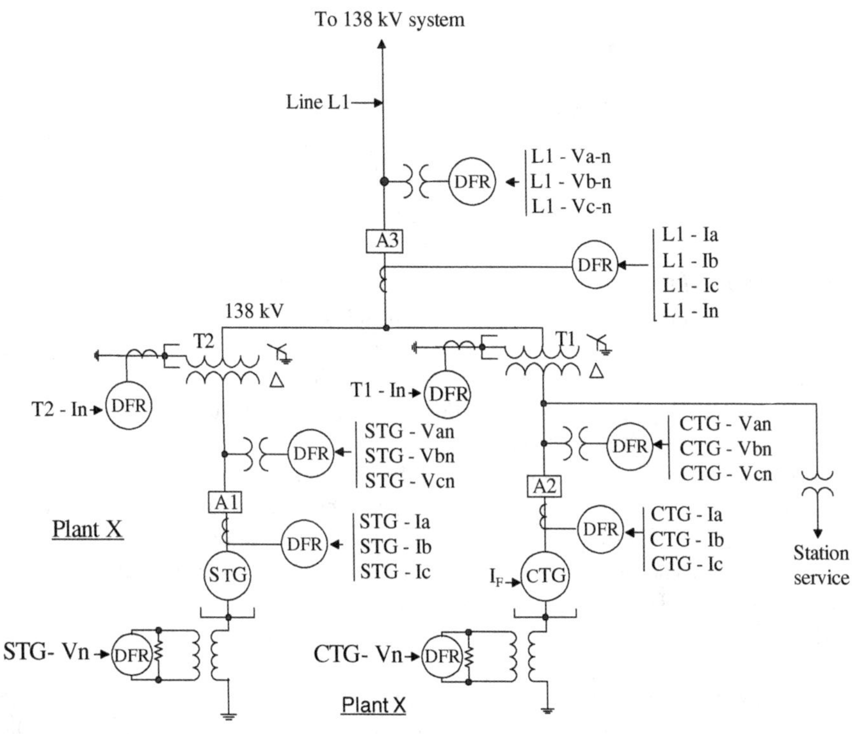

Fig. 8.11 One-line diagram showing DFR-monitored voltages and currents at plant X.

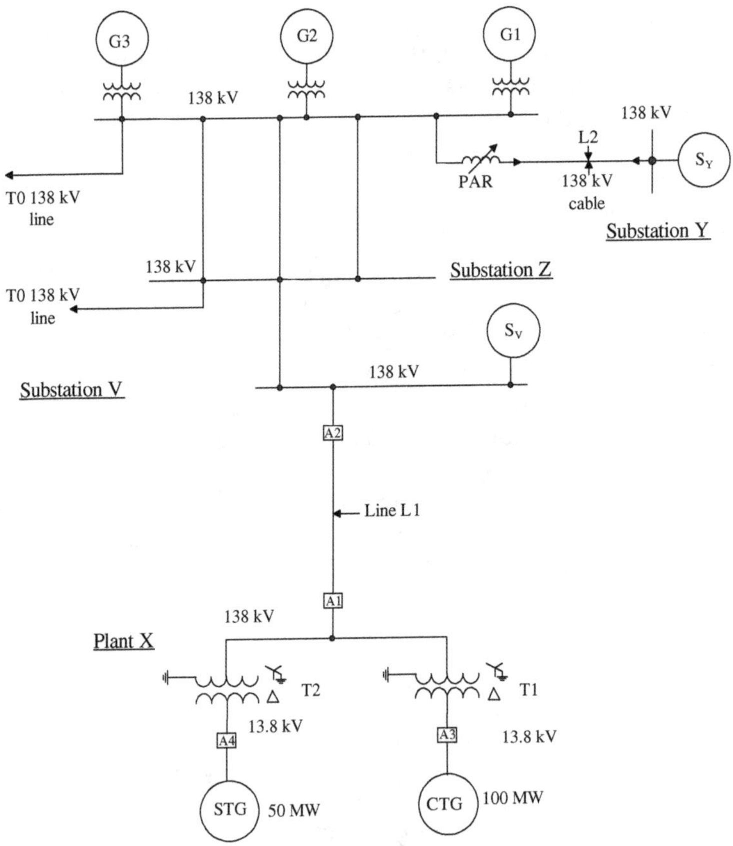

Fig. 8.12 System one-line diagram.

Description of the Incident A transient occurred on the power system during the energization of one of the 138-kV L2 cables at substation Y. The 10-mile cable shown in Fig. 8.12 serves as a tie-line between two different utility systems and is connected in series with a 300-MVA phase angle–regulating transformer. The switching transient apparently caused a reactive power swing, which resulted in ground current flow on line L1, which exceeded the setting of the ground element of the CB A1 solid-state breaker failure relaying. Operation of the fault detector ground element, coupled with a welded contact of one of the line relaying breaker failure initiation auxiliary relays, resulted in the tripping of CB A1 and thus the isolation and shutdown of the 150-MW plant. The switching transient coincided with the undesired CB A1 breaker failure operation. The DFR record shown in Fig. 8.13 illustrates the tripping of the CT turbine generator about 2.5 cycles after the occurrence of the switching transient.

Description of the Breaker Failure Relaying Scheme The 138-kV breaker failure protection shown in Fig. 8.14 is a solid-state (static) relay package. The relay logic is activated only when a fault occurs on the system, which will require tripping of

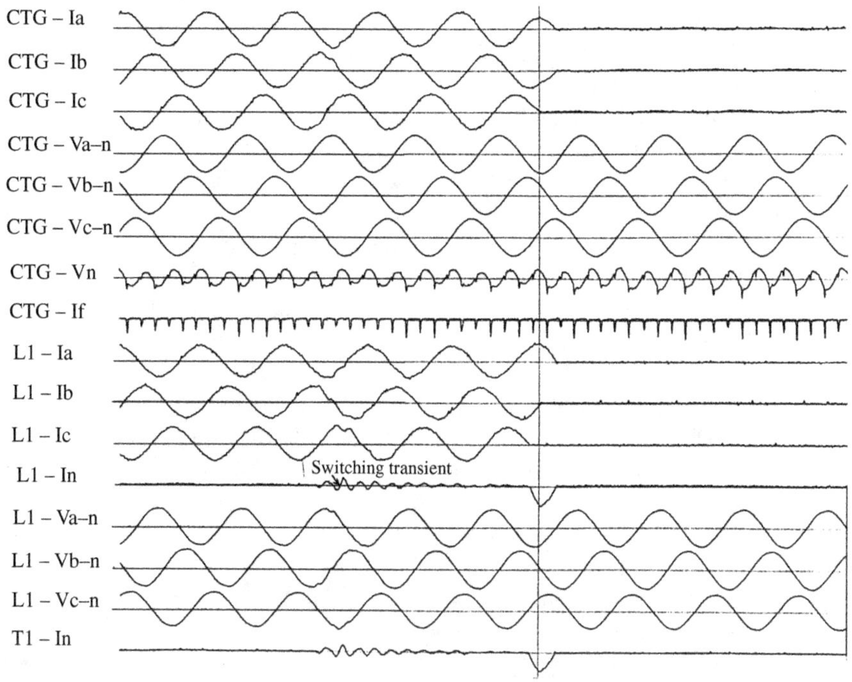

Fig. 8.13 CT generator and 138-kV line L1 currents and voltages, showing the plant trip and the switching transient.

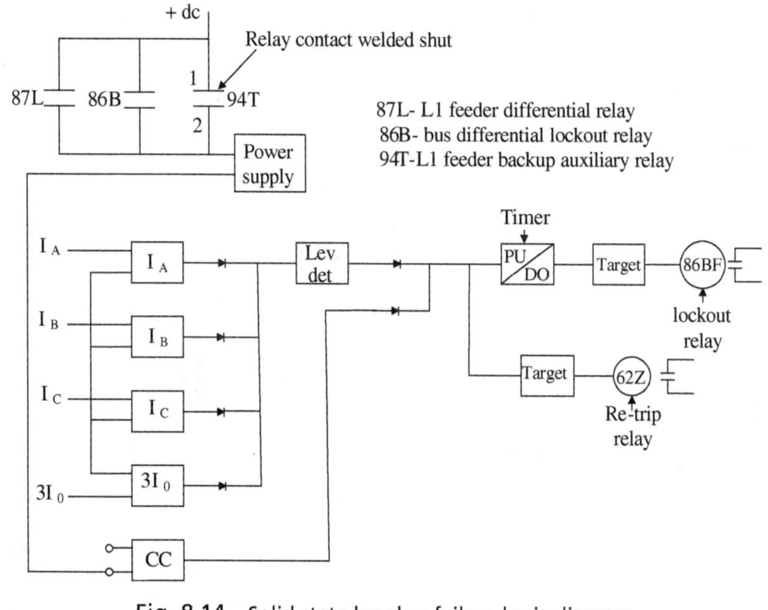

Fig. 8.14 Solid-state breaker failure logic diagram.

CB A1. The relay power supply is switched on to energize the relay by the operation of relays associated with the tripping of A1 via breaker failure initiation auxiliary relays 87L, 86B, and 94T. The re-trip relay 1T is energized by breaker failure initiation and is added to minimize the extent of breaker failure actions specifically during testing. The relay has an input from breaker auxiliary contacts 52a to initiate a breaker failure function for low-current conditions for faults cleared by a transformer sudden pressure relay or Buchholz or similar devices. These incipient faults may generate low fault currents, which may not be sufficient to pick up breaker failure fault detectors. The fault detectors are connected to a CT ratio of 1200 : 5 with the ground element set at 1 A, the phase element set at 5 A, and the breaker failure timer set at 167 ms (10 cycles).

Analysis of the DFR Records Figure 8.11 shows the power system one-line diagram and DFR-monitored voltages and currents. The digital fault recorder was triggered by the neutral residual current generated by interruption of the three-phase balanced generator output currents. The 120° angle separation between the phases generates a nonsinusoidal neutral residual current when individual phases are interrupting at zero-current crossings which were sufficient to trigger the DFR recording. The DFR record shown in Fig. 8.15 illustrates the 138-kV tie-line phase and neutral currents. The DFR reveals that phase C interrupts first at point c followed by phase B interruption at point b and finally, phase A interruption at point a. It was therefore coincident to record the early switching transient, which occurred a few cycles earlier and is shown in trace L1-In in Fig. 8.15.

The DFR record in Fig. 8.15 reveals that the frequency of this neutral current is about 300 Hz and lasted only about 40 ms (2.5 cycles). The magnitude of the neutral

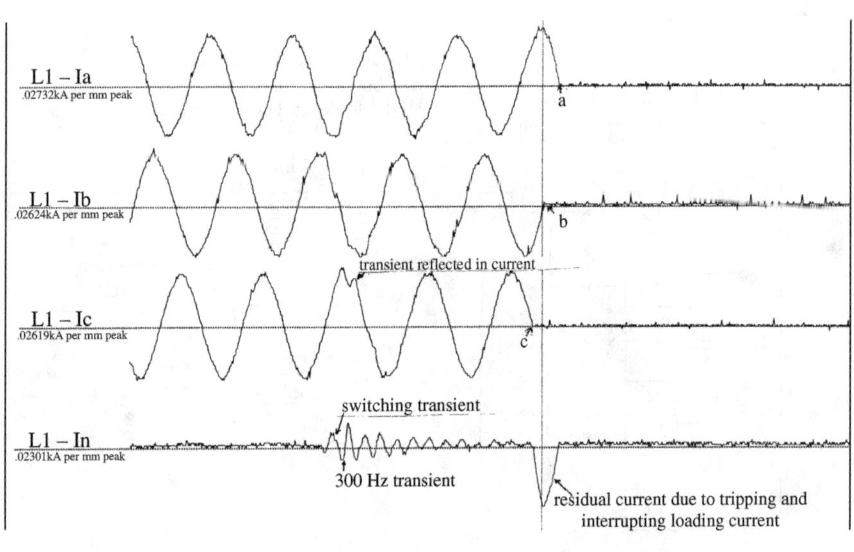

Fig. 8.15 138-kV line L1 currents prior to the plant trip, showing the switching transient.

current has apparently exceeded the threshold setting of the ground element of the solid-state CB A breaker failure relaying. The operation of the ground fault detector element for this transient current confirms that the element is not tuned to the 60-Hz (fundamental) current. Operation of the fault detector ground element, coupled with a welded contact of one of the line relaying breaker failure initiation auxiliary relays resulted in the tripping of CB A1 and thus the isolation and shutdown of the 150-MW plant. The total tripping time for the plant is about 2.75 cycles (46 ms). By factoring a two-cycle 138-kV breaker interruption time, this will imply that the total operating time of the re-trip function (ground fault detector element + auxiliary relay 1T) is less than 1 cycle. Figure 8.13 shows the CT unit generator currents and voltages, and Fig. 8.16 reveals the ST unit generator currents and voltages during the transient event. The tie-line voltages and currents are also shown in Figs. 8.13 and 8.16 as well as the none-60-Hz cable feeder neutral current that was generated, which triggered the DFR.

Troubleshooting the Incident Analysis of DFR records did not reveal a fault, so the plant was returned to service postulating incorrect operation of the CB A1 breaker failure relay. The DFR record for the incident was converted to COMTRADE data format and played back to test the breaker failure relay, after its isolation, using an automated test set. Due to the availability of sufficient backup protection for CB A1 failure event, the output of the breaker failure relay was isolated to troubleshoot the

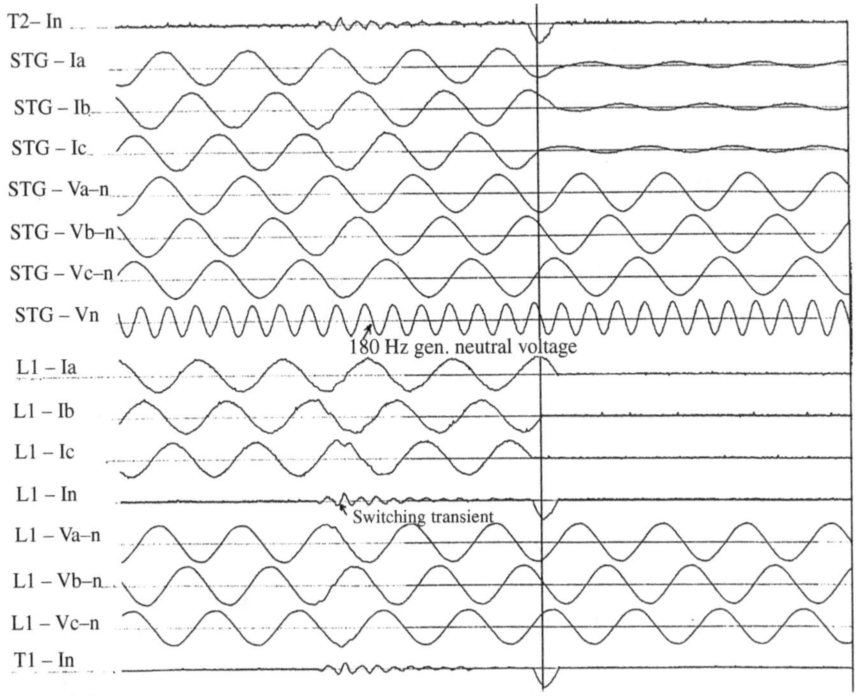

Fig. 8.16 ST generator and 138-kV line L1 currents, showing the switching transient.

undesired operation of the system. The playback of the incident confirmed that the breaker failure relay produced an output that is only present when the dc power is switched on, energizing the relay. The playback testing led to a review of the schematic and wiring drawings. The breaker failure relay terminals should have no "dc plus" during the nonfault condition. Direct current dc plus was observed at terminal 1 of device 94T, shown in Fig. 8.14 for one of the initiation contacts from the line backup protection system. The auxiliary relay contact was welded shut. To avoid tripping the plant after it was restored, the breaker failure relay was removed from service by isolating its tripping output.

Review of Breaker Failure Ground Fault Detector Setting The closed and open protection zones of CB A1 should first be defined to set the breaker failure fault detectors optimally. The end of the zone point will yield the minimum fault current required to set the breaker failure fault detectors. In addition, the setting margin should be included by taking system outages, which will result in lower fault currents. The setting of breaker failure fault detectors is normally done using a stub system condition (one end of the feeder is opened). This is justified on the basis that at any given time, one circuit breaker is only assumed to fail to clear a given system fault. Consideration should be given to examining fault currents during normal system condition with the two ends of the line closed. This will allow the breaker failure timer to start at the beginning of the fault (for some older schemes), thus avoiding longer clearing times that may occur if the fault detector is set on a system stub condition (one end of the faulted line opens first). As a corrective action, the present setting of the ground fault detector was reviewed and the following criteria applied:

1. The ground element should be set above the system loading unbalance. A value of 20% of maximum loading is normally used.
2. Sufficient margins should be established for minimum ground fault currents for normal and outage system conditions. Based on the criteria above and using a maximum loading of 175 MVA (730 A), a minimum ground fault current ($3I_0$) of 3250 A, and a CT ratio of 1200 : 5, the setting has to be greater than 146 A (0.2×730 A) and should provide about a fivefold pickup for a minimum fault. As a result, the present setting of the ground fault detector element was raised from 1 A to 3 A.

Power System Phenomena

1. Switching transient occurring when one of the 138-kV feeders is energized. Trace L1-In in Fig. 8.13 reveals the transient as it appeared in the neutral of the 138-kV plant feeder.
2. Appearance of nonsinusoidal neutral residual current during interruption of the three-phase balanced loading currents, as shown in trace L1-In in Fig. 8.15.

Corrective Actions

1. Temporarily remove the CB A1 breaker failure relay system from service and rely on the plant and 138-kV feeder backup protections to clear a breaker failure condition. The 138-kV line and generator backup relaying systems will provide a backup protection function that is carried out primarily using a CB A1 breaker failure relay. The additional delay in clearing times offered by the backup relays is acceptable. The line backup relays consist of phase and ground time overcurrent elements for line L1 and ground overcurrent relays in the GSU unit transformer neutrals. The generator backup protection consists of mho phase distance relay 21 and negative-sequence relay 46.
2. Place the A1 breaker failure in service during the next scheduled plant outage, by replacing the failed breaker failure initiation auxiliary relay 94T.
3. Review the breaker failure ground fault detector setting by performing a relay setting calculation to find the optimum set point.
4. Enhance the security of the CB A1 failure system by disabling the breaker failure re-trip function.
5. Review the DFR trigger setting for ground current to capture a significant system switching transient.

Lessons Learned

1. The breaker failure re-trip function enhances system security, especially during testing of relaying schemes; however, this feature can also cause undesired operation if certain system special conditions are met. The ground current during the switching transient lasted for only 3 cycles, which is less than the breaker failure timer setting of 10 cycles. Use of the re-trip feature for a single breaker configuration should be evaluated, especially since any undesired trip of breaker A1 will result in loss of the entire plant. The re-trip feature can provide good benefits when used in ring bus or breaker-and-a-half substations, where most elements are controlled by more than one circuit breaker. Therefore, for single breaker application similar to the system described for this case study, the re-trip feature should be disabled.
2. The circuit breaker failure ground fault detector should be set using a minimum fault current for faults only within the zone of protection for the respective circuit breaker.
3. The availability of COMTRADE format and fault record playback capability are useful in troubleshooting and assessing undesired operations. The system restoration phase following undesired operations can be enhanced and carried out faster by employing these tools.
4. The use of microprocessor relaying for breaker failure protection may eliminate some undesired operations when fundamental filters are used to calculate

phasors. The transient recorded on trace L1-In in Fig. 8.13 is about 300 Hz, which could be rejected by the 60-Hz discrete Fourier transform algorithm digital filter normally used for RMS calculations.

Case Study 8.2: 115-kV Dual Breaker Failures Resulting in the Loss of a 1000-MW Plant and Associated Substations

Abstract A transformer low-side fault began with an animal contact to a transformer bushing, then evolved to the transformer high-side bushing as a 115-kV three-phase fault. The fault was detected successfully by the transformer differential relaying system, which energized a lockout relay to trip and lock out 115-kV CBs A and A1. CB A1 failed to clear the fault, and its dual breaker failure relaying system detected the condition and attempted to isolate the failed A1 breaker by tripping and locking out all the 115-kV breakers, B, C1, D1, and E1, as well as tripping and locking out CB A. However, breaker B also failed to isolate the early failure of CB A1. This is now an event that is beyond power system design criteria. Normally, only one breaker has to be assumed to fail for any given event, not, as is the case here, two breakers failing in tandem (one after the other). In this case study we emphasize the importance of generator-phase backup devices 21 and 51V, as well as remote backup protection for substations around substation X, where the dual failure have taken place. Remote backup clearing systems are essential to clear extreme contingencies that occur in the case of either simultaneous breaker failure or total failure of the dual primary relaying systems.

Description of the Protection System As shown in Fig. 8.17, the transformer is protected by dual differential relaying systems, and each 115-kV breaker is equipped with a breaker failure relaying system (local backup). Each generator has protection against the generator stator phase and ground faults and abnormal system operating conditions. One of the key relaying elements for this disturbance is phase backup protection device 21, which is set to see the high-side 115-kV phase faults with a 25% margin and a timer setting at about 1 s. All the 115-kV lines are designed to have Z2 remote backup set to reach the next bus with a 24-cycle delay as well as ground time overcurrent elements.

Description of the Disturbance Figure 8.18 shows the system simple one-line diagram and the line L1 DFR-monitored voltages and transformer currents used to analyze the disturbance. The DFR record shown in Fig. 8.19 reveals that the disturbance started on the low-side 13.8-kV tertiary leads when an animal contacted phase a of the 13.8-kV bushing, causing a phase a-b-g fault (the phase b corner of the delta winding is grounded). The ionized cloud grew, causing the fault to evolve to a phase C-g fault on the 115-kV side, then to a phase C-B-g fault, and eventually to a high-side three-phase fault. Due to the large size of the substation, transformer T5 phase currents are not monitored directly by the DFR. Therefore, scattered phase

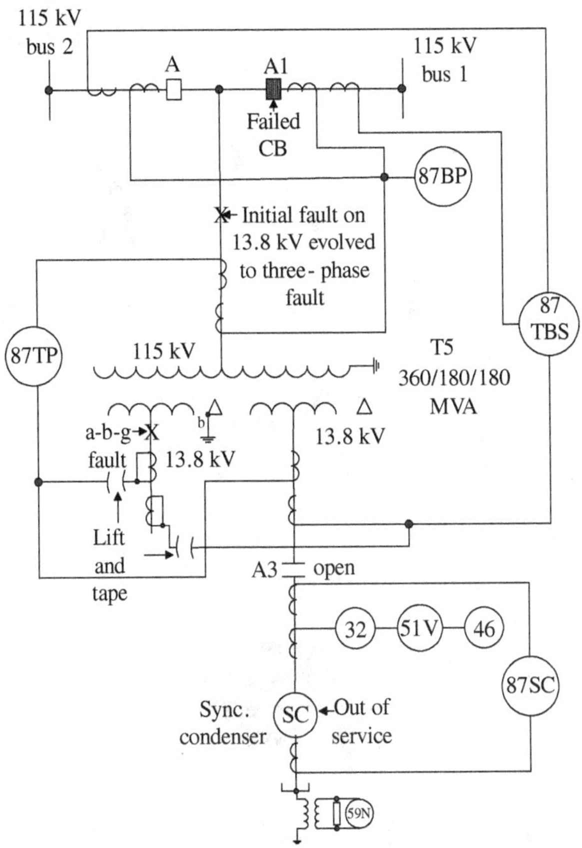

Fig. 8.17 Transformer T5 protection zones and the T5 low-side A-B-g fault, evolving to a high-side three-phase fault.

current traces from the remaining autotransformers, T1, T2, T3, and T4, are used for analysis of the disturbance.

The autotransformer has an embedded delta tertiary winding where the 13.8-kV bushings are connected only to the surge arresters (no other leads are connected to the bushings). In this type of application, one corner of the delta is grounded in accordance with ANSI standard requirements. The grounding of the delta b corner is normally done to reduce the likelihood of switching transient overvoltages and their associated insulation stresses on the transformer windings. The autotransformer has a delta that is not loaded, and therefore one of the corners of the delta is grounded, as shown in Fig. 8.20.

Classification of the Low-side 13.8-kV (Delta) Fault Using DFR Information from the 115-kV System Figure 8.21 shows the basic relationships between the transformer T5 high- and low-voltage sides for a phase a-b-g fault on the 13.8-kV delta side. Applying the transformer ampere-turns coupling principles, the low-side

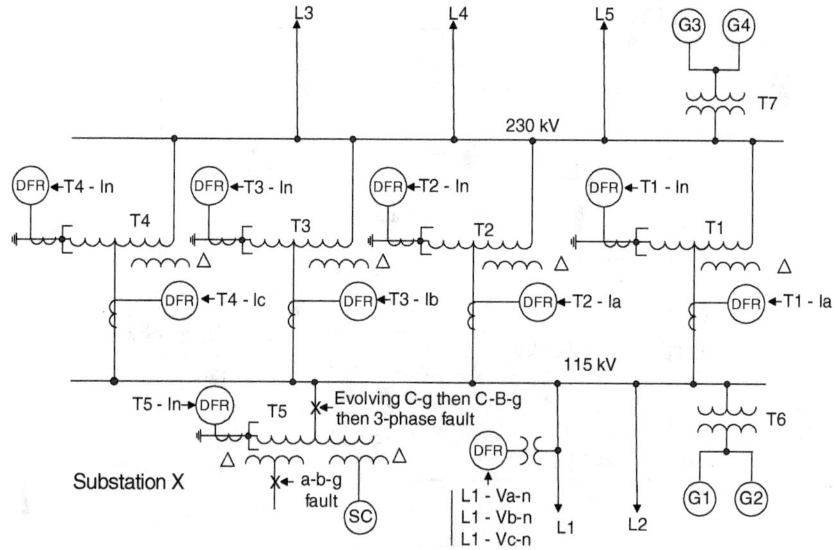

Fig. 8.18 One-line diagram showing DFR-monitored voltages and currents.

a-b-g fault results in current flows on all phases of the high-side 115-kV fault. Phase B current is equal to phase C current, and phase A will be equal and opposite to the sum of the flows on phases B and C. The transformer T5 high-side 115-kV currents are not monitored; therefore, scattered phase currents A, B, and C and neutral currents

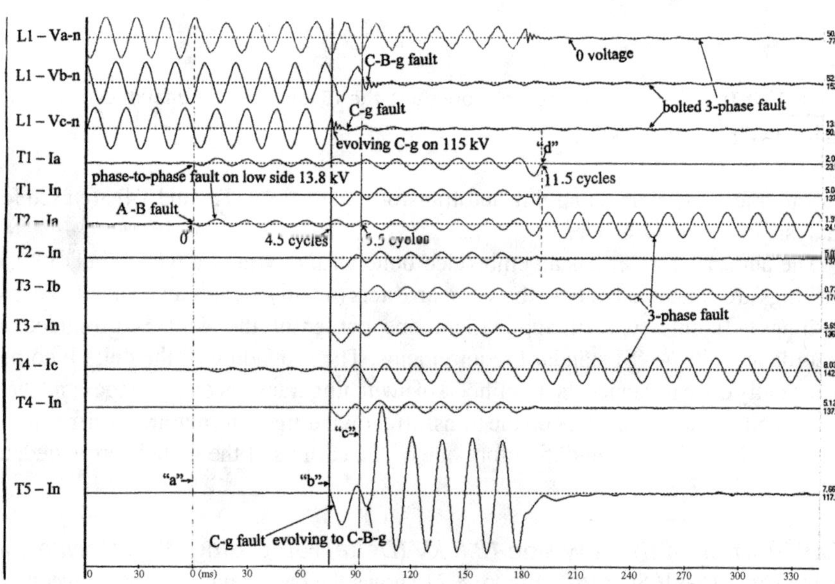

Fig. 8.19 DFR record showing line L1 voltages and transformer currents during the initial A-B fault and subsequent evolving faults.

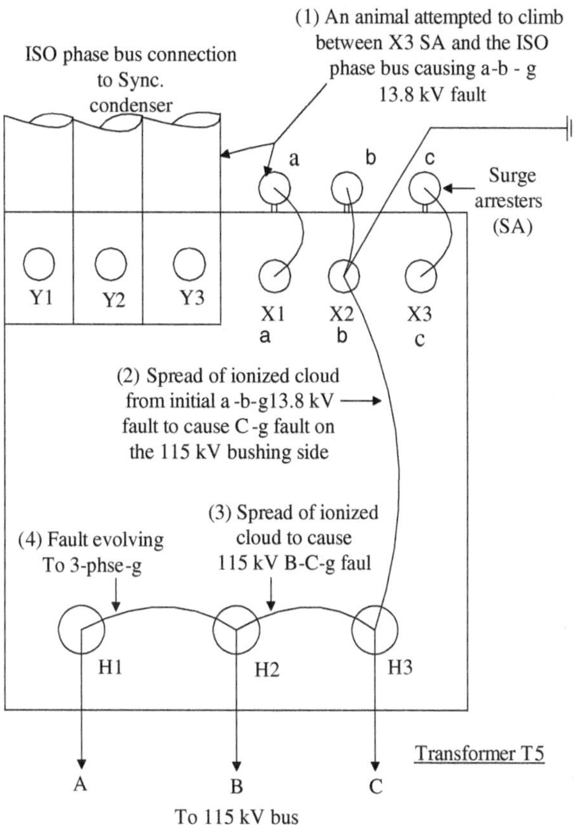

Fig. 8.20 Transformer T5 phasing diagram showing the starting fault and subsequent 115-kV evolving faults.

from the other autotransformers, T1, T2, T3, and T4, are used to confirm the analysis shown in Fig. 8.21 of the 13.8-kV fault caused by an animal climbing the transformer.

The transformer basic ampere-turns coupling principles can be verified using the DFR record shown in Fig. 8.19. By referring to the figure for the low-side a-b-g (the A-B fault on the DFR record) fault duration of 4.5 cycles, current traces T1-Ia and T2-Ia are opposite to current trace T4-Ic, while trace T3-Ib is very small and therefore difficult to read. In addition, trace T5-In reveals no neutral current flow for this duration. The high-side phase current flows in the DFR record will confirm the basic principles for current flows in Fig. 8.21, thus confirming the initial a-b-g fault shown in the T5 phasing diagram in Fig. 8.20.

Simultaneous Circuit Breaker Failures Circuit breaker A1 failed to clear the initial evolving fault at F2 and activated its breaker failure relaying. The CB A1 failure lockout relay operated to isolate CB A1 by tripping 115-kV breakers B, C1, D1,

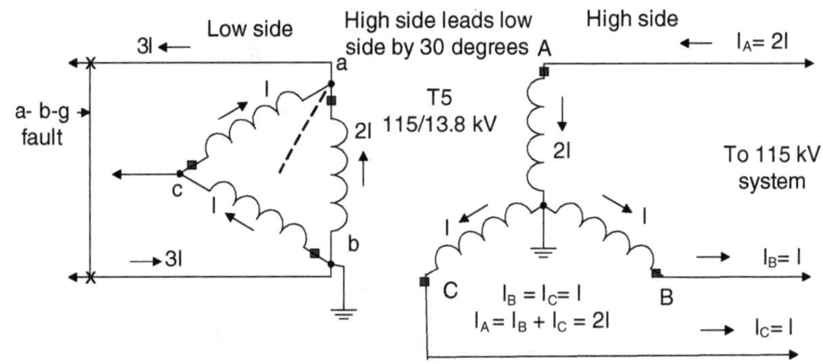

Fig. 8.21 Transformer T5 three-line A-C with 13.8-kV a-b-g fault location and the corresponding high-side fault currents.

and E1. Circuit breakers C1, D1, and E1 opened successfully. However, CB B failed to interrupt. Now we have two simultaneous CB failures, A1 and B, shown in Fig. 8.22. This event is now beyond normal power system operating criteria, and the CB B failure can only be cleared by tripping all local and remote sources feeding fault F2

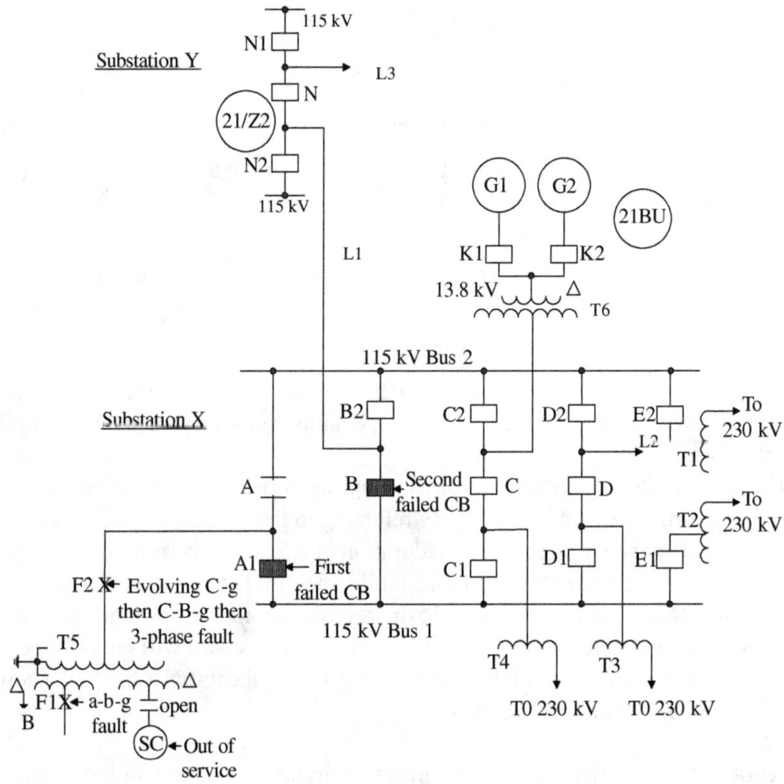

Fig. 8.22 One-line diagram showing 115-kV CB A1 failing to clear the fault and subsequent failure of 115-kV CB B.

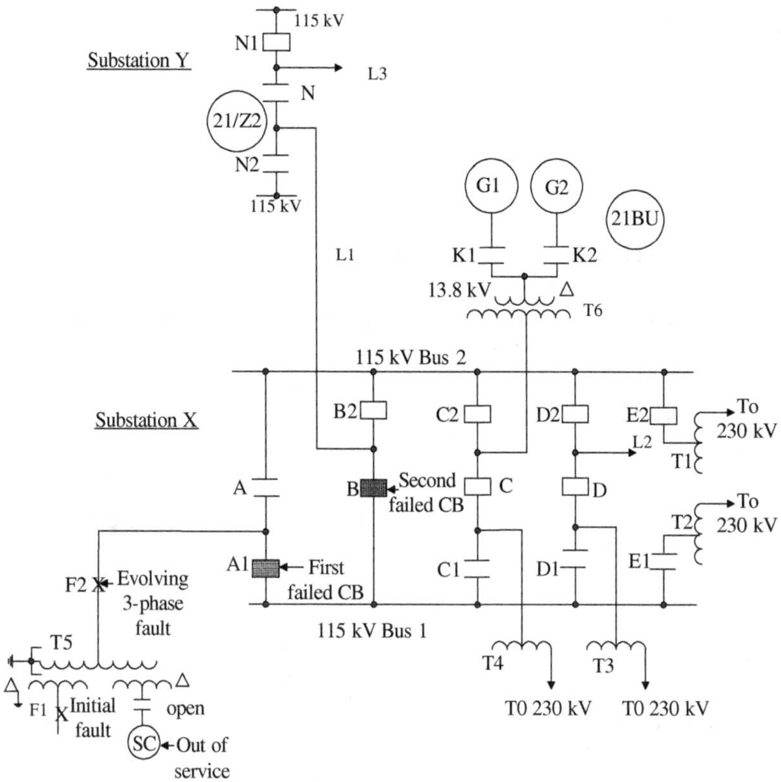

Fig. 8.23 One-line diagram confirming the total isolation of substation X and associated 230-kV substation and all connected generators.

via B. Generator remote phase backup relay 21 began to trip generating units in sequence in the timing range from 1 to 1.5 s. Figure 8.23 illustrates the total isolation of substation X and associated 230-kV substation and connected generators. Several remote backup sources operated from 2 to 3 s, and the entire event was fully cleared when CB B finally opened by itself 5 s after the start of the initial fault and saved the day.

Conclusions Using the basic analysis of the line-to-line fault on the low side and the derived currents on the high side, it can be concluded that the fault begun by an animal contacting phase a, causing a phase a-b-g fault (due to grounding of the delta corner at b). The ionized cloud grew and the fault evolved to a phase C-g fault on the 115-kV high-side bushing. The fault then evolved to include the 115-kV phase B bushing, causing a phase B-C-g fault and then eventually a three-phase fault.

Corrective Action Insulate the 13.8-kV connections to the bushings and surge arresters to eliminate faults caused by animals climbing and contacting the live 13.8-kV bushings.

Lessons Learned

1. The occurrence of simultaneous dual breaker failures is beyond the design criteria of power systems. It is assumed that only one circuit breaker will fail at any given clearing process of faults. It is therefore considered as an extreme contingency that can only be mitigated by system remote backup protection and good maintenance programs.
2. Generator phase backup protection device 21 or 51V plays an important role in tripping the generator out of service and thus stopping the generator fault current contribution during extreme system operating conditions such as in this case of simultaneous dual CB failures.
3. For line protection, zone 2 remote phase and ground elements as well as ground backup time overcurrent relaying also play an important role in removing lines contributing to the fault, similar to the function of generator relay 21 during extreme system contingency conditions.
4. Oil circuit breakers require a very comprehensive testing program to detect early breaker problems.
5. Testing of a three-phase fault with a breaker failure should be simulated even though it is not within the system testing criteria.
6. Remote backup must be applied to all lines, despite the application of breaker failure relaying locally for each breaker. The remote and local backup relaying systems supplement each other, especially during extreme system contingency conditions.
7. Use of a transformer numerical differential relay as a replacement for obsolete electromechanical relays will provide oscillography and fault records for high- and low-side transformer currents, thus simplifying the analysis of this type of disturbance.

Case Study 8.3: 230-kV Substation Outage Due to Circuit Breaker Problems During the Clearing of a Close-in Phase-to-Ground Fault

Abstract A phase-to-ground fault occurred on a 230-kV transmission line. Circuit breaker trip circuit problems resulted in complete isolation of the substation from the power system. Magnetic flux cancellation phenomena associated with a dual trip coil assembly are the main cause of substation isolation from the bulk power system. In this case study we describe the substation protection systems, events, and phenomena that led to isolation of the substation.

Description of the System and the Associated Protection Prior to the independent power producer (IPP) addition at plant W, substation X was operated as shown in Fig. 8.24, with the two diagonal breakers A1 and B2 normally operated open. As shown in Fig. 8.25, substation X was then modified to provide transmission access for the IPP 100-MW cogeneration facility at plant W. One new 230-kV circuit

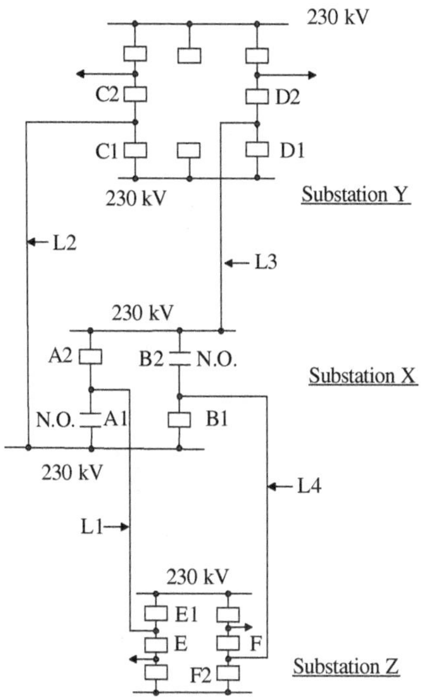

Fig. 8.24 One-line diagram for substation X prior to its upgrade to add an IPP plant.

breaker, a new 120-MVA, 240/120-kV autotransformer with a 13.8-kV tertiary, a new 115-kV circuit breaker, and 7.5 miles of a new 115-kV transmission line to plant W were added at the substation.

The addition of the IPP interconnection at substation X also forced the upgrade of existing protection systems to meet the bulk power system protection criteria. A second line relaying system needs to be added to all four 230-kV transmission lines, and breaker failure protection is now required for all existing and new circuit breakers. The 230-kV transmission lines will be protected by dual relaying systems: the existing numerical primary line relaying systems applied in a pilot scheme using directional comparison blocking over a power line carrier and a new numerical secondary line relaying system providing step-distance protection only. The secondary line relaying systems will be converted to a permissive overreaching transfer trip (POTT) system when the planned microwave communication becomes available at the substation.

The new 120-MVA autotransformer T1 is protected by dual differential relaying systems. The new 115-kV transmission line L5 is protected by a primary POTT system utilizing numerical relay and fiber optic communication and a secondary microprocessor-based current differential relay system utilizing fiber optic communication. An optical ground wire (OPGW) was included in the construction of the new 115-kV transmission line.

Two new 125-V dc battery systems were installed at substation X to provide dual dc systems. The new 230-kV and 115-kV circuit breakers were equipped with dual trip

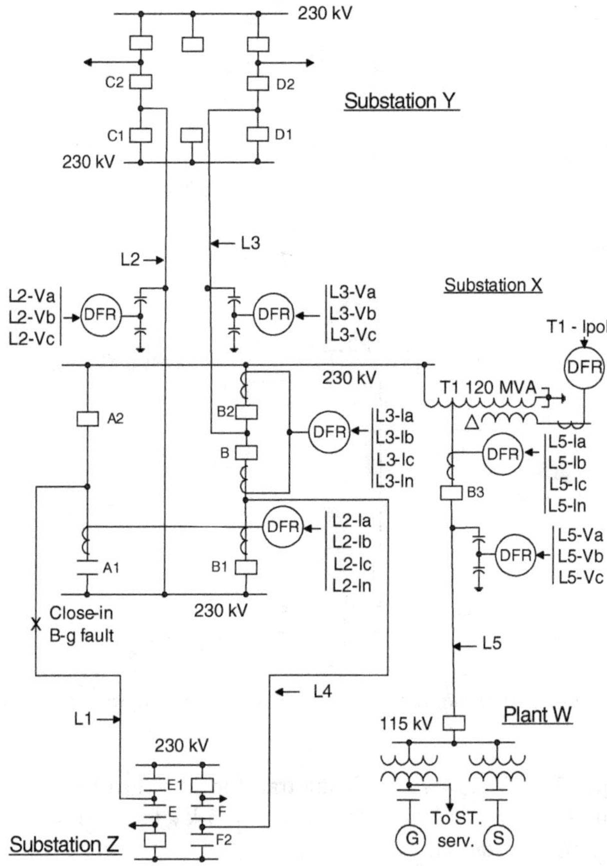

Fig. 8.25 One-line diagram showing DFR-monitored voltages and currents at substation X.

coils. Existing 230-kV circuit breakers were retrofitted with new dual trip coil assemblies, and new current transformers were added to the existing circuit breaker bushings to provide separate current transformers for the redundant line relaying systems.

Breaker failure relaying systems were added to all breakers at substation X. Normally, operation of the breaker failure relaying systems re-trips the failed breaker immediately, energizes a lockout relay to trip the local backup breakers, and initiates a direct transfer trip (DTT) to open the remote terminal breakers after an appropriate time delay when a new communication channel will become available. The 115-kV CB failure relaying will key the DTT system on the newly added OPGW fiber optic link.

Description of the Disturbance Figure 8.25 shows the system one-line diagram and DFR-monitored voltages and currents used for analysis of the disturbance. Based on an analysis of the DFR record shown in Fig. 8.26 and the substation X sequence-of-events system records, a phase B-g fault occurred on line L1. One cycle later, the

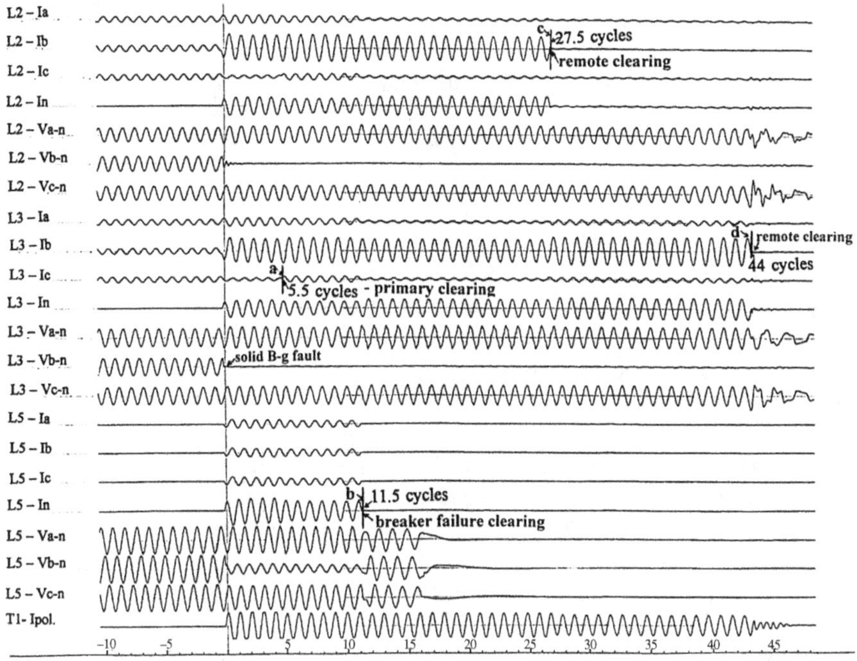

Fig. 8.26 Substation X DFR record showing voltages and currents during the ground fault with primary, local, and remote clearing times.

primary and secondary numerical relays of line L1 protection operated to energize the primary and secondary trip coils of CBs A1 and A2, respectively. Three cycles later (4 cycles from $t=0$), CB A1 at substation X opened. One and a half cycles later (5.5 cycles from $t=0$), CBs E and E1 at substation Z opened. In addition, CBs F and F2 at substation Z tripped undesirably (falsely). Eight and a half cycles from $t=0$, the CB A2 breaker failure relaying system timed out to energize its associated breaker failure lockout relay to trip CBs A1, B2, and B3. Three cycles later (11.5 cycles from $t=0$), 115-kV CB B3 for line L5 at substation X opened. Remote CBs C1 and C2 of line L2 at substation Y tripped after 27.5 cycles from $t=0$. Forty four cycles from $t=0$, remote 230-kV CBs D1 and D2 of line L3 at substation Y opened to remove the last source to the fault and isolate substation X from the bulk power system. The DFR record in Fig. 8.26 shows all these clearing steps with their marked time stamps in cycles.

Analysis of the Line-to-Ground Fault Figure 8.25 illustrates the DFR-monitored voltages and currents for lines L2, L3, and L5 that were used to analyze the fault that occurred on line L1. Figure 8.26 shows the entire DFR record of the initial phase B-to-ground fault with a primary clearing time of 5.5 cycles at point a, a breaker failure clearing time of 11.5 cycles at point b, a remote backup clearing time of 27.5 cycles at point c, and a remote backup clearing time of 44 cycles at point d. The fault clearing phases are summarized below.

Failure of Primary Fault Clearing The ground fault occurred when a strain insulator failed, causing the phase B line conductor between the A-frame and the line disconnect switch to snap to ground. The conductor fell onto the steel support for the disconnect switch, thus causing a solid B-g fault through connection to the substation ground mat. The fault was detected by both primary and secondary numerical line protection, which operated (at the same speed) to energize the primary and secondary trip coils for CBs A1 and A2 and to initiate breaker failure relaying for these breakers. Circuit breaker A2 failed to open, due to an error by the manufacturer during construction of the trip coil assembly. A magnetic flux cancellation phenomenon occurred and resulted in no movement of the tripping rod (mechanically, the breaker was never told to trip). The CB A2 breaker failure relaying system operated correctly to attempt to isolate the breaker by tripping and locking out 230-kV CBs A1 and B2 and 115-kV CB B3.

Failure of Local Backup (CB Failure Protection) to Clear the Fault The CB A2 breaker failure relaying system operated correctly in 7.5 cycles and energized both trip coils of CBs A1, B2, and B3 via a breaker failure lockout relay. The CB B2 dual trip coils were energized simultaneously by the breaker failure lockout relay. However, CB B2 failed to open, again due to improperly assembled dual trip coils and the resulting magnetic flux cancellation effect. As a result, the CB B2 tripping rod did not move (again, mechanically the breaker was never told to trip) and the initial B-g close-in fault on the line L1 circuit can be cleared from substations Y and Z only remotely.

Remote Backup Clearing Substation Z breakers E and E1 of the faulted line L1 tripped and opened successfully, and breakers F and F2 of line L4 tripped undesirably. As a result, the substation Z remote source to the fault was isolated. Since CB A2 failure relaying at substation X was unable to isolate the failed breaker, remote fault sources at substation Y continue to feed the initial fault. The zone 2 ground distance remote backup relaying system operated in 27.5 cycles to trip and isolate the L2 circuit at substation Y. The L3 circuit, which was the last source, was tripped by ground time overcurrent backup relaying in 44 cycles to clear the last source to the initial ground fault, thus isolating substation X from the power system.

Magnetic Flux Cancellation Phenomenon This phenomenon is explained in detail in Chapter 3. The existing trip coil assemblies for 230-kV circuit breakers A1, A2, B1, and B2 were modified to include dual trip coils. This modification was required due to the addition of dual line relaying systems fed from separate CTs and potential sources. The dual trip coils were assembled opposite each other on a common magnetic structure, inviting magnetic flux cancellation to take place when the coils are energized simultaneously with no movement of the armature. Figure 8.27 illustrates magnetic flux cancellation when dual trip coil assembly is used. The correct assembly is shown in Fig. 8.27(a), where magnetic flux addition will occur when both coils are energized simultaneously. Incorrect assembly, where no net magnetic flux is generated and hence there is no force to move the tripping rod, is shown in Fig. 8.27(b). When the initial fault occurred, the dual microprocessor-based line

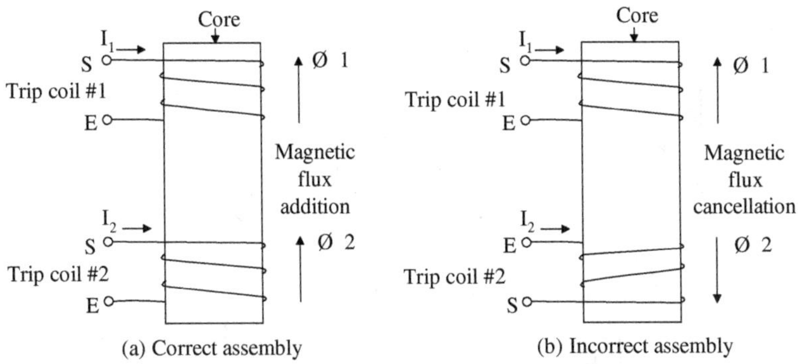

Fig. 8.27 Magnetic flux cancellation for dual trip coil assembly.

relaying systems detected the fault with a comparable speed of about 1 cycle. The auxiliary trip output relays apparently energized the dual trip coils at the same time, with relay 1 tripping trip coil TC1 and relay 2 tripping trip coil TC2. This resulted in a magnetic flux cancellation with no trip rod movement for CB A2 (the breaker was never told to move). The A2 failure relaying operated after expiration of the timer to clear the failed breaker from the system. The CB failure lockout relay was designed to energize both dual trip coils for CB B2. Again, magnetic flux cancellation for trip coil assembly produced no trip rod movement, and fault clearing had to be done by remote backup relaying. Three of the 230-kV existing oil circuit breakers A2, B1 and B2 were apparently placed in service with their trip coil assemblies wired in a flux cancellation mode, as shown in Fig. 8.28.

Lessons Learned

1. Addition to or modification of an existing protection system or associated equipment must be tested thoroughly in the field prior to placing the equipment in service. The entire event could have been confined to the faulted line had the magnetic flux cancellation phenomenon been detected prior to energizing the system.
2. The dual breaker trip coil assembly can be tested on the bench by paralleling the two coils through the connection of the individual positive leads as well as the individual negative leads of the two coils. The assembly should also be tested in service after installation by simultaneous energization of the coils.
3. The event highlighted the importance of remote backup protection system despite the local use of redundant relaying and breaker failure relaying systems.
4. Remote backup relaying is important and vital to limit cascading of system tripping when faults are not isolated, due to equipment problems or during extreme power system contingency conditions.
5. Careful analysis of system disturbances can provide basic and useful information and an explanation of power system phenomena.

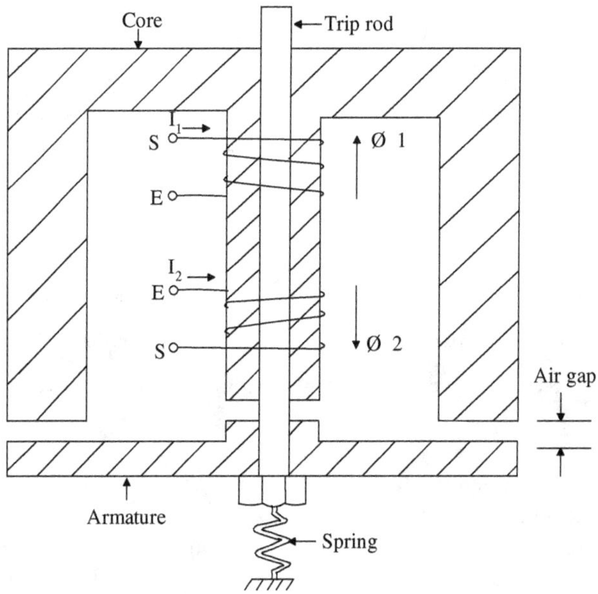

Fig. 8.28 Dual trip coil structure documenting incorrect assembly of the coils.

Case Study 8.4: Failure of a 230-kV Circuit Breaker Leading to Isolation of a 1000-MW Plant and Associated Substations

Abstract A 230-kV line surge arrester failed, causing a close-in C-g fault that was cleared successfully. A few cycles later, phase C of one of the 230-kV CBs flashed over to the tank, causing a C-g bus fault which was self-cleared from the system. The incident grew out of proportion, due to the undesired operation of a 230-kV breaker failure relaying system by backfeeding to the CT circuit, thus causing additional breaker tripping. The additional CB tripping resulted in the operation of a special protection system (SPS) which activated a 400-MW generation rejection and the net tripping of 31 CBs during the disturbance. In this case study we describe the substation protection and the events and phenomena that resulted in numerous tripping of breakers and isolation of generating units, as well as corrective actions.

Description of the Protection of the 230-kV Systems

Transmission Line L1 Protection Systems The 4-mile 230-kV circuit L1 is protected by dual relaying systems. As shown in Fig. 8.29, the A and B line protection systems are supplied from phase angle regulator (PAR) line-side CTs at substation Y. The A system consists of an electromechanical pilot wire relay that provides complete phase and ground protection applied over a privately owned hardwire. The B system is a permissive overreaching transfer trip scheme using electromechanical mho distance relaying interfacing to a dedicated hardwire communication link. PAR dual protection differential relaying zones are also shown in Fig. 8.29. The direct transfer trip system

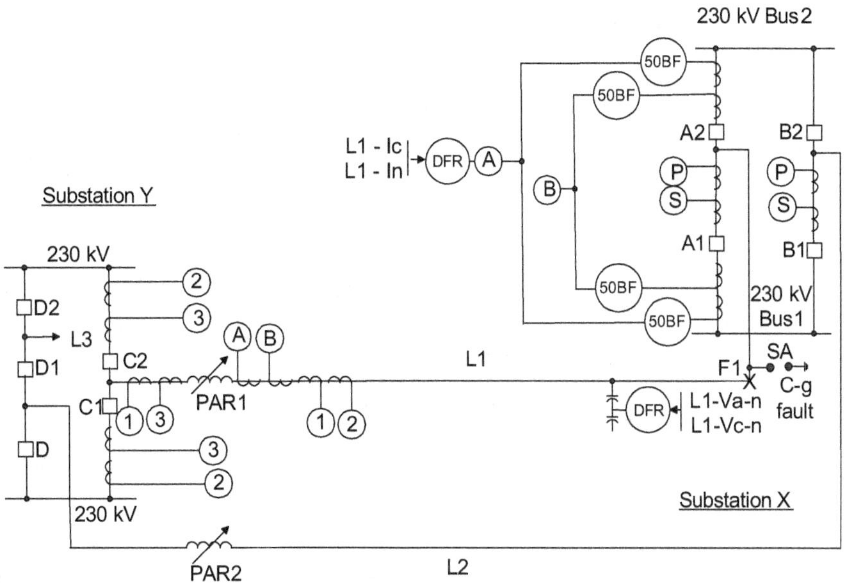

Fig. 8.29 One-line diagram showing the protection systems at substations X and Y and the DFR-monitored voltages and currents.

using the hardwire is keyed by the PAR relaying systems as well as breaker failure relaying to trip remote CBs.

Circuit Breaker Failure Protection Systems Substation X was originally designed to optimize the cost of running cables between the switchyard and the relay room, which is 1500 feet from the switchyard. For line relaying use, the CTs were paralleled at the entrance of the cable tunnel near the switchyard with a single four-wire conductor cable run to the relay room. Following the 1965 Northeast Blackout, a recommendation was made to add breaker failure (local backup) to the 230-kV bulk power system. To implement breaker failure relaying and to avoid running new cables due to the urgency of the implementation, instantaneous electromechanical fault detectors were added in the cable tunnel to be able to gain access to individual circuit breaker current. Figure 8.30 indicates the implementation of the breaker failure dc schematic. Although design criteria call for one relaying system for CB failure, dual identical systems (primary and secondary) are used to continue to implement complete redundancy of all types of bulk power relaying systems.

Bus Protection Systems The 230-kV buses are protected by dual relaying systems. Each system is employing the current differential relaying concept using high-impedance overvoltage relays. Each system also energizes lock out relays to trip and lock out all respective bus-associated circuit breakers.

Special Protection System for Generation Rejection The SPS associated with the 230-kV system is a generation rejection scheme designed to trip preselected generation

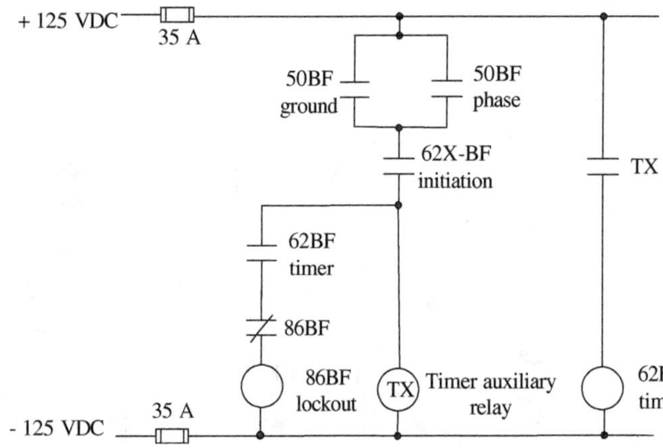

Fig. 8.30 Early design of breaker failure dc schematic for 230-kV CBs A1 and A2.

not directly involved in clearing a fault from the power system in response to certain contingencies or abnormal conditions. The generation rejection at substation X is activated to preserve system stability in the 230-kV transmission area. The SPS scheme is initiated either by breaker failure relaying operations, any autotransformer fault, any bus fault, or the loss of any group of transmission lines. One of the line groups is the combined loss of both L1 and L2 lines. The initial trip of L1 due to the C-g fault and the subsequent trapping of L2 at substation X due to the tripping of both 230-kV buses resulted in the loss of the two circuits and imitated the rejection of 400-MW generations.

Sequence of Events Figure 8.29 shows the system one-line diagram and DFR-monitored voltages and currents for the 230-kV line L1. The sequence of events as constructed from an analysis of the substation X 230-kV DFR record shown in Fig. 8.31 and sequence-of-events recorder is as follows. A phase C-g fault occurred on line L1 at F1, as shown in Fig. 8.29. The fault was caused by the failure of the line surge arrester at the X end of the circuit. From 1.5 to 2 cycles, the line L1 primary and secondary relaying systems operated to energize CBs A1 and A2 primary and secondary trip buses and to initiate primary and secondary breaker failure relaying for CBs A1 and A2, respectively. Three and a half cycles from $t=0$, CB A2 opened. CB A1 attempted unsuccessfully to interrupt the fault current at zero crossing at 4 and 4.5 cycles, as shown in Fig. 8.32. Circuit breakers C1 and C2 at substation Y tripped at 4.5 cycles. At 5 cycles, CB A1 opened and cleared the C-g fault from the 230-kV system.

At 9 cycles from $t=0$, the CB A1 phase C main interrupter failed, resulting in a flashover to the tank wall, causing C-g fault F2, as shown in Fig. 8.33. Six milliseconds later, the bus 1 phase C secondary differential relay operated to energize its associated lockout relay to trip and lock out all breakers connected to bus 1. Two milliseconds later, the bus 1 phase C primary differential relay operated to energize its associated lockout relay.

Three milliseconds later, the CB A2 primary breaker failure relaying system had an undesired operation which energized its lockout relay to clear the backup breakers,

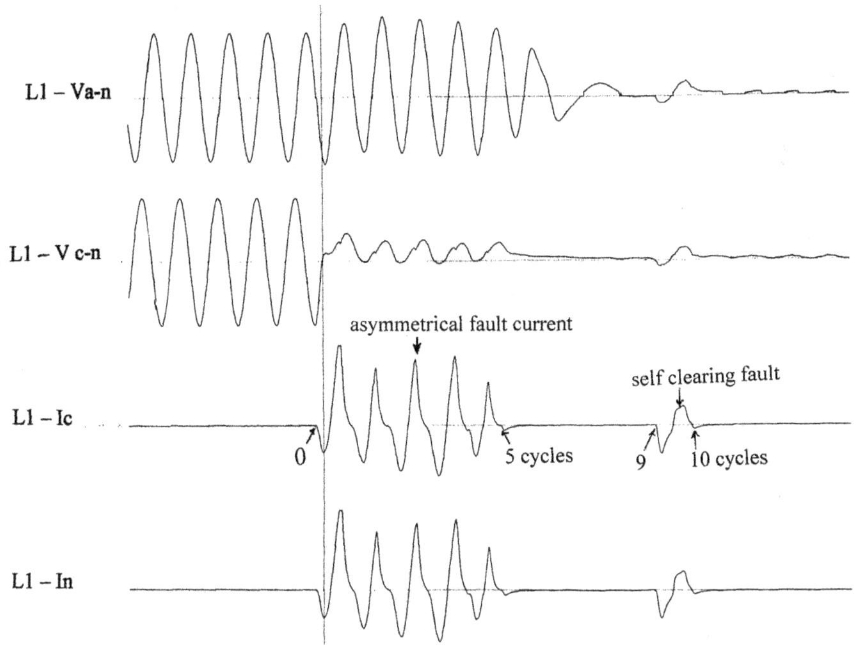

Fig. 8.31 DFR record showing line L1 voltages and currents during the initial L-g fault and the subsequent self-cleared fault.

including bus 2 (the CB A2 secondary breaker failure relay operated a few milliseconds later). Two milliseconds later, the CB A1 secondary breaker failure relaying system had an undesired operation which energized its lockout relay to clear the backup breakers (the CB A1 primary breaker failure relay operated few milliseconds later).

Ten cycles from $t = 0$, the phase C flashover inside breaker 2002 clears itself at current zero (the fault lasted for about 1 cycle), as shown in the DFR record in Fig. 8.31. Fourteen cycles from $t = 0$, 230-kV buses 1 and 2 are deenergized, as shown in Fig. 8.34. Two and a half cycles later a generation shedding scheme is initiated and generator breakers were tripped 3 cycles later. At this time a total of 31 high-voltage breakers were tripped and a total generation of 400 MW was shed.

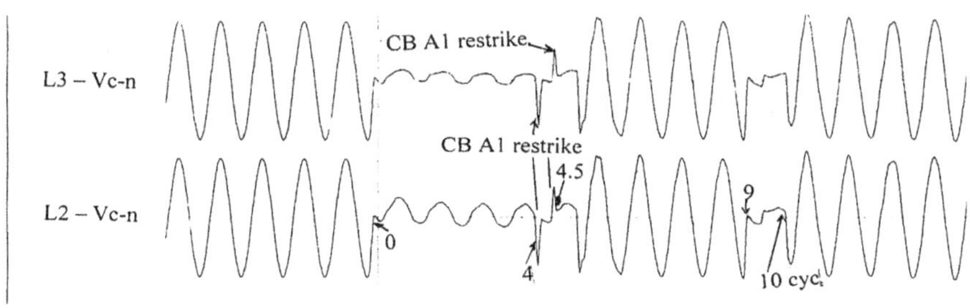

Fig. 8.32 DFR record with the faulted phase C-g voltage for lines L2 and L3 indicating CB A1 re-strikes at 4 and 4.5 cycles.

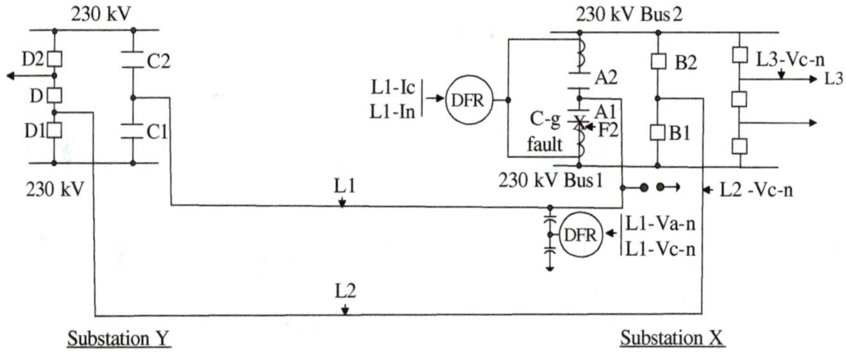

Fig. 8.33 Substation X configurations after clearing of the first fault and the subsequent C-g breaker fault.

Disturbance Analysis

Analysis of the Initial C-g Fault The initial C-to-ground fault F1 on the L1 circuit was caused by the failure of a conventional line surge arrester at the X end of the L1 line. Trace L1-Vc-n in Fig. 8.31 confirms that the fault was not solid-to-ground, due to the presence of fault resistances of arrester disks and breaker arcs. The gapped surge arrester apparently sparked-over under normal system conditions, which may have been caused by arrester gap deterioration. The arrester failed to reseal following the first current zero and conducted a fault current magnitude of over 24 kA for several cycles. Circuit breaker A2 at substation X tripped in 3.5 cycles, while CB A1 attempted to interrupt twice at zero crossings but was not successful. Finally, CB A1 tripped and cleared the fault. Circuit breakers C1 and C2 at the Y end of the line tripped in 4.5 cycles. The fault lasted for 5 cycles and was cleared by operation of the primary and secondary line relaying systems for the L1 circuit. As shown in trace L1-Ic in Fig. 8.31, the fault current magnitude for the last cycle before interruption was possibly lowered by an additional resistance introduced during CB A1 re-strikes.

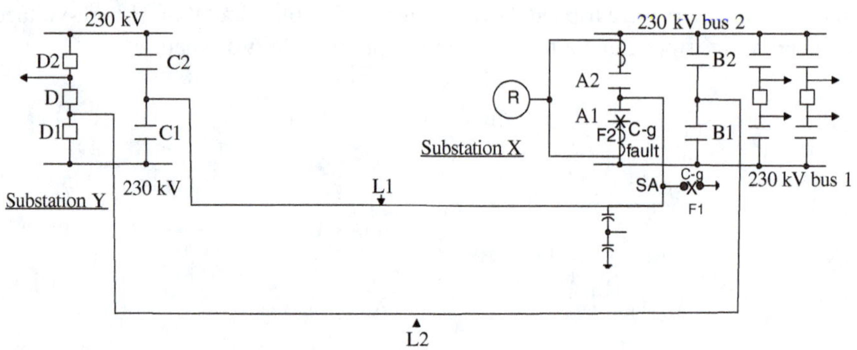

Fig. 8.34 Substation X configuration after clearing of the first and second faults and backfeeding to CB A2 breaker failure fault detectors.

Analysis of the Second C-g Fault The DFR record in Fig. 8.31 reveals that the second C-g fault, F2, occurred 4 cycles after successful interruption of the initial fault, F1, by tripping CB A1. The fault occurred inside CB A1 when the phase C main interrupter failed, resulting in the flashover to the tank wall at the CB bus side, as shown in Fig. 8.34. The CB fault was inside the overlapping zone areas of line and bus protection systems. However, the line relaying system was operated earlier and tripped CBs A1 and A2; therefore, the fault had to be cleared by the operation of either the bus 1 primary or secondary relaying system. In addition to bus differential relay operation and since the fault was inside CB A1, it was desirable to isolate the failed breaker. Consequently, the false operation of CB A1 breaker failure relaying systems for the fault was desirable. However, the incident grew out of proportion, due to the false (undesired) operation of CB A2 breaker failure relaying. The generation shedding scheme was initiated by the resulting line-end open condition for L2 and loss of the L1 circuit. A total of 400 MW was rejected by operation of the special relaying system, which energized lockout relays to trip the associated circuit breakers.

Analysis of the Breaker Failure Relaying Performance During the Fault The breaker failure relaying system was designed with solid-state timers, which were used primarily for Z2 and Z3 step-distance remote backup clearing applications. These were the only timers available at the time of the design. These timers have a dropout time of about 3 cycles, which is acceptable for a setting of 30 to 60 cycles for remote backup application. The poor performance of CB failure schemes during this disturbance is due to excessive delay in the dropout of timer auxiliary initiating relay TX and the breaker failure fault detectors shown in the dc schematic of Fig. 8.30. In addition, the fault detectors are located in a noncontrolled environment in the cable tunnel, subject to dust and humidity, which could degrade their performance. The CB A2 primary breaker failure indicated a TX dropout time of more than 34 ms and an initiation dropout time of more than 44 ms. From the analysis above it can be concluded that the timers never stopped after initiation by the TX coils. The occurrence of the C-g fault inside CB A1 9 cycles from $t = 0$ (4 cycles after successful interruption of the initial fault), which falls into a breaker failure total clearing time span of about 12 cycles, has presented a challenge to the performance of the breaker failure protection and its timing coordination margins.

Performance of Current Transformers in the Presence of Asymmetrical Currents As shown in the DFR record in Fig. 8.31 the secondary fault currents for the L1 circuit are not a faithful replica of the primary fault currents. The initial phase C-g fault has a large dc component with a slow decaying rate, due to the large X/R ratio for the generating plant associated with substation X. The dc component has far more influence than the ac fault current in producing severe saturation. Sizing the CT to have an adequate knee-point voltage with suitable allowance for a possible dc component and remanence can reduce the effect of saturation. The accuracy class C800 for CTs associated with CBs A1 and A2 was downgraded to C200 for the 800 : 5 connected tap. The L1 electromechanical line relaying, CT cable leads, CT winding, and CT lead resistances have a connected burden of about 2.2 Ω. For a symmetrical fault current of 22 kA, the impressed voltage across the CT is about 302 V, which is slightly below the knee point for the 800 : 5 tap. The combination of high fault current, high burden, and a low connected CT ratio have stressed the CT to be near its nonlinear region. In this incident, the initial asymmetrical fault current has caused CT partial saturation and left a remanence flux in the CT core. The only way to reduce remanence in CT cores that are presently in service is to demagnetize them by external means. In this incident the initial asymmetrical fault current had apparently caused a remanence in the CT core which was not removed. The occurrence of the second C-g fault inside CB A1 a few cycles after the initial fault stressed the CT and caused severe saturation, which contributed to the false operation of the CB A2 breaker failure relaying system.

Performance of the 230-kV Oil Circuit Breakers Circuit breaker A2 made a successful attempt to trip at 3.5 cycles. As shown in Fig. 8.32, CB A1 attempted without success to clear the fault at two zero crossings (at 4 and 4.5 cycles). The failure of CB A1 was not related directly to a re-strike; however, it may have been precipitated by the early re-strikes during the initial C-g fault. Compared with the earlier re-strikes, the higher short-circuit current during the initial C-g fault preceding these re-strikes could have contributed to the failure of breaker A1. The first inspection of CB A1 revealed no physical damage except for oil vented out of the breaker tank. Clearance was then taken and the CB oil was drained to prepare for inspection. The internal inspection revealed that the bus 1 interrupter assembly had failed and collapsed to the bottom of the breaker tank.

Breaker Failure Backfeeding Phenomenon This phenomenon is explained fully in Chapter 3. As indicated in the DFR record shown in Fig. 8.31, the initial phase C-g fault was cleared from the substation X end of line L1 in 5 cycles. No primary current was flowing through CB A2 when the CB A1 C phase failed to ground. Therefore, the undesired breaker failure relay operation for CB A2 was due to backfeeding to its breaker failure fault detector, resulting in secondary current flow that exceeded the relay ground element pickup of 2 A. Figure 8.35 shows the current flow through the CB A2 fault detector 50BF wiring circuit. CBs A1 and A2 both opened successfully and cleared the initial fault. However, when CB A1 faulted to ground, all fault current I_A was flowing from the breaker's bus-side CT to the line relaying. The flow of current I_L through burden Z_R will apply voltage V_R across the CB A2 current transformer. The voltage V_R, coupled with the fact that the initial fault current contained a large dc offset as shown in Fig. 8.31, had apparently stressed CB A2-associated breaker failure fault detector CT. As a result, a remanent magnetic flux could have been left. This residual flux, which degrades the quality of the CT, could have contributed to excessive current flow I_B through the fault detectors. This will force the CT associated with CB A2 to draw excessive magnetizing current flow I_B, enough to pick up the fault detectors and cause the breaker failure relaying scheme to operate.

Corrective Actions Breaker failure backfeeding can be mitigated by reducing the voltage imposed on the unenergized primary CT side by reducing secondary CT current and/or the connected CT burden. The reduction of the relay burden is

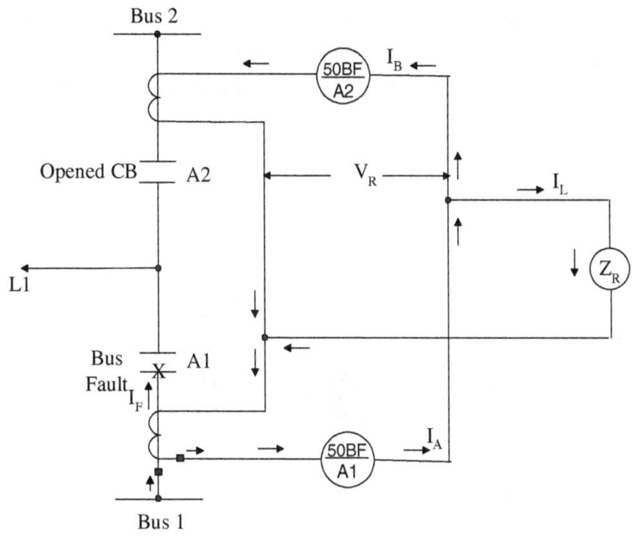

Fig. 8.35 Backfeeding to breaker A2 failure fault detector CT secondary circuit.

accomplished by replacing electromechanical relaying with a low-burden microprocessor-based (numerical) system. This replacement is normally justified, due to the obsolescence of electromechanical relaying and the advantages of new numerical relaying technology. Normally, a reduction in testing requirements, diagnostic capability, fault event records, and the reduced cost of numerical relaying justifies its implementation.

The complete solution to the problem is to provide replacement of the electromechanical and solid-state breaker failure components with an integrated microprocessor-based package. This plan will require adding new cables from the junction boxes in the tunnel to the relay room. The existing two identical primary and secondary relaying systems need to be replaced by two numerical breaker failure relaying systems preferably supplied by different manufacturers (to avoid common-mode failure).

Lessons Learned

1. The occurrence of asymmetrical faults followed by a second fault presents a challenge to CT accuracy performance, due to the presence of dc offset. This affects CT transient performance without demagnetization of the CT prior to the second fault.
2. Multiple faults occurring within the span of CB failure time also present a challenge to the scheme timing margin.
3. Secondary CT current backfeeding through breaker failure relaying for an opened breaker in a ring bus or breaker-and-a-half arrangement can cause additional breaker tripping and wider system disturbances.
4. Analysis of power system disturbances can provide basic and useful information for the protection systems. Periodic review of the existing electromechanical

relaying system should be conducted to check relay obsolescence or deterioration of their performance. Justification for electromechanical relaying replacement should be based on either the lack of spare parts or its degraded performance or reduced maintenance cost associated with the application of numerical relaying.

5. The availability of sequence-of-events recording (SER) systems at the substation level is essential. It should be capable of a fine timing resolution of 1 ms to provide a quick means to analyze disturbances. The SER records can supplement the DFR or numerical relaying oscillography records and speed up the process of developing the needed sequence of events. DFR can also provide a combination of analog and binary event monitoring to assist in system disturbance analysis.
6. The availability of RMS calculations as a DFR analytical tool provide an accurate and quick means of validating power system short-circuit modeling.

Case Study 8.5: Generator CB Failure During Automatic Synchronization of the Circuit Breaker

Abstract While in a pumping mode, a unit breaker developed a problem during unit synchronization to the system. The problem was attributed to the fact that the rack-and-pinion breaker mechanism did not complete its travel and close all three phases completely. This resulted in an out-of-step condition, with the unit oscillating three times against the system. In the first swing the unit was absorbing power from the system, and in the second swing the unit was giving power to the system. The breaker eventually failed, causing a phase A-C fault within the breaker. The fault was detected by the transformer differential relays, which activated lockout relays to isolate the unit and the failed breaker from the system. In this case study we describe the power system and associated protection and show DFR record analysis for the initial three unit swings and the subsequent phase A-C breaker fault. We also analyze the swing and corrective actions in detail.

Sequence of Events The sequence of events for 18-kV CB A1 failure during the automatic synchronization attempt of G/M1 unit while in the pumping mode was constructed from an analysis of the plant X DFR records and the SER printouts. Figure 8.36 shows the system one-line diagram and plant DFR-monitored voltages and currents used to analyze this disturbance. Figure 8.37 shows the unit protection overlapping zones and the protection systems for the unit and the generator step-up transformer.

Unit First Swing The machine had achieved 95% speed and was ready for synchronizing. The auto synchronization auxiliary relay operated to allow the auto synchronizer relay to close CB A1. One minute later, Fig. 8.38 reveals that phase B of CB A1 is still open, and unit G/M1 starts to swing against the system. The currents on phases A and C resemble a phase-to-phase fault, with a current magnitude of 36 kA; however, the machine terminal voltages do not support the theory of a phase-to-phase

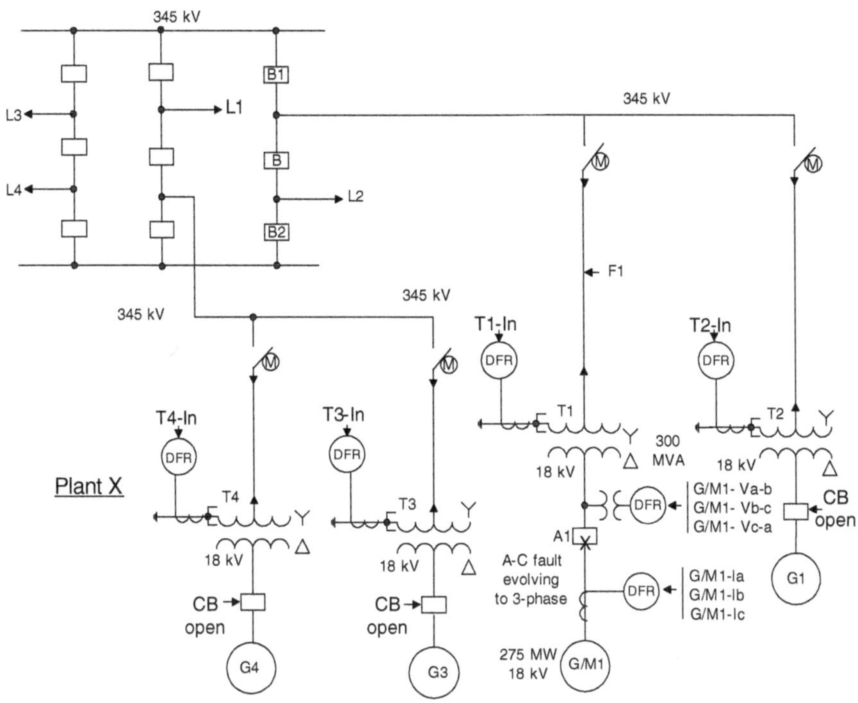

Fig. 8.36 One-line diagram showing DFR-monitored voltage and currents at substation X.

fault. Voltage $V_{a\text{-}b}$ was healthy and $V_{c\text{-}a}$ and $V_{b\text{-}c}$ show an initial voltage dip with slowly recoverable voltages. It is postulated that the phase A and C contacts of CB A1 have moved enough to allow the voltage difference across the contacts to arc over on phases A and C (similar to a unit out-of-phase closing or out-of-step condition).

This resulted in energizing phase A of the generator with an angle to the system. The machine began to draw power from the system, and the swing condition was supported by:

1. The nonoperation of both generator and transformer step-up overall differential relaying
2. The operation of the overcurrent 51M instantaneous units (produces targets only) on phases A and C
3. The operation of distance backup relaying device 21GB for the C-A and B-C units (the current duration was not sufficient for the 0.5-s. timer output)
4. Operation of fault detectors on the 345-kV line relaying phases A and C, indicating the flow of current during the unit swing to the system

Exactly 0.208 s later, unit G/M1 terminal voltages swung to normal values, and current flows on unit 1 phases A and C ceased (perfect interruption at current zero).

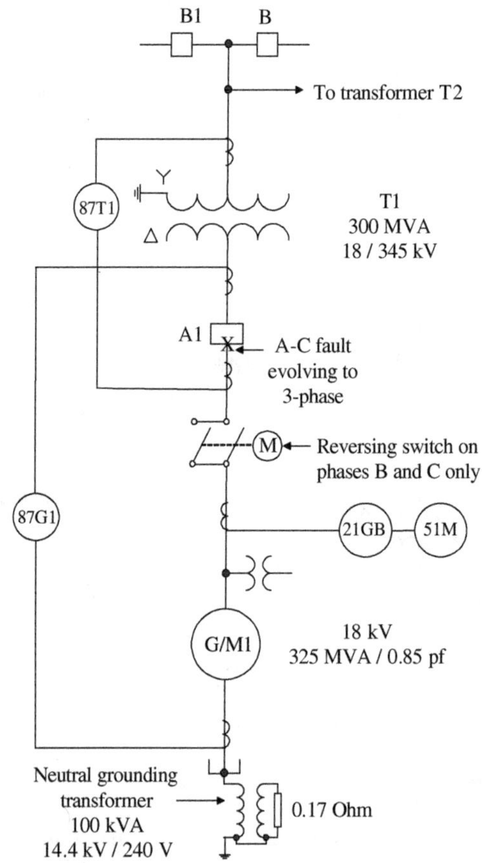

Fig. 8.37 Protection systems for the unit and the generator step-up transformer.

Time overcurrent relays 51M and distance relays 21GB reset early, due to the decreasing nature of the current values during and after the swing, which lasted for 12.5 cycles.

Since the unit was being started as a pump, the attempted synchronization would have occurred with the unit running at less than synchronous speed (slow scope). The combination of partially closed breaker and the unit running at less than synchronizing speed caused the unit to swing against the system (motor) and arc across the partially open breaker contacts via the preinsertion resistors. Figure 8.38 reveals that current trace G/M1-Ic is negative-going when trace G/M1-Vac is used as a reference with zero value going positive. During the swing period, power was flowing from the system toward the generator. The phase A and C currents shown in Fig. 8.38 started at about 36 kA. The current magnitudes were reduced to about 15 kA in 12.5 cycles at the interruption by the circuit breaker. The large inflow of current (36 kA) to the machine works to accelerate the unit, causing its angle to decrease, eventually

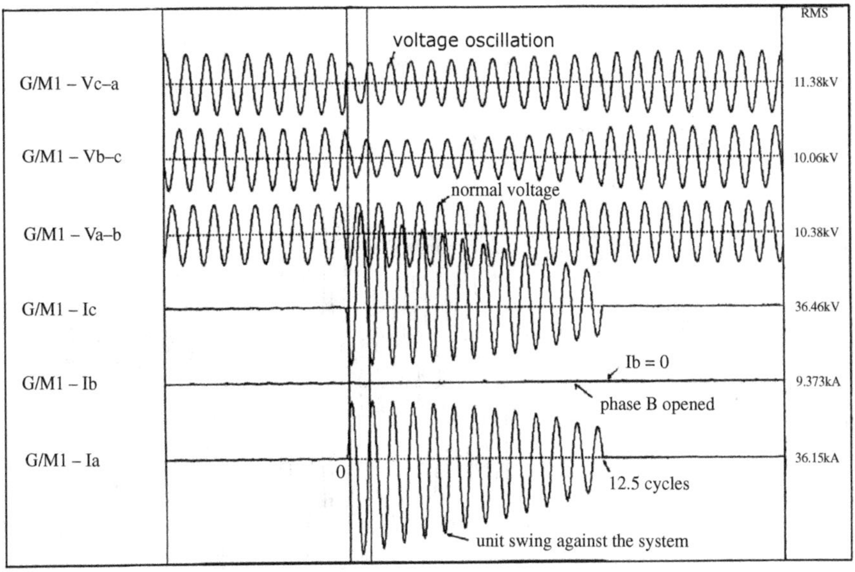

Fig. 8.38 Plant X DFR record showing G/M1 currents and voltages during the first synchronization attempt with RMS values at the first cycle.

reaching a value that will make the current sufficiently low for the partially open breaker contacts to interrupt at current zero, as shown in Fig. 8.38.

Unit Second Swing At this point the unit has been accelerated to above synchronous speed and once again, 0.5 s later, begins to swing against the system. The power flow is from the generator into the system. This can be verified from an analysis of Fig. 8.39, where current trace G/M1-Ic is positive-going when trace G/M1-Vac is used as a reference, with zero value going positive. This will confirm that the current has reversed polarity with respect to the machine voltage during the second swing, indicating power flow from the generator toward the system. The combination of the partially closed breaker and the unit at an accelerated speed caused the unit to swing against the system as a generator and arc across the partially open breaker contacts via the preinsertion resistors. The system begins deceleration of the unit to reach a current level low enough for the arc across the breaker contacts to extinguish at current zero. The swing lasted for 20.5 cycles.

Unit Third Swing and Subsequent Fault The decelerated unit once again pushes against the system, 0.5 s later, as it did at the first swing. Figure 8.40 reveals that current trace G/M1-Ic is negative-going when trace G/M1-Vac is used as a reference with the zero value going positive. This is similar to the first swing, indicating power flow from the system toward the generator. The unit swing lasted for 7.5 cycles, and before the unit can again be pulled back through the zero power angle, the heat and ionized gases from the arching contacts cause sufficient damage to initiate a phase A-C fault.

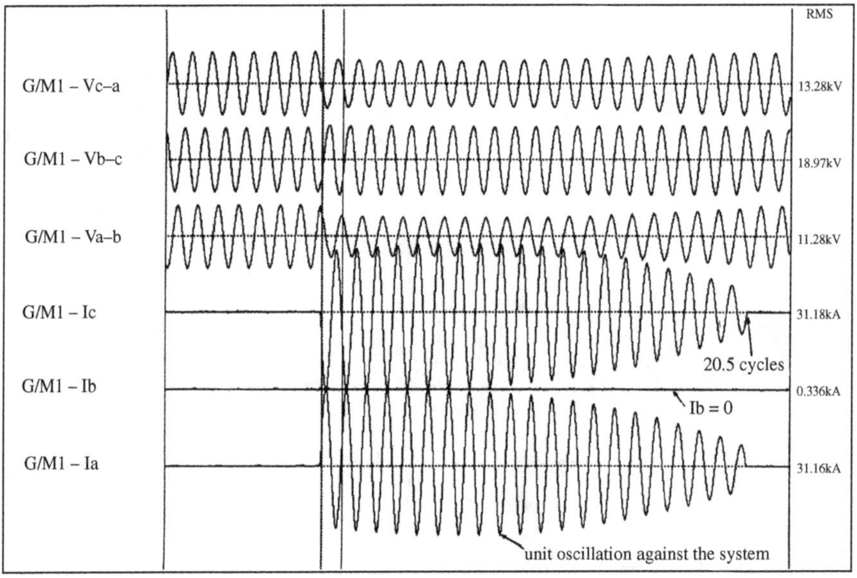

Fig. 8.39 Plant X DFR record showing G/M1 currents and voltages for the reappearance of the swing 0.5 s after the initial swing.

As shown in Figs. 8.40 and 8.41, the A-C fault then evolved to a three-phase fault where all three phases are faulted together across the arch shoot support insulators and through the I-beam that supports the insulators. The sequence of events is developed from an analysis of the DFR records and is as follows:

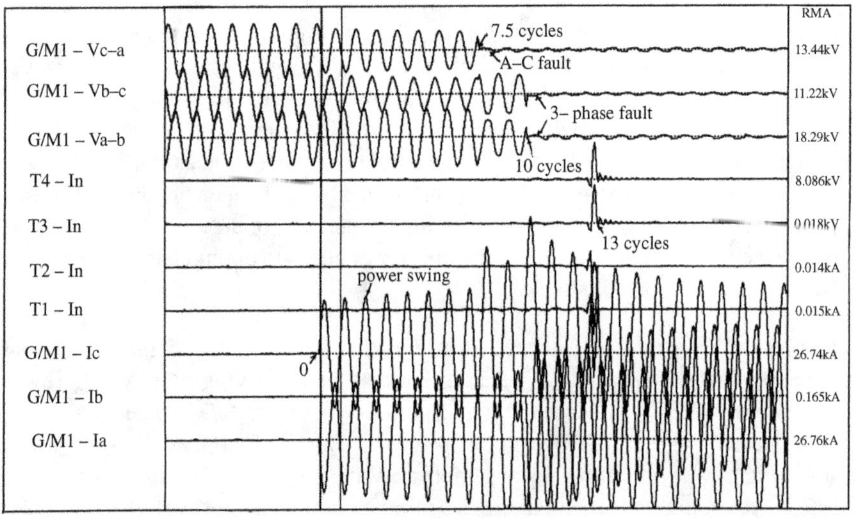

Fig. 8.40 Plant X DFR record showing G/M1 currents and voltages during the third swing followed by an A-C fault evolving into a three-phase fault.

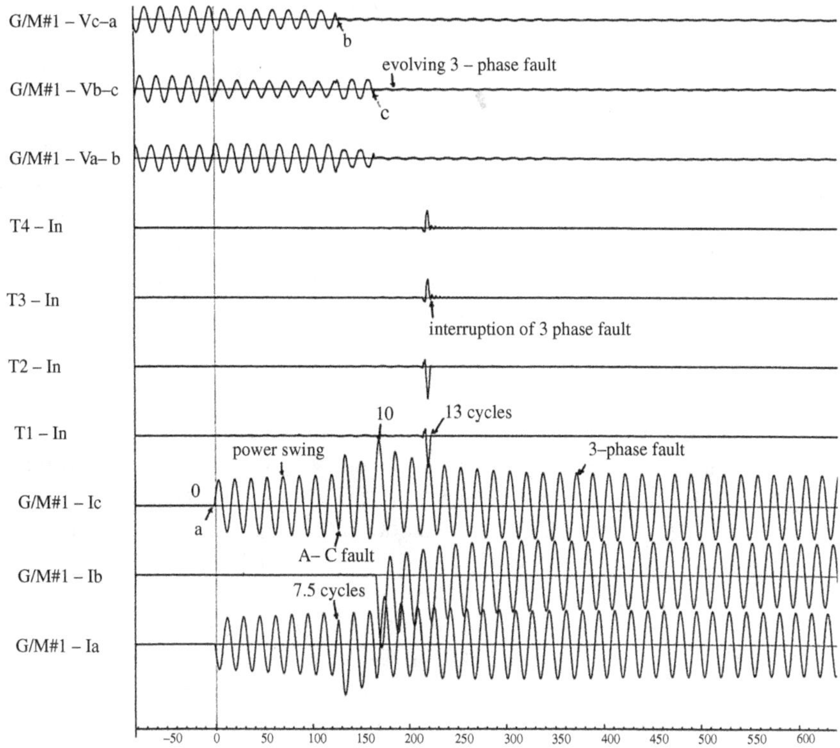

Fig. 8.41 Plant X DFR record showing G/M1 currents and voltages during the third swing, followed by an A-C fault evolving into a three-phase fault.

1. At $t = 0$, the third unit swing began and lasted for 7.5 cycles.
2. At $t = 0.125$ s (7.5 cycles) from $t = 0$, the swing evolved in a solid phase A-C fault around unit breaker A1. The phase c-a (V_{ca}) voltage dipped to zero, and the current contribution from the unit reached 42 kA.
3. At 0.167 s (10 cycles) from $t = 0$, the ionized cloud caused the phase A-C fault to evolve into a three-phase fault around the failed breaker. At the beginning of the three-phase fault, $I_a = 42.8$ kV, $I_b = 62.2$ kA, and $I_c = 55.9$ kA. The unit differential relays on phases A and C operated to energize the unit lockout relay 86ES to isolate the unit from the system and to initiate unit shutdown. The operation of phases A and C modified the targeting circuit, which prevented further targeting of phase B. Next, unit step-up transformer differential relay 87T2 phase C operated. Phase A operated 38 ms later, and phase B operated 9 ms following phase A.
4. At 0.216 s (13 cycles) from $t = 0$, the 345-kV breakers B and B1 tripped and cleared the circuit breaker fault from the system.
5. Unit decayed energy continued to feed the fault. The unit DFR record indicates that the unit decayed energy was still feeding the three-phase fault at the

failed breaker. Phase B of the unit is now feeding the ionized cloud for the three-phase fault.

Conclusions From an analysis of this operation, it appears that unit breaker A1 is the main cause of the problem. The three swings that the unit experienced were each separated by approximately 0.5 s, with a slip frequency of about 2 Hz. Calculated maximum out-of-phase currents and recorded DFR current levels confirm the postulated out-of-step condition. The protection systems performed correctly as designed during the entire sequence of events.

Corrective Actions

1. The analysis of events with line-to-line voltages is extremely difficult; therefore, the monitored voltages on all units were changed to phase-to-neutral. In addition, generator neutral voltages as well as generator field voltages on all units need to be monitored. Generator neutral voltage is the only analog signal that can be used to confirm any stator ground fault for high-resistance generator grounding system.
2. The failure of CB A1 has to be analyzed for possible enhancements of the protection and control systems (e.g., electrical sensing of unit breaker pole disagreement) at the plant.

Case Study 8.6: Circuit Breaker Re-strikes While Clearing Simultaneous Phase-to-Ground Faults on a 230-kV Double-Circuit Tower

Abstract Simultaneous B-g faults occurred on the two 230-kV circuits L1 and L2. The faults were cleared from the system successfully. However, oil circuit breakers associated with the fault clearing process at substation X re-struck. In, this case study we explain the breaker re-striking phenomena and show the two separate breaker re-strikes during clearing of the ground faults.

Description of the Disturbance A flashover to ground occurred on the two top insulators of a double-circuit 230-kV tower. The flashover, which probably occurred due to a lightning strike, created simultaneous B-g faults on the two 230-kV circuits L1 and L2, as illustrated in Fig. 8.42. Mechanical forces caused by high fault currents blew the L1 phase B insulator string, dropping the line on the lower arm, as shown in Fig. 8.43. The fault also damaged the insulator on circuit L2. Oil circuit breakers at substation X re-struck while clearing the ground faults on circuits L1 and L2. The fault was located at a tower where one of the static (ground or shield) wires was missing. During an ice storm that occurred earlier, one of the static wires for several towers failed, and the line was restored without repairing the failed wire.

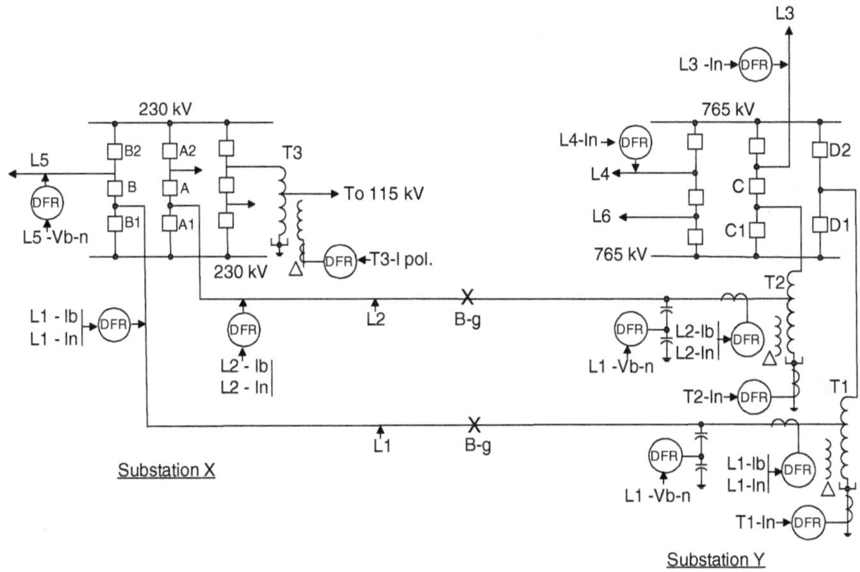

Fig. 8.42 One-line diagram showing DFR-monitored voltages and currents.

Sequence of Events Figure 8.42 shows the system one-line diagram and DFR-monitored voltages and currents for lines L1 and L2 at substations X and Y. The sequence of events is derived from an analysis of the DFR records in Fig. 8.44 from substation Y and in Fig. 8.45 from substation X, and the field report regarding the line insulator damage. The tower configuration shown in Fig. 8.43 is drawn based on field

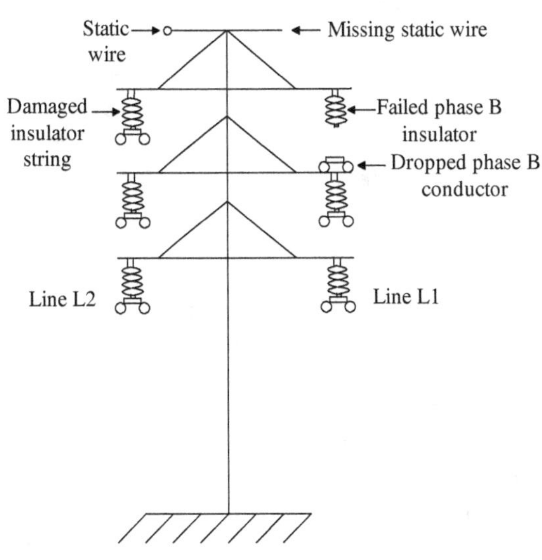

Fig. 8.43 Double-circuit tower with the extent of damage due to lightning hit shown.

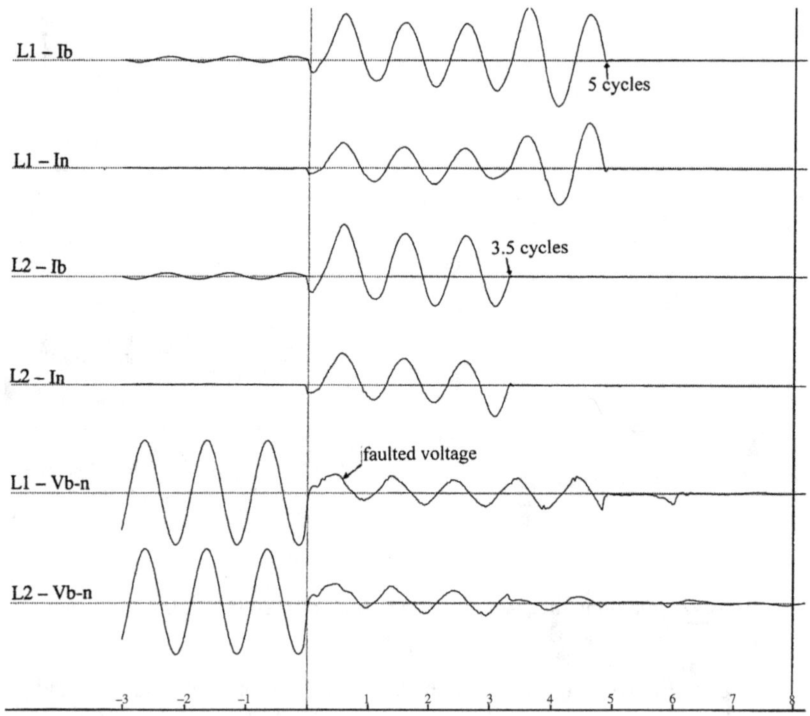

Fig. 8.44 Faulted voltages and currents at substation Y.

inspection of the tower damage. The sequence of events is derived from an analysis of Figs. 8.44 and 8.45.

1. *Initial simultaneous fault.* At time 0 (reference point), a simultaneous phase B-to-ground fault occurred on the L1 and L2 circuits.
2. *Clearing the line L2 B-g fault.* Three and a half cycles later, Fig. 8.44 reveals that CBs C1 and C at substation Y tripped and cleared the L2 ground fault from the system successfully. Figure 8.45 reveals that CBs A and A1 at substation X tripped and cleared the line L2 fault, also in 3.5 cycles. Four cycles from $t = 0$, current trace L2- Ib of Fig. 8.45 reveals that either CB A or A1 at substation X re-struck and started to feed the initial ground fault on L2 for an additional half cycle. Four and a half cycles from $t = 0$, the re-struck circuit breaker (A or A1) at substation X finally cleared the L2 B-g fault from the system successfully.
3. *Clearing the line L1 B-g fault.* Four cycles from $t = 0$, Fig. 8.45 reveals that CBs B and B1 at substation X tripped to clear the L1 ground fault. Four and a half cycles from $t = 0$, current trace L1-Ib in Fig. 8.45 reveals that either CB B or B1 at substation X re-struck and started to feed the initial ground fault on L1 circuit for an additional half-cycle. Five cycles from $t = 0$, the re-struck circuit breaker (B or B1) at substation X finally cleared the L1 B-g fault from the

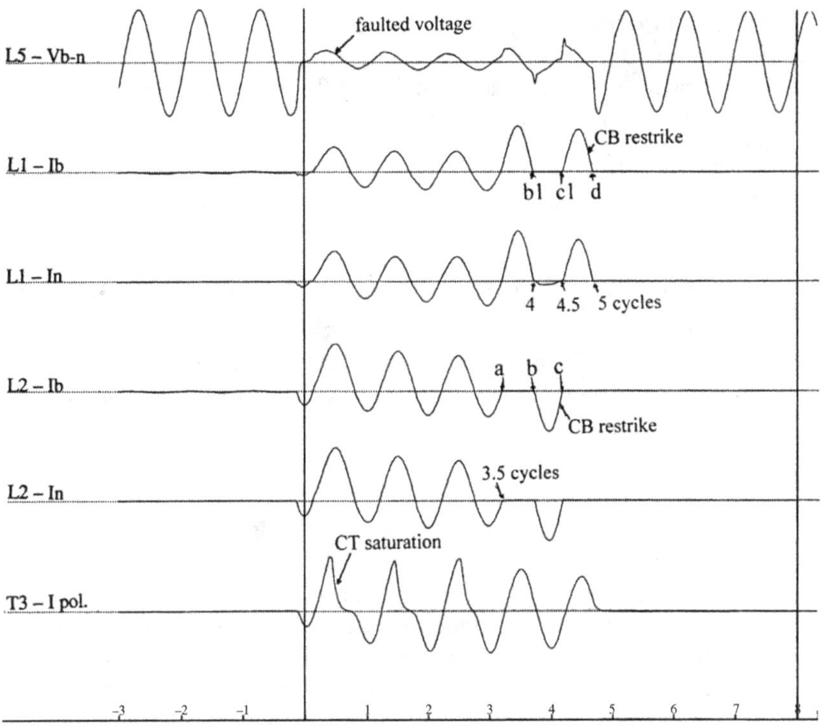

Fig. 8.45 Faulted voltages and currents at substation X.

substation. Figure 8.44 indicates that CBs D and D2 at substation Y also tripped and cleared the L1 B-g fault from the system successfully after 5 cycles.

Oil Circuit Breaker Re-striking Phenomenon The 230-kV oil circuit breakers experienced several re-strikes during the interruptions of fault currents. The breakers are equipped with grading resistors for the reduction of transient recovery voltage. However, these resistors are not very effective in coping with the rate of rise of the recovery voltage during the CB contact parting period.

Analysis of Circuit Breaker Re-strikes As shown in Fig. 8.45, CBs A and A1 at substation X interrupted the L2 ground fault in about 3.5 cycles at point a of trace L2-Ib. Half a cycle later, at point b on the same trace, one of the breakers re-struck and reenergized the initial fault for an additional 0.5 cycle. At point c the fault was finally cleared.

Similarly, CBs B and B1 interrupted the L1 fault at substation X in 4 cycles at point b1 of trace L1-Ib. A half cycle later, at point c1 on the same trace, one of the breakers (B1 or B) re-struck and reenergized the initial fault for an additional 0.5 cycle. At point d the fault was finally cleared. Reenergization of the faults at points b and c1 for an additional 0.5 cycle is known as a breaker re-striking phenomenon following the interruption of fault currents.

Power System Phenomena

1. Circuit breaker re-strikes, as shown in Fig. 8.45 for trace L1-Ib for line L1 and trace L2-Ib for line L2 at substation X
2. Current transformer saturation, as shown in trace T3-Ipol. in Fig. 8.45

Corrective Action

1. Repair the second failed shield wire for the double-circuit tower to provide additional lightning protection for the 230-kV circuit.
2. Inspect the four circuit breakers associated with the L1 and L2 circuits at substation X to look for any abnormality with the four oil circuit breakers.

Lessons Learned

1. Since the 230-kV oil circuit breakers are susceptible to re-strikes, monitoring of ground current for each breaker separately should help in identifying which of the line breakers has re-struck. For a breaker-and-a-half bus configuration, the line current CT input is formed by summing the individual breaker CT. Therefore, only an additional ground current input from the middle breaker is required for adequate monitoring of an individual circuit breaker. Some newer numerical line relaying is connected to individual breaker current and can therefore provide an oscillograph record for each circuit breaker to enhance individual breaker monitoring.
2. Circuit breakers should be inspected whenever re-striking phenomena are observed at the interruption zero-crossing point of fault currents.

Case Study 8.7: 345-kV Capacitor Bank Breaker Fault Coupled with an Additional Failure of a Dual SF6 Pressure 345-kV Breaker During the Clearing of the Fault

Abstract A 345-kV capacitor breaker, C1, failed, causing a C-g breaker fault, which falls in the overlapping protection zones for 345-kV bus 1 and capacitor bank 1. The bus 1 dual bus differential relaying (87B) detected the fault and attempted to isolate it by energizing the associated lockout relays (86). Subsequently, another 345-kV breaker, A1, also failed during the process of 345-kV bus C-g fault isolation. The second failed circuit breaker was detected by breaker failure relaying and isolated from the 765- and 345-kV sides. In this case study we describe the protection systems involved in the disturbance and the sequence of events, and analyze the C-g breaker fault and the second failed breaker.

Description of the Capacitor Bank and Associated Protection Systems Figure 8.46 shows the system one-line ac diagram and oscillograph- and

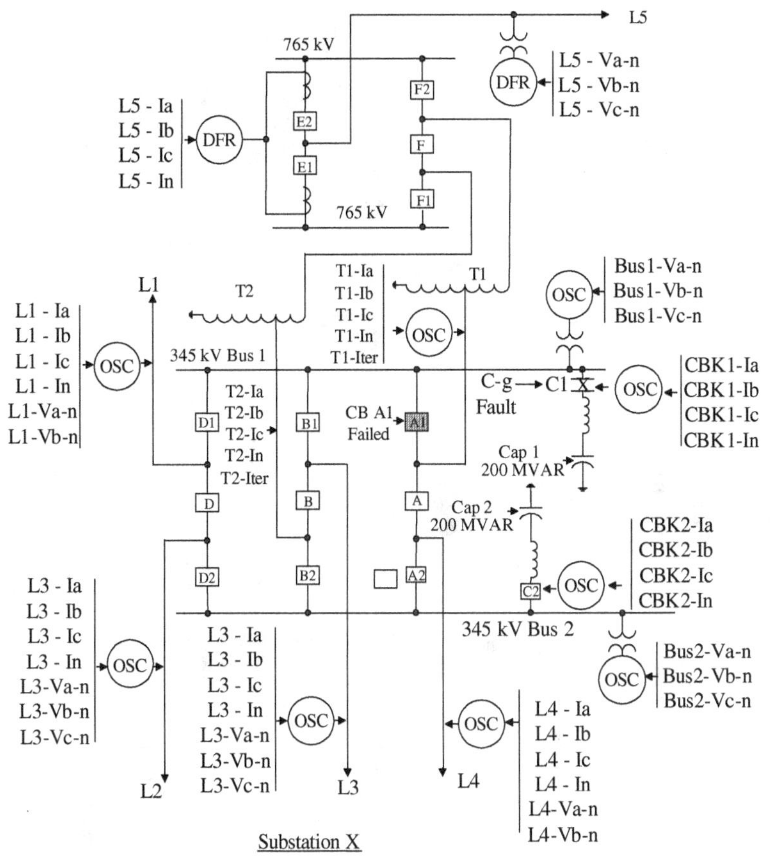

Fig. 8.46 One-line diagram showing the oscillograph-monitored voltages and currents.

DFR-monitored voltages and currents used for analysis of the breaker fault and the subsequent breaker failure. The 200-MVAR capacitor bank consists of two identical wye legs, each leg having 10 series and 11 parallel capacitor units. Each capacitor unit is rated as 300 kVAR, 19.92 kV and is protected by an external fuse. The capacitor bank protection system illustrated in Fig. 8.47 consists of two redundant 50/51 time overcurrent relays for phase and ground using diverse manufacturers. The capacitor bank overcurrent protection zone is overlapped with the 345-kV bus differential zone. Dual breaker failure relaying 50/62 is added for CB C1. The bank has dual-voltage balance relays to detect the loss of one or two cans. The first 60P system compares the individual phases of one leg against the other leg. The second system 64GS compares the two neutrals for unbalance caused by the loss of one, two, or more units from any of the wye legs. The relay systems use scaled voltages fed from voltage dividers tapping the bank, as shown in Fig. 8.47. The system alarms for the loss of one unit and trips for the loss of two units. In addition, individual capacitor units are protected by external fuses. The capacitor bank is also protected by two overvoltage

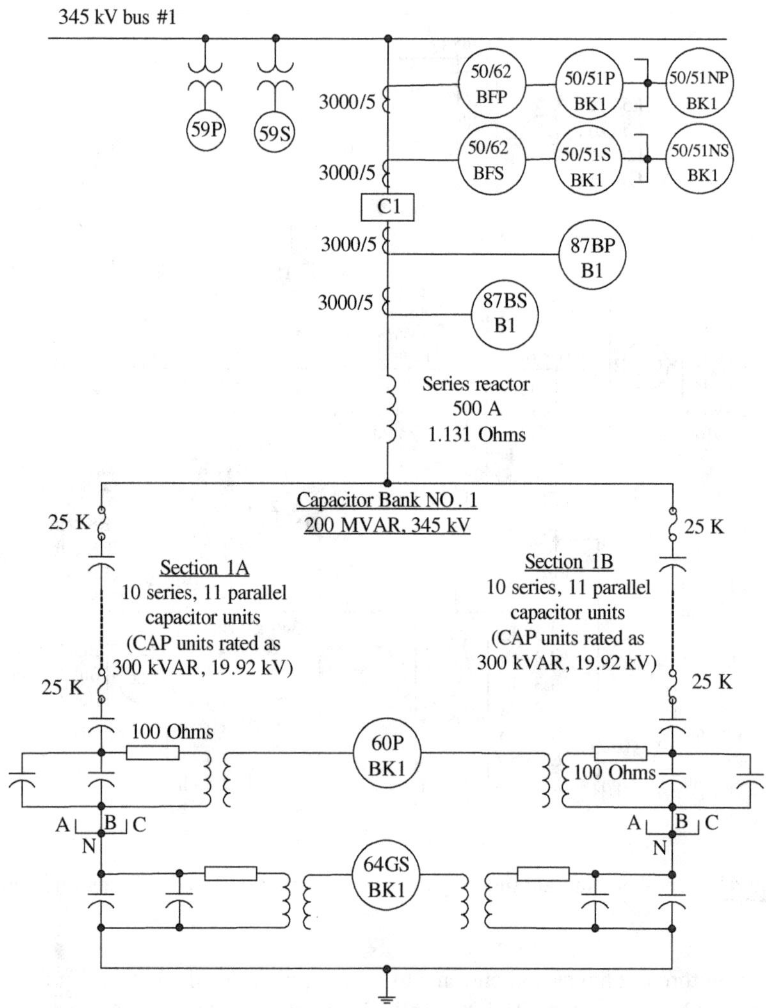

Fig. 8.47 Capacitor bank protection systems.

relays, 59P and 59S, for balanced overvoltage conditions above 110% of capacitor bank–rated voltage.

Circuit Breaker C1 Phase C-g Fault The C1 breaker re-struck 4 s after its successful opening via SCADA, momentarily connecting phase C of the capacitor bank to the system. As illustrated in Figs. 8.48 and 8.49, the breaker re-strike occurred near the voltage peak, and a damped high-frequency oscillatory capacitor current continued to flow on all phases for about 15 ms. Phase C of CB C1 eventually flashed over to ground away from the voltage peak point, causing the circuit breaker phase C-g fault current to be asymmetrical. The phase C voltage revealed that the flashover

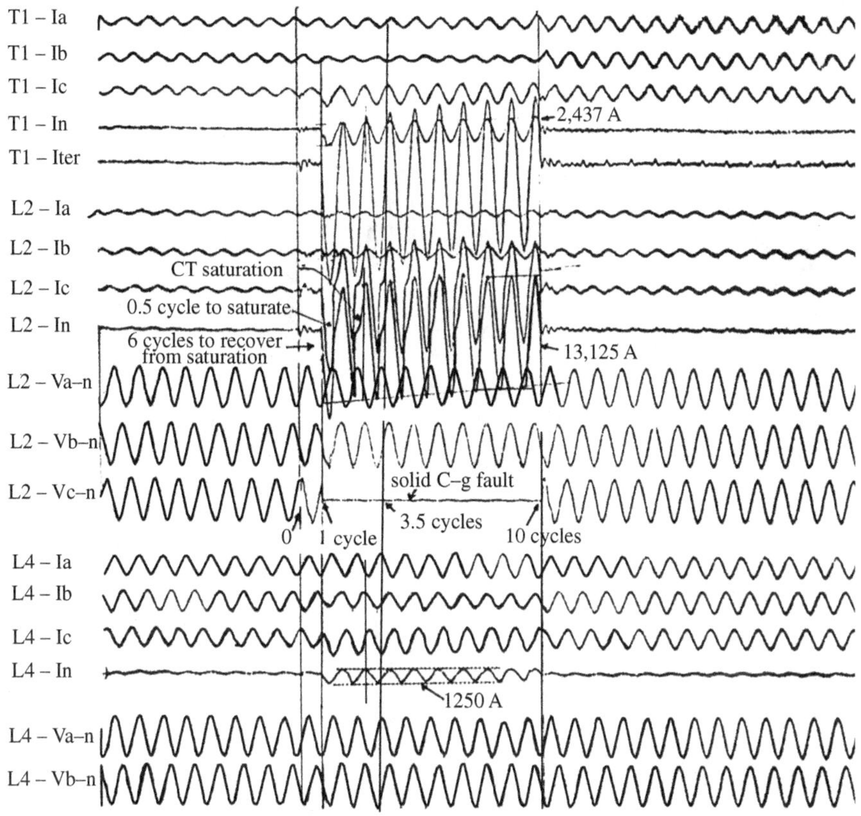

Fig. 8.48 Substation X 345-kV oscillogram showing 345-kV line L2 and L4 voltages and currents and T1 currents during the fault.

occurred prior to the voltage peak, resulting in an added dc component to the fault current. The 345-kV phase C voltage dipped to zero, indicating a solid phase-to-ground fault.

Protection System Reaction to the Circuit Breaker C-g Fault Solid-state medium-impedance primary bus differential relaying for the 345-kV bus 1 operated in 8 ms to energize the bus differential lockout relay. The CB C1 phase C-G fault falls in the overlapping protection zones of the 345-kV bus and the capacitor bank lead protection systems. The primary bus differential lockout relay operated to trip and lock out 345-kV CBs A1, B1, C1, and D1. In addition, lockout relay 86 initiated CBs A1, B1, C1, and D1, the primary breaker failure relaying systems. The electromechanical secondary high-impedance bus differential relaying for the 345-kV bus also operated to energize the secondary bus differential lockout relay. The relay operating time is about 20 ms compared to 8 ms for the solid-state primary system. The secondary bus differential lockout relay operated in 11 ms to energize

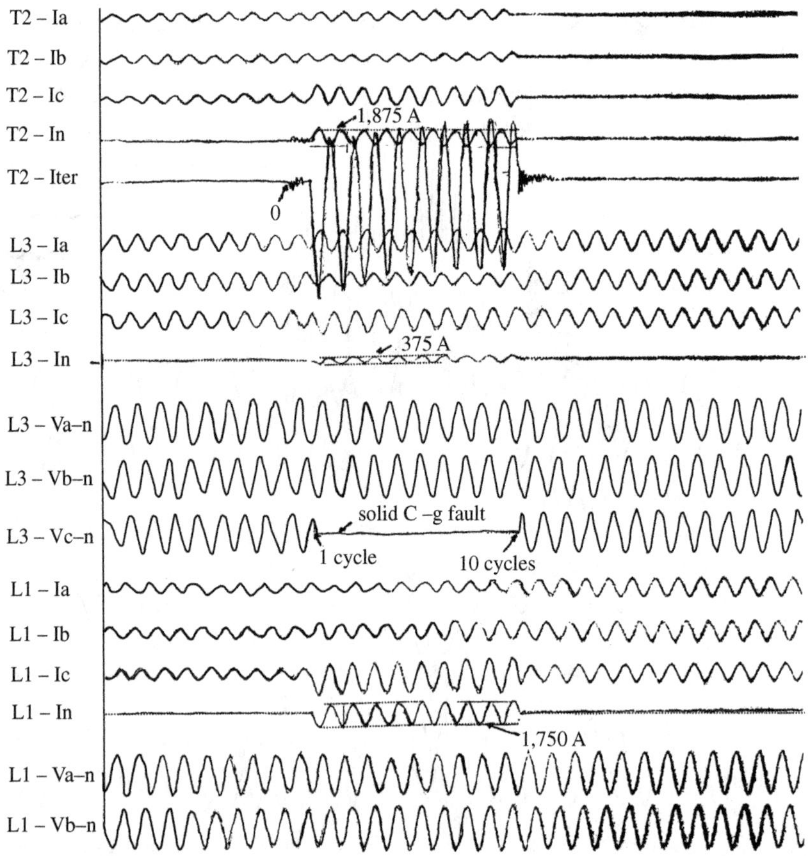

Fig. 8.49 Substation X 345-kV oscillogram showing asymmetrical fault current.

the secondary trip coils of CBs A1, B1, C1, and D1. In addition, lockout relay 86 initiated CBs A1, B1, C1, and D1, the secondary breaker failure relaying systems. Relay operating times were obtained from the sequence-of-events recorder, which has a 1-ms resolution time.

The capacitor bank 1 primary instantaneous protection operated in 21 ms to energize the capacitor bank 1 primary lockout relay to trip and lock out capacitor breaker C1. Capacitor bank 1 secondary instantaneous protection operated in 73 ms compared to 21 ms for the primary system to energize the capacitor bank 1 secondary lockout relay. The oscillogram shown in Fig. 8.50 reveals that the bus 1 phase A and B voltages are no longer present. This implies that CBs B1 and D1 have opened successfully to clear their associated bays and that phases A and B of CB A1 have opened successfully to clear the 345-kV bus 1. The total clearing time was about 2.5 cycles (41.7 ms) with a 1-cycle relaying system and 1.5 cycles for a CB interruption time for the fault currents. The oscillogram also shows that phase C of CB A1 was still conducting current and feeding (the only source) the initial fault.

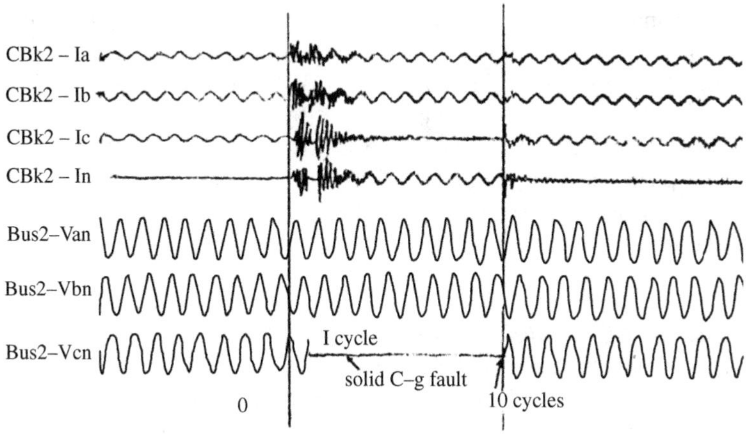

Fig. 8.50 Capacitor bank 2 out rush currents during the fault.

Breaker Failure Protection Operation CB A1 secondary and primary breaker failure protection systems were initiated by operation of the bus differential secondary and primary lockout relays, respectively. The CB A1 primary breaker failure system timed-out to energize the breaker failure primary lockout relay to isolate the failed circuit breaker A1. The CB A1 secondary breaker failure system also operated to energize the secondary CB failure lockout relay.

The 765-kV DFR record of Fig. 8.51 and the oscillograms in Figs. 8.48. 8.49, 8.50, and 8.52 show that the initial C-G fault was cleared from the system after 9 cycles. Fault clearing was accomplished by isolating the failed A1 breaker by tripping 345-kV CB A and 765-kV CBs F and F2. Total clearing time was about 9 cycles. The CB interruption time was about 25 ms.

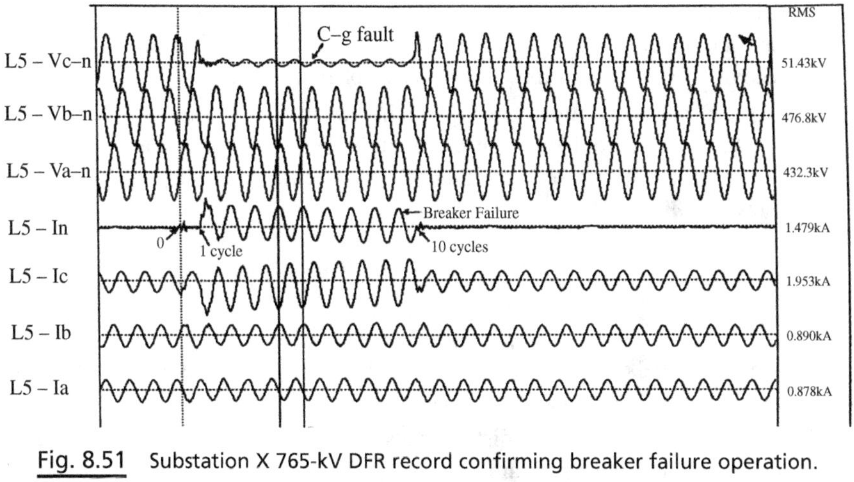

Fig. 8.51 Substation X 765-kV DFR record confirming breaker failure operation.

Oscillograms and DFR records indicated continuation of fault current flow on phase C of CB A1. The duration of fault current flow through phase C was about 6.9 cycles from the fault initiation time. CB A1 breaker failure lockout relay operated in 11 ms to isolate the failed breaker by tripping and locking out 345-kV CB A and 765-kV CBs F and F2. The CB A1 primary breaker failure system timed-out to energize the A1 breaker failure primary lockout relay to isolate the breaker. The scheme timed out in 7 cycles, and its setting of 7 cycles indicates perfect performance.

Performance of the Relay Systems During the CI Circuit Breaker Fault
Analysis of the disturbance confirmed that the initial failure of phase C of CB C1 resulted in a fault in the overlapping protection zone between 345-kV bus relaying and capacitor lead time overcurrent relaying. Both relaying systems operated correctly to clear the breaker fault. Subsequent failure of phase C of CB A1 was also detected correctly by breaker A1 failure relaying and cleared from the power system.

Performance of the Breaker Failure Relaying Scheme As illustrated in the fault records, the breaker failure system designed for the 345-kV system was tested with an actual CB failure event. The scheme performed perfectly to reset breaker failure schemes for CBs B1 and D1, which cleared breaker C1 fault successfully. In addition, the breaker failure scheme isolated the failed breaker A1 with an additional optimum clearing time of 6.5 cycles. The normal bus fault clearing time of 2.5 cycles for CBs B1 and D1 is considered optimum and is attributed to the fast operating time (8 ms) for the solid-state primary bus differential relaying.

Power System Phenomena

1. Current transformer saturation, as shown in trace L2-Ic in Fig. 8.48. The time to saturate the CT is 0.5 cycle from the initiation of the fault, and the CT is fully recovered from saturation after 6 cycles of fault current.
2. Capacitor bank 2 outrush phenomena into a close-in ground fault as shown in Fig. 8.50. Capacitor bank 1 outrush of high-frequency currents at the beginning of the fault, with phase C the faulted phase being the highest magnitude, are shown in Fig. 8.52.

Lessons Learned

1. A circuit breaker fault followed by another breaker failing to clear the initial fault is within the bulk power system protection criteria.
2. Analysis of power system disturbances can confirm protection system design and associated setting parameters.

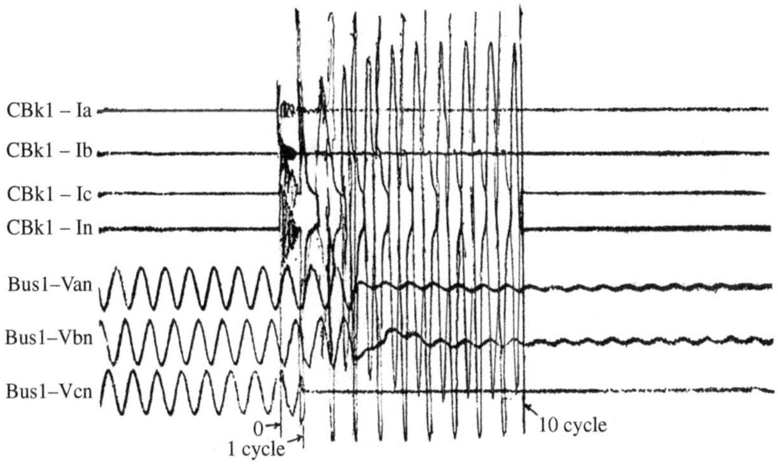

Fig. 8.52 Capacitor bank 1 fault currents and tripping of bus 1.

Case Study 8.8: Oil Circuit Breaker Failure Following the Clearing of a Failed 230-kV Surge Arrester

Abstract A 230-kV surge arrester failed, causing an A-g fault which was cleared from one end of the system in 4.5 cycles. Three and a half cycles later, a 230-kV circuit breaker arced-over to the breaker tank wall due to a breaker mechanical failure. The breaker fault was self-clearing and lasted only 0.5 cycle. The second fault, despite being of very short duration, was detected by the bus differential relays. The reoccurrence of the second fault energized the CB failure fault detectors, which coincided with a delay in the breaker failure circuit dropout, thus activating CB failure protection operation. Both the bus differential and breaker failure relaying operated their associated lockout relays to isolate the failed breaker. In this case study we describe the disturbance, oil circuit breaker problems, corrective actions, and lessons learned.

Description of the Protection Systems The protection systems shown in Fig. 8.53 that were involved in this disturbance are line relays, bus differential relays, and breaker failure relaying. Line L1 is protected by dual permissive transfer trip pilot relaying applied using diverse communication media, power line carriers, and leased telephone lines. Bus 1 is protected by a dual high-impedance electromechanical relaying system and breaker failure protection is employed using electromechanical fault detectors and solid-state timers.

Description of the Disturbance A phase A surge arrester associated with the 230-kV line L1 at substation X failed, causing an A-g fault. Figure 8.54 shows the system one-line diagram and DFR-monitored voltages and currents at substation X that are used to analyze the A-g fault. Analysis of the DFR record of Fig. 8.55 reveals

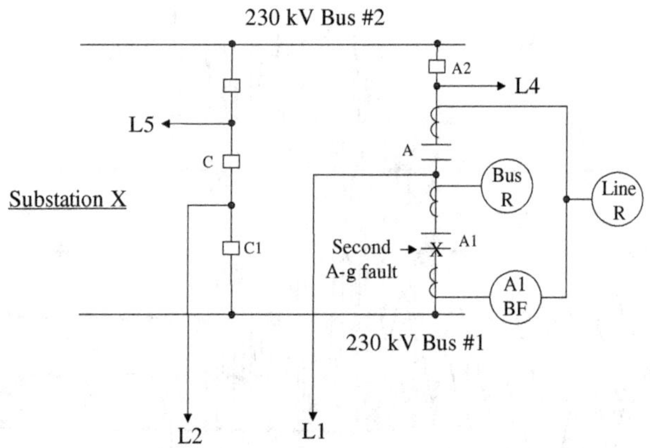

Fig. 8.53 One-line diagram showing the second A-g fault location and the relaying systems affected.

that the fault was cleared from substation X in 4.5 cycles and from substation Y in 7 cycles. The remote-end clearing of the line at substation Y can also be determined by analyzing the line voltage drop recorded at substation X. One-half cycle later, the phase A interrupter of circuit breaker A1 faulted internal to the tank wall, causing a

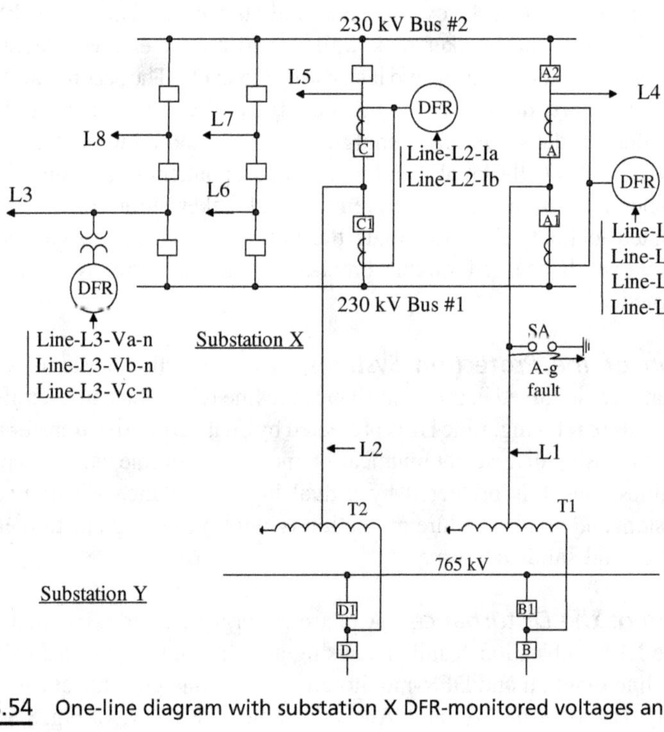

Fig. 8.54 One-line diagram with substation X DFR-monitored voltages and currents.

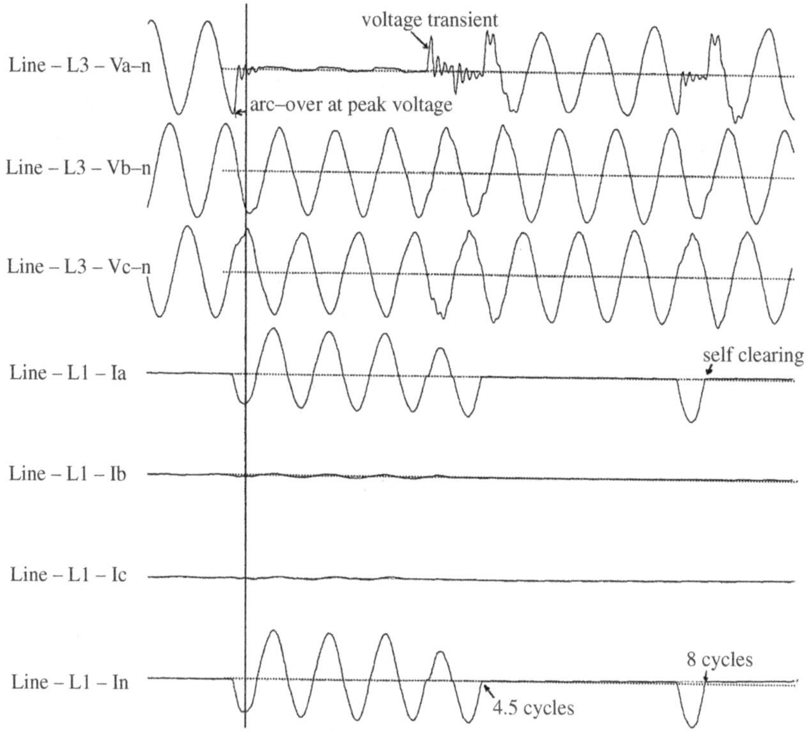

Fig. 8.55 Substation X DFR record showing L1 currents and L3 voltages during the A-g fault.

230-kV phase A-g bus fault. The fault lasted for only 0.5 cycle and was self-cleared from the system. The fault duration was sufficient to allow the bus 1 differential relay and CB A1 breaker failure relay to operate to isolate circuit breaker A1.

Analysis of the Initial A-g Line Fault As shown in Fig. 8.56, the initial phase A surge arrester failure incident point occurred at the phase A voltage peak, producing symmetric fault current with no dc offset. The close-in fault to substation X was detected by the Z1 relaying element at the substation X end of the line, and the fault was cleared in 4.5 cycles. Trace L2-In of the DFR record of Fig. 8.57 confirms that the substation Y end of line L1 cleared the fault in an additional 3 cycles by the permissive overreaching pilot element.

Analysis of the Second A-g Fault The second fault was self-clearing from the system with a duration of only 0.5 cycle. It is postulated that the live part of the oil circuit breaker interrupter failed, causing it to move and to arc-over to the breaker tank wall. This 0.5-cycle duration of fault current was sufficient to operate the bus differential relay and breaker failure fault detectors.

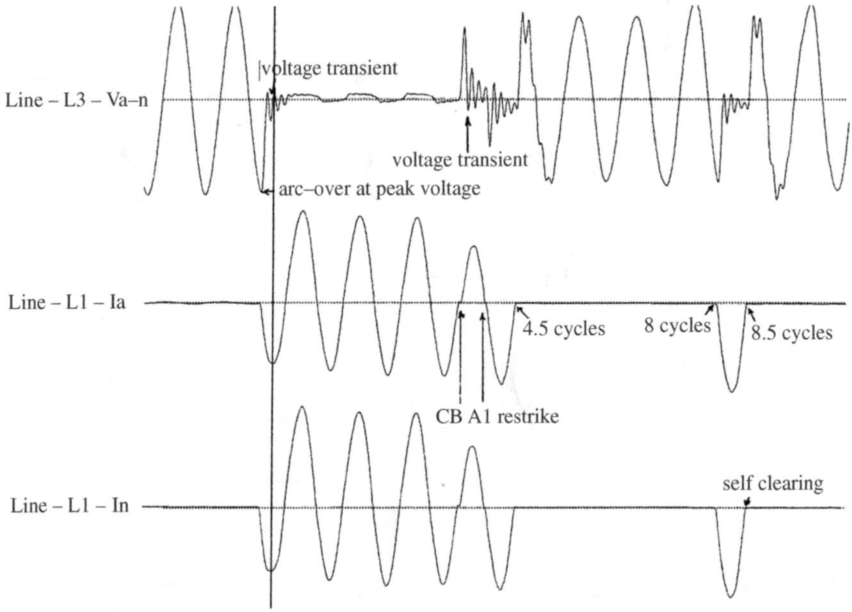

Fig. 8.56 DFR record showing CB A1 re-strikes and the self-clearing A-g fault.

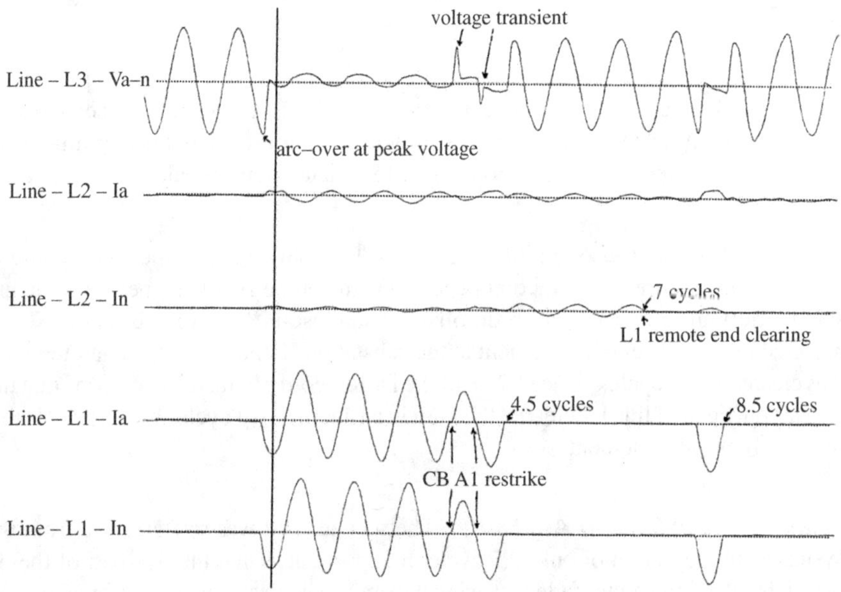

Fig. 8.57 Substation X DFR record showing L1 and L2 faulted phase and neutral currents and L3 voltage.

Analysis of the DFR Fault Record Figure 8.55 shows the DFR fault record for line L3 voltages and line L1 currents. The trace line-L3-Va-n reveals that the L3 voltage was near zero during both A-g faults. The phase A current trace line-L1-Ia confirms that either CB A or A1 attempted to clear the fault at two zero crossings (at 3.5 and 4 cycles), without success. The breakers finally interrupted the fault at 4.5 cycles. To summarize the sequence of events: At time zero, the surge arrestor failed. Four and a half cycles later, line L1 tripped and cleared the fault contribution from substation X. Seven cycles from $t = 0$, the substation Y end of line L1 tripped and cleared the fault. Eight cycles from $t = 0$, the A-g fault reappeared for a duration of 0.5 cycle and was self-cleared from the system.

Analysis of Circuit Breaker Failure Relaying This failure is similar in nature to the failure described in Case Study 8.4, where the failure was during the clearing of a C-g fault that was cleared successfully in 5 cycles, followed by a 1-cycle self-cleared breaker internal fault that occurred 9 cycles from time zero. The failure of CB A1 may not have been related directly to restrike phenomena; however, it may have been precipitated by the early re-strikes during the initial A-g fault. The higher short-circuit current of about 24 kA during the initial A-g fault preceding these re-strikes could have contributed to the failure of CB A1. The internal breaker inspection revealed that the bus 1 interrupter assembly had failed to ground.

Confirmation of the Short-Circuit Model Confirmation of the power system short-circuit model can be done by using fault type and fault location to run software packages to simulate faults and compare the study results with the recorded DFR RMS calculations. This case study reveals the fault incident point to be at the voltage peak, as shown in Fig. 8.56. This resulted in symmetrical fault current that contained no dc offset component. In addition, the faulted phase A voltage of the DFR record was near zero. This will imply that the surge arrester failure offers a very small fault resistance. The RMS measured DFR faulted phase voltage, as calculated by the DFR software, is 6.38 kV, compared to a normal phase-to-neutral voltage of 121.6 kV. The 6.38 kV voltage measured during the fault implies that the fault occurred through a small resistance of the surge arrestor. The DFR line L1 zero-sequence current measured for the close-in phase A-g fault, as stamped on the DFR record, is 25.7 kA. The simulated close-in line-to-ground fault in the short-circuit study for a fault resistance of 2.5 Ω is 27.4 kA.The resulting error between the zero-sequence current measured and that calculated is

$$\text{error} = \frac{27.4-25.7}{25.7} \times 100 = 6.6\%$$

Since the calculated value of 6.6% is less than 10%, confidence can be declared in the short-circuit model.

Power System Phenomena

1. Arc-over at voltage peak as the surge arrester failed, as shown in trace line-L3-Va-n in Fig. 8.56.

2. Breaker live contact arc-over to the breaker tank wall at the voltage peak as shown in trace line-L3-Va-n in Fig. 8.57.
3. Self-clearing fault as shown in trace line-L3-Ia in Fig. 8.56.
4. High-frequency transient that accompanies the breaker transient recovery voltage during an attempt to interrupt the fault current as shown in trace Line-L3-Va-n in Fig. 8.56.

Lessons Learned High-impedance bus differential relaying concepts using electromechanical relays permit optimum relay operation time of a few milliseconds. The 8-ms fault current was sufficient for the relay to operate. This confirms the high-speed operating time of the high-impedance electromechanical relay bus protection concept in the presence of ample fault current.

Case Study 8.9: Detection of a Remote Circuit Breaker Problem from Analysis of a Local Oscillogram Monitoring Line Currents and Voltages

Abstract Monitoring of three-phase voltages and three-phase and neutral currents of transmission lines by a DFR or an oscillograph can help to identify equipment problems at the remote end of the line. In this case study we illustrate how a remote breaker (located at a different substation) problem can be detected from an analysis of the local substation oscillograms.

Description of the Problem A scheduled maintenance condition required the removal from service of 345-kV line L2, shown in Fig. 8.58. The 345-kV line L2 was then taken out of service by opening associated breakers A and A2 at substation Y and C and C1 at substation Z. While attempting to trip CB A at substation Y, the breaker experienced a pole disagreement (discrepancy). This condition is normally detected by the wiring of a combination of the type a and b auxiliary switches shown in Fig. 8.60. The substation is designed to isolate circuit breakers experiencing pole disagreement conditions. Circuit breaker A was isolated locally by tripping CBs A1 and A2 and remotely by tripping CBs B and B1 at substation X via a direct transfer trip (DTT) system. Circuit breaker A1 at substation Y also experienced a pole disagreement condition, which resulted in additional tripping of the 345-kV bus at substation Y.

Analysis of the Oscillogram The substation X oscillograph was triggered by the presence of neutral current in the L1 circuit shown in Fig. 8.59. The L1 circuit was carrying about 480 MVA prior to the trip, which was initiated by the CB A pole disagreement protection feature at substation Y. Trace line L1-Ic of the oscillogram shown in Fig. 8.59 indicates a reduction of current magnitude at point a to a value of one-half the pre-trip loading current for a duration of 3.5 cycles. This reduction in phase C current resulted only in a neutral current ($3I_0$) that lasted until CBs B and B1 tripped at point b. In addition, trace line L1-Vb-n indicates that phase B voltage lasted for an additional 2.5 cycles longer than phases A and C, with the final circuit breaker

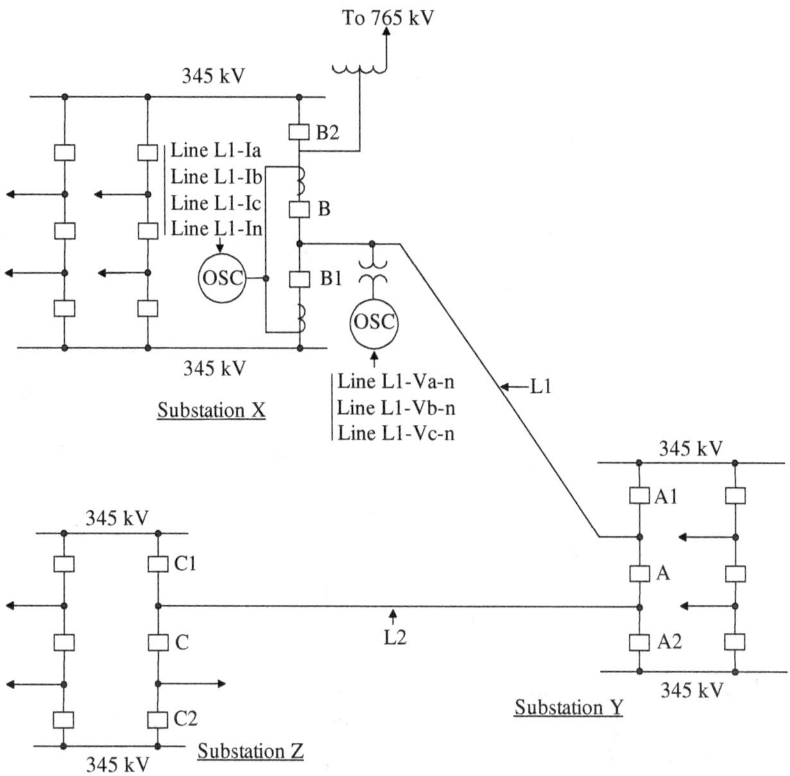

Fig. 8.58 One-line diagram showing oscillograph-monitored voltage and currents.

opening (line L1 dead) at point c. It can then be postulated that the reduction of phase C current and the prolonging of phase B voltage can be attributed to breaker pole disagreement problems at substation Y for CBs A and A1. The normal interruption of load currents as revealed by the traces line L1-Ia, line L1-Ib, and line L1-Ic implies

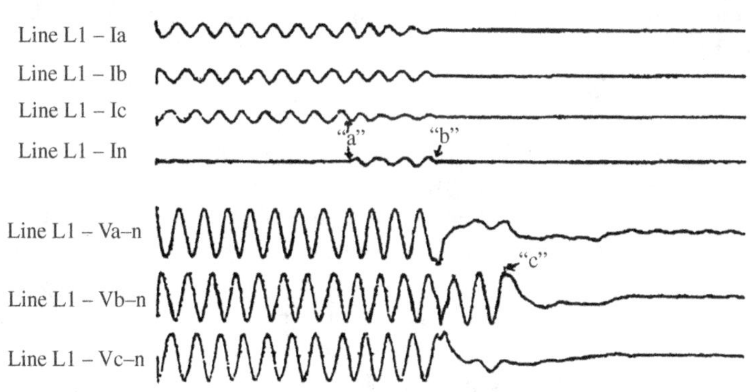

Fig. 8.59 Line L1 voltages and currents during breaker pole disagreement operation.

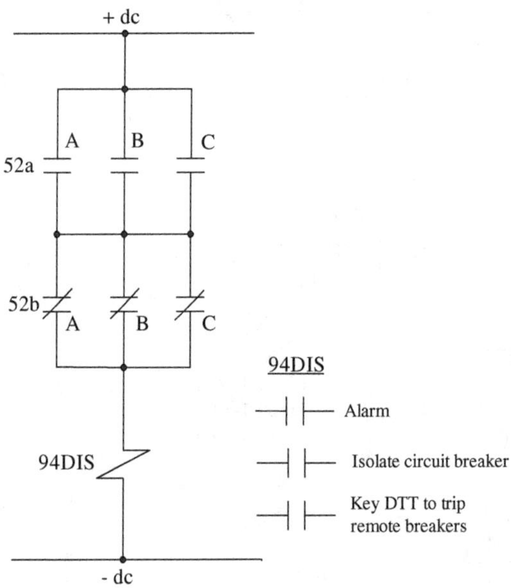

Fig. 8.60 Pole disagreement sensing circuit.

that CBs B and B1 tripped successfully upon receipt of the DTT signal from substation Y. The detection of pole disagreement conditions for independent pole mechanisms for high voltage and extrahigh voltage is done using a combination of type a and type b auxiliary switches, as shown in Fig. 8.60.

Case Study 8.10: Blackout of a 138-kV Load Area Due to a Primary Relay System Failure and the Lack of DC Control Power for the Secondary Relay System Circuit

Abstract A temporary B-g fault, caused by lightning, occurred on 138-kV line L1, shown in Fig. 8.61. This was followed by complete isolation of the 138-kV area around substation X from all adjacent line terminals. In this case study we describe the protection philosophy of line L1 and the reasons for complete substation isolation. In addition, we analyze the causes of the failure and lessons learned.

Description of the Line Protection The primary protection for the 138-kV line consists of an electromechanical pilot wire relay (current differential) for both phase and ground faults. The relay system uses a metallic pilot wire leased telephone circuit. The backup protection consists of step-distance relays for phase faults and a directional time overcurrent relay for ground faults. Breaker failure protection is also added to each of the 138-kV breakers in substation X and normally is initiated by line relay trip operations. Figure 8.62 shows the one-line ac diagram for the protection systems of line L1. The 138-kV relaying philosophy uses one set of CTs to supply the entire primary and backup relays. Figure 8.63 shows the dc elementary circuit for the

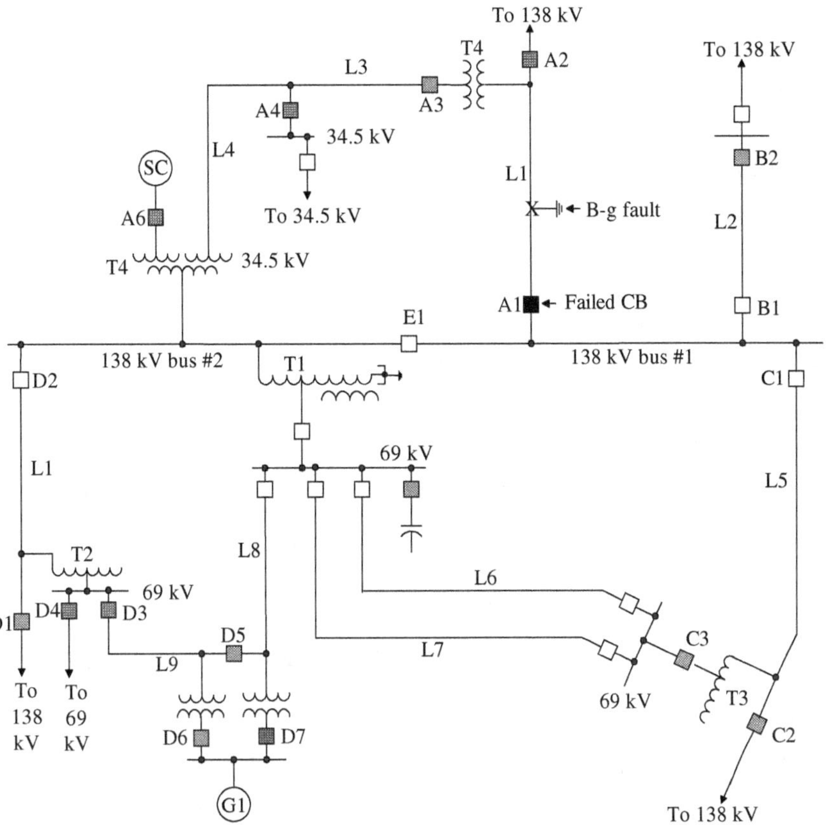

Fig. 8.61 One-line diagram for the 138-kV system with B-g fault on line L1.

line protection systems. One dc battery is used to supply all protection systems. Separate dc branches using separate breakers are utilized for the primary and backup protection systems. However, the breaker failure dc elementary circuit shares the dc supply with the backup line protection (a common dc breaker).

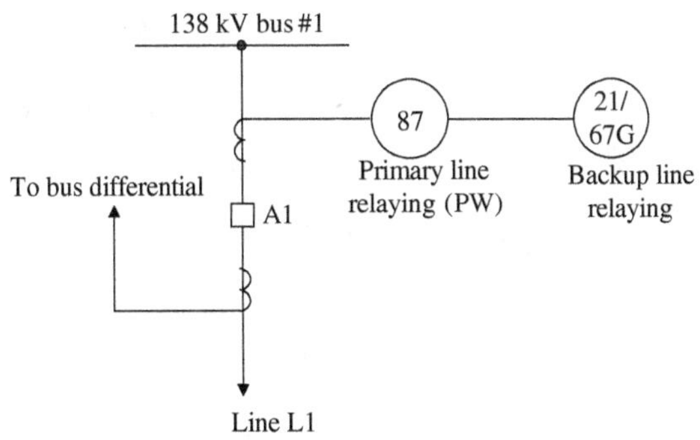

Fig. 8.62 Simplified relaying one-line diagram.

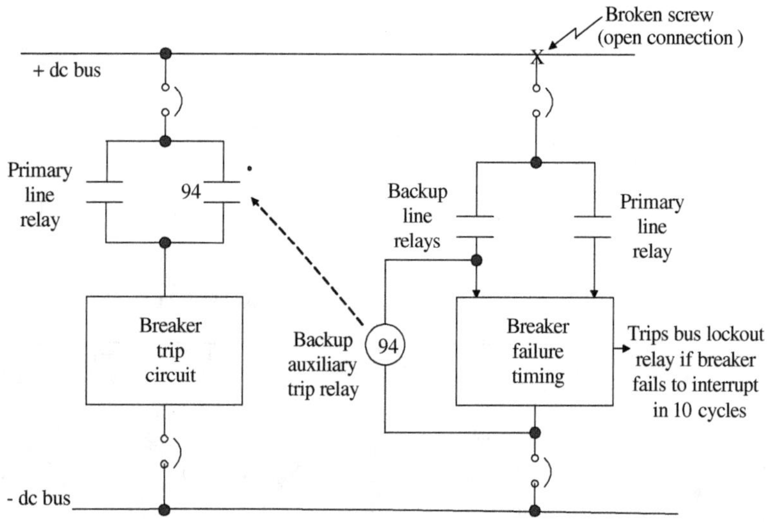

Fig. 8.63 Simplified dc schematic circuits.

Description of the Incident The ground fault originally caused by lightning was cleared from the system in approximately 2.5 s. The delayed fault clearing was due to the inability of the primary and backup line relays to trip the breaker. Therefore, it was necessary for the remote terminals of all other lines out of this substation to trip by remote backup relaying. Fortunately, the fault was cleared in time to prevent permanent line conductor damage. Automatic reclosing reenergized the station bus in 10 s to automatically restore the interrupted area to normal in less than a minute. However, some sensitive industrial customers were affected by the short-duration loss of the 138-kV area.

Causes of Line Primary and Backup Relay Failures As shown in Fig. 8.64, the pilot wire relay was disabled by a shorted diode in the bridge rectifier of the operating coil circuit. This condition rendered the relay incapable of operation. Referring to Fig. 8.63, the dc control voltage for the backup and breaker failure circuitry was lost due to a broken screw on the positive pole of the dc breaker located in the dc control cabinet, The head of the screw on the dc positive bus side of the dc breaker had popped off, causing an open connection for the dc circuit. The head of the screw was found lying on the bottom of the cabinet.

Analysis of the Causes of Line Primary and Backup Relay Failures The cause of the shorted diode in the pilot wire relay was not known. It can only be speculated that the diode failure may be attributed to a high-frequency surge originating from the pilot wire or other reasons. Other diode failures were experienced in similar relays. Unfortunately, a failed diode is not detectable except by means of a through relay operational routine checkout test. It was not known when the diode had become shorted, since the relay was checked more than a year prior to this incident. The cause

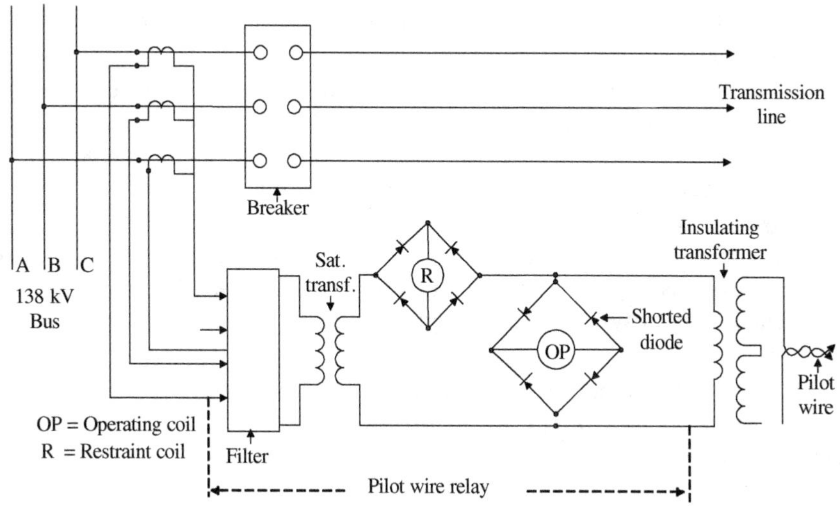

Fig. 8.64 Pilot wire primary relay failure.

of the broken screw seems to be excessive tightness coupled with thermal expansion, as it appears that the screw failed in tension about $\frac{1}{16}$ inch below the screw head. No signs of shear or of a manufacturing flaw in the screw material were noted.

Corrective Actions Relays to monitor loss of dc were added at the end of the runs of all relay dc circuits to alarm for fuse failure or lack of dc voltage at the unattended substation X.

Lessons Learned

1. The breaker failure dc circuit should be separated from the backup protection dc circuit. However, a dc separation from the backup relay circuit will not result in a different outcome for fault clearing in this incident, since both primary and backup relaying failed to operate and initiate CB failure protection.
2. Addition of alarm relays to monitor the dc circuits of backup and breaker failure circuits, especially in unattended stations, can enhance the dependability of the relaying systems.
3. In normal relaying schemes it is usually assumed that local backup relaying will take over if primary relaying fails. A dual failure is rare, but when it does occur, remote backup relaying must be counted on to clear the fault from the system. The main function of remote backup relaying is to protect the system during extreme power system conditions. This can occur when both primary relaying systems fail or during simultaneous breaker failure operations. Fortunately, remote backup worked well to clear the fault in this case study.
4. Due to the rareness of a dual relaying failure, this case study is worthy of being reported.

5. Remote backup relaying is vital for power systems despite the availability of local backup (breaker failure) relaying.
6. This case study emphasizes the importance of remote backup relaying to clear faults when a shorted or open CT cable or loss of the station battery can occur without due regard to redundancy and protection design separation criteria. This is valid for non-bulk power systems where redundant and separate relaying systems fed from separate CTs and dc batteries are not available.
7. Establishing and maintaining periodic routine trip testing and relay testing programs is important to ensure that protective relays, CT cables, and associated control systems are all in good operating condition.

Case Study 8.11: Installation of Two 345-kV Breakers in Series Within a Ring Substation Configuration to Mitigate the Loss of Critical Lines During Breaker Failure Events

Abstract For a ring bus configuration the installation of two breakers in series is done primarily to mitigate the loss of critical lines during breaker failure events. Figure 8.65 shows a conventional ring bus arrangement. For faults on either line L1 or L2 with CB A2 failing to clear, line elements L1 and L2 will both be tripped. For some power systems, the two-line loss may affect power flows between certain areas and may degrade system stability.

To avoid the impact of the loss of two critical circuits in system-sensitive power flow areas, an additional 345-kV breaker is added in series with an existing circuit breaker within the substation ring bus configuration. Figure 8.66 illustrates the installation of CB A5 in series with CB A2. A fault on line L1 with CB A2 failure will only result in the loss of line L1 by tripping breaker A5. Line L2 will remain in service supplying load via CB A3, as shown in Fig. 8.66.

Case Study 8.12: Design of Two 138-kV Circuit Breakers in Series to Fulfill the Need of Breaker Failure Protection

Abstract A 138-kV cable was tapped to connect a 50-MW combustion turbine (CT) unit. The cable is protected by a directional comparison pilot relaying scheme using a power line carrier (PLC) communication system. Due to unavailability of new

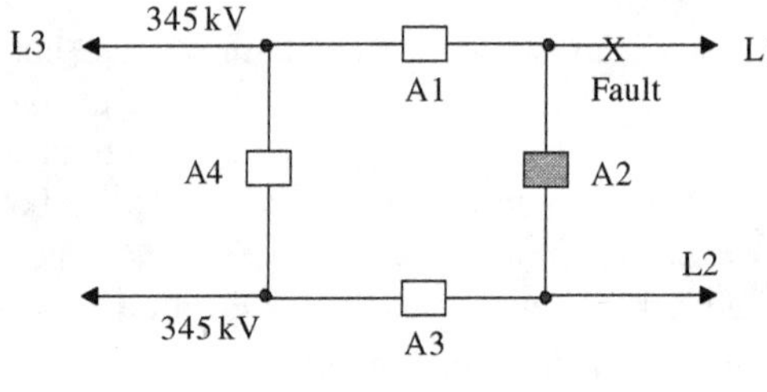

Fig. 8.65 Ring bus arrangements.

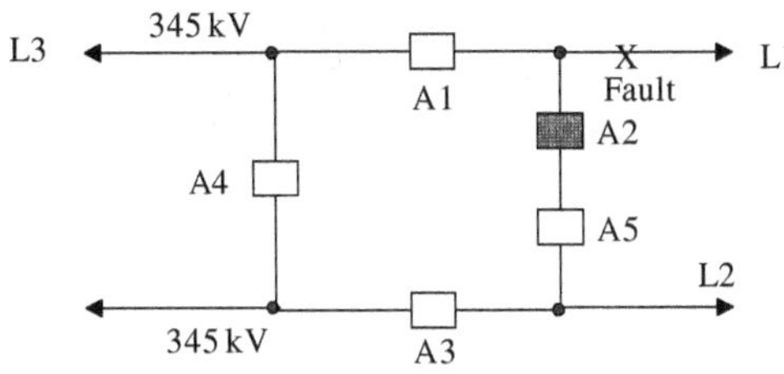

Fig. 8.66 Addition of one 345-kV breaker in series with an existing breaker.

communication media and the marginal performance of the existing PLC system over high-voltage cable, two 138-kV breakers are installed and connected in series to provide breaker failure protection function.

Description of the Facility Needed to Provide Breaker Failure Protection Function The addition of the plant converted the two terminal lines into a three-terminal line. A numerical distance relay was added to the line terminal at plant X to coordinate with the other terminals, Y and Z. The system one-line diagram with protection zones is shown in Fig. 8.67. Two 138-kV CBs A1 and A2 are installed in series to fulfill the breaker failure protection need at the plant facility.

Faults in the transformer overall differential zone and in the overlapping line protection zone will trip CB A1. In addition, all protection devices tripping breaker A1 will initiate its associated breaker failure relaying. This is accomplished by monitoring the fault current going through CB A1. Protective relay initiation and breaker failure fault detector operation will start the breaker failure timer, which is set at

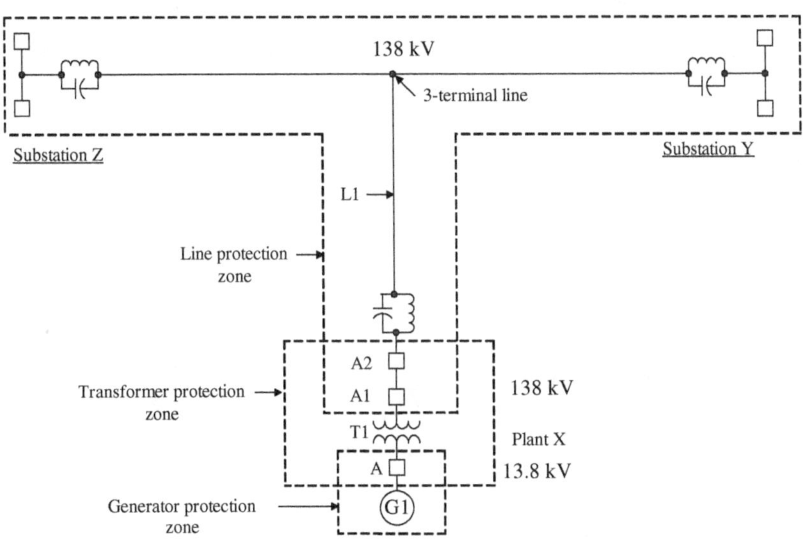

Fig. 8.67 System three-terminal one-line diagram with protection zones shown.

10 cycles. When the timer times-out, the breaker A1 failure lockout relay will trip and lock out breaker A2 and shut down CT plant X. Therefore, tripping and isolation of 138-kV remote circuit breakers at substations Y and Z by adding communication systems are not required.

REFERENCES

Blackburn, J. L. *Applied Protective Relaying*. Pittsburgh, PA: Westinghouse Electric Corporation, 1979.

Blackburn, J. L. *Protective Relaying Principles and Applications*. New York: Marcel Dekker, 1987.

Disturbance Monitoring and Reliability Requirements. NERC Reliability Standard PRC-002-02.

Elmore, W. A., Ed. *Protective Relaying Theory and Applications*. New York: Marcel Dekker, 2000.

Greenwood, A. *Electrical Transients in Power Systems*. New York: Wiley-Interscience, 1971.

Ibrahim, M. A. St. Lawrence FDR 230-kV substation disturbance of March 6, 1996. Presented at the Georgia Tech Protective Relaying Conference, April 30–May 2, 1998, Atlanta, GA, and at the 25th Annual Western Protective Relay Conference, October 26–28, 1998, Spokane, WA.

Ibrahim, M. A., and F. Stacom. Adirondack 230 kV substation outage of July 1, 1995. Presented at the Georgia Tech Protective Relaying Conference, May 1–3, 1996, Atlanta, GA, and at the 23rd Annual Western Protective Relay Conference, October 14–16, 1996, Spokane, WA.

IEEE Standard Common Format for Transient Data Exchange (COMTRAD) for Power Systems. IEEE Standard C37.111-1999, a revision of IEEE Standard C37.111-1991.

Khalifa, M. *High-Voltage Engineering*. New York: Marcel Dekker, 1990.

9

PROBLEMS

9.1. A fault occurred on the 345-kV line L1 shown in Fig. 9.1. The fault was caused by icing formation on the line insulator during an ice storm. The DFR record of substation X, shown in Fig. 9.2, documents the three-phase line currents, neutral current, and line-side three-phase-to-neutral voltages of line L1:

(a) Classify the fault type.

(b) Determine the fault incident point on the faulted phase voltage trace.

(c) Explain the fault mechanism scenario.

(d) Explain how a local DFR fault record can be used to determine the remote-end clearing time of a line fault.

(e) Determine the total clearing times of the local X and remote Y ends of the faulted line.

(f) Explain the nature of the varying current magnitudes for phase A current.

(g) Explain the increase in the phase A current few a cycles prior to interruption.

(h) Is any current reversal shown in the DFR record? Explain your answer.

Disturbance Analysis for Power Systems, First Edition. Mohamed A. Ibrahim.

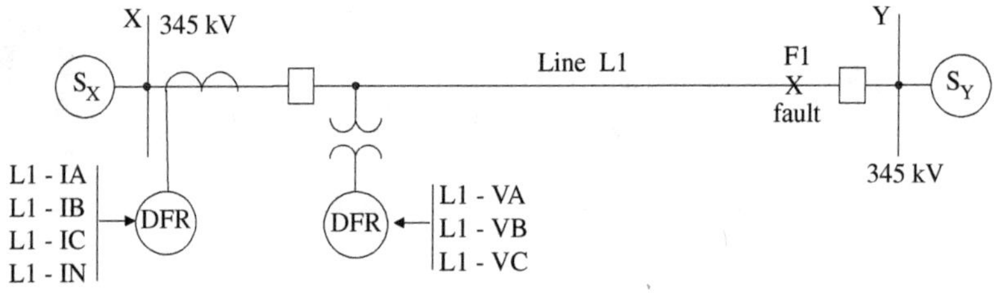

Fig. 9.1 One-line diagram showing DFR-monitored line voltages and currents.

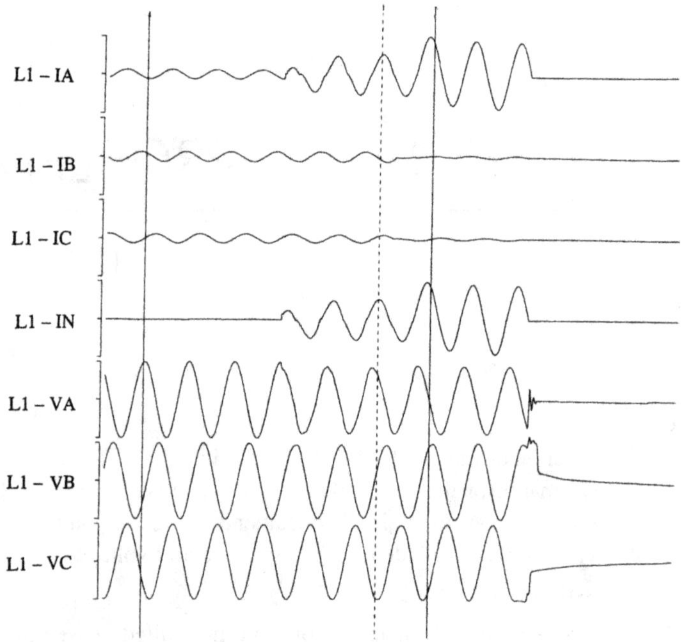

Fig. 9.2 Substation X DFR record for fault F1 on line L1.

(i) Does this operation exhibit a sequential clearing phenomenon? If yes, state the reasons behind sequential clearing, the effect on power system operation, and how this effect can be eliminated.

9.2. A fault occurred on feeder L1 of the power system shown in Fig. 9.3. The DFR record shown in Fig. 9.4 reflects the currents and voltages at the 138-kV side of the generator step-up transformer.

(a) Develop the sequence of events.

(b) Classify the fault type.

(c) Determine the fault incident point on the faulted phase voltage trace.

(d) Explain the fault mechanism scenario.

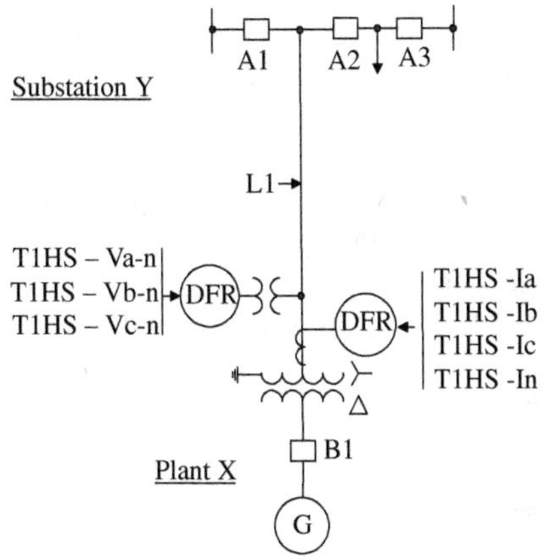

Fig. 9.3 One-line diagram showing DFR-monitored feeder voltages and currents.

(e) Determine the location of the fault based on an analysis of the 138-kV feeder voltages.

(f) Determine the total clearing times of the local generator plant end X and the remote substation Y ends.

(g) Explain the nature of the varying current magnitudes for the neutral current trace T1HS-In.

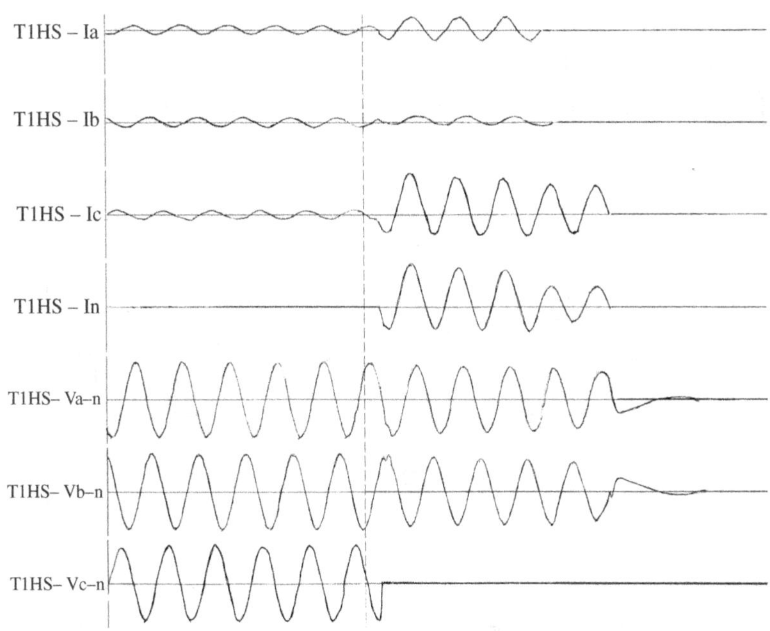

Fig. 9.4 Substation X DFR record showing GSU transformer T1 currents and voltages during the fault.

(h) Explain the decrease in the phase C current near interruption.

(i) Explain the increase in the phase A current trace T1HS-Ia

(j) Determine the fault resistance. Explain you answer.

9.3. A 230 kV B-g fault occurred on line L1 due to an insulator flashover, resulting in symmetrical currents, as shown in Fig. 9.5. The autotransformer tertiary voltage is 13.2 kV. Figure 9.6 shows a DFR record for all ground currents for the 230-kV lines L1, L2, L3, and L4, 115-kV line L5, and tertiary

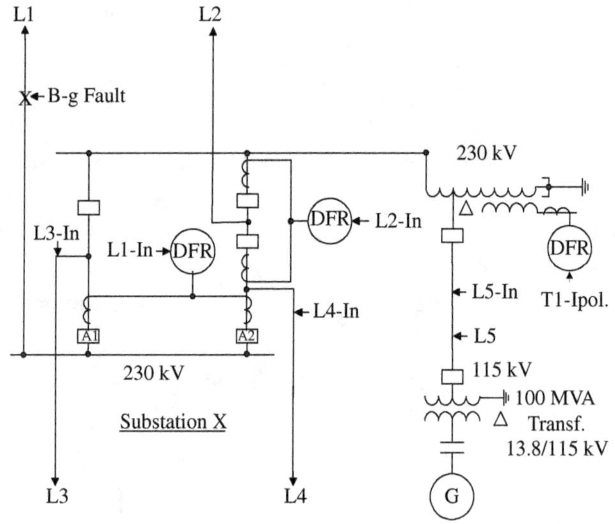

Fig. 9.5 One-line diagram showing DFR-monitored line currents during the fault.

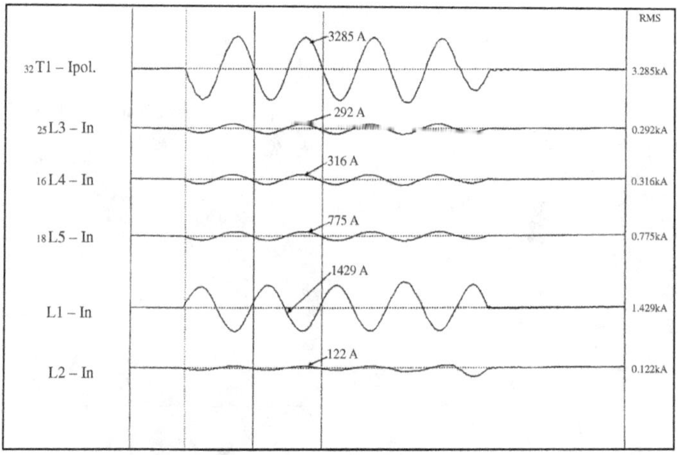

Fig. 9.6 Substation X DFR record showing RMS values for lines neutral currents and T1 tertiary current during the fault.

polarizing current with the RMS values calculated for the same timing window, as determined by the two one-cycle cursors marked on the record.

(a) Verify the accuracy of the DFR recording for the ground (neutral or $3I_0$) currents supplying the B-g fault. Shown in Fig. 9.6.

(b) Determine the clearing time at substation X.

(c) Determine the phase relationships for the currents feeding the B-g fault at substation X.

9.4. A fault occurred on the 230-kV line L1 shown in Fig. 9.7. The fault was caused by lightning. The DFR record of substation X shown in Fig. 9.8 documents the three-phase line currents, the neutral current, and the line-side three-phase neutral voltages of line L1 and the autotransformer 2 tertiary delta-connected winding. The DFR record of substation Y shown in Fig. 9.9 documents the three-phase line currents, the neutral current of L1, and the line-side three-phase voltages of line L1 and autotransformer 1 tertiary delta-connected winding.

(a) Classify the fault type.

(b) Determine the fault incident point on the faulted phase voltage trace.

(c) Explain the fault mechanism scenario.

(d) Explain why the fault current is not symmetric and contains partial dc offset.

(e) Determine the total clearing times of the line end at X and the line end at Y of the faulted line.

(f) Does this operation exhibit a sequential clearing phenomenon? If yes, state the reasons behind sequential clearing, the impact on power system operation, and how it can be eliminated.

(g) Explain the nature of the varying voltage magnitudes for phase A voltage on the DFR record at Y.

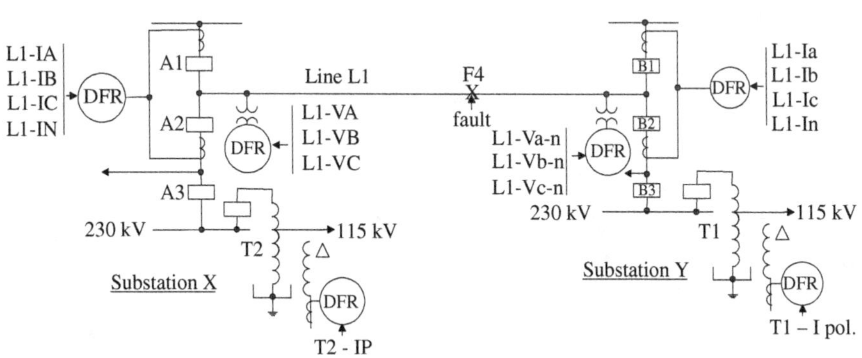

Fig. 9.7 One-line diagram showing DFR-monitored line voltages and currents.

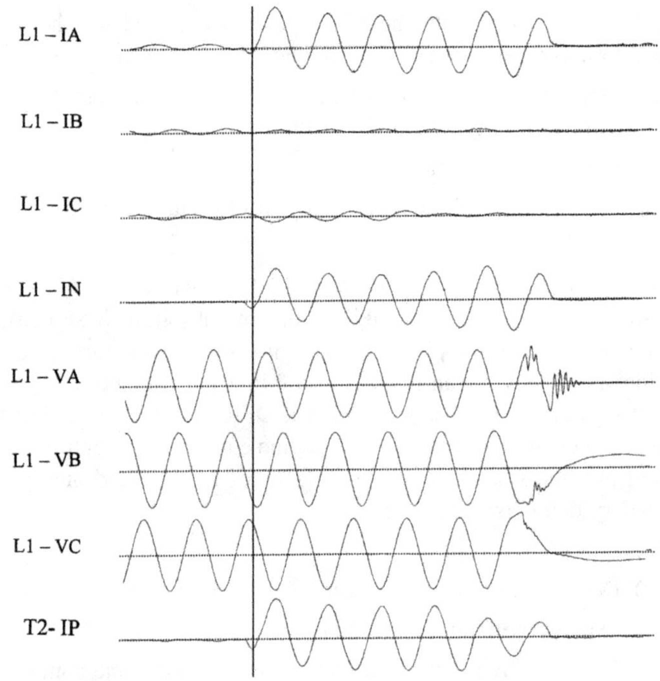

Fig. 9.8 Substation X DFR record for fault F4 on line L1.

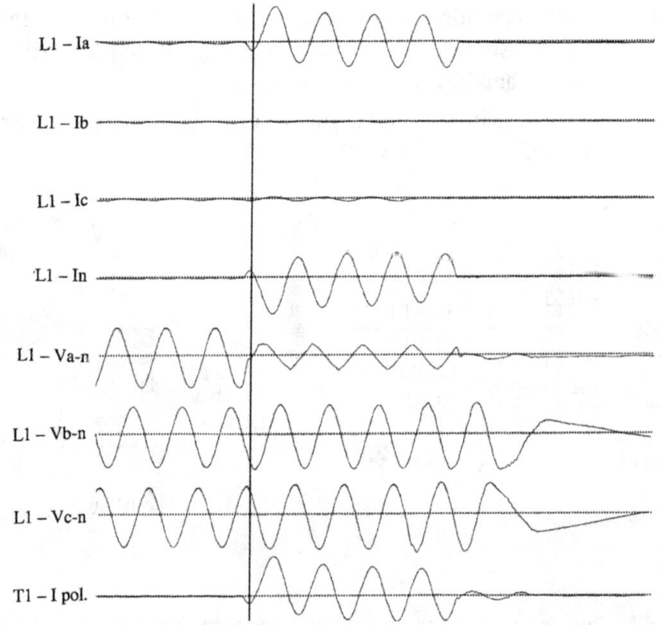

Fig. 9.9 Substation Y DFR record for fault F4 on line L1.

(h) Is there any current reversal shown in the DFR record? Explain to your answer.

(i) Explain why the phase A current and the neutral current of line L1 on DFR Y increases after 4 cycles from time zero.

(j) Can the sequence of events obtained from an analysis of the DFR records in Figs. 9.8 and 9.9 be obtained from an individual analysis of the DER of either substation X or substation Y? Explain your answer.

(k) Explain the nature of the varying current magnitudes for the tertiary current trace T2-IP.

9.5. A fault occurred on the 115-kV line L1 shown in Fig. 9.10, due to an apparent lightning strike. Figure 9.11 shows the DFR record of substation X, indicating the three-phase-to-neutral voltages and currents and neutral current of line L1.

(a) Classify the fault type, based on analysis of the DFR record. Support your answer.

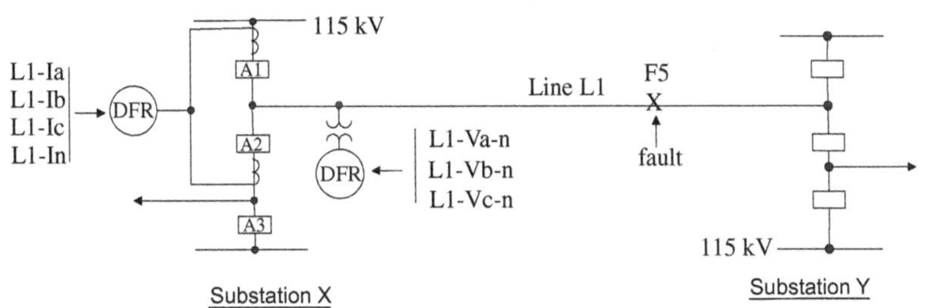

Fig. 9.10 One-line diagram showing DFR-monitored line voltages and currents.

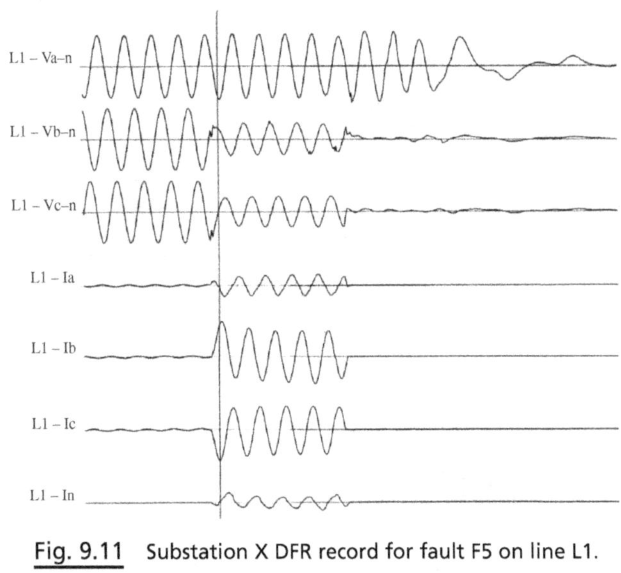

Fig. 9.11 Substation X DFR record for fault F5 on line L1.

(b) Determine the fault incident point on the faulted voltage traces.

(c) Determine the total clearing times for the local X and remote Y ends of the line.

(d) How can you deduce the remote clearing of the end Y of the line by analyzing the local DFR record at X?

(e) What is the pre-fault setting of the DFR in cycles? What is the estimated post-fault DFR setting in cycles? Is this an ideal setting? Explain your comments.

(f) Explain the nature of the pre-fault current shown in phases A, B, and C.

(g) Why there is no pre-fault current flow in the neutral trace L1-In?

9.6. A fault occurred on the 115-kV line L1 shown in Fig. 9.12, due to an apparent lightning strike. Figure 9.13 shows the DFR record of substation

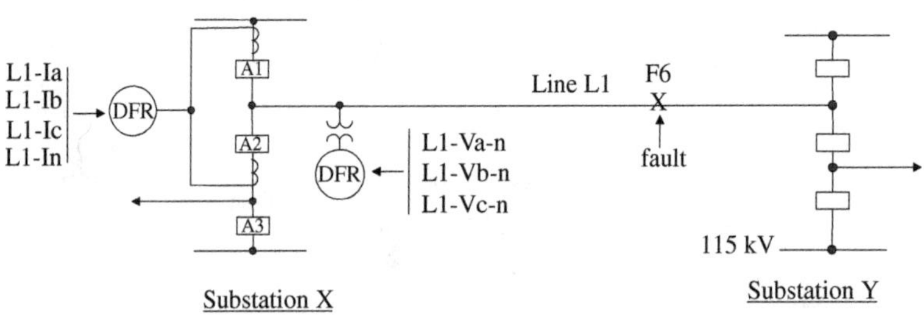

Fig. 9.12 One-line diagram showing DFR-monitored line voltages and currents.

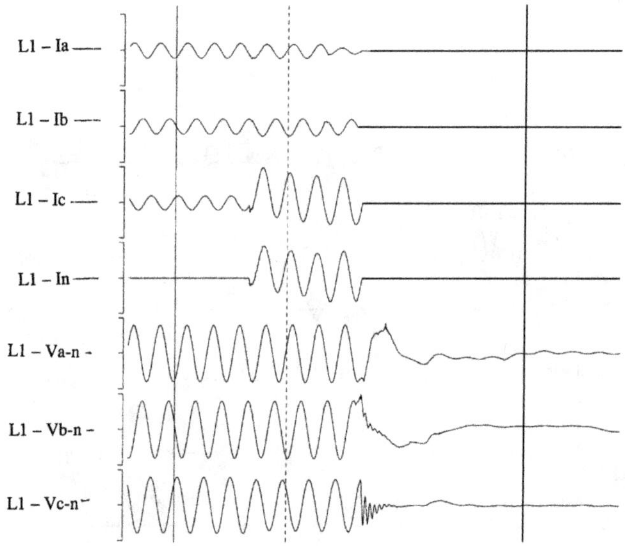

Fig. 9.13 Substation X DFR record showing line L1 currents and voltages during F6 fault.

X, indicating the three-phase-to-neutral voltages and currents and neutral current of line L1.

(a) Classify the fault type based on analysis of the DFR record. Support your answer.

(b) Determine the fault incident point on the faulted voltage traces.

(c) Determine the total clearing times for the local X and remote Y ends of the line.

(d) How can you deduce the remote clearing for the Y end of the line by analyzing the local DFR record at X?

(e) What is the nature of the pre-fault recorded currents on phases A, B, and C?

(f) Explain the nature of the transient appearing in the phase C voltage upon fault clearing.

9.7. A fault occurred on the 345-kV line L1 shown in Fig. 9.14. The DFR record of substation X shown in Figure 9.15 documents the three-phase line

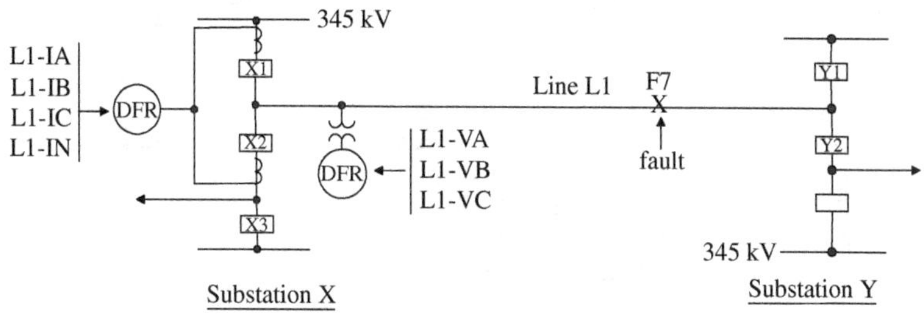

Fig. 9.14 One-line diagram showing DFR-monitored line voltages and currents.

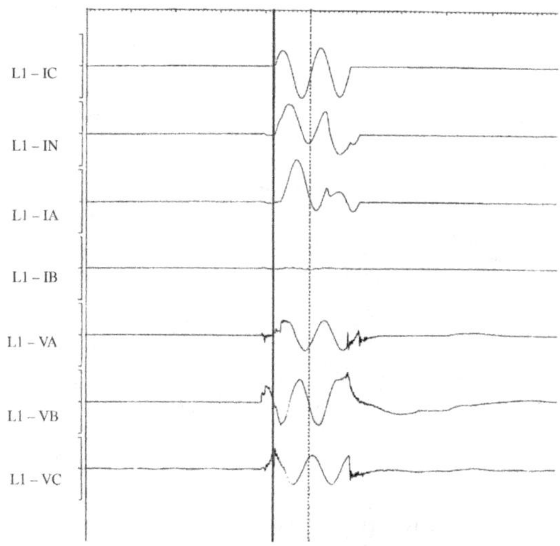

Fig. 9.15 Substation X DFR record for fault F7 on line L1.

currents, neutral current, and line-side three-phase-to-neutral voltages of line L1.

(a) Classify the fault type.

(b) Explain the fault mechanism scenario.

(c) Based on the pre-fault DFR data, determine the sequence of events for the system shown in Fig. 9.14.

(d) Determine the status (either closed or opened) of circuit breakers X1, X2, Y1, and Y2 prior to the occurrence of the fault and during and after fault clearing. CB X1 is designed to have automatic reclosing.

(e) Determine the total clearing time of the fault.

(f) Is the line L1 end at substation Y involved in the DFR shown in Fig. 9.15? Explain your answer.

9.8. A fault occurred on line L1 shown in Fig. 9.16. The DFR record of substation X shown in Fig. 9.17 documents the three-phase line currents, neutral current, and line-side three-phase-to-neutral voltages of line L1.

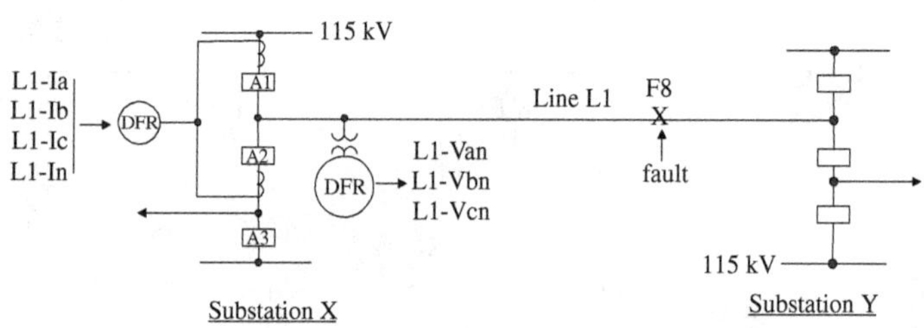

Fig. 9.16 One-line diagram showing DFR-monitored line voltages and currents.

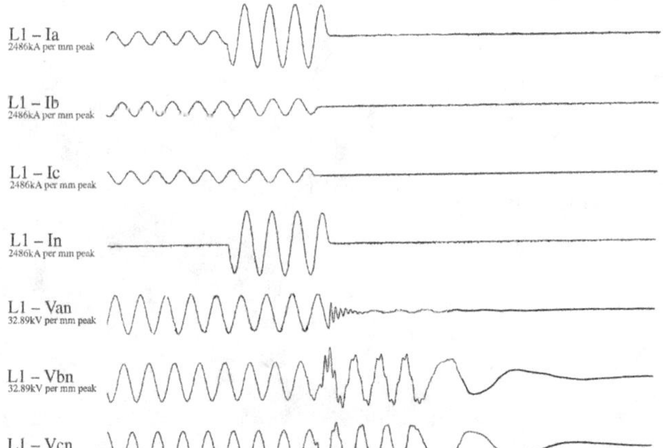

Fig. 9.17 Substation X DFR record showing line L1 currents and voltages during the F8 fault.

(a) Classify the fault type.

(b) Determine the fault incident point on the faulted-phase voltage trace.

(c) Explain the fault mechanism scenario.

(d) Explain how a local DFR fault record can be used to determine the remote-end clearing time of a line fault.

(e) Determine the total clearing times of the local X and remote Y ends of the faulted line.

(f) Explain the increase in voltages on phases b and c.

(g) Assess the magnitude of the fault resistance.

9.9. A fault occurred on 115-kV transmission line L1 shown in Fig. 9.18. The record DFR of substation X shown in Fig. 9.19 documents the three-phase line currents, line neutral current, and line-side three-phase-to-neutral voltages of line L1.

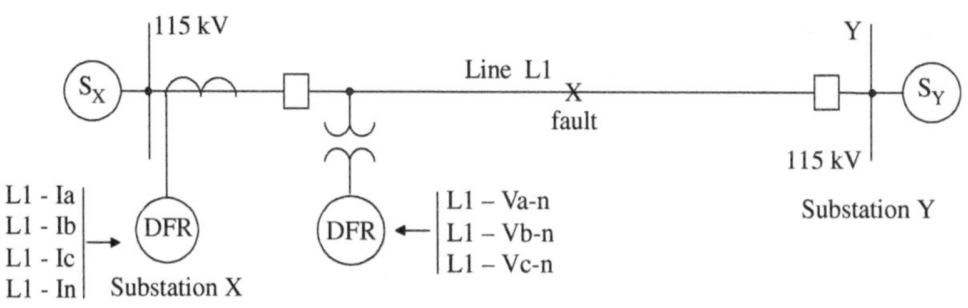

Fig. 9.18 One-line diagram showing DFR-monitored line voltages and currents.

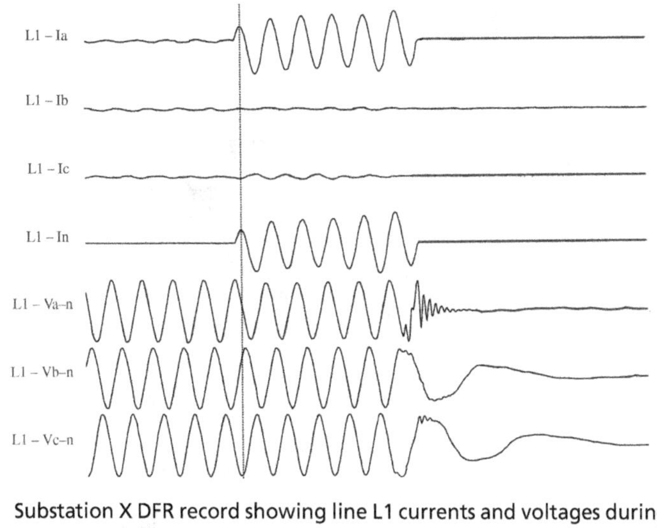

Fig. 9.19 Substation X DFR record showing line L1 currents and voltages during the fault.

(a) Determine the fault type.

(b) Determine the local clearing time at substation X.

(c) Determine the clearing time of the remote line end at Y.

(d) Analyzing the recorded three-phase voltages, can you tell if the fault was close to either substation X or substation Y?

9.10. A fault occurred on 115-kV transmission line L1 shown in Fig. 9.20. The DFR record of substation X shown in Fig. 9.21 documents the three-phase line currents, line neutral current, and line-side three-phase-to-neutral voltages of line L1.

(a) Determine the fault type.

(b) Determine the local clearing time at substation X.

(c) Determine the clearing time of the remote line end at Y.

(d) Analyzing the recorded three-phase voltages, can you tell if the fault was close to either substation X or substation Y?

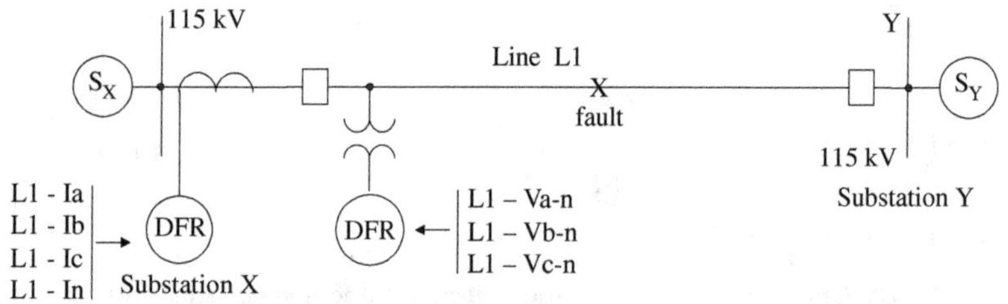

Fig. 9.20 One-line diagram showing DFR-monitored line voltages and currents.

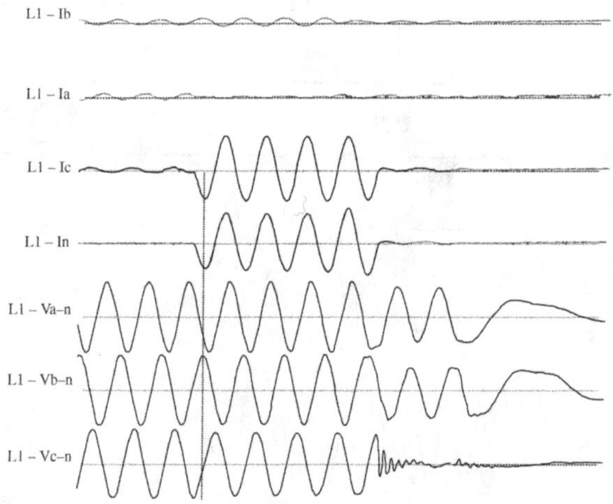

Fig. 9.21 Substation X DFR record showing line L1 currents and voltages during the fault.

(e) Assess the transient appearing on trace L1-Vc-n voltage 7.5 cycles from the beginning of the fault record. State the possible source of the signal. Estimate the frequency of the signal.

(f) Determine the fault incident point as related to the faulted phase voltage.

(g) Was the fault solid to ground? Explain your answer.

(h) What are the pre-fault and post-fault settings in cycles for the DFR record shown. Is this a good setting? Comment on your answer.

9.11. A fault occurred on the high-voltage system shown in Fig. 9.22. An oscillogram for line L1 was obtained at one of the substations and is shown in Fig. 9.23.

(a) Classify the system fault.

(b) Specify the duration of the fault.

(c) Is the fault current symmetric or asymmetric? Explain your answer.

(d) Was the fault internal or external to line L1? Explain your answer.

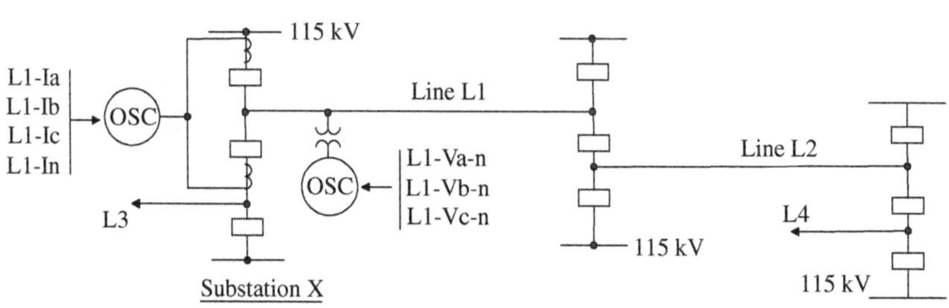

Fig. 9.22 One-line diagram showing oscillograph-monitored line voltages and currents.

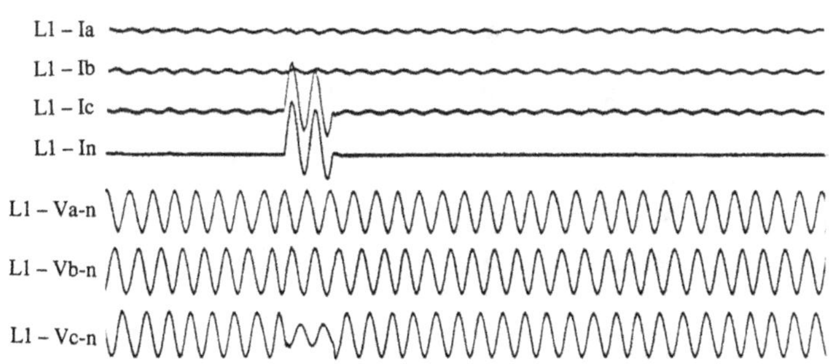

Fig. 9.23 Substation X oscillograph record showing the three-phase voltages and currents during the fault.

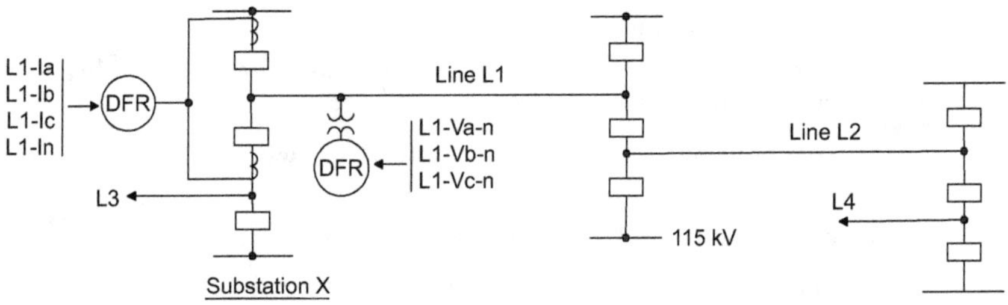

Fig. 9.24 One-line diagram showing DFR-monitored line voltages and currents.

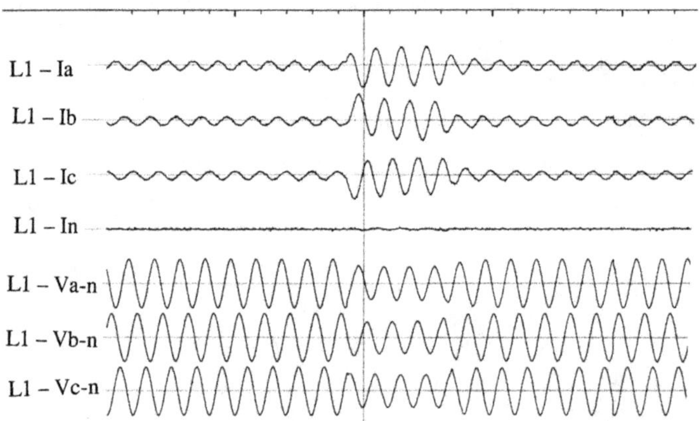

Fig. 9.25 Substation X DFR record showing line L1 voltages and currents during the fault.

9.12. A fault occurred in the high-voltage system shown in Fig. 9.24. The DFR record shown in Fig. 9.25 for line L1 was obtained at one of the substations.

(a) Classify the system fault.

(b) Specify the duration of the fault.

(c) Explain the nature of pre- and post-fault currents.

(d) Was the fault internal or external to line L1? Explain your answer.

(e) State the reason behind the different degree of asymmetrical fault currents for phases A, B, and C.

9.13. A fault occurred on the 345-kV line L1 shown in Fig. 9.26. The fault was caused by icing formation on the line insulator during an ice storm. The DFR record of substation X shown in Fig. 9.27 documents the three-phase line currents, neutral current, and line-side three-phase-to-neutral voltages of line L1.

(a) Classify the fault type.

(b) Determine the fault incident point on the faulted phase voltage trace.

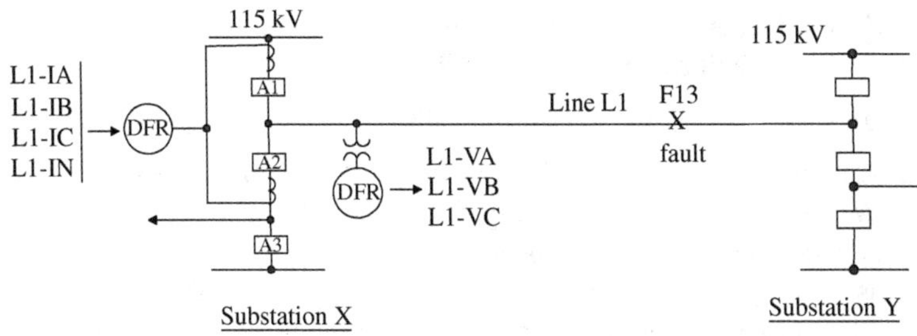

Fig. 9.26 One-line diagram showing DFR-monitored line voltages and currents.

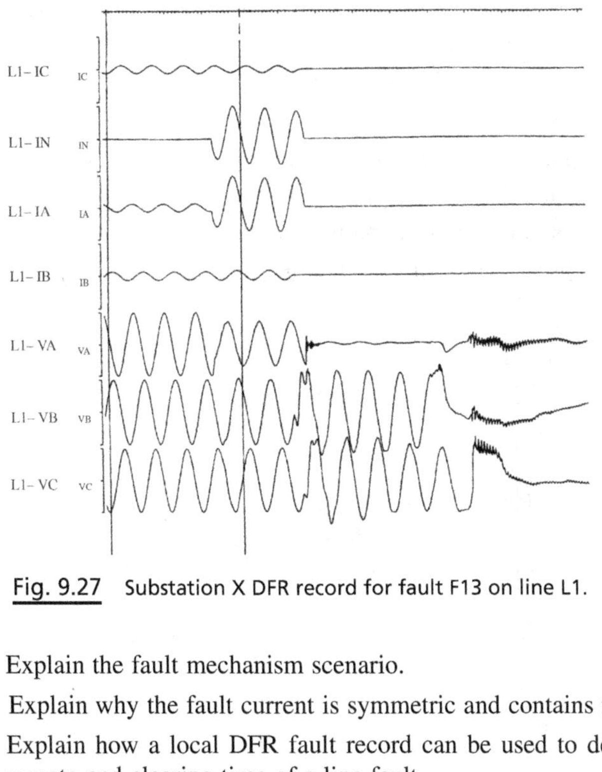

Fig. 9.27 Substation X DFR record for fault F13 on line L1.

(c) Explain the fault mechanism scenario.

(d) Explain why the fault current is symmetric and contains no dc offset.

(e) Explain how a local DFR fault record can be used to determine the remote-end clearing time of a line fault.

(f) Determine the total clearing times of the local X and remote Y ends of the faulted line.

(g) Explain why the phase B voltage has an increase in magnitude following the clearing of the line fault from substation X. Analyze and name this voltage rise phenomenon.

(h) Does this operation exhibit sequential clearing phenomena? If yes, state the reasons behind sequential clearing, the effect on power system operation, and how this effect can be eliminated.

(i) Explain the nature of the varying voltage magnitudes for phase A voltage. Is the fault resistance high or low? Explain your answer.

(j) Is any current reversal shown in the DFR record? Explain to your answer.

9.14. A combustion turbine (CT) plant is connected to the 138-kV system via transformer T1 as shown in Fig. 9.28. The CT unit was out of service, and generator breaker A1 was opened and racked out. The 138-kV system is used as a startup source and to power station service loads. The GSU transformer T1 is protected by dual numerical differential relays. One of the relays, shown in Fig. 9.28, operated and tripped 138-kV CB A4. The numerical differential relay fault record for the event is as follows:

Winding W1: $I_A = 2400$ A at 0; $I_B = 1200$ A at 180° lag; $I_C = 1200$ A at 180° lag

Winding W2: $I_a = 0$; $I_b = 0$; $I_c = 0$

Winding W3: $I_a = 0$; $I_b = 0$; $I_c = 0$

Given the phase currents for a numerical relay and the transformer phasor diagram shown in Fig. 9.29:

(a) Find the fault location that matches the relay event record.

(b) Classify the fault based on the relay information given.

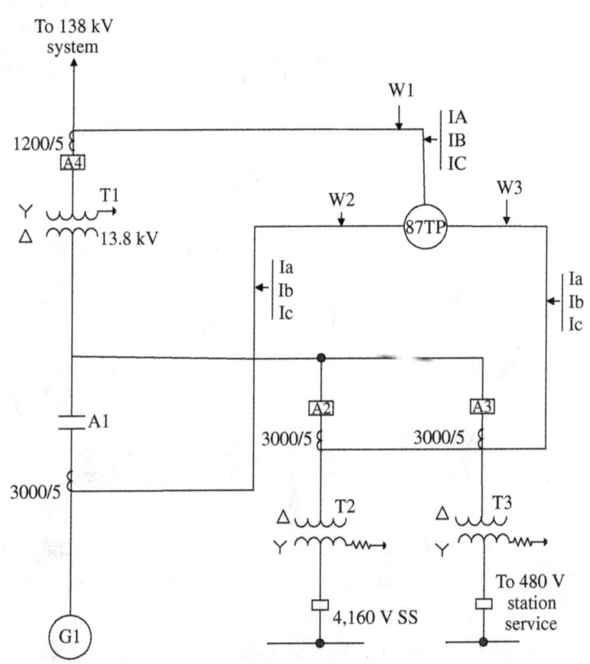

Fig. 9.28 One-line ac diagram showing T1 87TP numerical differential relay–monitored currents.

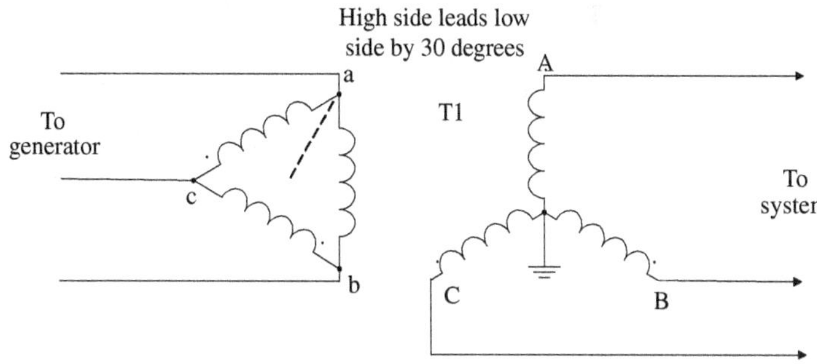

Fig. 9.29 Transformer T1 phasor diagram obtained from the transformer nameplate.

(c) Calculate the fault currents for the 13.8-kV delta currents and line currents for phases a, b, and c.

(d) Based on the results obtained in (c), demonstrate transformer T1 ampere-turns coupling theory for phases A, B, and C.

9.15. A 138-kV surge arrester failed, causing fault F15 at the top of transformer T2 of Fig. 9.30. The DFR record shown in Fig. 9.31 will be used to answer the following questions.

(a) Determine the sequence of events.

(b) Determine the type of fault.

(c) Describe any system phenomena that is part of the record.

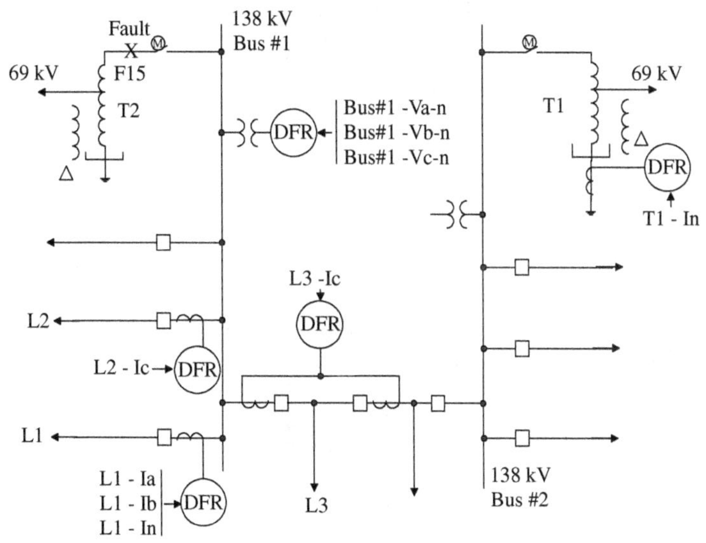

Fig. 9.30 One-line diagram showing DFR-monitored voltages and currents.

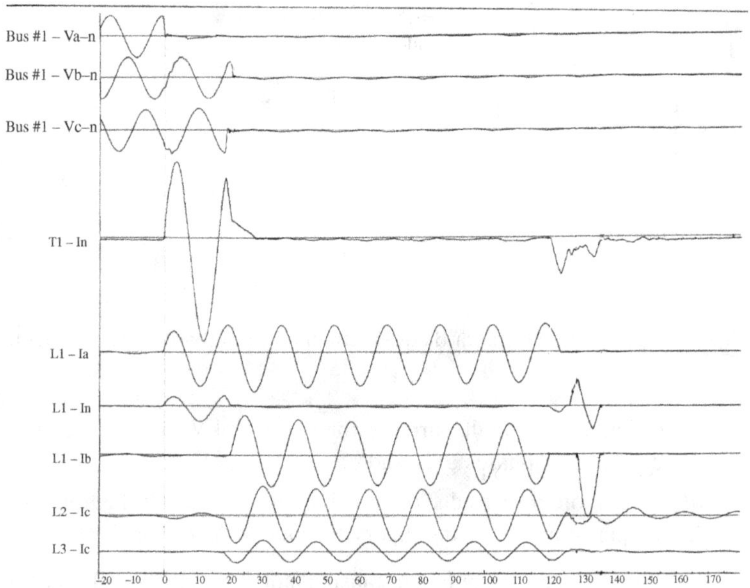

Fig. 9.31 DFR record showing currents and voltages that will be used for the fault F15 analysis.

9.16. A fault occurred on feeder L1 of the power system shown in Fig. 9.32. The DFR record shown in Fig. 9.33 reflects the currents and voltages at the 13.8-kV generating side of the GSU T1 transformer.

(a) Develop the sequence of events.

(b) Given the T1 phasor diagram in Fig. 9.34, classify the fault type.

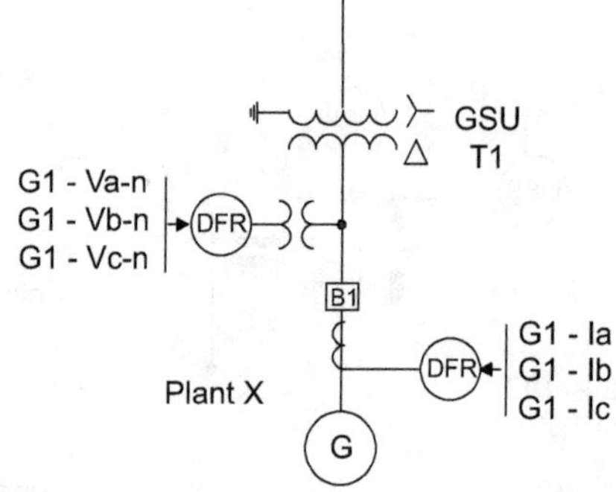

Fig. 9.32 One-line diagram showing DFR-monitored voltages and currents.

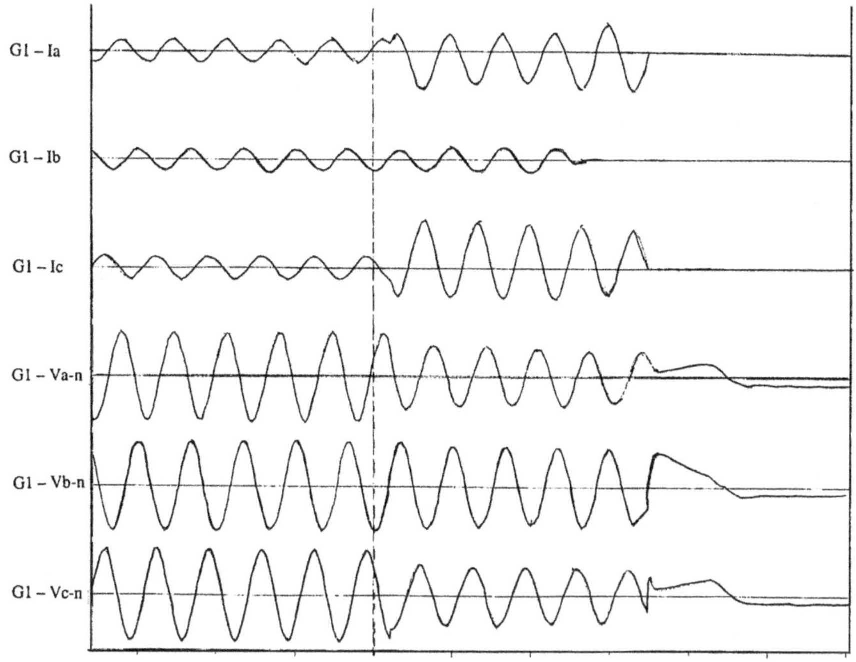

Fig. 9.33 Substation X DFR record showing generator G1 currents and voltages during the fault.

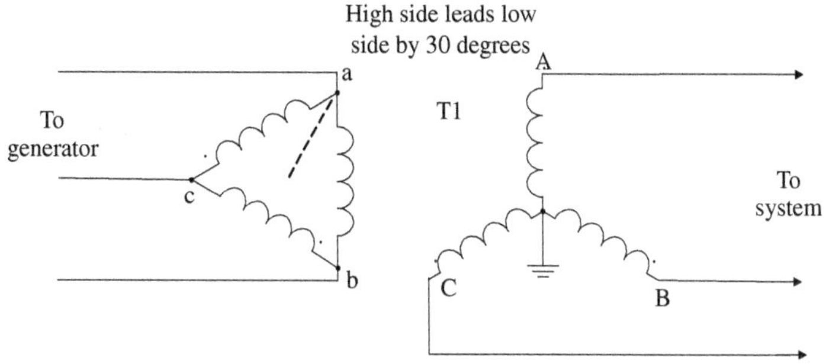

Fig. 9.34 Transformer T1 phasor diagram.

(c) Determine the fault incident point on the faulted phase voltage traces.

(d) Explain the fault mechanism scenario.

(e) Determine the total clearing times of the local generator plant end X and the remote substation Y ends.

(f) Classify the nature of the pre-fault currents.

9.17. For the system shown in Fig. 9.35, the DFR record shown in Fig. 9.36 represents the sequence of line restoration.

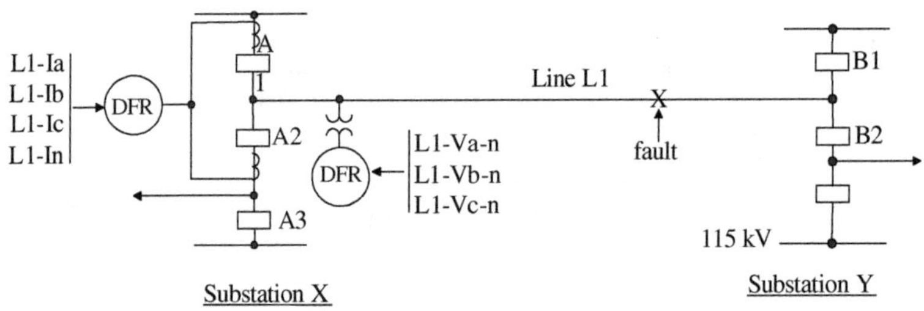

Fig. 9.35 One-line diagram showing DFR-monitored voltages and currents.

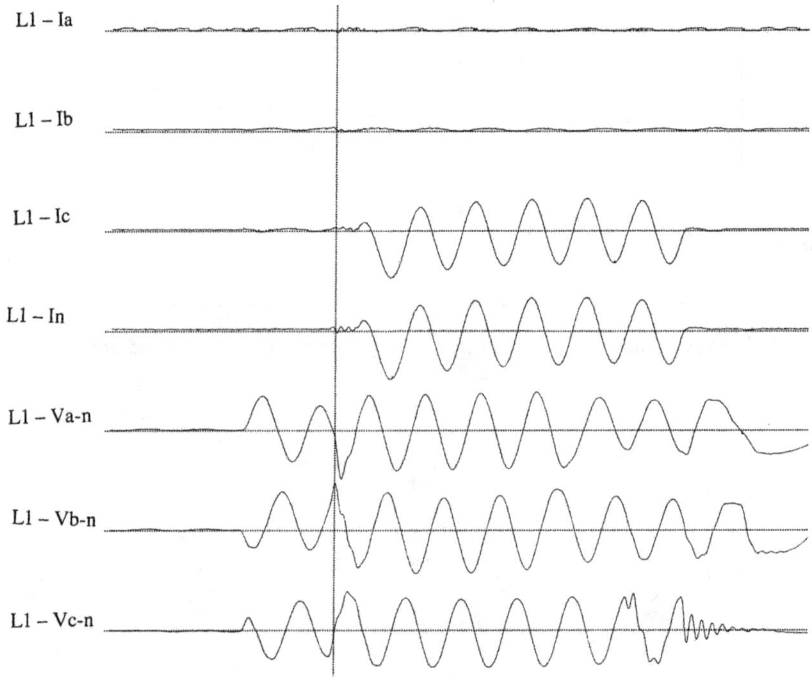

Fig. 9.36 Substation X DFR record showing line L1 voltages and currents during the fault.

(a) State the position of CBs A1 and A2 at substation X and B1 and B2 at substation Y in the pre-fault interval.

(b) Line L1 is designed to be restored automatically first from substation X. With this in mind, explain the sequence of events based on the recorded line voltages and currents.

(c) Determine the total clearing times.

(d) Determine the fault incident point using the faulted phase voltage.

(e) Explain the reason for the low voltage recorded for phases A, B, and C at the beginning of the DFR record for a duration of 1.5 cycles.

9.18. A fault occurred on feeder L1 of the power system shown in Fig. 9.37. The DFR record shown in Fig. 9.38 reflects the currents and voltages at the 138-kV side of the generator step-up transformer.

(a) State the positions of CBs A1 and A2 prior to occurrence of the fault.

(b) Explain the fault mechanism.

(c) Develop the sequence of events.

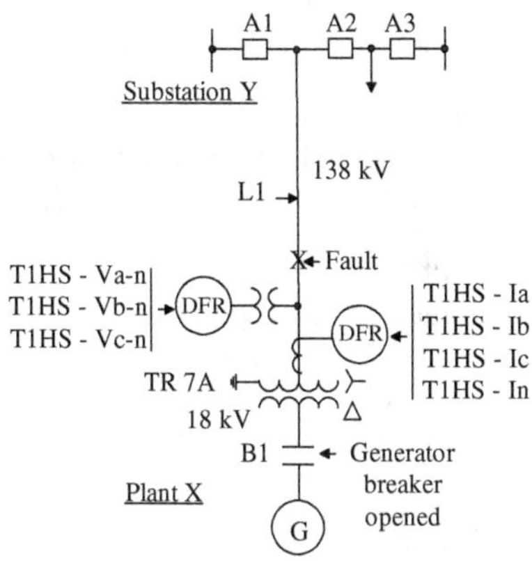

Fig. 9.37 One-line diagram showing feeder L1 DFR-monitored voltages and currents.

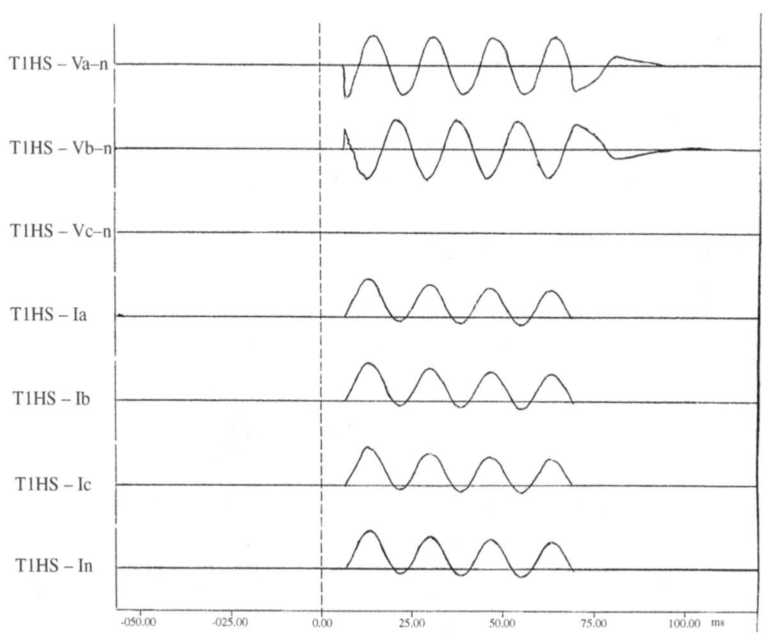

Fig. 9.38 Substation X DFR record showing feeder L1 currents and voltages during the fault.

(d) Classify the fault type.

(e) Determine the location of the fault based on an analysis of the 138-kV feeder voltages.

(f) Explain the nature of the current flows for phases A, B, and C and the neutral current traces.

(g) Explain the nature of currents during the fault and why they are equal on all phases.

(h) Determine the fault resistance. Explain you answer.

9.19. A line-to-ground fault occurred on line L1 with the DFR-monitored currents shown in Fig. 9.39. Using the DFR record shown in Fig. 9.40 for the voltages and currents for line L1:

(a) Classify the fault type.

(b) Determine the fault incident point on the faulted phase voltage trace.

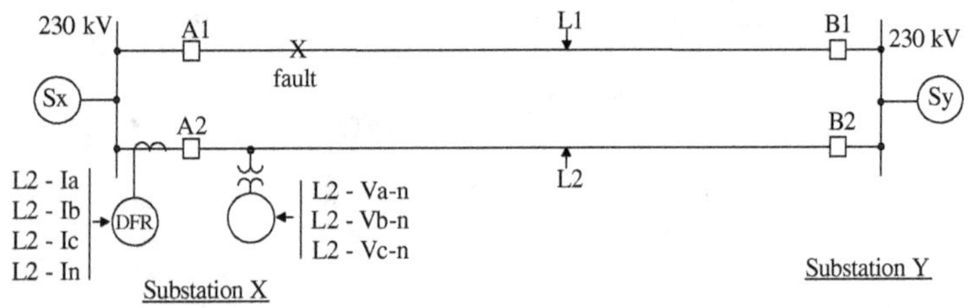

Fig. 9.39 One-line diagram showing DFR-monitored currents and voltages at substation X.

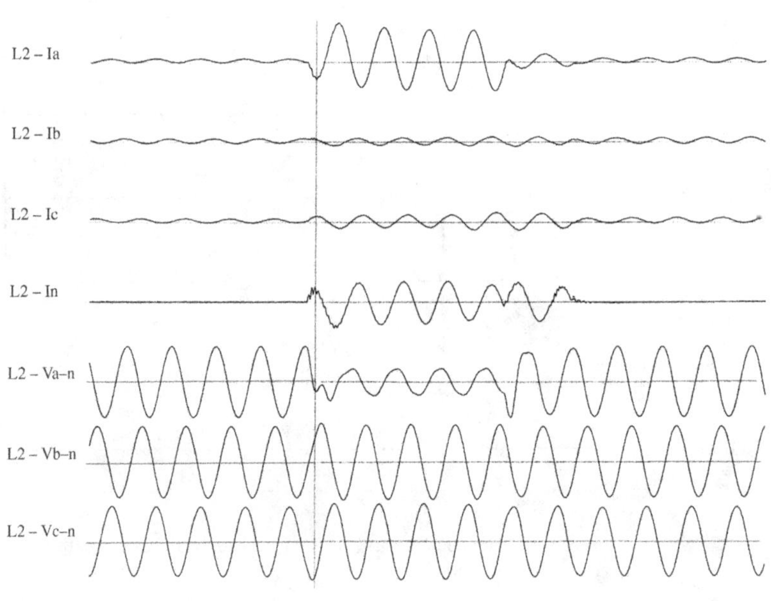

Fig. 9.40 Substation X DFR record showing feeder L2 currents and voltages during the fault.

(c) Determine the clearing time for CB A1 at substation X.

(d) Determine the clearing time for CB B1 at substation Y.

(e) Explain how the clearing times were obtained in (c) and (d).

(f) Explain the current directions using the simplified one-line diagram shown in Fig. 9.39.

9.20. A fault occurred on the 230-kV line L1 shown in Fig. 9.41. The fault was caused by icing formation on the line insulator during an ice storm. The DFR record of substation X shown in Fig. 9.42 documents the three-phase line currents, neutral current, and line-side three-phase-to-neutral voltages of line L1.

(a) Classify the fault type.

(b) Determine the fault incident point on the faulted phase voltage trace.

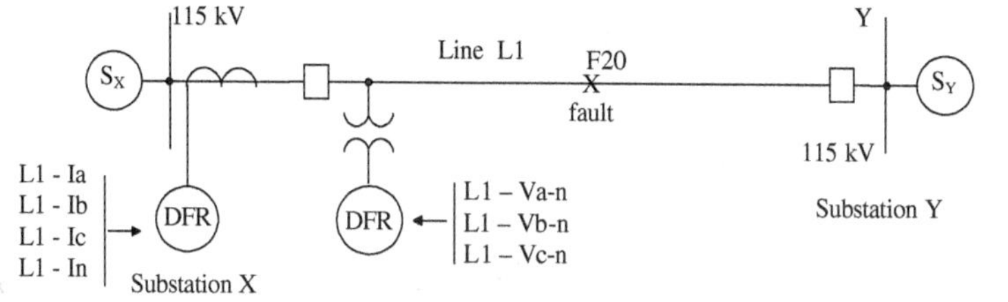

Fig. 9.41 One-line diagram showing DFR-monitored line voltages and currents.

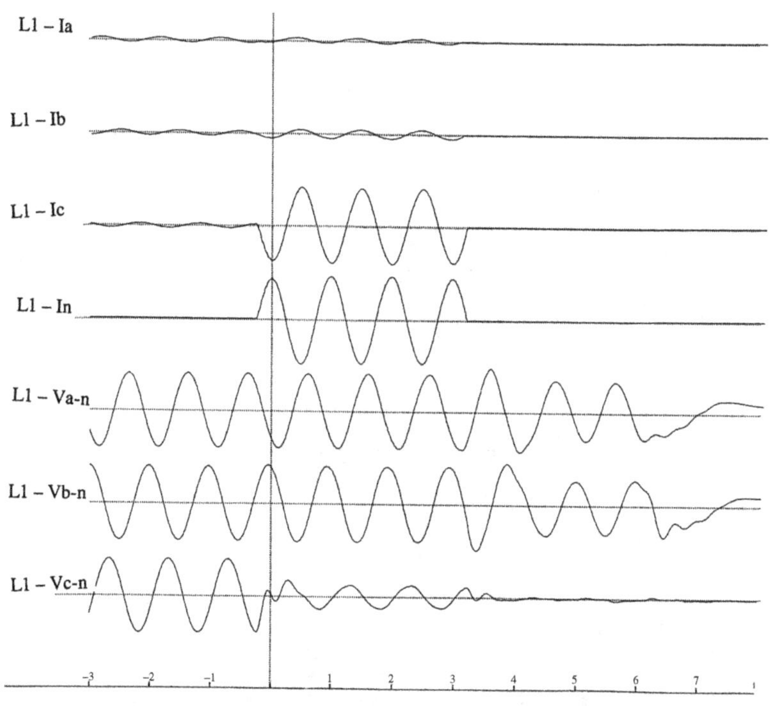

Fig. 9.42 Substation X DFR record for fault F20 on line L1.

(c) Explain the fault mechanism scenario.

(d) Explain how a local DFR fault record can be used to determine the remote-end clearing time of a line fault.

(e) Determine the total clearing times of the local X and remote Y ends of the faulted line.

9.21. A fault occurred on the 115-kV line L1 shown in Fig. 9.43. The fault was caused by a line insulator flashover during an ice storm. The DFR record of substation X shown in Fig. 9.44 documents the three-phase line currents, neutral current, and line-side three-phase-to-neutral voltages of line L1.

(a) Classify the fault type.

(b) Determine the fault incident point on the faulted phase voltage trace.

(c) Explain the fault mechanism scenario.

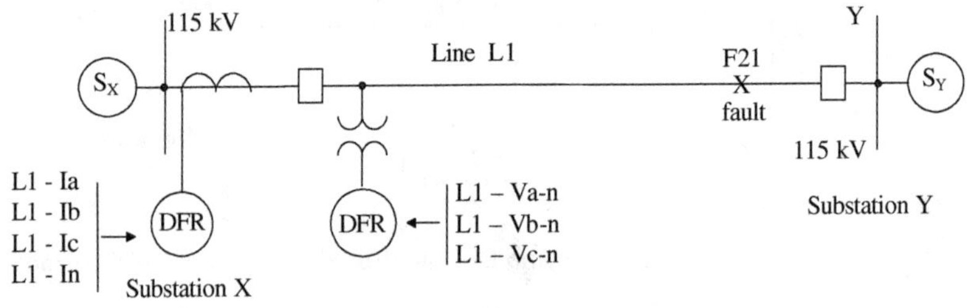

Fig. 9.43 One-line diagram showing DFR-monitored line voltages and currents.

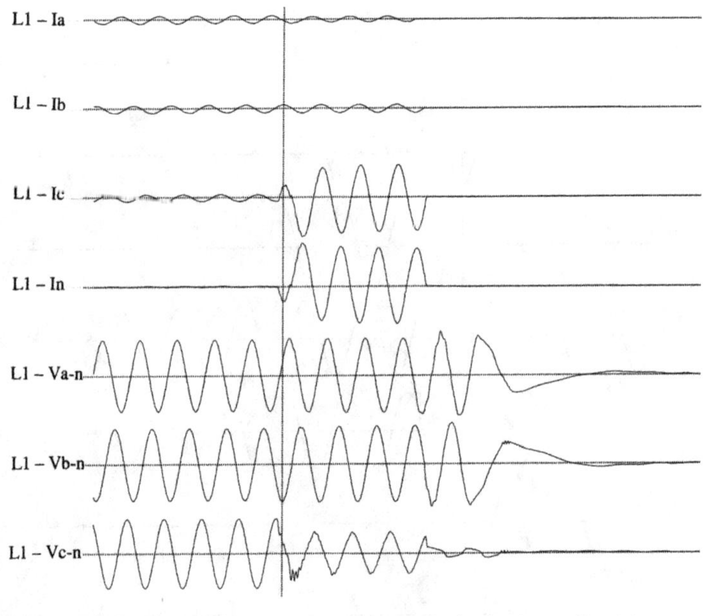

Fig. 9.44 Substation X DFR record for fault F21 on line L1.

(d) Explain how a local DFR fault record can be used to determine the remote-end clearing time of a line fault.

(e) Determine the total clearing times of the local X and remote Y ends of the faulted line.

9.22. A fault occurred on the 230-kV line L1 shown in Fig. 9.45. The fault was caused by a line insulator flashover. The DFR record of substation X shown in Fig. 9.46 documents the three-phase line currents, neutral current, and line-side three-phase-to-neutral voltages of line L1.

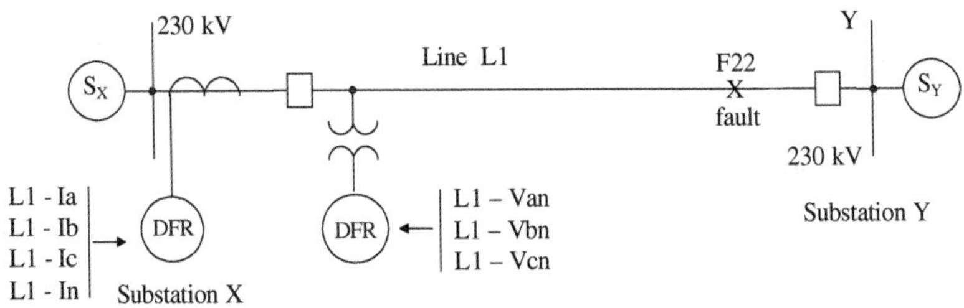

Fig. 9.45 One-line diagram showing DFR-monitored line voltages and currents.

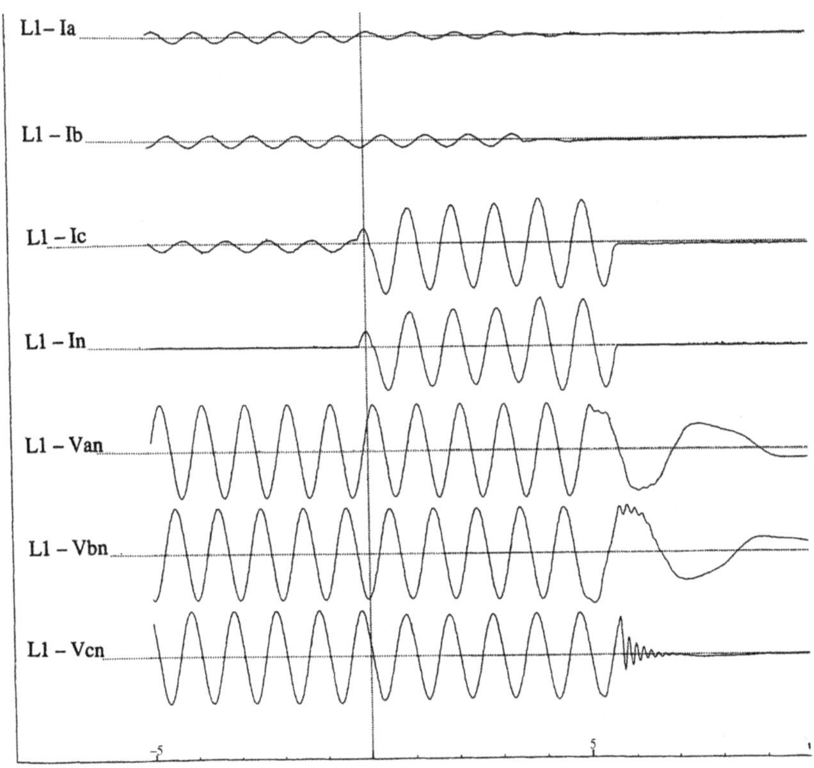

Fig. 9.46 Substation X DFR record for fault F22 on line L1.

(a) Classify the fault type.

(b) Determine the fault incident point on the faulted phase voltage trace.

(c) Explain the fault mechanism scenario.

(d) Explain how a local DFR fault record can be used to determine the remote-end clearing time of a line fault.

(e) Determine the total clearing times of the local X and remote Y ends of the faulted line.

(f) Explain the nature of the varying current magnitudes for the phase A current.

(g) Explain the increase in the phase A current a few cycles prior to interruption.

(h) Explain the disappearance of loading currents on phases A and B.

9.23. A line-to-ground fault occurred on line L1 with the DFR-monitored currents shown in Fig. 9.47. Using the DFR record shown in Fig. 9.48 for the neutral currents of lines L1 and L2:

(a) Determine the clearing time for CB A1 at substation X.

(b) Determine the clearing time for CB B1 at substation Y.

(c) Explain how the clearing times were obtained in (a) and (b).

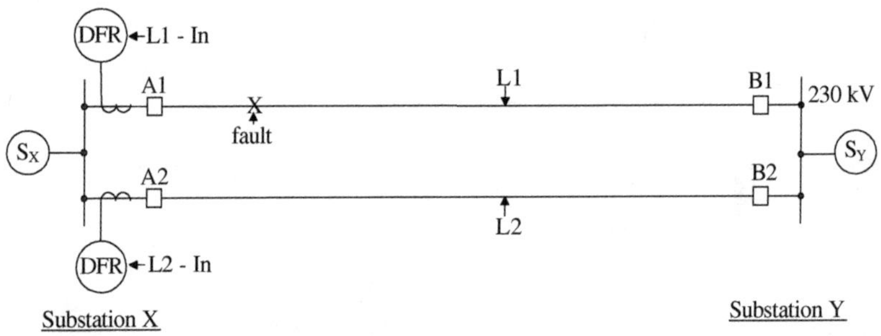

Fig. 9.47 One-line diagram showing DFR-monitored ground currents at substation X.

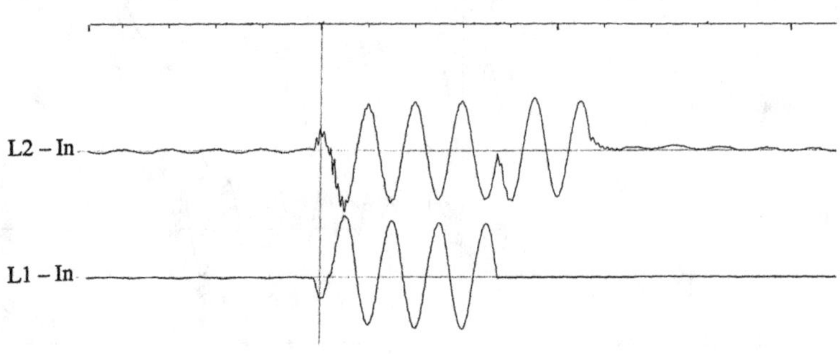

Fig. 9.48 Substation X DFR traces for the ground currents for L1 and L2.

(d) Explain the current directions using the simplified one-line diagram shown in Fig. 9.47.

9.24. A 230-kV A-g fault occurred on line L2, due to an insulator flashover resulting in symmetrical currents, as shown in Fig. 9.49. The autotransformer tertiary voltage is 13.2-kV. Figure 9.50 shows a DFR record for all ground currents for

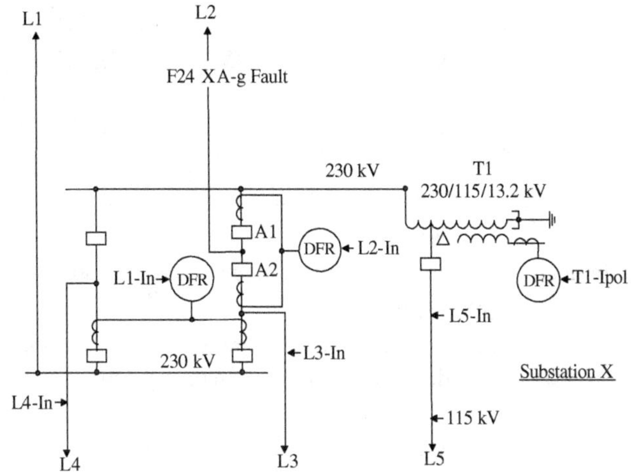

Fig. 9.49 One-line diagram showing DFR-monitored line currents during the A-g fault.

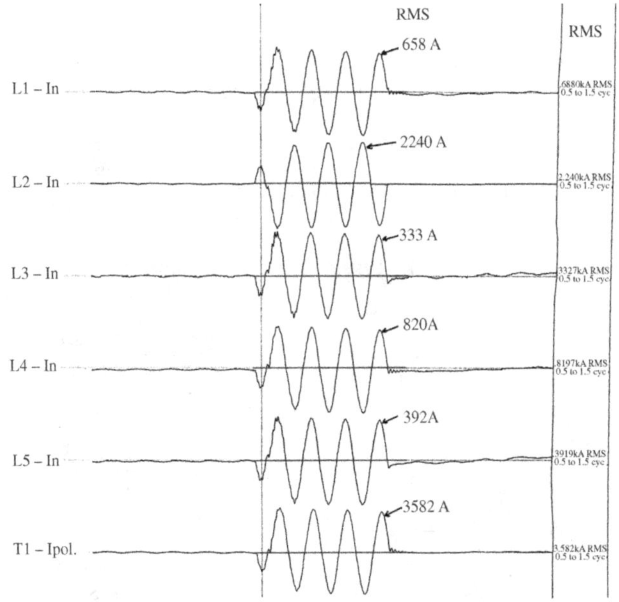

Fig. 9.50 Substation X DFR record showing the RMS values for line neutral currents and T1 tertiary current during the fault at F24.

230-kV lines L1, L2, L3, L4, 115-kV line L5, and the tertiary polarizing current with the RMS current values as calculated by the DFR software.

(a) Verify the accuracy of the DFR recording for the A-g fault shown.

(b) Determine the clearing time for the A-g fault at substation X.

(c) Using Fig. 9.49, determine the phase relationships for the currents feeding the A-g fault at substation X.

9.25. Figure 9.51 shows the system one-line ac diagram for transformer bank consisting of three single-phase 500-MVA 765/345/34.5-kV autotransformers with a delta tertiary. The standard taps for the current transformers for the 765-kV and 345- sides are shown in Fig. 9.52. The transformer is protected by dual transformer differential relays.

(a) Design (through calculation) the required single CT ratio required for the 34.5-kV winding.

(b) Indicate the ideal zones of protection on the attached one-line ac diagram. Explain your answer.

(c) What are the minimum and ideal number of relays that can provide protection redundancy and faster system restoration following a temporary

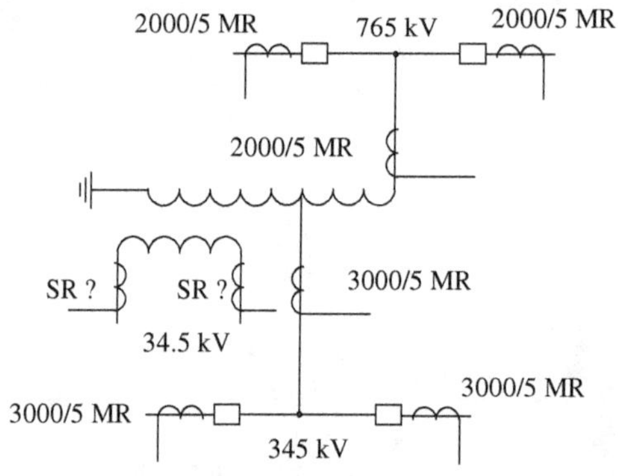

Fig. 9.51 System one-line ac diagram.

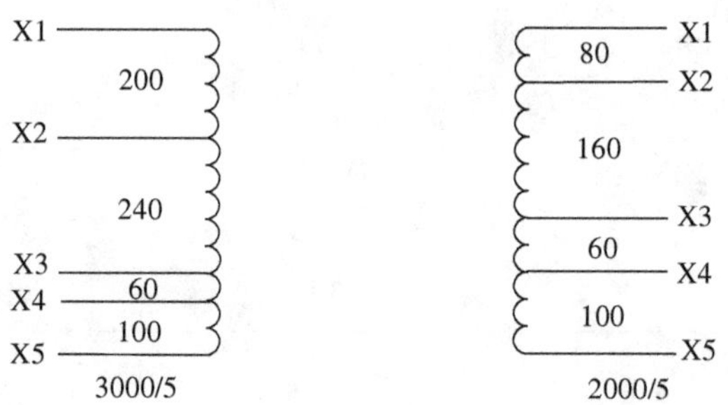

Fig. 9.52 Standard taps for busing current transformers.

fault on the 765-kV lead that connects the transformer to the substation? Explain your answer.

(d) If a numerical relay has a continuously adjustable tap range of 2.0 to 12.0 A and a software internal phase shift correction, select the CT ratios, CT connection, relay tap, and phase angle correction settings for each transformer side to balance the relay and achieve mismatch error of less than 5%. The transformer has a fixed 345-kV tap and five taps with 727-kV as the mid-tap voltage.

9.26. A combustion turbine (CT) plant is connected to the 115-kV system via transformer T2 as shown in Fig. 9.53. The CT unit was out of service and generator breaker A1 was opened and racked out. The 115-kV system is used as a startup source and to power station service loads. The GSU transformer T2 is protected by dual numerical differential relays. One of the relays, shown in Fig. 9.53, operated and tripped 115-kV circuit breaker A4. The numerical differential relay fault record for the event is as follows:

Winding W1: $I_A = 900$ A at 0; $I_B = 1800$ A at 180° lag; $I_C = 900$ A at 0°

Winding W2: $I_a = 0$; $I_b = 0$; $I_c = 0$

Winding W3: $I_a = 0$; $I_b = 0$; $I_c = 0$

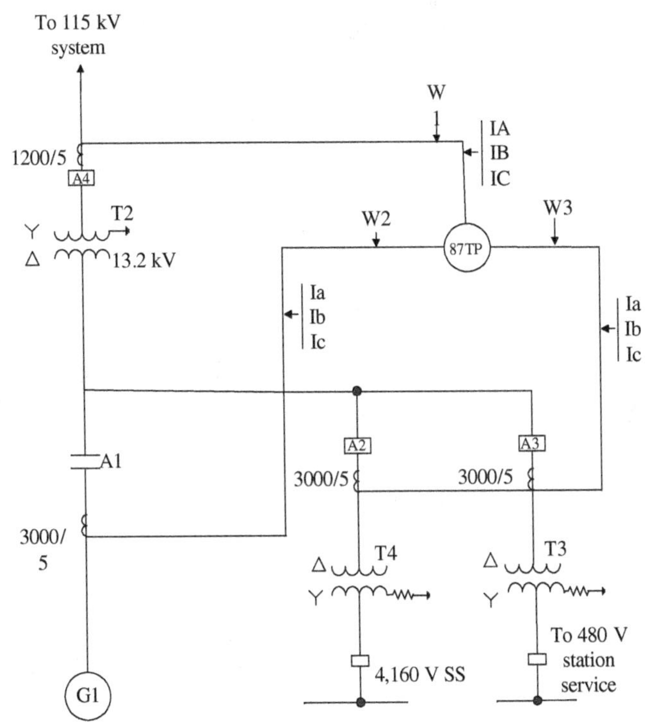

Fig. 9.53 One-line AC diagram showing the T2 87TP numerical differential relay–monitored currents.

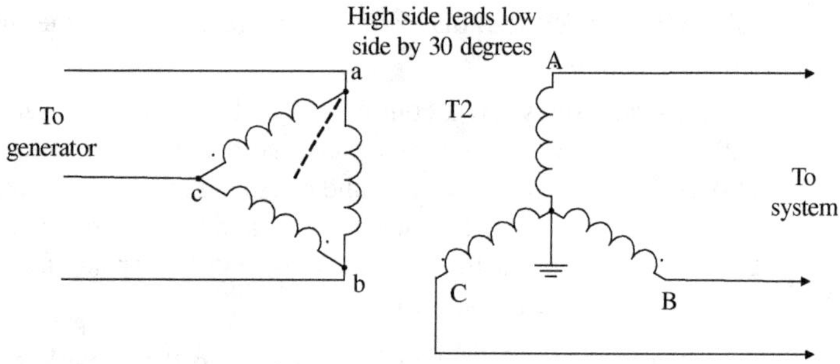

Fig. 9.54 Transformer T2 phasor diagram obtained from the transformer name plate.

Given the phase currents for a numerical relay and the transformer phasor diagram shown in Fig. 9.54.

(a) Find the fault location that matches the relay event record.

(b) Classify the fault based on the relay information given.

(c) Calculate the fault currents for the 13.2-kV delta currents and line currents for phases a, b, and c.

(d) Based on the results obtained in (c), demonstrate the transformer T2 ampere-turns coupling theory for phases A, B, and C.

INDEX